Lecture Notes in Bioinformatics 3500

Subseries of Lecture Notes in Computer Science

T0180640

Lecture Notes in Bioinformatics   3500

Edited by S. Istrail, P. Pevzner, and M. Waterman

Editorial Board: A. Apostolico, S. Brunak, M. Gelfand,
T. Lengauer, S. Miyano, G. Myers, M.-F. Sagot, D. Sankoff,
R. Shamir, T. Speed, M. Vingron, W. Wong

Subseries of Lecture Notes in Computer Science

Satoru Miyano   Jill Mesirov
Simon Kasif   Sorin Istrail
Pavel Pevzner   Michael Waterman (Eds.)

Research in Computational Molecular Biology

9th Annual International Conference, RECOMB 2005
Cambridge, MA, USA, May 14-18, 2005
Proceedings

 Springer

Volume Editors

Satoru Miyano
University of Tokyo
Human Genome Center, Institute of Medical Science, Tokyo, Japan
E-mail: miyano@ims.u-tokyo.ac.jp

Jill Mesirov
Broad Institute of MIT and Harvard, Cambridge, USA
E-mail: mesirov@broad.mit.edu

Simon Kasif
Boston University
Computational Genomics Laboratory, Boston, USA
E-mail: kasif@bu.edu

Sorin Istrail
California Institute of Technology, Pasadena, USA
E-mail: istrail@caltech.edu

Pavel Pevzner
University of California, San Diego, USA
E-mail: ppevzner@cs.ucsd.edu

Michael Waterman
University of Southern California, Los Angeles, USA
E-mail: msw@usc.edu

Library of Congress Control Number: Applied for

CR Subject Classification (1998): F.2.2, F.2, E.1, G.2, H.2.8, G.3, I.2, J.3

ISSN	0302-9743
ISBN-10	3-540-25866-3 Springer Berlin Heidelberg New York
ISBN-13	978-3-540-25866-7 Springer Berlin Heidelberg New York

Springer is a part of Springer Science+Business Media

springeronline.com

© Springer-Verlag Berlin Heidelberg 2005
Printed in Germany

Typesetting: Camera-ready by author, data conversion by Scientific Publishing Services, Chennai, India
Printed on acid-free paper   SPIN: 11415770   06/3142   5 4 3 2 1 0

Preface

This volume contains the papers presented at the 9th Annual International Conference on Research in Computational Molecular Biology (RECOMB 2005), which was held in Cambridge, Massachusetts, on May 14–18, 2005. The RECOMB conference series was started in 1997 by Sorin Istrail, Pavel Pevzner and Michael Waterman. The list of previous meetings is shown below in the section "Previous RECOMB Meetings." RECOMB 2005 was hosted by the Broad Institute of MIT and Harvard, and Boston University's Center for Advanced Genomic Technology, and was excellently organized by the Organizing Committee Co-chairs Jill Mesirov and Simon Kasif.

This year, 217 papers were submitted, of which the Program Committee selected 39 for presentation at the meeting and inclusion in this proceedings. Each submission was refereed by at least three members of the Program Committee. After the completion of the referees' reports, an extensive Web-based discussion took place for making decisions. From RECOMB 2005, the Steering Committee decided to publish the proceedings as a volume of *Lecture Notes in Bioinformatics* (LNBI) for which the founders of RECOMB are also the editors. The prominent volume number LNBI 3500 was assigned to this proceedings. The RECOMB conference series is closely associated with the *Journal of Computational Biology* which traditionally publishes special issues devoted to presenting full versions of selected conference papers. The RECOMB Program Committee consisted of 42 members, as listed on a separate page. I would like to thank the RECOMB 2005 Program Committee members for their dedication and hard work. Furthermore, I should thank the persons who assisted the Program Committee members by reviewing the submissions. In appreciation of their efforts, their names are given on a special page.

RECOMB 2005 invited eight keynote speakers: David Altshuler (Broad Institute of MIT and Harvard, Massachusetts General Hospital), Wolfgang Baumeister (Max Planck Institute for Biochemistry), James Collins (Boston University), Charles DeLisi (Boston University), Jonathan King (MIT), Eric Lander (Broad Institute of MIT and Harvard), Michael Levine (University of California at Berkeley), and Susan Lindquist (Whitehead Institute). The Stanislaw Ulam Memorial Computational Biology Lecture was given by Charles DeLisi. The Distinguished Biology Lecture was given by Jonathan King.

In addition to the keynote talks and the contributed talks, an important ingredient of the program was the lively poster session which was organized with tremendous efforts by the members of the RECOMB 2005 Organizing Committee, listed on a separate page.

I would like to thank the Organizing Committee for their contributions in making the RECOMB 2005 Conference a true Boston experience! Committee members are: Gary Benson (Boston University), Bonnie Berger (MIT),

Barb Bryant (Millennium Pharmaceuticals), Charlie Cantor (Boston University), George Church (Harvard University), Peter Clote (Boston College), Lenore Cowen (Tufts University), David Gifford (MIT), Jun Liu (Harvard University), Tommy Poggio (MIT), Aviv Regev (Harvard University), Brigitta Tadmor (MIT).

Especially, I would like to thank the Organizing Committee Co-chairs Jill Mesirov (Broad Institute of MIT and Harvard) and Simon Kasif (Boston University) for their strong leadership and guidance, as well as Gus Cervini and Lynn Vitiello (Broad Institute of MIT and Harvard) for their tireless efforts over the past 18 months in bringing this conference to Boston/Cambridge. Special thanks are due to Gus Cervini for his marvelous work even on the day of a huge snowstorm.

RECOMB 2005 would like to thank the institutions and corporations who provided financial support for the conference: Broad Institute of MIT and Harvard, Boston University's Center for Advanced Genomic Technology; Affymetrix, Apple, Biogen, EMC, Glaxo-SmithKline, Hewlett-Packard, IBM, Mathworks, Millennium Pharmaceuticals, NetApp, Novartis, Pfizer, Rosetta BioSoftware, Wyeth.

I thank the members of the Steering Committee for inviting me to be the Program Chair of RECOMB 2005 and for their support and encouragement when I carried out this task.

I would also like to thank Hideo Bannai, Seiya Imoto and Mizuho Wada (Human Genome Center, Univerisity of Tokyo) for their efforts in handling the submissions and editing the final versions for publication.

In closing, I would like to thank all the people who submitted papers and posters and those who attended RECOMB 2005 with enthusiasm.

May 2005   Satoru Miyano

Program Committee Members

Tatsuya Akutsu	Kyoto University, Japan
Alberto Apostolico	Univeristy of Padova, Italy and Purdue University, USA
Vineet Bafna	University of California, San Diego, USA
Serafim Batzoglou	Stanford University, USA
Philip E. Bourne	University of California, San Diego, USA
Alvis Brazma	EBI, UK
Siren Brunak	Technical University of Denmark, Denmark
Tim Ting Chen	University of Southern California, USA
Andy Clark	Cornell University, USA
Gordon M. Crippen	University of Michigan, USA
Eleazar Eskin	University of California, San Diego, USA
Nir Friedman	Hebrew University, Israel
Sridhar Hannenhalli	University of Pennsylvania, USA
Trey Ideker	University of California, San Diego, USA
Sorin Istrail	Caltech, USA
Tao Jiang	University of California, Riverside, USA
Richard M. Karp	University of California, Berkeley, USA
Simon Kasif	Boston University, USA
Giuseppe Lancia	University of Padua, Italy
Thomas Lengauer	Max-Planck-Institut für Informatik, Germany
Michal Linial	The Hebrew University in Jerusalem, Israel
Jill Mesirov	Broad Institute of MIT and Harvard, USA
Satoru Miyano	Chair, University of Tokyo, Japan
Richard Mott	Oxford University, UK
Gene Myers	University of California, Berkeley, US
Pavel A. Pevzner	University of California, San Diego, USA
Tzachi Pilpel	Weizmann Institute of Science, Israel
John Quackenbush	TIGR, USA
Aviv Regev	Harvard University, USA
Burkhard Rost	Columbia University, USA
Walter L. Ruzzo	University of Washington, USA
David Sankoff	University of Montreal, Canada
Ron Shamir	Tel Aviv University, Israel
Steve Skiena	State University of New York at Stony Brook, USA
Temple F. Smith	Boston University, USA
Terry Speed	University of California, Berkeley, USA
Martin Vingron	Max Planck Institute for Molecular Genetics, Germany
Michael Waterman	University of Southern California, USA
Haim J. Wolfson	Tel Aviv University, Israel
Wing Wong	Harvard University, USA
Ying Xu	University of Georgia, USA
Golan Yona	Cornell University, USA

Steering Committee Members

Sorin Istrail	RECOMB General Vice-Chair, Caltech, USA
Thomas Lengauer	Max-Planck-Institut für Informatik, Germany
Michal Linial	The Hebrew University of Jerusalem, Israel
Pavel A. Pevzner	RECOMB General Chair, University California, San Diego, USA
Ron Shamir	Tel Aviv University, Israel
Terence P. Speed	University of California, Berkeley, USA
Michael Waterman	RECOMB General Chair, University of Southern California, USA

Organizing Committee Members

Jill P. Mesirov	Co-chair (Broad Institute of MIT and Harvard)
Simon Kasif	Co-chair (Boston University)
Gary Benson	Boston University
Bonnie Berger	Massachusetts Institute of Technology
Barb Bryant	Millennium Pharmaceuticals
Charlie Cantor	Boston University
Gus Cervini	Broad Institute of MIT and Harvard)
George Church	Harvard University
Peter Clote	Boston College
Lenore Cowen	Tufts University
David Gifford	Massachusetts Institute of Technology
Jun Liu	Harvard University
Tommi Poggio	Massachusetts Institute of Technology
Aviv Regev	Harvard University
Fritz Roth	Harvard University
Brigitta Tadmor	Massachusetts Institute of Technology
Lynn Vitiello	Broad Institute of MIT and Harvard

Previous RECOMB Meetings

Date/Location	Hosting Institution	Program Chair	Organizing Chair
January 20–23, 1997 Santa Fe, NM, USA	Sandia National Lab	Michael Waterman	Sorin Istrail
March 22–25, 1998 New York, NY, USA	Mt. Sinai School of Medicine	Pavel Pevzner	Gary Benson
April 22–25, 1999 Lyon, France	INRIA	Sorin Istrail	Mireille Regnier
April 8–11, 2000 Tokyo, Japan	University of Tokyo	Ron Shamir	Satoru Miyano
April 22–25, 2001 Montréal, Canada	Université de Montréal	Thomas Lengauer	David Sankoff
April 18–21, 2002 Washington, DC, USA	Celera	Gene Myers	Sridhar Hannenhalli
April 10–13, 2003 Berlin, Germany	German Federal Ministry for Education & Research	Webb Miller	Martin Vingron
March 27–31, 2004 San Diego, USA	UC San Diego	Dan Gusfield	Philip E. Bourne

In addtion the Program Committee was greatly assisted by input from many people:

Tomohiro Ando
Iris Antes
Peter Arndt
George Asimenos
Vikas Bansal
Yoseph Barash
Aaron Birkland
Nikolaj Blom
Paola Bonizzoni
David Bryant
Jeremy Buhler
Tianxi Cai
Gal Chechik
Zhong Chen
Matteo Comin
Richard Coulson
Yan Cui
Vlado Danci
Sudeshna Das
Shailesh Date
Eugene Davydov

Milind Dawande
William Day
Michiel de Hoon
Arthur Delcher
Omkar Deshpande
Sharon Diskin
Chuong B. Do
Francisco Domingues
Riccardo Dondi
Pierre Donnes
Agostino Dovier
Gideon Dror
Steffen Durinck
Debojyoti Dutta
Nathan Edwards
Gal Elidan
Rani Elkon
Byron Ellis
Vince Emanuele
Anders Fausboll
Ari Frank

Jane Fridlyand
Iddo Friedberg
Carsten Friis
Tim Gardner
Olivier Gascuel
Irit Gat-Viks
Gary Glonek
Liang Goh
R. Gouveia-Oliveira
Apostol Gramada
Catherine Grasso
Clemens Groepl
Sam Gross
Steffen Grossmann
Jenny Gu
Brian Haas
Eran Halperin
Daniel Hanisch
Ke Hao
Boulos Harb
Tzvika Hartman

Anders Hinsby
Pengyu Hong
Katsuhisa Horimoto
Daniel Hsu
Haiyan Huang
Wolfgang Huber
Asa Ben Hur
Seiya Imoto
Ioannis Ioannidis
Rafael Irizarry
Ariel Jaimovich
Hanne Jarmer
Shane Jensen
Euna Jeong
Hongkai Ji
Yuting Jia
Hyun Min Kang
Ming-Yang Kao
Noam Kaplan
Tommy Kaplan
Ulas Karaoz
Uri Keich
Ryan Kelley
Wayne Kendal
Lars Kiemer
Gad Kimmel
Hironori Kitakaze
Satoshi Kobyashi
Oliver Kohlbacher
Mehmet Koyuturk
Chin-Jen Ku
Hiroyuki Kurata
Michal Lapidot
Zenia M. Larsen
Hyunju Lee
Christina Leslie
Haifeng Li
Jing Li
Lei Li
Ming Li
Xiaoman Li
Guohui Lin
Chaim Linhart
Yueyi Liu
Zhijie Liu

Ole Lund
Claus Lundegaard
Anne Mølgaard
Bin Ma
Bill Majoros
Craig Mak
Thomas Manke
Fenglou Mao
Osamu Maruyama
Hiroshi Matsuno
Kankainen Matti
Geoff McLachlan
Bud Mishra
Amir Mitchel
T.M. Murali
Iftach Nachman
Reiichiro Nakamichi
Naoki Nariai
H. Bjoern Nielsen
Mahesan Niranjan
Victor Olman
Sean O'Rourke
Paul S. Pang
Laxmi Parida
Dana Pe'er
Itsik Pe'er
Thomas N. Petersen
Ma Ping
Julia Ponomarenko
Liviu Popescu
Clint Potter
Steve Qin
Youxing Qu
Nusrat Rabbee
Jörg Rahnenführer
Gesine Reinert
Knut Reinert
Romeo Rizzi
Mark Robinson
Keith Robison
Wasinee Rungsarityotin
Gabriell Rustici
Yasubumi Sakakibara
Anastasia Samsonova
Oliver Sander

Simone Scalabrin
Thomas Schlitt
Russell Schwartz
Reut Shalgi
Roded Sharan
Itai Sharon
Kerby A. Shedden
Kirby Shedden
Noa Shefi
Tetsuo Shibuya
Ilya Shindyalov
Thomas Sicheritz-Ponten
Tobias Sing
Gordon Smyth
Lev Soinov
Ingolf Sommer
Rainer Spang
Eike Staub
Martin Steffen
Ulrike Stege
Israel Steinfeld
Nikola Stojanovic
Shih-Chieh Su
Zhengchang Su
Andreas Sundquist
Silpa Suthram
Amos Tanay
Haixu Tang
Martin Taylor
Elisabeth Tillier
Martin Tompa
Mercan Topkara
Umut Topkara
Nobuhisa Ueda
Igor Ulitsky
Dave Ussery
Gianluca Della Vedova
Stella Veretnik
Nicola Vitacolonna
Neils Volkmann
Yap Von Bing
Jeff Wall
Junwen Wang
Li-San Wang
Lusheng Wang

Todd Wareham

Atsuko Yamaguchi

John Zhang

Zasha Weinberg

Bo Yan

Qing Zhang

John Westbrook

Song Yang

Xuegong Zhang

Chris Workman

Chun Ye

Yu Zhang

Hongwei Wu

Lai Yinlei

Jie Zheng

Lei Xie

Ryo Yoshida

Qing Zhou

Dong Xu

Bin Yu

Chaya B.-Z. Zilberstein

Zohar Yakhini

Noah Zaitlen

Table of Contents

Keynote

Keynote

Keynote

Keynote

Keynote

Keynote

Efficient Algorithms for Detecting Signaling Pathways in Protein Interaction Networks

Jacob Scott[1], Trey Ideker[2], Richard M. Karp[3], and Roded Sharan[4]

[1] Computer Science Division, U. C. Berkeley,
387 Soda Hall, Berkeley, CA 94720
jhs@ocf.berkeley.edu
[2] Dept. of Bioengineering, U. C. San Diego,
9500 Gilman Drive, La Jolla, CA 92093
trey@bioeng.ucsd.edu
[3] International Computer Science Institute,
1947 Center St., Berkeley, CA 94704
karp@icsi.berkeley.edu
[4] School of Computer Science,
Tel-Aviv University, Tel-Aviv 69978, Israel
roded@cs.tau.ac.il

Abstract. The interpretation of large-scale protein network data depends on our ability to identify significant sub-structures in the data, a computationally intensive task. Here we adapt and extend efficient techniques for finding paths in graphs to the problem of identifying pathways in protein interaction networks. We present linear-time algorithms for finding paths in networks under several biologically-motivated constraints. We apply our methodology to search for protein pathways in the yeast protein-protein interaction network. We demonstrate that our algorithm is capable of reconstructing known signaling pathways and identifying functionally enriched paths in an unsupervised manner. The algorithm is very efficient, computing optimal paths of length 8 within minutes and paths of length 10 in less than two hours.

1 Introduction

A major challenge of post-genomic biology is to understand the complex networks of interacting genes, proteins and small molecules that give rise to biological form and function. Protein-protein interactions are crucial to the assembly of protein machinery and the formation of protein signaling cascades. Hence, the dissection of protein interaction networks has great potential to improve the understanding of cellular machinery and to assist in deciphering protein function.

The available knowledge about protein interactions in a single species can be represented as a *protein interaction graph*, whose vertices represent proteins and whose edges represent protein interactions; each edge can be assigned a weight, indicating the strength of evidence for the existence of the corresponding interaction. An important class of protein signaling cascades can be described as

S. Miyano et al. (Eds.): RECOMB 2005, LNBI 3500, pp. 1–13, 2005.

chains of interacting proteins, in which protein interactions enable each protein in the path to modify its successor so as to transmit biological information. Such structures correspond to simple paths in the protein interaction graph [1].

Steffen et al. [2] studied the problem of identifying pathways in a protein network. They applied an exhaustive search procedure to an unweighted interaction graph, considering all interactions equally reliable. To score the biological relevance of an identified path, they scored the tendency of its genes to have similar expression patterns. The approach was successful in detecting known signaling pathways in yeast. A related study that we have conducted [1] aimed at identifying pathways that are conserved across two species. The study employed a more efficient way of detecting simple paths in a graph that is based on finding acyclic orientations of the graph's edges.

The present work advances the methodology of searching for signaling cascades in two ways: first, by assigning well-founded reliability scores to protein-protein interactions, rather than putting all such interactions on the same footing; and second, by exploiting a powerful algorithmic technique by Alon et al. [3], called *color coding*, to find high-scoring paths efficiently. The color coding approach reduces the running time of the search algorithm by orders of magnitude compared to exhaustive search or to the faster acyclic orientation approach, thus enabling the search for longer paths. We also extend the color coding method to incorporate biologically-motivated constraints on the types of proteins that may occur in a path and the order of their occurrence, and to search for structures more general than paths, such as trees or two-terminal series-parallel graphs.

As evidence of the success of our approach we show that our method accurately recovers well-known MAP kinase and ubiquitin-ligation pathways, that many of the pathways we discover are enriched for known cellular functions, and that the pathways we find score higher than paths found in random networks obtained by shuffling the edges and weights of the original network while preserving vertex degrees.

The paper is organized as follows: Section 2 presents the path finding problem and describes the color coding approach. In Section 3 we develop biologically-motivated extensions of the color coding approach. Section 4 describes the estimation of protein interaction reliabilities and the path scoring methods used. Finally, Section 5 presents the applications of our method to yeast protein interaction data. For lack of space, some algorithmic details are omitted.

2   Finding Simple Paths: The Color Coding Technique

Alon et al. [3] devised a novel randomized algorithm, called *color coding*, for finding simple paths and simple cycles of a specified length k, within a given graph. In this section we describe this approach. Our presentation generalizes that in [3] in order to allow succinct description of biologically-motivated extensions of the basic technique.

Consider a weighted interaction graph in which each vertex is a protein and each edge (u, v) represents an experimentally observed interaction between proteins u and v, and is assigned a numerical value $p(u,v)$ representing the probability that u and v interact (computed as per Section 4 below). Each simple path in this graph can be assigned a *score* equal to the product of the values assigned to its edges. Among paths of a given length, those with the highest scores are plausible candidates for being identified as linear signal transduction pathways. Given a set I of possible start vertices we would like to find the highest-scoring paths from I to each vertex of the graph. In the case of signaling pathways I might be the set of all receptor proteins or a single protein of particular interest.

We begin by framing the problem mathematically. In order to work with an additive weight rather than a multiplicative one, we assign each edge (u,v) a weight $w(u,v) \equiv -\log p(u,v)$. We define the *weight* of a path as the sum of the weights of its edges, and the *length* of a path as the number of vertices it contains. Given an undirected weighted graph $G = (V, E, w)$ with n vertices, m edges and a set I of start vertices, we wish to find, for each vertex v, a minimum-weight simple path of length k that starts within I and ends at v. If no such simple path exists, our algorithm should report this fact.

For general k this problem is NP-hard, as the traveling-salesman problem is reducible to it. The difficulty of the problem stems from the restriction to simple paths; without this restriction the best path of length k is easily found. A standard dynamic programming algorithm for the problem is as follows. For each nonempty set $S \subseteq V$ of cardinality at most k, and each vertex $v \in S$, let $W(v, S)$ be the minimum weight of a simple path of length $|S|$ which starts at some vertex in I, visits each vertex in S, and ends at v. If no such path exists then $W(v, S) = \infty$. The following recurrence can be used to tabulate this function by generating the values $W(v, S)$ in increasing order of the cardinality of S:

$$W(v, S) = \min_{u \in S - \{v\}} W(u, S - \{v\}) + w(u, v), |S| > 1$$

where $W(v, \{v\}) = 0$ if $v \in I$ and ∞ otherwise.

The weight of the optimal path to v is the minimum of $W(v, S)$ over all pairs v, S such that $|S| = k$, and the vertices of the optimal path can be recovered successively in reverse order by a standard dynamic programming backtracking method. The running time of this algorithm is $O(kn^k)$ and its space requirement is $O(kn^k)$.

The idea of color coding is to assign each vertex a random color between 1 and k and, instead of searching for paths with distinct vertices, search for paths with distinct colors. The complexity of the dynamic programming algorithm is thereby greatly reduced, and the paths that are produced are necessarily simple. However, a path fails to be discovered if any two of its vertices receive the same color, so many random colorings need to be tried to ensure that the desired paths are not missed. The running time of the color coding algorithm is exponential in k and linear in m, and the storage requirement is exponential in k and linear in n. This method is superior when n is much larger than k, as is the case in our application, where typical values are $n = 6,000$ and $k = 8$.

The color coding algorithm requires repeated randomized trials. In each trial, every vertex $v \in V$ is independently assigned a color $c(v)$ drawn uniformly at random from the set $\{1, 2, ..., k\}$. Call a path *colorful* if it contains exactly one vertex of each color. We seek a minimum-weight colorful path from I to each vertex v. This problem can be solved using the following dynamic programming algorithm, which parallels the previous one: for each nonempty set $S \subseteq \{1, 2, ..., k\}$ and each vertex v such that $c(v) \in S$, let $W(v, S)$ be the minimum weight of a simple path of length $|S|$ that starts within I, visits a vertex of each color in S, and ends at v. This function can be tabulated using the following recurrence:

$$W(v, S) = \min_{u:c(u) \in (S - \{c(v)\})} W(u, S - \{c(v)\}) + w(u, v), |S| > 1$$

where $W(v, \{c(v)\}) = 0$ if $v \in I$ and ∞ otherwise.

The weight of a minimum-weight colorful path ending at v is $W(v, \{1, ..., k\})$. For each v, each trial yields a simple path of length k starting within I and ending at v, which is optimal among all the paths that are colorful under the random coloring in that trial. The running time of each trial is $O(2^k km)$ and the storage requirement is $O(2^k n)$. For any simple path P of length k, the probability that the vertices of P receive distinct colors in a given trial is $k!/k^k$, which is at least e^{-k} and is well approximated by $\sqrt{2\pi k}e^{-k}$. Thus, the chance that a trial yields an optimal path to v for our original problem is at least e^{-k}; for any $\epsilon \in (0, 1)$, the chance that the algorithm fails to find such a path in $e^k \ln \frac{1}{\epsilon}$ trials is at most ϵ. After $e^k \ln \frac{n}{\epsilon}$ trials the probability that there exists a vertex v for which an optimal path has not been found is at most ϵ.

3   Extensions of the Color Coding Method

In this section we present color-coding solutions to several biologically motivated extensions of the basic path-finding problem. These include: (1) constraining the set of proteins occurring in a path; (2) constraining the order of occurrence of the proteins in a path; and (3) finding pathway structures that are more general than simple paths.

3.1   Constraining the Set of Proteins

To ensure that a colorful path produced by our algorithm contains a particular protein, we can simply assign a color uniquely to that protein. By adding counters to the state set of the dynamic programming recurrence we can control the number of occurrences in the path of proteins from a specific family (e.g., proteins with a specific function). To enforce the constraint that our path must contain at least a and at most b proteins from a set T, we can define $W(v, S, c)$ as the minimum weight of a path of length $|S|$ ending at v that contains a vertex of each color in S and exactly c proteins from T. Here c ranges between 0 and b. This extension multiplies the storage requirement and running time of each trial by $b + 1$. Several counters can be added to enforce different constraints; each multiplies the time and storage requirement by a constant factor and does not affect the probability that the optimal path is colorful in any given trial.

3.2 Constraining the Order of Occurrence: Segmented Pathways

In many signaling pathways, the proteins occur in an inward order, from membrane proteins to nuclear proteins and transcription factors (see, e.g., Figure 2(a)). The color-coding method can be adapted to restrict attention to paths that respect such an ordering.

Unique Labeling. We restrict attention to simple paths that are the concatenation of t (possibly empty) ordered segments, where each segment contains proteins from a particular class (such as membrane proteins), and each protein is assigned to exactly one class. Subject to this restriction we seek, for each vertex v, a minimum-weight simple path of length k from some vertex in I to v. The segments are numbered successively in the order of their occurrence along the desired path. Depending on biological information (e.g., cellular component annotation), each protein u is assigned an integer label $L(u)$ which uniquely specifies the segment in which the protein may occur. We require that the labels of the proteins along the path form a monotonically nondecreasing sequence. Such a path is called *monotonic*.

As usual, in each trial we assign each vertex a color drawn uniformly at random from $\{1, 2, \ldots, k\}$. Since each vertex is restricted to a unique segment, the path will be simple provided that vertices in the same segment have different colors. For a vertex v and a subset of the colors $S \supseteq \{c(v)\}$, $W(v, S, k)$ is defined as the minimum weight of a simple monotonic path of length k from I to v, in which no two vertices with the same label have the same color, and the set of colors assigned to vertices with label $L(v)$ is S. We can tabulate this function using the following recurrence:

$$W(v, \{c(v)\}, l) = \min_{u:L(u)<L(v)} \min_{S} W(u, S, l-1) + w(u, v), l > 1$$

$$W(v, S, l) = \min_{u:L(u)=L(v),c(u)\in S-\{c(v)\}} W(u, S - \{c(v)\}, l-1) + w(u, v), 1 < |S| \le l$$

where $W(v, \{c(v)\}, 1) = 0$ if $v \in I$ and ∞ otherwise.

Each trial has a running time of $O(2^k km)$ and a storage requirement of $O(2^k kn)$. For any simple path P with at most h vertices in each segment, the probability that all vertices in each segment receive distinct colors is at least e^{-h}. Thus, the expected number of trials to discover an optimal segmented pathway with at most h proteins per segment is of order e^h, which is much smaller than e^k, the upper bound on expectation in the non-segmented case.

Interval Restrictions. It may be unrealistic to assume that every protein can be assigned *a priori* to a unique segment. Instead we can assume that, for each protein, there is a lower bound $L_1(u)$ and an upper bound $L_2(u)$ on the number of its segment. For example, if the successive segments correspond to membrane, cytoplasm, nucleus and transcription factor proteins, then a protein that is neither a membrane protein nor a transcription factor will have a lower bound of 2 and an upper bound of 3.

A path $(u_1, u_2, \ldots, u_k)$ is *consistent with the segmentation* if it is possible to assign to each protein u_i a segment number s_i such that the sequence of segment numbers along the path is monotone nondecreasing and, for each i, $L_1(u_i) \leq s_i \leq L_2(u_i)$. We can reformulate this condition as follows: for any path P, let $s(P)$ be the maximum, over all proteins u in P, of $L_1(u)$. Then the path $(u_1, u_2, \ldots, u_k)$ is consistent with the segmentation iff for all i, $L_2(u_i) \geq s(u_1, u_2, \ldots, u_{i-1})$.

Let each vertex u be assigned a color $c(u)$ drawn uniformly at random from $\{1, 2, \ldots, k\}$. For each vertex v, the color-coding method seeks a minimum-weight path of length k from I to v which is both colorful and consistent with the segmentation. Define $W(v, s, S)$, where $L_1(v) \leq s \leq L_2(v)$, as the minimum weight of a simple path P of length $|S|$ from I to v that is consistent with the segmentation, such that $s(P) = s$ and S is the set of colors assigned to the vertices in P. We obtain the following dynamic programming recurrence:

$$W(v, L_1(v), S) = \min_{u:c(u)\in(S-\{c(v)\})} \min_{s'\leq L_1(v)} W(u, s', S - \{c(v)\}) + w(u, v), |S| > 1$$

$$W(v, s, S) = \min_{u:c(u)\in(S-\{c(v)\})} W(u, s, S - \{c(v)\}) + w(u, v), L_1(v) < s \leq L_2(v), |S| > 1$$

where $W(v, L_1(v), \{c(v)\}) = 0$ if $v \in I$ and ∞ otherwise. The weight of a minimum-weight colorful path ending at v and consistent with the segmentation is $\min_s W(v, s, \{1, 2, \ldots, k\})$.

3.3   Finding More General Structures

In general, signaling pathways need not consist of a single path. For instance, the high osmolarity pathway in yeast starts with two separate chains that merge into a single path [2]. We shall demonstrate that the color-coding method can be used to find high-scoring signaling pathways with a more general structure. Our two principal examples are rooted trees, which are common when several pathway segments merge, and two-terminal series-parallel graphs, which capture parallel biological signaling pathways.

Rooted Trees. Let $G = (V, E)$ be a weighted graph with $I \subset V$, and let k be a positive integer. For each vertex v we wish to find a tree of minimum weight among all k-vertex subtrees in G that are rooted at v and in which every leaf is an element of I.

As usual, in each trial of the color coding method each vertex u is assigned a color drawn uniformly at random from $\{1, 2, \ldots, k\}$. For $v \in V$ and $\{c(v)\} \subseteq S \subseteq \{1, 2, \ldots, k\}$, let $W(v, S)$ be the minimum weight of a subtree with $|S|$ vertices that is rooted at v, contains a vertex of each color in S, and whose leaves lie in I. The following recurrence can be used to compute $W(v, S)$:

$$W(v, S) = \min\{ \min_{u:c(u)\in S-\{c(v)\}} W(u, S - \{c(v)\}) + w(u, v),$$
$$\min_{(S_1, S_2):S_1\cap S_2=\{c(v)\}, S_1\cup S_2=S} W(v, S_1) + W(v, S_2)\}$$

where $W(v, \{c(v)\}) = 0$ if $v \in I$ and ∞ otherwise. The running time for a trial is $O(3^k km)$ and the storage required is $O(2^k n)$.

Two-Terminal Series-Parallel Graphs. The definition of a two-terminal series-parallel graph (2SPG) is recursive:

[Base case] The graph with two vertices u and v connected by an edge is a 2SPG between terminals u and v.

[Series connection] If G_1 is a 2SPG between u and v, G_2 is a 2SPG between v and w, and G_1 and G_2 have no vertices in common except v, then $G_1 \cup G_2$ is a 2SPG between u and w.

[Parallel connection] If G_1 and G_2 are 2SPGs between u and v, and they have no vertices in common except u and v, then $G_1 \cup G_2$ is a 2SPG between u and v.

Our goal is to find, for each vertex v, a minimum-weight k-vertex 2SPG between some vertex in I and v. Let $W(u, v, S)$ be the minimum weight of a 2SPG between u and v with $|S|$ vertices in which the set of colors occurring is S. Then, following the recursive definition of a 2SPG we obtain:

$$W(u, v, S) = \min \{ \min_{w, S_1, S_2 : S_1 \cup S_2 = S, S_1 \cap S_2 = \{c(w)\}} W(u, w, S_1) + W(w, v, S_2),$$
$$\min_{T_1, T_2 : T_1 \cap T_2 = \{c(u), c(v)\}, T_1 \cup T_2 = S} W(u, v, T_1) + W(u, v, T_2)\}$$

where $W(u, v, \{c(u), c(v)\}) = w(u, v)$ for every edge (u, v). The execution time of a trial is $O(3^k kn^2)$ and the storage requirement is $O(2^k n^2)$.

4 Estimation of Interaction Reliabilities and Evaluation of Paths

Since experimental interaction data are notoriously noisy (see, e.g., [4, 5]), estimating and incorporating the reliability of the observed interactions in the path detection process are key to its success. Several authors have suggested methods for evaluating the reliabilities of protein interactions [5, 4, 6]. Here, we use a method we have previously developed [7], which is based on a logistic regression model. For completeness we describe it briefly in the sequel. We define the probability of a true interaction as a logistic function of three observed random variables on a pair of proteins: (1) the number of times an interaction between the proteins was experimentally observed; (2) the Pearson correlation coefficient of expression measurements for the corresponding genes (using 794 expression profiles obtained from Stanford Microarray Database [8]); and (3) the proteins' small world clustering coefficient [9], which is defined as the hypergeometric surprise for the overlap in the neighborhoods of two proteins.

According to the logistic distribution, the probability of a true interaction T_{uv} given the three input variables, $X = (X_1, X_2, X_3)$, is:

$$Pr(T_{uv}|X) = \frac{1}{1 + exp(-\beta_0 - \sum_{i=1}^{3} \beta_i X_i)}$$

where $\beta_0, \ldots, \beta_3$ are the parameters of the distribution. Given training data, one can optimize the distribution parameters so as to maximize the likelihood of the data. As positive examples we used the MIPS [10] interaction data, which is an accepted gold standard for yeast interactions. As negative examples, motivated by the large fraction of false positives in interaction data, we considered observed interactions chosen at random. We treated the chosen negative data as noisy indications that the corresponding interactions were false, and assigned those interactions a probability of 0.1397 for being true, where this value was optimized using cross-validation.

Denote the reliability of an edge (u, v) by $p(u, v)$. We use the estimated probabilities to assign weights to the interaction graph edges, where edge (u, v) is assigned the weight $-\log p(u, v)$. Under these assignments we seek minimum weight paths of specified lengths. We use two quality measures to evaluate the paths we compute: weight p-value and functional enrichment.

Given a path with weight w, its *weight p-value* is defined as the percent of top-scoring paths in random networks (computed using the same algorithm that is applied to the real network—see below) that have weight w or lower, where random networks are constructed by shuffling the edges and weights of the original network, preserving vertex degrees.

To evaluate the *functional enrichment* of a path P we associate its proteins with known Biological Processes using the Gene Ontology (GO) annotations [11]. We then compute the tendency of the proteins to have a common annotation using a method developed in [7]. The scoring is done as follows: define a protein to be *below* a GO term t, if it is associated with t or any other term that is a descendant of t in the GO hierarchy. For each GO term t with at least one protein assigned to it, we compute a hypergeometric p-value based on the following quantities: (1) the number of proteins in P that are below t; (2) the total number of proteins below t; (3) the number of proteins in P that are below all parents of t; and (4) the total number of proteins below all parents of t. The p-value is further Bonferroni corrected for multiple testing.

5   Application to the Yeast Protein Network

We implemented the color coding method for finding simple paths in a graph. The algorithm maintains a heap of the best paths found throughout the iterations and, thus, is able to report sub-optimal paths in addition to the optimal one. Table 1 presents a benchmark of its running time across a network with ~4,500 nodes and ~14,500 edges (the yeast network described below), when varying the desired path length, the probability of success and the size of the heap (number of required paths). The algorithm runs in minutes when searching for a path of length 8 with success probability of 99.9%, and in less than two hours when searching for a path of length 10. In comparison, the running time of our implementation of an exhaustive search approach was approximately 3-fold higher for length-8 paths (1009 seconds), and 14-fold higher for length-9 paths (17405 seconds).

Table 1. Running times of the path finding algorithm for different parameter settings

Path length	Success probability	#Paths	Time (sec)
10	99.9%	100	5613
9	99.9%	100	1241
8	99.9%	500	322
8	99.9%	300	297
8	99.9%	100	294
8	90%	100	99
8	80%	100	75
8	70%	100	61
8	50%	100	42
7	99.9%	100	86
6	99.9%	100	36

We applied our algorithm to search for pathways in the yeast protein interaction network. Protein-protein interaction data were obtained from the Database of Interacting Proteins [12] (February 2004 download) and contained 14,319 interactions among 4,389 proteins in yeast.

As a first test, we applied the algorithm to compute optimal paths of length 8 that start at a membrane protein (GO:0005886 or GO:0004872) and end at a transcription factor (GO:0030528). For every two possible endpoints we identified an optimal path between them (with the success probability set to 99.9%). We retained the 100 best paths of these (i.e., each of the paths had a distinct pair of endpoints) and for each of them evaluated its weight p-value and functional enrichment. The results are depicted in Figure 1. Clearly, both measures significantly exceed the random expectation. In particular, 60% of the identified paths had a significant Biological Process annotation ($p < 0.05$).

Next, we wished to test the utility of the algorithm in reconstructing known pathways in yeast. To this end, we concentrated on three MAPK signal transduction pathways that were also analyzed in [2]: pheromone response, filamentous growth and cell wall integrity. For each of the pathways we searched the net-

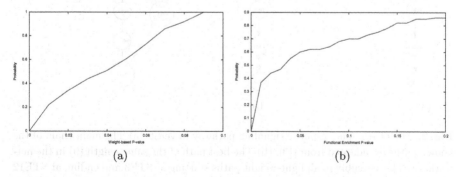

Fig. 1. Cumulative distributions of weight (a) and function (b) p-values. x-axis: p-value. y-axis: Percent of paths with p-value x or better

work for paths of lengths 6-10 using the pathway's endpoints to define the start
and end vertices. In all cases our results matched the known pathways well. We
describe these findings in detail below.

The pheromone response (mating type) pathway prepares the yeast cells for
mating by inducing polarized cell growth toward a mating partner, cell cycle
arrest in G1, and increased expression of proteins needed for cell adhesion, cell
fusion, and nuclear fusion. The main chain of this pathway (consisting of nine
proteins) is shown in Figure 2(a). In addition, proteins Bem1p, Rga1p, Cdc24p,
Far1p, Ste50p and Ste5p contribute to the operation of the pathway by interacting with proteins in the main chain.

Looking for the optimal path of length 9 in the yeast network yielded the path
depicted in Figure 2(b). This path mainly consists of proteins in the pheromone
response pathway, with the exception of Kss1p, which is a MAP kinase redundant to Fus3p, and Akr1p, which is a negative regulator of this pathway. The
occurrence of the latter protein is an artifact that arises because the direct link
between Ste3p and Ste4p is missing from the interaction data.

The aggregate of all the paths that the algorithm computed between Ste3p
and Ste12p, across a range of lengths (6-10), is depicted in Figure 2(c). All the

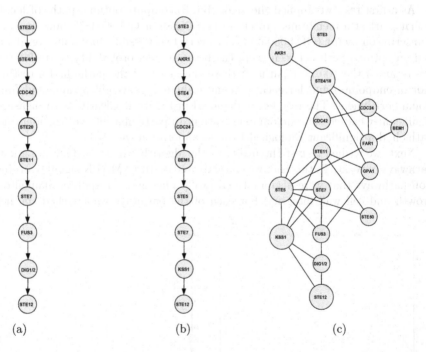

(a)   (b)   (c)

Fig. 2. The pheromone response signaling pathway in yeast. (a) The main chain of the
known pathway, adapted from [13]. (b) The best path of the same length (9) in the network. (c) The assembly of all light-weight paths starting at STE3 and ending at STE12
that were identified in the network. Nodes that occur in at least half of the paths are
drawn larger than the rest. Nodes that occur in less than 10% of the paths are omitted

proteins that we have identified are part of the pathway, except for Kss1p and Akr1p (see discussion above). A previous study by Steffen et al. reported similar results for this pathway [2]. In comparison to Figure 2(c), Steffen et al. identified three additional proteins (Sst2p, Mpt5p and Sph1p), which are related to the pathway, but are not part of the main chain. Steffen et al. failed to recover the true positive Cdc42p. Interestingly, this latter protein participates mainly in paths of length 9 and 10 in our computations (only two additional paths of length 8 contained this protein). Such long paths are very costly to compute using an exhaustive approach (about five hours for length-9 paths based on our benchmark).

The filamentous growth pathway is induced under stress conditions and causes yeast diploid cells to grow as filaments of connected cells. The pathway is depicted in Figure 3(a). Searching the network for the minimum-weight path of the same length as the known pathway (8), yielded the path shown in Figure 3(b), which largely matches the known pathway. The introduction of the proteins Cdc25p and Hsp82p is again an artifact that arises due to a missing link between Ras2p and Cdc42p in the network data.

The cell wall integrity pathway mediates cell cycle regulated cell wall synthesis. It is depicted in Figure 3(c). A search for the minimum-weight path of equal length starting at Ras2p and ending at Tec1p yielded the path shown in Figure 3(d). Again, the identified path matches the known pathway well. The only falsely detected protein, Rom2p, could be explained by the fact that the network does not contain a direct interaction between Mid2p and Rho1p.

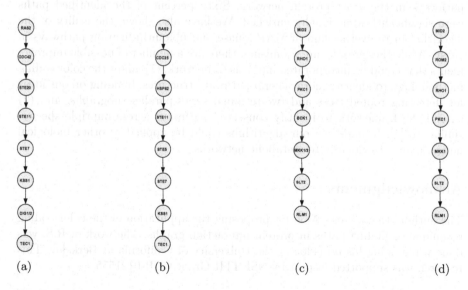

(a) (b) (c) (d)

Fig. 3. Search results for the filamentous growth and cell wall integrity pathways. (a) The known filamentous growth pathway, adapted from [13]. (b) The best path of length 8 between RAS2 and TEC1. (c) The known cell wall integrity pathway [13]. (d) The best path of length 7 between MID2 and RLM1

In addition, we used our algorithm to search for the high osmolarity MAPK pathway, starting at Sln1p and ending at Hog1p (leading to several transcription factors, including Mcm1p and Msn2/4p [13]). For this run, although we could recover the exact known pathway, it was only the 11th-scoring among the identified paths.

As a final test, we applied our algorithm to look for ubiquitin-ligation pathways by searching for paths of length 4-6 that start at a cullin (Cdc53p or Apc2p) and end at an F-box protein (Met30p, Cdc4p or Grr1p). For each pair of endpoints we output the best path for each specified length. To evaluate our success we computed the enrichment of the identified proteins within the GO category "ubiquitin-dependent protein catabolism" (GO:0006511). In total, 18 paths were computed, all of which were found to be highly enriched for this GO category ($p < 0.001$). A more careful examination of these paths revealed that they highly overlapped: In addition to their endpoints, these paths spanned four other proteins (Skp1p, Cdc34p, Hrt1p and Sgt1p), all of which are known ubiquitin-ligation proteins.

6   Conclusions

We have presented efficient algorithms for finding simple paths in graphs based on the color-coding technique, and several biologically-motivated extensions of this technique. We applied these algorithms to search for protein interaction pathways in the yeast protein network. Sixty percent of the identified paths were significantly functionally enriched. We have also shown the utility of the algorithm in recovering known MAP-kinase and ubiquitin-ligation pathways in yeast. While these results are promising, there are a number of possible improvements that could be incorporated into this framework: (1) adapt the color coding methodology to identify more general pathway structures, building on our ideas for detecting rooted trees and two-terminal series-parallel subgraphs; and (2) extend the framework to identify conserved pathways across multiple species, similar to [1]. In addition, our algorithms could be applied to other biological networks, most evidently to metabolic networks.

Acknowledgments

The authors thank Noga Alon for proposing the application of the color coding technique to finding paths in protein interaction graphs. The work of R.S. was done while doing his post-doc at the University of California at Berkeley. This research was supported in part by NSF ITR Grant CCR-0121555.

References

1. Kelley, B., Sharan, R., Karp, R., et al.: Conserved pathways within bacteria and yeast as revealed by global protein network alignment. Proc. Natl. Acad. Sci. USA **100** (2003) 11394–11399

2. Steffen, M., Petti, A., Aach, J., D'haeseleer, P., Church, G.: Automated modelling of signal transduction networks. BMC Bioinformatics **3** (2002) 34–44
3. Alon, N., Yuster, R., Zwick, U.: Color-coding. J. ACM **42** (1995) 844–856
4. Deng, M., Sun, F., Chen, T.: Assessment of the reliability of protein-protein interactions and protein function prediction. In: Proceedings of the Eighth Pacific Symposium on Biocomputing. (2003) 140–151
5. Bader, J., Chaudhuri, A., Rothberg, J., Chant, J.: Gaining confidence in high-throughput protein interaction networks. Nature Biotechnol. (2004) 78–85
6. von Mering, C., et al.: Comparative assessment of large-scale data sets of protein-protein interactions. Nature **417** (2002) 399–403
7. Sharan, R., Suthram, S., Kelley, R., Kuhn, T., McCuine, S., Uetz, P., Sittler, T., Karp, R., Ideker, T.: Conserved patterns of protein interaction in multiple species. Proc. Natl. Acad. Sci. USA (2005) in press.
8. Gollub, J., Ball, C., Binkley, G., Demeter, J., Finkelstein, D., Hebert, J., Hernandez-Boussard, T., Jin, H., Kaloper, M., Matese, J., et al.: The stanford microarray database: data access and quality assessment tools. Nucleic Acids Res. **31** (2003) 94–6
9. Goldberg, D., et al.: Assessing experimentally derived interactions in a small world. Proc. Natl. Acad. Sci. USA **100** (2003) 4372–6
10. Mewes, H., Amid, C., Arnold, R., Frishman, D., Guldener, U., Mannhaupt, G., Munsterkotter, M., Pagel, P., Strack, N., Stumpflen, V., et al.: MIPS: analysis and annotation of proteins from whole genomes. Nucleic Acids Res. **32** (2004) D41–4
11. The Gene Ontology Consortium: Gene ontology: Tool for the unification of biology. Nature Genetics **25** (2000) 25–9
12. Xenarios, I., et al.: DIP, the database of interacting proteins: a research tool for studying cellular networks of protein interactions. Nucleic Acids Res. **30** (2002) 303–305
13. Roberts, C., et al.: Signaling and circuitry of multiple MAPK pathways revealed by a matrix of global gene expression profiles. Science **287** (2000) 873–880

Towards an Integrated Protein-Protein Interaction Network

Ariel Jaimovich[1,2], Gal Elidan[3], Hanah Margalit[2,*], and Nir Friedman[1,*]

[1] School of Computer Science and Engineering,
The Hebrew University, Jerusalem, Israel
nir@cs.huji.ac.il
[2] Hadassah Medical School, The Hebrew University, Jerusalem, Israel
hanah@md.huji.ac.il
[3] Computer Science Department, Stanford University, Stanford, California, USA

Abstract. Protein-protein interactions play a major role in most cellular processes. Thus, the challenge of identifying the full repertoire of interacting proteins in the cell is of great importance, and has been addressed both experimentally and computationally. Today, large scale experimental studies of interacting proteins, while partial and noisy, allow us to characterize properties of interacting proteins and develop predictive algorithms. Most existing algorithms, however, ignore possible dependencies between interacting pairs, and predict them independently of one another. In this study, we present a computational approach that overcomes this drawback by predicting protein-protein interactions simultaneously. In addition, our approach allows us to integrate various protein attributes and explicitly account for uncertainty of assay measurements. Using the language of relational Markov Random Fields, we build a unified probabilistic model that includes all of these elements. We show how we can learn our model properties efficiently and then use it to predict all unobserved interactions simultaneously. Our results show that by modeling dependencies between interactions, as well as by taking into account protein attributes and measurement noise, we achieve a more accurate description of the protein interaction network. Furthermore, our approach allows us to gain new insights into the properties of interacting proteins.

1 Introduction

One of the main goals of molecular biology is to reveal the cellular networks underlying the functioning of a living cell. Proteins play a central role in these networks, mostly by interacting with other proteins. Deciphering the protein-protein interaction network is a crucial step in understanding the structure, function, and dynamics of cellular networks. The challenge of charting these protein-protein interactions is complicated by several factors. Foremost is the

---

* Corresponding authors.

S. Miyano et al. (Eds.): RECOMB 2005, LNBI 3500, pp. 14–30, 2005.

sheer number of interactions that have to be considered. In the budding yeast, for example, there are approximately 18,000,000 potential interactions between the roughly 6,000 proteins encoded in its genome. Of these, only a relatively small fraction occur in the cell [32, 27]. Another complication is due to the large variety of interaction types. These range from stable complexes that are present in most cellular states, to transient interactions that occur only under specific conditions (e.g. phosphorylation in response to an external stimulus).

Many studies in recent years address the challenge of constructing protein-protein interaction networks. Several experimental assays, such as *yeast two-hybrid* [31, 13] and *tandem affinity purification* [24] have facilitated high-through-put studies of protein-protein interactions on a genomic scale. Some computational approaches aim to detect functional relations between proteins, based on various data sources such as phylogenetic profiles [23] or mRNA expression [6]. Other computational assays try to detect physical protein-protein interactions by, for example, evaluating different combinations of specific domains in the sequences of the interacting proteins [26].

The various experimental and computational screens described above have different sources of error, and often identify markedly different subsets of the full interaction network. The small overlap between the interacting pairs identified by the different methods raises serious concerns about their robustness. Recently, in two separate works, von Mering *et al* [32] and Sprinzak *et al* [27] conducted a detailed analysis of the reliability of existing methods, only to discover that no single method provides a reasonable combination of sensitivity and recall. However, both studies suggest that interactions detected by two (or more) methods are much more reliable. These observations motivated later "meta" approaches that hypothesize about interactions by combining the predictions of computational methods, the observations of experimental assays, and other correlating information sources such as that of localization assays. These approaches use a variety of machine learning methods to provide a combined prediction, including support vector machines [1], naive Bayesian classifiers [14] and decision trees [36].

These methods, while offering a combined hypothesis, still ignore possible dependencies between different protein-protein interactions. In this paper, we argue that by explicitly modeling such dependencies in the model, we can leverage observations from varied sources to produce better *joint* predictions of the protein interaction network as a whole. As a concrete example, consider the budding yeast proteins Pre7 and Pre9. These proteins were predicted to be interacting by a computational assay [26]. However, according to a large-scale localization assay [11], the two proteins are *not* co-localized; Pre9 is observed in the cytoplasm and in the nucleus, while Pre7 is not observed in either of those compartments. Thus, a naive examination of this interaction alone (as in Figure 1a), might assign it a low probability. However, we can gain more confidence by looking at related interactions. For example, interactions of Pre5 and Pup3 with both Pre9 and Pre7 were reported by large scale assays [20, 26]; see Figure 1b. These observations suggest that these proteins might form a complex. Moreover, both Pre5 and Pup3 were found both in the nucleus and in the cytoplasm, implying

(a)   (b)

Fig. 1. Dependencies between interactions can be used to improve predictions. (a) Shows a possible interaction of two proteins (Pre7 and Pre9). Pre9 is localized in the cytoplasm and in the nucleus (dark blue) and Pre7 is not annotated to be in either one of those. This interaction was predicted by a computational assay [26] (shown by a dashed red line). When looking only at this evidence, we might assign to this interaction a low probability. (b) Introduces two new proteins which were also found to interact with Pre9 and Pre7 either by a computational assay [26] (shown again by a dashed red line) or experimental assays [20] (shown by a solid green line). By looking at this expanded picture, we can both hypothesize about the localization of Pre7 and increase the reliability of the interaction between Pre9 and Pre7

that Pre7 can possibly be localized in these two compartments, thus increasing our belief in the existence of an interaction between Pre9 and Pre7. Indeed, this intuition is confirmed by other interaction [9] and localization [17] assays. This example illustrates two types of inferences that our model can offer. First, certain patterns of interactions (*e.g.*, within complexes) might be more probable than others. Second, an observation on one interaction can provide information about the attributes of a protein (cellular localization in this example), which in turn can influence the likelihood of other interactions.

We present a unified probabilistic model for learning an integrated protein-protein interaction network. We build on the language of relational probabilistic models [8, 28] to explicitly define probabilistic dependencies between related protein-protein interactions, protein attributes, and observations regarding these entities. The use of probabilistic models also allows us to explicitly account for measurement noise of different assays. Propagation of evidence in our model allows interactions to influence one another as well as related protein attributes in complex ways. This in turn leads to better and more confident overall predictions. Using various proteomic data sources for the yeast *Saccharomyces cerevisiae* we show how our method can build on multiple weak observations to better predict the protein-protein interaction network.

2   A Probabilistic Protein-Protein Interaction Model

Our goal is to build a unified probabilistic model that can capture the integrative properties of the protein-protein interaction network that are demonstrated in the example of Figure 1. We map protein-protein interactions, interaction assays, and other protein attributes into random variables, and use the language of *Markov Random Fields* to represent the joint distribution over them. We now review Markov Random Field models, and the specific models we construct for modeling protein-protein interaction networks.

Markov Random Fields

Let $\mathcal{X} = \{X_1, \ldots, X_N\}$ be a finite set of random variables. A Markov Random Field over $\mathcal{X}$ describes a joint distribution by a set of potentials Ψ. Each potential $\psi_c \in \Psi$ defines a measure over a set of variables $\boldsymbol{X}_c \subseteq \mathcal{X}$. We call $\boldsymbol{X}_c$ the *scope* of ψ_c. The potential ψ_c quantifies local preferences about $\boldsymbol{X}_c$ by assigning a numerical value to each joint assignment of $\boldsymbol{X}_c$. Intuitively, the larger the value the more likely the assignment. The joint distribution is defined by combining the preferences of all potentials

$$P(\mathcal{X} = \boldsymbol{x}) = \frac{1}{Z} \prod_{c \in \mathcal{C}} e^{\psi_c(\boldsymbol{x}_c)} \tag{1}$$

where $\boldsymbol{x}_c$ refers to the projection of $\boldsymbol{x}$ onto the subset $\boldsymbol{X}_c$, and Z is a normalizing factor, often called the *partition function*, that ensures that P is a valid probability distribution. The above product form facilitates compact representation of the joint distribution, and in some cases efficient probabilistic computations.

Using this language to describe protein-protein interaction networks requires defining relevant random variables. A distribution over protein-protein interaction networks can be viewed as the joint distribution over binary random variables that denote interactions. Given a set of proteins $\mathcal{P} = \{p_i, \ldots, p_k\}$, an interaction network is described by interaction random variables I_{p_i,p_j} for each pair of proteins. The random variable I_{p_i,p_j} takes the value 1 if there is an interaction between the proteins p_i and p_j and 0 otherwise. Since this relationship is symmetric, we view I_{p_j,p_i} and I_{p_i,p_j} as two ways of naming the same random variable. Clearly, a joint distribution over all these interaction variables is equivalent to a distribution over possible interaction networks.

The simplest Markov Random Field model over the set of interaction variables has a univariate potential $\psi_{i,j}(I_{p_i,p_j})$ for each interaction variable. Each such potential captures the preference for the associated interaction. This model by itself is overly simplistic as it views interactions as independent from one another.

We can extend the model by introducing variables that denote protein attributes that can influence the probability of interactions. Here we consider cellular localization as an example of such an attribute. The intuition is clear: if two proteins interact, they have to be physically co-localized. As a protein may be present in multiple localizations, we model cellular localization by several indicator variables, L_{l,p_i} that denote whether the protein p_i is present in cellular localization $l \in \mathcal{L}$. We can now relate the localization variables for a pair of proteins with the corresponding interaction variable between them by introducing a potential $\psi_{l,i,j}(L_{l,p_i}, L_{l,p_j}, I_{p_i,p_j})$. Such a potential can capture preference for interactions between co-localized proteins. Note that in this case there is no importance to the order of p_i and p_j, and thus we require this potential to be symmetric around the role of p_i and p_j (we return to this issue in the context of learning). As with interaction variables, we might also have univariate potentials on each localization variable L_{l,p_j} that capture preferences over the localizations of specific proteins.

Assuming that $\mathcal{X}$ contains variables $\{I_{p_i,p_j}\}$ and $\{L_{l,p_i}\}$, we now have a Markov Random Field of the form:

$$P(\mathcal{X}) = \frac{1}{Z} \prod_{p_i,p_j \in \mathcal{P}} e^{\psi_{i,j}(I_{p_i,p_j})} \prod_{l \in \mathcal{L}, p_i \in \mathcal{P}} e^{\psi_{l,i}(L_{l,p_i})} \prod_{l \in \mathcal{L}, p_i,p_j \in \mathcal{P}} e^{\psi_{l,i,j}(I_{p_i,p_j}, L_{l,p_i}, L_{l,p_j})}$$

$$(2)$$

Noisy Sensor Models as Directed Potentials

The models we discussed so far make use of undirected potentials between variables. In many cases, however, a clear directional cause and effect relationship is known. In our domain, we do not observe protein interactions directly, but rather through experimental assays. We can explicitly discuss the noisy relations between an interaction and its assay readout within the model. For each interaction assay $a \in \mathcal{A}$ aimed towards evaluating the existence of an interaction between the proteins p_i and p_j, we define a binary random variable $IA^a_{p_i,p_j}$ (defined with the same logic as I_{p_i,p_j}).[1] It is natural to view the assay variable $IA^a_{p_i,p_j}$ as a noisy sensor of the real interaction I_{p_i,p_j}. In this case, we can use a *conditional distribution* $P(IA^a_{p_i,p_j} \mid I_{p_i,p_j})$ that captures the probability of the observation given the underlying state of the system. Conditional probabilities have several benefits. First, due to local normalization constraints, the number of free parameters of a conditional distribution is smaller (2 instead of 3 in this example). Second, since $P(IA^a_{p_i,p_j} = 0 \mid I_{p_i,p_j}) + P(IA^a_{p_i,p_j} = 1 \mid I_{p_i,p_j}) = 1$, such potentials do not contribute to the global partition function Z, which is typically hard to compute. Finally, the specific use of directed models will allow us to prune unobserved assay variables. Namely, if we do not observe $IA^a_{p_i,p_j}$, we can remove it from the model without changing the probability over interactions.

Probabilistic graphical models that combine directed and undirected relations are called *Chain Graphs* [2]. Here we examine a simplified version of Chain Graphs where a dependent variable associated with a conditional distribution (i.e., $IA^a_{p_i,p_j}$) is not involved with other potentials or conditional distributions. If we let $\mathcal{Y}$ denote the assay variables, then the joint distribution is factored as:

$$P(\mathcal{X}, \mathcal{Y}) = P(\mathcal{X}) \prod_{p_i,p_j \in \mathcal{P}, a \in \mathcal{A}} P(IA^a_{p_i,p_j} \mid I_{p_i,p_j}) \qquad (3)$$

where $P(\mathcal{X})$ is the Markov Random Field of Equation 2.

---

[1] Note that this random variable is not necessarily symmetric, since for some assays (*e.g.*, yeast two hybrid) $IA^a_{p_i,p_j}$ and $IA^a_{p_j,p_i}$ represent the results of two different experiments.

Template Markov Random Fields

Our aim is to construct a Markov Random Field over large scale protein-protein interaction networks. Using the model described above for this task is problematic in several respects. First, for the model with just univariate potentials over interaction variables, there is a unique parameter for each possible assignment of each possible interaction of protein pairs. The number of parameters is thus extremely large even for the simplest possible model (in the order of 6000^2 for the protein-protein interaction network of the budding yeast *S. cerevisiae*). Robustly estimating such a model from finite data is clearly impractical. Second, we want to apply the same "rules" (potentials) throughout the interaction network, regardless of the specific subset of proteins we happen to concentrate on. For example, we want the probabilistic relation between interaction (I_{p_i,p_j}) and localization (L_{l,p_i}, L_{l,p_j}), to be the same for all values of i and j.

We address these problems by using *template models*. These models are related to relational probabilistic models [8, 28] in that they specify a recipe with which a concrete Markov Random Field can be constructed for a specific set of proteins and localizations. This recipe is specified via *template potentials* that supply the numerical values to be reused. For example, rather then using a different potential $\psi_{l,i,j}$ for each protein pair p_i and p_j, we use a single potential ψ_l. This potential is used to relate an interaction variable I_{p_i,p_j} with its corresponding localization variables L_{l,p_i} and L_{l,p_j}, regardless of the specific choice of i and j. Thus, by reusing parameters, a template model facilitates a compact representation, and at the same time allows us to apply the same "rule" for similar relations between random variables.

The design of the template model defines what set of potentials are shared. For example, when considering the univariate potential over interactions, we can have a single template potential for all interactions $\psi(I_{p_i,p_j})$. On the other hand, when looking at the relation between localization and interaction, we can decide that for each value of l we have a different template potential for $\psi_l(L_{l,p_i})$. Thus, by choosing which templates to create we encapsulate the complexity of the model.

Protein-Protein Interaction Models

The discussion so far defined the basis for a simple template Markov Random Field for the protein-protein interaction network. The form given in Equation 3 relates protein interactions with multiple interaction assays and protein localizations. In this model the observed interaction assays are viewed as noisy sensors of the underlying interactions. Thus, we explicitly model experiment noise and allow the measurement to stochastically differ from the ground truth. For each type of assay we have a different conditional probability that reflects the particular noise characteristics of that assay. In addition, the basic model contains univariate template potential $\psi(I_{p_i,p_j})$ that is applied to each interaction variable. This potential captures the prior preferences for interaction (before we make any additional observations).

In this model, if we observe the interaction assay variables and the localization variables, then the posterior over interactions can be reformulated as an independent product of terms, each one involving I_{p_i,p_j}, its related assays, and the localization of p_i and p_j. Thus, the joint model can be viewed as a collection of independent models for each interaction. Each of these models is equivalent to a naive Bayes model (see, e.g., [14]). We call this the **basic** model (see Figure 2e).

We now consider two extensions to this basic model. The first extension relates to the localization random variables. Instead of using the experimental localization results to assign these variables, we can view these experimental results as noisy sensors of the true localization. To do so, we introduce localization assay random variables $LA_{l,p}$, which are observed, and relate each localization assay variable to its corresponding hidden ground truth variable using a conditional probability (Figure 2c). The parameters of this conditional probability depend on the type of assay and the specific cellular localization. For example, some localizations, such as "bud", are harder to detect as they represent a transient part of the cell cycle, while other localizations, such as "cytoplasm", are easier to detect since they are present in all stages of the cell's life and many proteins are permanently present in them. Allowing the model to infer the local-

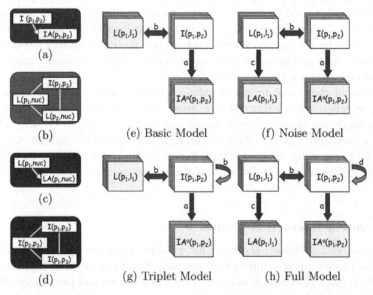

Fig. 2. Protein-protein interaction models. In all models, a plain box stands for a hidden variable, and a shadowed box represents an observed variable. The model consists of four classes of variables and four template potentials that relate them. (a) conditional probability of an interaction assay given the corresponding interaction; (b) potential between an interaction and the localization of the two proteins; (c) conditional probability of a localization assay given a corresponding localization; (d) potential between three related interacting pairs; (e)-(h) The four models we build and how they hold the variable classes and global relations between them

ization of a protein provides a way to create dependencies between interaction variables. For example, an observation of an interaction between p_i and p_j may change the belief in the localization of p_i and thereby influence the belief about the interaction between p_i and another protein, p_k. We use the name **Noise** model to refer to the basic model extended with localization assay variables (see Figure 2f).

The second extension to the basic model is to directly capture dependencies between interaction variables. We do so by introducing potentials over several interaction variables. The challenge is to design a potential that captures relevant dependencies in a concise manner. Here we consider dependencies between the three interactions among a triplet of proteins. More formally, we introduce a three variables potential $\psi_3(I_{p_i,p_j}, I_{p_i,p_k}, I_{p_j,p_k})$ (Figure 2d). This model is known in the social network literature as the *triad model* [7]. Such a triplet potential can capture properties such as preferences for (or against) adjacent edges, as well as transitive closure of adjacent edges. Given $\mathcal{P}$, the induced Markov Random Field has $\binom{|\mathcal{P}|}{3}$ potentials, all of which replicate the same parameters of the template potential. Note that this requires the potential to be ignorant of the order of its arguments (as we can "present" each triplet of edges in any order). Thus, the actual number of parameters for ψ_3 is four – one when all three edges are present, another for the case when two are present, and so on. We use the name **Triplet** model to refer to the basic model extended with these potentials (see Figure 2g). Finally, we use the name **Full** model to refer to the basic model with both the extensions of **Noise** and **Triplet** (see Figure 2h).

3   Inference and Learning

Inference

The task of inference is to compute queries about the joint distribution represented by a model. Two queries that will be relevant for us are computing the *likelihood* $P(e)$ of an assignment e to a set of variables E, and computing the *posterior probability* $P(X \mid e)$ over a set of variables X, given the evidence e. Specifically, in our model we are interested in the likelihood of the observations (interaction assays and localization assays) and the posterior probability of the hidden variables (interactions, localizations), given these observations. When reasoning about the interaction map over thousands of proteins, our method constructs models that involve tens or hundreds of thousands of variables. Performing inference in these models is a serious challenge.

In general, although exact inference algorithms are known, these are tractable only in networks with particular structures. The networks we construct here are not amenable to exact inference. Thus, we must resort to approximate inference (*e.g.*, [15]). In this work we rely on the simple and efficient *belief propagation* (BP) algorithm [21, 35] that iteratively *calibrates* beliefs by propagation of local messages between potentials with overlapping variables. Although this algorithm is not guaranteed to converge, empirical evidence shows that it often converges in general networks and provides reasonable estimates of marginal probabilities [21].

Learning

Our approach is data driven, meaning that we use real data to calibrate our model. Our aim is to estimate the parameters of the model from observations. Namely, we are given an observation e over some of the variables in the model and search for the parameters that give the best "explanation" for this observation. To do so we use the *maximum likelihood* principle and find a parameterization so that $\log P(e)$ is maximized.

We start with the case of *complete data* where the evidence assigns a value to all the variables in the domain. Recall that our model contains both undirected potentials and conditional probabilities. When learning from complete data, we can separately learn each type of parameters.

For the conditional probabilities, finding the maximum likelihood parameters is a relatively easy task. For example, to estimate the template parameter for $P(IA^a_{p_i,p_j} = 1 \mid I_{p_i,p_j} = 1)$ we simply count how many times we observed in e that both $IA^a_{p_i,p_j}$ and I_{p_i,p_j} equals one for some i and j, and normalize by the number of times we observe I_{p_i,p_j} equals one.

Finding the maximum likelihood parameters for undirected potentials is a much harder task. Although the likelihood function is concave, there is no closed form formula that returns the optimal parameters. A common heuristic is a gradient ascent search in the parameter space. To perform such a search we need to repeatedly evaluate both the likelihood and the partial derivatives of the likelihood with respect to each of the parameters. For an entry in a specific potential $\psi_c(\boldsymbol{x}_c)$, the gradient of the log-likelihood can be written as:

$$\frac{\partial \log P(e)}{\partial \psi_c(\boldsymbol{x}_c)} = \hat{P}(\boldsymbol{x}_c) - P(\boldsymbol{x}_c) \tag{4}$$

That is, it is equal to the difference between the the empirical count $\hat{P}(\boldsymbol{x}_c)$ of the event $\boldsymbol{x}_c$ and its probability $P(\boldsymbol{x}_c)$ according to the model (before we make observations) [4]. The first quantity is directly observed, and the later one requires inference.

Recall that in template models many potentials share the same parameters. Using the chain rule of partial derivatives, it is easy to see that if $\psi_c(\boldsymbol{x}_c) = \theta$ for all $c \in \mathcal{C}$, then the derivative of the shared parameter θ is $\frac{\partial \log P(e)}{\partial \theta} = \sum_{c \in \mathcal{C}} \frac{\partial \log P(e)}{\partial \psi_c(\boldsymbol{x}_c)}$. Thus, the derivatives with respect to the template parameters are aggregates of the derivatives of the corresponding entries in the potentials of the model. We can compute these derivatives by performing a single invocation of belief propagation to evaluate (an approximation to) all the terms that appear in the gradient of the likelihood. This invocation also computes an approximation to the likelihood itself. The computational bottleneck for learning is thus the multiple calls to the approximate inference procedure for estimating the likelihood and its gradients for different values of the parameters.

In the context of our models, we introduce additional constraints on learned template potentials. These constraints reduce the number of free parameters we need to learn, and ensure that the model captures the semantics we attribute to it. First, as discussed in Section 2, we require potentials to be symmetric about protein arguments. This implies that some entries in the template potential share a single parameter. We learn the shared parameter using the methods discussed above. Second, in some cases we force the potential to be indifferent to specific assignments. For example, when looking at the relation between I_{p_i,p_j}, L_{l,p_i} and L_{l,p_j}, we require that if $L_{l,p_i} = L_{l,p_j} = 0$ (i.e., both proteins are *not* in the cellular localization l), then the potential has no effect on I_{p_i,p_j}. This is done by fixing the value of the relevant potential entries to 0, and not changing it during parameter optimization.

In practice, learning is further complicated by the fact that our observations are incomplete — we do not observe all the variables in the model (*i.e.*, we do not observe the real localizations but only the assays). To deal with partial observations, we use the standard method of *Expectation-Maximization* (EM) [22]. The basic idea is that given the current parameters of the model, we can "guess" the unobserved values. We can then apply complete data techniques to the completed dataset to estimate new parameters. In our case, this procedure proceeds by iterating two steps until convergence.

- **E-step.** Infer (using Loopy Belief Propagation) the marginal probabilities of the random variables that are missing in the evidence given the evidence, and the current set of parameters. Use the resulting probabilities to estimate *expected* empirical counts for the relevant events.
- **M-step.** Maximize parameters using method for complete data, using the estimated counts instead of actual counts. In our case, this implies direct estimation of the conditional probabilities, and performing conjugate gradient search to find new parameters.

The theory of EM guarantees that in each iteration the likelihood increases until convergence to a local maximum [22].

4 Experimental Evaluation

In Section 2 we discussed a general framework for modeling protein-protein interactions and introduced four specific model variants that combine different aspects of the data. In this section, we evaluate the utility of these models in the context of the budding yeast *S. cerevisiae*. For this purpose we choose to use four data sources, each with different characteristics. The first is a large scale experimental assay for identifying interacting proteins by the yeast two hybrid method [31, 13]. The second is a large scale effort to curate direct experimental results from the literature about protein complexes [20]. The third is a collection of computational predictions based on correlated domain signatures learned from experimentally determined interacting pairs [26]. The fourth is a large scale experimental assay examining protein localization in the cell using GFP-tagged protein constructs [11]. Of the latter we regarded four cellular localizations (nucleus, cytoplasm, mitochondria, and ER).

In our models we have a random variable for each possible interaction and a random variable for each assay measuring such interaction. In addition, we have a random variable for each of the four possible localizations of each protein, and yet again another variable corresponding to each localization assay. A model for all 6000 proteins in the budding yeast includes close to 20, 000, 000 random variables. Such a model is too large to cope with using our current methods. Thus, we limit ourselves to a subset of the protein pairs, retaining both positive and negative examples. We construct this subset from the study of von Mering *et al* [32] who ranked 80, 000 protein-protein interactions according to their reliability based on multiple sources of evidence (including some that we do not examine here). From this ranking, we consider the 2000 highest ranked protein pairs as "true" interactions. These 2000 interactions involve 867 proteins. The selection of negative (non-interacting) pairs is more complex. There is no clear documentation of failure to find interactions, and so we consider pairs that do not appear in von Mering's ranking as non-interacting. Since the number of such non-interacting protein pairs is very large, we randomly selected pairs from the 867 proteins, and collected 2000 pairs that do not appear in von Mering's ranking as "true" non-interacting pairs. Thus, we have 4000 interactions, of these, half interacting and half non-interacting. For these entities, the **full** model involves approximately 17, 000 variables and 38, 000 potentials that share 37 parameters.

The main task is to learn the parameters of the model using the methods described in Section 3. To get an unbiased estimate of the quality of the predictions with these parameters, we want to test our predictions on interactions that were not used for learning the model parameters. We use a standard 4-fold cross validation technique, where in each iteration we learn the parameters using 1500 positive and 1500 negative interactions, and then test on 500 unseen interactions of each type. Cross validation in the relational setting is more subtle than learning with standard i.i.d. instances. In particular, when testing the predictions on the 1000 unseen interactions, we use both the parameters we learned from the interactions in the training set, and also the observations on these interactions. This simulates a real world scenario when we are given observations on some set of interactions, and are interested in predicting the remaining interactions, for which we have no direct observations.

To evaluate the performance of the different model elements, we compare the four models described in Section 2 (see Figure 2). Figure 3 compares the test set performance of these four models. The advantage of using an integrative model that allows propagation of influence between interactions and protein attributes is clear, as all three variants improve significantly over the baseline model. Adding the dependency between different interactions leads to a greater improvement than allowing noise in the localization data. We hypothesize that this potential allows for complex propagation of beliefs beyond the local region of a single protein in the interaction network. When both elements are combined, the full model reaches quite impressive results: above 85% true positive rate with just a 10% false positive rate. This is in contrast to the baseline model that achieves less than half of the true positive rate with the same amount of false positives.

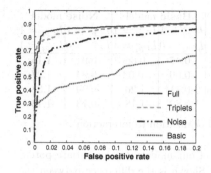

Fig. 3. Test performance of four models in a 4-fold cross validation experiment. Shown is the true positive vs. the false positive rates tradeoff for four models: **Basic** with just interaction, interaction assays, and localization variables; **Noise** that adds the localization assay variables; **Triplets** that adds a potential over three interactions to **Basic**; **Full** that combines both extensions

To evaluate the robustness of the results to the specific setup we used, we applied the learned parameters in additional settings. First, we chose a set of 4000 positive examples and 10000 negative examples and tested the quality of our predictions. Second, to deal with the concern that in real life we might have fewer observed interactions, we repeated our evaluations, but without using the evidence of the training interactions when making predictions on test interactions. In both cases the ROC curves are quite similar to Figure 3 with a slight decrease in sensitivity (especially in the second setting).

As an additional test, we repeated the original cross validation experiments with reshuffled localization data. As expected, the performance of the basic model decreased significantly. The performance of the full model, however, did not alter much. A possible explanation is that the training "adapted" the hidden localization variables to capture other dependencies between interactions. Indeed, the learned conditional probabilities in the model capture a weak relationship between the localization variables and the shuffled localization assays. This experiment demonstrates the expressive power of the model in capturing dependencies. It also reinforces the caution needed in interpreting what hidden variables represent.

We can gain better insight into the effect of adding a noisy sensor model for localization by examining the estimated parameters (Figure 4). As a concrete example, consider the potentials relating an interaction variable with the localization of the two relevant proteins in Figure 4b. In both models, when only one of the proteins is localized in the compartment, non-interaction is preferred, and if both proteins are co-localized, interaction is preferred. We see that smaller compartments, such as the mitochondria, provide stronger support for interaction. Furthermore, we can see that our noise model, allows us to be more confident in the localization attributes.

Another way of examining the effect of the noisy sensor is to compare the localization predictions made by our model with the original experimental observations. For example, out of 867 proteins in our experiment, 398 proteins are observed as nuclear [11]. Our model predicts that 482 proteins are nuclear. Of these, 389 proteins were observed as nuclear, 36 are nuclear according to YPD [3], 45 have other cellular localizations, and 22 have no known localization. We get similar results for other localizations. These numbers suggest that our

	Basic model	Noise model
Interaction	0	-0.02
Nucleus	-1.13	-0.91
Cytoplasm	-1.34	-1.13
Mitochondria	-1.96	-2.04
ER	-2.52	-2.52

(a) Univariate potentials

localization	Basic model $L_{l,p_i}=1$ $L_{l,p_j}=0$	Basic model $L_{l,p_i}=1$ $L_{l,p_j}=1$	Noise model $L_{l,p_i}=1$ $L_{l,p_j}=0$	Noise model $L_{l,p_i}=1$ $L_{l,p_j}=1$
Nucleus	-0.47	0.66	-0.91	1.15
Cytoplasm	-0.66	-0.02	-0.94	1.27
Mitochondria	-0.71	1.26	-0.99	1.38
ER	-0.82	1.18	-0.73	1.16

(b) Localization to interaction

Fig. 4. Examples of learned parameters in two of our models. (a) Univariate potential for interactions I_{p_i,p_j} and localization L_{l,p_i}. Shown is the difference between the potential values when the variable is set to 1 and when it is set to 0. (b) The potential between I_{p_i,p_j} and L_{l,p_i}, L_{l,p_j} for different localizations. Shown is the difference between the potential values when the interaction variable is set to 1 and when it is set to 0. As we can see, co-localization typically increases the probability of interaction, while disagreement on localization reduces it. In the **Noise** model, co-localization provides more support for interaction, especially in the nucleus and cytoplasm

model is able to correctly predict the localizations of many proteins, even when the experimental assay misses them.

To get a better sense of the performance of the model, we consider specific examples where the predictions of the full model differ from those of the basic model. Consider the unobserved interaction between the EBP2 and NUG1 proteins. These proteins are part of a large group of proteins involved in rRNA biogenesis and transport. Localization assays identify NUG1 in the nucleus, but do not report any localization for EBP2. The interaction between these two proteins was not observed in any of the three interaction assays included in our experiment, and thus considered unlikely by the basic model. In contrast, prop-

(a)   (b)

Fig. 5. Two examples demonstrating the difference between the predictions by our full model and those of the basic model. Solid lines denote observed interactions and a dashed line corresponds to an unknown one. Orange colored nodes represent proteins that are localized in the nucleus and blue colored ones represent proteins that are localized in the mitochondria. Uncolored nodes have no localization evidence. In (a), unlike the basic model, our full model correctly predicts that EBP2 is localized in the nucleus and that it interacts with NUG1. Similarly, in (b) we are able to correctly predict that MRPS9 is localized in the mitochondria and interacts with RSM25, that also interacts with MRPS28

agation of evidence in the full model effectively integrates information about interactions of both proteins with other rRNA processing proteins. We show a small fragment of this network in Figure 5a. In this example, the model is able to make use of the fact that several nuclear proteins interact with *both* EBP2 and NUG1, and thus predicts that EBP2 is also nuclear, and indeed interacts with NUG1. Importantly, these predictions are consistent with the cellular role of these proteins, and are supported by independent experimental assays [3, 32].

Another, more complex example involves the interactions between RSM25, MRPS9, and MRPS28. While there is no annotation of RSM25's cellular role, the other two proteins are known to be components of the mitochondrial ribosomal complex. Localization assays identify RSM25 and MRPS28 in the mitochondria, but do not report any observations about MRPS9. As in the previous example, neither of these interactions was tested by the assays in our experiment. As expected, the baseline model predicts that both interactions do not occur with a high probability. In contrast, by utilizing a fragment of our network shown in Figure 5b, our model predicts that MRPS9 is mitochondrial, and that both interactions occur. Importantly, these predictions are again supported by independent results [3, 32]. These predictions suggest that RSM25 is related to the ribosomal machinery of the mitochondria. Such an important insight could not be gained without using an integrated model such as the one presented here.

5 Discussion

In this paper we presented a general purpose framework for building integrative models of protein-protein interaction networks. Our main insight is that we should view this problem as a *relational learning problem*, where observations about different entities are not independent. We build on and extend tools from relational probabilistic models to combine multiple types of observations about protein attributes and protein-protein interactions in a unified model. We constructed a concrete model that takes into account interactions, interaction assays, localization of proteins in several compartments, and localization assays, as well as the relations between these entities. Our results demonstrate that modeling the dependencies between interactions leads to a significant improvement in predictions. We have also shown that including observations of protein properties, namely protein localization, and explicit modeling of noise in such observations, leads to further improvement in prediction. Finally, we have shown how evidence can propagate in the model in complex ways leading to novel hypothesis the can be easily interpreted.

Our approach builds on relational graphical models. These models exploit a template level description to induce a concrete model for a given set of entities and relations among these entities [8, 28]. In particular, our work is related to applications of these models to *link prediction* [10, 30]. In contrast to these works, the large number of unobserved random variables in the training data poses significant challenges for the learning algorithm. Our probabilistic model over network topology is also related to models devised in the literature of *social networks* [7]. Recently, other studies tried to incorporate global views of the interaction network when predicting interactions. For example, Iossifov *et al* [12] propose a method to describe properties of an interaction network topology when combining predictions from literature search and yeast two-hybrid data for a dataset of 83 proteins. Their model is similar to our **Triplet** model in that it combines a model of dependencies between interactions with the likelihood of independent observations about interactions. Their model of dependencies, however, focuses on the global distribution of node degrees in the network, rather than on local patterns of interactions. Other recent studies employ variants of Markov random fields to analyze protein interaction data. In these studies, however, the authors assumed that the interaction network is given and use it for other tasks, *e.g.*, predicting protein function [5, 18, 19] and clustering interacting co-expressed proteins [25]. In contrast to our model, these works can exploit the relative sparseness of the given interaction network to perform fast approximate inference.

Our emphasis here was on presenting the methodology and evaluating the utility of integrative models. These models can facilitate incorporation of additional data sources, potentially leading to improved predictions. The modeling framework allows us to easily extend the models to include other properties of both the interactions and the proteins, such as cellular processes or expression profiles, as well as different interaction assays. Moreover, we can consider additional dependencies that impact the global protein-protein interaction network. For example, a yeast two-hybrid experiment might be more successful for nuclear proteins and less successful for mitochondrial proteins. Thus, we would like to relate the cellular localization of a protein and the corresponding observation of a specific type of interaction assay, This can be easily achieved by incorporating a suitable template potential in the model. An exciting challenge is to learn which dependencies actually improve predictions. This can be done by methods of *feature induction* [4]. Such methods can also allow us to discover high-order dependencies between interactions and protein properties.

Extending our framework to more elaborate models and networks that consider a larger number of proteins poses several technical challenges. Approximate inference in larger networks is both computationally demanding and less accurate. Generalizations of the basic loopy belief propagation method (*e.g.*, [34]) as well as other related alternatives [16, 33], may improve both the accuracy and the convergence of the inference algorithm. Learning presents additional computational and statistical challenges. In terms of computation, the main bottleneck lies in multiple invocations of the inference procedure. One alternative is to utilize information learned efficiently from few samples to prune the search space when learning larger models. Recent results suggest that large margin discriminative training of Markov random fields can lead to a significant boost in prediction accuracy [29]. These methods, however, apply exclusively to fully observed training data. Extending these method to handle partially observable data needed for constructing protein-protein interaction networks is an important challenge.

Finding computational solutions to the problems discussed above is a necessary challenge on the way to a global and accurate protein-protein interaction model. Our ultimate goal is to be able to capture the essential dependencies between interactions, interaction attributes and protein attributes, and at the same time we want to be able to infer hidden entities. Such a probabilistic integrative model can elucidate the intricate details and general principals of protein-protein interaction networks.

Acknowledgments. We thank Aviv Regev, Noa Shefi, Einat Sprinzak, Ilan Wapinski, and the anonymous reviewers for useful comments on previous drafts of this paper. Part of this research was supported by grants from the Israeli Ministry of Science, the Isreal Science Foundation (ISF), European Union Grant QLRT-CT-2001-00015, and the National Institute of General Medical Sciences (NIGMS).

References

1. J. R. Bock and D. A. Gough. Predicting protein–protein interactions from primary structure. *Bioinformatics*, 17(5):455–460, 2001.
2. W. Buntine. Chain graphs for learning. In *Proc. Uncertainty in Art. Intel.*, p. 46–54. 1995.
3. M.C. Costanzo, *et al.* YPD, POMBEPD, and WORMPD: model organism volumes of the bioknowledge library, an integrated resource for protein information. *Nuc. Acids Res.*, 29:75–9, 2001.
4. S. Della Pietra, V. Della Pietra, and J. Lafferty. Inducing features of random fields. *IEEE Trans. on Pat. Anal. Mach. Intel.*, 19:380–393, 1997.
5. M. Deng, T. Chen, and F. Sun. An integrated probabilistic model for functional prediction of proteins. *J. Comput. Bio.*, 11:463–75, 2004.
6. M. B. Eisen, P. T. Spellman, P. O. Brown, and D. Botstein. Cluster analysis and display of genome-wide expression patterns. *Proc. Natl. Acad. Sci. USA*, 95:14863–8, 1998.
7. O. Frank and D. Strauss. Markov graphs. *J. Am. Stat. Assoc.*, 81, 1986.
8. N. Friedman, L. Getoor, D. Koller, and A. Pfeffer. Learning probabilistic relational models. In *Proc. Inte. Joint Conf. Art. Intel.*. 1999.
9. A. C. Gavin, *et al* . Functional organization of the yeast proteome by systematic analysis of protein complexes. *Nature*, 415(6868):141–147, Jan 2002.
10. L. Getoor, N. Friedman, D. Koller, and B. Taskar. Learning probabilistic models of relational structure. In *Int. Conf. Mach. Learning*. 2001.
11. W. Huh, *et al.* Global analysis of protein localization in budding yeast. *Nature*, 425:686 – 691, 2003.
12. I. Iossifov, M. Krauthammer, C. Friedman, V. Hatzivassiloglou, J.S. Bader, K.P. White, and A. Rzhetsky. Probabilistic inference of molecular networks from noisy data sources. *Bioinformatics*, 20:1205–13, 2004.
13. T. Ito, T. Chiba, R. Ozawa, M. Yoshida, M. Hattori, and Y. Sakaki. A comprehensive two-hybrid analysis to explore the yeast protein interactome. *Proc. Natl. Acad. Sci. USA*, 98:4569–4574, 2001.
14. R. Jansen, *et al.* A Bayesian networks approach for predicting protein-protein interactions from genomic data. *Science*, 302:449–453, 2003.

15. M. I. Jordan, ed. *Learning in Graphical Models*. Kluwer, 1998.

16. M. I. Jordan, Z. Ghahramani, T. Jaakkola, and L. K. Saul. An introduction to variational approximations methods for graphical models. In [15].

17. A. Kumar, *et al*. Subcellular localization of the yeast proteome. *Genes. Dev.*, 16:707–719, 2002.

18. M. Leone and A. Pagnani. Predicting protein functions with message passing algorithms. *Bioinformatics*, 21:239–247, 2005.

19. S. Letovsky and S. Kasif. Predicting protein function from protein/protein interaction data: a probabilistic approach. *Bioinformatics*, 19(Suppl 1):i97–204, 2003.

20. HW Mewes, J Hani, F Pfeiffer, and D Frishman. MIPS: a database for genomes and protein sequences. *Nuc. Acids Res.*, 26:33–37, 1998.

21. K. Murphy, Y. Weiss, and M. I. Jordan. Loopy belief propagation for approximate inference: An empirical study. In *Proc. Uncertainty in Art. Intel.*, 1999.

22. R. M. Neal and G. E. Hinton. A new view of the EM algorithm that justifies incremental and other variants. In [15].

23. M. Pellegrini, E. M. Marcotte, and T. O. Yeates. A fast algorithm for genome-wide analysis of proteins with repeated sequences. *Proteins*, 35:440–446, 1999.

24. G. Rigaut, A. Shevchenko, B. Rutz, M. Wilm, M. Mann, and B. Seraphin. A generic protein purification method for protein complex characterization and proteome exploration. *Nat. Biotech.*, 17:1030–1032, 1999.

25. E. Segal, H. Wang, and D. Koller. Discovering molecular pathways from protein interaction and gene expression data. *Bioinformatics*, 19(Suppl 1):i264–71, 2003.

26. E. Sprinzak and H. Margalit. Correlated sequence-signatures as markers of protein-protein interaction. *J. Mol. Biol.*, 311:681–692, 2001.

27. E. Sprinzak, S. Sattath, and H. Margalit. How reliable are experimental protein-protein interaction data? *J. Mol. Biol.*, 327:919–923, 2003.

28. B. Taskar, A. Pieter Abbeel, and D. Koller. Discriminative probabilistic models for relational data. In *Proc. Uncertainty in Art. Intel.*, p. 485–492, 2002.

29. B. Taskar, C. Guestrin, and D. Koller. Max-margin markov networks. In *Adv. Neu. Inf. Proc. Sys.*, 2003.

30. B. Taskar, M. F. Wong, P. Abbeel, and D. Koller. Link prediction in relational data. In *Adv. Neu. Inf. Proc. Sys.*, 2003.

31. P. Uetz, *et al*. A comprehensive analysis of protein-protein interactions in Saccharomyces cerevisiae. *Nature*, 403:623–627, 2000.

32. C. von Mering, R. Krause, B. Snel, M. Cornell, S. G. Oliver, S. Fields, and P. Bork. Comparative assessment of large-scale data sets of protein-protein interactions. *Nature*, 417:399–403, 2002.

33. M. J. Wainwright, T. Jaakkola, and A. S. Willsky. A new class of upper bounds on the log partition function. In *Proc. Uncertainty in Art. Intel.*, 2002.

34. J. Yedidia, W. Freeman, and Y. Weiss. Constructing free energy approximations and generalized belief propagation algorithms. TR-2002-35, Mitsubishi Electric Research Labs, 2002.

35. J. S. Yedidia, W. T. Freeman, and Y. Weiss. Generalized belief propagation. In *Adv. Neu. Inf. Proc. Sys.*, p. 689–695, 2000.

36. L. V. Zhang, S. L. Wong, O. D. King, and F. P. Roth. Predicting co-complexed protein pairs using genomic and proteomic data integration. *BMC Bioinformatics*, 5:38, 2004.

The Factor Graph Network Model for Biological Systems

Irit Gat-Viks, Amos Tanay, Daniela Raijman, and Ron Shamir

School of Computer Science,
Tel-Aviv University, Tel-Aviv 69978, Israel
{iritg, amos, rshamir}@tau.ac.il

Abstract. We introduce an extended computational framework for studying biological systems. Our approach combines formalization of existing qualitative models that are in wide but informal use today, with probabilistic modeling and integration of high throughput experimental data. Using our methods, it is possible to interpret genomewide measurements in the context of prior knowledge on the system, to assign statistical meaning to the accuracy of such knowledge and to learn refined models with improved fit to the experiments. Our model is represented as a probabilistic factor graph and the framework accommodates partial measurements of diverse biological elements. We develop methods for inference and learning in the model. We compare the performance of standard inference algorithms and tailor-made ones and show that hidden variables can be reliably inferred even in the presence of feedback loops and complex logic. We develop a formulation for the learning problem in our model which is based on deterministic hypothesis testing, and show how to derive p-values for learned model features. We test our methodology and algorithms on both simulated and real yeast data. In particular, we use our method to study the response of *S. cerevisiae* to hyper-osmotic shock, and explore uncharacterized logical relations between important regulators in the system.

1 Introduction

The integration of biological knowledge, high throughput data and computer algorithms into a coherent methodology that generates reliable and testable predictions is one of the major challenges in today's biology. The study of biological systems is carried out by characterizing mechanisms of biological regulation at all levels, using a wide variety of experimental techniques. Biologists are continuously refining models for the systems under study, but rarely formalize them mathematically. High-throughput techniques have revolutionized the way by which biological systems are explored by generating massive amounts of information on the genome-wide behavior of the system. Genome-wide datasets are subject to extensive computational analysis, but their integration into existing (informal) biological models is currently done almost exclusively manually. To rigorously integrate biological knowledge and high-throughput experiments,

S. Miyano et al. (Eds.): RECOMB 2005, LNBI 3500, pp. 31–47, 2005.
© Springer-Verlag Berlin Heidelberg 2005

one must develop computational methodologies that accommodate information from a broad variety of sources and forms, and handle highly complex systems and extensive datasets.

Recent studies on computational models for biological networks have attempted de-novo reconstruction of a network on genes (e.g., [6]), used prior knowledge on network topology (e.g., [9,11]), or combined transcription factor location and sequence data to learn a clustered model for the genome-wide behavior of the system [1,23,2]. Other studies built detailed models manually, utilizing existing biological knowledge [3,5] but lacked computational methods for model reassessment in light of additional evidence.

In this study we describe a new algorithmic framework for representing biological knowledge and integrating it with experimental data. Our methodology allows biologists to formalize their knowledge on a system as a coherent model, and then to use that model as the basis for computational analysis that predicts the system's behavior in various conditions. Most importantly, our framework allows the learning of a refined model with improved fit to the experimental data.

In previous works [25,8] we have introduced the notions of model refinement and expansion and studied it when applied to discrete deterministic models. Here we study these problems in the more general settings of probabilistic models. The probabilistic approach allows us to model uncertainty in prior biological knowledge, and to distinguish between regulatory relations that are known at high level of certainty and those that are only hypothesized. The probabilistic model also allows us to mix noisy continuous measurements with discrete regulatory logic. Unlike previous works that modeled gene networks as Bayesian networks (or dynamic Bayesian networks) on genes, our model expresses diverse biological entities (e.g., mRNAs, proteins, metabolites), and directly accommodates undelayed feedback loops which are essential in many biological systems. We formalize our model as a probabilistic factor graph [14], and show how it can be developed naturally from basic assumptions on the biochemical processes underlying the regulatory network and on the information we have on it.

Having established our methodology for probabilistic modeling, we develop algorithms for inferring the system's state given partial data. For example, we can infer the activity of proteins given gene expression data. We use inference algorithms as the basis for learning refined regulatory functions. We develop a formulation of the learning problem in our network model, which is based on deterministic hypothesis testing. Our approach to the learning of regulatory models uses regulatory features with clear biological meaning and allows the derivation of p-values for learned model features.

We tested the performance of our algorithms on simulated models and on two complex pathways in *S. cerevisiae*: the regulation of lysine biosynthesis, and the response to osmotic stress. In both cases our models successfully integrate prior knowledge and high throughput data and demonstrate improved performance compared to extant methods. In particular, our results suggest a novel model for regulation of genes coding for components of the HOG signaling pathway and robustly learn logical relations between central transcription factors downstream of the Hog1 kinase. Our results show that integration of prior biological knowledge with high throughput data is a key step toward making computational network analysis a practical part of the toolbox of the molecular biologist. For lack of space some proofs and details are omitted.

2   Modeling Prior Knowledge and Experimental Observations

In this section we present our probabilistic model for a biological regulatory network. The biological entities in the system under study are formulated as variables representing, e.g., mRNAs, proteins, metabolites and various stimulators. We assume that at a given condition, each of the entities attain a logical state, represented by an integer value of limited cardinality. We wish to study regulatory relations (or regulation functions) among variables. Such relations, for example, determine the level of a mRNA variable as a function of the levels of a set of transcription factor protein variables, or the level of a metabolite variable given the levels of other metabolites and of structural enzymes. Regulation functions approximate an underlying biochemical reaction whose exact parameterization is not known. Note that the regulatory process is stochastic at the single cell level, but the parameters of the reaction equations governing it are deterministic. Moreover, when we observe a large ensemble of cells in a high throughput experiment, we average millions of stochastic processes and in theory should obtain an almost deterministic outcome, or a superposition of several deterministic modes. Such deterministic outcome is obscured by significant experimental noise so a practical modeling strategy may assume deterministic logic and noisy observations. Given these notions, we wish to find a discrete, deterministic model that adequately approximates the system's reaction equations, in a way that can take into account the different sources of noise and uncertainty in the data.

In most studied biological systems, substantial prior knowledge on regulatory relations has accumulated. Such knowledge includes direct regulatory interactions, qualitative functional roles (activator/repressor), combinatorial switches, feedback loops, and more. Typically, that information is incomplete and of variable certainty. In order to optimally exploit it, we must model both the relations and their level of certainty. We do this by introducing a distribution on the regulation functions for each variable. This distribution may determine the regulation function with high probability if our prior knowledge is very strong. At the other extreme end, lack of information is modeled by uniform distribution over all possible regulation functions.

We formalize these notions as follows (see Figure 1). Let $X = \{X_1, ... X_n\}$ be a collection of biological variables. Let $S = \{0, 1, ..., k - 1\}$ be the set of logical *states* that each variable may attain. A *model state* s is an assignment of states to all the variables in X. Each variable X_i is regulated by a set of its *regulator* (or *parent*) variables $Pa_i = \{Pa_{i,1}, ..., Pa_{i,d_i}\} \subseteq X$. When addressing a particular regulation relation, the regulated variable is also called the *regulatee*.

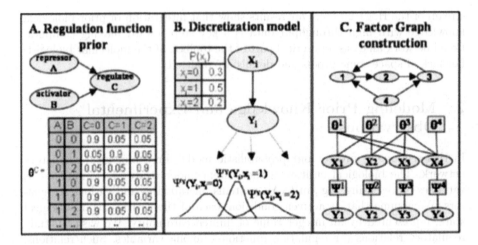

Fig. 1. An overview of the factor graph network model. A) Knowledge on the logical regulation functions is formalized as conditional probabilities. B) Continuous measurements and logical states are linked by joint discretizer distributions. C) A possibly cyclic network structure is transformed into a factor graph, using the regulation function priors and the discretizers' distributions

Lower case letters will indicate state assignments of the corresponding upper case variables. For example, given a model state s, x_i^s is the state of X_i, pa_i^s is the assignment of the set Pa_i. The *regulatory dependency graph* is a digraph $G_R = (X, A)$ representing direct dependencies, i.e., $(X_u, X_v) \in A$ iff $X_u \in Pa_v$ (G_R is sometimes called the wiring diagram of the model). The graph can contain cycles. The *regulation function prior* for a variable X_i is formulated as our belief that the variable attains a certain state given an assignment to its parents Pa_i. It is represented as in standard Bayesian networks, by the conditional probabilities θ^i:

$$\theta^i(X_i, Pa_i) = Pr(X_i|Pa_i) \tag{1}$$

Note that in current applications, the interpretation of the θ distributions is not as representing a stochastic process in which θ^i defines the conditional probability of X_i given its regulators. Instead, we assume that the true model deterministically determines X_i given its parents, but we are not sure which deterministic rule applies, and therefore what value X_i will attain. In the future, given refined understanding of regulatory switches, and measurements at the single cell level, the θ distributions may be applicable to describe the inherent stochasticity of some biological switches.

We wish to learn regulation functions from experimental data. In practice, biological experiments provide noisy observations on a subset of the variables in the system. The observations are continuous and we do not know in advance how to translate them into logical states. We thus introduce a set of real valued *sensor variables* $Y_1, .., Y_n$ and *discretizer distributions* $\psi^i(X_i, Y_i)$ that specify the joint distribution of a discrete logical state of X_i and the continuous observation on Y_i. In this work, we shall use mixtures of Gaussians (Figure 1b) to model ψ^i, but other formulations are also possible.

Our model is now represented as a probabilistic factor graph [14], defined by the joint distribution over logical (X) and sensor (Y) variables:

$$Pr_M(X, Y) = \frac{1}{Z} \prod_i \theta^i(X_i, Pa_i)\psi^i(X_i, Y_i) \tag{2}$$

Where Z is a normalization constant. We call this formulation a *factor graph network (FGN) model*. Factor graphs are widely used probabilistic graphical models that were originally applied to coding/decoding problems. Recently, factor graphs were used in computational biology, although in a different context [27].

When the dependency graph G_R is acyclic, our FGN model is equivalent to a Bayesian network on the variables X_i and Y_i, constructed using the edges of G_R and additional edges from each X_i to the corresponding Y_i. This can be easily seen from (2) by noting that in the acyclic case $Z = 1$ (the proof is as in Bayesian networks theory, e.g. [18]). When the model contains loops, the situation gets more complicated. For example, we note that according to the FGN model, $Pr_M(X_i|Pa_i)$ does not necessarily equals the original beliefs $\theta(X_i, Pa_i)$. We note that derivation of the FGN model from basic assumptions on deterministic approximations of the biological system and on our prior beliefs on them is possible, and will be described in a later publication.

We now outline several important extensions of our model that are not used in this study. In its simplest form, our model describes the steady state behavior of the system. Biological processes are ideally described as temporal processes, but when sampling rate is slow relative to the rate of the regulatory mechanisms, the steady state assumption can be invoked. Different regulatory processes operate on different time scales: In the typical high throughput experimental sampling rate, the steady state assumption may be highly adequate for metabolic pathways and post translational regulation and reasonable for transcriptional programs. For the models considered in this work we have combined interactions from all types and validated empirically (using, e.g., cross validation) that the steady state assumption still enables biologically meaningful results. Our model is unique in its handling of steady state feedback loops. It can be extended to handle slower temporal processes in a way analogous to the construction of dynamic Bayesian networks (DBN) [7, 24] from steady state Bayesian networks. As in DBNs, the algorithms for inference and learning can be naturally generalized from the steady state model to the dynamic model. Another possible extension is the consideration of other classes of regulation functions (for example, we can consider continuous or ranked function as in [26, 22, 12, 16]).

3   Inference

In this section we discuss the inference problem in the FGN model. Each experiment provides partial information on the value of model variables. Typically, a subset of the sensor real valued (Y) variables are observed in each experiment (for example, mRNA variables are determined in a gene expression experiment). The values of some logical (X) variables may also be determined by the experimenter (e.g., the nutritional condition is represented by the states of extra-cellular metabolite variables). Perturbations of certain regulation functions (e.g., by gene knockouts) are not discussed here for simplicity, but are easily incorporated by modifying the θ parameters at the appropriate conditions. The inference problem is defined with respect to an observation on a subset of the variables. The goal is to compute the posterior distribution of the unobserved (hidden) variables, i.e. the probability distribution of hidden variables values given the model and the observed data. For example, given a gene expression profile, we can compute the posterior distribution of protein variables or estimate the level of a certain metabolic intermediate. Solving instances of the inference problem is a pre-requisite for learning refined models.

Inference in graphical models is an NP hard problem [4] that was extensively studied. A popular class of algorithms [28] uses an approximation of the posterior distribution assuming certain decomposition over independent variables or clusters of variables. Algorithms from this class include loopy belief propagation (LBP), mean field, and their generalizations. They provide in many cases an effective way for estimating posteriors of small sets of variables. A different approach is needed if one is interested in *posterior modes* - complete system states with high probability mass. Modes provide an important insight into the system's state space, and cannot be directly computed from independent variable posteriors.

Posterior modes are particularly important when we study the behavior of systems with feedback loops. For example, the single variable posteriors in a positive feedback loop may contain no information even though the loop has two simple modes (all variables on or all variables off) that absorb all the probability mass. Still, when applicable, algorithms that compute independent posteriors perform well.

We studied the effects of our model's specific characteristics on the performance of several inference algorithms. We implemented a Gibbs sampler, the LBP algorithm, and a modes-based instantiation inference algorithm. Our **Gibbs sampler** is a naive MCMC algorithm [15] that performs a random walk over the space of model states, based on sampling from local distributions. In our model, sampling is done only for the X variables (unobserved sensors do not affect other posterior distributions). In Gibbs sampling, the posterior is estimated from the set of terminal trajectory states, so any query from the posterior is in principle possible (including single variable posteriors or complete modes). The **LBP** algorithm for the FGN model was implemented as described in [28]. The algorithm is a message passing procedure that is guaranteed to reach the exact solution for acyclic models. The algorithm, like others in the family, decomposes the posterior across variables or sets of variables, so we could only use it to obtain the posterior of small sets of variables. We also developed an instantiation-based inference algorithm that exploits the known dependency structure of the model and builds on ideas from [8]. The **modes instantiation (MI)** algorithm first builds a deterministic model by taking, for each variable, the maximum likelihood regulation function (using the prior θ^i and breaking ties arbitrarily). It then identifies a feedback set in G_R (see [8]) and enumerate all possible value assignments for variables in this set. Each feedback set configuration gives rise to a unique model state, which is used as the basis for further greedy optimization of the state likelihood. We add to the resulting set of high probability modes an additional small set of states derived using the Gibbs sampler. We now estimate the posterior as a mixture of the set of locally optimal and sampled model states, weighted by their likelihoods, where the partition function (our estimation of the Z parameter) equals the sum of likelihood over all of the states we consider. We thus map the posterior function landscape using a limited set of local optima.

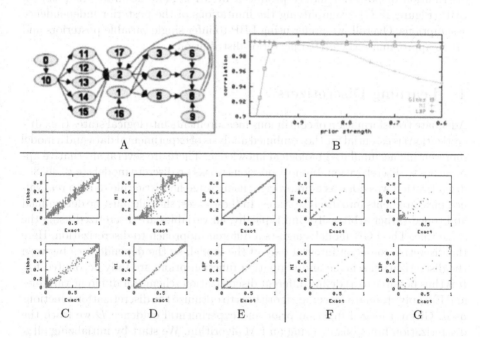

Fig. 2. Performance of different inference algorithms on a simulated model. A) The dependency graph G_R of the simulated model. B) Effect of prior strength on inference accuracy. Y axis: the correlation of approximated and exact single variable posteriors. X axis: prior strength (α). For strong priors, LBP and MI give a good approximation for the posterior, while the accuracy of the Gibbs sampler is low. As priors get weaker, the performance of MI deteriorates, indicating that the mixture of deterministic modes is a poor approximation for the posterior in these cases. C,D,E) Detailed correlation of algorithmic and exact single variable posteriors for $\alpha = 0.7$ (top) and $\alpha = 0.97$ (bottom). F,G)Detailed correlation of algorithmic and exact joint posteriors for $\alpha = 0.7$ (top) and $\alpha = 0.97$ (bottom). We see that MI outperforms LBP when comparing the joint posteriors modes

We tested the three inference algorithms on a simulated model (see Figure 2). We constructed simulated FGN models by adaptation of a deterministic model. We use a *prior strength* parameter α to construct θ functions that assign probability α for the anticipated deterministic function outcome and $\frac{1-\alpha}{k-1}$ to other values. For detailed description of the simulation, see our website (www.cs.tau.ac.il/ rshamir/fgn/). We explored the behavior of the different algorithms as a function of the prior strength (α) using the correct posterior as the reference. Models with α near one represent very good knowledge on the system under study. Models with α near $\frac{1}{k}$ represent complete lack of knowledge. Our analysis (see Figure 2) indicates that for estimation of single variable posteriors, LBP outperforms the other two algorithms (and also the mean field algorithm and a simple clustered variational algorithm [13], data not shown). Also, when the prior is strong, MI provides reasonable accuracy. However, the distribution of posterior modes produced by MI is more accurate than that of LBP (Figure 2F,G), exemplifying the limitations of the posterior independence assumptions. Overall, we prefer using LBP to infer single variable posteriors and MI to approximate the global posterior distribution.

4   Learning Discretizers

Adequate transformation of continuous measurements into logical states (i.e., discretization) is essential for the combined analysis of experimental data and a model representing accumulated biological knowledge. There are several alternative approaches to discretization. In most previous works on discrete models (e.g., [6, 8]), discretization was done as a preprocess, using some heuristic rule to map real valued measurements into discrete states. In these cases, the rule must be determined and tuned rather arbitrarily, and typically, all variables are discretized using the same rule. The FGN model suggests a different approach to discretization. Here the discretization is an integral part of the model, so the dependencies between the discretization scheme and regulation function priors are fully accounted for. It is thus possible to a) define different discretization scheme for different variables and b) apply standard learning algorithms to optimize the discretization functions used. Given a logical function prior and experimental evidence D we learn the discretization functions ψ^i using an EM algorithm. We start by initializing all ψ using any heuristic discretization scheme. On each EM iteration, we infer the posterior distributions for each of the variables X_i in each of the conditions, and then reestimate the ψ^i mixtures using these posteriors, by computing the Gaussians sufficient statistics $E(Y_i|X_i = j, D), V(Y_i|X_i = j, D)$. The new ψ^i distributions are used in the next iteration and the algorithm continues until convergence.

The FGN model thus provides a very flexible discretization scheme. In practice, this flexibility may lead to over-fitting and may decrease learnability. One can control such undesired effects by using the same few discretization schemes on all variables. As we shall see below, on real biological data, variable specific discretization outperforms global discretization using a single scheme, and is clearly more accurate than the standard preprocessing approach.

5   Learning Regulation Functions

Given an FGN model and experimental evidence, we wish to determine the optimal regulation function for each variable and provide statistical quantification of its robustness. The standard approach to learning in graphical models seeks parameters that maximize a likelihood or Bayesian score, possibly completing missing data using an EM algorithm. The applicability of this approach relies heavily on our ability to interpret the learned parameters in meaningful way. In many cases such interpretation is straightforward (e.g., learning the probabilities of each face in an unfair dice). In other cases, for example, when learning regulation functions in the FGN model, such interpretation is less obvious and should be carefully evaluated given the phenomenon we try to model. Specifically, if we assume the parameters of the logical factors in the FGN model represent our prior beliefs on the logical relations between variables, we may attempt to learn by confirming beliefs (deciding if a certain regulator assignment gives rise to a certain regulatee assignment) rather than finding optimal new beliefs. Given that currently most biological systems are only roughly characterized, and available experimental data provide estimated averages of the states of biological factors in them, this hypothesis driven approach may be a realistic alternative to parameters learning. In other cases (most importantly, when single cell measurements are available), we may prefer the standard approach. We assume throughout that the network topology is fixed and focus on learning regulation functions only. In principle, the methods introduced below can also be used to refine the topology (add and remove edges).

We focus on the regulation of some variable X_i given a fixed parents value assignment pa_i^s. Define h_j as the FGN model derived from M by setting $\theta(j, pa_i^s) = 1$ and $\theta(j', pa_i^s) = 0$ for $j' \neq j$. We define the learning problem in our model as selecting the maximum likelihood h_j. To that end we shall have to compute the likelihood of each h_j given the data, essentially solving the inference problem in the h_j model to compute the full probability of the data. Computing the full probability is a difficult problem, and as stated in section 3 we approximate it using the MI algorithm. We bound the likelihood of the data given h_j by summing the likelihoods of the posterior modes sampled by the MI algorithm (including modes computed from configurations of a feedback set in G_R and modes obtained using Gibbs sampling). While this is a very crude approximation, our empirical analysis shows it is still adequate (see below).

To assign statistical meaning to the learning procedure we use two methods: bootstrap and likelihood ratio testing. In the bootstrap method, we re-sample conditions from the original data D and reiterate the learning procedure. We count the number of times each h_j was selected as the maximum likelihood model and define the feature robustness as the fraction of times it was selected. Bootstrap is in widespread use in cases where sampling from the background distribution is impossible or very difficult. In our case, approximated sampling from $Pr(D|h_j)$ is possible given our representation of the posterior landscape as a mixture of modes. We can thus try to directly perform a likelihood ratio test and derive p-values for the learned features.

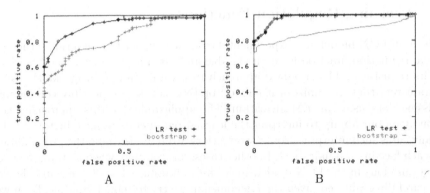

Fig. 3. Accuracy of learning regulation functions. Each figure is a ROC curve (X-axis: false positives rate Y-axis: true positives rate) for learning the functions in a simulated model using bootstrap and likelihood ratio test for determining the significance of learned features. Results are shown for learning from 15 (A) and 80 (B) conditions. The accuracy of the likelihood ratio test method is consistently higher

We fix j and define $H_1 : h_j, H_0 : \cup_{k \neq j} h_k$ and $\lambda = \frac{max_{h_i \in H_0 \cup H_1} Pr(D|h_i)}{max_{h_i \in H_0} Pr(D|h_i)}$. In order to compute a p-value for the observed statistic λ, we have to estimate the distribution of λ given H_0. To that end we generate samples from the distribution $Pr(D|H_0)$, compute the corresponding λ's and reconstruct the distribution. When sampling datasets D, we take into account both the model (defined by H_0) and the properties of the original dataset. We do this as follows: for each of the conditions in the original dataset, we fix all observations of X variables (these correspond to the experimental conditions) and all model perturbations. We then apply the MI algorithm with the H_0 model, and compute the set of posterior modes matching these experimental conditions, without using any of the observations on Y variables. We then generate a sample by a) selecting a mode from the mixture, and b) generating observations on Y variables using the model discretizer distributions ψ.

We analyzed the performance of the bootstrap and likelihood ratio test methods by learning features in a simulated model (see our website for details). Figure 3 shows ROC curves for learning in the simulated model using 15 and 80 conditions. We see consistently better accuracy when using the likelihood ratio tests, probably due to better resolution of features that are nearly ambiguous given the data. While bootstrap has the advantage of not assuming an approximation to the global posterior, it is less accurate when the posterior can be reasonably approximated.

6   Results on Biological Data

In order to test the applicability of our methods to real biological systems, we constructed models of two important yeast pathways, the lysine intake and biosynthesis pathway, and the Hog1 MAPK pathway, which mediates the yeast response to osmotic stress. For each of the models, we performed extensive literature survey in order to construct the initial model of the system (for the lysine system, our previously developed deterministic model [8] was the main source). The HOG model is outlined in Figure 6. The lysine model contained 140 variables and the HOG model contained 50 variables. We used prior strength $\alpha = 0.9$ in all reported experiments. We collected published experimental data on each of the models. The data consisted of 23 conditions (cf. [8]) for the lysine model and 129 conditions for the HOG model [17]. Differential measurements from cDNA microarrays were transformed into absolute values as described in [8].

6.1   Learning Discretization

The FGN model couples continuous measurements and discrete states via the discretizer distributions ψ^i. We tested our ability to learn the functions ψ^i by performing cross validation using gene expression data for the HOG and lysine models.

We used cross validation to compare three alternatives: (A) single common predefined mixture of Gaussians. (B) using the EM algorithm described in Section 4 to learn a single common maximum likelihood ψ distribution (C) applying an unconstrained EM to learn variable specific ψ^i-s.

Cross validation was done as follows. For each condition, we used one of the above methods to learn the ψ distributions, using all data excluding that condition. We then iterated over all the model's variables. For each variable v, we hid its observation in the omitted condition, and inferred its posterior distribution using the trained ψ's. Finally, we computed the likelihood of v's observation given the posterior.

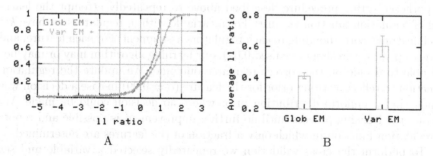

A   B

Fig. 4. Learning discretization distributions. Cross validation results for alternative methods for estimating the discretization functions ψ^i in the HOG model. Glob EM - optimized single common discretization function. Var EM - optimized variable specific discretization. A) Cumulative distribution of log likelihood (ll) ratios comparing each of the two discretization methods to the global preprocessed discretization scheme. B) Average ll ratios for the two methods. Bars indicate the predicted standard deviation of the averages

Figure 4 shows the results of the cross validation on the HOG model. We present the distribution and the average log likelihood ratio of each of the methods B and C to the predefined discretization (method A). This comparison allows us to view the results in terms of the generalization capabilities of the optimized discretizers: likelihood ratios smaller than 0 represent cases where the refined discretization resulted in overfitting, ratios larger than 0 represent successful generalizations. We conclude that for 80% of the cases, incorporating the discretization into the model significantly improves performance. Moreover, variable specific discretization outperforms the optimized global discretization scheme. Similar results were obtained for the lysine model.

6.2   Learning Regulation Functions

We used cross validation in the lysine model to confirm the capability of our method to learn real regulation functions from real data and to compare its performance to the deterministic and to a naive Bayesian approaches. The deterministic model approach [8] learns a deterministic regulation function by optimizing a least squares score. It assumes a prior model that is 100% certain and solves the deterministic analog of the inference problem to enable the learning of a regulation function from partial observations. To allow comparison of the deterministic model with the current one, we transformed its discrete predictions into continuous distributions using predefined Gaussians. The same discretizers were used in the other two models, in order to ensure that differences in model performance were not due to the discretization. In the naive Bayesian approach, we assume the topology of a Bayesian network over the observed variables (the mRNAs in our case) is given, and we learn the conditional probabilities of each variable separately given its regulators using complete data. The learning problem in this case is trivially solved by building a frequency table. Learning in the FGN model was done given the probabilistic function priors θ^i. We used the hypothesis testing procedure described above to repeatedly attempt the learning of regulation function features. For a variable with m regulators we have k^m such features, corresponding to each regulators assignment. For each feature, and given a p-value threshold (we used 0.01), our learning algorithm may or may not be able to decide on the correct regulatee outcome. We update the regulation function to reflect a strong prior for the feature ($\alpha = 0.99$) in case a decision was made, and a uniform distribution prior where no decision could be made. We iterate the learning process until no further improvement is possible and report a regulation function in which only a fraction of the features are determined.

To perform the cross validation we repeatedly selected a variable and set its prior θ^i to the uniform distribution. We removed one condition from the dataset, learned the variable's regulation function and used it to compute the posterior of the variable, given the omitted condition without the observation for the test variable. Figure 5 depicts the log likelihood ratio distribution for the three methods (compared to a uniform prior model). We see that the FGN model improves over the other two methods. Detailed examination of the distribution reveals that the probabilistic model makes half as many erroneous predictions

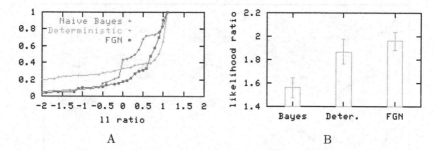

Fig. 5. Learning regulation functions. Cumulative distributions (A) and averages (B) of the log likelihood (ll) ratio for cross validation in the lysine model using three methods for learning regulation functions: A naive Bayesian method, assuming the network topology, a deterministic learning scheme as in [8], and learning using the FGN model. Bars indicate the predicted standard deviation of the averages

(negative log likelihood ratios) as does its deterministic counterpart, probably due to its ability to tolerate mistakes in the prior model. Both the deterministic and probabilistic methods make good use of the additional knowledge, formalized into the model logic, to obtain better results than the naive, topology based approach.

6.3   Biological Analysis of the HOG Model

The response of yeast to hyper-osmotic stress is mediated through parallel MAPK signaling pathways, the multi target kinase Hog1, and an array of transcription factors that coordinate a complex process of adaptation by transient growth repression and modifications to glycerol metabolism, membrane structure and more [10]. We have constructed an FGN model that represents known regulatory relations in the HOG system (Figure 6A) and used it to study the transcriptional program following treatment by variable levels of KCl [17]. The data we used contained observations of all mRNA variables in the model and assignments of fixed values for logical variables describing experimental conditions (e.g., Turgor pressure). We used the MI inference algorithm to estimate the states of all logical variables in the model and applied the model to test the accuracy of the prior logic distributions modeled from the literature. We summarize the model predictions in the *discrepancy matrix* shown in Figure 6B. The discrepancy matrix shows the correspondence between model predictions and experimental observations at the level of a single variable under a single condition. Essentially, the discrepancy matrix is the result of a leave-one-out cross validation procedure. To generate it, we examine each sensor variable Y_i in each condition. We infer the posterior distribution of Y_i given the observations on all other variables and compute the expected value and the probability of Y_i observation. We present the difference between the predicted values and the observations in a color coded matrix.

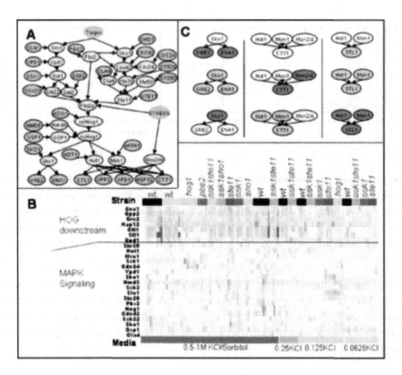

Fig. 6. Analyzing yeast response to hyper-osmotic shock. mRNA variable names are capitalized, protein variables names appear in initial capital letters. Stimulator variables appear as unoutlined ovals. A) Topology of the HOG FGN model used in this study. B) The discrepancy matrix for the HOG model and data from O'Rourke et al. Columns correspond to mRNA variables and rows to experimental conditions, see text for details. Green (Red) cells indicate observations that are smaller (bigger) than the expected prediction. Color intensity is proportional to minus the log likelihood of the observation given the inferred variable posteriors. C) examples of model features learned by the FGN methodology. We show logical relations that were learned with significant p-values. Each graph depicts the regulation of one regulatee given the particular states of its regulators. Variable states are indicated as node colors: white - 0, pink - 1, red - 2

The discrepancy matrix reveals several important discrepancies between the current model for osmo-regulation and the microarray experiments we analyzed. We discuss here briefly two major trends. The first trend affects a group of genes coding for proteins participating in the MAPK signaling cascade (*SSK1, SHO1, STE20, PBS2, CDC42, HOG1* and more). These genes are repressed during the peak of the osmoregulation program (10-30 minutes after treatment with 0.5M KCl, around 60 minutes in the 1M KCl treatment). This repression is not reported in the current literature. We hypothesize that as part of the adaptation to high levels of osmotic pressure, yeasts may reduce the Hog1 signaling cascade sensitivity, by slowing down the production of some central components in it.

A second group of discrepancies involves genes that are targets of the Hog1 downstream regulators Sko1, Hot1, Msn1 and Msn2,4 [19, 21, 20]. In many cases, the literature does not specify the logical relations among the regulators and each of their regulatees, and this lack of knowledge is manifested as discrepancies.

We thus used our model learning machinery to refine the regulatory logic for several model variables that are known to be affected by Hog1-downstream regulators. Figure 6c shows examples of logical relations we learned. First, we were able to learn the known repressive role of Sko1 in the regulation of *GRE2* and *ENA1* [19]. We learned three model features that associated the mRNA variables of these two genes with the opposite state of the inferred Sko1 regulator state. Note that the expression of the *SKO1* gene during osmotic stress is static, and the correct regulation function could only be learned given the inferred Sko1 activities. These inferred activities take into account, in addition to the mRNA measurements, the entire model and its regulatory functions. A second example for a variable we learned regulation for was *STL1*. The regulation of *STL1* is reported to be completely dependent on Hot1 and Msn1 [20], but it is not clear what are the logical relations among them. Our results show that although these two regulators have a positive effect on *STL1* expression, the gene can be induced even when both regulators lack any activity. We can thus hypothesize that a third factor is involved in *STL1* regulation. A third, more complex regulation function, associates the Hog1 specific regulators Hot1, Msn1 and the general stress factor Msn2/4 into a single program controlling several genes (we model just four representatives of a larger regulon: *GPP2*, *GPD1*, *HSP12* and *CTT1* [21]). Our results indicate that the two signaling pathways (the HOG cascade and the general stress pathway) act in parallel, and each of the branches can induce the regulon in the absence of activity from the other.

Acknowledgment

We thank Nir Friedman, Dana Pe'er and the anonymous referees for helpful comments. IGV was supported by a Colton fellowship. AT was supported in part by a scholarship in Complexity Science from the Yeshaia Horvitz Association. DR was supported by a summer student fellowship from the Weizmann Institute of Science. RS holds the Raymond and Beverly Sackler Chair for Bioinformatics at Tel Aviv University, and was supported in by the Israel Science Foundation (Grant 309/02) and by the EMI-CD project that is funded by the European Commission within its FP6 Programme, under the thematic area "Life sciences, genomics and biotechnology for health", contract number LSHG-CT-2003-503269. "The information in this document is provided as is and no guarantee or warranty is given that the information is fit for any particular purpose. The user thereof uses the information at its sole risk and liability."

References

1. Z. Bar-Joseph, G.K. Gerber, T.I. Lee, N.J. Rinaldi, J.Y. Yoo, F. Robert, D.B. Gordon, E. Fraenkel, T.S. Jaakkola, R.A. Young, and D.K. Gifford. Computational discovery of gene modules and regulatory networks. *Nature Biotechnology*, 21:1337–1342, 2003.

2. M.A. Beer and S. Tavazoie. Predicting gene expression from sequence. *Cell*, 117:185–198, 2004.

3. K.C. Chen et al. Kinetic analysis of a molecular model of the budding yeast cell cycle. *Mol Biol Cell*, 11:369–91, 2000.

4. G. Cooper. The computational complexity of probabilistic inference using Bayesian belief networks. *Artificial Intelligence*, 42:393–405, 1990.

5. M.W. Covert, E.M. Knight, J.L. Reed, M.J. Herrgard, and B.O. Palsson. Integrating high-throughput and computational data elucidates bacterial networks. *Nature*, 429:92–96, 2004.

6. N. Friedman, M. Linial, I. Nachman, and D. Pe'er. Using Bayesian networks to analyze expression data. *J. Comp. Biol.*, 7:601–620, 2000.

7. N. Friedman, K. Murphy, and S. Russell. Learning the structure of dynamic probabilistic networks. In *Proc. 14th Conference on Uncertainty in Artificial Intelligence.*, pages 139–147, 1998.

8. I. Gat-Viks, A. Tanay, and R. Shamir. Modeling and analysis of heterogeneous regulation in biological networks. In *Proceedings of the first Annual RECOMB Satellite Workshop on Regulatory Genomics*, pages 21–31, 2004. Also J. Comput Biol in press.

9. A. Hartemink, D. Gifford, T. Jaakkola, and R. Young. Combining location and expression data for principled discovery of genetic regulatory networks. In *Proceedings of the 2002 Pacific Symposioum in Biocomputing (PSB 02)*, pages 437–449, 2002.

10. S Hohmann. Osmotic stress signaling and osmoadaptation in yeasts. *Microbiol Mol Biol Rev.*, 66(2):300–72, 2002.

11. S. Imoto, T. Higuchi, T. Goto, K. Tashiro, S. Kuhara, and S. Miyano. Combining microarrays and biological knowledge for estimating gene networks via Bayesian networks. *J. Bioinform. Comput. Biol.*, 2:77–98, 2004.

12. S. Imoto, S. Kim, T. Goto, S. Aburatani, K. Tashiro, S. Kuhara, and S. Miyano. Bayesian network and nonparametric heteroscedastic regression for nonlinear modeling of genetic network. *J. Bioinform. Comput. Biol.*, 1:231–252, 2004.

13. T.S. Jaakkola. Tutorial on variational approximation methods. In D. Saad and M. Opper, editors, *Advanced Mean Field Methods - Theory and Practice*, pages 129–160. MIT Press, 2001.

14. F.R. Kschischang, B.J. Frey, and Loeliger H. Factor graphs and the sum-product algorithm. *IEEE Transactions on Information Theory*, 47:498–519, 2001.

15. D. J. C. MacKay. Introduction to Monte Carlo methods. In M. I. Jordan, editor, *Learning in Graphical Models*, pages 175–204. Kluwer Academic Press, 1998.

16. I. Nachman, A. Regev, and N. Friedman. Inferring quantitative models of regulatory networks from expression data. *Bioinformatics*, 20:248–256, 2004.

17. S.M. O'Rourke and I. Herskowitz. Unique and redundant roles for hog mapk pathway components as revealed by whole-genome expression analysis. *Mol Biol Cell.*, 15(2):532–42, 2004.

18. J. Pearl. *Probabilistic Reasoning in intelligent systems*. Morgan Kaufmann publishers, inc, 1988.

19. M. Proft and R. Serrano. Repressors and upstream repressing sequences of the stress-regulated ena1 gene in saccharomyces cerevisiae: bzip protein sko1p confers hog-dependent osmotic regulation. *Mol Biol Cell.*, 19:537–46, 1999.

20. M. Rep, M. Krantz, J.M. Thevelein, and S. Hohmann. The transcriptional response of saccharomyces cerevisiae to osmotic shock. hot1p and msn2p/msn4p are required for the induction of subsets of high osmolarity glycerol pathway-dependent genes. *J. Biol. Chem,* 275:8290–8300, 2000.

21. M. Rep, V. Reiser, U. Holzmller, J.M. Thevelein, S. Hohmann, G. Ammerer, and H. Ruis. Osmotic stress-induced gene expression in saccharomyces cerevisiae requires msn1p and the novel nuclear factor hot1p. *Mol. Cell. Biol,* 19:5474–5485, 1999.

22. M. Ronen, R. Rosenberg, B. Shraiman, and U. Alon. Assigning numbers to the arrows: Parameterizing a gene regulation network by using accurate expression kinetics. *Proceedings of the National Academy of Science USA,* 99:10555–10560, 2002.

23. E. Segal, M. Shapira, A. Regev, D. Pe'er, D. Botstein, D. Koller, and N. Friedman. Module networks: identifying regulatory modules and their condition-specific regulators from gene expression data. *Nat Genet.,* 34(2):166–76, 2003.

24. V.A. Smith, E.D. Jarvis, and A.J. Hartemink. Evaluating functional network inference using simulations of complex biological systems. *Bioinformatics,* 18:216–224, 2002.

25. A. Tanay and R. Shamir. Computational expansion of genetic networks. *Bioinformatics,* 17:S270–S278, 2001.

26. A. Tanay and R. Shamir. Modeling transcription programs: inferring binding site activity and dose-response model optimization. *J. Comp. Biol.,* 11:357 – 375, 2004.

27. C.H. Yeang, T. Ideker, and T. Jaakkola. Physical network models. *J Comput Biol.,* 11(2-3):243–62, 2004.

28. S. Yedidia, WT. Freeman, and Y. Weiss. Constructing free energy approximations and generalized belief propagation algorithms. Technical Report TR-2004-040, Mitsubishi electric resaerch laboratories, 2004.

Pairwise Local Alignment of Protein Interaction Networks Guided by Models of Evolution*

Mehmet Koyutürk, Ananth Grama, and Wojciech Szpankowski

Dept. of Computer Sciences, Purdue University, West Lafayette, IN 47907
{koyuturk, ayg, spa}@cs.purdue.edu

Abstract. With ever increasing amount of available data on protein-protein interaction (PPI) networks and research revealing that these networks evolve at a modular level, discovery of conserved patterns in these networks becomes an important problem. Recent algorithms on aligning PPI networks target simplified structures such as conserved pathways to render these problems computationally tractable. However, since conserved structures that are parts of functional modules and protein complexes generally correspond to dense subnets of the network, algorithms that are able to extract conserved patterns in terms of general graphs are necessary. With this motivation, we focus here on discovering protein sets that induce subnets that are highly conserved in the interactome of a pair of species. For this purpose, we develop a framework that formally defines the pairwise local alignment problem for PPI networks, models the problem as a graph optimization problem, and presents fast algorithms for this problem. In order to capture the underlying biological processes correctly, we base our framework on duplication/divergence models that focus on understanding the evolution of PPI networks. Experimental results from an implementation of the proposed framework show that our algorithm is able to discover conserved interaction patterns very effectively (in terms of accuracies and computational cost). While we focus on pairwise local alignment of PPI networks in this paper, the proposed algorithm can be easily adapted to finding matches for a subnet query in a database of PPI networks.

1   Introduction

Increasing availability of experimental data relating to biological sequences, coupled with efficient tools such as BLAST and CLUSTAL have contributed to fundamental understanding of a variety of biological processes [1, 2]. These tools are used for discovering common subsequences and motifs, which convey functional, structural, and evolutionary information. Recent developments in molecular biology have resulted in a new generation of experimental data that bear relationships and interactions between biomolecules [3]. An important class of molecular

---

* This research was supported in part by NIH Grant R01 GM068959-01 and NSF Grant CCR-0208709.

S. Miyano et al. (Eds.): RECOMB 2005, LNBI 3500, pp. 48–65, 2005.

interaction data is in the form of protein-protein interaction (PPI) networks, which provide the experimental basis for understanding modular organization of cells, as well as useful information for predicting the biological function of individual proteins [4]. High throughput screening methods such as two-hybrid analysis [5], mass spectrometry [6], and TAP [7] provide large amounts of data on these networks.

As revealed by recent studies, PPI networks evolve at a modular level [8] and consequently, understanding of conserved substructures through alignment of these networks can provide basic insights into a variety of biochemical processes. However, although vast amounts of high-quality data is becoming available, efficient network analysis counterparts to BLAST and CLUSTAL are not readily available for such abstractions. As is the case with sequences, key problems on graphs derived from biomolecular interactions include aligning multiple graphs [9], finding frequently occurring subgraphs in a collection of graphs [10], discovering highly conserved subgraphs in a pair of graphs, and finding good matches for a subgraph in a database of graphs [11]. In this paper, we specifically focus on discovering highly conserved subnets in a pair of PPI networks. With the expectation that conserved subnets will be parts of complexes and modules, we base our model on the discovery of two subsets of proteins from each PPI network such that the induced subnets are highly conserved.

Based on the understanding of the structure of PPI networks that are available for several species, theoretical models that focus on understanding the evolution of protein interactions have been developed. Among these, the duplication/divergence model has been shown to be successful in explaining the power-law nature of PPI networks [12]. In order to capture the underlying biological processes correctly, we base our framework on duplication/divergence models through definition of duplications, matches, and mismatches in a graph-theoretic framework. We then reduce the resulting alignment problem to a graph optimization problem and propose efficient heuristics to solve this problem. Experimental results based on an implementation of our framework show that the proposed algorithm is able to discover conserved interaction patterns very effectively. The proposed algorithm can be also adapted to finding matches for a subnet query in a database of PPI networks.

2   Related Work

As the amount of cell signaling data increases rapidly, there have been various efforts aimed at developing methods for comparative network analysis. In a relatively early study, Dandekar et al. [13] comprehensively align glycolysis metabolic pathways through comparison of biochemical data, analysis of elementary modes, and comparative genome analysis, identifying iso-enzymes, several potential pharmacological targets, and organism-specific adaptations. While such efforts demonstrate the potential of interaction alignment in understanding cellular processes, these analyses are largely manual, motivating the need for automated alignment tools.

As partially complete interactomes of several species become available, researchers have explored the problem of identifying conserved topological motifs in different species [8, 14]. These studies reveal that many topological motifs are significantly conserved within and across species and proteins that are organized in cohesive patterns tend to be conserved to a higher degree. A publicly available tool, PathBLAST, adopts the ideas in sequence alignment to PPI networks to discover conserved protein pathways across species [11]. By restricting the alignment to pathways, *i.e.*, linear chains of interacting proteins, this algorithm renders the alignment problem tractable, while preserving the biological implication of discovered patterns.

Since the local alignment of PPI networks for patterns in the form of general graphs leads to computationally intractable problems, tools based on simplified models are generally useful. However, as functional modules and protein complexes are likely to be conserved across species [8], algorithms for aligning general graphs are required for understanding conservation of such functional units. In a recent study, Sharan et al. [15] have proposed probabilistic models and algorithms for identifying conserved modules and complexes through cross-species network comparison. Similar to their approach, we develop a framework for aligning PPI networks to discover subsets of proteins in each species such that the subgraphs induced by these sets are highly conserved. In contrast to existing methods, our framework relies on theoretical models that focus on understanding the evolution of protein interaction networks.

3   Theoretical Models for Evolution of PPI Networks

There have been a number of studies aimed at understanding the general structure of PPI networks. It has been shown that these networks are power-law graphs, *i.e.*, the relative frequency of proteins that interact with k proteins is proportional to $k^{-\gamma}$, where γ is a network-specific parameter [16]. In order to explain this power-law nature, Barábasi and Albert have proposed [16] a network growth model based on preferential attachment, which is able to generate networks with degree distribution similar to PPI networks. According to this model, networks expand continuously by addition of new nodes and these new nodes prefer to attach to well-connected nodes when joining the network. Observing that older proteins are better connected, Eisenberg and Levanon [17] explain the evolutionary mechanisms behind such preference by the strength of selective pressure on maintaining connectivity of strongly connected proteins and creating proteins to interact with them. Furthermore, in a relevant study, it is observed that the interactions between groups of proteins that are temporally close in the course of evolution are likely to be conserved, suggesting synergistic selection during network evolution [18].

A common model of evolution that explains preferential attachment and power-law nature of PPI networks is the duplication/divergence model that is based on gene duplications [12, 19, 20, 21]. According to this model, when a gene is duplicated in the genome, the node corresponding to the product of this gene is also du-

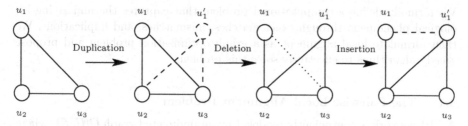

Fig. 1. Duplication/divergence model for evolution of PPI networks. Starting with three interactions between three proteins, protein u_1 is duplicated to add u_1' into the network together with its interactions (dashed circle and lines). Then, u_1 loses its interaction with u_3 (dotted line). Finally, an interaction between u_1 and u_1' is added to the network (dashed line)

plicated together with its interactions. An example of protein duplication is shown in Figure 1. A protein loses many aspects of its functions rapidly after being duplicated. This translates into divergence of duplicated (paralogous) proteins in the interactome through deletion and insertion of interactions. Deletion of an interaction in a PPI network implies the elimination of an existing interaction between two proteins due to structural and/or functional changes. Similarly, insertion of an interaction into a PPI network implies the emergence of a new interaction between two non-interacting proteins, caused by mutations that change protein surfaces. Examples of insertion and deletion of interactions are also illustrated in Figure 1. If a deletion or insertion is related to a recently duplicated protein, it is said to be correlated; otherwise, it is uncorrelated [19]. Since newly duplicated proteins are more tolerant to interaction loss because of redundancy, correlated deletions are generally more probable than insertions and uncorrelated deletions [12]. Since the elimination of interactions is related to sequence-level mutations, one can expect a positive correlation between similarity of interaction profiles and sequence similarity for paralogous proteins [20]. It is also theoretically shown that network growth models based on node duplications generate power-law distributions [22].

In order to accurately identify and interpret conservation of interactions, complexes, and modules across species, we base our framework for the local alignment of PPI networks on duplication/divergence models. While searching for highly conserved groups of interactions, we evaluate mismatched interactions and paralogous proteins in light of the duplication/divergence model. Introducing the concepts of match (conservation), mismatch (emergence or elimination) and duplication, which are in accordance with widely accepted models of evolution, we are able to discover alignments that also allow speculation about the structure of the network in the common ancestor.

4   Pairwise Local Alignment of PPI Networks

In light of the theoretical models of evolution of PPI networks, we develop a generic framework for the comparison of PPI networks in two different species.

We formally define a computational problem that captures the underlying biological phenomena through exact matches, mismatches, and duplications. We then formulate local alignment as a graph optimization problem and propose greedy algorithms to effectively solve this problem.

4.1   The Pairwise Local Alignment Problem

A PPI network is conveniently modeled by an undirected graph $G(U, E)$, where U denotes the set of proteins and $uu' \in E$ denotes an interaction between proteins $u \in U$ and $u' \in U$. For pairwise alignment of PPI networks, we are given two PPI networks belonging to two different species, denoted by $G(U, E)$ and $H(V, F)$. The homology between a pair of proteins is quantified by a similarity measure that is defined as a function $S : (U \cup V) \times (U \cup V) \to \Re$. For any $u, v \in U \cup V$, $S(u, v)$ measures the degree of confidence in u and v being orthologous if they belong to different species and paralogous if they belong to the same species. We assume that similarity scores are non-negative, where $S(u, v) = 0$ indicates that u and v cannot be considered as potential orthologs or paralogs. In this respect, S is expected to be highly sparse, i.e., each protein is expected to have only a few potential orthologs or paralogs. We discuss the reliability of possible choices for assessing protein similarity in detail in Section 4.4.

For PPI networks $G(U, E)$ and $H(V, F)$, a *protein subset pair* $P = \{\tilde{U}, \tilde{V}\}$ is defined as a pair of protein subsets $\tilde{U} \subseteq U$ and $\tilde{V} \subseteq V$. Any protein subset pair P induces a local alignment $\mathcal{A}(G, H, S, P) = \{\mathcal{M}, \mathcal{N}, \mathcal{D}\}$ of G and H with respect to S, characterized by a set of duplications $\mathcal{D}$, a set of matches $\mathcal{M}$, and a set of mismatches $\mathcal{N}$. The biological analog of a *duplication* is the duplication of a gene in the course of evolution. Each duplication is associated with a penalty, since duplicated proteins tend to diverge in terms of their interaction profiles in the long term [20]. A *match* corresponds to a conserved interaction between two orthologous protein pairs, which is rewarded by a match score that reflects our confidence in both protein pairs being orthologous. A *mismatch*, on the other hand, is the lack of an interaction in the PPI network of one of the species between a pair of proteins whose orthologs interact in the other species. A mismatch may correspond to the emergence (insertion) of a new interaction or the elimination (deletion) of a previously existing interaction in one of the species after the split, or an experimental error. Thus, mismatches are also penalized to account for the divergence from the common ancestor. We provide formal definitions for these three concepts to construct a basis for the formulation of local alignment as an optimization problem.

Definition 1. Local Alignment of PPI networks. *Given protein interaction networks $G(U, E)$, $H(V, F)$, and a pairwise similarity function S defined over the union of their protein sets $U \cup V$, any protein subset pair $P = (\tilde{U}, \tilde{V})$ induces a local alignment $\mathcal{A}(G, V, S, P) = \{\mathcal{M}, \mathcal{N}, \mathcal{D}\}$, where*

$$\mathcal{M} = \{u, u' \in \tilde{U}, v, v' \in \tilde{V} : S(u, v) > 0, S(u', v') > 0, uu' \in E, vv' \in F\} \quad (1)$$

$$\mathcal{N} = \{u, u' \in \tilde{U}, v, v' \in \tilde{V} : S(u, v) > 0, S(u', v') > 0, uu' \in E, vv' \notin F\}$$
$$\cup \{u, u' \in \tilde{U}, v, v' \in \tilde{V} : S(u, v) > 0, S(u', v') > 0, uu' \notin E, vv' \in F\} \quad (2)$$

$$\mathcal{D} = \{u, u' \in \tilde{U} : S(u, u') > 0\} \cup \{v, v' \in \tilde{V} : S(v, v') > 0\} \quad (3)$$

Each match $M \in \mathcal{M}$ is associated with a score $\mu(M)$. Each mismatch $N \in \mathcal{N}$ and each duplication $D \in \mathcal{D}$ are associated with penalties $\nu(N)$ and $\delta(D)$, respectively.

The score of alignment $\mathcal{A}(G, H, S, P) = \{\mathcal{M}, \mathcal{N}, \mathcal{D}\}$ is defined as:

$$\sigma(\mathcal{A}) = \sum_{M \in \mathcal{M}} \mu(M) - \sum_{N \in \mathcal{N}} \nu(N) - \sum_{D \in \mathcal{D}} \delta(D). \quad (4)$$

We aim to find local alignments with locally maximal score (drawing an analogy to sequence alignment [23], *high-scoring subgraph pairs*). This definition of the local alignment problem provides a general framework for the comparison of PPI networks, without explicitly formulating match scores, mismatch, and duplication penalties. These functions can be selected and their relative contributions can be tuned based on theoretical models and experimental observations to effectively synchronize with the underlying evolutionary process. Clearly, an appropriate basis for deriving these functions is the similarity score function S. We discuss possible choices for scoring functions in detail in Section 4.4.

A sample instance of the pairwise local alignment problem is shown in Figure 2(a). Consider the alignment induced by the protein subset pair $\tilde{U} = \{u_1, u_2, u_3, u_4\}$ and $\tilde{V} = \{v_1, v_2, v_3\}$, shown in Figure 2(b). The only duplication in this

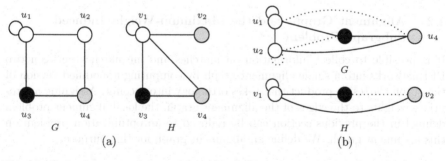

Fig. 2. (a) An instance of the pairwise local alignment problem. The proteins that have non-zero similarity scores (*i.e.*, are potentially orthologous), are colored the same. Note that S does not necessarily induce a disjoint grouping of proteins in practice. (b) A local alignment induced by the protein subset pair $\{u_1, u_2, u_3, u_4\}$ and $\{v_1, v_2, v_3\}$. Ortholog and paralog proteins are vertically aligned. Existing interactions are shown by solid lines, missing interactions that have an existing ortholog counterpart are shown by dotted lines. Solid interactions between two aligned proteins in separate species correspond to a match, one solid one dotted interaction between two aligned proteins in separate species correspond to a mismatch. Proteins in the same species that are on the same vertical line correspond to duplications

alignment is (u_1, u_2). If this alignment is chosen to be a "good" one, then, based on the existence of this duplication in the alignment, if $S(u_2, v_1) < S(u_1, v_1)$, we can speculate that u_1 and v_1 have evolved from the same gene in the common ancestor, while u_2 is an in-paralog that emerged from duplication of u_1 after split. The match set consists of interaction pairs (u_1u_1, v_1v_1), (u_1u_2, v_1v_1), (u_1u_3, v_1v_3), and (u_2u_4, v_1v_2). Observe that v_1 is mapped to both u_1 and u_2 in the context of different interactions. This is associated with the functional divergence of u_1 and u_2 after duplication. Moreover, the self-interaction of v_2 in H is mapped to an interaction between paralogous proteins in G. The mismatch set is composed of (u_1u_4, v_1v_2), (u_2u_2, v_1v_1), (u_2u_3, v_1v_3), and (u_3u_4, v_3v_2). The interaction u_3u_4 in G is left unmatched by this alignment, since the only possible pair of proteins in $\tilde{V}$ that are orthologous to these two proteins are v_3 and v_2, which do not interact in H. One conclusion that can be derived from this alignment is the elimination or emergence of this interaction in one of the species after the split. The indirect path between v_3 and v_2 through v_1 may also serve as a basis for the tolerability of the loss of this interaction. We can also simply attribute this observation to experimental noise. However, if we include v_4 in $\tilde{V}$ as well, then the induced alignment is able to match u_3u_4 and v_3v_4. This will strengthen the probability that this interaction existed in the common ancestor. However, v_4 comes at the price of another duplication since it is paralogous to v_2. This example illustrates the challenge of correctly matching proteins to their orthologs in order to reveal the maximum amount of reliable information about the conservation of interaction patterns. Our model translates this problem into a trade-off between mismatches and duplications, favoring selection of duplicate proteins that have not quite diverged in the alignment.

4.2   Alignment Graphs and the Maximum-Weight Induced Subgraph Problem

It is possible to collect information on matches and mismatches between two PPI networks into a single alignment graph by computing a modified version of the graph Cartesian product that takes orthology into account. Assigning appropriate weights to the edges of the alignment graph, the local alignment problem defined in the previous section can be reduced to an optimization problem on this alignment graph. We define an alignment graph for this purpose.

Definition 2. Alignment Graph. *For a pair of PPI networks $G(U, E)$, $H(V, F)$, and protein similarity function S, the corresponding weighted alignment graph $\mathbf{G}(\mathbf{V}, \mathbf{E})$ is computed as follows:*

$$\mathbf{V} = \{\mathbf{v} = \{u, v\} : u \in U, v \in V \text{ and } S(u, v) > 0\}. \tag{5}$$

In other words, we have a node in the alignment graph for each pair of ortholog proteins. Each edge $\mathbf{vv'} \in \mathbf{E}$, where $\mathbf{v} = \{u, v\}$ and $\mathbf{v'} = \{u', v'\}$, is assigned weight

$$w(\mathbf{vv'}) = \mu(uu', vv') - \nu(uu', vv') - \delta(u, u') - \delta(v, v'). \tag{6}$$

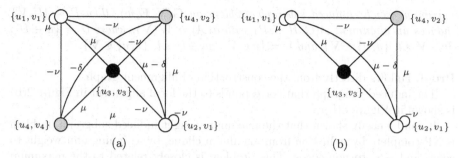

Fig. 3. (a) Alignment graph corresponding to the instance of Fig. 2(a). Note that match scores, mismatch and duplication penalties are functions of incident nodes, which is not explicitly shown in the figure for simplicity. (b) Subgraph induced by node set $\tilde{\mathbf{V}} = \{\{u_1, v_1\}, \{u_2, v_1\}, \{u_3, v_3\}, \{u_4, v_2\}\}$, which corresponds to the alignment shown in Fig. 2(b)

Here, $\mu(uu', vv') = 0$ if $(uu', vv') \notin \mathcal{M}$, and similarly for mismatch and duplication penalties.

Consider the PPI networks in Figure 2(a). To construct the corresponding alignment graph, we first compute the product of these two PPI networks to obtain five nodes that correspond to five ortholog protein pairs. We then put an edge between two nodes of this graph if the corresponding proteins interact in both networks (*match edge*), interact in only one of the networks (*mismatch edge*), or at least one of them is paralogous (*duplication edge*), resulting in the alignment graph of Figure 3(a). Note that the weights assigned to these edges, which are shown in the figure, are not constant, but are functions of their incident nodes. Observe that the edge between $\{u_1, v_1\}$ and $\{u_2, v_1\}$ acts a match and duplication edge at the same time, allowing analysis of the conservation of self-interactions of duplicated proteins.

The weighted alignment graph is conceptually similar to the orthology graph of Sharan et al. [15]. However, instead of accounting for similarity of proteins through node weights, we encapsulate the orthology information in edge weights, which also allows consideration of duplications effectively. This construction of the alignment graph allows us to formulate the alignment problem as a graph optimization problem defined below.

Definition 3. Maximum Weight Induced Subgraph Problem. *Given graph* $\mathbf{G}(\mathbf{V}, \mathbf{E})$ *and a constant* ϵ, *find a subset of nodes,* $\tilde{\mathbf{V}} \in \mathbf{V}$ *such that the sum of the weights of the edges in the subgraph induced by* $\tilde{\mathbf{V}}$ *is at least* ϵ, *i.e.,* $W(\tilde{\mathbf{V}}) = \sum_{\mathbf{v}, \mathbf{v}' \in \tilde{\mathbf{V}}} w(\mathbf{v}\mathbf{v}') \geq \epsilon$.

Not surprisingly, this problem is equivalent to the local alignment of PPI networks defined in the previous section, as formally stated in the following theorem:

Theorem 1. *Given PPI networks* G, H, *and a protein similarity function* S, *let* $\mathbf{G}(\mathbf{V}, \mathbf{E}, w)$ *be the corresponding alignment graph. If* $\tilde{\mathbf{V}}$ *is a solution to the* *maximum weight induced subgraph problem on* $\mathbf{G}(\mathbf{V}, \mathbf{E}, w)$, *then* $P = \{\tilde{U}, \tilde{V}\}$ *induces an alignment* $\mathcal{A}(G, H, S, P)$ *with* $\sigma(\mathcal{A}) = W(\hat{\mathbf{V}})$, *where* $\tilde{U} = \{u \in U : \exists v \in V \ s.t. \ \{u, v\} \in \hat{\mathbf{V}}\}$ *and* $\tilde{V} = \{v \in V : \exists u \in U \ s.t. \ \{u, v\} \in \hat{\mathbf{V}}\}$.

Proof. Follows directly from the construction of alignment graph.

The induced subgraph that corresponds to the local alignment in Figure 2(b) is shown in Figure 3(b).

It can be easily shown that the maximum-weight induced subgraph problem is NP-complete by reduction from maximum clique, by assigning unit weight to edges and $-\infty$ to non-edges. This problem is closely related to the maximum edge subgraph [24] and maximum dispersion problems [25] that are also NP-complete. Although the positive weight restriction on these problems limits the application of existing algorithms to the maximum weight induced subgraph problem, the nature of the conservation of PPI networks makes a simple greedy heuristic quite effective for the local alignment of PPI networks.

4.3   A Greedy Heuristic for Local Alignment of Protein Interaction Networks

In terms of protein interactions, functional modules and protein complexes are densely connected while being separable from other modules, *i.e.*, a protein in a particular module interacts with most proteins in the same module either directly or through a common module hub, while it is only loosely connected to the rest of the network [26]. Since analysis of conserved motifs reveals that proteins in highly connected motifs are more likely to be conserved suggesting that such dense motifs are parts of functional modules [8], high-scoring local alignments are likely to correspond to functional modules. Therefore, in the alignment graph, we can expect that proteins that belong to a conserved module will induce heavy subgraphs, while being loosely connected to other parts of the graph. This observation leads to a greedy algorithm that can be expected to work well for the solution of the maximum weight induced subgraph problem on the alignment graph of two PPI networks. Indeed, similar approaches are shown to perform well in discovering conserved or dense subnets in PPI networks [15, 27]. By seeding a growing subgraph with a protein that has a large number of conserved interactions and small number of mismatched interactions (*i.e.*, a *conserved hub*) and adding proteins that share conserved interactions with this graph one by one, it is possible to discover a group of proteins with a set of dense interactions that are conserved, likely being part of a functional module.

A sketch of the greedy algorithm for finding a single conserved subgraph on the alignment graph is shown in Figure 4. This algorithm grows a subgraph, which is of locally maximal total weight. To find all non-redundant "good" alignments, we start with the entire alignment graph and find a maximal subgraph. If this subgraph is statistically significant according to the reference model described in Section 4.5, we record the alignment that corresponds to this subgraph and mark its nodes. We repeat this process by allowing only unmarked nodes to be chosen as seed until no subgraph with positive weight can be found. Re-

procedure GREEDYMAWISH(**G**)
> **Input G**(**V**, **E**, w): Alignment graph
> **Output** $\tilde{\mathbf{V}}$: Subset of selected nodes
> $g(\mathbf{v})$: Gain of adding **v** into $\tilde{\mathbf{V}}$
> W: Total subgraph weight
1 for each $\mathbf{v} \in \mathbf{V}$ do
2      $g(\mathbf{v}) \leftarrow w(\mathbf{vv})$
3      $w(\mathbf{v}) = \sum_{\mathbf{vv'} \in \mathbf{E}} w(\mathbf{vv'})$
4 $\tilde{\mathbf{V}} \leftarrow \emptyset$, $W \leftarrow 0$
5 $\tilde{\mathbf{v}} \leftarrow \text{argmax}_{\mathbf{v} \in \mathbf{V}} w(\mathbf{v})$
6 repeat
7      $\tilde{\mathbf{V}} \leftarrow \tilde{\mathbf{V}} \cup \{\tilde{\mathbf{v}}\}$
8      $W \leftarrow W + g(\tilde{\mathbf{v}})$
9   for each $\mathbf{v} \in (\mathbf{V} \setminus \tilde{\mathbf{V}})$ s.t. $\tilde{\mathbf{v}}\mathbf{v} \in \mathbf{E}$ do
10        $g(\mathbf{v}) \leftarrow g(\mathbf{v}) + w(\tilde{\mathbf{v}}\mathbf{v})$
11      $\tilde{\mathbf{v}} \leftarrow \text{argmax}_{\mathbf{v} \in (\mathbf{V} \setminus \tilde{\mathbf{V}})} g(\mathbf{v})$
12 until $g(\mathbf{v}) \leq 0$
13 if $W > 0$
14   then return $\tilde{\mathbf{V}}$
15   else return $\emptyset$

Fig. 4. Greedy algorithm for finding a set of nodes that induces a subgraph of maximal total weight on the alignment graph

stricting the seed to only non-aligned nodes avoids redundancy while allowing discovery of overlapping alignments. Finally, we rank all subgraphs based on their significance and report the corresponding alignments. A loose bound on the worst-case running time of this algorithm is $O(|\mathbf{V}||\mathbf{E}|)$, since each alignment takes $O(|\mathbf{E}|)$ time and each node can be the seed at most once. Assuming that the number of orthologs for each protein is bounded by a constant, the size of the alignment graph is linear in the total size of the input networks.

4.4 Selection of Model Components

In order for the discovered PPI network alignments to be biologically meaningful, selection of the underlying similarity function and the models for scoring and penalizing matches, mismatches, and duplications is crucial, as in the case of sequences.

Similarity Function. Since proteins that are involved in a common functional module, or more generally, proteins that interact with each other, show local sequence similarities, care must be taken while employing pairwise sequence alignment as a measure of potential orthology between proteins. Furthermore, while aligning two PPI networks and interpreting the alignment, only duplications that correspond to proteins that are duplicated after the split of species are of interest. Such protein pairs are called in-paralogs, while the others are called out-paralogs [28]. Unfortunately, distinguishing between in-paralogs and out-paralogs is not trivial. Therefore, we assign similarity scores to protein pairs conservatively by detecting orthologs and in-paralogs using a dedicated algorithm, INPARANOID [28], which is developed for finding disjoint ortholog clusters in two species. Each ortholog cluster discovered by this algorithm is characterized by two *main orthologs*, one from each species, and possibly several other in-paralogs from both species. The main orthologs are assigned a confidence value of 1.0, while the in-paralogs are assigned confidence scores based on their relative similarity to the main ortholog in their own species. We define the similarity between two proteins u and v as

$$S(u,v) = confidence(u) \times confidence(v). \tag{7}$$

This provides a normalized similarity function that takes values in the interval $[0,1]$ and quantifies the confidence in the two proteins being orthologous or paralogous.

Scores and Penalties. *Match score.* A match is scored positively in an alignment to reward a conserved interaction. Therefore, the score represents the similarity between the two interactions that are matched. Since the degree of conservation in the two ortholog protein pairs involved in the matched interactions need not be the same, it is appropriate to conservatively assign the minimum of the similarities at the two ends of the matching interaction to obtain:

$$\mu(uu',vv') = \bar{\mu}S(uu',vv'), \tag{8}$$

where $S(uu',vv') = \min\{S(u,v), S(u',v')\}$ and $\bar{\mu}$ is a pre-determined parameter specifying the relative weight of a match in the total alignment score. While we use this definition of $S(uu',vv')$ in our implementation, $S(u,v) \times S(u',v')$ provides a reliable measure of similarity between the two protein pairs as well.

Mismatch penalty. Similar to match score, mismatch penalty is defined as:

$$\nu(uu',vv') = \bar{\nu}S(uu',vv'), \tag{9}$$

where $\bar{\nu}$ is the relative weight of a mismatch. With this penalty function, each lost interaction of a duplicate protein is penalized to reflect the divergence of duplicate proteins.

Duplication penalty. Duplications are penalized to account for the divergence of the proteins after duplication. Sequence similarity provides a crude approximation to the age of duplication and likelihood of being paralogs [21]. Hence, duplication penalty is defined as:

$$\delta(u,u') = \bar{\delta}(d - S(u,u')), \tag{10}$$

where $\bar{\delta}$ is the relative weight of a duplication and $d \geq \max_{u,u' \in U} S(u,u')$ is a parameter that determines the extent of penalizing duplications. Considering the similarity function of (7), setting $d = 1.0$ results in no penalty for duplicates that are paralogous to the main ortholog with 100% confidence.

4.5   Statistical Significance

To evaluate the statistical significance of discovered high-scoring alignments, we compare them with a reference model generated by a random source. In the reference model, it is assumed that the interaction networks that belong to the two species are independent from each other as well as the protein sequences. To accurately capture the power-law nature of PPI networks, we assume that the interactions are generated randomly from a distribution characterized by a given degree sequence. The probability $q_{uu'}$ of observing an interaction between two proteins $u, u' \in U$ for the degree sequence derived from G can be estimated by a Monte Carlo algorithm that repeatedly swaps the incident nodes of randomly chosen edges [15]. On the other hand, we assume that the sequences are generated by a memoryless source, such that $u \in U$ and $v \in V$ are orthologous with probability p. Similarly, $u, u' \in U$ and $v, v' \in V$ are paralogous with probability p_U and p_V, respectively. Since the similarity function of (7) provides a measure of the probability of true homology between a given pair of proteins, we estimate p by $\frac{\sum_{u \in U, v \in V} S(u,v)}{|U||V|}$. Hence, $E[S(u,v)] = p$ for $u \in U, v \in V$. The probabilities of paralogy are estimated similarly.

In the reference model, the expected value of the score of an alignment induced by $\tilde{\mathbf{V}} \subseteq \mathbf{V}$ is

$$E[W(\tilde{\mathbf{V}})] = \sum_{\mathbf{v},\mathbf{v}' \in \tilde{\mathbf{V}}} E[w(\mathbf{v}\mathbf{v}')],$$

where

$$E[w(\mathbf{v}\mathbf{v}')] = \bar{\mu}p^2 q_{uu'}q_{vv'} - \bar{\nu}p^2(q_{uu'}(1 - q_{vv'}) + (1 - q_{uu'})q_{vv'}) \\ - \bar{\delta}(p_U(1 - p_U) + p_V(1 - p_V)) \tag{11}$$

is the expected weight of an edge in the alignment graph. Moreover, with the simplifying assumption of independence between interactions, we have $Var[W(\tilde{\mathbf{V}})] = \sum_{\mathbf{v},\mathbf{v}' \in \tilde{\mathbf{V}}} Var[w(\mathbf{v}\mathbf{v}')]$, enabling us to compute the z-score to evaluate the statistical significance of each discovered high-scoring alignment, under the normal approximation that we assume to hold.

4.6   Extensions to the Model

Accounting for Experimental Error. PPI networks obtained from high-throughput screening are prone to errors in terms of both false negatives and positives [4]. While the proposed framework can be used to detect experimental errors through cross-species comparison to a certain extent, experimental noise can also degrade the performance of the alignment algorithm. In other words, mismatches should be penalized for lost interactions during evolution, not for experimental false negatives. To account for such errors while analyzing interaction networks, several methods have been developed to quantify the likelihood of an interaction or complex co-membership between proteins [29, 30, 31]. Given the prior probability distribution for protein interactions and set of observed interactions, these methods compute the posterior probability of interactions based on Bayesian models. Hence, PPI networks can be modeled by weighted graphs to account for experimental error more accurately.

While the network alignment framework introduced in Section 4.1 assumes that interactions are represented by unweighted edges, it can be easily generalized to a weighted graph model as follows. Assuming that weight ϖ_{uv} represents the posterior probability of interaction between u and v, we can define match score and mismatch penalty in terms of their expected values derived from these posterior probabilities. Therefore, for any $u, u' \in U$ and $v, v' \in V$, we have

$$\mu(uu', vv') = \bar{\mu}S(uu', vv')\varpi_{uu'}\varpi_{vv'} \tag{12}$$

$$\nu(uu', vv') = \bar{\nu}S(uu', vv')(\varpi_{uu'}(1 - \varpi_{vv'}) + (1 - \varpi_{uu'})\varpi_{vv'}). \tag{13}$$

Note that match and mismatch sets are not necessarily disjoint here in contrast to the unweighted graph model, which is indeed a special case of this model.

Tuning Model Components and Parameters. *Sequence similarity.* A more flexible approach for assessing similarity between proteins is direct employment of sequence alignment scores. In PathBLAST [32], the similarity between two proteins is defined as the log-likelihood ratio for homology, *i.e.*, $S(u, v) = \log(p(u, v)/\bar{p})$, where $p(u, v)$ is the probability of true homology between u and v given the BLAST E value of their alignment and $\bar{p}$ is the expected value of p over all proteins in the PPI networks being aligned. To avoid consideration of similarities that do not result from orthology, it is necessary to set cut-off values on the significance of alignments [32, 20].

Shortest-path mismatch model. Since proteins that are linked by a short alternative path are more likely to tolerate losing their interaction, mismatch penalty can be improved using a shortest-path mismatch model, defined as:

$$\nu(uu', vv') = \bar{\nu}S(uu', vv')(\max\{\Delta(u, u'), \Delta(v, v')\} - 1), \tag{14}$$

where $\Delta(u, u')$ is the length of the shortest path between proteins u and u'. While this model is likely to improve the alignment algorithm, it is computationally expensive since it requires solution of the all pairs shortest path problem on both PPI networks.

Linear duplication model. The alignment graph model enforces each duplicate pair in an alignment to be penalized. For example, if an alignment contains n paralogous proteins in one species, $\binom{n}{2}$ duplications are penalized to account for each duplicate pair. However, in the evolutionary process, each paralogous protein is the result of a single duplication, *i.e.*, n paralogous proteins are created in only $n - 1$ duplications. Therefore, we refer to the current model as *quadratic duplication model*, since the number of penalties is a quadratic function of number of duplications. While this might be desirable as being more restrictive on duplications, to be more consistent with the underlying biological processes, it can be replaced by a *linear duplication model*. In this model, each duplicate protein is penalized only once, based on its similarity with the paralog that is most similar to itself.

5   Experimental Results

In this section, we present local alignment results to illustrate the effectiveness of the proposed framework and the underlying algorithm on interaction data retrieved from the DIP protein interaction database [33]. We align the PPI networks of two mammalians that are available in the database; *Homo sapiens* (Hsapi) and *Mus musculus* (Mmusc). As of October 2004, the Hsapi PPI network contains 1369 interactions among 1065 proteins while Mmusc PPI network contains 286 interactions among 329 proteins. Running INPARANOID on this set of 1351 proteins, we discover 237 ortholog clusters. Based on the similarity function induced by these clusters, we construct an alignment graph that consists of 273 nodes and 1233 edges. The alignment graph contains 305 matched interactions, 205 mismatched interactions in Hsapi, 149 mismatched interactions in Mmusc, 536 duplications in Hsapi, and 384 duplications in Mmusc. We then compute local alignments using the algorithm of Section 4.3 on this graph. By trying alternate settings for the relative weights of match score and mismatch, duplication penalties, we identify 54 non-redundant alignments, 15 of which contain at least 3 proteins on each network. Note that construction of alignment graph and discovery of local alignments on this graph takes only a few milliseconds.

A conserved subnet of DNA-dependent transcription regulation that is found to be statistically significant (z-score=18.1) is shown in Figure 5. The subnet is composed of three major common functional groups, namely transcription factors and coactivators PO22, PO31, OCT1, TIF2, OBF1, steroid hormone receptors GCR, ANDR, ESR1, PRGR, GRIP1, THB1, and high mobility proteins HMG1 and HMG2. Indeed, it is known that HMG1 and HMG2 are co-regulatory proteins that increase the DNA binding and transcriptional activity of the steroid hormone class of receptors in mammalian cells [34]. All proteins in this subnet are localized in nucleus, with mobility proteins particularly localizing in condensed chromosome. This subnet contains 17 matching interactions between 15 proteins. Two interactions of TIF2 (transcriptional intermediary factor 2) that exist in human are missing in mouse. If we increase the relative weight of mismatch penalties in the alignment score, the alignment does not contain TIF2 any more, providing a perfect match of 16 interactions.

The subnet that is part of transforming growth factor beta receptor signaling pathway, which is significantly conserved (z-score=19.9) in human and mouse PPI networks is shown in Figure 6. This subnet contains 8 matching

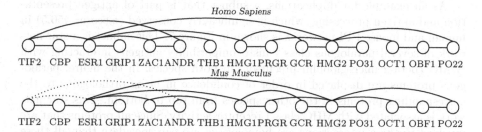

Fig. 5. A conserved subnet that is part of DNA-dependent transcription regulation in human and mouse PPI networks. Ortholog proteins are vertically aligned. Existing interactions are shown by solid edges, missing interactions that have an existing orthologous counterpart in the other species are shown by dotted edges

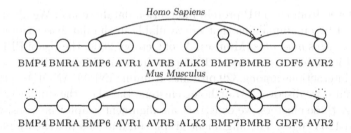

Fig. 6. A conserved subnet that is part of transforming growth factor beta receptor signaling pathway in human and mouse PPI networks

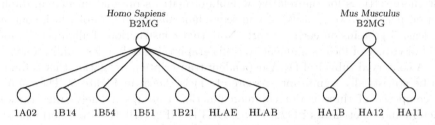

Fig. 7. A conserved subnet that is part of antigen presentation and antigen processing in human and mouse PPI networks. Homologous proteins are horizontally aligned. Paralogous proteins in a species are shown from left to right in the order of confidence in being orthologous to the respective proteins in the other species

interactions among 10 proteins. It is composed of two separate subnets that are connected through the interaction of their hubs, namely BMP6 (bone morphogenetic protein 6 precursor) and BMRB (activin receptor-like kinase 6 precursor). All proteins in this subnet have the common function of transforming growth factor beta receptor activity and are localized in the membrane. Note that self-interactions of three proteins in this subnet that exist in human PPI network are missing in mouse and one self-interaction that exists in mouse is missing in human.

As an example for duplications, a subnet that is part of antigen presentation and antigen processing, which is significantly conserved (z-score=456.5) in human and mouse PPI networks is shown in Figure 7. This subnet is a star network of several paralogous class I histocompatibility antigens interacting with B2MG (beta-2 microglobulin precursor) in both species. In the figure, paralogous proteins are displayed in order of confidence in being orthologous to the corresponding proteins in the other species from top to bottom. This star network is associated with MHC class I receptor activity. Since all proteins that are involved in these interactions are homologous, we can speculate that all these interactions have evolved from a single common interaction. Note that such patterns are found only with the help of the duplication concept in the alignment model. Neither a pathway alignment algorithm, nor an algorithm that tries to match each protein with exactly one ortholog in the other species will be able to detect such conserved patterns. Indeed, this subnet can only be discovered when the duplication coefficient is small ($\bar{\delta} \leq 0.12\bar{\mu}$).

6 Concluding Remarks and Ongoing Work

This paper presents a framework for local alignment of protein interaction networks that is guided by theoretical models of evolution of these networks. The model is based on discovering sets of proteins that induce conserved subnets with the expectation that these proteins will constitute a part of protein complexes or functional models, which are expected to be conserved together. A preliminary implementation of the proposed algorithm reveals that this framework is quite successful in uncovering conserved substructures in protein interaction data.

We are currently working on a comprehensive implementation of the proposed framework that allows adaptation of several models for assessing protein similarities and scoring/penalizing matches, mismatches and duplications. Furthermore, we are working on a rigorous analysis of distribution of the alignment score, which will enable more reliable assessment of statistical significance. Once these enhancements are completed, the proposed framework will be established as a tool for pairwise alignment of PPI networks, that will be publicly available through a web interface. The framework will also be generalized to the search of input queries in the form of subnets in a database of PPI networks. Using this tool researchers will be able to find conserved counterparts of newly discovered complexes or modules in several species.

Acknowledgments

The authors would like to thank Prof. Shankar Subramaniam of UCSD for many valuable discussions.

References

1. Altschul, S.F., Madden, T.L., Schffer, A.A., J. Zhang, Z.Z., Miller, W., Lipman, D.J.: Gapped BLAST and PSI-BLAST: a new generation of protein database search programs. Nuc. Acids Res. **25** (1997) 3389–3402
2. Thompson, J.D., Higgins, D.G., Gibson, T.J.: CLUSTAL-W: improving the sensitivity of progressive multiple sequence alignment through sequence weighting, position-specific gap penalties and weight matrix choice. Nuc. Acids Res. **22** (1994) 4673–4680
3. Hartwell, L.H., Hopfield, J.J., Leibler, S., Murray, A.W.: From molecular to modular cell biology. Nature **402** (1999) C47–C51
4. Titz, B., Schlesner, M., Uetz, P.: What do we learn from high-throughput protein interaction data? Exp. Rev. Prot. **1** (2004) 111–121
5. Ito, T., Chiba, T., Ozawa, R., Yoshida, M., Hattori, M., Sakaki, Y.: A comprehensive two-hybrid analysis to explore the yeast protein interactome. PNAS **98** (2001) 4569–4574

6. Ho, Y. et al.: Systematic identification of protein complexes in Saccharomyces cerevisae by mass spectrometry. Nature **415** (2002) 180–183
7. Gavin, A.C. et al.: Functional organization of the yeast proteome by systematic analysis of protein complexes. Nature **415** (2002) 141–147
8. Wuchty, S., Oltvai, Z.N., Barabási, A.L.: Evolutionary conservation of motif constituents in the yeast protein interaction network. Nature Gen. **35** (2003) 176–179
9. Tohsato, Y., Matsuda, H., Hashimoto, A.: A multiple alignment algorithm for metabolic pathway analysis using enzyme hierarchy. In: 8th Intl. Conf. Intel. Sys. Mol. Bio. (ISMB'00). (2000) 376–383
10. Koyutürk, M., Grama, A., Szpankowski, W.: An efficient algorithm for detecting frequent subgraphs in biological networks. In: Bioinformatics Suppl. 12th Intl. Conf. Intel. Sys. Mol. Bio. (ISMB'04). (2004) i200–i207
11. Kelley, B.P., Yuan, B., Lewitter, F., Sharan, R., Stockwell, B.R., Ideker, T.: Path-BLAST: a tool for aligment of protein interaction networks. Nuc. Acids Res. **32** (2004) W83–W88
12. Vázquez, A., Flammini, A., Maritan, A., Vespignani, A.: Modeling of protein interaction netwokrs. ComPlexUs **1** (2003) 38–44
13. Dandekar, T., Schuster, S., Snel, B., Huynen, M., Bork, P.: Pathway alignment: application to the comparative analysis of glycolytic enzymes. Biochem. J **343** (1999) 115–124
14. Lotem, E.Y., Sattath, S., Kashtan, N., Itzkovitz, S., Milo, R., Pinter, R.Y., Alon, U., Margalit, H.: Network motifs in integrated cellular networks of transcription-regulation and protein-protein interaction. PNAS **101** (2004) 5934–5939
15. Sharan, R., Ideker, T., Kelley, B.P., Shamir, R., Karp, R.M.: Identification of protein complexes by comparative analysis of yeast and bacterial protein interaction data. In: 8th Intl. Conf. Res. Comp. Mol. Bio. (RECOMB'04). (2004) 282–289
16. Barabási, A., Albert, R.: Emergence of scaling in random networks. Science **286** (1999) 509–512
17. Eisenberg, E., Levanon, Y.: Preferential attachment in the protein network evolution. Phys. Rev. Let. **91** (2003) 138701
18. Qin, H., Lu, H.H.S., Wu, W.B., Li, W.: Evolution of the yeast protein interaction network. PNAS **100** (2003) 12820–12824
19. Pastor-Satorras, R., Smith, E., Solé, R.V.: Evolving protein interaction networks through gene duplication. J Theo. Bio. **222** (2003) 199–210
20. Wagner, A.: The yeast protein interaction network evolves rapidly and contains few redundant duplicate genes. Mol. Bio. Evol. **18** (2001) 1283–1292
21. Wagner, A.: How the global structure of protein interaction networks evolves. Proc. R. Soc. Lond. Biol. Sci. **270** (2003) 457–466
22. Chung, F., Lu, L., Dewey, T.G., Galas, D.J.: Duplication models for biological networks. J Comp. Bio. **10** (2003) 677–687
23. Smith, T.F., Waterman, M.S.: Identification of common molecular subsequences. J Mol. Bio. **147** (1981) 195–197
24. Feige, U., Peleg, D., Kortsarz, G.: The dense k-subgraph problem. Algorithmica **29** (2001) 410–421
25. Hassin, R., Rubinstein, S., Tamir, A.: Approximation algorithms for maximum dispersion. Oper. Res. Let. **21** (1997) 133–137
26. Tornow, S., Mewes, H.W.: Functional modules by relating protein interaction networks and gene expression. Nuc. Acids Res. **31** (2003) 6283–6289
27. Bader, J.S.: Greedily building protein networks with confidence. Bioinformatics **19** (2003) 1869–1874

28. Remm, M., Storm, C.E.V., Sonnhammer, E.L.L.: Automatic clustering of orthologs and in-paralogs from pairwise species comparisons. J Mol. Bio. **314** (2001) 1041–1052

29. Jansen, R., Yu, H., Greenbaum, D., Kluger, Y., Krogan, N.J., Chung, S., Emili, A., Snyder, M., Greenblatt, J,F., Gerstein, M.: A Bayesian networks approach for predicting protein-protein interactions from genomic data. Science **302** (2003) 449–453

30. Ashtana, S., King, O.D., Gibbons, F.D., Roth, F.P.: Predicting protein complex membership using probabilistic network reliability. Genome Research **14** (2004) 1170–1175

31. Gilchrist, M.A., Salter, L.A., Wagner, A.: A statistical framework for combining and interpreting proteomic datasets. Bioinformatics **20** (2003) 689–700

32. Kelley, B.P., Sharan, R., Karp, R.M., Sittler, T., Root, D.E., Stockwell, B.R., Ideker, T.: Conserved pathways withing bacteria and yeast as revealed by global protein network alignment. PNAS **100** (2003) 11394–11399

33. Xenarios, I., Salwinski, L., Duan, X.J., Higney, P., Kim, S., Eisenberg, D.: DIP: The Database of Interacting Proteins. A research tool for studying cellular networks of protein interactions. Nuc. Acids Res. **30** (2002) 303–305

34. Boonyaratanakornkit, V. et al.: High-mobility group chromatin proteins 1 and 2 functionally interact with steroid hormone receptors to enhance their DNA binding in vitro and transcriptional activity in mammalian cells. Mol. Cell. Bio. **18** (1998) 4471–4488

Finding Novel Transcripts in High-Resolution Genome-Wide Microarray Data Using the GenRate Model

Brendan J. Frey[1], Quaid D. Morris[1,2],
Mark Robinson[1,2], and Timothy R. Hughes[2]

[1] Elec. and Comp. Eng., Univ. of Toronto,
Toronto ON M5S 3G4, Canada
http://www.psi.toronto.edu
[2] Banting and Best Dep. Med. Res., Univ. of Toronto,
Toronto ON M5G 1L6, Canada
http://hugheslab.med.utoronto.ca

Abstract. Genome-wide microarray designs containing millions to tens of millions of probes will soon become available for a variety of mammals, including mouse and human. These "tiling arrays" can potentially lead to significant advances in science and medicine, *e.g.*, by indicating new genes and alternative primary and secondary transcripts. While bottom-up pattern matching techniques (*e.g.*, hierarchical clustering) can be used to find gene structures in tiling data, we believe the many interacting hidden variables and complex noise patterns more naturally lead to an analysis based on generative models. We describe a generative model of tiling data and show how the iterative sum-product algorithm can be used to infer hybridization noise, probe sensitivity, new transcripts and alternative transcripts. We apply our method, called GenRate, to a new exon tiling data set from mouse chromosome 4 and show that it makes significantly more predictions than a previously described hierarchical clustering method at the same false positive rate. GenRate correctly predicts many known genes, and also predicts new gene structures. As new problems arise, additional hidden variables can be incorporated into the model in a principled fashion, so we believe that GenRate will prove to be a useful tool in the new era of genome-wide tiling microarray analysis.

1 Introduction

One of the most important current problems in molecular biology is the development of techniques for building libraries of genes and gene variants for organisms, and in particular higher mammals such as mice and humans. While an analysis of genomic nucleotide sequence data can be used to build such a library (c.f. [30]), it is the mRNA molecules that are transcribed from genomic DNA ("transcripts") that directly or indirectly constitute the library of functional elements. In fact, the many complex mechanisms that influence transcription of genomic DNA into mRNAs produces a set of functional transcripts that is

S. Miyano et al. (Eds.): RECOMB 2005, LNBI 3500, pp. 66–82, 2005.

much richer than can be currently explained by analyzing genomic DNA alone. This richness is due to many mechanisms including transcription of non-protein-coding mRNA molecules that are nonetheless functional (c.f. [2]), tissue-specific transcriptional activity, alternative transcription of genomic DNA (*e.g.*, alternative start/stop transcription sites), and alternative post-transcriptional splicing of mRNA molecules (c.f. [3, 5, 4, 9]). Instead of attempting to understand these variants by studying genomic DNA alone, microarrays can be used to directly study the rich library of functional transcriptional variants. Previously, we used microarrays to study subsets of variants, including non-protein-coding mRNAs [6] and alternative-splicing variants [8]. Here, we describe a technique called "GenRate", which applies the max-product algorithm in a graphical model to perform a genome-wide analysis of high-resolution (many probes "per gene") microarray data.

In 2001, Shoemaker *et al.* demonstrated for the first time how DNA microarrays can be used to validate and refine predicted transcripts in portions of human chromosome 22q, using 8,183 exon probes [21]. By "tiling" the genome with probes, patterns of expression can be used to discover expressed elements. In the past 3 years, the use of microarrays for the discovery of expressed elements in genomes has increased with improvements in density, flexibility, and accessibility of the technology. Two complementary tiling strategies have emerged. In the first, the genome is tiled using candidate elements (*e.g.*, exons, ORFs, genes, RNAs), each of which is identified computationally and is represented one or a few times on the array [19, 21, 11, 14]. In the second, the entire genome sequence is comprehensively tiled, *e.g.*, overlapping oligonucleotides encompassing both strands are printed on arrays, such that all possible expressed sequences are represented [21, 13, 20, 23, 14, 15, 31]. Genome-wide tiling data using both approaches is currently becoming available [32, 1].

The above tiling approaches, as well as independent analysis by other methods [17, 11] have indicated that a substantially higher proportion of genomes are expressed than are currently annotated. The most recent genome-wide mammalian survey by Bertone *et al.* is based on expression in a single tissue (liver) and claims to have found 10,595 novel transcribed sequences over and above genes detected by other methods. However, since microarray data is noisy and since probe sensitivity and cross-hybridization noise can vary tremendously from one probe to another (a factor of 40 is not unusual), it is quite difficult to control the false detection rate using only one tissue. Most protein-coding transcripts are composed of multiple exons and most mammalian genes vary in expression between tissues, so tissue-dependent co-regulation of probes that are nearby in the genome provides evidence of a transcriptional unit [21]. We will refer to such a co-regulated transcriptional unit as a "CoReg".

Microarrays do not inherently provide information regarding the length of the RNA or DNA molecules detected, nor do they inherently reveal whether features designed to detect adjacent features on the chromosome are in fact detecting the same transcript. mRNAs, which account for the largest proportion of transcribed sequence in a genome, present a particular challenge. mRNAs are composed only of spliced exons, often separated in the genome (and in the primary transcript) by thousands to tens of thousands of bases of intronic sequence. Each gene may have a variety of transcript variants, *e.g.*, due to alternative splicing [3, 4] and exons that are conserved across species (*e.g.* human and mouse) often undergo species-specific splicing [9]. Identifying the exons that comprise individual transcripts from genome- or exon-tiling data is not a trivial task, since falsely-predicted exons, overlapping features, transcript variants, and poor-quality measurements can confound assumptions based on simple correlation of magnitude or co-variation of expression.

Heuristics that group nearby probes using intensity of expression or co-regulation across experimental conditions can be used to approach this problem [21, 13, 23, 14, 32]. In [21], correlations between the expression patterns of nearby probes are used to merge probes into CoRegs. A merge step takes place if the correlation exceeds 0.5, but not if the number of non-merged probes between the two candidate probes is greater than 5. In [13], the density of the activity map of RNA transcription is used to verify putative exons. In [32], a single tissue is studied and two probes are merged if their intensities are in the top 90th percentile and if they are within 250nt of each other. In [14], principal components analysis (PCA) is first applied to probes within a window. Then, the distribution of PCA-based Mahalanobis distances of probes are compared with the distribution of distances for known intron probes, and each probe is merged into a CoReg if the distance of the probe to a selection of the PCA subspaces is low.

While the above techniques have been quite helpful in analyzing microarray data, an important disadvantage of the techniques is that they do not directly model various sources of noise and the noisy relationships between variables. For example, a highly-sensitive probe will indicate the presence of transcript, even if the true abundance is negligible. A poorly-designed probe will cross-hybridize to many other transcripts, again misleadingly indicating the presence of the transcript for which the probe was designed. By not optimizing a global scoring function derived from a model of the various processes, these techniques tend to make greedy local decisions that are not globally optimal. For example, while the assignment of a probe to a CoReg may be locally optimal, this decision removes the probe from consideration in other CoRegs, so the decision may not be globally optimal. Further, because these techniques do not clearly identify the probabilistic relationships between relevant hidden variables (*e.g.*, gene start/stop sites), it is not straightforward to modify them to account for new hidden variables or new data types. Also, because the separation between modeling assumptions and the optimization technique is not clear, it is difficult to improve performance in a principled manner.

Inspired by recent successes in using graphical probability models (*e.g.*, Bayesian networks) to analyze microarray data (c.f. [22, 24]), we have developed a generative probability model which jointly accounts for the stochastic nature of the arrangement of exons in genomic DNA, the stochastic nature of transcript expression, and the properties of probe sensitivity and noise in microarray data.

Inference in this model balances different sources of probabilistic evidence and makes a globally-optimal set of decisions for CoRegs, *after* combining probabilistic evidence. In contrast to our recent work [25] where exact inference is tractable, the model described in this paper accounts for more complex gene structures, such as alternative splicing isoforms, so exact inference is computationally burdensome. We describe how iterative application of the sum-product algorithm (a.k.a. "loopy belief propagation") can be used for efficient probabilistic inference. We compare the performance of our technique with a bottom-up threshold-based hierarchical clustering method [21], and we find that at low false positive rates, GenRate finds at least five times more exons. We also present new results showing that out of many novel mouse gene structures predicted by GenRate, the 9 highest-scoring structures that we tested are *all* confirmed by RT-PCR sequencing experiments.

2   Microarray Data

The microarray data set, a portion of which is shown in Fig. 1, is a subset of a full-genome data set to be described and released elsewhere [1]. Briefly, exons were predicted from repeat-masked mouse draft genome sequence (Build 28) using five different exon-prediction programs. Once Build 33 became available, we mapped the putative exons to the new genome. (While this data is based on putative *exons*, GenRate can be applied to any sequence-based expression data set, including genome tiling data.) A total of 48,966 non-overlapping putative exons were contained on chromosome 4 in Build 33. One 60-mer oligonucleotide probe for each exon was selected using conventional procedures, such that its binding free energy for the corresponding putative exon was as low as possible compared to its binding free energy with sequence elsewhere in the genome, taking into account other constraints on probe design. (For simplicity, we assume each probe has a unique position in the genome.) Arrays designs were submitted

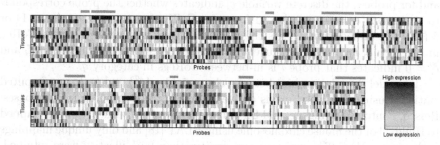

Fig. 1. A small fraction of our data set for chromosome 4, consisting of an expression measurement for each of 12 mouse tissue pools and 48,966 60-mer probes for repeat-masked putative exons arranged according to their order in Build 33 of the genome. Some "CoRegs" (co-regulated transcriptional units) were labeled by hand and are shown with green bars

to Agilent Technologies (Palo Alto, California) for array production. Twelve diverse samples were hybridized to the arrays, each consisting of a pool of cDNA from poly-A selected mRNA from mouse tissues (37 tissues total were represented). The pools were designed to maximize the diversity of genes expressed between the pools, without diluting them beyond detection limits [7]. Scanned microarray images were quantitated with GenePix (Axon Instruments), complex noise structures (spatial trends, blobs, smudges) were removed from the images using our spatial detrending algorithm [10], and each set of 12 pool-specific images was calibrated using the VSN algorithm [26] (using a set of one hundred "housekeeping" genes represented on every slide). For each of the 48,966 probes, the 12 values were then normalized to have intensities ranging from 0 to 1.

3   Generative Model

Our model accounts for the expression data by identifying a large number of CoRegs, each of which spans a certain number of probes. Each probe within a CoReg may correspond to an exon that is part of the CoReg or an intron that is not part of the CoReg. The probes for the tiling data are indexed by i and the probes are ordered according to their positions in the genome. Denote the expression vector for probe i by x_i, which contains the levels of expression of probe i across M experimental conditions. In our data, there are $M = 12$ tissue pools. To account for alternative primary and secondary transcripts, we allow CoRegs to overlap. So, we assume that when the genome sequence data is scanned in order, if a probe corresponds to an exon, the exon belongs to one of a small number of CoRegs that are currently active. Exons that take part in multiple transcripts are identified in a post-processing stage. This model enables multiple concurrent CoRegs to account for alternative splicing isoforms.

For concreteness, in this paper we assume that at most two CoRegs may be concurrently active, but the extension to a larger number is straightforward. So, each CoReg can be placed into one of two categories (labeled $q = 1$ and $q = 2$) and for probe i, the discrete variable e_i indicates whether the probe corresponds to an intron ($e_i = 0$) or an exon from the CoReg from category 1 ($e_i = 1$) or 2 ($e_i = 2$). At position i, ℓ_i^q is the remaining length (in probes) of the CoReg in category q, including the current probe. The maximum length is $\ell_i^q = L$ and $\ell_i^q = 0$ indicates that probe i is in-between CoRegs in category q.

To model the relationships between the variables $\{\ell_i^q\}$ and $\{e_i\}$, we computed statistics using confirmed exons derived from four cDNA and EST databases: Refseq, Fantom II, Unigene, and Ensembl. The database sequences were mapped to Build 33 of the mouse chromosome using BLAT [18] and only unique mappings with greater than 95% coverage and greater than 90% identity were retained. Probes whose chromosomal location fell within the boundaries of a mapped exon were taken to be confirmed. We model the lengths of CoRegs using a geometric distribution, with parameter $\lambda = 0.05$, which was estimated using cDNA genes. Importantly, there is a significant computational advantage in using the memoryless geometric distribution. Using cDNA genes to select the length prior will introduce a bias during inference. However, we found in our experiments that the effect of this bias is small. In particular, the results are robust to up to one order of magnitude in variation of λ.

The "control knob" that we use to vary the number of CoRegs that GenRate finds is κ, the *a priori* probability of starting a CoReg at an arbitrarily chosen position. Combining the above distributions, and recalling that $\ell_i^q = 0$ indicates position i is in-between CoRegs in category q, we have

$$P(\ell_i^q | \ell_{i-1}^q \in \{0,1\}) = \begin{cases} 1 - \kappa & \text{if } \ell_i^q = 0 \\ \kappa(0.05e^{-0.05\ell_i^q}) & \text{if } \ell_i^1 > 0, \end{cases}$$

$$P(\ell_i^q | \ell_{i-1}^q \in \{2,\dots,L\}) = [\ell_i^q = \ell_{i-1}^q - 1],$$

where square brackets indicate Iverson's notation, *i.e.*, $[True] = 1$ and $[False] = 0$. Both 0 and 1 are included in the condition "$\ell_{i-1} \in \{0,1\}$", because a new CoReg may start directly after the previous CoReg has finished. The term $\kappa(0.05e^{-0.05\ell_i^q})$ is the probability of starting a CoReg with length ℓ_i^1.

From genes in the cDNA databases, we found that within individual genes, probes are introns with probability $\epsilon = 0.3$. Depending on whether one or two CoRegs are active, we use a multinomial approximation to the probability that a probe is an exon:

$$P(e_i = 0 | \ell_i^1, \ell_i^2) = \epsilon^{[\ell_i^1 > 0] + [\ell_i^2 > 0]},$$

$$P(e_i = 1 | \ell_i^1, \ell_i^2) = [\ell_i^1 > 0](1 - \epsilon)(\frac{1 + \epsilon}{2})^{[\ell_i^2 > 0]},$$

$$P(e_i = 2 | \ell_i^1, \ell_i^2) = [\ell_i^2 > 0](1 - \epsilon)(\frac{1 + \epsilon}{2})^{[\ell_i^1 > 0]},$$

where again square brackets indicate Iverson's notation. Note that under this model, $P(e_i = 0 | \ell_i^1 > 0, \ell_i^2 > 0) = \epsilon^2$, $P(e_i = 1 | \ell_i^1 > 0, \ell_i^2 = 0) = 1 - \epsilon$ and $P(e_i = 1 | \ell_i^1 > 0, \ell_i^2 > 0) = (1 - \epsilon^2)/2$.

The similarity between the expression profiles belonging to the same CoReg is accounted for by a prototype expression vector. Each CoReg has a unique, hidden index variable and the prototype expression vector for CoReg j is μ_j. We denote the index of the CoReg at probe i in category q by c_i^q.

Different probes may have different sensitivities (for a variety of reasons, including free energy of binding), so we assume that each expression profile belonging to a CoReg is similar to a scaled version of the prototype. Since probe sensitivity is not tissue-specific, we use the same scaling factor for all M tissues. Also, different probes will be offset by different amounts (*e.g.*, due to different average amounts of cross-hybridization), so we include a tissue-independent additive variable for each probe. The expression profile x_i for an exon (where $e_i > 0$) is equal to the corresponding prototype $\mu_{c_i^{e_i}}$, plus isotropic Gaussian noise, we have

$$P(x_i | e_i = q, c_i^1, c_i^2, a_i, \{\mu_j\}) = \prod_{m=1}^{M} \frac{1}{\sqrt{2\pi a_{i3}^2}} e^{-(x_{im} - [a_{i1}\mu_{c_i^q,m} + a_{i2}])^2 / 2a_{i3}^2},$$

where a_{i1}, a_{i2} and a_{i3} are the scale, offset and isotropic noise variance for probe i, collectively referred to as a_i. In the *a priori* distribution $P(a_i)$ over these variables, the scale is assumed to be uniformly distributed in $[1/30, 30]$, which corresponds to a liberal assumption about the range of sensitivities of the probes. The offsets are assumed to be uniform in $[-0.5, 0.5]$ and the variance is assumed to be uniform in $[0, 1]$. While these assumptions are simplistic, we find they are sufficient for obtaining high-precision predictions, as described below.

We assume that the expression profiles for false exons are independent of the identify of the CoReg. While this assumption is also simplistic and should be further researched, it simplifies the model and leads to good results, so we make it for now. Thus, the false exon profiles are modeled using a background expression profile distribution:

$$P(x_i | e_i = 0, c_i^1, c_i^2, a_i, \{\mu_j\}) = P_0(x_i)$$

Since the background distribution doesn't depend on c_i^1, c_i^2, a_i or $\{\mu\}$, we also write it as $P(x_i | e_i = 0)$. We obtained this background model by training a mixture of 100 Gaussians on the entire, unordered set of expression profiles using a robust split-and-merge training procedure, and then including a component that is uniform over the range of expression profiles.

The Bayesian network in Fig. 2 shows the dependencies between the random variables in this generative model. Often, when drawing Bayesian networks, the parameters (prototypes) are not shown. We include the prototypes in the Bayesian network to highlight that they induce long-range dependencies in the model. For example, if a learning algorithm uses too many prototypes to model CoRegs (gene structures) in the first part of the chromosome, not enough will be left to model the remainder of the chromosome. So, during learning, prototypes must somehow be distributed in a fair fashion across the chromosome. We address this problem in the next section.

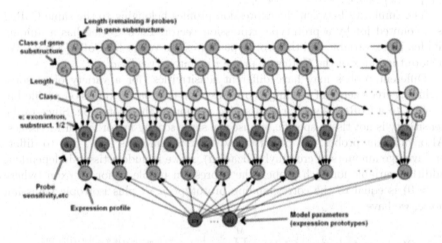

Fig. 2. A Bayesian network showing the variables and parameters in GenRate

Combining the structure of the Bayesian network with the conditional distributions described above, we can write the joint distribution as follows:

$$P(\{x_i\}, \{a_i\}, \{e_i\}, \{c_i^1\}, \{c_i^2\}, \{\ell_i^1\}, \{\ell_i^2\}, \{\mu_j\}) =$$

$$\left(\prod_{i=1}^{N} P(x_i | e_i, c_i^1, c_i^2, a_i, \{\mu_j\}) P(a_i) P(e_i | \ell_i^1, \ell_i^2) \right.$$

$$\left. \left(\prod_{q=1}^{2} P(c_i^q | c_{i-1}^q, \ell_{i-1}^q) P(\ell_i^q | \ell_{i-1}^q) \right) \right) \prod_{j=1}^{G} P(\mu_j),$$

where the appropriate initial conditions are obtained by setting $P(c_1^q | c_0^q, \ell_0^q) = [c_1^q = 1]$ and $P(\ell_1^q | \ell_0^q) \propto (1 - \lambda)^{\ell_1^q} \lambda, \ell_q^1 = 1, \ldots, L$. The constant of proportionality normalizes the distribution (if ℓ were not bounded from above by L, the distribution not require normalization). Whenever a gene terminates, c_i^q is incremented in anticipation of modeling the next gene, so $P(c_i^q = n | c_{i-1}^q = n, \ell_{i-1}^q) = 1$ if $\ell_{i-1}^q > 1$ and $P(c_i^q = n + 1 | c_{i-1}^q = n, \ell_{i-1}^q) = 1$ if $\ell_{i-1}^q = 1$. We assume the prototypes are distributed according to the background model: $P(\mu_j) = P_0(\mu_j)$.

4   Inference and Learning

Exact inference of the variables *and* parameters in the above model is computationally intractable. Given the model parameters, the model has a chain-type structure, so a standard approach is to use the EM algorithm [27]. EM iterates between performing exact inference for the variables in the chain while holding the parameters constant, and then updating the parameters based on sufficient statistics computed during inference in the chain. However, the EM algorithm fails spectacularly on this problem, because it gets stuck in local minima where prototypes are used to model weakly-evidenced gene patterns in one part of the chromosome, at the cost of not modeling gene patterns elsewhere in the chromosome. In fact, the EM algorithm in long hidden Markov models is known to be extremely sensitive to initial conditions and tends to find poor local minima caused by suboptimal parsings of the long data sequence [33].

To circumvent the problem of poor local minima, we devised a computationally efficient scheme for finding good solutions in a discrete subspace of the parameter space, which can then be finely tuned using the EM algorithm. In our scheme, the prototypes are represented using examples from the data set (in a manner akin to using data points as cluster centers in "k-centers clustering"). In the original model, the prototype for x_i is derived from nearby expression patterns, corresponding to nearby exons in the genomic DNA. Thus, if x_i is part of a CoReg, there is likely another x nearby that is a good representative of the profile for the CoReg. In the new representation, we replace each pair of variables ℓ_i^q and c_i^q with a variable r_i^q that encodes the *relative location* of the prototype for x_i. r_i^q gives the distance, in indices, from x_i to the prototype x_j for the CoReg that x_i belongs to, *i.e.* $r_i^q = j - i$. For example, $r_i^q = -1$ indicates that the profile immediately preceding x_i is the prototype for the gene to which x_i belongs. We limit the range of r_i^q to $-W, \ldots, W$, where W is a "window length". Within a CoReg, r_i^q decrements, indicating that the relative position of the prototype always decreases. We set aside a particular value of r_i^q, $r_0 = W + 1$, to account for the situation where i is in-between CoRegs.

Note that in this representation, the start of a CoReg corresponds to the condition $r_i^q = r_0$ and $r_{i+1}^q \neq r_0$, while the end of a CoReg corresponds to the condition $r_i^q \neq r_0$ and $r_{i+1}^q = r_0$. If a new CoReg starts directly after the previous CoReg, the boundary corresponds to the condition $r_{i+1}^q > r_i^q$. Since the r-variables are sufficient for describing CoReg boundaries, in fact, the ℓ variables need not be represented in the model. So, the new representation contains variables $\{r_i^q\}$, $\{e_i\}$, $\{a_i\}$, and $\{x_i\}$. If x_i is an exon in category q $(e_i = q)$, the conditional distribution of x_i is

$$
\prod_{m=1}^{M} \frac{1}{\sqrt{2\pi a_{i3}^2}} e^{-(x_{im} - [a_{i1}x_{i+r_i^q, m} + a_{i2}])^2 / 2a_{i3}^2},
$$

except if x_i is the prototype for the gene $(r_i^q = 0)$, in which case the distribution of x_i is $P_0(x_i)$. This model cannot be expressed as a Bayesian network, because constructing a Bayesian network from the above form of conditional distribution would create directed cycles. However, it can be expressed as a factor graph [28] or a directed factor graph [25].

The above model is a product of a Markov chain on $\{r_i^1\}$ and another Markov chain on $\{r_i^2\}$, coupled together by the switch e_i, which determines which chain is used to model the current expression vector, x_i. By combining the state spaces of the two Markov chains, exact inference can be performed using the forward-backward algorithm or the Viterbi algorithm. However, the combined state space has $4W^2$ states, where $2W$ is the maximum width of a CoReg, in probes. To enable our algorithm to find long CoRegs, we set $W = 100$, so the number of states in the combined chain would be $40,000$, making exact application of the Viterbi algorithm too slow. Instead, we apply the iterative sum-product algorithm to perform inference in the pair of coupled chains [28]. In each iteration, the algorithm performs a forward-backward pass in one chain, propagates probabilistic evidence across to the other chain, and then performs a forward-backward pass in the other chain.

We implemented the above inference algorithm in MATLAB, and for a given value of κ, our implementation takes approximately 10 minutes on a 2.4GHz PC to process the 48,966 probes and 12 tissue pools in chromosome 4 (with $W = 100$). The only free parameter in the model is κ, which sets the statistical significance of the genes found by GenRate.

5   Discussion of Computational Results

Fig. 3 shows a snapshot of the GenRate view screen that contains interesting examples of CoRegs found by GenRate. After we set the sensitivity control, κ,

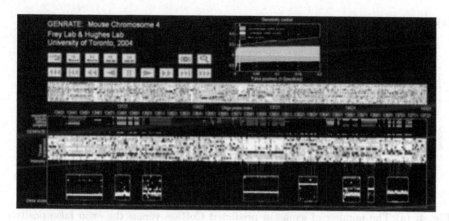

Fig. 3. The GenRate program (implemented in MATLAB) shows the genomic expression data and predicted CoRegs for a given false positive rate. The new genes, known genes and extensions of known genes that are found by GenRate are identified by shaded blocks, each of which indicates that the corresponding exon is included in the gene. Genes in cDNA databases (Ensembl, Fantom II, RefSeq, Unigene) are also shown. Each box at the bottom of the screen corresponds to a predicted gene structure and contains the normalized profiles for exons determined to be part of the gene. The corresponding raw profiles are connected to the box by lines. The score of each gene is printed below the corresponding box

to achieve a false positive rate of 1%, as described below, GenRate found 9,332 exons comprising 712 CoRegs. To determine how many of these predictions are new, we extracted confirmed genes derived from four cDNA and EST databases: Refseq, Fantom II, Unigene, and Ensembl. The database sequences were mapped to Build 33 of the mouse chromosome using BLAT and only unique mappings with greater than 95% coverage and greater than 90% identity were retained. Probes whose chromosomal location fell within the boundaries of a mapped exon were taken to be confirmed.

An important motivation for approaching this problem using a probability model is that the model should be capable of balancing probabilistic evidence provided by the expression data and the genomic exon arrangements. For example, there are several expression profiles that occur frequently in the data (in particular, profiles where activity in a single tissue pool dominates). If two of these profiles are found adjacent to each other in the data, should they be labeled as a gene? Obviously not, since this event occurs with high probability, *even if the putative exons are arranged in random order.*

To test the statistical significance of the results obtained by GenRate, we constructed a new version of the chromosome 4 data set, where the order of the columns (probes) is randomly permuted. For each value of κ in a range of values, we applied GenRate to the original data and the permuted data, and measured the number of positives and the number of false positives.

(a)   (b)

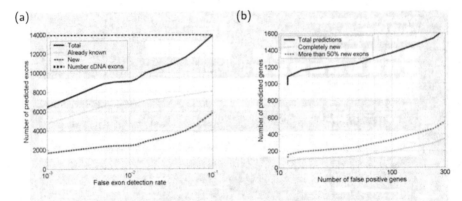

Fig. 4. (a) The number of exons in predicted CoRegs versus the exon false positive rate. The 3 curves correspond to the total number of predicted exons, the number of exons that are not in cDNA databases, and the number of known exons. The dash-dot line shows the number of exons in known genes from chromosome 4 (according to cDNA databases). (b) The number of predicted CoRegs versus the number of false positive predictions. The 3 curves correspond to the total number of predicted CoRegs, the number genes that are completely new (all exons within each structure are new), and the number of genes that contain at least 50% new exons

Fig. 4a shows the number of exons in CoRegs predicted by GenRate versus the false positive rate. Fig. 4b shows the number of predicted CoRegs versus the number of false positives. These curves demonstrate that GenRate is able to find CoRegs and associated exons with high precision. At an exon false positive rate of 1%, GenRate identifies 9,118 exons, 2,416 of which do not appear in known genes in cDNA databases, and GenRate identifies 65% of the exons in known genes in cDNA databases. This last number is a reasonable estimate of the proportion of genes that are expected to be expressed in the tissue pools represented in the data set. In Fig. 4b, when κ is set so GenRate finds 20 false positive CoRegs, GenRate identifies approximately 1,280 CoRegs, 209 of which contain at least 50% new exons, and 107 of which have no overlap with genes in cDNA databases.

Interestingly, the genes found by GenRate tend to be longer than genes in cDNA databases, as shown in Fig. 5a. While some of this effect can be accounted for by the fact that GenRate tends to find longer transcripts because they have higher statistical significance than short transcripts (*e.g.*, those containing 1 or 2 exons), there are two other explanations that should be considered. First, neighboring genes that are co-regulated may be identified by GenRate as belonging to a single transcript. We found that 23% of pairs of neighboring genes in the RefSeq cDNA database that were both detected by GenRate were identified as a single CoReg by GenRate. However, it is possible that in many of these cases the neighboring pair of "genes" in the cDNA database are in fact a single gene and that GenRate is correctly merging the predictions together. This possibility is consistent with the latest revision of the human genome, which shows that the number of genes is significantly lower than previously predicted [30], so that average gene length is longer than previously thought. In fact, as described below, we have shown using overlapping RT-PCR experiments that the longest, highest-scoring CoRegs identified by GenRate that consist of multiple cDNA "genes" exist as single, long transcripts.

5.1 Comparison to Hierarchical Clustering

A previously-described technique for assembling CoRegs from microarray tiling data consists of recursively merging pairs of probes into clusters, based on the correlation between the corresponding expression profiles and the distance between the probes in the genome [21]. In particular, if the correlation exceeds a threshold θ_1 *and* the genomic distance is less than another threshold θ_2, the probes are merged. In Fig. 5b, we compare the sensitivities of GenRate and this

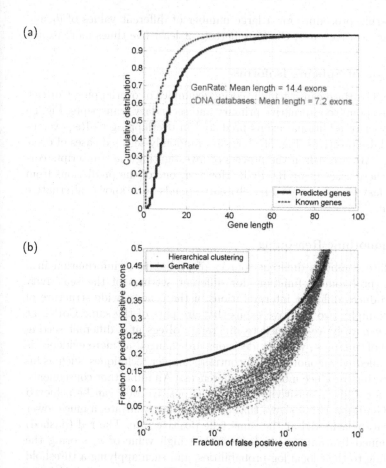

Fig. 5. (a) Cumulative distributions of gene lengths (in exons) for genes in cDNA databases and genes found by GenRate at an exon false positive rate of 1%. (b) A comparison between GenRate and correlation-based hierarchical clustering

(a)   (b)

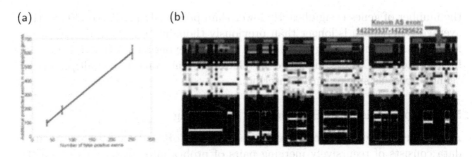

Fig. 6. (a) Additional exons predicted when overlapping CoRegs are taken into account. (b) Alternative splicing isoforms predicted by GenRate. The indicated structure corresponds to a known splicing event

recursive clustering procedure for a large number of different values of θ_1 and θ_2. For low false positive rates, GenRate detects at least five times more exons.

5.2   Detection of Splicing Isoforms

After inference, GenRate can list high-scoring examples of overlapping CoRegs, which may correspond to alternative primary and secondary transcripts. Fig. 6a shows the number of additional exons predicted in overlapping CoRegs versus the number of false positives. Fig. 6b shows six randomly-selected cases of overlapping CoRegs. We are still in the process of investigating the transcripts corresponding to these cases using RT-PCR. However, one of the predictions from GenRate (the last one shown in Fig. 6b) corresponds to a known alternative splicing isoform.

5.3   Non-monotonic Reasoning

Because GenRate combines different sources of probabilistic information in a global scoring (probability) function, for different settings of the sensitivity κ, GenRate produces different interpretations of the genome-wide structure of CoRegs. For example, two putative exons that are part of the same CoReg at one setting of κ may be re-assigned to different CoRegs at a different setting of κ. This type of inference, whereby decisions are changed as more evidence is considered, is called non-monotonic. (In contrast, simpler techniques, such as hierarchical clustering, produce monotonic inferences.) An important consequence of this is that for a given sensitivity, a low false positive rate can be achieved by re-running GenRate. Fig. 7 shows that by re-running GenRate, a much lower false positive rate is achieved at the same true positive rate. The red (dashed) curve was obtained by running GenRate with a high value of κ, scoring the CoRegs according to their local log-probabilities, and then applying a threshold to the scores to produce a predicted set of CoRegs. This was repeated with the randomly permuted data to obtain the plot of detected CoRegs versus false positives. The blue (solid) curve was produced by running GenRate with different

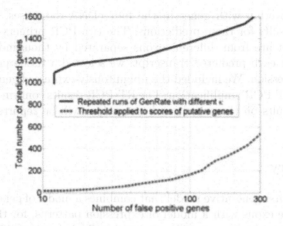

Fig. 7. When GenRate is re-run with a different sensitivity, κ, it may re-assign putative exons to other CoRegs. Compared to running GenRate once and using CoReg scores to rank CoRegs (red, dashed curve), running GenRate multiple times (blue, solid curve) leads to a significantly lower false positive rate at a given sensitivity

values of κ and retaining all predicted CoRegs (not applying a score threshold). By re-running GenRate, a much lower false positive rate is achieved for the same detection rate.

6   RT-PCR Experiments

Using RT-PCR, we have verified nine of the novel, high-scoring transcripts predicted by GenRate. In three cases, we selected predicted CoRegs that had high

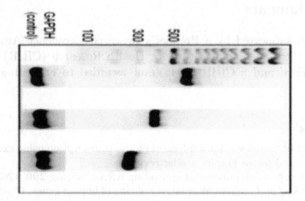

Fig. 8. RT-PCR results for three new transcripts identified by GenRate. The horizontal axis corresponds to the weight of the RT-PCR product and the darkness of each band corresponds to the amount of product with that weight

scores and no overlap with genes in the four cDNA databases. Fig. 8 shows the RT-PCR results for these predictions. The two PCR primers for each predicted transcript are from different exons separated by thousands of bases in the genome. For each predicted transcript, we selected a tissue pool with high microarray expression. We included the ubiquitously-expressed gene GAPDH to ensure proper RT-PCR amplification. The RT-PCR results confirm the predicted transcripts. Results on the other novel transcripts will be reported in another article [1].

7   Summary

GenRate is the first generative model that combines a model of genomic arrangement of putative exons with a model of expression patterns, for the purpose of discovering CoRegs in genome-wide tiling data. By balancing different sources of uncertainty, GenRate is able to achieve a significantly lower false positive rate than correlation-based hierarchical clustering methods. Applied to our microarray data, GenRate identifies many novel CoRegs with a low false-positive rate. We confirmed three of the predicted transcripts using RT-PCR experiments, and were able to recover known alternative splicing events and predict some new ones, albeit with high false positive rate. We have recently completed a genome-wide analysis of novel transcripts and this work has led us to a surprising conclusion, reported in [1], which appears to contradict recent results obtained by other researchers using microarrays to detect novel transcripts.

Because GenRate is based on a principled probability model, additional hidden variables can be incorporated in a straight-forward fashion. We believe GenRate will be a useful tool for analyzing other types of genome-wide tiling data, such as whole-genome tiling arrays.

Acknowledgments

This work was supported by a Premier's Research Excellence Award to Frey, a grant from the Canadian Institute for Health Research (CIHR) awarded to Hughes and Frey, and a CIHR NET grant awarded to Frey, Hughes and M. Escobar.

References

1. Frey B. J. *et al.* Genome-wide analysis of mouse transcription using exon-resolution microarrays and factor graphs. Under review.
2. Storz G. An expanding universe of noncoding RNAs. *Science* **296**, 1260-1263, 2002.
3. Mironov, AA *et al.* Frequent alternative splicing of human genes. *Genome Research* **9**, 1999.
4. Maniatis, T *et al.* Alternative pre-mRNA splicing and proteome expansion in metazoans. *Nature* **418**, 2002.
5. Burge C *et al.* Splicing of precursors to mRNAs by the spliceosomes. 2 edn, New York, Cold Spring Harbor Laboratory Press.

6. Peng WT *et al.* A panoramic view of yeast noncoding RNA processing. *Cell* **113:7**, 919-933, June 2003.

7. Zhang W *et al.* The functional landscape of mouse gene expression. *J. Biol.* **3:21**, Epub Dec 2004.

8. Pan Q. *et al.* Revealing global regulatory features of mammalian alternative splicing using a quantitative microarray platform. *Mol. Cell* **16**, 929-941, December 2004.

9. Pan Q. *et al.* Alternative splicing of conserved exons is frequently species-specific in human and mouse. *Trends in Genet.* **21:2**, 73-77, Feb 2005.

10. Shai O *et al.* Spatial bias removal in microarray images. Univ. Toronto TR PSI-2003-21. 2003.

11. Hild M *et al.* An integrated gene annotation and transcriptional profiling approach towards the full gene content of the Drosophila genome. *Genome Biol.* 2003;5(1):R3. Epub 2003 Dec 22.

12. Hughes TR *et al.* Expression profiling using microarrays fabricated by an ink-jet oligonucleotide synthesizer. *Nat. Biotechnol.* 2001 Apr;19(4):342-7.

13. Kapranov P *et al.* Large-scale transcriptional activity in chromosomes 21 and 22. *Science* 2002 May 3;296(5569):916-9.

14. Schadt EE *et al.* A comprehensive transcript index of the human genome generated using microarrays and computational approaches. *Genome Biology* 5 (2004).

15. Stolc V *et al.* A gene expression map for the euchromatic genome of drosophila melanogaster. To appear in *Science* (2004).

16. Nuwaysir EF *et al.* Gene expression analysis using oligonucleotide arrays produced by maskless photolithography. *Gen. Res.* 2002 Nov;12(11):1749-55.

17. FANTOM Consortium: RIKEN Genome Exploration Research Group Phase I & II Team. Analysis of the mouse transcriptome based on functional annotation of 60,770 full-length cDNAs. Okazaki Y *et al. Nature.* 2002 Dec 5;420(6915):563-73.

18. Kent WJ. BLAT–the BLAST-like alignment tool. *Genome Res.* 2002 Apr;12(4):656-64.

19. Pen SG *et al.* Mining the human genome using microarrays of open reading frames. *Nat. Genet.* 2000 Nov;26(3):315-8.

20. Rinn JL *et al.* The transcriptional activity of human chromosome 22. *Genes Dev.* 2003 Feb 15;17(4):529-40.

21. Shoemaker DD *et al.* Experimental annotation of the human genome using microarray technology. *Nature* 2001 Feb 15;409(6822):922-7.

22. Friedman N *et al.* Using Bayesian networks to analyze expression data. *Jour Comp Biology* **7**, 2000.

23. Yamada K *et al.* Empirical Analysis of Transcriptional Activity in the *Arabidopsis* Genome. *Science* **302**, 2003.

24. Segal E *et al.* Genome-wide discovery of transcriptional modules from DNA sequence and gene expression. *Bioinformatics* **19**, 2003.

25. Frey BJ *et al.* GenRate: A generative model that finds and scores new genes and exons in genomic microarray data. *Proc. PSB*, Jan. 2005.

26. Huber W *et al.* Variance stabilization applied to microarray data calibration and to quantification of differential expression. *Bioinformatics* **18**, 2002.

27. Dempster AP *et al.* Maximum likelihood from incomplete data via the EM algorithm. *Proc. Royal Stat. Soc.* **B-39**, 1977.

28. Kschischang FR *et al.* Factor graphs and the sum-product algorithm. *IEEE Trans. Infor. Theory* **47**, 2001.

29. Frey BJ. Extending factor graphs so as to unify directed and undirected graphical models. *Proc. UAI*, Aug. 2003.

30. International Human Genome Sequencing Consortium. Finishing the euchromatic sequence of the human genome. *Nature* **431**, 2004.
31. Berman P *et al.* Fast optimal genome tiling with applications to microarray design and homology search. *JCB* **11**, 766-785, 2004.
32. Bertone P *et al.* Global identification of human transcribed sequences with genome tiling arrays. *Science* **24**, Dec, 2004.
33. Ostendorf M. *IEEE Trans. Speech & Audio Proc.* **4**:360, 1996.

Efficient Calculation of Interval Scores for DNA Copy Number Data Analysis

Doron Lipson[1], Yonatan Aumann[2], Amir Ben-Dor[3],
Nathan Linial[4], and Zohar Yakhini[1,3]

[1] Computer Science Dept., Technion, Haifa
[2] Computer Science Dept., Bar-Ilan University, Ramat Gan
[3] Agilent Laboratories
[4] Computer Science Dept., Hebrew University of Jerusalem

Abstract. *Background.* DNA amplifications and deletions characterize cancer genome and are often related to disease evolution. Microarray based techniques for measuring these DNA copy-number changes use fluorescence ratios at arrayed DNA elements (BACs, cDNA or oligonucleotides) to provide signals at high resolution, in terms of genomic locations. These data are then further analyzed to map aberrations and boundaries and identify biologically significant structures.

Methods. We develop a statistical framework that enables the casting of several DNA copy number data analysis questions as optimization problems over real valued vectors of signals. The simplest form of the optimization problem seeks to maximize $\varphi(I) = \sum v_i / \sqrt{|I|}$ over all subintervals I in the input vector. We present and prove a linear time approximation scheme for this problem. Namely, a process with time complexity $O\left(n\epsilon^{-2}\right)$ that outputs an interval for which $\varphi(I)$ is at least Opt/$\alpha(\epsilon)$, where Opt is the actual optimum and $\alpha(\epsilon) \to 1$ as $\epsilon \to 0$. We further develop practical implementations that improve the performance of the naive quadratic approach by orders of magnitude. We discuss properties of optimal intervals and how they apply to the algorithm performance.

Examples. We benchmark our algorithms on synthetic as well as publicly available DNA copy number data. We demonstrate the use of these methods for identifying aberrations in single samples as well as common alterations in fixed sets and subsets of breast cancer samples.

1 Introduction

Alterations in DNA copy number are characteristic of many cancer types and are thought to drive some cancer pathogenesis processes. These alterations include large chromosomal gains and losses as well as smaller scale amplifications and deletions. Because of their role in cancer development, regions of chromosomal instability are useful for elucidating other components of the process. For example, since genomic instability can trigger the over expression or activation of oncogenes and the silencing of tumor suppressors, mapping regions of common

S. Miyano et al. (Eds.): RECOMB 2005, LNBI 3500, pp. 83–100, 2005.
© Springer-Verlag Berlin Heidelberg 2005

genomic aberrations has been used to discover cancer related genes. Understanding genome aberrations is important for both the basic understanding of cancer and for diagnosis and clinical practice.

Alterations in DNA copy number have been initially measured using local fluorescence in situ hybridization-based techniques. These evolved to a genome wide technique called Comparative Genomic Hybridization (CGH, see [11]), now commonly used for the identification of chromosomal alterations in cancer [13, 1]. In this genome-wide cytogenetic method differentially labeled tumor and normal DNA are co-hybridized to normal metaphases. Ratios between the two labels allow the detection of chromosomal amplifications and deletions of regions that may harbor oncogenes and tumor suppressor genes. Classical CGH has, however, a limited resolution (10-20 Mbp). With such low resolution it is impossible to predict the borders of the chromosomal changes or to identify changes in copy numbers of single genes and small genomic regions. In a more advanced method termed array CGH (aCGH), tumor and normal DNA are co-hybridized to a microarray of thousands of genomic clones of BAC, cDNA or oligonucleotide probes [16, 18, 10, 6, 4, 3, 2]. The use of aCGH allows the determination of changes in DNA copy number of relatively small chromosomal regions. Using oligonucleotides arrays the resolution can, in theory, be finer than single genes.

To fully realize the advantages associated with the emerging high resolution technologies, practitioners need appropriate efficient data analysis methods. A common first step in analyzing DNA copy number (DCN) data consists of identifying aberrant (amplified or deleted) regions in each individual sample. Indeed, current literature on analyzing DCN data describes several approaches to this task, based on a variety of optimization techniques. Hupe et al [15] develop a methodology for automatic detection of breakpoints and aberrant regions, based on Adaptive Weight Smoothing, a segmentation technique that fits a piecewise constant function to an input function. The fit is based on maximizing the likelihood of the (observed) function given the piecewise constant model, penalized for the number of transitions. The penalty weight is a parameter of the process. Sebat et al. [6], in a pioneering paper, study DCN variations that naturally occur in normal populations. Using an HMM based approach they compare signals for two individuals and seek intervals of 4 or more probes in which DCNs are likely to be different. A common shortcoming of many other current approaches is lack of principled optimization criteria that drive the method. Even when a figure of merit forms the mathematical basis of the process such as in [6], convergence of the optimization process is not guaranteed.

Further steps in analyzing DCN data include the automatic elucidation of more complex structures. A central task involves the discovery of common aberrations, either in a fixed set of studied samples or in an unsupervised mode, where we search for intervals that are aberrant in some significant subset of the samples. There are no formal treatments of this problem in the literature but most studies do report common aberrations and their locations. The relationship of DCN variation and expression levels of genes that reside in the aberrant region is of considerable interest. By measuring DNA copy numbers and mRNA expression levels on the same set of samples we gain access to the relationship of copy number alterations to how they are manifested in altering expression profiles. In [17] the authors used (metaphase slides) CGH to identify large scale amplifications in 23 metastatic colon cancer samples. For each identified large amplified region they compared the median expression levels of genes that reside there (2146 genes total), in the samples where amplification was detected, to the median expression levels of these genes in 9 normal control colon samples. A 2-fold over-expression was found in 81 of these genes. No quantitative statistical assessment of the results is given. In [19] a more decisive observation is reported. For breast cancer samples the authors establish a strong global correlation between copy number changes and expression level variation. Hyman et al. [10] report similar findings. The statistics used by both latter studies is based on simulations and takes into account single gene correlations but not local regional effects. In Section 2.3 we show how our general framework enables us to efficiently compute correlations between gene expression vectors and DCN vectors of genomic intervals, greatly extending the scope of the analysis.

In summary, current literature on CGH data analysis does not address several important aspects of DCN data analysis and addresses others in an informal mathematical setup. In particular, the methods described in current literature do not directly address the tasks of identifying aberrations that are common to a significant subset of a study sample set. In Section 2 we present methods that optimize a clear, statistically motivated, score function for genomic intervals. The analysis then reports all high scoring intervals as candidate aberrant regions. Our algorithmic approach yields performance guarantees as described in Sections 3 and 4. The methods can be used to automatically map aberration boundaries as well as to identify aberrations in subsets and correlations with gene expression, all within the same mathematical framework. Actual results from analyzing DCN data are described in Section 5.

Our approach is based on finding intervals of consistent high or low signals within an ordered set of signals, coming form measuring a set of genomic locations and considered in their genomic order. This principle motivates us to assign scores to intervals I of signals. The scores are designed to reflect the statistical significance of the observed consistency of high or low signals. These interval scores are useful in many levels of the analysis of DCN data. By using adequately defined statistical scores we transform the task of suggesting significant common aberrations as well as other tasks to optimizing segment scores in real valued vectors. Segment scores are also useful in interpreting other types of genomic data such as LOD scores in genetic analysis [14].

The computational problem of optimizing interval scores for vectors of real numbers is related to segmentation problems, widely used in time series analysis as well as in image processing [9, 8].

2   Interval Scores for CGH

In this section we formally define interval scores for identifying aberrant chromosomal intervals using DCN data. In Section 2.1 we define a basic score that is used to identify aberrations in a single sample. In Sections 2.2 and 2.3 we extend the score to accommodate data from multiple samples as well as joint DCN and gene-expression data.

2.1   Aberrant Intervals in Single Samples

Detection of chromosomal aberrations in a single sample is performed for each chromosome separately. Let $V = (v_1, \ldots, v_n)$ denote a vector of DCN data for one chromosome (or chromosome arm) of a single sample, where v_i denotes the (normalized) data for the i-th probe along the chromosome. The underlying model of chromosomal instabilities suggests that amplification and deletion events typically span several probes along the chromosome. Therefore, if the target chromosome contains amplification or deletion events then we expect to see many consecutive positive entries in V (amplification), or many consecutive negative entries (deletion). On the other hand, if the target chromosome is normal (no aberration), we expect no localized effects. Intuitively, we look for intervals (sets of consecutive probes) where signal sums are significantly larger or significantly smaller than expected at random. To formalize this intuition we assume (null model) that there is no aberration present in the target DNA, and therefore the variation in V represents only the noise of the measurement.

Assuming that the measurement noise along the chromosome is independent for distinct probes and normally distributed[1], let μ and σ denote the mean and standard deviation of the normal genomic data (typically, after normalization $\mu = 0$). Given an interval I spanning $|I|$ probes, let

$$\varphi^{sig}(I) = \sum_{i \in I} \frac{(v_i - \mu)}{\sigma \sqrt{|I|}}.$$ (1)

Under the null model, $\varphi^{sig}(I)$ has a Normal$(0,1)$ distribution, for any I. Thus, We can use $\varphi^{sig}(I)$ (which does not depend on probe density) to assess the statistical significance of values in I using, for example, the following large deviation bound [7]:

$$\text{Prob}(|\varphi^{sig}(I)| > z) \approx \frac{1}{\sqrt{2\pi}} \cdot \frac{1}{z} e^{-\frac{1}{2}z^2}.$$ (2)

Given a vector of measured DCN data V, we therefore seek all intervals I with $\varphi^{sig}(I)$ exceeding a certain threshold. Setting the threshold to avoid false positives we report all these intervals as putative aberrations.

2.2   Aberrant Intervals in Multiple Samples

When DCN data from multiple samples is available it may, of course, be processed one sample at a time. However, it is possible to take advantage of the additional data to increase the significance of the located aberrations by searching

---

[1] The normality assumption of the noise can be somewhat relaxed as the distribution of average noise for large intervals will, in any event, be close to normal (central limit theorem).

for *common* aberrations. More important than the statistical advantage, common aberrations are indicative of genomic alterations selected for in the tumor development process and may therefore be of higher biological relevance.

Given a set of samples S, and a matrix $V = \{v_{s,i}\}$ of CGH values where $v_{s,i}$ denotes the (normalized) data for the i-th probe in sample $s \in S$, identification of common aberrations may be done in one of two modes. In the *fixed set* mode we search for genomic intervals that are significantly aberrant (either amplified or deleted) in all samples in S. In the *class discovery* mode we search for genomic intervals I for which there exists a subset of the samples $C \subseteq S$ such that I is significantly aberrant only in the samples within C.

Fixed Set of Samples. A simple variant of the single-sample score (1) allows accommodation of multiple samples. Given an interval I, let

$$\varphi_S^{sig}(I) = \sum_{s \in S} \sum_{i \in I} \frac{(v_{s,i} - \mu)}{\sigma\sqrt{|I| \cdot |S|}}. \tag{3}$$

Although this score will indeed indicate whether I contains a significant aberration within the samples in S, it does not have the ability to discern between an *uncommon* aberration that is highly manifested in only one sample, and a *common* aberration that has a more moderate effect on all samples. In order to focus on common aberrations, we employ a *robust* variant of the score: Given some threshold τ^+, we create a binary dataset: $B = \{b_{s,i}\}$ where $b_{s,i} = 1$ if $v_{s,i} > \tau^+$ and $b_{s,i} = 0$ otherwise. Assuming (null model) that the appearance of 1s in B is independent for distinct probes and samples we expect the number of positive values in the submatrix defined by $I \times S$ to be Binom(n, p) distributed with $n = |I| \cdot |S|$ and $p = \text{Prob}(v_{s,i} > \tau^+) = \sum_{s \in S} \sum_{i=1}^{n} \frac{b_{s,i}}{|B|}$. The significance of k 1s in $I \times S$ can be assessed by the binomial tail probability:

$$\sum_{i=k}^{n} \binom{n}{i} p^i (1-p)^{(n-i)}. \tag{4}$$

For algorithmic convenience we utilize the Normal approximation of the above [5]. Namely, for each probe i we define the score of an interval I as in (3):

$$\varphi_S^{rob}(I) = \sum_{s \in S} \sum_{i \in I} \frac{(b_{s,i} - \nu)}{\rho\sqrt{|I| \cdot |S|}}, \tag{5}$$

where $\nu = p$ and $\rho = \sqrt{p(1-p)}$ are the mean and standard deviation of the variables $b_{s,i}$. A high score $\varphi_S^{rob}(I)$ is indicative of a common amplification in I. A similar score indicates deletions using a negative threshold τ^-.

Class Discovery. In the mode of class discovery we search a genomic interval I and a subset $C \subseteq S$ such that I is significantly aberrant on the samples within C. Formally, we search for a pair (I, C) that maximizes the score:

$$\varphi^{sig}(I, C) = \sum_{s \in C} \sum_{i \in I} \frac{(v_{s,i} - \mu)}{\sigma\sqrt{|I| \cdot |C|}}. \tag{6}$$

A *robust* form of this score is:

$$\varphi^{rob}(I,C) = \sum_{s \in C} \sum_{i \in I} \frac{(b_{s,i} - \nu)}{\rho \sqrt{|I| \cdot |C|}}. \tag{7}$$

2.3   Regional Correlation to Gene Expression

In previous work [12] we introduced the Regional Correlation Score as a measure of correlation between the expression levels pattern of a gene and an aberration in or close to its genomic locus. For any given gene g, with a known genomic location, the goal is to find whether there is an aberration in its chromosome that potentially effects the transcription levels of g. Formally, we are given a vector e of expression level measurements of g over a set samples S, and a set of vectors $V = (v_1, ..., v_n)$ of the same length corresponding to genomically ordered probes on the same chromosome, where each vector contains DCN values over the same set of samples S. For a genomic interval $I \subset [1, ..., n]$ we define a regional correlation score:

$$\varphi^{cor}(I,e) = \frac{\sum_{i \in I} r(e, v_i)}{\sqrt{|I|}}, \tag{8}$$

where $r(e, v_i)$ is some correlation score (e.g. Pearson correlation) between the vectors e and v_i. We are interested in determining whether there is some interval I for which $\varphi^{cor}(I,e)$ is significantly high. Note that for this decision problem it is sufficient to use an approximation process, such as described in Section 3.

2.4   A Tight Upper Bound for the Number of Optimal Intervals

In general, we are interested in finding the interval with the maximal score. A natural question is thus what is the maximal possible number of intervals with this score. The following theorem, proof of which appears in Appendix A, provides a tight bound on this number.

Theorem 1. *For any of the above scores, there can be n and at most n maximal intervals.*

3   Approximation Scheme

In the previous section we described interval scores arising from several different motivations related to the analysis DCN data. Despite the varying settings, the form of the interval scores in the different cases is similar. We are interested in finding the interval with maximal score. Clearly, this can be done by exhaustive search, checking all possible intervals. However, even for the single sample case this would take $\Theta(n^2)$ steps, which rapidly becomes time consuming, as the number of measured genomic loci grows to tens of thousands. Moreover, even for $n \approx 10^4$, $\Theta(n^2)$ does not allow for interactive data analysis, which is called for by practitioners (over 1 minute for 10K probes on 40 samples, see Section 5.1). For the class discovery case, a näive solution would require an exponential $\Theta(n^2 2^{|S|})$ number of steps. Thus, we seek more efficient algorithms. In this section we present a linear time approximation scheme. Then, based on this approximation scheme, we show how to efficiently find the actual optimal interval.

3.1   Fixed Sample Set

Note that the single sample case is a specific case of the fixed multiple-sample case ($|S| = 1$), on which we concentrate. For this case there are two interval scores defined in Section 2.2. The optimization problem for both these scores, as well as the score for regional correlations of Section 2.3, can all be cast as a general optimization problem. Let $W = (w_1, \ldots, w_n)$ be a sequence of numbers. For an interval I define

$$\varphi(I) = \frac{\sum_{i \in I} w_i}{\sqrt{|I|}} \tag{9}$$

Setting $w_i = \frac{\sum_{s \in S}(v_{s,i} - \mu)}{\sigma \sqrt{|S|}}$ gives $\varphi(I) = \varphi_S^{sig}(I)$; setting $w_i = \frac{\sum_{s \in S}(b_{s,i} - \nu)}{\rho \sqrt{|S|}}$ gives $\varphi(I) = \varphi_S^{rob}(I)$; and setting $w_i = r(e, v_i)$ gives $\varphi(I) = \varphi^{cor}(I, e)$. Thus, we focus on the problem of optimizing $\varphi(I)$ for a general sequence W.

A Geometric Family of Intervals. For an interval $I = [x, y]$ define $sum(I) = \sum_{i=x}^{y} w_i$. Fix $\epsilon > 0$. We define a family $\mathcal{I}$ of intervals of increasing lengths, the *geometric family*, as follows. For integral j, let $k_j = (1 + \epsilon)^j$ and $\Delta_j = \epsilon k_j$. For $j = 0, \ldots, \log_{(1+\epsilon)} n$, let

$$\mathcal{I}(j) = \left\{ [i\Delta_j, i\Delta_j + k_j - 1] : 0 \leq i \leq \frac{n - k_j}{\Delta_j} \right\} \tag{10}$$

In words, $\mathcal{I}(j)$ consists of intervals of size k_j evenly spaced Δ_j apart. Set $\mathcal{I} = \cup_{j=0}^{\log_{(1+\epsilon)} n} \mathcal{I}(j)$. The following lemma shows that any interval $I \subseteq [1..n]$ contains an interval of $\mathcal{I}$ that has "almost" the same size.

Lemma 1. *Let I be an interval, and J - the leftmost longest interval of $\mathcal{I}$ fully contained in I. Then $\frac{|I| - |J|}{|I|} \leq 2\epsilon + \epsilon^2$.*

Proof. Let j be such that $|J| = k_j$. In $\mathcal{I}$ there is an interval of size k_{j+1} every Δ_{j+1} steps. Thus, since there are no intervals of size k_{j+1} contained in I, it must be that $|I| < k_{j+1} + \Delta_{j+1}$. Therefore

$$|I| - |J| < k_{j+1} + \Delta_{j+1} - k_j = (1+\epsilon)k_j + \epsilon(1+\epsilon)k_j - k_j = (2\epsilon + \epsilon^2)k_j \leq (2\epsilon + \epsilon^2)|I| \quad \square$$

The Approximation Algorithm for a Fixed Set. The approximation algorithm simply computes scores for all $J \in \mathcal{I}$ and outputs the highest scoring one:

Algorithm 1 Approximation Algorithm - Fixed Sample Case

Input: Sequence $W = \{w_i\}$.
Output: Interval J with score approximating the optimal score.

$sum([1, 0]) = 0$
for $j = 1$ to n $sum([1, j]) = sum([1, j-1]) + w_j$
Foreach $J = [x, y] \in \mathcal{I}$ $\varphi(J) = \dfrac{sum([1, y]) - sum([1, x-1])}{|I|^{1/2}}$
output $J_{max} = \text{argmax}_{J \in \mathcal{I}}\{\varphi(J)\}$

The approximation guarantee of the algorithm is based on the following lemma:

Lemma 2. *For $\epsilon \leq 1/5$ the following holds. Let I^* be an interval with the optimal score and let J be the leftmost longest interval of $\mathcal{I}$ contained in I. Then $\varphi(J) \geq \varphi(I^*)/\alpha$, with $\alpha = (1 - \sqrt{2\epsilon(2 + \epsilon)})^{-1}$.*

Proof. Set $M^* = \varphi(I^*)$. Assume by contradiction that $\varphi(J) < M^*/\alpha$. Denote $J = [u, v]$ and $I^* = [x, y]$. Define $A = [x, u-1]$ and $B = [v+1, y]$ (the segments of I^* protruding beyond J to the left and right). We have,

$$M^* = \varphi(I^*) = \frac{sum(A) + sum(J) + sum(B)}{\sqrt{|I^*|}}$$

$$= \frac{sum(A)}{\sqrt{|A|}} \cdot \frac{\sqrt{|A|}}{\sqrt{|I^*|}} + \frac{sum(J)}{\sqrt{|J|}} \cdot \frac{\sqrt{|J|}}{\sqrt{|I^*|}} + \frac{sum(B)}{\sqrt{|B|}} \cdot \frac{\sqrt{|B|}}{\sqrt{|I^*|}}$$

$$= \varphi(A) \cdot \frac{\sqrt{|A|}}{\sqrt{|I^*|}} + \varphi(J) \cdot \frac{\sqrt{|J|}}{\sqrt{|I^*|}} + \varphi(B) \cdot \frac{\sqrt{|B|}}{\sqrt{|I^*|}}$$

$$\leq M^* \frac{\sqrt{|A|}}{\sqrt{|I^*|}} + \varphi(J) \frac{\sqrt{|J|}}{\sqrt{|I^*|}} + M^* \frac{\sqrt{|B|}}{\sqrt{|I^*|}} \tag{11}$$

$$\leq \left(\frac{\sqrt{|A|} + \sqrt{|B|}}{\sqrt{|I^*|}} \right) M^* + \varphi(J) \leq \sqrt{2 \cdot \frac{|A| + |B|}{|I^*|}} \cdot M^* + \varphi(J) \tag{12}$$

$$\leq \sqrt{2(2\epsilon + \epsilon^2)} \cdot M^* + \varphi(J), \tag{13}$$

$$< \sqrt{2(2\epsilon + \epsilon^2)} \cdot M^* + M^*/\alpha = M^*, \tag{14}$$

in contradiction. In the above, (11) follows from the optimality of M^*; (12) follows from arithmetic-geometric means inequality; (13) follows from Lemma 1; and (14) by the contradiction assumption and by the definition of α. $\square$

Thus, since the optimal interval must contain at least one interval of $\mathcal{I}$, we get,

Theorem 2. *For $\epsilon \leq 1/5$, Algorithm 1 provides an $\alpha(\epsilon) = (1 - \sqrt{2\epsilon(2 + \epsilon)})^{-1}$ approximation to the maximal score.*

Note that $\alpha(\epsilon) \to 1$ as $\epsilon \to 0$. Hence, the above constitutes an approximation scheme.

Complexity. The complexity of the algorithm is determined by the number of intervals in $\mathcal{I}$. For each j, the intervals of $\mathcal{I}_j$ are Δ_j apart. Thus, $|\mathcal{I}_j| \leq \frac{n}{\Delta_j} = \frac{n}{\epsilon(1+\epsilon)^j}$. Hence, the total complexity of the algorithm is:

$$|\mathcal{I}| \leq \sum_{j=0}^{\log_{(1+\epsilon)} n} \frac{n}{\epsilon}(1+\epsilon)^{-j} \leq \frac{n}{\epsilon}\sum_{j=0}^{\infty}(1+\epsilon)^{-j} = \epsilon^{-2}n = O(n\epsilon^{-2})$$

3.2   Class Discovery

Consider the problem of optimizing the scores $\varphi^{sig}(I,C)$ and $\varphi^{rob}(I,C)$. Similar to the fixed sample case, both problems can be cast as an instance of the general optimization problem. For an interval I and $C \subseteq S$ let:

$$\varphi(I,C) = \frac{\sum_{i \in I, s \in C} w_{s,i}}{\sqrt{|I| \cdot |C|}}, \quad opt(I) = \max_{C \subseteq S} \varphi(I,C)$$

Note that $\max_{I,C}\{\varphi(I,C)\} = \max_I\{opt(I)\}$.

Computing $opt(I)$. We now show how to efficiently compute $opt(I)$, without actually checking all possible subsets $C \subseteq S$. The key idea is the following. Note that for any C, $\varphi(I,C) = \sum_{s \in C} sum_s(I)/(|C||I|)^{1/2}$. Thus, for a fixed $|C| = k$, $\varphi(I,C)$ is maximized by taking k s's with the largest $sum_s(I)$. Thus, we need only sort the samples by this order, and consider the $|S|$ possible sizes, which is done in $O(|S| \log |S|)$. A description of the algorithm is provided in Appendix B.

The Approximation Algorithm for Class Discovery. Recall the geometric family of intervals $\mathcal{I}$, as defined in Section 3.1. For the approximation algorithm, for each $J \in \mathcal{I}$ compute $opt(J)$. Let $J_{\max}$ be the interval with the largest $opt(J)$. Using the Algorithm 4 find C for which $opt(J)$ is obtained, and output the pair J, C. The approximation ratio obtained for the class discovery case is identical to that of the fixed case, and the analysis is also similar.

Theorem 3. *For any $\epsilon \leq 1/5$, the above algorithm provides an $\alpha(\epsilon) = (1 - \sqrt{2\epsilon(2+\epsilon)})^{-1}$ approximation to the maximal score.*

The proof, which is essentially identical to that of Theorem 2, is omitted. Again, since $\alpha(\epsilon)$ approaches 1 as ϵ approaches 0, the above constitutes an approximation scheme.

Complexity. Computing $\varphi([1,j],s)$ for $j = 1, \ldots, n$, takes $O(|S|n)$ steps. For each $J \in \mathcal{I}$ computing $opt(J)$ is $O(|S| \log |S|)$. There are $O(n\epsilon^{-2})$ intervals in $\mathcal{I}$. Thus, the total number of steps for the algorithm is $O(n|S| \log |S|\epsilon^{-2})$.

4   Finding the Optimal Interval

In the previous section we showed how to *approximate* the optimal score. We now show how to find the absolute optimal score and interval. We present two algorithms. First, we present the *LookAhead* algorithm. Then, we present the GFA (*Geometric Family Algorithm*) which is based on a combination of the *LookAhead* algorithm, and the approximation algorithm described above. We note that for both algorithms we cannot prove that their worst case performance is better than $O(n^2)$, but in Section 5 we show that in practice *LookAhead* run in $O(n^{1.5})$ and GFA runs in linear time.

4.1   LookAhead Algorithm

Consider the fixed sample case. The algorithm operates by considering all possible interval starting points, in sequence. For each starting point i, we check the intervals with ending point in increasing distance from i. The basic idea is to try and not consider all possible ending point, but rather to skip some that will clearly not provide the optimum. Assume that we are given two parameters t and m, where t is a lower bound on the optimal score ($t \leq \max_I \varphi(I)$) and m is an upper bound on the value of any single element ($m \geq \max_i w_i$). Assume that we have just considered an interval of length k: $I = [i, ..., i + k - 1]$ with $\sigma = sum(I)$, and $\varphi(I) = \frac{\sigma}{\sqrt{k}}$. After checking the interval I, an exhaustive algorithm would continue to the next ending point, and check the interval $I_1 = [i, ..., i + k]$. However, this might not always be necessary. If σ is sufficiently small and k is sufficiently large then I_1 may stand no chance of obtaining $\varphi(I_1) > t$. For any x, setting $I_x = [i, ..., i + k + x]$ (skipping the next $(x - 1)$ intervals) an upper bound on the score of I_x is given by $\varphi(I_x) \leq \frac{\sigma + mx}{\sqrt{k + x}}$. Thus, $\varphi(I_x)$ has a chance of surpassing t only if $t \leq \frac{\sigma + mx}{\sqrt{k + x}}$. Solving for x, we obtain $x \geq (t^2 - 2m\sigma + \sqrt{4k - 4m\sigma + t^2})/2m^2$. Thus, the next ending point to consider is $i + h - 1$, where $h = \lceil k + x \rceil$.

The improvement in efficiency depends on the number of ending points that are skipped. This number, in turn, depends on the tightness of the two bounds t and m. Initially, t may be set to $t = \max_i w_i$. As we proceed t is replaced by the maximal score encountered so far, gradually improving performance.

m provides an upper bound on the values of single elements in W. In the description above, we used the *global* maximum $m = \max_i w_i$. However, using this bound may limit the usefulness of the LookAhead approach, since even a single high value in the data will severely limit the skip size. Thus, we use a *local* approach, where we bound the value of the single elements within a window of size κ. The most efficient definition of local bounds is the *slope maximum*: For a window size κ, for all $1 \leq i < n$, pre-calculate, $f_i = \max_{i < j \leq i + \kappa} \frac{\sum_{\ell = i + 1}^{j} w_\ell}{j - i}$. Although f_i is not an upper bound on the value of elements within the κ-window following i, the value of $f_i x$ is indeed an upper bound on $\sum_{j = i + 1}^{i + x} w_j$ within the κ-window, maintaining the correctness of the approach. Note that the skip size must now be limited by the window size: $h = \min(\lceil k + x \rceil, \kappa)$. Thus, larger values of κ give better performance although preprocessing limits the practical window size to $O(\sqrt{n})$.

The psuedocode in Algorithm 2 summarizes this variant of the LookAhead algorithm.

Algorithm 2 LookAhead algorithm - fixed sample case

Input: Sequence W, window size κ.	$maxScore = 0$
Output: The maximum-scoring interval I.	**for** $i = 1$ to n **do**
	$\quad \sigma = w_i, k = 1$
	$\quad$ **while** $i + k - 1 < n$ **do**
Preprocessing	$\quad\quad x = \lceil [(t^2 - 2m\sigma + \sqrt{4k - 4m\sigma + t^2})/2m^2] \rceil \;(\ast)$
$t = \max_i w_i, I = [\operatorname{argmax}_i w_i]$	$\quad\quad k = k + \min(\kappa, x)$
$sum[1,0] = 0;$	$\quad\quad \sigma = sum([1, i+k]) - sum([1, i-1])$
for $i = 1$ to n **do**	$\quad\quad score = \frac{\sigma}{\sqrt{k}}$
$\quad f_i = \max_{i < j \le i+\kappa} \frac{\sum_{\ell=i+1}^{j} w_\ell}{j - i}$	$\quad\quad$ **if** $score > maxScore$ **then**
$\quad sum[1,i] = sum([1, i-1]) +$	$\quad\quad\quad maxScore = score, I = [i, i+k-1]$
$\quad\quad w_i$	

A simple variant of the LookAhead algorithm allows to limit the search for the maximal scoring interval only to intervals starting within a fixed zone Z_1 and ending with another fixed zone Z_2. This is obtained by limiting the two loops in the algorithm to the required ranges. The GFA algorithm, described shortly, uses this variant of the algorithm.

Class Discovery. Application of the LookAhead heuristic to the class discovery mode is more complex, since the size of the subset C optimizing the score of each interval may vary. The following variation accommodates this difficulty, albeit at reduced algorithmic efficiency. Given the matrix W over a samples set S, create a vector $d = d_i$ as follows. For each $1 \le i \le n$ order the set $\{w_{s,i} : s \in S\}$ in decreasing order: $w_{s_1,i} \ge \ldots \ge w_{s_{|S|},i}$. Set $d_i = \max_{j:1 \le j \le |S|} \frac{\sum_{\ell=1}^{j} w_{s_\ell,i}}{\sqrt{j}}$. The vector d is used only in the preprocessing step, to compute the slope values f_i. At the main loop of the algorithm, the score of a specific interval $[i, i+k-1]$ is computed from the original data matrix W.

The correctness of the variant follows from the observation that if the subset C_1 maximizes the score of interval I_1 and subset C_2 maximizes the score of an extended interval I_2 (i.e. $I_1 \subset I_2$), setting $x = |I_2| - |I_1|$, then:

$$opt(I_2) = \frac{\sum_{s \in C_2} \sum_{i \in I_2} w_{s,i}}{\sqrt{|I_2| \cdot |C_2|}} \le \frac{\sum_{s \in C_2} \sum_{i \in I_1} w_{s,i} + f_i x \sqrt{|C_2|}}{\sqrt{|I_2| \cdot |C_2|}} \quad (15)$$

$$\le \frac{\sum_{s \in C_1} \sum_{i \in I_1} w_{s,i}}{\sqrt{|I_1| \cdot |C_1|}} + \frac{f_i x \sqrt{|C_2|}}{\sqrt{|I_2| \cdot |C_2|}} = opt(I_1) + \frac{f_i x}{\sqrt{|I_2|}} \quad (16)$$

$$= opt(I_1) + \frac{f_i x}{\sqrt{|I_1| + x}}$$

which can then be solved for x in the step marked by $(\ast)$ in Algorithm 2.

4.2   Geometric Family Algorithm (GFA)

GFA is based on a combination of the approximation algorithm of Section 3, which is used to zero-in on "candidate zones", and the *LookAhead* algorithm, which is then used to search within these candidate zones.

Specifically, let M be the maximum score of the intervals in $\mathcal{I}$, and let M^* be the maximal score of all intervals. Consider an interval $J \in \mathcal{I}$ with $\varphi(J) < M/\alpha \leq M^*/\alpha$. By Lemma 2, if I^* is the optimal interval then J cannot be the leftmost largest interval of $\mathcal{I}$ contained in I^*. Thus, when searching for the optimal interval, we need not consider any interval for which J is the leftmost largest interval. For each interval J we define the *cover zone* of J to be those intervals I for which J is the leftmost largest interval. Specifically, for $J = [x, y]$ such that $|J| = k_j$, let L-COV$(J) = x - \Delta_j + 1$ and R-COV$(J) = x + k_{j+1} - 2$. The *cover zone* of J is COVER$(J) = \{I = [u, v] : u \in [\text{L-COV}(J), x], v \in [y, \text{R-COV}(J)]\}$. In GFA we concentrate only on intervals J with $\varphi(J) \geq M/\alpha$, and search COVER(J) for the optimal interval using the *LookAhead* algorithm. If several intervals overlap, then we combine their cover zones, as described in Algorithm 3.

Algorithm 3 Finding the Optimal Interval - Fixed Sample Case

Input: $W = \{w_i\}_{i=1}^n$.
Output: The maximum-scoring interval $I_{\max}$.

forall $J \in \mathcal{I}$ compute $\varphi(J)$
$M = \max_{J \in \mathcal{I}}\{\varphi(J)\}$, $\quad U = \{J \in \mathcal{I} : \varphi(J) > M/\alpha\}$
while $U \neq \emptyset$ **do**
$\quad$ $J' = \text{argmax}_{J \in U}\{\varphi(J)\}$, $\quad U' = \{J \in U : J \cap J' \neq \emptyset\}$
$\quad$ L-COV$(U') = \min\{\text{L-COV}(J) : J \in U'\}$ and R-COV$(U') = \max\{\text{R-COV}(J) : J \in U'\}$
$\quad$ $J_\cap = \cap_{J \in U'} J$. Denote $J_\cap = [\text{l-}J_\cap, \text{r-}J_\cap]$
$\quad$ Run *LookAhead* to find the max score interval among the intervals starting in $[\text{L-COV}(U'), \text{l-}J_\cap]$ and ending in $[\text{r-}J_\cap, \text{R-COV}(U')]$.
$\quad$ Denote the optimal by $I_{opt}(U')$.
$\quad$ $U = U - U'$
$I_{\max} = \text{argmax}_{I_{opt}(U')}\{\varphi(I_{opt}(U'))\}$
return $I_{\max}$

Class Discovery. The algorithm for the class discovery case is identical to that of the fixed sample case except that instead of using $\varphi(J)$ we use $opt(J)$, and using Algorithm 4.

4.3   Finding Multiple Aberrations

In the previous section we showed how to find the aberration with the highest score. In many cases, we want to find the k most significant aberrations, for some fixed k, or to find all aberrations with score beyond some threshold t.

Fixed Sample. For the fixed sample case, finding multiple aberrations is obtained by first finding the top scoring aberration, and then recursing on the remaining left and right intervals. We may also find *nested* aberrations by recursing within the interval of aberration. We note that when using the GFA algorithm, in the recursion we need not recompute the scores of intervals in the geometric family $\mathcal{I}$. Rather, we compute these scores only once. Then, in each recursive step we find the max score within each region, and rerun the internal search (using *LookAhead*) within the candidate cover zones.

Class Discovery. For the class discovery case, we must first clearly define the desirable output; specifically, how to handle overlaps between the high scoring rectangles ($I \times C$). If no overlaps are permitted then a simple greedy procedure may be used to output all non-overlapping rectangles, ordered by decreasing score. If overlaps are permitted, then it is necessary to define how much overlap is permitted and of what type. We defer this issue to further research.

5   Performance Benchmarking and Examples

In this section we present results of applying the scores and methods described in the previous sections to several datasets. In Section 5.1 we benchmark the performance of both the LookAhead and GFA algorithms on synthetic datasets and on data from breast cancer cell-lines [10] and breast tumor data [19]. In Section 5.2 we demonstrate results of applying the multiple-sample methods to DCN data from breast cancer samples, using both the fixed sample mode and the class discovery mode.

5.1   Single Samples

We benchmark the performance of the exhaustive, LookAhead and GFA approaches on synthetic data, generated as follows. A vector $V = \{v_i\}$ is created by independently drawing n values from the Normal(0,1) distribution. A synthetic amplification is "planted" in a randomly placed interval I. The length of I is drawn randomly from the distribution Geom(0.01)+10, and the amplitude of the amplification is drawn uniformly in the range $[0.5, 10]$. Synthetic data was created for four different values of n - 1,000, 5,000, 10,000, and 50,000. In addition, we applied the algorithms to six different biological DCN vectors from [19] and [10]. Benchmarking results of finding the maximum scoring interval are summarized in Table 1. Note that the data from different chromosomes were concatenated for each biological sample to produce significantly long benchmark instances. The running times of the exhaustive, LookAhead and GFA algorithms exhibit growth rates of $O(n^2)$, $O(n^{1.5})$ and $O(n)$, respectively.

Figure 1 depicts high scoring intervals identified in chromosome-17, based on aCGH data from a breast cancer cell line sample (MDA-MB-453 – 7723 probes, from [2]). Here, all intervals with scores > 4 were identified, including nested aberrations. Running time of the full search dropped from 1.1 secs per sample with the exhaustive search, to 0.22 secs using the recursive LookAhead heuristic.

Table 1. Benchmarking results of the Exhaustive, LookAhead and GFA algorithms on synthetic vectors of varying lengths (Rnd1-Rnd4), and six biological DCN vectors from [19] (Pol) and [10] (Hym). Running times are in seconds; simulations performed on a 0.8GHz Pentium III. For Hym and Pol data from all chromosomes were concatenated to produce significantly long benchmarks. LookAhead was run with $\kappa = \sqrt{n}$, GFA with $\epsilon = 0.1$. Linear regression to log-log plots suggests that running times of the Exhaustive, LookAhead and GFA algorithms are $O(n^2)$, $O(n^{1.5})$ and $O(n)$, respectively

	Rnd1	Rnd2	Rnd3	Rnd4	Pol	Hym	
n	1,000	5,000	10,000	50,000	6,095	11,994	
# of instances	1000	200	100	100	3	3	
Exhaustive	0.0190	0.467	1.877	57.924	0.688	2.687	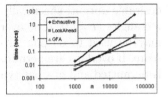
LookAhead	0.0045	0.044	0.120	1.450	0.053	0.143	
GFA	0.0098	0.047	0.093	0.495	0.079	0.125	

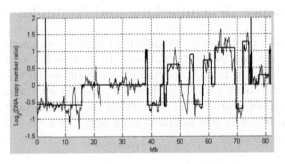

Fig. 1. Significant alterations in breast cancer cell line sample MDA-MB-453, chromosome 17 (7723 probes, from [2]). The thin line indicates the raw data (1Mb moving average window). Overlaid thick lined step function denotes the identified aberrations, where intervals above the x-axis denote amplifications with score $\varphi_S^{sig}(I) > 4$ and intervals below the x-axis – deletions with score $\varphi_S^{sig}(I) < -4$. The relative y-position of each interval indicates the average signal in the altered interval

5.2   Multiple Samples

Fixed Sample Set. We searched for common alterations in the two different breast cancer datasets. We used φ^{rob} to detect common aberrations in a set of 14 breast cancer cell line samples (data from [10]). Figure 2a depicts the chromosomal map of common aberrations that were identified with a score of $|\varphi_S^{rob}(I)| > 4.5$. No intervals with scores of this magnitude were located when the analysis was repeated on random data.

Class Discovery. We used φ^{rob} to detect classes of common alterations in 41 breast tumor samples (data from [19]). A large number of significant common alterations were identified in this dataset, with significantly high intervals scores, $\varphi^{rob}(I, C) > 12$. Again, no intervals with scores of this magnitude were located

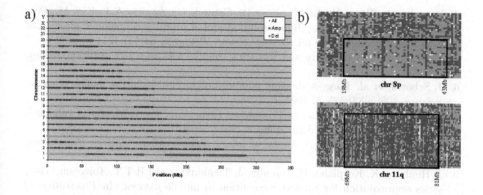

Fig. 2. a) Common alterations detected in 14 breast cancer cell line samples (data from [10]), in *fixed set* mode. All alterations with score $|\varphi_S^{rob}(I)| > 4.5$ are depicted - amplifications by red marks, and deletions by green marks. Dark blue marks denote all probe positions. b) Common deletion in 8p ($\varphi^{rob}(I, C) = 23.6$) and common amplification in 11q ($\varphi^{rob}(I, C) = 22.8$) identified in two subsets of 41 breast cancer samples in *class discovery* mode, $\tau = 0$ (data from [19]). X-axis denotes chromosomal position. Samples are arbitrarily ordered in y-axis to preserve a different class structure for each alteration. Red positions denote positive values and green – negative values

when the analysis was repeated on randomly permuted data. Some large aberrations were identified, including alterations affecting entire chromosomal arms as described in the literature [19]. Specifically, amplifications in 1q, 8q, 17q, and 20q, and deletions in 1p, 3p, 8p, and 13q, were identified, as well as numerous additional smaller alterations. Some of these regions contain known oncogenes (e.g. MYC (8q24), ERBB2 (17q12), CCND1 (11q13) and ZNF217 (20q13)) and tumor suppressor genes (e.g. RB (13q14), TP53 (17p13), BRCA1 (17q21) and BRCA2 (13q13)). Figure 2 depicts two significant common alterations that were identified in 8p (deletion) and 11q (amplification). An interesting aspect of the problem, which we did not attempt to address here, is the separation and visualization of different located aberrations, many of which contain significant intersections.

References

1. B.R. Balsara and J.R. Testa. Chromosomal imbalances in human lung cancer. *Oncogene*, 21(45):6877–83, 2002.
2. M.T. Barrett, A. Scheffer, A. Ben-Dor, N. Sampas, D. Lipson, R. Kincaid, P. Tsang, B. Curry, K. Baird, P.S. Meltzer, Z. Yakhini, L. Bruhn, and S. Laderman. Comparative genomic hybridization using oligonucleotide microarrays and total genomic DNA. *PNAS*, 101(51):17765–70, 2004.
3. G. Bignell, J. Huang, J. Greshock, S. Watt, A. Butler, S. West, Grigorova M., K. Jones, W. Wei, M. Stratton, P. Futreal, B. Weber, M. Shapero, and R. Wooster. High-resolution analysis of DNA copy number using oligonucleotide microarrays. *Genome Research*, 14(2):287–95, 2004.

4. Brennan C., Zhang Y., Leo C., Feng B., Cauwels C., Aguirre A.J., Kim M., Protopopov A., and Chin L. High-resolution global profiling of genomic alterations with long oligonucleotide microarray. *Cancer Research*, 64(14):4744–8, 2004.
5. M.H. DeGroot. *Probability and Statistics*, chapter 5.7, page 275. Addison-Wesley, 1989.
6. J. Sebat J. et al. Large-scale copy number polymorphism in the human genome. *Science*, 305(5683):525–8, 2004.
7. W. Feller. *An Introduction to Probability Theory and Its Applications*, volume I, chapter VII.6, page 193. John Wiley & Sons, 1970.
8. K.S. Fu and J.K. Mui. A survey of image segmentation. *Pattern Recognition*, 13(1):3–16, 1981.
9. J. Himberg, K. Korpiaho, H. Mannila, J. Tikanmki, , and H.T.T. Toivonen. Time series segmentation for context recognition in mobile devices. In *Proceedings of the 2001 IEEE International Conference on Data Mining (ICDM 2001)*, pages 203–210, 2001.
10. E. Hyman, P. Kauraniemi, S. Hautaniemi, M. Wolf, S. Mousses, E. Rozenblum, M. Ringner, G. Sauter, O. Monni, A. Elkahloun, O.P. Kallioniemi, and A. Kallioniemi. Impact of DNA amplification on gene expression patterns in breast cancer. *Cancer Research*, 62:6240–5, 2002.
11. O.P. Kallioniemi, A. Kallioniemi, D. Sudar, D. Rutovitz, J.W. Gray, F. Waldman, and D. Pinkel. Comparative genomic hybridization: a rapid new method for detecting and mapping DNA amplification in tumors. *Semin Cancer Biol*, 4(1):41–46, 1993.
12. D. Lipson, A. Ben-Dor, E. Dehan, and Z. Yakhini. Joint analysis of DNA copy numbers and expression levels. In *Proceedings of WABI 04*, LNCS 3240, page 135. Springer-Verlag, 2004.
13. F. Mertens, B. Johansson, M. Hoglund, and F. Mitelman. Chromosomal imbalance maps of malignant solid tumors: a cytogenetic survey of 3185 neoplasms. *Can. Res.*, 57:2765–80, 1997.
14. M. Morley, C. Molony, T. Weber, J. Devlin, K. Ewens, R. Spielman, and V. Cheung. Genetic analysis of genome-wide variation in human gene expression. *Nature*, 430(7001):743–7, 2004.
15. Hupe P., Stransky N., Thiery J.P., Radvanyi F., and Barillot E. Analysis of array CGH data: from signal ratio to gain and loss of DNA regions. *Bioinformatics*, 2004. (Epub ahead of print).
16. D. Pinkel, R. Segraves, D. Sudar, S. Clark, I. Poole, D. Kowbel, C. Collins, W.L. Kuo, C. Chen, Y. Zhai, S.H. Dairkee, B.M. Ljung, J.W. Gray, and D.G. Albertson. High resolution analysis of DNA copy number variation using comparative genomic hybridization to microarrays. *Nature Genetics*, 20(2):207–211, 1998.
17. P. Platzer, M.B. Upender, K. Wilson, J. Willis, J. Lutterbaugh, A. Nosrati, J.K. Willson, D. Mack, T. Ried, and S. Markowitz. Silence of chromosomal amplifications in colon cancer. *Cancer Research*, 62(4):1134–8, 2002.
18. J.R. Pollack, C.M. Perou, A.A. Alizadeh, M.B. Eisen, A. Pergamenschikov, C.F. Williams, S.S. Jeffrey, D. Botstein, and P.O. Brown. Genome-wide analysis of DNA copy-number changes using cDNA microarrays. *Nature Genetics*, 23(1):41–6, 1999.
19. J.R. Pollack, T. Sorlie, C.M. Perou, C.A. Rees, S.S. Jeffrey, P.E. Lonning, R. Tibshirani, D. Botstein, A. Borresen-Dale, and P.O. Brown. Microarray analysis reveals a major direct role of DNA copy number alteration in the transcriptional program of human breast tumors. *PNAS*, 99(20):12963–8, 2002.

A   Proof of Theorem 1

Let $f : \mathcal{R} \to \mathcal{R}$ be a strictly concave function (that is $tf(x) + (1 - t)f(y) < f(tx + (1 - t)y)$ for all $0 < t < 1$). We define the score of an interval I with respect to f, denoted $S_f(I)$, to be the sum of I's entries divided by $f(|I|)$. In particular, for $f(x) = \sqrt{x}$, $S(I)$ coincides with the interval score function we use throughout this paper.

For a given vector of real numbers, of length n, let $m > 0$ denote the maximal interval score obtained (among all $\binom{n}{2}$ intervals). We call an interval *optimal* if its score equals m. The following theorem bounds the number of optimal intervals:

Theorem 4. *There can be at most n optimal intervals.*

Claim. For all $x, y > z > 0$ the following inequality holds: $f(x) + f(y) > f(z) + f(x + y - z)$

Proof. Define $t = (x - z)/(x + y - 2z)$. Note that

$$x = t(x + y - z) + (1 - t)z, \quad \text{and} \quad y = tz + (1 - t)(x + y - z).$$

As f is strictly concave, and $0 < t < 1$

$$tf(x + y - z) + (1 - t)f(z) < f(t(x + y - z) + (1 - t)(z)) = f(x)$$
$$tf(z) + (1 - t)f(x + y - z) < f(t(z) + (1 - t)(x + y - z)) = f(y)$$

Summing the above two inequalities, we get

$$f(x + y + z) + f(z) < f(x) + f(y)$$

Claim. If two optimal intervals I and J intersect each other than either I properly contains J or vise versa.

Claim. If two optimal intervals I and J are disjoint, there must be at least one point between them which is not included in either. Otherwise, consider the interval $K = I; J$ representing their union.

Let $\mathcal{I}$ be a collection of optimal intervals. For each $x \in \{1, \ldots, n\}$ we denote by $\mathcal{I}(x)$ the set set of optimal intervals that contains x.

Claim. All intervals in $\mathcal{I}(x)$ have different sizes.

Proof. Any two optimal intervals that contains x intersect. By Claim A one of them properly contains the other. Thus, they have different size.

We define for each x the smallest interval that contains x, denoted by $T(x)$. Note that by the previous claim, there can be only one minimal interval that contains x. To complete the upper bound proof, we show now that the mapping $T : \{1, \ldots, n\} \leftarrow \mathcal{I}$ is onto.

Lemma 3. *For each optimal interval I there exist a point x such that $T(x) = I$.*

Proof. Let I be an optimal interval, $I = [i, \leftarrow, j]$. We consider two cases

- *The leftmost point i, of I, is not included in any interval $J \subset I$. By Claim A all the intervals that contains i must contain I, and thus I is the smallest interval that contains i. Therefore $T(i) = I$.*
- *Otherwise, let J denote the largest interval that is properly contained in I and $i \in J$. let k denote the rightmost point in J. We show now that $T(k+1) = I$. First note that since J is properly contained in I, and J contains the leftmost point of I, J cannot contain the rightmost point of I. Therefore $k + 1 \in I$. Assume, towards contradiction, that there exist a shorter interval I' that contains $k + 1$. As I' intersect I and it is shorter, it must be (Claim A) properly contained in I. Let $i' \geq i$ denote the leftmost point of I'. We consider three case*
  - *$i' = i$. In this case $i \in I'$, and $J \subset I'$ a contradiction to the definition of J.*
  - *$i < i' \leq k$. In this we contradict Claim A as*
    $$k \in J \cap I', i \in J \setminus I', k+1 \in I' \setminus J.$$
  - *$i' = k + 1$. In this case, $k + 1$ is the leftmost point of I', while k is the leftmost point of J, a contradiction to Claim A.*

Therefore, there I is the smallest interval containing $k+1$, and thus $T(k+1) = I$.

We conclude

Theorem 5. *There can be at most n optimal intervals*

This bound is, in fact, tight as can be seen by considering any prefix of $v_1 = 1, ... v_i = f(i) - f(i-1), ...$

B   Algorithm for Computing $opt(I)$

For a Matrix W and an interval I, the following algorithm efficiently computes $opt(I)$. Note that we assume that for each s and j, $sum_s([1, j])$ is already known. This can be done in a preprocessing phase, for all s and j in $O(n|S|)$.

Algorithm 4 Computing $opt(I)$

Input: Matrix $W = \{w_{s,i}\}$, Interval $I = [x, y]$.
Output: $opt(I)$.

Foreach $s \in S$ compute $sum_s(I) = sum_s([1, y]) - sum_s([1, x-1])$
Sort the values of $sum_s(I)$ in decreasing order. Let $s_1, \ldots, s_m$ be the order of the s's.
$sum(0) = 0$
for $\ell = 1$ to m **do**
    $sum(\ell) = sum(\ell - 1) + sum(I)_{s_\ell}$
Output $opt(I) = \max_\ell \{sum(\ell)/\sqrt{\ell}\}$

A Regulatory Network Controlling Drosophila Development

Dmitri Papatsenko and Mike Levine

Department of Molecular and Cell Biology, Division of Genetics and Development,
University of California, Berkeley, CA 94720, USA

We have used a combination of genetics, microarray assays, bioinformatics methods and experimental perturbation to determine the genomic regulatory program underlying the early development of the fruitfly, Drosophila melanogaster. The process of gastrulation is initiated by a sequence-specific transcription factor called Dorsal. The Dorsal protein is distributed in a broad concentration gradient across the dorsal-ventral axis of the early embryo, with peak levels in ventral regions and progressively lower levels in more dorsal regions. The Dorsal gradient controls gastrulation by regulating a variety of target genes in a concentration-dependent fashion. Microarray assays have identified at least 50 genes that display localized expression across the dorsal-ventral axis and something like 30 of these genes are directly regulated by the Dorsal gradient. That is, these genes contain nearby enhancers with essential Dorsal binding sites. Most of the enhancers have been identified and experimentally validated. Computational analysis suggests that the threshold readout of the Dorsal gradient depends on the quality of the best Dorsal binding sites contained in the enhancer. Enhancers with Dorsal binding sites containing a poor match to the consensus sequence are activated by high levels of the gradient, while those enhancers with good matches to the consensus are regulated by intermediate or low levels of the gradient.

Most of the Dorsal target enhancers are associated with genes that encode either regulatory proteins or components of cell signaling pathways. Linking the encoded proteins to specific Dorsal target enhancers produces a network-like architecture that defines the genomic program for gastrulation. The analysis of the network, or circuit diagram, suggests a number of testable hypotheses that are made possible only by the computational visualization of this complex process. Most of the initial regulatory interactions that are seen involve transcriptional repressors and the formation of different boundaries of gene expression (corresponding to the primary embryonic tissues). Later, the Dorsal network directs the localized activation of the Notch signaling pathway, which in turn establishes a stable positive feedback loop generated by EGF signaling. Some of he enhancers that are activated by EGF signaling also correspond to direct targets of the Dorsal gradient. The transition from Dorsal to EGF essentially "locks down" stable patterns of expression that are only transiently specified by Dorsal. We are currently testing the possibility that aspects of the Dorsal signaling network can be circumvented by the directed activation of EGF signaling in specific regions of the embryo.

S. Miyano et al. (Eds.): RECOMB 2005, LNBI 3500, p. 101, 2005.

Yeast Cells as a Discovery Platform for Neurodegenerative Disease

Susan Lindquist, Ernest Fraenkel, Tiago Outeiro, Aaron Gitler,
Julie Su, Anil Cashikar, and Smitha Jagadish

Whitehead Institute for Biomedical Research,
Cambridge, Massachusetts, USA
Lindquist_admin@wi.mit.edu

Protein conformational changes govern most processes in cell biology. Because
these phenomena are universal, we use a variety of model systems to study them,
from yeast to mice. Recently we've been particularly interested in proteins that
undergo self-perpetuating changes in conformation, such as prions. Surprisingly,
our studies in yeast suggested that a cytoplasmic form of the mammalian prion,
PrP, might be involved in disease. Mouse models we constructed to test this
possibility established that cytoplasmic PrP is sufficient to cause neurodegener-
ative disease. This led us to establishing other models of protein-folding diseases
in yeast, with the aim of learning about the causes of misfolding, the nature of
its cyto-toxic effects, and mechanisms that might ameliorate them. A variety of
high throughput methods are being used to exploit these. Computational methods are being employed to integrate data from different platforms. Results with
alpha synuclein (whose misfolding is directly related to Parkinson's disease) and
Huntingtin (whose misfolding is directly related to Huntingtin's disease) will be
discussed.

S. Miyano et al. (Eds.): RECOMB 2005, LNBI 3500, p. 102, 2005.
© Springer-Verlag Berlin Heidelberg 2005

RIBRA–An Error-Tolerant Algorithm for the NMR Backbone Assignment Problem[*]

Kuen-Pin Wu[1], Jia-Ming Chang[1], Jun-Bo Chen[1],
Chi-Fon Chang[2], Wen-Jin Wu[3], Tai-Huang Huang[2,3],
Ting-Yi Sung[1], and Wen-Lian Hsu[1]

[1] Institute of Information Science, Academia Sinica, Taiwan
{kpw, jmchang, philip, tsung, hsu}@iis.sinica.edu.tw
[2] Genomics Research Center, Academia Sinica, Taiwan
[3] Institute of Biomedical Sciences, Academia Sinica, Taiwan
{cfchang, bmthh, winston}@ibms.sinica.edu.tw

Abstract. We develop an iterative relaxation algorithm, called RIBRA, for NMR protein backbone assignment. RIBRA applies nearest neighbor and weighted maximum independent set algorithms to solve the problem. To deal with noisy NMR spectral data, RIBRA is executed in an iterative fashion based on the quality of spectral peaks. We first produce spin system pairs using the spectral data without missing peaks, then the data group with one missing peak, and finally, the data group with two missing peaks. We test RIBRA on two real NMR datasets: hb-SBD and hbLBD, and perfect BMRB data (with 902 proteins) and four synthetic BMRB data which simulate four kinds of errors. The accuracy of RIBRA on hbSBD and hbLBD are 91.4% and 83.6%, respectively. The average accuracy of RIBRA on perfect BMRB datasets is 98.28%, and 98.28%, 95.61%, 98.16% and 96.28% on four kinds of synthetic datasets, respectively.

1 Introduction

Nuclear magnetic resonance (NMR) spectroscopy and X-ray crystallography are the only two methods that can determine three-dimensional structures of proteins to atomic level. NMR spectroscopy has the additional power that it can also study the dynamics of proteins and screen for interacting partners such as drug screening in solution, also to the atomic/residue level. However, before any of the detailed studies can be carried out, sequence specific backbone resonance assignments must be completed. Multi-dimensional NMR spectra contain cross-peaks, which contain resonance frequencies (*chemical shifts*) and correlation information. The cross-peak represents a covalent-bond linkage (COSY type) or a spatial relation (NOESY type) among a set of nuclei depending on the types of NMR

---

[*] This work is supported in part by the thematic program of Academia Sinica under Grant AS91IIS1PP.

experiments performed. Different kinds of NMR experiments provide different partial resonance information of residues so that biologists can decide which experiments should be performed to best suit their needs. For example, the two dimensional HSQC experiment concerns whether there is a covalent bond between two atoms N and H^N; if there is, then a corresponding peak appears in the spectrum and its coordinate is composed of the chemical shifts of the two atoms.

Once cross-peaks are available, we assign chemical shifts to the corresponding atoms within residues. This important stage is known as *resonance assignment* and is the main topic in this paper. Cross-peaks are usually extracted according to an intensity threshold. If the threshold is set too high, peaks with low intensities will be ignored and become missing peaks (false negatives). On the other hand, if it is set too low, noisy peaks with high intensity will be regarded as real peaks and become false positives. Even if spectral data contains no false negatives or false positives, there are still some other problems. The same kind of atoms may be located in similar environments. In such cases, these atoms may have nearly the same chemical shifts and be represented as a single cross-peak in a spectrum; namely, more than one peak may cluster together as a single point in the spectrum and become indistinguishable. Another problem is experimental error. Theoretically, each atom in a residue has a unique chemical shift. However, different NMR experiments may generate slightly different chemical shifts for an atom due to differences in experimental conditions such as slight change in temperature or difference in digital resolution. Ambiguity arises when the results of different experiments are cross-referenced. In summary, there are mainly four kinds of problems in real spectral data: 1) false negatives, 2) false positives, 3) clustered peaks and 4) experimental errors. As these four kinds of data problems mix together in real data, resonance assignment becomes very challenging. For example, with respect to a residue, we cannot easily distinguish whether there is a missing peak, or the residue is a Glycine, which intrinsically has no C^β atom.

To overcome the above problems, researchers usually perform the following five procedures [22]:

1. filter peaks and relate resonances from different spectra (filtering and referencing)
2. group resonances into spin systems (grouping). A *spin system* contains the chemical shifts of atoms within a residue.
3. identify the amino acid types of spin systems (typing)
4. find and link sequential spin systems into segments (linking) and
5. map spin-system segments onto the primary sequence (mapping).

Note that different researcher might perform the above procedures in different order. Many works have been done to carry out part of these five procedures [3, 4, 5, 9, 10, 11, 13, 14, 15, 16, 19, 21, 23, 24, 25, 26]. Most models assume that the spin system grouping is already given (thus, eluding the possible ambiguity problem in grouping). Some model (such as the constrained bipartite graph approach of [26]) further prohibits linking ambiguity. Some other works perform all these five procedures [1, 2, 7, 17, 18, 20, 27]. The reader is referred to [22] for a thorough survey.

The contribution of this paper is twofold. First, we design an error-tolerant algorithm, RIBRA (Relaxation and Iterative Backbone Resonance Assignment) based on iterative relaxation, to solve the backbone resonance assignment problem (*backbone assignment* for short) with good precision and recall. We use HSQC, CBCANH and CBCA(CO)NH spectral data to assign chemical shifts to atoms N, H^N, C^α and C^β along the backbone of a target protein. We use the rules of TATAPRO II [2] to perform typing. Instead of performing the procedures as separate tasks, RIBRA adopts two operations RGT and LM, in which RGT does mixed referencing, grouping and typing, and LM does mixed linking and mapping. Based on data quality, we apply a relaxation technique to perform RGT and LM iteratively for noise filtering, and gradually carry out the entire backbone assignment.

An important source of NMR datasets is the BioMagResBank[1]. Since the data in BMRB is normally error-free, researchers need to generate noises to test whether their algorithms can cope with these problems. However, in the past, these noises are generated in an ad-hoc fashion. Our second contribution is to create comprehensive synthetic datasets that reflect the following potential problems: false positive, false negative and experimental errors. The original BMRB datasets can largely be regarded as "perfect standard datasets." Note, however, there is no need to synthesize datasets containing clustered peaks since they are intrinsically embedded in the original BMRB data. These synthetic BMRB datasets are designed to serve as benchmark datasets for testing any future assignment algorithms.

We use real experimental data from Academia Sinica, perfect BMRB data and synthetic BMRB data to evaluate RIBRA. Two real datasets are substrate binding domain of BCKD (hbSBD) and lipoic acid bearing domain of BCKD (hbLBD [8]). Each of them contains more than 50% false positives and false negatives. Define the *precision* and *recall* of an assignment as follows:

$$\text{precision} = \frac{\text{number of correctly assigned amino acids}}{\text{number of assigned amino acids}} \times 100\% \qquad (1)$$

$$\text{recall} = \frac{\text{number of correctly assigned amino acids}}{\text{number of amino acids with known answers}} \times 100\% \qquad (2)$$

Compared with the best manual solution, the precision and recall of RIBRA on the first dataset hbSBD are 91.43% and 76.19%, respectively; and those on the second dataset hbLBD are 83.58% and 70.00%, respectively. Such a performance is regarded as quite satisfactory in practice in the sense that the additional human postprocessing effort is quite minimal. We also test RIBRA on 902 perfect datasets in BMRB. The average precision and recall on these datasets are 98.28% and 92.33%, respectively. From these perfect datasets, we generate four kinds of synthetic datasets each with one type of errors from false positives, false

---

[1] BMRB, http://www.bmrb.wisc.edu/index.html

negatives, grouping errors and linking errors. The average precision of RIBRA on false positive, false negative, grouping errors and linking errors datasets are 98.28%, 95.61%, 98.16% and 96.28%, respectively; the average recall of RIBRA on false positive, false negative, grouping errors and linking errors dataset are 92.35%, 77.36%, 88.57% and 89.15%, respectively.

The remainder of this paper is organized as follows. Section 2 describes the RIBRA algorithm. Section 3 presents experimental results and analysis. Finally, conclusions are summarized in Section 4.

2   RIBRA

The input of RIBRA is HSQC, CBCANH and CBCA(CO)NH spectral data. We first introduce the basic idea of forming spin systems using these three spectral data. To map spin-system segments onto the target protein sequence, we model it as a graph optimization problem, and provide a solution based on a heuristic maximum independent set algorithm. Finally, we introduce the iterative relaxation technique used by RIBRA to perform backbone assignment.

Our goal is to assign N, H^N, C^α and C^β chemical shifts along the backbone of target proteins. Figure 1a shows two consecutive residues, the $(i-1)$-th and the i-th residues, where only atoms along the backbone are depicted; we use R to denote all atoms of a side chain. For the i-th residue, the HSQC experiment

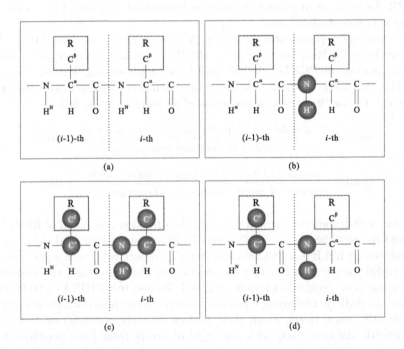

Fig. 1. Different NMR experiments on two consecutive residues. The detected atoms with available chemical shifts are marked in black

detects H_i^N and N_i chemical shifts (see Figure 1b). It generates one peak of the form $(H_i^N, N_i, +)$, where the first two elements are chemical shifts of H_i^N and N_i, respectively. The "+" sign is used to denote that there is a positive peak intensity.

For the i-th residue, the CBCANH experiment detects H_i^N, N_i, C_{i-1}^α, C_{i-1}^β, C_i^α and C_i^β chemical shifts (see Figure 1c). It generates four peaks of the form $(H_i^N, N_i, C_?, +/-)$, where the first three elements are chemical shifts of H_i^N, N_i and C, respectively. The question mark of the third element is used to indicate that we do not know whether the carbon is located in the $(i-1)$-th residue or the i-th residue. The fourth element is the peak intensity in which C^α has a positive value and C^β has a negative one. Two of the four peaks are associated with the α and β carbons of the $(i-1)$-th residue, and the other two peaks are associated with those of the i-th residue.

For the i-th residue, the CBCA(CO)NH experiment detects H_i^N, N_i, C_{i-1}^α and C_{i-1}^β chemical shifts (see Figure 1d). It generates two peaks of the form $(H_i^N, N_i, C_{i-1}^?, +)$, where the first three elements are chemical shifts of H_i^N, N_i and C, respectively. The question mark of the third element indicates that it is not known whether the carbon is C^α or C^β, since the intensity of this experiment is always positive. Cross-referencing the HSQC, CBCANH and CBCA(CO)NH peaks for the i-th residue, we can generate two consecutive spin systems. That is, we use HSQC to select a spin system of a residue, say the i-th residue, use CBCA(CO)NH to distinguish carbons of the $(i-1)$-th and the i-th residues, and use the "+" and "−" signs in CBCANH to distinguish α and β carbons. These groups of seven peaks are called a *spin system group* in the rest of the paper.

2.1   *RGT* Operation: Referencing, Grouping and Typing

In practice, perfect dataset is not available due to experimental errors and biases. So using an exact match algorithm is not feasible for comparing different experimental data. We design an operation RGT to group related peaks from different experiments, to generate two consecutive spin systems (or *spin system pair*), and to determine the amino acid types of the two spin systems. These three procedures, referencing, grouping and typing are all mixed together in RGT. Initially, we plot all HSQC, CBCANH and CBCA(CO)NH peaks onto an H^N-N plane (see Figure 2). Ideally, for each residue, there should be seven peaks mapped to a single point in the plane since they share the same H^N and N chemical shifts (i.e., these seven points coincide completely): one HSQC peak, two CBCA(CO)NH peaks and four CBCANH peaks. In reality, they usually do not coincide, but are clustered nearby. Since HSQC is more reliable than the other two experiments, we use its peaks as bases to identify different clusters. For each HSQC point in the plane, RGT finds the closest four CBCANH points and two CBCA(CO)NH points. If the distances between these points are within a certain threshold, RGT regards them as peaks associated with a specific residue, and forms the corresponding spin system pairs. If there are more

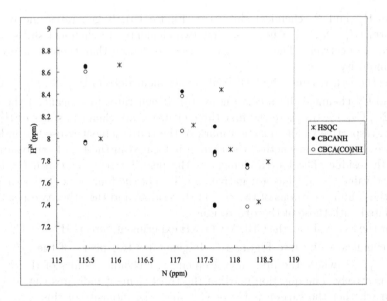

Fig. 2. An H^N-N plane. The unit of chemical shifts is ppm (parts per million). Peaks with the same H^N and N chemical shifts are normally close to each other

than six peaks close enough to a HSQC peak, we generate all legal combinations (two CBCA(CO)NH peaks and four CBCANH peaks) to represent possible spin system pairs. Once spin system pairs are generated, we determine their residue types according to the ranges given by TATAPRO II [2]; see Table 1. According to Table 1, some typed spin systems are associated with a set of possible residues rather than a unique one. If a generated spin system cannot be typed according to the table, it is deleted. The typed spin systems are basic units to perform

Table 1. Amino acid types based on carbon chemical shift characteristics [2]. Since Glysine has only a proton on its side chain, it has no C^β chemical shift. Proline intrinsically has no peaks appearing in NMR spectra, so it has no C^α or C^β chemical shifts. Cys^{red} and Cys^{oxd} represent reduced Cystein and oxidized Cystein, respectively

Carbon chemical shift	Amino acid
Absence of C^β	Gly
$14 < C^\beta < 24$	Ala
$56 < C^\beta < 67$	Ser
$24 < C^\beta < 36$ and $C^\alpha < 64$	Lys, Arg, Gln, Glu, His, Trp, Cys^{red}, Val and Met
$24 < C^\beta < 36$ and $C^\alpha \geq 64$	Val
$36 < C^\beta < 52$ and $C^\alpha < 64$	Asp, Asn, Phe, Tyr, Cys^{oxd}, Ile and Leu
$36 < C^\beta < 52$ and $C^\alpha \geq 64$	Ile
–	Pro
$C^\beta > 67$	Thr

linking and mapping. Note that RGT only performs referencing, grouping and typing, and it does not handle false negatives or false positives. False positives are handled by the LM operation (Section 2.3); false negatives are handled by the iterative relaxation technique in Sections 2.4 and 2.5.

2.2 LM Operation: Linking and Mapping

Given a set of typed spin system pairs, we try to link them to form longer spin-system segments that can be mapped onto the target protein sequence. A typed spin system pair generated by RGT can be regarded as a segment of length 2. Initially, all segments are placed in possible positions with respective to the target sequence according to Table 1. A segment may be placed in more than one position. Two typed spin systems are matched only if they satisfy a set of predefined thresholds. Any two segments (of length ≥ 2) can be linked to form a longer one if their overlapped typed spin systems are matched. Since segments have already been placed in all possible positions, there is no need to check all segment-pair combinations; it suffices to check consecutive segment pairs. Note that for each typed spin system pair/segment there may be more than one candidate to link to. In the LM operation, we generate all possible linked segments to prevent false negatives; this may generate false positives, which will be handled in the mapping mechanism below.

If linked segment cannot be extended further and their lengths are greater than 3, some of them are then mapped to a target protein sequence. Even though segments can be mapped to the target sequence, there are some potential problems to consider. First, due to false positive of peaks, a segment may contain fake spin systems and should not be mapped to the target sequence; this causes more than one segment to be mapped to overlapped regions of the target protein, which is a contradiction (see Figure 3a). Second, since more than one residue may be associated with a typed spin system, a segment can probably be mapped to more than one position in the target sequence (see Figure 3b). Third, since we generate all possible segments, a spin system may be contained in more than one segment, and those segments cannot be mapped to the target protein sequence simultaneously. In fact, at most one of them can be correctly mapped (see Figure 3c). To resolve these mapping problems, we model them as a graph optimization problem.

Let $G(V, E)$ be an undirected graph, where V is a set of nodes and E a set of edges. Each node in V represents a mapping from a segment to the target sequence. If a segment can be mapped to $n \geq 2$ positions, there will be n nodes to represent the n mappings. There is an edge between two nodes if there is a conflict between them. Two nodes are in conflict if 1) they share the same spin system(s), or 2) they are mapped to overlapped regions in the target protein sequence. Our goal is to map segments to the target protein containing as many residues as possible without conflict. This can be formulated as the following weighted maximum independent set problem:

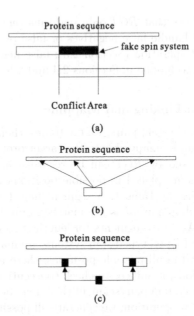

Fig. 3. Problems in mapping. (a) Two segments overlap because one of them contains a fake spin system. (b) A segment maps to more than one position in the sequence. (c) Two segments that are farther apart could contain an identical spin system

Given an undirected graph $G(V, E)$, find a set $S \subseteq V$ such that

1. *For any two distinct nodes p, $q \in S$, $(p, q) \notin E$.*
2. *The weighted sum of all nodes in S is greater than or equal to that of any other subset of V satisfying (1).*

The weight $w(v)$ of a node v is defined as follows:

$$w(v) = \frac{|v| + \sum_{x \in v} \frac{1}{N(x)}}{fre(v)} \tag{3}$$

where $|v|$ is the length of v in terms of mapped segment, x is a spin system of v, $N(x)$ is the number of spin systems having the same H^N and N chemical shifts as x, and $fre(v)$ is the number of positions on the target sequence to which v may map. This weight function satisfies

1. longer segments have higher weights;
2. segments that are more specific to the target sequence have higher weights;
3. segments containing less ambiguous spin systems have higher weights.

Nodes not included in S are regarded as false positives. This graph optimization problem is known to be NP-hard [12]. Although there are algorithms for solving small size problems (such as 200 nodes) in reasonable amount of time,

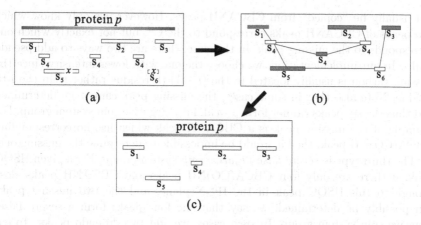

Fig. 4. Use a maximum independent set algorithm to map segments. (a) There are six segments $S_1 \dots S_6$ to be mapped to the target protein p. Among them, S_4 can be mapped to three positions of p; S_4 may overlap S_1, S_2 and S_5; S_3 overlaps S_6; both S_5 and S_6 contain the same spin system x. (b) The corresponding undirected graph, in which nodes are represented by horizontal segments. The shaded nodes form the maximum weighted independent set S of the graph. (c) The final mapping

there are good heuristics with much better performance. In the *LM* operation, we adopt a modified heuristic proposed in [6]. The modified heuristic generates several independent sets (rather than a single maximum one) to include more candidates in the future extending and linking. *LM* is operated on the longest n segments to generate S; in our implementation, n is 100. Figure 4 is an example of our mapping approach.

2.3   Classification of Spin System Group Data

To handle false negatives, we distinguish three types of spin system groups based on their data quality. The first type is *perfect spin system groups*. For a given HSQC peak, if we can find two CBCA(CO)NH peaks and four CBCANH peaks close enough to the peak in the H^N-N plane, we say that these seven peaks form a perfect spin system group, and we can easily generate a corresponding spin system pair.

The second type is *weak false-negative spin system groups*. For a given HSQC peak, if there are only five CBCA(CO)NH peaks and CBCANH peaks that are close enough to this HSQC peak in the H^N-N plane and the missing one can possibly be determined, we say that these six peaks form a weak false-negative spin system group. In such cases, we add a "pseudo peak" to enable cross-referencing and generate a corresponding spin system pair. Note that there may be more than one way to add the pseudo peak. For example, if only one CBCA(CO)NH peak (H^N, N, C, +) is available, we compare it with four CBCANH peaks to determine whether it corresponds to a C^α or a C^β. Suppose it corresponds to a C^α. Then we know that the peak containing the C^β information is missing, which can usually be "copied" from CBCANH peaks. However, we only know which two peaks of CBCANH peaks correspond to the C^β, but not exactly which one corresponds to the missing peak. In this case, there are two ways to add pseudo peaks. In our implementation, we choose the one with lower intensity since the missing carbon is usually located in the $(i-1)$-th residue rather than the i-th residue. Note also that in some cases, the missing peak cannot be determined and thus the six peaks cannot form a weak false-negative spin system group. For example, if the missing peak is a CBCANH peak which has no corresponding CBCA(CO)NH peak, then it would be impossible to determine the missing one.

The third type is *severe false-negative spin system groups*. For a given HSQC peak, if there are only four CBCA(CO)NH peaks and CBCANH peaks close enough to this HSQC peak in the H^N-N plane, and the two missing peaks can possibly be determined, we say that the four peaks form a severe false-negative spin system group. In such cases, we add two "pseudo peaks" to enable cross-referencing and generate a corresponding spin system pair; we may cross-reference CBCANH and CBCA(CO)NH peaks to recognize which peaks are missing and should be added. Similarly, the way to add the pseudo peaks is not unique, and we generate some possible candidates to create spin system pairs. In some cases, the two missing peaks cannot be determined and the five peaks cannot form a severe false negative spin system group.

2.4   Overview of RIBRA

RIBRA uses a relaxation approach to perform backbone assignment in an iterative fashion. The relaxation approach is controlled by the data quality. We first use peaks with top-level quality (perfect spin system groups) to generate a partial assignment of high confidence, and then use peaks with middle-level (weak false-negative spin system group) and low-level quality (severe false-negative spin system groups) to iteratively make more assignment. The general steps of RIBRA is as follows:

1. Extract peaks that can form perfect spin system groups.
   (a) Apply *RGT* to generate typed spin system pairs.
   (b) Apply *LM* on the generated typed spin system pairs to carry out assignment.
   (c) Delete used spin systems and related peaks.
2. Extract peaks that can form weak false-negative spin system groups.
   (a) Add required pseudo peaks.
   (b) Apply *RGT* to generate typed spin system pairs.
   (c) Apply *LM* on the newly generated typed spin system pairs and existing segments to modify the assignment generated in Step 1.b.
   (d) Delete used spin systems and related peaks.
3. Extract peaks that can form severe false-negative spin system groups.
   (a) Add required pseudo peaks.
   (b) Apply *RGT* to generate typed spin system pairs.
   (c) Apply *LM* on the newly generated typed spin system pairs and existing segments to modify the assignment generated in Step 2.c.

In Steps 2 and 3, we may add new spin-system segments or extend those contained in the previous alignment. These segments can be used to modify the alignment by applying the weighted maximum independent set algorithm.

3   Implementation

We implement RIBRA as a C++ executable file and test it on three kinds of data: BMRB perfect datasets, synthetic datasets based on BMRB data, and real wet-lab datasets. We evaluate RIBRA by calculating its recall and precision on different datasets. Note that Prolines are not taken into account because they have no NMR peaks in HSQC, CBCANH and CBCA(CO)NH experiments.

3.1   Experimental Results on BMRB Perfect Datasets

We download the full BMRB datasets with 3129 proteins on September 10, 2004. Proteins satisfying the following two conditions are chosen for our experiments:

1. $50 \leq$ protein length ≤ 400;
2. Each protein has at least half of the amino acids whose chemical shifts are available;

In the end, we have collected 902 proteins for testing. The average length of proteins in the collection is 128.12, and the average number of amino acids with available chemical shifts is 110.87. For each protein, we generate the corresponding HSQC, CBCANH and CBCA(CO)NH peaks for testing. The average precision and recall of RIBRA on these datasets are 98.28% and 92.33%, respectively.

3.2   Experimental Results on Synthetic Datasets

Synthetic Dataset Construction. We generate synthetic testing datasets to simulate real world noises. Three modifications on the 902 proteins used in Section 3.1 are considered: false negatives, false positives and errors. Since the problem of clustered peaks is an intrinsic property of NMR data, we do not synthesize such datasets.

To generate false negative datasets, we create missing peaks by randomly deleting peaks. Since we use HSQC peaks as bases, we delete only CBCANH and CBCA(CO)NH peaks. For the i-th residue, the peaks associated with C_{i-1}^{α} and C_{i-1}^{β} in CBCANH are more likely missing than those associated with C_i^{α} and C_i^{β} due to their longer distances to H_i^N and N_i. And they are usually missing together. Note that we do not delete C_i^{α} and C_i^{β} peaks in CBCANH spectra since their peak intensities are usually high enough. Therefore, we delete peaks as follows:

1. Delete both C_{i-1}^{α} and C_{i-1}^{β} peaks in CBCANH randomly with probability 0.06
2. Delete the C_{i-1}^{α} peak in CBCANH randomly with probability 0.02

3. Delete the C_{i-1}^{β} peak in CBCANH randomly with probability 0.02
4. Delete the C_{i-1}^{α} peak in CBCA(CO)NH randomly with probability 0.05
5. Delete the C_{i-1}^{β} peak in CBCA(CO)NH randomly with probability 0.05

We select deleted peaks based on a uniform distribution.

To generate false positive datasets, we generate fake peaks to simulate false positives. Noisy peaks are usually randomly distributed. Hence, we randomly generate fake peaks with carbon chemical shifts between 10-70. We follow the rules below to generate fake CBCANH peaks:

1. Randomly generate 5% of C_i^{α} and C_i^{β} peaks, respectively, to the CBCANH experiment.
2. Randomly generate 5% of carbon peaks with chemical shifts between 10-70 to the CBCA(CO)NH experiment.

We consider data errors arising from different experimental conditions. They are classified into *grouping errors* and *linking errors*:

- When grouping related peaks to form a spin system pair, one has to match chemical shifts of the CBCANH and CBCA(CO)NH peaks. The grouping error complicates this matching since we do not know whether two peaks with similar chemical shifts are actually two peaks or identical. We simulate such error as follows. When we generate two CBCA(CO)NH peaks for the i-th residue, alter the chemical shifts of H^N, N, C_{i-1}^{α} and C_{i-1}^{β} off their originals by ± 0.06, ± 0.8, ± 0.2 and ± 0.4 ppm, respectively. Furthermore, assume these chemical shift differences follow normal distributions with mean 0 and different standard deviations of 0.0024, 0.32, 0.08 and 0.16 ppm for H^N, H, α carbon and β carbon, respectively.
- When linking two consecutive segments, we match their overlapped spin systems. The match cannot be exact due to linking errors of C^{α} and C^{β} chemical shifts. We simulate such error as follows. When we generate four CBCANH peaks for the i-th residue, we alter the chemical shifts of C_{i-1}^{α} and C_{i-1}^{β} off their originals by ± 0.2 and ± 0.4 ppm, respectively. Furthermore, these chemical shift differences all follow normal distributions with mean 0 and different standard deviations of 0.08 and 0.16 ppm for α carbon and β carbon, respectively.

Experimental Results. The precision and recall of RIBRA on false positive, false negative, linking errors and grouping errors datasets are listed in Table 2.

Among all 902 proteins with linking errors, the PACES [11] has also tested 21 of them with a similar parameter setting (which is the only test on synthetic data that PACES performed in their paper). PACES can handle only 20 of them and gets 95.24% precision and 86.48% recall in average. RIBRA can handle all 21 proteins and gets 95.05% precision and 87.81% recall in average. Keep in mind that the final results of PACES are further edited by human experts, whereas RIBRA is fully automatic. Moreover, PACES assumes that the spin systems are already given.

Table 2. Experimental results of RIBRA on synthetic datasets

# of proteins	902	
# of available residues	100,001	
Testing Cases	Precision	**Recall**
False positive	98.28%	92.35%
False negative	95.61%	77.36%
Grouping errors	98.16%	88.57%
Linking errors	96.28%	89.15%

Table 3. The attributes of hbSBD and hbLBD datasets

Datasets	hbSBD	hbLBD
# of amino acids (A.A.)	53	85
# of A.A. manually assigned by biologists	42	80
# of HSQC peaks	58	78
# of CBCA(CO)NH peaks	258	271
# of CBCANH peaks	224	620
False positive (CBCA(CO)NH)	67.4%	41.0%
False positive (CBCANH)	25.0%	48.4%

3.3 Experimental Results on Real NMR Spectral Data

Real NMR spectra normally are mingled with false negatives, false positives, clustered peaks and errors. We test RIBRA on two datasets that correspond to substrate binding domain of BCKD (hbSBD) and lipoic acid bearing domain of BCKD (hbLBD), respectively. The detailed properties of hbSBD and hbLBD are listed in Table 3. The precision and recall of RIBRA on the hbSBD dataset are 91.43% and 76.19%, respectively; and those on the hbLBD dataset are 83.58% and 70.00%, respectively.

4 Conclusion

Many backbone assignment approaches have been developed. They usually perform well on perfect data but cannot achieve good accuracy and recall simultaneously on real wet-lab spectral data containing missing peaks and noises. RIBRA resolves these problems in two ways by: 1) using weighted independent set algorithm to deal with false positives; and 2) using an iterative approach to deal with false negatives in a relaxation fashion. RIBRA uses the best quality NMR peaks to generate basic spin-system segments that serve as building blocks. Then use less confident peaks to extend previous results. The experimental results show that this approach achieves a high degree of precision.

In addition, we have also created four synthesized datasets from BMRB that can be used to perform comprehensive verification on automatic backbone assignment algorithms in the future. They contain false negatives, false positives, linking errors and grouping errors respectively to simulate real NMR spectral data.

References

1. Atreya, H.S., Sahu, S.C., Chary, K.V.R. and Govil, G.: A tracked approach for automated NMR assignments in proteins (TATAPRO). J. Biomol. Nmr. **17** (2000) 125-136
2. Atreya, H.S., Chary, K.V.R. and Govil, G.: Automated NMR assignments of proteins for high throughput structure determination: TATAPRO II. Curr. Sci. India. **83** (2002) 1372-1376
3. Bailey-Kellogg, C., Widge, A., Kelley, J.J., Berardi, M.J., Bushweller, J.H. and Donald, B.R.: The NOESY Jigsaw: Automated protein secondary structure and main-chain assignment from sparse, unassigned NMR data. J. Comput. Biol. **7** (2000) 537-558
4. Bailey-Kellogg, C., Chainraj, S. and Pandurangan, G.: A Random Graph Approach to NMR Sequential Assignment. The Eighth Annual International Conference on Research in Computational Molecular Biology (RECOMB'04). (2004) 58-67
5. Bartels, C., Guntert, P., Billeter, M. and Wuthrich, K.: GARANT–A general algorithm for resonance assignment of multidimensional nuclear magnetic resonance spectra. J. Comput. Chem. **18** (1997) 139-149
6. Boppana, R. and Halldorsson, M.M.: Approximating Maximum Independent Sets by Excluding Subgraphs. Bit. **32** (1992) 180-196
7. Buchler, N.E.G., Zuiderweg, E.R.P., Wang, H. and Goldstein, R.A.: Protein heteronuclear NMR assignments using mean-field simulated annealing. J. Magn. Reson. **125** (1997) 34-42
8. Chang, C.F., Chou, H.T., Chuang, J.L., Chuang, D.T. and Huang, T.H.: Solution structure and dynamics of the lipoic acid-bearing domain of human mitochondrial branched-chain alpha-keto acid dehydrogenase complex. J. Biol. Chem. **277** (2002) 15865-15873
9. Chen, Z.Z., Jiang, T., Lin, G.H., Wen, J.J., Xu, D. and Xu, Y.: Improved approximation algorithms for NMR spectral peak assignment. Lect. Notes. Comput. Sc. **2452** (2002) 82-96
10. Chen, Z.Z., Jiang, T., Lin, G.H., Wen, J.J., Xu, D., Xu, J.B. and Xu, Y.: Approximation algorithms for NMR spectral peak assignment. Theor. Comput. Sci. **299** (2003) 211-229
11. Coggins, B.E. and Zhou, P.: PACES: Protein sequential assignment by computer-assisted exhaustive search. J. Biomol. Nmr. **26** (2003) 93-111
12. Garey, M.R. and Johnson, D.S.: Computer and Intractability: A guide to the Theory of NP-completeness. New York: W.H. Freeman and Co.(1979)
13. Guntert, P., Salzmann, M., Braun, D. and Wuthrich, K.: Sequence-specific NMR assignment of proteins by global fragment mapping with the program MAPPER. J. Biomol. Nmr. **18** (2000) 129-137
14. Hitchens, T.K., Lukin, J.A., Zhan, Y.P., McCallum, S.A. and Rule, G.S.: MONTE: An automated Monte Carlo based approach to nuclear magnetic resonance assignment of proteins. J. Biomol. Nmr. **25** (2003) 1-9

15. Hyberts, S.G. and Wagner, G.: IBIS–A tool for automated sequential assignment of protein spectra from triple resonance experiments. J. Biomol. Nmr. **26** (2003) 335-344.

16. Langmead, C.J., Yan, A., Lilien, R., Wang, L. and Donald, B.R.: Large a polynomial-time nuclear vector replacement algorithm for automated NMR resonance assignments. The Seventh Annual International Conference on Research in Computational Molecular Biology (RECOMB'03). (2003)

17. Leutner, M., Gschwind, R.M., Liermann, J., Schwarz, C., Gemmecker, G. and Kessler, H.: Automated backbone assignment of labeled proteins using the threshold accepting algorithm. J. Biomol. Nmr. **11** (1998) 31-43

18. Li, K.B. and Sanctuary, B.C.: Automated resonance assignment of proteins using heteronuclear 3D NMR. Backbone spin systems extraction and creation of polypeptides. J. Chem. Inf. Comp. Sci. **37** (1997) 359-366

19. Lin, G.H., Xu, D., Chen, Z.Z., Jiang, T., Wen, J.J. and Xu, Y.: An efficient branch-and-bound algorithm for the assignment of protein backbone NMR peaks. Proceedings of the First IEEE Computer Society Bioinformatics Conference (CSB'02). (2002) 165-174

20. Lukin, J.A., Gove, A.P., Talukdar, S.N. and Ho, C.: Automated probabilistic method for assigning backbone resonances of (13C, 15N)-labeled proteins. J. Biomol. NMR. **9** (1997) 151-166

21. Malmodin, D., Papavoine, C.H.M. and Billeter, M.: Fully automated sequence-specific resonance assignments of heteronuclear protein spectra. J. Biomol. Nmr. **27** (2003) 69-79

22. Moseley, H.N.B. and Montelione, G.T.: Automated analysis of NMR assignments and structures for proteins. Curr. Opin. Struc. Biol. **9** (1999) 635-642

23. Ou, H.D., Lai, H.C., Serber, Z. and Dotsch, V.: Efficient identification of amino acid types for fast protein backbone assignments. J. Biomol. Nmr. **21** (2001) 269-273

24. Slupsky, C.M., Boyko, R.F., Booth, V.K. and Sykes, B.D.: Smartnotebook: A semi-automated approach to protein sequential NMR resonance assignments. J. Biomol. Nmr. **27** (2003) 313-321

25. Wang, X., Xu, D., Slupsky, C.M. and Lin, G.H.: Automated Protein NMR Resonance Assignments. Proceedings of the Second IEEE Computer Society Bioinformatics Conference (CSB'03). (2003) 197-208

26. Xu, Y., Xu, D., Kim, D., Olman, V., Razumovskaya, J. and Jiang, T.: Automated assignment of backbone NMR peaks using constrained bipartite matching. Comput. Sci. Eng. **4** (2002) 50-62

27. Zimmerman, D.E., Kulikowski, C.A., Huang, Y.P., Feng, W.Q., Tashiro, M., Shimotakahara, S., Chien, C.Y., Powers, R. and Montelione, G.T.: Automated analysis of protein NMR assignments using methods from artificial intelligence. J. Mol. Biol. **269** (1997) 592-610

Avoiding Local Optima in Single Particle Reconstruction

Marshall Bern[1], Jindong Chen[1], and Hao Chi Wong[1,2]

[1] Palo Alto Research Center,
3333 Coyote Hill Rd., Palo Alto, CA 94304, USA
[2] Departamento de Ciência da Computação,
UFMG, Belo Horizonte, MG, Brazil
{bern, jchen, hcwong}@parc.com

Abstract. In single-particle reconstruction, a 3D structure is reconstructed from a large number of randomly oriented 2D projections, using
techniques related to computed tomography. Unlike in computed tomography, however, the orientations of the projections must be estimated at
the same time as the 3D structure, and hence the reconstruction process
can be error-prone, converging to an incorrect local optimum rather than
the true 3D structure. In this paper, we discuss and further develop a
maximum-likelihood approach to reconstruction, and demonstrate that
this approach can help avoid incorrect local optima for both 2D and 3D
reconstructions.

1   Introduction

Cryo-Electron Microscopy (cryo-EM) uses a transmission electron microscope to
acquire 2D projections of a specimen preserved in vitreous ice. A 3D electron
density map can then be reconstructed from the 2D projections computationally.
In "single-particle" cryo-EM, the specimen consists of many ostensibly identical copies of randomly oriented particles, and the reconstruction process must
estimate the unknown orientations at the same time that it estimates the 3D
structure. (Other types of cryo-EM specimens, such as 2D crystals or helical
filaments, exhibit naturally ordered arrangements that greatly simplify the orientation problem.)

Cryo-EM is emerging as an important technique in structural biology, because it can determine 3D structures without the need for crystallization. It is
especially well-suited to the study of large molecular complexes, which are vi-
tally important but very difficult to study by other means [1]. Single particle
cryo-EM [14] has been used quite extensively to study the ribosome, achieving
about 10 Å resolution with 20,000–50,000 images. Cryo-EM can also be used in
conjunction with x-ray crystallography, with x-ray data supplying atomic resolution (3 Å or better) for subunits whose overall arrangement in natural conditions
is determined by cryo-EM. A recent demonstration [20] of the tremendous potential of such a combined approach is the atomic-resolution structure of an
11-subunit bacterial flagellar filament (a helical specimen).

S. Miyano et al. (Eds.): RECOMB 2005, LNBI 3500, pp. 118–132, 2005.

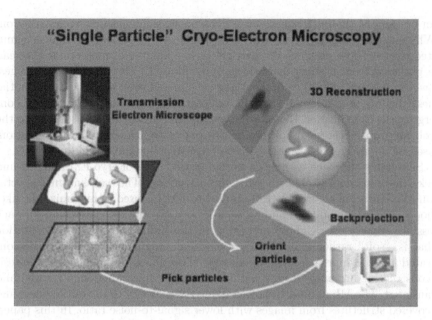

Fig. 1. The processing pipeline for single-particle cryo-EM includes particle picking, followed by iterations of orientation estimation and 3D reconstruction. (Figure adapted from [17])

The central issue in single particle reconstruction is the chicken-and-egg problem of determining image orientations without knowing the particle shape, and determining particle shape without knowing the orientations. The solution developed over the last 20 years [5, 18] is an iterative process that starts from an initial guess of the shape, perhaps as simple as an ellipsoid, and then iteratively aligns images to the current shape and reconstructs a new shape from the aligned images, as shown in Figure 1. In the standard approach [15], (sometimes called "angular reconstitution" [18]), each particle image contributes equally to the reconstruction, with a single orientation set by the image's best alignment, typically determined by maximum correlation. The standard approach—implemented in three major software packages, IMAGIC [18], SPIDER [6], and EMAN [10]—is remarkably successful, giving correct structures from images that to the human eye appear to be almost pure noise. Yet occasionally, depending upon the initial guess, the standard approach gives an incorrect structure, close to a fixed point for the iteration but nowhere near the true structure. See Figure 8(c).

What can be done to prevent such a disaster? One possible solution is a better initial guess. If the particle has strongly preferred orientations, as many do, then it is possible with the "random conical tilt" approach [5] to obtain a set of projections with known relative orientations. Similarly, cryo-electron tomography [9] images the very same particle from a number of known orientations. In either case the result is a low-resolution (say 30–40 Å) 3D reconstruction that can be used as the initial guess for a subsequent single-particle reconstruction. (Why not use tomography for the entire problem? The answer is that accumulated electron dose destroys the particle, so the number of projections that can be gathered from a single particle is quite limited.) Obtaining tilted images, however, requires special equipment and set-up, and even with the better initial guess there is no guarantee that the orientation/reconstruction cycle will converge to the right answer. Hence we would also like an algorithmic solution to the problem of incorrect local optima. In this paper we investigate such a solution, based on an approach to single-particle reconstruction due to Sigworth [16].

Sigworth's approach assumes a probabilistic model of cryo-EM imaging and seeks the maximum-likelihood (ML) reconstruction. The well-known Expectation Maximization algorithm (EM applied to EM!) is used to maximize the likelihood [3]; this amounts to adding each image into the reconstruction for each possible orientation weighted according to the probability of that orientation (that is, "soft" rather than "hard" orientation assignments). Sigworth demonstrated his ML approach for an analogous 2D problem of independent interest: reconstructing an image from a number of noisy copies, randomly rotated and translated. The ML approach showed reduced sensitivity to the initial guess and recovered structures from images with lower signal-to-noise ratio. In this paper, we demonstrate that the ML approach can also avoid incorrect local optima, first with synthetic 2D examples (Section 3), and then with real 2D and 3D data (Sections 4 and 5).

2   Reconstruction Algorithms

Following Sigworth [16], we describe reconstruction algorithms for an analogous 2D problem, rather than for the full 3D problem. The analogy between the two problems is not perfect, because the 2D problem does not involve projections (integrals of the density along parallel viewing lines) of the original structure; nevertheless both include the central chicken-and-egg problem. The 2D problem arises as a subproblem in 3D reconstruction in each of the major software packages [5, 6, 10].

Let the particle images—the data—be denoted $D_1, D_2, \ldots, D_n$. Each image D_i is assumed to be a noisy copy of the same original image Φ, rotated and translated by some unknown amounts. We are assuming that the D_i's have been located by a particle picking process that approximately locates the center of each particle, but leaves the rotation and the precise translation unknown [16]. Let $\theta_i = (\alpha_i, x_i, y_i)$ denote the *alignment parameters* for image D_i, with α_i the rotation angle and (x_i, y_i) the translation vector. That is,

$$D_i = z \cdot (-\theta_i(\Phi)) + (1 - z) \cdot G_i, \qquad (1)$$

where G_i is the noise, z is a constant with $0 \leq z \leq 1$, and the notation $(-\theta_i(\Phi))$ means Φ transformed by the rotation and translation $-\theta_i()$ (so that the alignment parameters $\theta_i()$ return D_i to its original orientation). Assuming Φ and G_i have the same variance, $z^2/(1-z)^2$ gives the signal-to-noise ratio (SNR), defined as the ratio of signal variance to noise variance [5]. When we refer to images as 95% noise we mean that $1 - z = 0.95$.

2.1   The Standard Method

The standard method [12, 15] attempts to directly determine the alignment parameters. Given the reconstruction $\Phi^{(k)}$ after k iterations, this method computes a number of *reference images*, denoted $R_1^{(k)}, R_2^{(k)}, \ldots, R_m^{(k)}$, all possible rotations and translations of $\Phi^{(k)}$ quantized to some pre-determined level of precision.[1] To determine the alignment parameters for image D_i, the standard method finds the j that maximizes the cross-correlation $D_i \cdot R_j^{(k)}$ and sets $\theta_i^{(k)}$ to be equal to the rotation and translation values of the maximizing $R_j^{(k)}$. As usual the cross-correlation is the inner product of the images, treated as vectors.

Once the alignment parameters are determined, then the next reconstruction is simply the aligned average of the data images,

$$\Phi^{(k+1)} = \frac{1}{n} \sum_{i=1}^{n} \theta_i^{(k)}(D_i). \tag{2}$$

For continuous values of the alignment parameters, this iteration converges to a local maximum of the Euclidean norm of the reconstruction $\|\Phi\|_2$, in other words the strongest-contrast reconstruction [12, 16]. With quantized alignment parameters, however, the iteration does not necessarily converge but rather reaches a small limit cycle. Limit cycles are not usually reached in real biological projects, which use 1000's of data images, 1000's of reference images, and fewer than 50 iterations, but they are indeed observable in our small 2D experiments (Figure 4).

2.2   Maximum-Likelihood Method

For low-SNR images, random correlation peaks cause mistakes in the alignment parameters, and the standard method can give poor results. Sigworth's method [16] never actually determines the alignment parameters, but rather treats them as hidden random variables with probability distributions over possible values, and then attempts to maximize the likelihood of the data images D_i as a function of the reconstruction and the parameters of the probability distributions. Sigworth derives his algorithm, a version of the Baum-Welch or EM algorithm [3], assuming that the data arises from a specific generative model: additive Gaussian pixel noise (independent and identically distributed), normally distributed image translations, and uniformly distributed image rotations. In this section, we rederive the ML approach in a more realistic setting. In Section 2.3 we discuss the key step of the algorithm in more detail. Then in Section 2.4 we develop three variations on the ML approach that address some of the practical issues.

---

[1] In practice, references are produced only for different rotations, and translations are determined by convolution using the Fast Fourier Transform.

The likelihood is the probability that the observed data set $D_1, D_2, \ldots, D_n$ arises from a given model. Our model assumes an original image Φ and alignment parameters $\theta_1, \theta_2, \ldots, \theta_n$ for the data images. Assume that the alignment parameters, rather than being continuous random variables as they are in real life, are discrete variables that take exactly the same values as the rotation angles and translations of the reference images $R_1, R_2, \ldots, R_m$. Assume that the data images $D_1, D_2, \ldots, D_n$ are produced by a random process that first selects a reference R_j $(= R_j(\Phi))$ according to any probability distribution and then sets D_i to be a copy of R_j with added noise. This *mixture model* is more realistic than Sigworth's classical parametric model, because cryo-EM target particles usually have preferred orientations and are not uniformly distributed over 3D rotations. Let π_j denote the probability of selecting reference R_j, with $\sum_j \pi_j = 1$. For the model just described, the likelihood is a function of Φ (through the reference copies $\{R_j\}$) and $\Pi = (\pi_1, \pi_2, \ldots, \pi_m)$. Assuming independence of the random variables associated with the individual data images, we can express the logarithm of the likelihood (log likelihood) $\mathcal{L}(\Phi, \Pi)$ as a sum of the log likelihoods of the individual images.

$$\mathcal{L}(\Phi, \Pi) = \sum_{i=1}^{n} \log \left(\sum_{j=1}^{m} \pi_j P[D_i \mid R_j] \right). \tag{3}$$

The expression in equation (3) is difficult to maximize directly, due to the sum of terms inside the logarithm. So we introduce unobserved "latent" variables δ_{ij}, with $\delta_{ij} = 1$ if D_i came from R_j and $\delta_{ij} = 0$ otherwise. (See [7], pp. 236–241, for a similar development of the EM algorithm.) This trick allows the sum to be pulled outside the logarithm, and the log likelihood becomes

$$\mathcal{L}(\Phi, \Delta) = \sum_{i=1}^{n} \sum_{j=1}^{m} \delta_{ij} \log P[D_i \mid R_j]. \tag{4}$$

The EM algorithm given in Figure 2 is the standard way to maximize a likelihood expression of this form [7]. In its "E-step" (Steps 2.1 and 2.2), this algorithm makes a soft assignment to δ_{ij}; that is, rather than using a value in $\{0, 1\}$ for δ_{ij}, the algorithm uses a fractional value $\gamma_{ij}^{(k)}$, the expectation of δ_{ij} given the current reconstruction $\Phi^{(k)}$ and references $R_j^{(k)}$. Rather remarkably, this algorithm for maximizing $\mathcal{L}(\Phi, \Delta)$ also monotonically increases the original log likelihood $\mathcal{L}(\Phi, \Pi)$ until it reaches a local maximum [7]; indeed the two log likelihoods agree at a joint maximum.

2.3   Estimating the Image Probabilities

The algorithm as given in Figure 2 is not fully specified. We must give a way to compute $\gamma_{ij}^{(k)}$, which is an estimate of the probability that image D_i is a noisy copy of reference $R_j^{(k)}$. One possibility is simply to set $\gamma_{ij}^{(k)} = 1$ if $R_j^{(k)}$ maximizes $D_i \cdot R_j^{(k)}$ and $\gamma_{ij}^{(k)} = 0$ otherwise. With these crude estimates of $\gamma_{ij}^{(k)}$, the EM

1. Compute references $R_1^{(k)}, R_2^{(k)}, \ldots, R_n^{(k)}$ from the current reconstruction $\Phi^{(k)}$
2. **for** each data image D_i, $i = 1, 2, \ldots, n$ **do**
   2.1. **for** $j = 1, 2, \ldots, m$, compute $\gamma_{ij}^{(k)}$, an estimate of probability $P[D_i \mid R_j^{(k)}]$
   2.2. Normalize $\gamma_{ij}^{(k)}$ values so they sum to 1, that is, $\gamma_{ij}^{(k)} \leftarrow \gamma_{ij}^{(k)} / \left(\sum_j \gamma_{ij}^{(k)}\right)$
   2.3. **for** $j = 1, 2, \ldots, m$, add D_i to the next reconstruction $\Phi^{(k+1)}$
       with orientation j and weight $\gamma_{ij}^{(k)}$
3. $k \leftarrow k + 1$ and go to 1.

Fig. 2. The EM algorithm applied to single particle reconstruction

algorithm reduces to the standard method of Section 2.1. A more principled estimate of $\gamma_{ij}^{(k)}$ assumes a probabilistic model of image formation. Sigworth's model [16] assumes independent Gaussian noise with variance σ^2 in the image pixels, and hence

$$ P[D_i \mid R_j^{(k)}] \propto \exp\left(-\sum_p \frac{(D_i(p) - R_j^{(k)}(p))^2}{\sigma^2} \right), \qquad (5) $$

where $D_i(p)$ is the p-th pixel of D_i, and $R_j^{(k)}(p)$ is the p-th pixel of $R_j^{(k)}$. Since multiplying all the $\gamma_{ij}^{(k)}$'s by the same constant does not change the reconstruction, we can simply set $\gamma_{ij}^{(k)}$ in Step 2.1 to the expression on the right in equation (5). To do so, we must also have an estimate of the noise variance σ^2. In our synthetic-data experiments, we simply plugged in the true value for σ^2. In attempting to add an EM option to EMAN, however, Ludtke (personal communication) found that setting a value for σ^2 that works for real data images was the major stumbling block. The exponential in equation (5) makes $\gamma_{ij}^{(k)}$ quite sensitive to incorrect σ^2 and to deviations of the real data from the imaging model. The problem is not estimating the noise variance, which is relatively easy, but rather the assumption of independent pixels. Another issue is numerical underflow, as $\gamma_{ij}^{(k)}$ values are small, something like e^{-2400} for our synthetic images, so calculations must be done with logarithms. Assuming these difficulties can be overcome, a well-known shortcut to computing the expression in equation (5) is to use the cross-correlation $D_i \cdot R_j^{(k)}$ and the relation

$$ -\sum_p (D_i(p) - R_j^{(k)}(p))^2 = 2D_i \cdot R_j^{(k)} - \sum_p (D_i(p))^2 - \sum_p (R_j^{(k)}(p))^2. \qquad (6) $$

Step 2.2 also has an impact on the image probabilities. This step sets $\sum_j \gamma_{ij}^{(k)}$ equal to one, which has the consequence that each D_i contributes equally to the reconstruction $\Phi^{(k+1)}$. Sigworth's version of the EM algorithm omits this step, because it is derived from a continuous probabilistic model, rather than from a discrete mixture model that assumes that one of the reference orientations is indeed correct. Step 2.2 turns out to be essential. Due to the huge range of $\gamma_{ij}^{(k)}$ values, if Step 2.2 is omitted a small subset of data images—those that happen to be closest to reference image angles—will dominate the reconstruction.

We can imagine interesting ways to modify Step 2.2; for example, if we had an independent measurement of the probability $p_i^{(k)}$ that D_i is indeed a good data image, we could normalize the $\gamma_{ij}^{(k)}$ values to sum to $p_i^{(k)}$ rather than to one. Such a variable normalization could be useful, as cryo-EM particle images vary quite substantially in quality (see Section 4), and the quality of an image cannot be easily assessed *a priori*.

2.4   Three Variations

In this section, we describe three variations of the EM algorithm. The first variation, which we call EM-BASIC, is essentially the same as Sigworth's original algorithm using Equations (5) and (6). The original algorithm, however, is very slow (over an hour for a reconstruction from 200 images), so EM-BASIC incorporates some practical tricks that speed up the processing by a factor of 20, yet give reconstructions almost visually indistinguishable from those of the original algorithm. An obvious speed-up is to round the smallest $\gamma_{ij}^{(k)}$ values to zero, and thus skip Step 2.3 for most i, j pairs. We rounded $\gamma_{ij}^{(k)} < 0.001$ to zero. A further speed-up, found empirically, is to set all except the single largest of the $\gamma_{ij}^{(k)}$ values corresponding to the same rotation angle to zero (before the normalization in Step 2.2), and then to low-pass filter D_i (Gaussian filter with standard deviation $\sigma_F = 1.0$ pixels) before adding it into the reconstruction with orientation $\gamma_{ij}^{(k)}$ to approximate the effect of using multiple translations. This speed-up works because $\gamma_{ij}^{(k)}$ tends to vary in a smooth and predictable way over small translations, but in a much less predictable way over rotations.

Running time is a significant issue in cryo-EM reconstruction, as large-scale reconstructions (say 20,000 images) can take more than a day on a compute cluster. Indeed, large-scale reconstruction almost always includes speed-ups to the standard method, for example, only testing a data image D_i against reference images $R_j^{(k)}$ for orientations that D_i had matched fairly well in previous iterations.

The second variation is called EM-MULTI for multiresolution. EM-MULTI is the same as EM-BASIC, with the addition that it low-pass filters data images heavily in early iterations, and gradually reduces this filtering in later iterations. The standard deviation of the filter is $\sigma_F = \max\{1.0, 12/(k+1)\}$ pixels, where k is the iteration number, running from 1 to 20 in our experiments. Multiresolution optimization is common in pose estimation problems in computer vision [2]. Low-pass filtering can change the SNR of the images, because the signal (the particle) has more low-spatial-frequency content than the noise, so for fair comparison, all our algorithms (including our implementation of the standard method) low-pass filtered images with $\sigma_F = 1.0$, the final value used by EM-MULTI.

Our third variation, EM-ROBUST, addresses the problem reported by Ludtke, that is, the difficulty of robustly estimating $P[D_i \mid R_j^{(k)}]$ for real data images. EM-ROBUST is the same as EM-MULTI, but for each i, EM-ROBUST blindly sets $\gamma_{ij}^{(k)} = \frac{1}{2}$ for the j maximizing $D_i \cdot R_j^{(k)}$, and $\gamma_{ij}^{(k)} = \frac{1}{4}$ for the second best j, and $\gamma_{ij}^{(k)} = \frac{1}{8}$ for the third best, and so forth. Thus EM-ROBUST does not even attempt to map image correlations to probabilities. EM-ROBUST could be used in the case that the matching scores for data and reference images cannot be interpreted as probabilities. For example, EMAN uses a two-stage approach that would be difficult to treat as probabilities: it first matches data images to projections $R_j^{(k)}$ and then reclusters data images, as in the k-means algorithm, by matching them to the average of the data images that matched each $R_j^{(k)}$. In Section 6, we further discuss the problem of setting the $\gamma_{ij}^{(k)}$'s. We are not advocating EM-ROBUST as the solution, but rather we show that even a "deliberately dumb" algorithm such as EM-ROBUST can give reasonable results.

3   Experiments with Synthetic 2D Data

For our 2D experiments we generated synthetic data, grayscale images with values 0–255 and size 73×73 pixels. We used uniformly distributed rotation angles α_i, and normally distributed x- and y-translations x_i and y_i with zero mean and $\sigma = 2.0$ pixels; we used bilinear interpolation for subpixel accuracy. The pixel noise, denoted G_i in equation (1), was independent Gaussian noise with zero mean and $\sigma = 36$ gray levels. We imposed periodic boundary conditions (wrap-around), so that a a pixel moved off the right boundary reappeared on the left, and similarly for top and bottom; wrap-around neatly handles the issue of filling in missing pixels [16]. We did not quantize the rotation angle; hence our data images were not exact matches to any of the reference images, giving a more realistic test of the EM algorithm than was done previously [16].

In all experiments, we used references with rotations of $0°, 5°, 10°, \ldots, 355°$. Each reference image is the average image over three rotations in order to represent "patches" of orientations. For example, the reference image for $10°$ is the average of the rotations at $5°$, $15°$, and two copies of $10°$. Rotations every $5°$ may seem rather coarse, but it would take over 1000 reference images to cover the sphere of views of a 3D particle this densely (not counting planar rotations as separate orientations), and hence this level of quantization is realistic for most biological studies.

We used one additional data image, with α_i, x_i and y_i all equal to zero, as the initial guess $\Phi^{(0)}$. After each iteration, we linearly transformed the grayscale of the reconstruction $\Phi^{(k)}$ so that it had mean 128 and standard deviation 36. This normalization helps with some numerical issues and also enhances the contrast in the reconstructions, making them easier to compare by eye.

Figure 3 shows some examples of synthetic data. Our synthetic particles were deliberately chosen to pose difficult reconstruction problems. For example, the particle shown in Figure 3 has no rotational symmetry, yet is close to a particle—

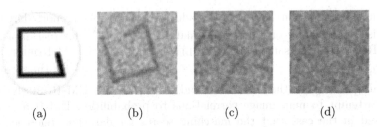

(a)   (b)   (c)   (d)

Fig. 3. Synthetic images: (a) the original particle, (b) with 80% noise, (c) with 90% noise, and (d) with 95% noise, the value used in most of our experiments. The particle shape was chosen to have a likely incorrect local optimum: a square. We used periodic boundary conditions (wrap-around) as seen in (c), to avoid edge effects

a square—with 90° rotational symmetry. Incorrect orientations, approximately 90° or 180° away from the correct rotation, can easily lead the reconstruction algorithm to converge to a square.

Results are given in Figures 4–6. STANDARD is the standard algorithm of Section 2.1, which uses maximum correlation to give a single hard orientation assignment to each data image. We found that all the algorithms including STANDARD perform well with high-SNR images, images that are at most 92% noise.[2] As the noise level grows larger, STANDARD is always the first algorithm to fail, with the various versions of the EM algorithm giving reasonable results with somewhat higher noise percentages. Typically STANDARD fails with noise percentage at about 94% (SNR = 0.052) and the EM algorithms fail with noise percentage at about 96% (SNR = 0.023). We found that results were surprisingly insensitive to initial guess: reconstructions that failed with a data image as the initial guess always failed with the original black-and-white particle as the initial guess as well. Because of its hard orientation assignments, STANDARD also fails in a qualitatively different way from the EM algorithms: it often falls into a noticeable limit cycle with an unstable reconstruction as shown in Figure 4. The EM algorithms reach more stable reconstructions with only slight variations from iteration to iteration.

The noise percentage failure thresholds for the various EM algorithms are typically close, differing by less than 1%. Generally EM-MULTI is the best algorithm. For example, Figure 5(e) not only looks better than Figure 5(d), but also achieves 1.1% larger value on an objective measure of reconstruction quality, the sum of maximum image correlation scores: $S^{(k)} = \sum_i \max_j D_i \cdot R_j^{(k)}$ for $k = 20$. EM-ROBUST falls between the other two EM algorithms, as its $S^{(20)}$ score is 0.6% better than that of EM-BASIC.

---

[2] For black-and-white particles, 92% noise corresponds to SNR of $\sigma_S^2/\sigma_N^2 = (0.08 \times 128)^2/(0.92 \times 36)^2 \approx 0.095$, where σ_S^2 is signal variance and σ_N^2 is noise variance. Similarly 95% noise corresponds to SNR = 0.035, and 96% corresponds to SNR = 0.023. Realistic SNR values for cryo-EM are generally in the range from 0.01 to 0.5, depending upon the particle and the defocus.

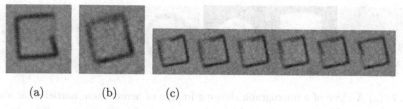

(a)   (b)   (c)

Fig. 4. The success of the STANDARD algorithm depends upon the signal-to-noise ratio. (a) With 93% noise (SNR = 0.072) the algorithm gives a correct reconstruction. (b) With 95% noise (SNR = 0.035) the algorithm falls into an incorrect local optimum. (c) With 94% noise (SNR = 0.051) the algorithm finds a limit cycle with a rotating particle that alternately resembles reconstructions (a) and (b). All reconstructions were made using 200 data images and 20 iterations, but reconstructions changed little after the first 10–12 iterations, except in the case of the limit cycle

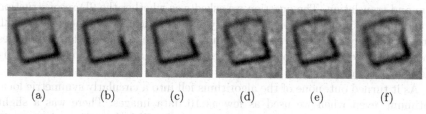

(a)   (b)   (c)   (d)   (e)   (f)

Fig. 5. The max-likelihood approach is more resistant to local optima. (a) EM-BASIC is just starting to fail at 95% noise (SNR = 0.035), but (b) EM-MULTI and (c) EM-ROBUST give correct reconstructions. All these results (from 200 images and 20 iterations) are significantly better than Figure 4(b). (d) EM-BASIC fails at 96% noise (SNR = 0.023), but (e) EM-MULTI is still correct and (f) EM-ROBUST is borderline

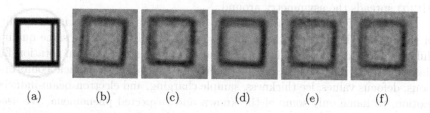

(a)   (b)   (c)   (d)   (e)   (f)

Fig. 6. (a) This particle is an attempt to fool EM-MULTI and EM-ROBUST, which will blur away the doubled edge until late iterations. Yet (b) EM-BASIC, (c) EM-MULTI, and (d) EM-ROBUST achieve fairly similar results at 94% noise. They all start to fail at 95% noise; shown are (e) EM-BASIC and (f) EM-MULTI

4   Real 2D Data

We also applied the algorithms described above to a real 2D data set. The particle is keyhole limpet hemocyanin [11], a didecamer cylindrical "cage" about 400 Å long. This particle shows strongly preferred orientations; for our 2D re-

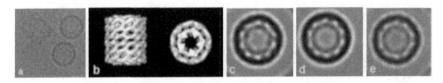

Fig. 7. (a) A piece of a micrograph showing images of hemocyanin particles (at about 1 μm defocus). (b) An isosurface (side and top views) of a 3D reconstruction of hemocyanin made from about 1100 particle images. (c) A 2D reconstruction of the top view made by STANDARD from 250 images. (d) A 2D reconstruction made by EM-MULTI from the same 250 images. (e) A 2D reconstruction made by EM-MULTI from the 25 "best" images

construction experiment we used 250 automatically selected, 100×100 pixel, "top" views [19] with the same nominal defocus, ostensibly identical up to rotation and translation. The question we asked was whether the 2D reconstruction would find the 5-fold rotational symmetry of the particle, or whether it would fall into a circularly symmetric local optimum. As above, we used reference images every 5°. For σ^2 in Equation (5) we used the image variance, in effect treating the image as independent pixels of pure noise.

As it turned out, none of the algorithms fell into a circularly symmetric local optimum, even when we used as few as 10 data images. There was a slight difference, however, between STANDARD and the EM algorithms. As seen in Figure 7(c) and (d), STANDARD gives a less symmetric solution, with some sides of the knobby, white, inner ring more clearly dumbbell-shaped than others. (The strongest dumbbell side, bottom left in Figure 7(c), also "rotates" as in Figure 4.) We believe that many of the top views are slightly tipped, and STANDARD, with its single orientation per image, tends to align the images by tip, whereas EM-MULTI spreads the asymmetry around.

A major difference between synthetic and real data is variation in the quality of the particle images. Synthetic images are essentially all of the same quality, differing only in pseudorandom noise and in how well the random rotation fits the references. Real images, on the other hand, vary due to particle conformations, defocus values, ice thickness, sample charging, and electron-beam-induced motion, to name only some of the known and suspected phenomena. We used a delete-half jackknife [4] to estimate the variance of reconstruction quality, as measured by the sum of maximum correlations, $S^{(10)} = \sum_i \max_j D_i \cdot R_j^{(10)}$, over all the images. (For speed we ran only 10 iterations.) We computed 20 reconstructions, each time leaving out a randomly chosen half of the images. For 200 synthetic images (those used in Figure 5), $S^{(10)}$ for the delete-half subsamples ranged from 97.2% to 99.1% of the $S^{(10)}$ achieved by the entire data set. For 250 real images, $S^{(10)}$ for the delete-half subsamples ranged from 95.4% to 101.1%.

We also ran an experiment to see if we could sense the high-quality images. The algorithm used to make Figure 7(e) considers all the images, but at each iteration chooses only the best 10% (by maximum correlations) to go into the reconstruction. The best 10% of the real images give a reconstruction with $S^{(10)}$

that is 97.1% of the $S^{(10)}$ given by all 250 images; whereas, the same procedure with the synthetic images gave $S^{(10)}$ that is 95.5% of the $S^{(10)}$ given by all the images. Interestingly, the "best" 50% of the real images gave $S^{(10)}$ with 99.4% of the $S^{(10)}$ given by all the images, which seems quite good until compared with the 101.1% achieved by the best delete-half subsample. This experiment suggests that maximum correlation gives only a weak assessment of image quality.

5   Real 3D Data

Finally we applied our ideas to a 3D reconstruction problem from a real biological study, shown in Figure 8. The particle is p97 AAA ATPase [13], a 6-fold symmetric particle about 150 Å across. A large p97 data set converges to the correct optimum (Figure 8(b)), whereas a small data set converges to an incorrect local optimum (Figure 8(c)). The incorrect optimum seems to occur more often with EMAN than with SPIDER, perhaps due to EMAN's reclassification of image orientations, which tends to accelerate convergence.

We modified EMAN in an attempt to correct the problem. Actually implementing some version of the ML algorithm would have required extensive modifications to EMAN, so we tried an approximation instead: we added a random component to EMAN's matching score for each image-orientation pair. Thus suboptimal asssignments occasionally win by chance, as in a Monte Carlo sampling algorithm. If each particle orientation is represented by a number of images in the data set, these randomized assignments approximate the soft assignments of the ML algorithm, but with rather arbitrary "probabilities" as in EM-ROBUST. Specifically, at each iteration k, for $k = 1, 2, \ldots, 20$, for each image i, we added a random component, uniform in the range $[0, (1/4^k) \cdot MaxCorr_i(k)]$ to each correlation score $D_i \cdot R_j^{(k)}$, where $MaxCorr_i(k) = \max_j \{D_i \cdot R_j^{(k)}\}$ is the maximum correlation score for image i at iteration k.

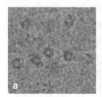

Fig. 8. (a) A piece of a micrograph showing images of p97 AAA ATPase particles (the dark blobs). (b) An isosurface of a correct reconstruction of p97 made from about 4000 particle images. (c) An incorrect reconstruction made from 617 images, using EMAN's default initial guess, an extruded shape. What went wrong is that some of the "top" and "bottom" views (hexagonal rings) of the particle were mistaken for "side" views. (d) A better reconstruction using EMAN with the same initial guess, along with multiresolution refinement and "annealing" of image orientations. (e) A "best-possible" reconstruction from the 617 images using the reconstruction from (b) as the initial guess

The size of the random component is reduced in later iterations, so that the best assignment becomes relatively better as the reconstruction grows more reliable. This reduction step, akin to simulated annealing [8], enables the sampling algorithm to reach a stable reconstruction. Reduction is unnecessary for stability in the 2D versions of the EM algorithm, because they use continuous weights for the soft assignments.

Our modified version of EMAN also includes the multiresolution idea of EM-MULTI. We blur images very heavily at first and gradually reduce the amount of blurring in later iterations. Both Monte Carlo sampling with annealing and multiresolution seem to be necessary to avoid the local optimum for the 617-image p97 data set, as either one alone gives a reconstruction more closely resembling Figure 8(c) than (d).

We specified C_6 symmetry in EMAN, so the symmetry of the reconstructions in Figure 8 is imposed rather than discovered as in Figure 7. Following EMAN's guidelines, we used 49 reference orientations—not counting planar rotations as separate orientations—for all the reconstructions from 617 images. The 617 images are ones picked automatically by a model-based particle picker we developed [19]. Figure 8(d) shows the results from our modified EMAN.

6   Discussion

In this paper we have confronted the most vexing problem of single-particle reconstruction: convergence to incorrect local optima. In experiments with synthetic 2D data, we found that variations of Sigworth's ML approach can indeed avoid incorrect local optima at lower signal-to-noise ratios than the standard approach. We have left open the question of how to estimate the probabilities $\gamma_{ij}^{(k)}$ in a robust and effective way. Accurate estimates may require good models of both the signal (projections of the target particle) and the noise, including the dependencies between neighboring pixels, which are largely due to the point spread, or contrast transfer function (CTF), of the imaging system. Such detailed models could be obtained, at least in principle; for example, the CTF is often estimated using the power spectrum of an entire micrograph. It would be more convenient, however, if a very simple method, such as that used in EM-ROBUST, could give reliable results. Here we are imagining an empirically determined schedule of probabilities, rather than the exponentially decreasing sequence used by EM-ROBUST. Such a method would have to be trained on a variety of real data sets, so it is beyond the scope of the present work.

Although the 2D experimentation described in this paper involved very simple signal and noise models, it apparently served as a reasonable guide for algorithm development, as the multiresolution idea, first implemented in EM-MULTI, proved effective on a real 3D data set when incorporated into EMAN. We now believe that the key to solving the problem of incorrect local optima is to slow convergence and allow the initial iterations to explore the space of reconstructions. Besides multiresolution, other techniques from areas such as machine learning and computer vision, could find useful application in cryo-EM.

Acknowledgements

We thank Fred Sigworth and Steve Ludtke for numerous valuable discussions. The p97 and hemocyanin data used here was collected at the National Resource for Automated Molecular Microscopy which is supported by the National Institutes of Health through the National Center for Research Resources (P41 grant RR17573).

References

1. A. Abbott. The society of proteins. *Nature* 417, 27 June 2002, 894–896.
2. L.G. Brown. A survey of image registration techniques. *ACM Computing Surveys*, 24 (1992), 325–376.
3. A. Dempster, N. Laird, and D. Rubin. Maximum likelihood from incomplete data via the EM algorithm (with discussion). *J. Royal Statistical Soc. B* 39 (1977), 1–38.
4. B. Efron. *The Jackknife, the Bootstrap and Other Resampling Plans*. CBMS-NSF Regional Conference Series in Applied Mathematics #38, SIAM, 1982.
5. J. Frank. *Three-dimensional electron microscopy of macromolecular assemblies*. Academic Press, 1996.
6. J. Frank, M. Radermacher, P. Penczek, J. Zhu, Y. Li, M. Ladjadj, and A. Leith. SPIDER and WEB: processing and visualization of images in 3D electron microscopy and related fields. *J. Struct. Biol.* 116 (1996), 190–199. http://www.wadsworth.org/spider_doc/spider/docs/
7. T. Hastie, R. Tibshirani, and J. Friedman. *The Elements of Statistical Learning: Data Mining, Inference, and Prediction*. Springer, 2001.
8. S. Kirkpatrick, C.D. Gelatt, and M.P. Vecchi. Optimization by simulated annealing. *Science* 220 (1983), 671–680.
9. A.J. Koster, R. Grimm, D. Typke, R. Hegerl, A. Stoschek, J. Walz, and W. Baumeister. Perspectives of molecular and cellular electron tomography. *J. Struct. Biology* 120 (1997), 276–308.
10. S.J. Ludtke, P.R. Baldwin, and W. Chiu. EMAN: Semiautomated software for high-resolution single-particle reconstructions. *J. Struct. Biol.* 128 (1999), 82–97. http://ncmi.bcm.tmc.edu/homes/stevel/EMAN/doc/home.html
11. E.V. Orlova, P. Dube, J.R. Harris, E. Beckman, F. Zemlin, and J. Markl. Structure of keyhole limpet hemocyanin type 1 (KLH1) at 15 Å resolution by electron cryomicroscopy and angular reconstitution. *J. Molecular Biology* 271 (1997) 417–437.
12. P. Penczek, M. Radermacher, and J. Frank. Three-dimensional reconstruction of single particles embedded in ice. *Ultramicroscopy* 40 (1992), 33–53.
13. I. Rouiller, B. DeLaBarre, A.P. May, W.I. Weis, A.T. Brunger, R. Milligan, E.M. Wilson-Kubalek. Conformational changes of the multifunction p97 AAA ATPase during its ATPase cycle. *Nature Struct. Biol.* 9 (2002), 950–957.
14. J. Ruprecht and J. Nield. Determining the structure of biological macromolecules by transmission electron microscopy, single particle analysis and 3D reconstruction. *Progress in Biophysics and Molecular Biology*, 75 (2001), 121–164.
15. W.O. Saxton and J. Frank. Motif detection in quantum noise-limited electron micrographs by cross-correlation. *Ultramicroscopy* 2 (1977), 219–227.
16. F.J. Sigworth. A maximum-likelihood approach to single-particle image refinement. *J. Struct. Biology* 122 (1998), 328–339.

17. H.   Stark.   3D   Electron   Cryomicroscopy.   http://www.mpibpc.gwdg.de/
    abteilungen/103/index.html
18. M. van Heel, *et al.* Single-particle electron microscopy: towards atomic resolution.
    Quarterly Reviews of Biophysics 33 (2000), 307–369.
19. H.C. Wong, J. Chen, F. Mouche, I. Rouiller, and M. Bern. Model-based particle
    picking for cryo-electron microscopy. *J. Struct. Biology* 145 (2004), 157–167.
20. K. Yonekura, S. Maki-Yonekura, and K. Namba. Complete atomic model of the
    bacterial flagellar filament by electron cryomicroscopy. *Nature* 424 (2003), 643–650.

A High-Throughput Approach for Associating microRNAs with Their Activity Conditions

Chaya Ben-Zaken Zilberstein[1], Michal Ziv-Ukelson[1],
Ron Y. Pinter[1], and Zohar Yakhini[12]

[1] Dept. of Computer Science, Technion - Israel Institute of Technology,
Haifa 32000, Israel
{chaya,michalz,pinter}@cs.technion.ac.il
[2] Agilent Technologies, Tel Aviv, Israel
zohar_yakhini@agilent.com

Abstract. Plant microRNAs (miRNAs) are short RNA sequences that bind to target mRNAs and change their expression levels by influencing their stabilities and marking them for cleavage. We present a high throughput approach for associating between microRNAs and conditions in which they act, using novel statistical and algorithmic measures. Our new prototype tool, *miRNAXpress*, computes a (binary) matrix T denoting the potential targets of microRNAs. Then, using T and an additional predefined matrix X indicating expression of genes under various conditions, it produces a new matrix that predicts associations between microRNAs and the conditions in which they act.

The computational intensive part of *miRNAXpress* is the calculation of T. We provide a hybridization search algorithm which given a query microRNA, a text mRNA, and a predefined energy cutoff threshold, finds and reports all targets (putative binding sites) of the query in the text with binding energy below the predefined threshold. In order to speed it up, we utilize the sparsity of the search space without sacrificing the optimality of the results. Consequently, the time complexity of the search algorithm is almost linear in the size of a sparse set of locations where base-pairs are stacked at a height of three or more.

We employed our tool to conduct a study, using the plant *Arabidopsis thaliana* as our model organism. By applying *miRNAXpress* to 98 microRNAs and 380 conditions, some biologically interesting and statistically strong relations were discovered.

Further details, including figures and pseudo-code, can be found at:
http://www.cs.technion.ac.il/~michalz/LinearRNA.ps

1 Introduction

Genes in plants may be expressed in specific locations (*e.g.* leaf-specific genes), at specific times (*e.g.* seedling-specific genes), or in response to environmental stimuli (*e.g.* light-responsive genes). The cellular expression levels of genes are largely influenced by transcription rates as well as by the degradation rates of their mRNAs (messenger RNA). Plant microRNAs (miRNAs) are non coding RNA molecules (of size $\approx$ 22 nucleotides) that regulate gene expression in plants by moderating mRNA degradation rates. Plant microRNAs bind to mRNAs and

S. Miyano et al. (Eds.): RECOMB 2005, LNBI 3500, pp. 133–151, 2005.

mark them for cleavage [10, 12, 23]. For example, *Arabidopsis thaliana* microRNA 39 interacts with mRNAs of several transcription factors to reduce their expressions and to direct the plant development process.

In this paper we propose a high throughput approach for associating between plant microRNAs and conditions in which they act, given a set of microRNAs and a set of conditions. A condition can be, for example, a certain tissue in which the gene expression is to be measured or exposure of the plant to a long darkness period. Note that there are other factors which influence gene expression, including transcription factors which regulate molecular transcription rates and proteins which mediate mRNA degradations, but they are beyond the scope of this study.

Our framework for associating a microRNA and a specific condition can be formalized as follows. First note that since there is a correspondence between genes and mRNAs, we use the term "mRNA" to represent the corresponding gene. Suppose that mRNAs t_1, t_2 and t_3 are targets of microRNA Y, yet other mRNAs t_4, t_5 and t_6 clearly do not bind to Y. Furthermore, assuming there is evidence that t_4, t_5 and t_6 are highly expressed under some given condition, yet t_1, t_2 and t_3 are expressed at a low level under the very same condition, then it may be possible to statistically assert that microRNA Y is active under this condition, contributing to the degradation and low expressions of t_1, t_2 and t_3.

Given a set of p candidate microRNAs $\alpha_1 \ldots \alpha_p$, another set of q conditions $\beta_1 \ldots \beta_q$, and a third set of r genes $\gamma_1 \ldots \gamma_r$, the association between microRNAs and their predicted activity conditions can be formally captured in an *Association Matrix*.

Definition 1. *The* **Association Matrix** A *is a* $q \times p$ *matrix of real numbers such that* $A[i, j]$ *reflects the association of microRNA* α_j *and condition* β_i.

Figure 3 contains highlight entries from the association matrix computed using our method by the experimental setup described in Section 4.

The matrix A is computed by combining information (operating on the rows and columns) from two pre-computed matrices: the Expressions Matrix X and the Targets Matrix T, defined as follows.

Definition 2. *The* **Expressions Matrix** X *is a matrix of real numbers such that* $X[i, j]$, *for* $i = 1 \ldots q$ *and* $j = 1 \ldots r$, *represents the expression of gene* γ_j *under condition* β_i.

(This is a standard representation of the results of expression profiling studies).

Definition 3. *The* **Targets Matrix** T *is a binary matrix where* $T[i, j] = 1$ *means that gene* γ_i *is a predicted target of microRNA* α_j.

In our approach $A[i, j]$ is computed by performing an operation between row X_i of the *Expressions Matrix* and column T_j of the *Targets Matrix*, that is $A[i, j] = F(X_i, T_j)$. We focus on the following instance of an association operation F:

Definition 4. *The* **Association Operation** F: *Let* $\bar{z}$ *denote the binary complement of* z. $F(X_i, T_j)$ *returns a p-value which is the result of a statistical test* *comparing two sets of numbers, $\{X[i,\ell] \times T[\ell,j]\}_{l=1}^{r}$ and $\{X[i,\ell] \times \overline{T[\ell,j]}\}_{l=1}^{r}$, and assessing whether they are significantly different (whether one set contains significantly lower values than the other set).*

The operation F reflects the intuition explained above: it measures the differential expression between potential targets ($\{X[i,l] \times T[l,j]\}_{l=1}^{r}$) of a microRNA and non targets ($\{X[i,l] \times \overline{T[l,j]}\}_{l=1}^{r}$). Thus, $A[i,j]$ is set to the *p-value* which quantifies the association between a microRNA and a condition. This *p-value* is the result of a statistical test intended to reject the hypothesis that the atypical low expression of the targets is due to factors other than quick degradation induced by the microRNA activity. Clearly, the lower the *p-value*, the higher the assumed likelihood that the expression of the targets is influenced by the microRNA activity; see Figure 3.

The *Expressions Matrix X* contains expression values experimentally measured in a micro-array study. A wealth of such information has already been collected by various laboratories and is now available to the public. Naturally, the construction of this matrix does not involve any significant computational task.

However, such is not the case with the *Targets Matrix T*. In order to calculate $T[i,j]$ one needs to asses whether mRNA i is a target of microRNA j in vivo. This can be done computationally, as will be discussed in the next section.

1.1 Computing the Target Matrix T: Approach and Background

In traditional sequence analysis, one tries to assess the likelihood of a hypothesis, for example, whether similarity between sequences is due to common ancestry or continuity of functional constraints or a chance occurrence. Here, however, we aim to assess the likelihood of an actual physical interaction between two molecular species, one is a short microRNA and the other a full length mRNA. The most intuitive measure for this interaction is the free energy of a microRNA/target hybrid. Therefore, we compute the free energy of the best duplex formed between the two. This duplex may contain mismatches, bulges and interior loops, since although the complementarity between plant microRNAs and their targets is usually quite high, it is not perfect. For example, the complementarity between the sequence of DCL1 mRNA and the sequence of miR162 is not perfect, and contains a bulge nucleotide. (DCL1 is subject to negative feedback regulation of miR162 which marks it for cleavage [26].) Another example is miR-JAW with 4-5 mismatches to the mRNAs of several transcription factors [16].

Moreover, even though our study here focuses on plant microRNAs, the tool we developed is general enough to apply to animal microRNAs as well. A recent study by Yekta *et al.* [27] has shown that plants are not unique in using microRNAs to regulate mRNA target cleavage. This study demonstrated that natural metazoan microRNAs direct mRNA cleavage as well. Since, in animals, microRNAs often display limited complementarity to their targets, designing our algorithms to allow mismatches, bulges and loops of various size is important in order to keep them general enough to support animal studies as well. Thus, the core of our approach is an efficient algorithm for computing the optimal free energy of a duplex between a microRNA and an mRNA target, using scoring rules derived from the nearest neighbor thermodynamic approach [28, 13]. Performing this calculation for the entire set of microRNAs and genes of interest yields the targets matrix T.

Work to date on microRNA target prediction consists of methods that either rely solely on edit distance [19], or combine sequence similarity information with secondary structure prediction by energy minimization. The combination is achieved via a two-stage approach [4, 17, 22]: During the first stage potential binding sites are identified by searching for near-prefect base complementarity to the 5'-end of the microRNAs (or some heuristically set "nucleus" of base pairs). For each such match a candidate consisting of a stretch of about twice the microRNA size from the target is extracted, anchored at the highly matched nucleus. In the second stage, the secondary structure is computed, for the candidates suggested by the first stage, by applying the standard folding program *Mfold* by Zuker *et al.* [28] to the concatenation of potential binding site and microRNA. If this results in a score above a predefined threshold then a microRNA/mRNA target relation is determined.

Note that there are clear drawbacks to this approach. The first drawback is that the sequences have to be concatenated with a short linker sequence that can lead to artifacts in the prediction, as explicitly demonstrated by [18]. The second problem is that mRNA or microRNA self-structure can occur. However, this self-binding should not be allowed, for the following reason: microRNAs function as guides for RISC (RNA-induced silencing complex) to cleave mRNA, and thus the whole microRNA/mRNA duplex is incorporated into the RISC complex [3]. Base pairing between target necleotides or microRNA nucleotides would result in a structure that would prevent this incorporation due to steric interference. Moreover, the microRNA is so short that it is natural to assume that no two loops cross and that no multiple loops are formed. The third drawback is that for prediction of multiple bindings in one target, as in the case of animal microRNAs, the appropriate potential binding sites have to be cut out and folded separately. Furthermore, these algorithms rely on heuristic parameters such as the size of the required "nucleus" of base pairs or the predefined limit on the number of target nucleotides to participate in the duplex. All these could result in false negatives.

In Rehmsmeier *et al.* 2004 [18] a program is described that directly predicts multiple potential binding sites of microRNAs in large target mRNAs. In general, the program finds the energetically most favorable bindings of a small RNA to a large RNA. Self-structures are not allowed in this program. However, this program uses brute-force dynamic programming, whose time complexity would naively be $O(m^2 n^2)$, where m is the size of the microRNA and n the size of the target. The program is sped up by heuristically restricting the size of the allowed gap to 15.

1.2   Our Results

miRNAXpress is a novel tool for associating between microRNAs and the conditions in which they act. The program is composed of two main modules that work in tandem to compute the desired output.

The first component is a microRNA/mRNA *target prediction engine* that, given a query microRNA, a text mRNA and a predefined energy cutoff threshold, finds and reports all targets (putative binding sites) of the query in the text with binding energy below the predefined threshold. The process does not allow self-structures. The target prediction engine is based on an efficient algorithm that exploits the sparsity of the search space without sacrificing the optimality of the results (no heuristics are used). The time complexity of the algorithm is almost linear with the size s of a sparse set of stacked base pair locations. It is based on the approaches of Eppstein, Galil, Giancarlo, and Italiano [7] and Miller and Myers [14, 15], and is extended to an algorithm which utilizes the score-bounding as well as the convexity of the loop-cost function to speed up the search. Further reduction of the complexity is shown for the target prediction decision problem with discrete scores. The algorithm is described in Section 2.

The second component takes a pre-defined *Expression Matrix* (based on a gene set and a condition set which are given as input to the program) and a *Targets Matrix*, which is computed by the *target prediction engine* component (given the gene set and a set of input microRNAs), and applies the *Associating Operation F* to the two matrices. The resulting *Association Matrix* is the output of the program. The formation function is described in Section 3.

Note that the prediction of the targets is the heavier part of the above process in terms of computational complexity. The efficient approach to this part facil-

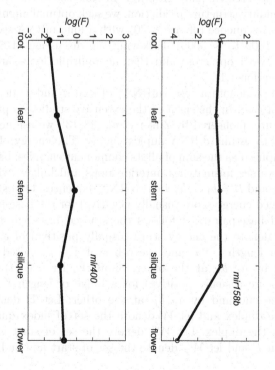

Fig. 1. Tissue specific microRNAs

itates the entire process. Also note that the algorithm described in this paper avoids the artifacts associated with the concatenation of the microRNAs to the targets, which was a drawback of previous target prediction algorithms which applied a two-stage approach.

We employed *miRNAXpress* to conduct a study, using the *A. thaliana* plant as our study subject, as follows. An *Expression matrix* was assembled from 5800 genes and 380 conditions. These conditions included various tissues, hormonal treatments, ionic and cationic stress, and pathogens. A *Targets Matrix* was then constructed by testing the *AHP Existence*(See Definition 6.2) of each one of 98 previously discovered *A. thaliana* microRNAs in each of the 5800 mRNAs. Some of the associations established as a result of our study are discussed in Section 4. These are plant microRNA α_i such that their association with specific condition β_j got a significant *p-value*.

As an additional result, this study led to the discovery of some tissue specific microRNAs. These microRNAs yielded a significant *p-value* only in specific tissues of the plant (see Figure 1).

2   The Target Prediction Engine

The problem of target prediction is similar in essence to that of finding RNA secondary structure (see *e.g.*[20]). However, in contrast to the traditional problem of RNA secondary structure prediction, we seek optimal alignments between a short microRNA sequence (of size < 30) with a long mRNA sequence (of size varying from ~ 1000 to ~ 3000). Following the discussion of Section 1.1, we assume that no two loops cross and that no multiple loops are allowed (see explanation in Section 1.1).

Therefore, let T denote a text (mRNA) of size n and P denote a pattern (microRNA) of size m. In the spirit of the extensive studies on predicting RNA secondary structure, pioneered by Zuker *et al.* [28, 13], we use nearest-neighbor thermodynamics to estimate RNA duplex energy. The energy of a duplex will therefore be computed as the sum of all its component energies: base-pairs which create stability, versus mismatches, interior-loops and bulges, which reduce it. Let $A = a_1 \ldots a_n$ and $B = b_1 \ldots b_m$ be two RNA sequences. Let $es(i-1, i, j-1, j)$ denote a predefined energy score (usually negative) for the base-pair (i, j) (such that the stacked base-pair (i, j) follows the stacked base-pair $(i - 1, j - 1)$). Let $w(i', j', i, j)$ denote the energy score (usually positive) of an interior loop with one side of length $i - i'$ and the other of $j - j'$, closed by pair a_i, b_j at one end and by $a_{i'}, b_{j'}$ at the other. Similarly, let $w'(i', i)$ denote a predefined energy score (usually positive) for a bulge of length $i' - i$, closed by pair a_i, b_j at one end and by $a_{i'}, b_{j'}$ at the other. Let $\mathcal{D}$ denote the set of base pairs in the duplex and let $\mathcal{W}$ denote the set of index quartets defining interior loops in the duplex. Let $\mathcal{W}_P$ denote the set of start and end index-pairs of bulges in P and let $\mathcal{W}_T$ denote the set of start and end index-pairs of bulges in T.

Definition 5. *Given two hybridized RNA sequences, R_1 and R_2, which form a duplex, the* **Duplex Hybridization Energy Score** *of R_1 and R_2, denoted $DHES[R_1, R_2]$, is*

$$DHES[R_1, R_2] = \sum_{\forall(i,j)\in\mathcal{D}} es(i-1,i,j-1,j) + \sum_{\forall(i',j',i,j)\in\mathcal{W}} w(i',j',i,j)$$

$$+ \sum_{\forall(i,i')\in\mathcal{W}'_P} w'(i',i) + \sum_{\forall(i,i')\in\mathcal{W}'_T} w'(j',j)$$

In this section we address a new and challenging problem, based on RNA secondary structure prediction: finding all approximate helix bindings of a short RNA sequence (microRNA) in a long one (mRNA), given a threshold bound on the allowed score. This challenge is formally defined as the following search and decision problems.

Definition 6. Approximate Hybridization Prediction (AHP):
1. The Search Problem:

The **AHP Search** *problem is, given a predefined score threshold e, to find all the approximate predicted hybridization sites of P in T, where the DHES score of the duplex formed by P and the corresponding putative binding site in T is at most e.*

2. The Decision (Existence) Problem:

The **AHP Existence** *problem is, given a predefined score threshold e, to predict whether or not there exists a hybridization site of P in T where the DHES score of the duplex formed by P and the corresponding putative binding site in T is at most e.*

Note that casting AHP as a search problem is applicable to target discovery in animals, where microRNAs display complementarity to multiple sites in a single mRNA. In plants, however, microRNAs display complementarity to a single site only, and therefore applying the decision version of the problem is more appropriate in this case.

The thermodynamic parameters used by our target prediction engine are experimentally derived and are similar to those used by *Mfold Version 3* for RNA folding [28], scaled to accommodate the relevant plant and animal conditions. The loop energy cost w sums up three terms: a purely entropic term that depends on the loop size, the terminal stacking energies for the mismatched base pairs adjacent to both closing base pairs, and an asymmetric loop penalty for non-symmetric interior loops. In both w and w' (bulge energy cost) the loop destabilizing energy grows logarithmically with the total size of the loops, *i.e.* $i - i' + j - j' - 2$ for interior loops, $i - i'$ for bulges in P and $j - j' - 1$ for bulges in T.

Thus, the energy calculation can be modeled by the following dynamic programming equations, based on the formalization of Waterman and Smith [25]:

$$D[i, j] = \min\{D[i-1, j-1] + es(i-1, i, j-1, j), H[i, j], V[i, j], E[i, j]\} \quad (1)$$

where

$$V[i,j] = \min_{0<i'<i} \{D[i',j-1] + w'(i',i)\} \tag{2}$$

$$H[i,j] = \min_{0<j'<j} \{D[i-1,j'] + w'(j',j)\} \tag{3}$$

$$E[i,j] = \min_{0<i'<i,0<j'<j} \{D[i',j'] + w(i',j',i,j)\} \tag{4}$$

Note that the heavier part of the computation of Equation 1 is the $E[i,j]$ term, as computed in Equation 4. For the sake of simplicity we will therefore assume throughout this section that recurrences 2 and 3 are simplified instances of Equation 4 and focus our explanations on the handling of Equation 4. Also, from now on we will use the term "arc" when referring to both bulges and loops.

The naive dynamic programming algorithm solves recurrence 1 for sequences of length n in $O(n^4)$ time. A lower complexity of $O(n^3)$ can be achieved with no assumptions on loop destabilizing functions [25]. Eppstein *et al.* [5] considered loop destabilizing functions satisfying certain convexity or concavity conditions, and developed an $O(n^2 \log^2 n)$ algorithm for this case. This was later improved to $O(n^2 \log n)$ [1], and finally to $O(n^2 \alpha(n))$ (where α is the inverse Ackerman function) for logarithmically growing destabilizing functions [11].

2.1   Sparsification

For the parameters and cost functions that are standard for base stacking and loops, one can show that the energy cost of breaking a loop will be more than the energy benefit from a single base pair or two stacked base pairs. Thus, we ignore base pairs that cannot be stacked without gaps at height of 3 or more.

This insight can be used to greatly reduce the number of possible pairs, yielding a sparsification of the dynamic programming matrix for P versus T. Instead of nm entries we drop to about $nm/64$ (the probability of a complementary triplet), without sacrificing the optimality of the results. (In fact in our data the number of relevant base pairs was analyzed to be about one percent on average.) Thus the computation and minimization needs to be taken only over positions (i,j) which end a triplet of complementary base-pairs, taking into account wobble pairs *i.e.* G-U pairs as well. Let S be the set of such positions, to be denoted "points" in the rest of this paper, and let $s = |S|$. Note that the set S can be computed in $O(s + n \log \Sigma)$ time, where Σ is the size of the alphabet (four nucleotides in this case), using standard string matching techniques (suffix trees). The effects of sparsity on the alignment of RNA has been studied, in a unifying framework, by Eppstein *et al.* [6,7]. With Johnson's data structure [7] and a special implementation of the binary search, an $O(n + s \log s \log min(s, n^2/s))$ bound can be obtained. For simple destabilizing functions, it reduces to $O(n + s \log s \log \log min(s, n^2/s))$. Larmore and Schieber [11] have improved one of the algorithms of Eppstein *et al.* [7] to $O(n + s \log min(s, n^2/s))$ for concave w and $O(n + s\alpha \log min(s, n^2/s))$ for convex w. However their algorithm uses matrix searching techniques, which lead to a high constant factor in the time bound. Therefore, we chose the algorithm of [7] as the basis for our target prediction engine, to be described in the next section.

2.2   The AHP Algorithm (for Both the Search and Decision Problems)

Following Definition 6.2, the *AHP* algorithm searches for potential bindings sites of P in T that may form a duplex with P of score below a predefined *DHES* cutoff threshold e. Our cutoff threshold is a constant fraction (specifically 0.85) of the optimal energy possible for the microRNA/target duplex (this optimal energy is calculated by aligning the microRNA with its exact complement sequence).

Observation 1. *The score threshold e imposes a lower bound on the number of microRNA nucleotides that must hybridize in order to achieve the threshold score, and therefore an upper bound on the number of deletions allowed from the microRNA in the sought alignment.*

For example, assume that matches contribute -2 to the free energy and consider a microRNA of size 10 and a threshold of -18 (that is, -18 is an upper bound on the minimal energy allowed for an accepted duplex). In that case the number of allowed deletions is 1, because more deletions would result in less than 9 matches, yielding a hybridization energy that is higher than -18. For each microRNA, knowing the content of its nucleotides, we can easily derive the minimal number of matches necessary to achieve the required threshold.

Therefore, let k denote a pre-defined bound, based on the *DHES* threshold e and on the maximum number of deletions from P allowed in aligning P and T. Note that k is immediately a bound on the maximal size of an allowed gap in P. Also note the asymmetric nature of the gap binding: the acceptance threshold score does not impose a limit on the number of deletions from the long sequence T.

2.2.1   A Division of the *DP* Table Based on the Gap Bound in P

In this section we demonstrate how the k bound on gaps in P, as imposed by the pre-defined score threshold, can be used to restrict the search space and speed up the search, under the sparse dynamic programming model, without sacrificing the optimality of the scores. The dynamic programming table is divided to m/k slices of k rows each. To each block we apply a divide-and-conquer recursion on the points of S which are included in its rows. The invariant is that at each level of the recursion the gaps in P are confined to a size that is twice the size allowed in the previous level. For each level of the recursion, having t rows in the subproblem of that level, we partition the rows into two blocks, the first consisting of the first top $t/2$ rows of the block and the second consisting of the bottom $t/2$ rows of the block.

Definition 7. *Given two points $(i', j'), (i, j) \in S$ such that $i > i'$ and $j > j'$, the* **arc-contribution** *of point (i', j') to point (i, j) is the contribution of the score term suggested by an optimal alignment that consists of an arc from (i', j') to (i, j) to the score minimum computation of $E[i, j]$ in Equation 4.*

The score minima for the points of each sub-block are computed as follows:

1. Recursively solve the problem for the top sub-block.
2. Compute the arc-contribution of the points in the top sub-block to the points in the bottom sub-block.
3. Recursively solve the problem on the bottom sub-block.

Note that since the gap size in P does not exceed k, all arcs are confined to a maximum of two consecutive k-blocks (i.e. blocks of k rows each). Therefore, the k-blocks are processed in the following order. First the divide and conquer computation is applied to the first block. Then the arc-contribution of the first block to the second block is computed. Only then is the divide and conquer computation applied to the second block. Now the arc-contribution of the second block to the third block can be computed, followed by a computation of the third block, and so on: for each pair of consecutive blocks the bottom block in the pair is only computed after the top one has been computed and the arc-contribution of the top block to the bottom block has already been resolved.

Lemma 1. *Imposing the block division as above will not sacrifice the optimality of the scores.*

Proof. *No false negatives:* All correct arcs of sizes up to $2k$ will be considered by the algorithm, and this set includes of course all potential arcs of size up to k. *No false positives:* The block division allows the consideration of arcs of sizes up to $2k$ (two consecutive blocks). However, since the k bound is derived from the score cutoff threshold e, we know that all hybridizations with gaps in P that are greater than k will by definition be ruled out by the search algorithm. □

2.2.2   Computing the Arc-Contribution of One Set of Points to Another

In this section we describe a point-traversal order which will allow us to efficiently compute the arc contribution of a top sub-block to a consecutive bottom sub-block, following the partitioning of the dynamic programming table as described in Section 2.2.1. This point-traversal order will support dynamic minimization and is based on the approach described in [7]. We say that point (i', j') *precedes* point (i, j), denoted by $(i', j') \prec (i, j)$, if and only if $i' < i$ and $j' < j$. The *diagonal index* of a point (i, j) is defined as $i + j$. Let d_k be the set of points (i', j') in S whose diagonal index is k.

Let S_1 be the set of points on the top sub-block, and S_2 be the set of points on the bottom sub-block in a given arc-contribution computation. Points in S_1 and S_2 are processed in order of their column indices first and only then by their row indices. Within a given column we first process the points of S_2, and then the points of S_1.

Claim 1. *If the points of $S_1 \cup S_2$ are scanned in the order described above, then:*
1. When a point $p \in S_2$ is reached, all points $p' \in S_1$ such that $p' \prec p$, and only those points, have already been traversed and analyzed.
2. The diagonal index of any point $p \in S_2$ that has not yet been scanned is greater than the diagonal index of any of the points already scanned in S_1.

Proof.

1. Clearly $i > i'$ since S_2 is the lower part of the block, and we also impose $j > j'$.

2. Consider any pair of points $\{(i', j') \in S_1, (i, j) \in S_2\}$ such that point (i', j') has already been scanned and processed in S_1 and point (i, j) has not yet been scanned in S_2. Clearly $i > i'$ due to the block separation. Also $j' > j$ due to the column-scanning order. Therefore, the indices of the diagonals corresponding to the two points follow $i + j > i' + j'$. $\square$

Note that the recurrence of Equation 4 has the property that the function w depends only on the differences between the two diagonals defined by its four argument points. Thus, during the computation of $E[i, j]$, the source points (i', j') can be grouped into sets of points with identical diagonal numbers, such that each set will contribute only one term to the minimization formulated in Equation 4. For the set of source points $(i', j'); i' + j' = x$ on a given diagonal x, we have only to consider the minimum among the $D[i', j']$ entries in d_x. Furthermore, by Claim 1, Equation 4 could be reformulated in terms of diagonal indices as follows. Let $x = i' + j'$, $y = i + j$, and let x_max denote a dynamic variable which, at any moment of the point traversal, stores a value which is greater by one than that of the largest diagonal in S_1 which was scanned so far. This yields

$$E'[y] = \min_{x < x\_max} \{D'[x] + w(x, y)\} \quad for\ x\_max \le y \le n + t, \tag{5}$$

Using the point-traversal order described above, the arc-contribution of points in S_1 to points in S_2 can be computed as a minimization problem with dynamically changing input values, based on Equation 5, as follows.

1. A point (i', j') in S_1 is processed by performing the operation of decreasing $D'[x]$ to $min(D'[x], D[i', j'])$.

2. A point (i, j) in S_2 is processed by performing the operation of computing $E'[y]$.

2.2.3   Using the Convexity and Simplicity of w to Compute $E'[y]$ Efficiently

Note that, for the application at hand, it is customary to use the *ln* function for quantifying the growth of w and w' when increasing the loop/bulge size. Also note that the *ln* function is convex.

Definition 8. *A weight function w is convex w.r.t. the y minima computation if $\forall\ x < x' < y < y'$, $w(x, y') - w(x, y) \le w(x', y') - w(x', y)$.*

The computation of $E'[y]$ is viewed as a competition among a sparse set of $|S_1|$ candidate diagonals from the range $0, 1 \ldots x\_max$ for the minimum in Equation 5, to be computed for a sparse set of target diagonals in the range $x\_max + 1 \ldots n + t$. In the next Lemma we will show that the competing source diagonals divide the target diagonals into intervals, to be denoted *leadership intervals*. In each such interval only one of the source diagonals from S_1 will yield the minima values for Equation 5.

Lemma 2. *Given a function w which is convex with respect to the minima of Equation 5.*

1. *For any x, y and x', with $x < x'$, if $D[x] + w(x, y) \leq D[x'] + w(x', y)$, then for all $y' > y$, $D[x] + w(x, y') \leq D[x'] + w(x', y')$.*

2. *Conversely, if $D[x] + w(x, y) > D[x'] + w(x', y)$, then for all $x\_max < y' < y$, $D[x] + w(x, y') > D[x'] + w(x', y')$.*

Proof.

1. By the definition of convexity, $w(x, y') + w(x', y) \leq w(x, y) + w(x', y')$. Subtracting $w(x, y') + w(x', y') + D[x'] - D[x]$ from both sides and rearranging yields $(D[x] + w(x, y)) - (D[x'] + w(x, y')) \leq ((D[x] + w(x', y)) - (D[x'] + w(x', y'))$. But by assumption $((D[x] + w(x, y)) - (D[x'] + w(x, y'))$ is positive, and therefore $(D[x] + w(x', y)) - (D[y'] + w(x', y'))$ must also be positive and the first statement holds.

2. Similar to the former statement. The only difference is that here we encounter an interval $x \leq y < x'$ where $w(x', y)$ is not defined and therefore x rather than x' would yield the diagonal minima in this interval. However, by Claim 1.2, at the stage when $E(i, j)$ is computed this interval is no longer relevant to the minimization computation.   □

We point out that a similar lemma was stated and proven for the *concave* case in [7]. The *convex* case, however, is more complicated, since $w(x', y)$ is undefined for any $x \leq y < x'$. Therefore, x rather than x' yields the diagonal minima in this interval. This imposes a fragmentation of the leadership-intervals which could strongly affect the efficiency of the algorithm. However, we resolve this problem for applications which follow the divide and conquer approach and traversal order described in Section 2.2.1, as follows: By Lemma 2.2 at the stage when $E(i, j)$ is computed, the fragmented leadership-intervals, all of which by definition fall to the left of x_max, are no longer relevant to the minimization computation and are therefore discarded.

Conclusion 1. *At any given point in the traversal of $S_1 \cup S_2$, the values of $D[x]$ supplying the minima for the positions of $E[y]$ partition the possible indices of y, $x\_max < y < n + t$, into a sequence of "leadership intervals". If $x' < x$ and if y is in the interval in which $D[x] + w(x, y)$ is best, and y' is in the interval in which $D[x'] + w(x', y)$ is best, then $y' > y$. Also, if both x' and x have active leadership-intervals, then $D[x] > D[x']$.*

Based on Conclusion 1, the algorithm maintains in a data structure, a subset of candidates which satisfies the property that $E'[y]$ depends only on these candidates. Diagonals carrying points from S_1, whose potential arc-contribution term is no longer a candidate to yield the minimal E' (as computed by Equation 5) for some future point in S_2, are discarded from this subset. Intervals corresponding to target diagonals that are smaller than the dynamically growing x_max are also discarded. The interested reader is referred to the candidate list algorithms of [5, 7, 8, 9, 14] which also utilize convexity/concavity properties.

Definition 9. *Given x and x', $x < x' \leq n+t$, **Intersect**(x, x') is the minimal index y, $x' < y \leq n+t$, such that $D[x]+w(x, y) \leq D[x']+w(x', y)$. Intersect$(x, x')$ is ∞ if there is no such index y.*

By Conclusion 1, the source diagonal numbers giving the minima for E', computed for increasing target diagonal numbers, are nonincreasing when w is convex. Furthermore, for log functions $Intersect(b, a)$ can be computed in constant time. A candidate diagonal x is *live* if it supplies the minimum for some $E'[y]$. The algorithm maintains live diagonals and their leadership intervals in which these diagonals give the minimum using a priority queue data structure. Computing $E'[y]$ then reduces to looking up which leadership interval contains y. Decreasing $D'[x]$ involves updating the interval structure by deleting some neighboring live diagonals and finally a constant time $Intersect$ computation at each end.

2.3   The Complexity of Sparse *AHP*

Theorem 1. *The algorithm described in this section computes sparse AHP with logarithmically growing destabilizing functions in time $O(s \log \log s)$.*

Proof. During a preprocessing stage, the set S of stacked triplets can be computed in $O(s+n \log \Sigma)$ time, where Σ is the size of the alphabet (four nucleotides in this case), using standard string matching techniques (suffix trees).

Within each level of the recursion, we will need the points of each set to be sorted by their column indices and only then by their row indices. To achieve that we initially bucket-sort all points, and then at each level of the recursion perform a pass through the sorted list to divide it into the two sets. Thus the order we need is achieved at a linear cost per level of the recursion. We also need a data structure to efficiently support the following two operations:

1. Compute the value of $E'[y]$ for some $y \in S$.
2. Decrease the value of $D'[x]$ for some $x \in S$.

Note that Operation 2 involving one value of $D'[x]$ may simultaneously change $E'[y]$ for many y's. A candidate list can be implemented to maintain the live diagonal candidates and support union, find and split operations on their leadership intervals. If a balanced search tree is used for implementing this list the amortized time per operation is $O(\log n)$. The bound can be improved to $O(\log \log n)$ with van Emde Boas's data structure [24]. Note that the time for each data structure operation can be taken to be $O(\log s)$ or $O(\log \log s)$ rather than $O(\log n)$ or $O(\log \log n)$. This is because only diagonals of the dynamic programming matrix that actually contain some of the s points in the sparse problem need to be considered. The flat trees can be set up at $O(s)$ preprocessing time and then be reused at different levels of the recursion. Thus the arc-contribution of S_1 to S_2 can be computed in $O(s \log \log s)$ time. Multiplying by the number of levels of recursion, and adding the work spent on arc-connecting consecutive k-block pairs, the total time is $O(n + s \log k \log \log s) \stackrel{k \leq s}{=} O(s \log \log s)$. □

Theorem 2. *The sparse AHP Decision problem with logarithmically growing destabilizing functions and discrete DHES cost can be computed in time $O(s)$.*

Proof. A score function f is said to be discrete only if there exists some constant r such that every element of f is some integral multiple of r. Let $min\_triplet$ denote the minimal cost of a triplet of base pairs. When computing AHP Existence, the algorithm immediately halts once an e-scoring duplex is found. Therefore, the range of scores does not exceed $e - min\_triplet$. Let $c = (e - min\_triplet)/r$. Clearly, at any given moment during the execution of the algorithm there can only be c different $DHES$ scores which are shared by all candidate diagonals from S_1. By Conclusion 1, for any two diagonal candidates (x', x) from S_1 such that $x' < x$, if $G[x'] = G[x]$, where $G[x]$ denotes the maximal score of a point from S_1 on diagonal x, then only x is a live candidate. Therefore, the total number of live candidates in a given moment is c and the time complexity of the algorithm in this case is therefore $O(s \log \log c) \overset{c \ is \ constant}{=} O(s)$.   □

3   Computing the Association Operation F

The association operator (see Definition 4 in Section 1), $F(X_i, T_j)$, represents the probability that microRNA j is active under condition i. It is set to the confidence level by which we reject the hypothesis that $\{X[i,k] \cdot T[k,j]\}_{k=1}^r$ and $\{X[i,k] \cdot \overline{T[k,j]}\}_{k=1}^r$ were sampled from the same distributions. We use the following theorem to calculate a statistic t for the difference between $\{X[i,k] \cdot T[k,j]\}_{k=1}^r$ and $\{X[i,k] \cdot \overline{T[k,j]}\}_{k=1}^r$ and then set $F(X_i, T_j)$ to $\Phi(t)$, the corresponding level of significance associated with t.

Theorem 3 [2]. *Let x_1 and x_2 be two observations and let n_1 and n_2 be their sizes respectively. Let $\overline{x}$ denote the mean of sample x, and SD_x its standard deviation.*
Let $SD_x = \sqrt{\frac{SD_{x_1}^2(n_1-1)+SD_{x_2}^2(n_2-1)}{n_1+n_2-2}}$. If x_1 and x_2 were sampled from the same normal population and assuming that $n_1 + n_2 \geq 30$ then $t = \frac{\overline{x_1}-\overline{x_2}}{SD\sqrt{1/n_1+1/n_2}}$ has an approximate standard normal distribution.

4   A Study Associating *A. Thaliana* microRNA with Various Activation Conditions

Different expression patterns are part of how a plant grows, develops, and adapts to environmental changes. In this section we use our computational approach to shed light on the contributions of various microRNAs to these expression patterns in *A. thaliana*. The Expression Matrix X is based on 380 conditions and 5800 genes. The expressions were measured using a two dye microarray assay where the condition sample was labeled red and the reference sample was labeled green. The red/green ratio presents the abundance of the gene's mRNA in the condition, in comparison to the reference. A low red/green ratio can indicate that the mRNA is more degraded in the condition than in the reference, and vice versa for high ratios. We used the average values of $\log_2$ of the normalized red/green ratio. The reference samples changed between the conditions and are indicated at the relevant positions. The microarray data for the 380 different conditions was retrieved from the *TAIR* database, http://www.Arabidopsis.org/.

The T matrix is based on 98 previously discovered microRNAs. These were retrieved form the RFAM database, http://www.sanger.ac.uk/Software/Rfam/, and were discovered either experimentally (by isolating, reverse transcribing, cloning, and sequencing small cellular mRNAs) or computationally (by seeking microRNA precursors conserved between *A. thaliana* and other related organisms [19]). The mRNA sequences of all 5800 *A. thaliana* genes were retrieved from the *TAIR* database. To construct the T matrix we used the untranslated regions (UTRs) of these molecules (since UTRs control mRNAs stabilities).

Figure 3 describes some relations corresponding to significant entries of the Association Matrix A, and Figure 2 includes the corresponding predicted targets. The results presented in Figure 3 were selected based on their *p-values*. From all the microRNAs with *p-value* better than 10^{-6} we chose 6 for further investigation. The symbol $\overline{x}_{targets}$ in Figure 3 represents the average expression of the set of the potential microRNA targets, *i.e.* $\{X[i,\ell] \times T[\ell,j]\}_{l=1}^{r}/\{T[\ell,j]\}_{l=1}^{r}$, and $\overline{x}_{others}$ is the average expression of all other genes, *i.e.* $\{X[i,l] \times \overline{T[l,j]}\}_{l=1}^{r}/\{\overline{T[l,j]}\}_{l=1}^{r}$. Since we use relative expressions, if $\overline{x}_{targets} < \overline{x}_{others}$ we hypothesize that the microRNA is active at the condition and not in the reference. On the other hand, if $\overline{x}_{targets} > \overline{x}_{others}$ we hypothesize that the microRNA is silenced at the condition and is active at the reference. Both cases can result in a significant *p-value* as represented by column $A[i,j]$. Moreover, to asses both cases one naturally needs to use a two tail test such as the t-test we used which is described in Section 3. Note that negative values of $\overline{x}_{targets}$ and $\overline{x}_{others}$ reflect higher expression in the reference compared to the condition, since we are using the $\log_2$ of the red/green ratio.

$A[i,j]$ is a *p-value* for whether the expression of the targets is distinguished from the expression of all other genes. For this *p-value* to be correct the expression levels of the genes should be normally distributed (see Section 3). Although in practice this is the case for most microarray measurements, we further asserted our *p-values* by conducting the following shuffling simulations. We first randomly shuffle the mRNA sequences with respect to the genes and then computed a new T matrix. Using this matrix and the E matrix we computed the A matrix again. Repeating this process 1000 times allows us to estimate the *false/positive* rate of our approach. For example, if the lowest number computed for an entry $A[i,j]$ in all 1000 different assays is x, we conclude that the null probability for obtaining a value $A[i,j] \leq x$ is smaller than 10^{-3}. The smallest and average numbers computed for the relevant entries are also presented in Figure 3 (see min $A[i,j]_{shuffled}$ and average $A[i,j]_{shuffled}$, respectively). Note that the *p-values* computed for the microRNAs are significantly better than these numbers, strengthening the power of our approach and the significance of our results.

microRNA	Target list	Hybrid-ization energy	GO term
mir168a	AT3G26420	-43.2	RNA binding
	AT5G60800	-41.7	Metal ion binding
miR159C	AT4G08320	-35.90	Endomembrane system
	AT3G26420	-36.30	RNA binding
	AT4G02520	-36.70	Glutathione transferase
	AT3G10770	-41.40	Glutathione transferase
	AT1G78880	-37.30	Unknown
	AT4G01810	-36.10	Photoreceptor activity
	AT1G09570	-37.30	Photoreceptor activity
	AT1G73060	-36.80	Unknown
	AT4G22890	-39.60	Chloroplast protein
	AT4G24230	-38.40	Acyl-CoA binding
	AT5G40760	-36.00	Unknown
	AT2G24940	-36.90	Electron transport
mir399C	AT4G24470	-37.50	Transcription factor
	AT1G51200	-38.10	Electron transporter
	AT3G13460	-36.10	Unknown
	AT3G07390	-36.80	Response to Auxin
	AAT3G27260	-36.30	DNA binding
mir395	AT3G26420	-38.20	RNA binding
	AT2G31800	-38.70	Kinase activity
	AT1G55255	-36.80	Zinc ion binding
	AT1G78080	-36.20	Transcription factor
	AT5G61170	-36.50	Ribosomal RNA
	AT1G73060	-42.20	Chloroplast protein
	AT5G27650	-36.60	Unknown
	AT5G35360	-36.10	Chloroplast protein
	AT2G36400	-36.70	Chemokine receptor
	AT3G52800	-36.80	Zink ion binding
	AT4G24230	-36.90	Endomembrane system
	AT1G60730	-36.10	Kcl channel
	AT2G43970	-37.90	RNA binding
	AT3G18210	-36.60	Unknown
	AT5G10860	-38.40	Acetolactate synthase
	AT1G08990	-37.50	Endomembrane system
	AT5G54300	-36.20	Sodium ion transport
	AT5G15320	-38.50	Endomembrane system
	AT3G50960	-36.60	Unknown

Fig. 2. microRNAs and their potentialtargets

Our high throughput approach to computing A enables us to focus on the more interesting associations, *e.g.:*

1. miR159C was found to be more active when *nph4* mutants were subjected to phototropic stimulation than when wild-type (non-mutant) cells did. A phototropic stimulation is the process of exposing a plant to light after a long period of darkness. In this experiment, for example, plants were grown in the dark for 2.5 days and then exposed to blue light for one hour. NPH4 is a regulatory protein which is crucial for the normal response of the plant to a phototropic stimulation. Our findings indicate that one derivative of this protein action

microRNA	Condition	$A[i,j]$	$\bar{x}_{others}$	$\bar{x}_{targets}$	min $A[i,j]_{shuffled}$	average $A[i,j]_{shuffled}$
miR168A	Xanthophyll mutant vs wildtype	10^{-22}	0.03	-1.82	0.009	0.48
miR168A	Arabidopsis leaves inoculated in bacterial phatogen vs non non inaculated leaves	10^{-6}	0.01	-1.76	0.045	0.47
miR168A	Shoot exposed to light after 1hr darkness vs unexposed shoot	10^{-9}	1	0.41	0.25	0.5
miR156A	Defective DST-mediated mRNA degradation pathway vs wildtype	10^{-14}	22	1.23	0.05	0.52
miR159C	Phototropic stimulation of NPH4 mutants vs phototropic stimulation of wildtype	10^{-11}	0.07	0.62	-0.85	0.1
miR399A	Roots treated with Auxin compare to untreated roots.	10^{-11}	-0.85	0.07	0.1	0.55

Fig. 3. MicroRNAs and associated conditions. For Bonferroni corrected p-values, entries of the table should be multiplied by 10^4

might be miR159C, which was found to be more active in *nph4* mutants than in wild-types, possibly indicating NPH4 pathway silencing of miR159C in response to light stimulus in wild-types. Indeed, many of the predicted targets (as in Figure 2) of miR159C are involved in light processes, *e.g.* belong to endomembrane systems, or have photoreceptor activity, or code for chloroplast protein. The loss of the phototropic response in the mutant might be partly due to the loss of suppression of miR159C by NPH4. Intriguingly, miR159C seems to be active also in flowers that do not respond to phototropic stimulations. Moreover, since flowers do not perform photosynthesis they do not need most of miR159C's target products.

2. We found mir168A to be active in response to bacterial pathogen inoculation. This is a condition in which a virus virulent protein is introduced into the plant cells. Supporting evidence for our findings is the fact that AT5G60800, one of mir168A targets (Figure 2), is a metal binder. Metals have been shown to facilitate viral attacks in plants [21]. Our findings, therefore, suggest that part of the virus strategy for attacking the cells is in elevating metal levels by activating mir168A and eliminating the metal binder, AT5G60800.

The very same microRNA is also activated in Xanthophyll cycle mutants. Xanthophyll cycle is a photoprotection mechanism. We propose that AT5G60800 is also related to this observation, since it is capable of binding free radicals which are typical to light exposure. We propose that in normal cells mir168A is deactivated by the Xanthophyll cycle in order to elevate the levels of AT5G60800 and reduce light damage.

3. miR399A was identified to be silenced under Auxin response. Concurrently, one of miR399A predicted targets, AT3G07390, is an Auxin responsive gene. This gene should be necessary for the cell's response to that hormone.

It is interesting to note that our findings give rise to hypotheses about microRNA regulation in the positive direction (conditions that induce microRNA activity) as well as in the negative direction (pathways that reduce microRNA activity).

Lastly, we looked for tissue specific microRNAs (see Figure 1). We focused on the subset of rows in A corresponding to flowers, stems, siliques, leaf, and root tissue conditions. Within this subset of selected rows we chose columns with only one significant entry. The microRNAs corresponding to such columns are potentially specific to the tissue represented by this row.

Acknowledgements. We thank Eleazar Eskin for fruitful discussions and Uri Levy for contributing code. The research of M. Z.-U. was supported in part by the Aly Kaufman Post Doctoral Fellowship.

References

1. A. Aggarawal and J. Park. Notes on searching in multidimensional monotone arrays. *Proc. 29th IEEE Symp. on Foundations of Computer Science*, pages 497–512, 1988.
2. C. Chatfield. *Statistics for technology, a course in applied statistics, Sci. Papreback, 1970.*
3. D.V. Dugas and B. Bartel. MicroRNA regulation of gene expression in plants. *Curr. Opin. Plant Biol.*, 7:512–520, 2004.
4. A.J. Enright et al. MicroRNA targets in drosophila. *Genome Biol.*, 5(1), 12 2003.
5. D. Eppstein, Z. Galil, and R. Giancarlo. Speeding up dynamic programming. *Proc. 29th IEEE Symp. on Foundations of Computer Science*, pages 488–296, 1988.
6. D. Eppstein, Z. Galil, R. Giancarlo, and G.F. Italiano. Sparse dynamic programming I: Linear cost functions. *JACM*, 39:519–545, 1992.
7. D. Eppstein, Z. Galil, R. Giancarlo, and G.F. Italiano. Sparse dynamic programming II: Concave and convex cost functions. *JACM*, 39:519–545, 1992.
8. Z. Galil and R. Giancarlo. Speeding up dynamic programming with applications to molecular biology. *Theoretical Computer Science*, 64:107–118, 1989.
9. D.S. Hirshberg and L.L. Larmore. The least weight subsequence problem. *SIAM J. Compt.*, 16(4):628–638, 1987.
10. K.D. Kasschau et al. P1/HC-Pro, a viral suppressor of RNA silencing, interferes with Arabidopsis development and miRNA function. *Dev. Cell*, 4:205–217, 2003.
11. L. Larmore and B. Schieber. On-line dynamic programming with applications to the prediction of RNA secondary structure. *J. Algorithms*, 12(3):490–515, 1991.
12. C. Llave et al. Cleavage of scarecrow-like mRNA targets directed by a class of Arabidopsis miRNA. *Science*, 23:2053–2056, 2002.
13. D.H. Mathews, J. Sabina, M. Zuker, and D.H. Turner. Expanded sequence dependence of thermodynamic parameters improves prediction of RNA secondary structure. *J. Mol. Biol.*, 288:911–940, 1999.
14. W. Miller and E. Myers. Sequence comparison with concave weighting functions. *Bull. of Mathematical Biology*, 50(2):97–120, 1988.

15. E. Myers and W. Miller. Chaining multiple-alignment fragments in sub-quadratic time. *ACM-SIAM Symposium on Discrete Algorithms*, pages 1–10, 1995.

16. J.F. Palatnik et al. Control of leaf morphogenesis by microRNAs. *Nature*, 425:257–263, 2003.

17. N. Rajewsky and N.C. Socci. Computational identification of microRNA targets. *Genome Biology*, 5, 2004.

18. M. Rehmsmeier, P. Steffen, M. Hochsmann, and R. Giegerich. Fast and effective prediction of microRNA/target duplexes. *RNA*, 10:1507–1517, 2004.

19. M.W. Rhoades et al. Prediction of plant microRNA targets. *Cell*, 23:513–520, 2002.

20. J. Setubal and J. Meidanis. Introduction to computational molecular biology. 1997.

21. A. Shevchenko et al. Plant virus infection development as affected by heavy metal stress. *Archives of Phytopathology and Plant Protection*, 23:139–146, 2004.

22. A. Stark et al. Identification of Drosophila microRNA targets. *PLoS. Biol.*, 1(3), 2003.

23. G. Tang et al. Framework for RNA silencing in plants. *Genes Dev.*, 17:49–63, 2003.

24. P. van Emde Boas, R. Kaas, and E. Zijlstra. Design and implementation of an effcient priority queue. *Mathematical Systems Theory*, 10:99–127, 1977.

25. M.S. Waterman and T.F. Smith. Rapid dynamic programming algorithms for RNA secondary structure. *Adv. Appl. Math.*, 7:455–464, 1986.

26. Z. Xie et al. Negative feedback regulation of dicer-like1 in Arabidopsis by microRNA-guided mRNA degradation. *Curr. Biol.*, 13:784–789, 2003.

27. S. Yekta. MicroRNA-directed cleavage of HOXB8 mRNA. *Science*, 304:594–596, 2004.

28. M. Zuker. Mfold web server for nucleic acid folding and hybridization prediction. *Nucleic Acids Res.*, 31(13):3406–3415, 2003.

RNA-RNA Interaction Prediction and Antisense RNA Target Search

Can Alkan[1,2], Emre Karakoç[2], Joseph H. Nadeau[3],
S. Cenk Şahinalp[2], and Kaizhong Zhang[4]

[1] Department of EECS, Case Western Reserve University,
Cleveland, OH 44106, USA
[2] School of Computing Science, Simon Fraser University,
Burnaby, BC V5A 1S6, Canada
[3] Department of Genetics, Case Western Reserve University,
Cleveland, OH 44106, USA
[4] Department of Computer Science, University of Western Ontario,
London, ON N6A 5B7, Canada**

Abstract. Recent studies demonstrating the existence of special non-coding "antisense" RNAs used in post-transcriptional gene regulation have received considerable attention. These RNAs are synthesized naturally to control gene expression in C.elegans, Drosophila and other organisms; they are known to regulate plasmid copy numbers in E.coli as well. Small RNAs have also been artificially constructed to knock-out genes of interest in humans and other organisms for the purpose of finding out more about their functions.

Although there are a number of algorithms for predicting the secondary structure of a *single* RNA molecule, no such algorithm exists for reliably predicting the *joint* secondary structure of two interacting RNA molecules, or measuring the stability of such a joint structure. In this paper, we describe the RNA-RNA interaction prediction (RIP) problem between an antisense RNA and its target mRNA and develop efficient algorithms to solve it. Our algorithms minimize the joint free-energy between the two RNA molecules under a number of energy models with growing complexity. Because the computational resources needed by our most accurate approach is prohibitive for long RNA molecules, we also describe how to speed up our techniques through a number of heuristic approaches while experimentally maintaining the original accuracy. Equipped with this fast approach, we apply our method to discover targets for any given antisense RNA in the associated genome sequence.

1 Introduction

Recent studies on both prokaryotic and eukaryotic cells have demonstrated the existence of small RNAs that are used in post-transcriptional gene regulation. These small RNAs usually bind to their target mRNAs to prohibit their

---

** The authors are listed in alphabetical order.

S. Miyano et al. (Eds.): RECOMB 2005, LNBI 3500, pp. 152–171, 2005.

translation and, in effect, down-regulate the expression levels of corresponding genes [21]. Other mechanisms for down-regulating gene expression through small RNAs are described in [11].

Regulatory RNAs provide a subclass of the antisense RNA family; other antisense RNAs include snoRNAs, snRNAs, gRNAs, stRNAs, that are naturally used for rRNA modification, RNA editing, mRNA splicing, developmental regulation, and plasmid copy-number regulation. Antisense RNAs are also artificially synthesized for studying specific gene functions since they can knock out targeted genes [21].

Since the first reports on natural antisense RNAs that regulate gene expression (e.g. in C.elegans [19]), there has been substantial interest to better understand how antisense RNAs interact with target mRNAs. In this paper, we describe a new framework and corresponding algorithms for predicting the secondary structure of two interacting RNA molecules by means of free energy minimization. We then evaluate how well our algorithms predict known secondary structures of naturally interacting RNA molecules recently observed in E.coli [21].

Figure 2a shows the natural joint structure of interacting RNA molecules CopA and CopT. Similarly, Figure 2b shows the natural joint structure of interacting RNA molecules OxyS and fhlA.

In Figures 3a and 4a we present the same interactions in a more illustrative manner: here blue links represent *internal bonds* whereas red links represent *external bonds* between bonded bases. Our predictions of the joint structure of these two RNA molecule pairs are also given in Figures 3 and 4.

There are a number of algorithmic tools for predicting the secondary structure of a *single* RNA molecule [23, 24, 9, 20]; there are also several algorithms to compute "similarity" or "alignment" between two *non-interacting* RNA molecules[14, 4, 22]. However, there are only a few studies related to the problem of predicting the secondary structure formed by two RNA molecules: The HyTher package by the SantaLucia Lab[18] predicts the hybridization thermodynamics of a given duplex given the two strands; it does not aim to minimize the joint free energy or predict the secondary structure of the interacting RNA strands. The *Pairfold* program [2] aims to predict the secondary structure of two interacting RNA sequences by simply concatenating two RNA strands and performing a secondary structure prediction as if there is only one strand, using the *Mfold* algorithm (for folding a single strand [23, 9]). Because *Mfold* avoids pseudoknots, possible topologies that can be predicted by *PairFold* are very limited; e.g. it can not predict any "kissing" hairpin loops, which are essential to joint structure prediction of two RNA sequences. In principle, *PairFold* can employ the *pknots* method of Rivas and Eddy [20] which can predict certain types of pseudoknots. However the pseudoknot types allowed by *pknots* (as per the characterization in [3]) do not capture any non-trivial kissing loop complex such as the ones explored in this paper. Thus even by employing *pknots*, the *PairFold* approach would not be able to predict the joint structure of interacting RNA molecules of interest.

We recently became aware of a new paper [17] which was published after the submission of our paper. This paper descibes the *IRIS* software tool/algorithm which aims to solve the joint structure prediction problem. *IRIS* is based on a very simple energy model, almost identical to the *basepair energy model*, which we describe as a warmup exercise. As we observed in our own experiments, this approach does not provide good results: the only known natural joint RNA structure examined by [17] is the OxyS-fhlA pair; on this example, the predicted structure by *IRIS* is quite different from the naturally occuring structure.

1.1   Preliminaries and Contributions

In this paper we introduce the general RNA-RNA Interaction Prediction (RIP) Problem. Given two RNA sequences S and R (e.g. an antisense RNA and its target), RIP problem asks to predict their joint secondary structure. A joint secondary structure between S and R is a set of "pairings" where each nucleotide of S and R is paired with at most one other nucleotide, either from S or R.

Let the ith nucleotide of an RNA sequence S be denoted by $S[i]$ and the substring of S extending from $S[i]$ to $S[j]$ denoted by $S[i, j]$. As a notational convenience, let $S[k, k]$ denote $S[k]$, $S[i, i-1]$ denote an empty sequence and $S[i, i-1]^r$ denote the reverse of $S[i-1, i]$. In the rest of the paper, we assume that $S[1]$ denotes the $5'$ end of S and $R[1]$ denotes the $3'$ end of R.

We compute the joint structure between S and R through minimizing their *total free energy* which is, in general, a function of (stacked) pairs of bases as well as the topology of the joint structure. In this paper we consider three models for computing the free energy of the joint structure of interacting RNA sequences.

1. We first use the sum of free energies of individual Watson-Crick base pairs as a crude approximation to the total joint free energy. This *basepair* energy model is quite similar to that used in [15] for predicting the structure of a single RNA molecule. Although the basepair energy model is known to be inaccurate, it provides a good starting point for our further explorations.

2. Our second free energy model is based mostly on *stacked pair* energies given in [10], which provide the main contribution to the energy model employed by the *Mfold* program for pseudoknot free single RNA structure prediction. Unfortunately there is very little thermodynamic information on pseudoknots or kissing loops in the literature. Thus we employ the approach used by Rivas and Eddy [20] to differentiate the thermodynamic parameters of "external" bonds from the "internal" bonds by multiplying the external parameters with a *weight* slightly smaller than 1. This *stacked pair* energy model turns out to be quite accurate, especially in predicting the joint structure of shorter (≤ 150 bases) RNA molecule pairs.

3. The final energy model enriches the above models by summing up the free energies of various types of internal loops and stacked pairs as per [23, 10] as well as the weighted free energies of externally interacting ("kissing") loops. This model, which will be referred to as the *loop* energy model, appears to be more accurate especially for longer (≥ 150 bases) RNA molecules.

Although we allow arbitrary loops to form kissing pairs, we impose the following constraints on the topology of a joint structure between RNA sequences. First, a joint structure can have no *internal pseudoknots*; i.e., if $S[i]$ bonds with $S[j]$ then no $S[i']$ for $i < i' < j$ can bond with any $S[j']$ for $j < j'$. The same property will be satisfied by the nucleotides of R as well. Second, a joint structure can not have any *external pseudoknots*; i.e., if $S[i]$ bonds with $R[j]$ then no $S[i']$ for $i' > i$ can bond with any $R[j']$ for $j' < j$.

These assumptions are satisfied by all examples of complex RNA-RNA interactions we have encountered in the literature. Furthermore allowing arbitrary pseudoknots in the secondary structure of even a single RNA molecule makes the energy minimization problem NP-hard [1]. In fact we prove in Section 2 that the RIP problem is NP-hard for each one of our energy models, even when no internal or external pseudoknots are allowed. This necessitates the addition of one more natural constraint on the topology of the joint secondary structure prediction, which is again satisfied by all known joint structures in the literature. Under this constraint we then show how to obtain efficient algorithms to minimize the free energy of the joint structure under all three energy models and test the accuracy of the algorithms on known joint structures. We finally apply our structure prediction techniques to compute target mRNA sequences to any given small RNA molecule.

2   RIP Problem for Both Basepair and Stacked Pair Energy Models Is NP-Complete

We start our discussion by showing that the RIP problem is NP-Complete under both the basepair and the stacked pair energy models.

Theorem 1. *RIP problem under the Basepair Energy Model is NP-Complete.*

Proof. The NP-Completeness of RIP is established through a reduction from the longest common subsequence of multiple binary strings (mLCS) which is a known NP-Complete problem. Our proof is an extension to the one in [1] for the single RNA secondary structure prediction problem with pseudoknots.

The decision version of the mLCS problem is as follows: Given a set of *binary* strings $L = \{S_1, S_2, \cdots, S_m\}, (|S_1| = \cdots = |S_m| = n)$ and an integer k, decide whether there exists a sequence C of length k which is a subsequence of each S_i. Here we assume that m is an odd number; if it is even, we simply add a new string $S_{m+1} = S_m$ to L.

From an instance of mLCS, we first construct two "RNA" sequences S and R, using an extended nucleotide alphabet $\Sigma^e = \{a, b, c, d, e, f, u, w, x, y, z\}$. (The NP-hardness proof for the -more interesting- stacked pair energy model below uses the standard RNA nucleotide alphabet $\{A, C, G, U\}$.)

Let v^j denote the string formed by concatenating j copies of character v and let $\bar{v}$ denote the *complementary residue* of v. In our extended alphabet we set $\bar{x} = w$, $\bar{y} = z$, $\bar{a} = b$, $\bar{c} = d$, and $\bar{e} = f$. Given a string T, we denote by $\bar{T}$ its *reverse complement*.

For $i = 1, \cdots, m$, we construct strings D_i and E_i as follows. Note that we set $s_{i,j}$ to x if the jth character of string S_i is 0; if it is 1, $s_{i,j}$ is set to be y.

$$
\begin{aligned}
D_i &= a\ s_{i,1}\ a\ s_{i,2}\ a\ \cdots\ a\ s_{i,n}\ a, & \text{if } i \text{ is odd;} \\
D_i &= a\ \overline{s_{i,n}}\ a\ \overline{s_{i,n-1}}\ a\ \cdots\ a\ \overline{s_{i,1}}\ a, & \text{if } i \text{ is even;} \\
E_i &= b\ \overline{s_{i,1}}\ b\ \overline{s_{i,2}}\ b\ \cdots\ b\ \overline{s_{i,n}}\ b, & \text{if } i \text{ is odd;} \\
E_i &= b\ s_{i,n}\ b\ s_{i,n-1}\ b\ \cdots\ b\ s_{i,1}\ b, & \text{if } i \text{ is even.}
\end{aligned}
$$

We now construct the RNA sequences S and R as follows.

$$S = u^k, D_1, c^1, D_2, D_3, d^1, c^2, D_4, D_5, d^2 \cdots c^{(m-1)/2}, D_{m-1}, D_m, d^{(m-1)/2}$$
$$R = e^1, E_1, E_2, f^1, e^2, E_3, E_4, f^2 \cdots e^{(m-1)/2}, E_{m-2}, E_{m-1}, f^{(m-1)/2}, E_m, u^k$$

Note that the lengths of S and R are polynomial with the total size of all sequences $S_1 \ldots S_m$.

We now set the energy function for bonded nucleotides pairs. The bond between each nucleotide with its complement has a free energy of -1.0. The bond between u with x, y, z, w also has a free energy of -1.0. For other bonds between nucleotide pairs, the free energy is 0.0.

In the basepair energy model, the free energy of the overall structure is defined to be the sum of the free energies of all bonded pairs of nucleotides. Thus, according to the above setting, each nucleotide other than u will tend to get bonded with their complementary nucleotides, and u will tend to get bonded with any of x, y, z, w and vice versa. We call such bondings *valid* bondings. The free energy of the joint structure is minimized when the number of valid bondings between nucleotide pairs is maximized.

We now show that there exists a common subsequence of length k among $S_1, \ldots, S_m$ if and only if there exists a joint secondary structure of S and R where *every* nucleotide forms a valid bonding.

Suppose that $S_1 \ldots S_m$ have a common subsequence C of length k; we can construct a secondary structure of S and R where every nucleotide forms a valid bonding as follows.

- For each i, form a bond between the ith a in S with the ith b in R.
- For each i, bond the substring c^i to the substring d^i in S and bond the substring e^i to the substring f^i in R.
- For each string $S_i \in L$ there is a corresponding substring D_i in S and E_i (which is the complement of D_i) in R. Consider for each S_i the sequence that remains when the common subsequence C is deleted out; denote this sequence by C'. Bond each nucleotide in D_i that corresponds to a character in C' to its corresponding complementary nucleotide in E_i.
- All that remains in S and R are those nucleotides that correspond to the common subsequence C in each string S_i. There is also the substring u^k at the left end of S and another substring of the form u^k at the right end of R. Bond the u^k block in S to the unbonded nucleotides (that correspond to C) in D_1. For all $1 \le i \le (m-1)/2$, bond the unbonded nucleotides in E_{2i-1} to those in E_{2i}. Similarly bond the unbonded nucleotides in D_{2i} to those in D_{2i+1}. Finally bond the unbonded nucleotides in E_m to the u^k block in R.

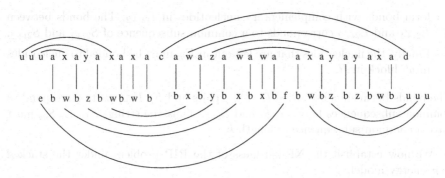

Fig. 1. Sample RIP solution for mLCS problem on $S_1 = \{xyxx\}, S_2 = \{xxyx\}, S_3 = \{xyyx\}$. The mLCS is determined with the internal bondings, here it is xyx

The reader can easily verify that our construction establishes a valid bonding for all nucleotides in S and R. The process of constructing S and R and establishing the bonds described above is demonstrated in Figure 1. Here $L = \{s_1 = xyxx,\ s_2 = xxyx,\ s_3 = xyyx\}$.

Now we show that if there is a joint secondary structure between S and R where every nucleotide forms a valid bonding, then there is a common subsequence of strings $S_1, S_2, \cdots, S_m$ of length k.

- Nucleotides a and b are complementary and do not form bonds with u. S only has as and R only has bs. If all as and bs form valid bonds, the ith a *must* form a bond with the ith b.
- Nucleotides c, d only occur in S and only form valid bonds with each other. Because we do not allow internal pseudoknots, each c^i block will be bonded with the d^i block. Similarly, nucleotides e, f only occur in R and only form valid bonds with each other. Again, because there are no internal pseudoknots, each e^i block will be bonded with the f^i block.
- The above bondings necessitate that nucleotides of the u^k block in S must bond with those in D_1 and nucleotides of the u^k block in R must bond with those in E_m. The remaining nucleotides of D_1 must bond with corresponding nucleotides in E_1 and the remaining nucleotides of E_m must bond with corresponding nucleotides in D_m.
- The nucleotides that are left in E_1 are the nucleotides that correspond to those in D_1 which have been bonded to u^k block - they must be bonded to complementary nucleotides in E_2. The bonds between E_1 and E_2 corresponds to a common subsequence of S_1 and S_2 of size k.
- Inductively, for $i = 1 \ldots (m-1)/2$, the nucleotides left out in E_{2i} must form bonds with corresponding nucleotides in D_{2i}. The ones that are left out in D_{2i} must form bonds with complementary nucleotides in D_{2i+1}. The bonds between D_{2i} and D_{2i+1} corresponds to a common subsequence of S_{2i} and S_{2i+1}.
- Similarly, the nucleotides left out in D_{2i+1} must form bonds with corresponding nucleotides in E_{2i+1}. The ones that are left out in E_{2i+1} must

form bonds with complementary nucleotides in E_{2i+2}. The bonds between E_{2i+1} and E_{2i+2} corresponds to a common subsequence of S_{2i+1} and S_{2i+2}.

- Finally, the nucleotides that are left out in E_m must be bonded to nucleotides in u^k block in R.

The bonds between consecutive D_i, D_{i+1} pairs and E_i, E_{i+1} pairs correspond to common subsequences between S_i and S_{i+1}. Thus the strings $S_1, \ldots, S_m$ must have a common subsequence of length k.

We now establish the NP-hardness of the RIP problem under the stacked pair energy model.

Theorem 2. *RIP problem under the Stacked Pair Energy Model is NP-Complete.*

Proof. The proof is through an indirect reduction from the mLCS problem as per Theorem 1. Consider the reduction of the mLCS problem to the RIP problem under the basepair energy model. Given sequences S and R that were obtained as a result of this reduction, we construct two new RNA sequences S' and R' from the standard nucleotide alphabet by replacing each character in S and R with quadruplets of nucleotides as follows: $a \leftarrow CCGU$, $b \leftarrow GGCU$, $c \leftarrow GCCU$, $d \leftarrow CGGU$, $e \leftarrow CGCU$, $f \leftarrow GCGU$, $u \leftarrow AAAU$, $x \leftarrow ACAU$, $z \leftarrow CACU$, $y \leftarrow AGAU$, $w \leftarrow GAGU$.

We now determine the energy function for stacked pairs of nucleotides. The free energy of the following stacked pairs are all set to -0.5:
$(A-A, A-C), (A-A, C-A), (A-A, A-G), (A-A, G-A), (A-C, A-A), (A-C, C-A), (A-G, A-A), (A-G, G-A), (C-A, A-A), (C-A, A-C), (C-G, C-G), (C-G, G-C), (G-A, A-A), (G-A, A-G), (G-C, C-G), (G-C, G-C)$. For other bondings between nucleotides, the free energy is set to 0.0. Thus bonding U with any nucleotide will not reduce the free energy of the joint structure.

In the stacked pair energy model, the free energy of the overall structure is defined to be the sum of the free energies of all stacked pairs of bonded nucleotides. The reader can verify that above setting of stacked pair energies ensure that the bonds between the characters of S and R presented in Theorem 1 will be preserved between S' and R'. (e.g. a bond between a and b has free energy -1.0. Because a corresponds to $CCGU$ and b corresponds to $GGCU$, the stacked pairs obtained will be $(C-G, C-G)$ and $(C-G, G-C)$ each with free energy -0.5. The total free energy will thus be -1.0.)

2.1   Additional Topological Constraints on Joint Structures

The hardness of the RIP problem under both basepair and stacked pair energy models necessitate one more constraint on the topology of the interaction between two RNA molecules. Based on our observations of known joint structures of RNA molecule pairs in Figure 2, we impose the following constraint (which is satisfied by all known structures in the literature). Let $S[i]$ be bonded with $S[j]$ and $R[i']$ be bonded with $R[j']$. Then exactly one of the following must be satisfied:

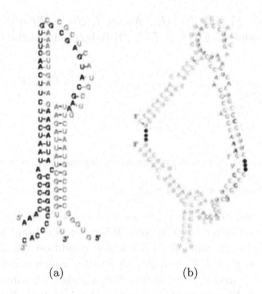

(a)   (b)

Fig. 2. (a) Natural joint structure between small RNA molecules CopA (antisense) and CopT (its target) in *E.Coli*. (b) Natural joint structure between small RNA molecules fhlA (target[black]) and OxyS (antisense[red]) in *E.Coli*

1. There are no $i < k < j$ and $i' < k' < j'$ such that $S[k]$ bonds with $R[k']$.
2. For all $i < k < j$, if $S[k]$ bonds with some $R[k']$ then $i' < k' < j'$.
3. For all $i' < k' < j'$, if $R[k']$ bonds with some $S[k]$ then $i < k < j$.

The condition simply states that if two "substructures" $S[i,j]$ and $R[i',j']$ interact, then one must "subsume" the other. A joint structure of two RNA sequences S and R is considered to be *valid* if all above conditions are satisfied.

3   Structure Prediction in the Basepair Energy Model

The basepair energy model approximates the free energy of the joint structure between interacting RNA molecules as the sum of the free energies of bonded nucleotide pairs. We denote the Watson-Crick free energy of a bond between nucleotides x and y by $e(x,y)$ if they are on the same RNA strand (this is called an *internal bond*) and by $e'(x,y)$ if they are on different strands (this is called an *external bond*). Although in our experiments we set $e' = e$, our formulation also allows to differentiate these two energy functions. Below, we obtain a valid pairing between the nucleotides of S and R that minimizes the free energy of their joint structure through the computation of $E\left(S[i,j], R[i',j']\right)$ the free energy between interacting RNA strands $S[i,j]$ and $R[i',j']$ for all $i < j$ and $i' < k'$. Clearly E gives the overall free energy between S and R when

$i = i' = 1$ and $j = |S|$ and $j' = |R|$. We set $E(S[i,i], R[i',i'])$ to $e'(S[i], R[i'])$ and compute the value of $E(S[i,j], R[i',j'])$ inductively as the minimum of the following:

1. $\min_{i-1 \leq k \leq j; i'-1 \leq k' \leq j'' : (k \neq i-1 \ or \ k' \neq i'-1),(k \neq j \ or \ k' \neq j')} E(S[i,k], R[i',k']) + E(S[k+1,j]), R[k'+1,j'])$.
2. $E(S[i+1, j-1], R[i',j']) + e(S[i], S[j])$.
3. $E(S[i,j], R[i'+1, j'-1]) + e(R[i'], R[j'])$.

Lemma 1. *The above dynamic programming formulation is correct.*

Proof. There are two cases to deal with:

1. Consider the case that either $S[i]$ or $S[j]$ or $R[i']$ or $R[j']$ bonds with a nucleotide on the other RNA strand. Wlog, let $S[i]$ bond with $R[h']$; then either (i) $R[i']$ bonds with $R[j']$ for which condition (3) will be satisfied, or (ii) $i' = h'$ so that $R[i']$ bonds with $S[i]$ for which condition (1) will be satisfied for $k = i$ and $k' = i'$, or (iii) $R[i']$ bonds with some $R[\ell']$ for which condition (1) will be satisfied for some "breakpoint" $S[k], R[k']$, for $i \leq k \leq j$ and $i' \leq k' \leq j'$ such that $S[i,k]$ interacts only with $R[i',k']$ and $S[k+1,j]$ interacts only with $R[k'+1,j']$.
2. If the above condition is not satisfied then wlog we can assume that $S[i]$ bonds with $S[h]$ and $R[i']$ bonds with $R[h']$. If for no $\ell > h$, $S[\ell]$ interacts with any $R[\ell']$ for $\ell' > h'$ then condition (1) will be satisfied with $k = h$ and either $k' = i' - 1$ or $k' = j' + 1$. If for no $\ell < h$, $S[\ell]$ interacts with any $R[\ell']$ for $\ell' < h'$ then condition (1) will be satisfied again with $k = h$ and either $k' = i' - 1$ or $k' = j' + 1$. The possibility of none of these two cases hold is excluded by our topological constraints described earlier.

Lemma 2. *The table E can be computed in time $O(|S|^3 \cdot |R|^3)$ and in space $O(|S|^2 \cdot |R|^2)$.*

3.1   Testing the Basepair Energy Model

We tested the basepair energy model on naturally occurring joint structures of interacting RNA molecule pairs CopA-CopT and OxyS-fhlA. Our results are given in Figures 3b and 4b. Perhaps not surprisingly, the predicted joint structures by the Basepair Energy Model is quite different from the natural secondary structures (Figures 3a and 4a). Observe that in natural joint structures, internal or external bonds usually form stacked pairs; i.e., a bond $S[i] - S[j]$ usually implies bonds $S[i+1] - S[j-1]$ and $S[i-1] - S[j+1]$. Similarly a bond $S[i] - R[i']$ usually implies bonds $S[i+1] - R[i'+1]$ and $S[i-1] - R[i'-1]$. Furthermore, in natural joint structures unbonded nucleotides seem to form uninterrupted sequences rather than being scattered around.

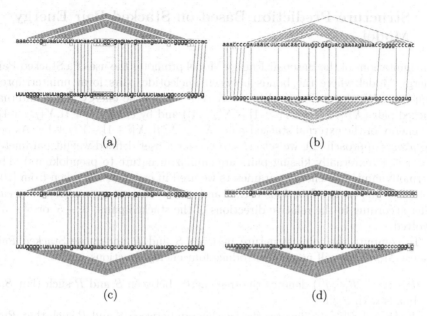

Fig. 3. (a) Known joint structure of CopA and CopT, (b) as predicted by Basepair Energy Model, (c) as predicted by Stacked Pair Energy Model, (d) as predicted by Loop Energy Model

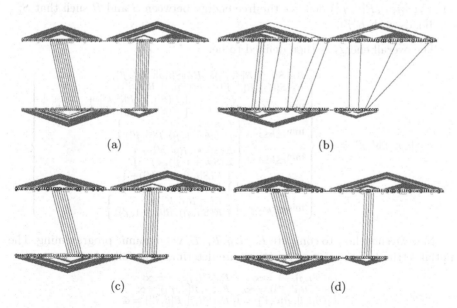

Fig. 4. (a) Known joint structure of OxyS and fhlA, (b) as predicted by Basepair Energy Model, (c) as predicted by Stacked Pair Energy Model, (d) as predicted by Loop Energy Model

4   Structure Prediction Based on Stacked Pair Energy Model

The limitations of the Basepair Energy Model promotes the use of a Stacked Pair Energy Model where the bonds between nucleotide pairs form uninterrupted sequences. We denote by $ee(X[i, i + 1], X[j - 1, j])$ the energy of the internal stacked pair $(X[i] - X[j], X[i + 1] - X[j - 1])$ and by $ee'(X[i, i + 1], Y[j, j + 1])$ the energy of the external stacked pair $(X[i] - Y[j], X[i + 1] - Y[j + 1])$. As per the *pknots* approach [20] we set $ee' = \sigma \cdot ee$ for a user defined weight parameter $0 < \sigma \leq 1$ (externally kissing pairs are similar in nature to pseudoknots). The thermodynamic free energy parameters we used in our tests are taken from [10]. Note that the energy functions $ee(.,.)$ and $ee'(.,.)$ are not symmetric; they can differ according to the relative directions of the stacked pairs $(3' - 5'$ or $5' - 3')$ involved.

To compute the joint structure between S and R under the Stacked Pair Energy Model we will need to introduce four energy functions.

1. $E_S(S[i, j], R[i', j'])$ denotes the free energy between S and R such that $S[i]$ bonds with $S[j]$.
2. $E_R(S[i, j], R[i', j'])$ denotes the free energy between S and R such that $R[i']$ bonds with $R[j']$.
3. $E_l(S[i, j], R[i', j'])$ denotes the free energy between S and R such that $S[i]$ bonds with $R[i']$.
4. $E_r(S[i, j], R[i', j'])$ denotes the free energy between S and R such that $S[j]$ bonds with $R[j']$.

The overall energy is then defined to be:

$$E(S[i, j], R[i', j']) = \min \left\{ \begin{array}{l} E_S(S[i, j], R[i', j']), \quad E_R(S[i, j], R[i', j']), \\ E_r(S[i, j], R[i', j']), \quad E_l(S[i, j], R[i', j']), \\ \min_{i \leq k \leq j-1; i' \leq k' \leq j'-1} \left\{ \begin{array}{l} E(S[i, k], R[i', k'])+ \\ E(S[k+1, j]), R[k'+1, j']) \end{array} \right\} \\ \min_{i \leq k \leq j-1} \left\{ \begin{array}{l} E(S[i, k], -)+ \\ E(S[k+1, j]), R[i', j']) \end{array} \right\} \\ \min_{i \leq k \leq j-1} \left\{ \begin{array}{l} E(S[i, k], R[i', j'])+ \\ E(S[k+1, j]), -) \end{array} \right\} \\ \min_{i' \leq k' \leq j'-1} \left\{ \begin{array}{l} E(S[i, j], R[i', k'])+ \\ E(-, R[k'+1, j']) \end{array} \right\} \\ \min_{i' \leq k' \leq j'-1} \left\{ \begin{array}{l} E(-, R[i', k'])+ \\ E(S[i, j]), R[k'+1, j']) \end{array} \right\} \end{array} \right\}.$$

Now we show how to compute E_S, E_R, E_r, E_l via dynamic programming. The initial settings of the energy functions are determined as follows.

$$\begin{array}{ll} E_l(S[i, j], -) = \infty & E_r(S[i, j], -) = \infty \\ E_l(-, R[i', j']) = \infty & E_r(-, R[i', j']) = \infty \\ E_l(S[i, i], R[i', i']) = 0 & E_r(S[i, i], R[i', i']) = 0 \end{array}$$

$$E_S(S[i, i], -) = \infty \qquad E_R(-, R[i', i']) = \infty$$

Now we give the complete description of the dynamic programming formulation. Note that because sequence R is assumed to be in $3' - 5'$ direction, reversing the stacked pairs involved is required for the correct use of ee function in E_R.

$$E_l(S[i,j], R[i',j']) = \min \left\{ \begin{array}{l} E_l(S[i+1,j], R[i'+1,j']) + ee'(S[i,i+1], R[i',i'+1]), \\ E(S[i+1,j], R[i'+1,j']) \end{array} \right\}$$

$$E_r(S[i,j], R[i',j']) = \min \left\{ \begin{array}{l} E_r(S[i,j-1], R[i',j'-1]) + ee'(S[j-1,j], R[j'-1,j']), \\ E(S[i,j-1], R[i',j'-1]) \end{array} \right\}$$

$$E_S(S[i,j], R[i',j']) = \min \left\{ \begin{array}{l} E_S(S[i+1,j-1], R[i',j']) + ee(S[i,i+1], S[j-1,j]), \\ E(S[i+1,j-1], R[i',j']) \end{array} \right\}$$

$$E_R(S[i,j], R[i',j']) = \min \left\{ \begin{array}{l} E_R(S[i,j], R[i'+1,j'-1]) + ee(R[j'-1,j']^r, R[i',i'+1]^r), \\ E(S[i,j], R[i'+1,j'-1]) \end{array} \right\}$$

The following lemmas follow from the DP formulation above and their proofs are left for the full version for the paper.

Lemma 3. *The above dynamic programming formulation is correct.*

Lemma 4. *The tables E_S, E_R, E_l, E_r and the overall energy table E can be computed in time $O(|S|^3 \cdot |R|^3)$ and in space $O(|S|^2 \cdot |R|^2)$.*

As will be discussed below, the Stacked Pair Energy Model formulation works very well with the joint structure prediction problems considered in this paper. However this formulation does not necessarily aim to cluster gaps in uninterrupted sequences, as observed in natural joint structures. Thus, we also provide a more general formulation for the Stacked Pair Energy Model, that employs an "affine" cost model for the gaps involved. Also considered in this formulation are penalties for switching from internal to external bonds (and vice versa). This general formulation does not necessarily improve our predictions for the joint structures considered in this paper; however it could be useful for other examples and thus is provided in the Appendix.

4.1 Testing Stacked Pair Energy Model

The Stacked Pair Energy Model as defined above has only one user defined parameter (as per [20]), σ, which is the ratio between the free energies of internal and external stacked pairs. Unfortunately no miracle prescription for determining the right value of σ is available (see for example [20]). It is possible to approximately determine the value for σ by closely inspecting the natural joint structure of CopA-CopT pair (Figure 3a). CopA and CopT sequences are (almost) perfectly complimentary to each other, thus they can, in principle, form a stable duplex structure that would prevent any internal bonding pairs, leaving out just *one* nucleotide unbonded. However, as one can observe from Figure 3a this does not happen. The ratio between the length of the external bonding sequences in the joint structure and that of the internal bonding sequences implies that $\sigma \in [0.7, 0.8]$.

Under these observations we tested our algorithm that implements the Stacked Pair Energy Model with $\sigma \in [0.7, 0.8]$. The secondary structures predicted by our algorithm on CopA-CopT and OxyS-fhlA pairs are given in Figures 3c and 4c. As one can observe, there are only very slight differences between the natural joint structure and the predicted joint structure of the RNA pairs. For example, the predicted joint structure of OxyS-fhlA pair (Figure 4c) has 53 internal-bonds, 14 external-bonds, and 23 unbonded nucleotides. In *all* aspects, these figures are superior to the natural joint structure of the pair (Figure 4a), which has 50 internal-bonds, 16 external-bonds, and 25 unbonded nucleotides. Because the external bond scores are smaller than internal ones, under *any selection of* $\sigma < 1$ the prediction of our algorithm results in a higher score/lower free energy than that implied by the natural joint structure of OxyS-fhlA pair. Nevertheless, the differences between the natural structures and the predicted ones are very small implying that the Stacked Pair Energy Model can be used as the central tool of our RNA target prediction algorithm.

5   Structure Prediction Based on Loop Energy Model

The structure prediction algorithm to find the optimal joint structure between two RNA molecules based on Stacked Pair Energy Model requires substantial resources in terms of running time and memory. On a Sun Fire v20z server with 16GB RAM and AMD Opteron 2.2GHz processor, the running time for predicting the joint secondary structure of OxyS-fhlA pair is 15 minutes; this could be prohibitive for predicting the targets of sufficiently long RNA molecules. In this section we make a number of observations on the natural joint structures of RNA molecule pairs for speeding up our approach through heuristic shortcuts - without losing its (experimental) predictive power.

An interesting observation is that the (predicted) self structures are mostly preserved in the joint secondary structures. In fact, external interactions only occur between pairs of predicted hairpins. Thus it may be be sufficient to compute the joint structure of two RNA sequences by simply computing the set of loop pairs that form bonds to minimize the total joint free energy.

The above observation prompts an alternative, simpler approach which is described below. This new approach maintains that each RNA sequence will tend to preserve much of its original secondary structure after interacting with the other RNA sequence, which is achieved by means of preserving what we call "independent subsequences" that form hairpins. More formally:

Definition 1. *Independent Subsequences:*

Given an RNA sequence R and its secondary structure, the substring $R(i, j)$ is an independent subsequence of R if it satisfies the following conditions.

 – $R[i]$ is bonded with $R[j]$.
 – $j - i \leq \kappa$ for some user specified length κ.

– There exists no $i' < i$ and $j' > j$ such that $R[i']$ is bonded with $R[j']$ and $j' - i' \leq \kappa$. (This condition prohibits overlaps between independent subsequences).

It is possible to compute the (locations of) independent sequences of a given RNA molecule, from its secondary structure predicted by *Mfold*, through a simple greedy algorithm as follows.

1. Let IS be the set of independent subsequences in R; initially we set $IS = \emptyset$.
2. Starting from the first nucleotide of R find the first nucleotide $R[i]$ which bonds with another nucleotide $R[j]$, $(j > i)$.
3. If $j - i \leq \kappa$ then update $IS = IS \cup R[i, j]$ and move to $R[j + 1]$. Else move to $R[i + 1]$.
4. Repeat Step 2.

The proofs of the following are quite easy to obtain and hence are not given.

Lemma 5. *The above algorithm finds the correct independent subsequences.*

Lemma 6. *Given the secondary structure of an RNA sequence R, its independent subsequences can be computed in $O(|R|)$ time via the above algorithm.*

5.1 Computing the Interactions Between Independent Subsequences

In our new model, the external bondings between nucleotide pairs will be permitted among the independent subsequences of the two RNA sequences S and R, predicted by *Mfold*. Below we show how to compute the external bonds between such nucleotides which minimize the total free energy in the interacting RNA sequences.

From this point on we will treat each RNA molecules as an (ordered) set of independent subsequences (IS), where each IS is indeed a string of nucleotides. The ith IS of an RNA molecule S is denoted by $S_{IS}[i]$. The sequence of ISs between $S_{IS}[i]$ and $S_{IS}[j]$ are thus denoted as $S_{IS}[i, j]$.

We calculate the joint structure between R and S by minimizing the total free energy of their ISs via means of establishing bonds between their nucleotides as follows. Let the minimum free energy of the joint secondary structure of the two ISs $S_{IS}[i]$ and $R_{IS}[j]$ be $e_{IS}(i, j)$. The value of $e_{IS}(i, j)$ can be computed via the algorithm we described in Section 4.

The minimum joint free energy between the consecutive sets of ISs of R and S is calculated once $e_{IS}(i, j)$ is computed for all i, j. Let n and m denote the number of ISs in S and R respectively. Now let $E(S_{IS}[i], R_{IS}[j]) = E[i, j]$ be the smallest free energy of the interacting independent subsequence lists $S_{IS}[1, i]$ and $R_{IS}[1, j]$ (which satisfy the distance constraint) provided that $S_{IS}[i]$ and $R_{IS}[j]$ interact with each other.

Before we show how to compute the values of $E[i, j]$, we make one final observation on the OxyS-fhlA pair that the "distance" between two interacting subsequences in OxyS appears to be be very close to that in fhlA. This may be due to the limited flexibility of "root stems" that support the independent subsequences when they interact with each other. In order to ensure that the predictions made by our algorithm satisfy such limitations we impose restrictions on the "distances" between interacting independent subsequences as follows.

Definition 2. *Let $S_{IS}[i]$ and $S_{IS}[j]$ be two independent subsequences in a given RNA sequence S. The distance between $S_{IS}[i]$ and $S_{IS}[j]$, denoted $d(S_{IS}[i], S_{IS}[j])$ is defined as the number of nucleotides $S[k]$ that do not lie between a bonded pair of nucleotides $S[h]$ and $S[h']$ that are both located between $S_{IS}[i]$ and $S_{IS}[j]$.*

The above definition simply ignores all nucleotides that lie in the independent subsequences between $S_{IS}[i]$ and $S_{IS}[i']$ regardless of their lengths. Our algorithm ensures that if $S_{IS}[i] - R_{IS}[j]$ and $S_{IS}[i'] - R_{IS}[j']$ are pairs of consecutive independent subsequences that interact with each other and if $d(S_{IS}[i], S_{IS}[i']) \geq d(R_{IS}[j], R_{IS}[j'])$ then $d(S_{IS}[i], S_{IS}[i']) \leq (1 + \epsilon) \cdot d(R_{IS}[j], R_{IS}[j']) + \delta$; here $\epsilon < 1$ and $\delta > 0$ are user defined constants.

The value of $E[i, j]$ can be computed through dynamic programming as follows.

$$E[i, j] = \min_{i' < i, j' < j \mid d(S_{IS}[i'], S_{IS}[i]) \leq (1+\epsilon) \cdot d(R_{IS}[j'], R_{IS}[j]) + \delta} \left(\begin{array}{l} E[i', j'] + e_{IS}(i, j) + \\ \sum_{i' < i'' < i} e_{IS}(i'', 0) + \\ \sum_{j' < j'' < j} e_{IS}(0, j''). \end{array} \right)$$

Here $e_{IS}(i'', 0)$ and $e_{IS}(0, j'')$ denote the free energy of independent subsequences $S_{IS}[i'']$ and $R_{IS}[j'']$ respectively. The overall free energy of the interacting independent subsequence sets of R and S is thus:

$$\min_{\forall i, j} E[i, j] + \sum_{i < i'} e_{IS}(i', 0) + \sum_{j < j'} e_{IS}(0, j').$$

The following lemmas are easy to verify and are left to the full version of the paper.

Lemma 7. *The above dynamic programming formulation correctly computes $E[i, j]$.*

Lemma 8. *Given $e_{IS}(i, j)$ for all i, j, the values of E can be computed in time $O(n^3 \cdot m^3)$. As the computation of $e_{IS}(i, j)$ takes $O(\kappa^6)$ time the overall running time of the algorithm is $O(n \cdot m \cdot \kappa^6 + n^3 \cdot m^3)$.*

Because $n \leq |S|/\kappa$ and $m \leq |R|/\kappa$ the worst case running time of this algorithm is $O(|S| \cdot |R| \cdot \kappa^4 + |S|^3 \cdot |R|^3/\kappa^6)$. This is substantially faster than our earlier approach requiring $O(|S|^3 \cdot |R|^3)$ time. In fact this version can predict the joint structure of the OxyS-fhlA pair in 5 seconds, improving our earlier approach by a factor of 180.

5.2   Testing the Loop Energy Model

We tested our third model on the interacting RNA pairs CopA-CopT and OxyS-fhlA, with the same σ values we used in Stacked Pair Energy Model: $\sigma \in [0.7, 0.8]$.

Joint structure predictions obtained by Loop Energy Model are given in Figures 3d for CopA-CopT pair, and 4d for OxyS-fhlA pair. Although there is a slight loss in the prediction quality in CopA-CopT pair with respect to the Stacked Pair Energy Model prediction (Figure 3c), the "kissing" hairpin sequence is predicted correctly. In our OxyS-fhlA test, notice that the predictions obtained by the Loop Energy Model and Stacked Pair Energy Model are even more similar. Furthermore, careful observation shows that the total free energy in the predicted structure is still better than the natural joint structure (Figure 4a).

6   Target Prediction for Antisense RNAs

An important byproduct of our algorithms for the RIP problem is the ability to search for target sequences for specific antisense RNA molecules in whole genomic and plasmid sequences. Because of the time and space constraints, the Stacked Pair Energy Model is not efficient when searching through large sequences. Therefore, in our target prediction approach is based on Loop Energy Model. Our search strategy employs the following steps:

1. First, we need to find the "candidate" target sequences from a given genome sequence (or plasmid) that is known to include the target. This is achieved via using the cDNA annotation available for genomic sequences. To compute the potential mRNA each such cDNA is extended towards $5'$ and $3'$ UTR ends as follows.
   (a) Each cDNA is extended up to l_1 nucleotides at its $5'$ UTR, and by l_2 nucleotides at its $3'$ UTR, where l_1 and l_2 are user defined parameters. (In our experiments we set $l_1 = 250$ and $l_2 = 25$).
   (b) Then each "extended" cDNA sequence is trimmed from both ends via a dynamic programming routine in order to compute its subsequence which has the lowest "energy density" (this will be the subsequence of the extended cDNA sequence whose secondary structure is most stable.) We predict the resulting mRNA of each such cDNA as its trimmed extension. The details of this mRNA prediction approach (from a given cDNA) is of independent interest and is left to the full version of the paper.
2. We then run our joint secondary structure prediction algorithm based on Loop Energy Model to determine if there are any external bonds formed between each candidate target sequence and the antisense RNA sequence under the following constraints. (1) At least one IS in the candidate target sequence which lies before the start codon (i.e. AUG) should interact with an independent subsequence in the query sequence. We impose this constraint in order to capture the ribosome binding site interactions. (2) All predicted interactions between pairs of ISs should include at least ξ uninterrupted bonds for some user specified constant ξ. We impose this constraint to favor long uninterrupted external bonds, since ribosomes are capable of breaking shorter interactions. (3) At least two pairs of independent sequences must be interacting with each other.

We tested the above approach on both RNA-RNA interactions that we considered in the paper.

(1) We first searched the target mRNA sequences for CopA in the R1 plasmid sequence in *E.coli*. It is known that CopA regulates the copy number of R1 plasmid by binding to the CopT sequence which is a part of the 125Kb long plasmid [6, 21]. Our program needed about 40 hours on a PC equipped with 2 Ghz Pentium IV processor and 1 GB of main memory to detect all targets of the CopA sequence on the complete R1 plasmid. Out of the 141 potential mRNA segments obtained from the annotated cDNA sequences it returned only the correct target CopT as a potential target.

(2) We then used our program to detect the target mRNA sequences of the OxyS antisense RNA on a 130Kb long portion of E.coli genome that included the known target fhlA [16]. Out of the 100 potential mRNA segments obtained from the annotated cDNA sequences, our program returned 9 hits including the known target fhlA.

Notice that the joint structure between CopA and CopT are much more stable than that between OxyS and fhlA (the former one has a half-life of about an hour where as the latter one has a half-life of only a couple of minutes). It is possible that OxyS may have other targets in the E.coli genome with which it may establish unstable joint structures, not strong enough to make impact. We are aiming to test whether such *in silico* interactions between OxyS and its 8 unknown targets (i.e. those predicted by our program) actually take place *in vitro* or *in vivo* in the future.

Acknowledgments. We would like to thank the anonymous referees for pointing out the *IRIS* software package to our attention.

References

1. Akutsu T., Dynamic programming algorithms for RNA secondary structure prediction with pseudoknots, *Discrete Applied Mathematics*, 104:45-62, 2000.
2. Andronescu M., Aguirre-Hernandes R., Condon A., and Hoos H., RNAsoft: a suite of RNA secondary structure prediction and design software tools, *Nucleic Acids Research*, 31(13):3416-3422, 2003.
3. Condon A., Davy B., Rastegari B., Zhao S., Tarrant F., Classifying RNA pseudoknotted structures. *Theoretical Computer Science*, 320(1): 35-50, 2004.
4. Collins G., Le S., Zhang, K., A new algorithm for computing similarity between RNA structures, *Proc. 5th Joint Conf. on Information Science*, Atlantic City, NJ, vol. 2, pp. 761-765, March, 2000.
5. Kim, C.-H., and Tinoco Jr., I. A Retroviral RNA Kissing Complex Containing Only Two G-C Base Pairs, *Proc.Nat.Acad.Sci.* USA 97 pp. 9396, 2000.
6. Kolb, F.A., Engdahl, H.M, Slagter-Jager, J.G., Ehresmann, B., Ehresmann, C., Westhof, E., Wagner, E.G.H., and Romby, P., Progression of a loop-loop complex to a four-way junction is crucial for the activity of a regulatory antisense RNA, *EMBO Journal*, 19(21):5905-5915, 2000.

7. Lagos-Quintana, M., Rauhut, R., Lendeckel, W., and Tuschl, T., Identification of novel genes coding for small expressed RNAs, *Science*, 294:853-857, 2001.
8. Lau, N.C., Lim, L.P., Weinstein, E.G., Bartel, D.P., An abundant class of tiny RNAs with probable regulatory roles in Caenorhabditis elegans, *Science*, 294:858-862, 2001.
9. Lyngso, R.B., Zuker, M., and Pedersen, C.N.S. Fast evaluation of internal loops in RNA secondary structure prediction, *Bioinformatics*, 15:440-445, 1999.
10. Mathews, D., Sabina, J., Zuker, M. and Turner, D., Expanded sequence dependence of thermodynamic parameters improves prediction of RNA secondary structure. *Journal of Molecular Biology*, 288:911-940, 1999.
11. McManus, Michael T., and Sharp, Phillip A., Gene silencing in mammals by small interfering RNAs, *Nature Reviews Genetics*, 10:737-747, 2002.
12. Moss, Eric G., RNA interference: It's a small RNA world, *Current Biology*, 11:R772-R775, 2001.
13. Moss, Eric G., MicroRNAs: Hidden in the Genome, *Current Biology*, 12:R138-R140, 2002.
14. Notredame, C., O'Brien, E.A., and Higgins, D.G., RAGA: RNA sequence alignment by genetic algorithm, *Nucleic Acids Research*, 25(22):4570-4580, 1997.
15. Nussinov, R. and Jacobson, A. Fast algorithm for predicting the secondary structure of single stranded RNA, *PNAS*, 77:6309-6313, 1980.
16. NCBI web site, `http://www.ncbi.nlm.nih.gov`
17. Pervouchine, Dmitri D. IRIS: Intermolecular RNA Interaction Search, *15th Int. Conf. Genome Informatics*, 2004.
18. Peyret N. and SantaLucia, J., HYTHER^{TM} version 1.0, `http://ozone2.chem.wayne.edu/Hyther/hythermenu.html`, Wayne State University.
19. Reinhart, B.J., Slack, F.J., Basson, M., Pasquinelli, A.E., Bettinger, J.C., Rougvie, A.E., Horvitz, H.R., and Ruvkun, G., The 21-nucleotide let-7 RNA regulates developmental timing in Caenorhabditis elegans, *Nature*, 403:901-906, 2000.
20. Rivas, E., and Eddy, S.R., A dynamic programming algorithm for RNA structure prediction including pseudoknots, *J Mol Biol.*, 285(5):2053-68, 1999.
21. Wagner, E.G.H., and Flardh, K., Antisense RNAs everywhere?, *TRENDS in Genetics*, 18(5):223-226, 2002.
22. Zhang, K., Wang, L., and Ma, B., Computing similarity between RNA structures, *Theoretical Computer Sciences*, 276(1-2):111-132, 2002. March, 2000.
23. Zuker,M. and Stiegler,P., Optimal computer folding of large RNA sequences using thermodynamics and auxiliary information. *Nucleic Acids Res.*, 9:133148, 1981.
24. Zuker, Michael, On finding all suboptimal foldings of an RNA molecule, *Science*, 244:48-52, 1989.

A   A More General Stacked Pair Energy Formulation

Our more general formulation of the Stacked Pair Energy Model adds two more energy functions e and e', and two penalty parameters g and G. This necessitates the use of four additional energy tables $E_{S,l}, E_{S,r}, E_{R,l}, E_{R,r}$ to the set (E_S, E_R, E_l, E_r) already used in Section 4:

1. $E_{S,l}(S[i,j], R[i',j'])$ denotes the free energy between S and R such that $S[i]$ remains unbonded.
2. $E_{S,r}(S[i,j], R[i',j'])$ denotes the free energy between S and R such that $S[j]$ remains unbonded.
3. $E_{R,l}(S[i,j], R[i',j'])$ denotes the free energy between S and R such that $R[i']$ remains unbonded.
4. $E_{R,r}(S[i,j], R[i',j'])$ denotes the free energy between S and R such that $R[j']$ remains unbonded.

The addition of four more parameters (and four new degrees of freedom) makes this approach more adjustable to specific properties of the input RNA strands.

In addition to the stacked pair energies, this formulation also considers the free energies of an internally and externally bonded individual nucleotide pairs denoted $e(X[i], Y[j])$ and $e'(X[i], X[j])$ respectively. For further generality, this formulation induces an additive penalty for switching between the two types of bonds. More specifically, the energy function has an additive penalty g to any nucleotide $X[k]$ (X could be S or R), if (i) $X[k]$ is bonded with $X[j]$ however $X[k+1]$ is not bonded with $X[j-1]$, (ii) $X[k]$ is bonded with $X[j]$ however $X[k-1]$ is not bonded with $X[j+1]$, (iii) $X[k]$ is bonded with $Y[k']$ however $X[k+1]$ is not bonded with $Y[k'+1]$, (iv) $X[k]$ is bonded with $Y[k']$ however $X[k-1]$ is not bonded with $Y[k'-1]$. For unbonded nucleotides $X[k]$ another additive penalty G is charged if (i) $X[k+1]$ is bonded, (ii) $X[k-1]$ is bonded. The gap penalties are also added to the first and last nucleotides of X - this is only for avoiding further complexity in the dynamic programming formulation and does not affect the energy minimization process or the resulting structure prediction.

Our new energy formulation is as follows.

$$E(S[i,j], R[i',j']) = \min \left\{ \begin{array}{l} E_S(S[i,j], R[i',j']) + 2g, \quad E_R(S[i,j], R[i',j']) + 2g, \\ E_r(S[i,j], R[i',j']) + 2g, \quad E_l(S[i,j], R[i',j']) + 2g, \\ E_{S,l}(S[i,j], R[i',j']) + G, \quad E_{S,r}(S[i,j], R[i',j']) + G, \\ E_{R,l}(S[i,j], R[i',j']) + G, \quad E_{R,r}(S[i,j], R[i',j']) + G, \\ \min_{i \le k \le j-1; i' \le k' \le j'-1} \left\{ \begin{array}{l} E(S[i,k], R[i',k']) + \\ E(S[k+1,j]), R[k'+1,j']) \end{array} \right\} \\ \min_{i \le k \le j-1} \left\{ \begin{array}{l} E(S[i,k], -) + \\ E(S[k+1,j]), R[i',j']) \end{array} \right\} \\ \min_{i \le k \le j-1} \left\{ \begin{array}{l} E(S[i,k], R[i',j']) + \\ E(S[k+1,j]), -) \end{array} \right\} \\ \min_{i' \le k' \le j'-1} \left\{ \begin{array}{l} E(S[i,j], R[i',k']) + \\ E(-, R[k'+1,j']) \end{array} \right\} \\ \min_{i' \le k' \le j'-1} \left\{ \begin{array}{l} E(-, R[i',k']) + \\ E(S[i,j]), R[k'+1,j']) \end{array} \right\} \end{array} \right\}$$

These tables need to be initialized as follows.

$E_l(S[i,j], -) = \infty$ $E_r(S[i,j], -) = \infty$
$E_l(-, R[i',j']) = \infty$ $E_r(-, R[i',j']) = \infty$
$E_l(S[i,i], R[i',i']) = e'(S[i], R[i']) + 2g$ $E_r(S[i,i], R[i',i']) = e'(S[i], R[i']) + 2g$

$E_S(S[i,i], -) = \infty$ $E_R(-, R[i',i']) = \infty$
$E_{S,l}(S[i,i], -) = e(S[i], -) + G$ $E_{R,l}(-, R[i',i']) = e(-, R[i']) + G$
$E_{S,r}(S[i,i], -) = e(S[i], -) + G$ $E_{R,r}(-, R[i',i']) = e(-, R[i']) + G$

Here is the complete description of the dynamic programming formulation.

$$E_l(S[i,j], R[i',j']) = \quad e'(S[i], R[i'])+$$
$$\min \left\{ \begin{array}{c} E_l(S[i+1,j], R[i'+1,j']) + ee'(S[i,i+1], R[i',i'+1]), \\ E(S[i+1,j], R[i'+1,j']) + 2g \end{array} \right\}$$

$$E_r(S[i,j], R[i',j']) = \quad e'(S[j], R[j'])+$$
$$\min \left\{ \begin{array}{c} E_r(S[i,j-1], R[i',j'-1]) + ee'(S[j-1,j], R[j'-1,j']), \\ E(S[i,j-1], R[i',j'-1]) + 2g \end{array} \right\}$$

$$E_S(S[i,j], R[i',j']) = \quad e(S[i], S[j])+$$
$$\min \left\{ \begin{array}{c} E_S(S[i+1,j-1], R[i',j']) + ee(S[i,i+1], S[j-1,j]), \\ E(S[i+1,j-1], R[i',j']) + 2g \end{array} \right\}$$

$$E_R(S[i,j], R[i',j']) = \quad e(R[i'], R[j'])+$$
$$\min \left\{ \begin{array}{c} E_R(S[i,j], R[i'+1,j'-1]) + ee(R[j'-1,j']^r, R[i',i'+1]^r), \\ E(S[i,j], R[i'+1,j'-1]) + 2g \end{array} \right\}$$

$$E_{S,l}(S[i,j], R[i',j']) = \quad \min \left\{ \begin{array}{c} E_{S,l}(S[i+1,j], R[i',j']) + e(S[i],-), \\ E(S[i+1,j], R[i',j']) + e(S[i],-) + G \end{array} \right\}$$

$$E_{S,r}(S[i,j], R[i',j']) = \quad \min \left\{ \begin{array}{c} E_{S,r}(S[i,j-1], R[i',j']) + e(-,S[j]), \\ E(S[i,j-1], R[i',j']) + e(-,S[j]) + G \end{array} \right\}$$

$$E_{R,l}(S[i,j], R[i',j']) = \quad \min \left\{ \begin{array}{c} E_{R,l}(S[i,j], R[i'+1,j']) + e(R[i'],-), \\ E(S[i,j], R[i'+1,j']) + e(R[i'],-) + G \end{array} \right\}$$

$$E_{R,r}(S[i,j], R[i',j']) = \quad \min \left\{ \begin{array}{c} E_{S,r}(S[i,j], R[i',j'-1]) + e(-,R[j']), \\ E(S[i,j], R[i',j'-1]) + e(-,R[j']) + G \end{array} \right\}$$

Consensus Folding of Unaligned RNA Sequences Revisited

Vineet Bafna[1], Haixu Tang[2], and Shaojie Zhang[1]

[1] Dept. of Computer Science and Engineering,
University of California, San Diego, La Jolla, CA 92093
{vbafna, shzhang}@cs.ucsd.edu
[2] School of Informatics and Center for Genomics and Bioinformatics,
Indiana University, Bloomington, IN 47408
hatang@indiana.edu

Abstract. As one of the earliest problems in computational biology, RNA secondary structure prediction (sometimes referred to as "RNA folding") problem has attracted attention again, thanking to the recent discoveries of many novel non-coding RNA molecules. The two common approaches to this problem are *de novo* prediction of RNA secondary structure based on energy minimization and "consensus folding" approach (computing the common secondary structure for a set of unaligned RNA sequences). Consensus folding algorithms work well when the correct seed alignment is part of the input to the problem. However, seed alignment itself is a challenging problem for diverged RNA families.

In this paper, we propose a novel framework to predict the common secondary structure for unaligned RNA sequences. By matching putative stacks in RNA sequences, we make use of both primary sequence information and thermodynamic stability for prediction at the same time. We show that our method can predict the correct common RNA secondary structures even when we are only given a limited number of unaligned RNA sequences, and it outperforms current algorithms in sensitivity and accuracy.

1 Introduction

With the recent discovery of novel non-coding RNA (*ncRNA*) families, RNA is rapidly gaining importance as a molecule of interest [1, 2]. Recent article puts the number of human genes down to about $20000 - 25000$ [3]. By comparison, even the worm *C. elegans* has around $19,500$ genes. On the other hand, expression activity has been detected on a much larger portion of human genome [4]. It is likely that many of these novel transcripts are ncRNA genes, which may carry on many unknown cellular functions. Like proteins, RNA structures are more important for function than their sequences: RNA with similar function often have similar structures but distinct primary sequences. Therefore, understanding the structures of these RNA molecules will help elucidate their functions. Consider the recent exciting discovery of riboswitches [5, 6] as an example. These control

S. Miyano et al. (Eds.): RECOMB 2005, LNBI 3500, pp. 172–187, 2005.

elements, with a conserved secondary structure, are located in the untranslated regions of genes coding for proteins that are involved in variant metabolite (nucleic acids, amino acids etc.) synthesis pathways. The riboswitches can turn off the expression of their downstream genes by binding to certain metabolites, subsequently changing their secondary structures and blocking the translation initiation.

While there is a resurgence in interest in ncRNA, the problem of RNA secondary structure prediction has been extensively studied since the 70s. The key idea is that to stabilize its structure, distant base-pairs in the single stranded RNA molecule must form hydrogen bonds. There are two distinct approaches to predict RNA secondary structure. The RNA folding approach, initiated by Tinoco et al. [7], assigns free energies to the components of RNA secondary structure, and then computes the RNA secondary structure with the minimum energy. Dynamic programming algorithms have been developed to compute minimum energy secondary structures [8, 9, 10, 11, 12], and implemented in software packages such as MFOLD [13] and ViennaRNA [14]. However, RNA folding via energy minimization has its shortcomings. First, fold prediction depends critically upon correct values of the energy parameters, as shown by Jaeger et al. [15], which are hard to obtain experimentally. Also, RNA folding in a real cell is mediated by interactions with other molecules, and the absence of knowledge of these interactions may cause mis-folding *in silico*. Pavesi et al. [16] tried to alleviate this problem by comparing minimum energy structures of a set of RNA sequences from the same family to determine conserved secondary structure. However, it is unclear how the misprediction of secondary structure for a single RNA sequence can affect the accuracy of this approach.

A different approach attempts to resolve these shortcomings by using evolutionary conservation of structure as the basis for structure prediction. It needs as input multiple RNA sequences from an RNA family, which have common secondary structures. Since for divergent sequences, the mutations in base-pairing regions must be compensated in the complementary base to preserve structure, the presence of multiple covarying mutations is a strong signal for base-pairing. In fact most RNA sequences are selected more for maintenance of the structure than conservation of primary sequence. If the sequence similarity between the given RNA sequences is appropriate, one can first align these sequences using multiple sequence alignment algorithm and then figure out the potential base pairs in RNA secondary structures by looking at regions with a high number of compensating mutations. Levitt successfully derived the theoretical tRNA secondary structure using this approach, which was largely confirmed by crystallography [17], and various other structures have been determined through such analysis. Computer programs were implemented later on to achieve this goal automatically [18].

However, aligning multiple and divergent RNA sequences so as to preserve their conserved structures is not easy, because many compensatory mutations decrease the overall sequence similarity. For unaligned sequences, one must compute the structure and alignment simultaneously. Sankoff proposed an algorithm that can simultaneously align RNA sequences and find the optimal common fold [19, 20, 21]. However, the complexity of this algorithm is $O(l^6)$, where l is the length of RNA sequences, too high to be practical even for two sequences. The complexity can be reduced to $O(l^4)$ [22], but only when RNA has no multi-loop structure. Eddy and Durbin, and other groups [23, 24, 25] pioneered the approach of modeling RNA sequences using stochastic context free grammars. The rules of the SCFG allow for position dependent scoring of distant base-pairs and primary sequence conservation, and also allow automated estimation of model parameters from unaligned sequences using EM. However, in practice, the extensive divergence of RNA sequences makes it hard to reconstruct structure and alignment with perfect accuracy, and the covariance models work best when supplied with good seed alignments. Much recent work has focused on improving fold prediction for aligned sequences [18, 26].

In our approach to this well-researched problem, we are motivated by the idea of constraining allowed folds to make it more likely to reach the final correct structure. This idea has been used with good success in aligning divergent genomic sequences. For diverged DNA sequences, there is not enough signals for probabilistic models such as HMMs to be effective without prior information (in the form of seed alignments). The recent methods [27] identify anchors corresponding to highly conserved orthologous regions, and use these to constrain the multiple alignments. This approach has been used in RNA as well. Waterman [28] pioneered this with a statistical approach for choosing conserved stem-loops within a pair of fixed size windows in a set of unaligned RNA sequences. Ji and Stormo [29] extended this idea considerably. Starting with putative stem-loops, they remove all but the ones conserved in a global sequence alignment of two sequences. These are further culled to retain only those that are present in every sequence in the family. Perriquet et al. [30] proposed a different anchoring approach to solving the same problem by first determining anchor regions that are highly conserved in given RNA sequences and then seeking a set of conserved stems crossing the same anchor regions that have minimum folding energy. Both methods use primary sequence conservation extensively, and limit the variability in the length of loop regions, which may lead to accurate but relatively few predicted anchor stacks for diverged families. On the contrary, Bouthinon and Soldano [31], Davydov and Batzoglou [32] each proposed algorithms to select conserved base-pairs among the given RNA sequences based solely on the conservation of the structure that they form, considering neither their sequence similarity nor the thermodynamic stability of this structure. As a result, these methods may risk selecting wrong base-pairs when only a limited number of RNA sequences are given.

In this paper, we describe a method, RNAscf (RNA Stack based Consensus Folding), for predicting the consensus fold of an RNA family, given unaligned sequences. Our method is based on the notion of finding structurally conserved anchors, and an iterative extension constrained by the anchors. With relatively few parameters, and limited training, the method outperforms other competing methods (See RESULTS), detecting 88% of true stacks (sensitivity), and overlapping with a correct stack in 93% of all predictions. The method establishes the validity of this new paradigm for computing consensus structure in RNA. We also discuss extensions based on an iterative refinement of the predicted structure. (See DISCUSSION).

2   Approach

As shown in Figure 1(a), the secondary structure of an RNA has a tree like shape. Assume that there is a dummy base-pair between 0 and $n + 1$. Define a *loop* as a set of indices $i_1 \leq i_2 \ldots \leq i_k$ such that for all j, either $i_{j+1} = i_j + 1$, or (i_j, i_{j+1}) form a base-pair. Further, if for some j, j', $(i_j, i_{j'})$ form a base-pair, then $|j - j'| = 1$. It can be seen that the structure can be decomposed uniquely into a set of loops, and the loops can be classified as *hairpin* (containing only one base-pair), a *stem-loop* (two base-pairs with no unpaired bases), an *interior loop/bulge* (two base-pairs with unpaired bases), and a *multi-loop* ($k > 2$ base-pairs).

The stability of the RNA structure is determined predominantly by *stacks* of consecutive stem-loops. The stacks are stabilized by hydrogen bonds between the base-pairs, and in general, the longer a stack region, the more energetically favorable it is. Each stack corresponds to a pair of sub-strings. These pairs are typically non-interleaving. While interleaved stacks, or *pseudo-knots* (such as the pair (f, f'), and (h, h') in Figure 1(a)) do occur, they are relatively less common and ignored here. In our approach for finding anchors, we ignore individual base-pairs and work with a slightly generalized notion of a stack that includes unpaired bases. Configurations of stacks that are conserved in multiple sequences will be the anchors in determining consensus structures.

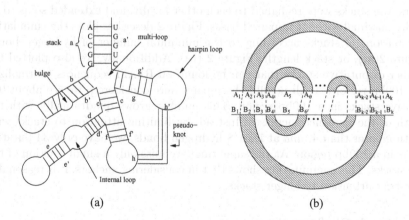

(a)   (b)

Fig. 1. (a) An RNA secondary structure structure with various structural elements including stacked stem-loops, bulges, hairpins, and multi-loops. (b) Two stack configurations match to each other for both unpaired regions and paired regions

2.1   Predicting Putative Stacks

The thermodynamic stability of a stack is proportional to the number of hydrogen bonds between the base-pairs in the stack. Any pair of strings can be aligned (with gaps) so as to optimize the energy of the paired bases. Therefore, given an RNA string A, we construct a local alignment of $A[1, \ldots, n]$ with $A[n, \ldots, 1]$. Let $\delta_h(i, j)$ be the score (number of h-bonds) in an $(A[i], A[j])$ base-pairing. Thus base-pairing of G-C, A-T, G-U, is scored 3, 2, and 1 respectively. Let $S[i, j]$ be optimum score for a stack with left end-point i, and right end-point j. Then

$$S[i, j] = \max \begin{cases} S[i+1, j-1] + \delta_h(i, j) \; (\delta_h[i, j] > 0) \\ S[i+1, j] + g \\ S[i, j-1] + g \\ 0 \end{cases} \tag{1}$$

where g is a gap penalty. In our implementation, we modify this basic approach to include affine gap costs. We select each (i, j) for which $S[i, j]$ is greater than some threshold. In order to avoid predicting overlapping stacks, we sort the stacks by decreasing score values. Each time a stack is picked, all base-pairs in it are excluded. While straightforward, this is an effective procedure. Intuitively the probability of finding a base-pair at random is much higher than the probability of finding a high-scoring stack. Waterman showed that finding a k-long stack within certain window size in many given sequences (even not all) can be significant [28]. Also, most real base-pairs in correct structures should be stabilized by multiple stacked base-pairs, implying that limiting consideration to high-scoring stacks does not result in many false negatives. To test this, we computed statistics on the seed-alignments in the RFAM database [33]. All stacks from seed alignments of 379 families (9559 sequences) were selected. The seed alignment contains known representative members of the family, which is hand-curated and is annotated with structural information. To correct for annotation errors, the stacks were realigned to each other locally, and extended so as to be locally maximal without unpaired bases. Figure 2 describes plots the cumulative distributions of stacks according to the minimum number of hydrogen bonds (Figure 2(a)), or stack length (Figure 2 (b)). Additionally, we also plotted the number of putative stacks that can be found on RFAM sequences (normalized to a 100bp region). If all possible base-pairs were considered, we see about 900 putative stacks in a 100bp region. This number grows quadratically with the length of the sequence increases. Instead, by limiting attention to stacks with length greater than 4, *and* at least 8 hydrogen bonds, we have only 34 putative stacks in a 100bp region. At the same time, we miss only a small fraction of the true stacks. This computation shows that in considering anchors, it is reasonable to restrict attention to longer stacks.

2.2   Stack Configurations

A putative stack of length 4 can still be found by chance. Ji and Stormo [29] and Perriquet et al. [34] both use sequence similarity to further select stacks. We propose to select a set of stacks instead of one at a time. We evaluate the

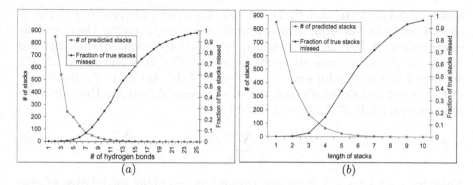

(a)   (b)

Fig. 2. Statistics of the stacks in Rfam database. (a) One line shows the fraction of annotated stacks which would be missed out by using different cutoff number of hydrogen bonds. The other line shows the number of predicted stacks per 100 based all Rfam seed sequences for each cutoff number of hydrogen bonds. (b) One line shows the fraction of annotated stacks which would be missed out by using different cutoff length of the stacks. The other line shows the number of predicted stacks per 100 bases for each cutoff length of stacks. We also can combine two cutoffs together - the length of the stacks and the number of hydrogen bonds- to achieve better performance

set of stacks by both the stability (free energy) of the structure they form and the sequence similarity computed based on these common stacks as anchors. To describe evaluation function of the selected set, we must define a notion of *configuration* of stacks. Note that two stacks P_1 and P_2 may have one of the following relations [31]: (1) P_1 and P_2 are interleaving; (2) P_1 is enclosed within P_2 (denoted by $P_1 <_I P_2$); (3) P_1 is juxtaposed to P_2 so that its right end-point is before the left end-point of P_2 ($P_1 <_p P_2$). An RNA structure R_A on A is a collection of non-interleaving stacks on A. The stacks have a partial order ($<_I \cup <_p$) defined on them. The free energy of the structure is the sum of free energies of each of the constituent hairpins, stem-loops, interior loops, and multi-loops. Here, we work with the notion of generalized stacks instead of stem-loops. The energy of a stack P denoted by $\mathcal{E}_s(P)$ is as described in equation 1. The energies of loops $\mathcal{E}_h$ (hairpin), $\mathcal{E}_b$ (bulge), $\mathcal{E}_m$ (multi-loops), are a function of the length of the sequence and set as described in Jaeger et al. [15].

Define a *consensus* structure $\mathcal{P}(A, B)$ as a pair of structures R_A and R_B on A and B respectively, with a one-one correspondence between stacks in R_A and R_B such that the corresponding stacks in the two structures maintain identical partial order relationships. We define the free energy $\phi(\mathcal{P}(A, B))$ of the consensus structure similar to the energy of the individual structures. For each pair of corresponding stacks, or loops, the maximum of the two energy values is chosen.

Given a consensus $\mathcal{P}(A, B)$, the sequences A and B can be aligned to be consistent with $\mathcal{P}(A, B)$ (see Figure 1(c)), so that the sequences in stacks are aligned to each other, and likewise for the sequences in the unpaired regions. This alignment partitions sequence A and B into alternating stack and non-stack regions $A_1, A_2, \ldots, A_k$, and $B_1, B_2, \ldots, B_k$. Each pair of sequences (A_i, B_i), is aligned optimally. We define such an alignment as a *configuration*. The cost of the configuration $(A, B, \mathcal{P}(A, B))$ is defined as a function of sequence similarity and the energy of the consensus structure. Denote $i \in \mathcal{P}(A, B)$ if and only if (A_i, B_i) are paired in a stack with some (A_j, B_j) in $\mathcal{P}(A, B)$. Let $\mathcal{S}(A_i, B_i)$ denote the cost of an optimal global alignment of subsequences A_i and B_i. The *cost* of the configuration $(A, B, \mathcal{P}(A, B))$ is denoted by

$$M(\mathcal{P}(A, B)) = w_1 \Phi(\mathcal{P}(A, B)) + w_2 \sum_{i \in \mathcal{P}(A, B)} \mathcal{S}(A_i, B_i) + w_3 \sum_{i \notin \mathcal{P}(A, B)} \mathcal{S}(A_i, B_i) \quad (2)$$

where $w_1 + w_2 + w_3 = 1$, represent parameters describing the relative weights to the free energy of the configuration, sequence similarity in stack regions, and sequence similarity within loop regions. Ideally, these weights should be adjusted according to the number and divergence of the given sequences. However, in the tests through this paper, we use an identical set of weights ($w_1 = 0.84$, $w_2 = 0.06$, and $w_3 = 0.1$). For a given pair, we compute consensus structures of minimum cost.

The definition of a configuration of stacks for a pair of sequences also extends to multiple sequences: A *configuration* $\mathcal{P}(A_1, A_2, \ldots, A_s)$ is a collection of s RNA structures $\{P^{A_1}, P^{A_2}, \ldots, P^{A_s}\}$, one for each sequence with the following property: For each pair of structures, there is a one-one correspondence between the stacks that are consistent with the partial orders $<_l$ and $<_p$. A configuration with l stacks partitons each sequence A_i into $2l + 1$ blocks denoted $A_{i,1}, A_{i,2}, \ldots, A_{i,2l+1}$, where each block $A_{*,j}$ is either a stack in the configuration ($j \in \mathcal{P}$), or part of the loop region. We modify equation 2 to describe the cost of the configuration $\mathcal{P}(A_1, A_2, \ldots, A_s)$, as follows:

$$M(\mathcal{P}(A_1, \ldots, A_s)) = w_1 \Phi(\mathcal{P}(A_1, \ldots, A_s)) + w_2 \sum_{j \in \mathcal{P}} \mathcal{S} \begin{pmatrix} A_{1,j}, \\ A_{2,j}, \\ \ldots, \\ A_{s,j} \end{pmatrix} + w_3 \sum_{j \notin \mathcal{P}} \mathcal{S} \begin{pmatrix} A_{1,j}, \\ A_{2,j}, \\ \ldots, \\ A_{s,j} \end{pmatrix}$$

$$(3)$$

Here the function $\mathcal{S}$ computes the score of a multiple alignment. The *RNA stack based consensus folding problem* can be described formally: given s RNA sequences, compute a minimum cost *stack configuration*. In the following section, we describe algorithms for computing optimal configurations.

3   Stack Based Consensus Folding

3.1   Computing Optimal Stack Configuration in Two RNA Sequences

We use dynamic programming to compute an optimal configuration. The algorithm is similar to prior work [35, 20] with an important difference being that stacks (instead of individual base-pairs) are now used. Given sequences

A, B, we compute all potential stacks in them, using the algorithm from Section 2.1. Assume these two sequences have m and n stacks respectively. Let $\mathcal{P}^A = P_1^A, P_2^A, ..., P_m^A$ and $\mathcal{P}^B = P_1^B, P_2^B, ..., P_n^B$ denote the stacks, ordered according to increasing values of the right-most base-pair. Denote the index of the first and last base pair of a stack P as P_b, P_e, and the length as P_l. Define the following terms:

Seq(P^A): The sub-sequence covered by the stack P^A, given by $A[P_b^A \ldots P_b^A + P_l^A - 1]$, and $A[P_e^A - P_l^A + 1 \ldots P_e^A]$.

Loop(P^A): The sub-sequence covered by by the first and last positions of the stack P^A after excluding the bases in Seq(P^A). In other words, the sequence $A[P_b^A + P_l^A \ldots P_e^A - P_l^A]$.

Loop($P^A, P^{A'}$): If $P^{A'}$ is enclosed within P^A, then the loop region corresponds to the sequence in between the two stacks (i.e., the subsequences $A[P_b^A + P_l^A \ldots P_b^{A'} - 1]$, and $A[P_e^{A'} + 1 \ldots P_e^A - P_l^A]$). If $P^{A'}$ is to the left of P^A, the loop region corresponds to $A[P_e^{A'} + 1 \ldots P_b^A - 1]$. Otherwise, the term is undefined.

$M(P^A, P^B)$: The cost of an optimum configuration of A and B over all consensus structures, given that stacks P^A and P^B are in the consensus structure and aligned to each other.

Clearly, it is sufficient to compute $M[P^A, P^B]$ for all pairs in $\mathcal{P}^A \times \mathcal{P}^B$, which would need $O(m^2 n^2)$ time. In computing $M[P^A, P^B]$, we have 3 choices for the subsequences Loop(P^A), and Loop(P^B), as they could either form a hairpin, an interior loop/bulge, or a multi-loop. Therefore,

$$M[P^A, P^B] = M_s[P^A, P^B] + \min \left\{ \begin{array}{ll} M_h[P^A, P^B], & \text{(* hairpin loop *)} \\ M_b[P^A, P^B], & \text{(* interior loop/bulge *)} \\ M_m[P^A, P^B] & \text{(* multi-loop *)} \end{array} \right\} \quad (4)$$

here, $M_s(P^A, P^B)$ is the score matching stacks P^A and P^B, based on sequence and structure conservation, and can be computed by

$$M_s[P^A, P^B] = w_1 \max \left\{ \begin{array}{l} \mathcal{E}_s(P^A), \\ \mathcal{E}_s(P^B) \end{array} \right\} + w_2 \mathcal{S} \left(\begin{array}{l} \text{Seq}(P^A) \\ \text{Seq}(P^B) \end{array} \right) \quad (5)$$

$M_h[P^A, P^B]$ is the score of the loop regions of P^A and P^B given that no other matched stack pair is included by P^A and P^B i.e. these regions form matched hairpin loops.

$$M_h[P^A, P^B] = w_1 \max \left\{ \begin{array}{l} \mathcal{E}_h(|\text{Loop}(P^A)|), \\ \mathcal{E}_h(|\text{Loop}(P^B)|) \end{array} \right\} + w_3 \mathcal{S} \left(\begin{array}{l} \text{Loop}(P^A), \\ \text{Loop}(P^B) \end{array} \right) \quad (6)$$

$M_b[P^A, P^B]$ represents the matching score when P^A, and P^B are followed by an interior loop, or bulge. Consider all stacks P^x, P^y that are enclosed by P^A, and P^B, respectively. Then, $M_b(P^A, P^B)$ is the minimum free energy of any matching of P^x, P^y.

$$M_b[P^A, P^B] = \min_{\substack{P^x <_I P^A, \\ P^y <_I P^B}} \left\{ w_1 \max \left\{ \begin{array}{l} \mathcal{E}_b(|\text{Loop}(P^x, P^A)|), \\ \mathcal{E}_b(|\text{Loop}(P^y, P^B)|) \end{array} \right\} + w_3 \mathcal{S} \left(\begin{array}{l} \text{Loop}(P^x, P^A), \\ \text{Loop}(P^y, P^B) \end{array} \right) + M[P^x, P^y] \right\} \quad (7)$$

For the multi-loop case, we need to define some additional terms. A sequence of stacks $P_1, P_2, \ldots$ form a *chain* if $P_1 <_p P_2 <_p \ldots$. $M_m[P^A, P^B]$ represents the matching score between P^A, and P^B, given that there is a pair of chains included by P^A and P^B that form the multiloop. Let $P_1^A, P_2^A, \ldots$ (respectively $P_1^B, P_2^B, \ldots$) denote stacks enclosed by P^A (P^B, respectively) and ordered according to increasing values of the last coordinate. Denote $P_{i_1}^A \in F(P_{i_2}^A)$ if $P_{i_1}^A <_p P_{i_2}^A$ and there is no stack P_j^A such that $P_{i_1}^A <_p P_j^A <_p P_{i_2}^A$. Then

$$M_m[P^A, P^B] = \min_{i,j} \{ M_c[P_i^A, P_j^A] + \max \left\{ \begin{array}{l} \mathcal{E}_m(|\text{Loop}(P^i, P^A)|), \\ \mathcal{E}_m(|\text{Loop}(P^j, P^B)|) \end{array} \right\} \quad (8)$$

Here, $M_c[P_i^A, P_j^B]$ is defined as the minimum energy of a chain that ends at P_i^A, and P_j^B, and begins at some $P_{i'}^A <_p P_i^A$, and $P_{j'}^B <_p P_j^B$. (The end conditions are added for efficiency reasons). Then,

$$M_c[P_i^A, P_j^B] = \min_{\substack{P_x^A \in F(P_i^A) \\ P_y^A \in F(P_i^B)}} \left\{ \begin{array}{l} M_c[P_x^A, P_y^B] + M_o[P_x^A, P_i^A; P_y^B, P_j^B] + M[P_i^A, P_j^B], \\ M_c[P_i^A, P_j^B] + \mathcal{E}_m(|\text{Loop}(P_x^A, P_i^A)| + |\text{Seq}(P_i^A)|), \\ M_c[P_x^A, P_j^B] + \mathcal{E}_m(|\text{Loop}(P_y^B, P_j^B)| + |\text{Seq}(P_j^B)|) \end{array} \right\}$$

$$(9)$$

where $M_o[P_x^A, P_i^A; P_y^B, P_j^B]$ is the minimum free energy of the matching between the loops (P_x^A, P_i^A) and (P_y^B, P_j^B),

$$M_o[P_x^A, P_i^A; P_y^B, P_j^B] = w_1 \max \left\{ \begin{array}{l} \mathcal{E}_b(|\text{Loop}(P_x^A, P_i^A)|), \\ \mathcal{E}_b(|\text{Loop}(P_y^B, P_j^B)|) \end{array} \right\} + w_3 \mathcal{S} \left(\begin{array}{l} \text{Loop}(P_x^A, P_i^A), \\ \text{Loop}(P_y^B, P_j^B) \end{array} \right)$$

$$(10)$$

3.2 Consensus Fold Computation for Multiple RNA Sequences

The optimal configuration of a randomly chosen pair of sequences from a family already shows high sensitivity (data not shown). It is likely that an optimal configuration of structures conserved in diverse multiple sequences will be very accurate. Recall the cost of the configuration as Equation (3), where $\mathcal{S}$ denotes the score of a multiple alignment of the block. Clearly, the problem of computing optimal configuration is hard, given the discussion for the pairwise case. Here, we use a heuristic principle based on the notion of a star-alignment, with a seed configuration chosen from an optimal configuration of a random pair of sequences. To understand why our approach should work, we describe a back of the envelope calculation.

Consider a stack x of length k from the seed structure defined on sequence A_1 that is in fact incorrect (does not overlap with a true stack). For x to be retained in the final anchored configuration, it must match with stacks in a large fraction of the other sequences. Let p be the probability that a random pair of bases can form base-pairs. Given an interval (i,j) in some sequence, the probability that (i,j) is the end of a stack of length k is p^k. Then, the probability that x is matched up with some random set of base-pairs defined by the end-points (i,j) is no more than p^k, even after ignoring sequence similarity. However, x cannot be matched to any other arbitrary stack. As we also score for primary sequence conservation, and the match should maintain the partial order of the configuration, (i,j) and x must be 'similarly situated'. To model this, we introduce a parameter w. Define w as the number of distinct pairs (i,j) such that x is allowed to match to. Then, the probability that x finds a match by chance is given by $p_x = 1 - (1 - p^k)^w$. Allowing a more flexible definition, we say that x is f-conserved in the configuration if it finds a match in at least $(1 - f)s$ of the s sequences.

$$Pr[x \text{ is } f\text{-conserved}] = P_c(x) = \sum_{l \geq (1-f)s} \binom{s}{l} p_x^l (1 - p_x)^{s-l} \qquad (11)$$

Fix some parameters as follows. Let $p = \frac{3}{8}$ (corresponding to G-C, G-U, A-U), $f = 0.7$. The probability of getting an incorrect conserved stack depends critically on the parameters w. If w is too large, there is a high probability of getting random stacks to match up ($P_c(x) = 0.91$, when $k = 4, w = 80$ and $s = 20$). This effect might be offset by increasing k ($P_c(x) = 1.8e - 6$, when $k = 5, w = 80$ and $s = 20$), but then we risk losing many true (smaller size) stacks, which may cause incorrect pairs to be matched. The effect is also offset (to a less degree) by increasing the number of sequences, but that may not always be possible. The reason our approach works is because the choice of a conserved configuration restricts the possible stacks that x can match, effectively keeping the value of w low.

Before describing the approach, we must first modify the formulation to allow stacks to be *partially conserved*, and therefore, absent from some sequences. For $0 < f \leq 1$, define an *f-configuration* as a configuration with the following property: for every set of s corresponding stacks (one from each sequence), at most $(1 - f)s$ can be absent. In Figure 3, we describe a procedure for computing an anchor configuration.

The anchor configuration consists of stacks that optimize the cost of the configuration, and are conserved across the family. Thus, the stacks are highly likely to be correct. However, the procedure might also miss some true stacks due to a high initial value of k, and requirement of conservation. To increase sensitivity, we now search for less conserved, and shorter stacks. However, the new stacks are forced to be consistent with the anchor configuration.

Recall from the definition of loops in Section 2.1 that all unpaired bases can be uniquely assigned to a loop. Additionally, a stack does not interleave with the anchor configuration if and only if it is defined on unpaired bases within a single loop. This forms the basis of the final procedure RNASCF. See Figure 4.

Procedure COMPUTEANCHORCONFIGURATION(k, f)

1. Pick a pair of sequences (A, B) at random from the set $\mathcal{R}$ of RNA sequences.
2. Compute putative stacks $\mathcal{P}^A$ from A, and $\mathcal{P}^B$ from B with minimum length k. k is chosen according to the lengths of sequences in $\mathcal{R}$, and typically $k = 4$.
3. Compute the optimum configuration. Reduce $\mathcal{P}^A$ to retain only the stacks from the optimum configuration. Denote $\mathcal{P}'^A$ as the reduced set.
4. For each sequence $R \in \mathcal{R}$, compute the optimum pairwise configuration of (A, R) using the reduced set $\mathcal{P}'^A$. Denote $M[(A, B), \mathcal{R}]$ as the sum of the configuration costs.
5. Recompute Steps 1-4 for various random choices of (A, B), and pick the pair (A, B) with minimum configuration cost $M[(A, B), \mathcal{R}]$.
6. Retain only the stacks in $\mathcal{P}'^A$ that appear in $1 - f$ fraction of the sequences in $\mathcal{R}$. Denote the subset as $\mathcal{P}''^A$. Output $\mathcal{P}''^A$ as the *anchored structure* of $\mathcal{R}$.

Fig. 3. Procedure for computing anchor configuration

Procedure RNAscf(k, f)

1. $\mathcal{P} = \text{ComputeAnchorConfiguration}(k, f)$
2. In each sequence, partition the unpaired bases according to their loop region in $\mathcal{P}$.
3. For every loop region that has a minimum number of unpaired bases, predict additional putative stacks with $k' < k$. Each 'arm' of the stack is constrained to have contiguous base-pairs.
4. For each stack in the optimal configuration that was not present in every member of the family, recompute the alignment with the additional putative stacks to retrieve less conserved stacks.
5. For each set of loop regions and potential stacks, recurse using RNAscf(k', f) to compute additional stacks in the loop regions.

Fig. 4. The procedure RNAscf for computing Consensus Folds

3.3   Implementation Details

The program is implemented in C, and is available upon request. In default setting, we limit ourselves to two iterations. For the first iteration, we choose the default parameters as $k = 4, h = 8$ (k is the minimum length and h is the minimum number of hydrogen bonds in the putative stacks) and $g = 0$ (no unpaired bases allowed in stacks). For the second iteration, the default settings are changed to $k = 3, h = 6$.

4   Results and Discussion

To test the performace of RNAscf, we chose a set of 12 RNA families from the Rfam database [33]. 20 sequences were chosen for each family, except for CRE (RF00220) and glmS (RF00234) for which we chose available 10 sequences respectively. Stacks were retrieved from the annotated structures for each of these sequences. In all, there are 953 stacks. We chose 3 other programs to compare the performance of RNAscf, choosing the best representative of different method-ologies: RNAfold, which is an implementation of energy based minimization [15] from the Vienna package; COVE, is an implementation of covariance model [23], and comRNA [29], which is based on computing anchors in multiple sequence align-ment. Only comRNA predicts stacks explicitly. COVE and RNAfold do not explicitly predict stacks, but most of their base-pairs appear in stacks. For best results, we ignored unstacked base-pairs (with unpaired bases on either side) for RNAfold and COVE. Larger stack length cutoffs were also tried, but this choice gave the best balance of sensitivity and accuracy. *Sensitivity* is defined as the fraction of true stacks that overlapped with predicted stacks. A sensitivity of 1 would im-ply that all true stacks overlapped with some predicted stacks. Correspondingly *accuracy* is the fraction of predicted stacks that overlapped with a true stack. As COVE expects aligned sequences, we aligned the sequences using ClustalW [36]. The alignment was used to train the Covariance model, and the model was then used to align sequences, and predict structure. We also ran COVE on unaligned sequences, but the performance in that case was inferior to the performance on ClustalW aligned sequences.

Figure 5 shows the plots of the sensitivity and specificity of all programs on the test (detailed numbers are shown in Table 1). As can be seen in the tables, RNAscf is at the top or near the top in every family, and maintains high sensitivity and accuracy throughout, with an average accuracy of 0.884, and average sensitivity of 0.926. Only comRNA shows consistently high accuracy because it predicts very few stacks (leading to low-sensitivity) that are well-conserved in both sequence and structure. COVE occasionally shows very poor sensitivity, possibly because of incorrect seed alignment. RNAfold predicts many

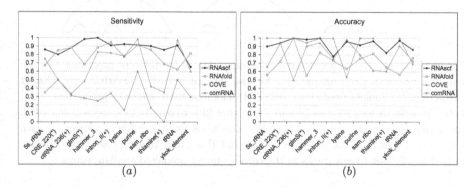

(a)   (b)

Fig. 5. Sensitivity and accuracy of RNA secondary structure prediction on 12 RNA families. The default parameters for RNAscf are f=0.7 and k=4 and h=8 for the first iteration. RNAfold is run under default parameters. COVE is run under the default parameters, using the multiple alignment from ClustalW as input. comRNA is run under the recommended parameter (p=0.7, s=0.56). (*) There are only 10 sequences in these families. (+) RNAscf is run under k=3 and h=6 on these families, due to their small size

Table 1. A complete list of the comparison of sensitivity and accuracy of RNA secondary structure prediction on 12 RNA families shown in Figure 5

Name (Rfam_id)	Sensitivity				Accuracy			
	RNAscf	RNAfold	Cove	comRNA	RNAscf	RNAfold	Cove	comRNA
5s_rRNA (RF00001)	0.86	0.67	0.75	0.35	0.9	0.558	0.658	1
Rhino_CRE (RF00220)	0.8	0.85	0.5	0.5	0.941	0.72	0.941	1
ctRNA_pGA1 (RF00236)	0.882	0.882	0.333	0.313	1	1	0.5	1
glmS(RF00234)	0.983	0.683	0.483	0.283	0.983	0.552	0.906	0.944
Hammerhead_3 (RF00008)	1	0.883	0.833	0.25	1	0.828	0.943	1
Intron_gpII (RF00029)	0.91	0.946	0.821	0.339	0.782	0.731	0.754	1
Lysine (RF00168)	0.925	0.775	0.783	0.142	0.958	0.633	0.984	0.531
Purine (RF00167)	0.917	0.917	0.983	0.6	0.917	0.753	0.797	1
Sam_riboswitch (RF00162)	0.903	0.858	0.425	0.168	0.966	0.813	0.613	1
Thiamine (RF00059)	0.858	0.690	0.354	0	0.824	0.654	0.606	-
tRNA (RF00005)	0.912	0.625	0.975	0.5	0.973	0.567	0.910	1
ykok (RF00380)	0.656	0.817	0.606	0.3	0.863	0.762	0.727	0.692
Average	0.884	0.8	0.654	0.312	0.926	0.714	0.778	0.924

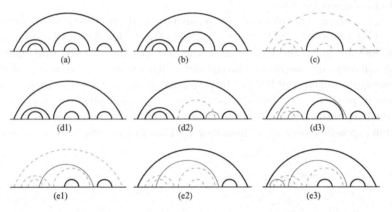

Fig. 6. A comparison of predicted stack configurations by different programs. (a) The true consensus stacks configuration for the sam riboswitch (RF00162). (b) RNAscf prediction. (c) comRNA prediction. (d1)-(d3) The first three RNAfold predictions. (e1)-(e3) The first three COVE predictions. Note that RNAfold and COVE are not limited to predicting conserved stacks, and therefore give potentially a different answer for each sequence. Thick line, dashed (green) line and thin (red) line represent true stacks, missed stacks and wrong stacks in the corresponding predicted configurations

stacks, and therefore has good sensitivity, but the extra predictions lead to a loss of accuracy. While our method shows robust performance for a limited number of given RNA sequences, its performance improves when the number of the given sequences increase.

Finally, we emphasize that even though sometimes we cannot predict all the stacks in all the given sequences, the consensus structure obtained by RNAscf is always the right configuration; the prediction errors in a few input sequences are usually due to an incorrect stack that is very close to a correct one. Since this cannot be quantified, we use one example to demonstrate that the minor prediction error in a few given sequences does not affect the prediction of the common structure. Figure 6 shows the predicted configuration of the four programs on the SAM riboswitch. Clearly, RNAFold can predict the correct configuration in some of the RNA sequences, but make the wrong prediction on the others. This is not surprising, because it analyzes each RNA sequence separately and doesn't presume they have the common structure. However, it will be difficult to derive their common structure based on these results. comRNA tends to miss many real stacks, although the ones it predicts are often correct. COVE predicted some correct stacks but it may miss some correct stacks and also predict some wrong ones. Similar results were seen for all families.

In conclusion, RNAscf establishes the principle that anchored stacks selection based on seed configurations, and prediction of consensus structure subject to anchored constraints is a valid approach to RNA structure prediction. Our future work will be aimed at correcting errors by using a stochastic iterative scheme such as Gibbs sampling [37]. In each step, we will remove a stack from the consensus structure, and add a new stack sampled from possibilities that are consistent with the remaining configuration, and weighted according to the energy. Early experiments have shown the promise of such refinement. An expended version of this paper is available upon request.

References

1. Eddy, S.: Non-coding RNA genes and the modern RNA world. Nat Rev Genet **2** (2001) 919–929
2. Storz, G.: An expanding universe of noncoding RNAs. Science **296** (2002) 1260–63
3. International Human Genome Sequencing Consortium: Finishing the euchromatic sequence of the human genome. Nature **431** (2004) 931–945
4. Kampa, D., et al.: Novel RNAs identified from an in-depth analysis of the transcriptome of human chromosomes 21 and 22. Genome Res **14** (2004) 331–342
5. Nahvi, A., Sudarshan, N., Ebert, M., Zou, X., Brown, K., Breaker, R.: Genetic control by a metabolite binding mRNA. Chemical Biology **9** (2003) 1043–1049
6. Vitreschak, A., et al.: Riboswitches: the oldest mechanism for the regulation of gene expression? Trends in Genetics **20** (2003) 44–50
7. Tinoco, I., Uhlenbeck, O., Levine, M.: Estimation of secondary structure in ribonucleic acids. Nature **230** (1971) 362–367
8. Nussinov, R., Jacobson, A.: Fast algorithm for predicting the secondary structure of single-stranded RNA. Proc Natl Acad Sci U S A **77** (1980) 6309–6313
9. Nussinov, R., Pieczenik, G., Griggs, J., Kleitman, D.: Algorithms for loop matchings. SIAM J. Appl. Math. **35** (1978) 68–82
10. Smith, T., Waterman, M.: RNA Secondary structure. Math. Biosci. **42** (1978) 257–266
11. Waterman, M.: Secondary structure of single stranded nucleic acids. Adv. Math. Suppl. Stud. **I** (1978) 167–212

12. Zuker, M., Sankoff, D.: RNA secondary structure and their prediction. Bull. Math. Biol. **46** (1984) 591–621
13. Zuker, M.: Prediction of RNA secondary structure by energy minimization. Methods Mol Biol **25** (1994) 267–294
14. Hofacker, I.: Vienna RNA secondary structure server. Nucl Acids Res **31** (2003) 3429–3431
15. Jaeger, J., Turner, D., Zuker, M.: Improved predictions of secondary structures for RNA. Proc Natl Acad Sci U S A **86** (1989) 7706–7710
16. Pavesi, G., Mauri, G., Stefani, M., Pesole, G.: RNAProfile: an algorithm for finding conserved secondary structure motifs in unaligned RNA sequences. Nucl Acids Res **32** (2004) 3258–3269
17. Levitt, M.: Detailed molecular model for transfer ribonucleic acid. Nature **224** (1969) 759–763
18. Hofacker, I., Fekete, M., Stadler, P.: Secondary structure prediction for aligned RNA sequences. J Mol Biol **319** (2002) 1059–1066
19. Gorodkin, J., Stricklin, S., Stormo, G.: Discovering common stem-loop motifs in unaligned RNA sequences. Nucl Acids Res **29** (2001) 2135–2144
20. Sankoff, D.: Simultaneous solution of the RNA folding, alignment and protosequence problems. Siam J. Appl. Math. **45** (1985) 810–825
21. Mathews, D., Turner, D.: Dynalign: an algorithm for finding the secondary structure common to two RNA sequences. J Mol Biol **317** (2002) 191–203
22. Gorodkin, J., Heyer, L., Stormo, G.: Finding the most significant common sequence and structure motifs in a set of RNA sequences. Nucl Acids Res **25** (1997) 3724–32
23. Eddy, S., Durbin, R.: RNA sequence analysis using covariance models. Nucl Acids Res **22** (1994) 2079–2088
24. Sakakibara, Y., et al.: Recent methods for RNA modeling using Stochastic Context Free Grammars. Combinatorial Pattern Matching **807** (1994)
25. Knudsen, B., Hein, J.: Pfold: RNA secondary structure prediction using stochastic context-free grammars. Nucl Acids Res **31** (2003) 3423–3428
26. Knight, R., Birmingham, A., Yarus, M.: BayesFold: rational 2 degrees folds that combine thermodynamic, covariation, and chemical data for aligned RNA sequences. RNA **10** (2004) 1323–1336
27. Bray, N., Pachter, L.: MAVID: Constrained Ancestral Alignment of Multiple Sequences. Genome Res. **14** (2004) 693–699
28. Waterman, M.: Consensus methods for fodling single-stranded nucleic acids. Mathematical methods for DNA Sequences (1989) 185–224
29. Ji, Y., Xu, X., Stormo, G.: A graph theoretical approach for predicting common RNA secondary structure motifs including pseudoknots in unaligned sequences. Bioinformatics **20** (2004) 1591–1602
30. Perriquet, O., Touzet, H., Dauchet, M.: Finding the common structure shared by two homologous RNAs. Bioinformatics **19** (2003) 108–116
31. Bouthinon, D., Soldano, H.: A new method to predict the consensus secondary structure of a set of unaligned RNA sequences. Bioinformatics **15** (1999) 785–798
32. Davydov, E., Batzoglou, S.: A computational model for rna multiple structural alignment. Combinatorial Pattern Matching (2004)
33. Griffiths-Jones, S., Bateman, A., Marshall, M., Khanna, A., Eddy, S.: Rfam: an RNA family database. Nucl Acids Res **31** (2003) 439–441

34. Touzet, H., Perriquet, O.: CARNAC: folding families of related RNAs. Nucl Acids Res **32** (2004) 142–145
35. Bafna, V., Muthukrishnan, S., Ravi, R.: Computing similarity between RNA strings. Combinatorial Pattern Matching **937** (1995) 1–14
36. Thompson, J., Higgins, D., Gibson, T.: CLUSTAL W: improving the sensitivity of progressive multiple sequence alignment through sequence weighting, position-specific gap penalties and weight matrix choice. Nucl Acids Res **22** (1994) 4673–4680
37. Lawrence, C., et al.: Detecting subtle sequence signals: a Gibbs sampling strategy for multiple alignment. Science **262** (1993) 208–214

Discovery and Annotation of Genetic Modules

Charles DeLisi

Boston University
delisi@bu.edu

Biological complexity, and the complexity of a cell in particular, scales only weakly with the number of components. Instead, a cell's ability to process information, to respond and adapt to an environment in fugue, is directly related to the combinatorially large number of ways genes can be selected and modulated to express its phenotypic repertoire. Viewed in this way it is clear that a cell lacks a fixed network structure, but instead has the potential to form, subject to physical chemical and structurally determined constraints, an extremely large number of environmentally selected networks of genes and proteins with shared components.

With time frozen, i.e. environment fixed and in the steady state, virtually any protein network will be hierarchically organized. Sets of tightly regulated proteins will form modules (e.g. co-regulated components of a pathway) and these will be organized into higher order modules, sharing sub modules, and so on, somewhat reminiscent of an integrated circuit. We can reasonably expect that many, if not all, modules will participate in more than one process, and that the selection frequency will be scale free, some modules being selected very frequently, others only rarely. Identifying and characterizing such modules and determining the rules governing their organization is one of cell biology's central challenges in the coming decades. The task is daunting because most genomes have not yet been fully parsed, and for those stretches of DNA with precisely annotated start and stop codons, only about 40.

I will discuss recent developments in evolutionary-based and other computational methods for assigning unannotated genes to known functional modules, for determining the conditions under which they are active, for uncovering new modules, and for identifying sets of co-regulated genes and the transcription factors that regulate them.

S. Miyano et al. (Eds.): RECOMB 2005, LNBI 3500, p. 188, 2005.
© Springer-Verlag Berlin Heidelberg 2005

Efficient q-Gram Filters for Finding All ϵ-Matches over a Given Length

Kim R. Rasmussen[1], Jens Stoye[2], and Eugene W. Myers[3]

[1] International NRW Graduate School in Bioinformatics and Genome Research,
Center of Biotechnology, Universität Bielefeld,
33594 Bielefeld, Germany
kim.rasmussen@cebitec.uni-bielefeld.de

[2] Technische Fakultät, Universität Bielefeld,
33594 Bielefeld, Germany
stoye@techfak.uni-bielefeld.de

[3] Div. of Computer Science, UC Berkeley,
Berkeley, CA 94720-1776, USA
gene@eecs.berkeley.edu

Abstract. Fast and exact comparison of large genomic sequences remains a challenging task in biosequence analysis. We consider the problem of finding all ϵ-matches between two sequences, i.e. all local alignments over a given length with an error rate of at most ϵ. We study this problem theoretically, giving an efficient q-gram filter for solving it. Two applications of the filter are also discussed, in particular genomic sequence assembly and BLAST-like sequence comparison. Our results show that the method is 25 times faster than BLAST, while not being heuristic.

1 Introduction

Searching a biological sequence database for sequences similar to a given query sequence is a problem of fundamental importance in bioinformatics. Of particular interest is the case where sequences are considered similar if they have a local alignment that scores above a given threshold. The first algorithm to solve this problem is due to Smith and Waterman [19], whose names have become synonymous with the search. A Smith-Waterman search is guaranteed to find all local alignments scoring above a given threshold, and is therefore commonly referred to as a *full-sensitivity* search. Unfortunately, the algorithm has quadratic time complexity and spends most of its running time verifying that there are in fact *no* alignments of interest.

This inefficiency led to the development of *heuristics* such as the very popular FASTA [17] and BLAST [1, 2] programs. The latter is based on the assumption that biologically interesting alignments must contain at least one pair of highly similar substrings, called a *seed*. Aided by a preprocessed query sequence, the BLAST algorithm efficiently locates and extends each seed to a local alignment containing the seed. However, it is important to note that the regions of the search space thusly disregarded, can actually contain a match.

S. Miyano et al. (Eds.): RECOMB 2005, LNBI 3500, pp. 189–203, 2005.

In contrast, a *filter* is an algorithm that rapidly and stringently eliminates a large part of a search space from consideration. That is, unlike a heuristic, it guarantees not to eliminate a region containing a match. Full sensitivity can therefore be obtained by applying a full sensitivity algorithm on the unfiltered regions. Effective filtration under scoring measures typical for protein searches is very difficult [15]. It is somewhat easier for applications on DNA, when matches of high identity ($\geq 90\%$) are sought. The first filtration algorithms for this problem appeared in the early 90's and include the works of Ukkonen [20], Chang and Lawler [6] and Myers [13].

Typical for these algorithms is the *preprocessing* of the query sequence or the sequence database in order to accelerate searches. In particular, after building an *index*, such as a suffix tree or array (see e.g. [8]) or a simple list of locations of q-grams (i.e. strings of length q), it can be used for any number of searches. This is important for many applications, where the typically large sequence database remains unchanged over a high number of usually shorter queries. Myers [13] was the first to deliver an asymptotic improvement using such preprocessing of the sequence database. This setting – a filter using an index of the sequence database, aimed for applications on DNA – is the focus of this paper.

Related Work. The program QUASAR [3] is the closest precursor to the work presented in this paper and is itself a refinement of the earlier q-gram algorithm of Ukkonen [20]. It uses a suffix array to retrieve the positions of any given q-gram in the target sequence. At query time the target sequence is logically split into blocks. As a sliding window proceeds over the query sequence, the number of q-grams co-occurring in the window and each block is determined. Blocks with less than a certain threshold of q-grams in common with the query sequence are eliminated, and BLAST is run on the query sequence and the remaining blocks to find the reported alignments.

The heuristics SSAHA [16] and BLAT [10] use a q-gram index of the target sequence, containing however only the positions of the non-overlapping q-grams. This reduces the index size and the expected number of q-gram seeds by a factor of q, but at the cost of a loss in sensitivity. At query time the set of matching q-gram positions between the query and target sequences is collected and sorted by diagonal. A linear scan locates stretches of hits on identical diagonals. These stretches are then sorted by target sequence position. Another linear scan identifies contiguous stretches of matching positions in the target sequence. These stretches are extended in traditional BLAST-style to produce the final alignments. We note that the idea of diagonal sorting actually first appeared in FASTA [17].

The use of sample indexing is taken even further in FLASH [5]. Based on a probabilistic model, randomly chosen discontiguous patterns are indexed in multiple highly redundant indexes. Consequently, the index is very large, requiring approximately 18 Gb for a 100 Mb nucleotide database. However, it was shown that this approach yielded high sensitivity in practice. The use of discontiguous, or *gapped*, seed patterns has more recently been refined in PATTERNHUNTER [12] and [4], giving fast and sensitive heuristic searches. The gapped-seed idea is orthogonal to the filtration method we employ here and could conceivable be layered on, but we do not address it in this paper.

Contributions. In this work we consider the problem of detecting local alignments under the unit cost measure or Levenshtein distance. That is, the distance between two strings is the number of insertions, deletions, and substitutions in the alignment between them. An absolute threshold on the number of differences in a local alignment is inappropriate as the lengths of the aligned strings are unconstrained. Normalizing by dividing by the length of the aligned strings, we seek instead local alignments where the *error rate* is at most $\epsilon > 0$. In other words, we seek all ϵ-*matches* for small ϵ.

The main contribution of this paper is an efficient filter for identifying regions of the implied edit matrix that are guaranteed to overlap with possible ϵ-matches. The filter is much more selective than QUASAR and Ukkonen's early work, while operating at comparable or superior speeds depending on parameter settings. Moreover, it finds all matches over a given length, a criterion not considered nor met by the earlier q-gram filters. It is thus a very effective, full-sensitivity filter for DNA searches.

The organization of the paper is as follows. In Section 2 we formalize the problem definition and present a filter criterion that identifies regions of the query and target sequences that may contain an ϵ-match. Section 3 gives an efficient algorithm for realizing the filter. Section 4 describes several applications of the basic filter and Sect. 5 presents experimental results for these applications.

2 q-Gram Filters for ϵ-Matches

2.1   Problem Definition

Given a string A over a finite alphabet Σ, $|A|$ is the length of A, $A[i]$ refers to the ith character of A, and $A[i,j]$ is the substring of A that starts with the ith character and ends with the jth. A substring of length $q > 0$ of A is a q-*gram* of A. Let ε denote the empty string. An *alignment* L of strings A and B is a sequence $(\alpha_1 \to \beta_1, \ldots, \alpha_\ell \to \beta_\ell)$ of *edit operations* (i.e., insertions $\varepsilon \to \beta$, deletions $\alpha \to \varepsilon$, and substitutions $\alpha \to \beta$ of single character substrings) such that $A = \alpha_1 \ldots \alpha_\ell$ and $B = \beta_1 \ldots \beta_\ell$. We denote the number of edit operations $\alpha \to \beta$, $\alpha \neq \beta$ in an alignment L by $\delta(L)$. The (*unit cost*) *edit distance* between A and B is then defined as $dist_\delta(A, B) := \min\{\delta(L) \mid L \text{ is an alignment of } A \text{ and } B\}$. It is well known that the edit distance can be calculated in quadratic time using dynamic programming. An $(|A| + 1) \times (|B| + 1)$ *edit matrix* E_δ is tabulated such that $E_\delta(i,j) := dist_\delta(A[1,i], B[1,j])$. Then, $E_\delta(|A|,|B|) = dist_\delta(A, B)$.

The problem we consider is that of finding ϵ-matches. The normalized relative distance, or *error rate*, is defined as the edit distance divided by the length of the query substring involved in the local alignment. An ϵ-*match* is then a local alignment with an error rate of at most ϵ. More precisely, our problem is defined as follows. Given a *target* string A and a *query* string B, a minimum match length n_0 and a maximum error rate $\epsilon > 0$, find all ϵ-matches (α, β)

where α and β are substrings of A and B, respectively, such that $|\beta| \geq n_0$ and $dist_\delta(\alpha, \beta) \leq \lfloor \epsilon|\beta| \rfloor$.

2.2   Filters

Our goal is to devise an efficient filter for identifying the regions between A and B that may contain an ϵ-match. We derive our idea for the filter from the q-gram method, which is based on the observation that the substrings of an approximate match must have a certain number of q-grams in common [9]. Define a q-hit as a pair (i, j) such that $A[i, i+q-1] = B[j, j+q-1]$. The basic q-gram method then works as follows. First, find all q-hits between the query and target strings. Second, identify regions between the strings that have 'enough' hits. Such candidate regions are subsequently subject to a closer examination.

We now show that all q-hits in an ϵ-match occur in a well-defined region of the edit matrix. We first consider ϵ-matches of length n_0 and then extend to ϵ-matches of length n_0 or greater.

Finding ϵ-Matches of Length n_0. Let an $n \times e$ *parallelogram* of the edit matrix be a set of entries on $n + 1$ consecutive columns and $e + 1$ consecutive diagonals. The *A-projection* p_A of an $n \times e$ parallelogram is the substring of A between the last row of the first column and the first row of the last column, implying $|p_A| = n - e$. Similarly, the *B-projection* p_B of an $n \times e$ parallelogram is the substring of B between the first and the last column of the parallelogram, with $|p_B| = n$. A q-hit (i, j) between A and B corresponds to a sequence of $q+1$ consecutive entries along the diagonal $j - i$ of the edit matrix. We say that a q-hit is *contained* in a given parallelogram if its entries are a subset of those of the parallelogram. Figure 1 illustrates.

An ϵ-match (α, β) of length n between A and B with at most $e = \lfloor \epsilon n \rfloor$ differences relates to an $n \times e$ parallelogram by the following lemma.

Lemma 1. *Let α and β be substrings of A and B, respectively, s.t. $|\beta| = n$ and $dist_\delta(\alpha, \beta) \leq e$. Then, there exists an $n \times e$ parallelogram such that (a) it contains at least $T(n, q, e) := (n+1) - q(e+1)$ q-hits, (b) its B-projection is β, and (c) its A-projection is contained in α.*

Proof. The proof is straightforward, c.f. Fig. 1.   □

Hence, regions between A and B that can hold an ϵ-match of length n_0 can be found as follows. First, count the number of q-hits in each $n_0 \times \lfloor \epsilon n_0 \rfloor$ parallelogram. Second, identify parallelograms that contain at least $T(n_0, q, \lfloor \epsilon n_0 \rfloor)$ q-hits. Then, for each such parallelogram there can be an ϵ-match (α, β) where α is intersected by p_A and $\beta = p_B$.

Finding ϵ-Matches of Length n_0 or Greater. We now consider how to find all ϵ-matches of length n_0 or greater. To solve this problem our idea is to look for the existence of a $w \times e$ parallelogram whose projections intersect α and β, respectively, and which we can guarantee to contain at least τ q-hits. For a

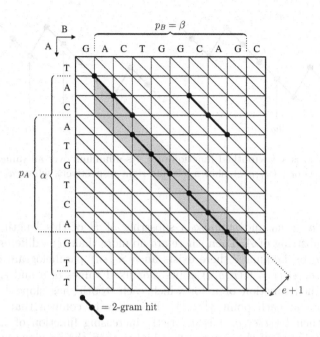

Fig. 1. An 8×2-parallelogram in the edit graph between the sequences $A =$ TACATGTCAGTT and $B =$ GACTGGCAGC. For $q = 2$, there are four q-hits within the parallelogram and $dist_\delta(\alpha, \beta) = 2$

given choice of the parameters q, ϵ and n_0, the following lemma guarantees the existence of such a parallelogram. Moreover, it provides its dimensions w and e, and the q-hit threshold τ.

Lemma 2. *Let β denote a substring of B of length n_0 or greater that has an ϵ-match to a substring α of A. Let $U(n, q, \epsilon) := (n+1) - q(\lfloor \epsilon n \rfloor + 1)$ and assume that the q-gram size q and the threshold τ have been chosen such that*

$$q < \lceil 1/\epsilon \rceil \quad and \quad \tau \leq min\{U(n_0, q, \epsilon), U(n_1, q, \epsilon)\}, \tag{1}$$

where $n_1 = \lceil (\lfloor \epsilon n_0 \rfloor + 1)/\epsilon \rceil$. Then, there is guaranteed to exist a $w \times e$ parallelogram containing at least τ q-hits whose projections intersect α and β, where

$$w = (\tau - 1) + q(e + 1) \quad and \quad e = \left\lfloor \frac{2(\tau - 1) + (q - 1)}{1/\epsilon - q} \right\rfloor . \tag{2}$$

Further, if $|\beta| \leq w$ then the B-projection of the $w \times e$ parallelogram contains β, otherwise it is a substring of β.

Proof. The outline of the proof is as follows. First, we determine the lower bound τ on the number of q-hits contained in the $n \times \lfloor \epsilon n \rfloor$ parallelogram of an ϵ-match of length $n \geq n_0$. Then, we argue that there exists a $w \times e$ parallelogram that contains at least τ q-hits. Finally, we determine the dimensions w and e of such a parallelogram over all values $n \geq n_0$.

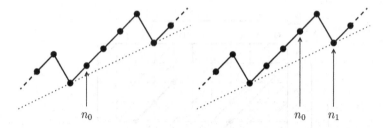

Fig. 2. $U(n, q, \epsilon)$ is a sawtoothed function of n. Its minimum over all values $n \geq n_0$ is either $U(n_0, q, \epsilon)$ or $U(n_1, q, \epsilon)$, shown left and right respectively, where n_1 is the next tooth value

Consider $n \geq n_0$ and suppose a substring of B of this length has an ϵ-match to a substring of A. Then, there are no more than $\lfloor \epsilon n \rfloor$ differences in the match and so by Lemma 1 there exists an $n \times \lfloor \epsilon n \rfloor$ parallelogram containing at least $T(n, q, \lfloor \epsilon n \rfloor) = U(n, q, \epsilon)$ q-hits. For fixed values of q and ϵ, $U(n, q, \epsilon)$ is a saw-toothed function of n for which each 'tooth' has slope 1, dropping to a minimum at each point $\{\lceil i/\epsilon \rceil\}_i$. It is easy to confirm that as long as $q < \lceil 1/\epsilon \rceil$, then $U(\lceil i/\epsilon \rceil, q, \epsilon)$ is a strictly increasing function of i, i.e., each successive minimum of the saw-tooth is higher than the previous one. It thus follows that the smallest value of $U(n, q, \epsilon)$ over all values $n \geq n_0$ is either $U(n_0, q, \epsilon)$ or the value $U(n_1, q, \epsilon)$ at the next tooth value $n_1 = \lceil (\lfloor \epsilon n_0 \rfloor + 1)/\epsilon \rceil$, c.f. Fig. 2. Therefore, if one chooses $\tau \leq \min\{U(n_0, q, \epsilon), U(n_1, q, \epsilon)\}$ there will always be at least τ q-hits in an $n \times \lfloor \epsilon n \rfloor$ parallelogram for $n \geq n_0$.

The $U(n, q, \epsilon) \geq \tau$ hits within an $n \times \lfloor \epsilon n \rfloor$ parallelogram are interspersed between up to $\lfloor \epsilon n \rfloor$ differences. For any n, the question is what is the largest e such that for every consecutive sequence of e differences there are less than τ q-hits in the e spaces below them? Some thought reveals that a way to maximize e is to cluster the q-hits into groups of $\tau - 1$ and to then spread these as far apart between the $\lfloor \epsilon n \rfloor$ differences, and in this case $e = \left\lfloor \frac{\lfloor \epsilon n \rfloor}{\lceil U(n,q,\epsilon)/(\tau-1) \rceil - 1} \right\rfloor$. Furthermore, it follows that one will have at least τ hits in a $w \times e$ parallelogram for $w = (\tau - 1) + q(e + 1)$.

It thus remains to find the largest value of e (and its associated w) over all values of $n \geq n_0$. First observe that one need only consider the minimum points $\{\lceil i/\epsilon \rceil\}_i$ of the sawtooth as for any given tooth, $U(n, q, \epsilon)$ increases while $\lfloor \epsilon n \rfloor$ stays the same. Therefore we seek

$$e = \max_{i \geq \lceil \epsilon n_0 \rceil} \left\{ \left\lfloor \frac{i}{\lceil U(\lceil i/\epsilon \rceil, q, \epsilon)/(\tau - 1) \rceil - 1} \right\rfloor \right\} . \tag{3}$$

By the definition of the ceiling function and the fact that $\lceil i/\epsilon \rceil$ is the only fractional part of the denominator it follows that this equals

$$e = \max_{i \geq \lceil \epsilon n_0 \rceil} \left\{ \left\lfloor \frac{i}{\lceil U(i/\epsilon, q, \epsilon)/(\tau - 1) \rceil - 1} \right\rfloor \right\} . \tag{4}$$

For all values of i for which $\lceil U(i/\epsilon, q, \epsilon)/(\tau - 1)\rceil$ is the same value, say m, the largest value of i will give the largest value of e. But this value is

$$\max\{j : U(j/\epsilon, q, \epsilon) \leq m(\tau - 1)\} = \left\lfloor \frac{m(\tau - 1) + (q - 1)}{1/\epsilon - q} \right\rfloor . \tag{5}$$

Moreover, since $U(n, q, \epsilon) \geq \tau$ for all $n \geq n_0$ we are in effect considering all $m \geq 2$. Thus,

$$e = \max_{m \geq 2}\left\{ \left\lfloor \frac{\left\lfloor \frac{m(\tau-1)+(q-1)}{1/\epsilon-q} \right\rfloor}{m - 1} \right\rfloor \right\} = \left\lfloor \max_{m \geq 2}\left\{ \frac{m(\tau - 1) + (q - 1)}{(m - 1)(1/\epsilon - q)} \right\} \right\rfloor , \tag{6}$$

which is clearly decreasing in m and therefore has its maximum at $m = 2$.   □

Feasible Values of n_0 Given τ. For a given choice of the parameters ϵ and q, there is a set of pairs (n_0, τ) such that for any ϵ-match a $w \times e$ parallelogram exists that contains this match. In Lemma 2, we give the set of feasible τ for a given n_0. We now compute the set of feasible n_0 for a given choice of τ.

Corollary 1. *If $n_0 \geq q \left\lceil \frac{\tau+q-1}{1/\epsilon-q} \right\rceil + \tau - 1$ then a $w \times e$ parallelogram, as defined by Lemma 2, exists.*

Proof. Consider a given choice of τ, q, and ϵ. We seek the value n_0 for which $U(n, q, \epsilon) \geq \tau$ for all $n \geq n_0$. First we find the smallest tooth point $n_1 = \lceil d_1/\epsilon\rceil$ whose value $U(n_1, q, \epsilon)$ is not less than τ. That is we seek the minimum d such that $\lceil d/\epsilon\rceil + 1 - q(d + 1) \geq \tau$. Performing a bit of algebra, we get $d_1 = \min\left\{d : d \geq \frac{\tau+q-1}{1/\epsilon-q}\right\} = \left\lceil \frac{\tau+q-1}{1/\epsilon-q} \right\rceil$. So, n_0 occurs in the previous tooth and satisfies $n_0 + 1 - q((d_1 - 1) + 1) = \tau$. Solving for n_0 gives the result.   □

Table 1 illustrates the complex relationships between the parameters of the filter, as calculated by Lemma 2 and Corollary 1. Moreover, the values give an indication of the very good selectivity of the filter.

Table 1. Filter parameters for $\epsilon = 0.05$, by Lemma 2 and Corollary 1, respectively

	$q = 7$			$q = 9$			$q = 11$				$q = 11$								
n_0	30	50	100	30	50	100	30	50	100	τ	7	8	9	10	11	12	13	14	15
w	44	71	128	48	77	136	40	71	133	n_0	28	29	41	42	43	44	45	46	47
e	3	5	9	3	5	9	2	4	8	w	39	40	52	53	54	55	67	68	69
τ	17	30	59	13	24	47	8	17	35	e	2	2	3	3	3	3	4	4	4

3   An Efficient Algorithm

We now describe an efficient algorithm for finding all $w \times e$ parallelograms for ϵ-matches of length n_0 or greater between the strings A and B.

3.1   Preprocessing

In the preprocessing step we construct a q-gram index for the target sequence A. The index consists of two tables. The *occurrence table* is a concatenation of the lists $L(G) := \{i \mid A[i, i+q-1] = G\}$ for all q-grams $G \in \Sigma^q$ in A, and the *lookup table* is an array indexed by the natural integer encoding of G to base $|\Sigma|$, giving the start of each list in the occurrence table.

3.2   Finding $w \times e$ Parallelograms

The $w \times e$ parallelograms that contain at least τ q-hits can be found trivially using a sliding window. The implied edit matrix is split into all overlapping *bins* of $e + 1$ adjacent diagonals. At any time, each bin counts the number of q-hits contained in the $w \times e$ parallelogram defined by the intersection of the diagonals of the bin and the rows of the sliding window $W_j = B[j, j+w-1]$. As the sliding window proceeds to W_{j+1}, the bin counters are updated to reflect the changes caused by the q-grams leaving and entering the window. If a bin counter reaches τ, the corresponding parallelogram is reported; overlapping parallelograms are trivially merged on the fly.

Improving space requirements. The number of bin counters is reduced by searching for $w \times (e + \Delta)$ parallelograms, where $\Delta > 0$. We associate each bin counter with $e + \Delta + 1$ adjacent diagonals and let successive bins overlap by e diagonals. This is sufficient as the τ q-hits cannot be spread over more than $e+1$ diagonals. In total, only $\left\lceil \frac{|A|-e-\Delta}{\Delta+1} \right\rceil$ bin counters are required. A good choice for Δ is 2^z, where $z \in \mathbb{N}$ and $2^z > e$. Bin indices are then calculated with fast bit-operations.

Improving running time. We reduce the considering of each q-hit from twice to once by use of two observations. First, two q-hits that are more than $w - q$ apart (counted as the difference between their starting positions in B) cannot both be in the same $w \times e$ parallelogram. Secondly, the τ q-hits in an ϵ-match cannot occur in a string shorter than $q + \tau - 1$. Hence, we relax the search to finding $w' \times e$ parallelograms, where $w' \geq q + \tau - 1$, and update the bins as follows. For each bin we keep track of the minimum and maximum B-position of the contained q-hits, min and max respectively. The number of q-hits in a bin is counted until a q-hit (i, j) is found such that $j - w + q > max$. If the bin counter has reached the threshold τ, we report the matching $(max - min + q) \times e$ parallelogram. We then reset the bin counter and set $min = max = j$ as the current q-hit is counted. Algorithm 1 shows the bin updating step in pseudocode.

The improved approach for updating bins has a few subtle points worth mentioning. Unless care is taken, it may return too short parallelograms; in fact as short as q. This can happen when a very dense cluster of q-hits falls into a bin. In particular, single-character repeat runs are a source of such dense hit clusters. To alleviate the problem, one possibility would be to extend the reporting criterion such that also the validity of the parallelogram length is checked. A

Algorithm 1: `UpdateBin`(r, j, d)

Input : Bin record r; q-hit position j in sequence B; and offset bin diagonal d.
Output: Empty or singleton parallelogram set P.

1 $P \leftarrow \emptyset$
2 **if** $j - w + q > r.max$ **then**
3   **if** $r.count \geq \tau$ **then**
4        $p.left \leftarrow |A| - d$
5        $p.top \leftarrow r.max + q$
6        $p.bottom \leftarrow r.min$
7        $P \leftarrow \{ p \}$
8     $r.count \leftarrow 0$

9 **if** $r.count = 0$ **then**
10    $r.min \leftarrow j$

11 **if** $r.max < j$ **then**
12    $r.max \leftarrow j$
13    $r.count \leftarrow r.count + 1$

14 **return** P

---

more elegant solution is, however, for each bin to count all q-hits with identical B positions as one hit only. Another point is that parallelograms can be generated which are not in accordance with the filter criterion. That is, they do not contain at least τ q-hits within every window of length w. In the worst case, this happens when a bin receives one q-hit exactly every $w-q+1$ positions in B. Although this is very unlikely to occur in practice, other likewise unfortunate hit distributions can cause the generation of similar strictly invalid filter parallelograms. However, it should be remarked that when searching for local alignments in biological sequences the regions triggering such parallelograms are often of great interest anyway. Algorithm 2 shows the pseudo-code for the main loop of our filtration algorithm.

Summing up, the specificity of our improved approach is slightly lower than that of the simple sliding window approach. This is largely due to the use of larger bins, but also because of the slight risk of producing parallelograms that do not strictly adhere to the filter criterion. On the other hand, our approach improves the time and space requirements considerably.

3.3 Complexity

The q-gram index is constructed in $\mathcal{O}(|A| + |\Sigma|^q)$ time. Each occurrence list is found in $\mathcal{O}(1)$ time, but the length can be linear in $|A|$. The worst case time for the filter is therefore $\mathcal{O}(|A| \cdot |B|)$. If we assume random strings of uniformly i.i.d. characters, the expected length of each occurrence list is $|A| \cdot |\Sigma|^{-q}$. Hence, under this assumption the filter requires $\mathcal{O}(|B| + |A| \cdot |B| \cdot |\Sigma|^{-q})$ expected time.

The space complexity is dominated by the q-gram index, which requires $|A|+|\Sigma|^q$ integers. Bins and parallelograms are each represented using 3 integers,

---

Algorithm 2: Filter for identifying parallelograms for ϵ-matches

---

Input : Query B; q-gram index I for target A; parameters w, e, τ; and $\Delta = 2^z$
Output: Set of parallelograms P

1 Allocate and initialize array of bin records *Bins*
2 $P \leftarrow \emptyset$
3 **for** $j \leftarrow 0$ *to* $|B| - q$ **do**
4       $G \leftarrow B[j, j + q - 1]$
5       $L(G) \leftarrow$ lookup occurrence list for G in I
6   **foreach** $i \in L(G)$ **do**
7           $d \leftarrow |A| + j - i$
8           $b_0 \leftarrow d \gg_{bit} z$
9           $b_m \leftarrow b_0 \bmod |Bins|$
10          $P \leftarrow P \cup \texttt{UpdateBin}(Bins[b_m], j, b_0 \ll_{bit} z)$
11   **if** $(d \mathbin{\&_{bit}} (\Delta - 1)) < e$ **then**
12              $b_m \leftarrow (b_m + |Bins| - 1) \bmod |Bins|$
13              $P \leftarrow P \cup \texttt{UpdateBin}(Bins[b_m], j, (b_0 - 1) \ll_{bit} z)$
14   **if** $(j - e) \bmod (\Delta - 1) = 0$ **then**
15              $b_0 \leftarrow (j - e) \gg_{bit} z$
16              $b_m \leftarrow b_0 \bmod |Bins|$
17   /* *CheckAndResetBin is similar to lines 3–8 of* UpdateBin */
18              $P \leftarrow P \cup \texttt{CheckAndResetBin}(Bins[b_m], j, b_0 \ll_{bit} z)$

19 $P \leftarrow P \cup \{$ remaining parallelograms in *Bins* $\}$

---

so with a bin size of $e + 2^z$ we need $3 \cdot 2^{-z}|A|$ integers for all bins, and $3p$ integers to return p parallelograms. In total, the filter requires $(3 \cdot 2^{-z} + 1)|A| + |\Sigma|^q + 3p$ integers. Thus, for 32-bit integers and parameters $|A| = 3 \cdot 10^9$, $q = 11$ and $z = 3$, the filter requires only 5.5 bytes per input character and $12p$ bytes for the result.

4   Applications

This section covers two applications of the basic q-gram filter. We first consider its use in an overlapper for sequence assembly and then for general purpose BLAST-like alignment searching.

4.1   Sequence Assembly

When building a typical DNA assembler, one faces the problem of comparing a collection of $600 - 1000$ bp fragments against each other in search of overlaps, typically say over 50 bp long at 95 % or greater identity. The filter is ideal for this application in that the error rate is low and the requirement is to find all matches under a given percent difference and over a lower length limit. Typically there are thousands or millions of reads $f_1, f_2, \ldots, f_n$, in total $\sum_{i=1}^{n} |f_i|$ bases. The reads are concatenated together to make a single large string A that our filter is run over, with an auxiliary table $map[k] = \sum_{i=1}^{k} |f_i|$ giving the start of each read in A. Note that we are comparing A against itself so we carefully modify the filter to ignore hits on or below the diagonal of the implied edit matrix.

In essence after running the filter, all we need to do is identify the pairs of reads that intersect parallelograms and then check each pair for a proper overlap. This is simply a matter of mapping base positions in A to read positions through the inverse of map. We used a quick sort of all parallelograms in one dimension and then an insertion sort in the second dimension as the sort buckets are expected to be small. Multiple hits to a given read pair are merged during the insertion sort of a bucket and a parallelogram is required to have more than 5 base pairs in a sequence as often a legitimate parallelogram in one read pair will extend slightly into the next read due to the concatenation. For each read pair, we keep track of the maximum and minimum diagonal of the edit matrix between their sequences that is covered by a parallelogram so that the check for an overlap need only perform dynamic programming within a band consistent with these diagonals and the maximum error rate. The dynamic programming itself is done with a bit-vector acceleration method by Myers [14].

4.2 BLAST-Like Searching

Another application of the filter is BLAST-like alignment searching. In this setting, we use ϵ-matches as seeds and extend these into longer alignments by dynamic programming. After running the filter, we identify the possible ϵ-matches in each parallelogram by chaining of the contained q-hits. Our approach uses a simple variation of sparse dynamic programming [7] and the fact that an ϵ-match must contain at least τ q-hits, which are separated by no more than e differences.

We define the partial ordering relation $\ll$ on a set of q-hits as follows. Let $h = (i, j)$ and $h' = (i', j')$ denote q-hits and define $diag(h) := j - i$. Let $dist_\infty(h, h')$ refer to the Chebychev distance $\max\{|i' - i - q|, |j' - j - q|\}$ between the ending and starting points of h and h', respectively. Then, $h \ll h'$ if and only if (1) $dist_\infty(h, h') \le e$, and (2a) $diag(h) \ne diag(h')$ and $i + q \le i'$ and $j + q \le j'$, or (2b) $diag(h) = diag(h')$ and $i < i'$. A chain is then a sequence of q-hits $\langle h_1, h_2, \ldots, h_l \rangle$ where $h_i \ll h_{i+1}$ for all $1 \le i < l$. If we assign the score q to each q-hit and use $dist_\infty$ for the penalty of connecting successive q-hits, the score of a chain C is thus given by $chain(C) := q \cdot l - \sum_{i=1}^{l-1} dist_\infty(h_i, h_{i+1})$. Denoting the maximum score over all chains ending in q-hit h' by $chain(h')$, the recurrence relation $chain(h') := \max\{0, \max_{h \ll h'}\{chain(h) - dist_\infty(h, h')\} + q$ immediately gives the basis for the algorithm.

The chaining requires one sweep over each parallelogram. The q-hits are found column-wise by lookup in a hash table over the q-grams in the sliding window that is defined on A by the first and the last rows of the current column. A balanced search structure D maintains the q-hits found on the previous $q + e + 1$ columns, ordered by diagonal number and starting position in B. For each q-hit h' in the current column, D is searched for q-hits in the diagonal range $[i-j-e, i-j+e]$. In every diagonal in this range, the nearest chain (w.r.t. position in B) ending in q-hit h, such that $h \ll h'$, is candidate for chaining with h'. Of all such candidate chains, the maximum scoring is chosen. If multiple candidate chains reach the maximum score, we choose the closest (w.r.t. diagonal). Then, the new chain ending in h' is inserted in D.

Chains shorter than τ cannot be part of an ϵ-match and they are therefore immediately disposed. Otherwise, the chain is rescored by finding the optimal gap position between successive q-hits. To avoid reporting multiple only slightly differing chains, we partition the rescored chains into equivalence classes by their first q-hit. That is, two chains belong to the same equivalence class if and only if they begin with the same q-hit. Only the maximum scoring chain for each equivalence class is extended by gapped X-drop extension [21]. We initiate the extension procedure at the end and beginning of the first and last q-hits of the chain, respectively.

5 Experimental Results

In this section we describe some experimental results for the two applications.

5.1 Sequence Assembly

With a gigabyte of memory we can reasonably solve 60 Mbp by 60 Mbp comparisons for 50 bp overlaps at less than 5 % difference in roughly 90 seconds on an Apple PowerBook G4 laptop. Larger problems are solved by partitioning the data set into 60 Mbp segments and solving either serially or in parallel all the necessary pairwise comparisons of segments. On the same laptop, the 1.8 Gbp data set for *D. melanogaster* can be compared in a total of 18 CPU hours. With more memory, larger q-grams can be used and larger segments can be accomodated in a single run. For example, on an Intel Itanium II with 16 Gb memory we can compute the same overlaps for *D. melanogaster* in under two hours.

5.2 EST Clustering

We compare the performance of our BLAST-like alignment searching application with that of BLAST and the Smith-Waterman algorithm. The Smith-Waterman implementation that we use is SSEARCH [18], which is part of the FASTA package [24]. The BLASTN version is 2.2.9 from NCBI [22]. Our BLAST-like alignment searching application is implemented in SWIFT, available at [25].

The setup resembles that of [11]. Briefly, we select EST sequences from two species and perform a full-sensitivity, all-against-all comparison. All sequence pairs where the best Smith-Waterman local alignment scores above a given threshold are recorded. The all-against-all comparison is then repeated using BLAST and SWIFT. For each, the sensitivity, or recall ratio, is determined as follows. Suppose the full-sensitivity search finds p pairs with a best local alignment score of s. If p' of the p pairs are found with score at least $\frac{s}{2}$ by BLAST or SWIFT, the ratio $\frac{p'}{p}$ is the sensitivity for alignment score s. Neither BLAST or SWIFT attempt to compute the optimal alignment for found homologies and we therefore consider a sequence pair recalled if its local alignment score is not lower than $\frac{s}{2}$. We note that any other threshold ratio of s can equally well be used.

Table 2. Running times for EST all-against-all comparison. The time for the database formatting and preprocessing in BLAST (3 s) and SWIFT (12 s) is not included

	SWIFT			BLAST	SSEARCH
(ϵ, n_0)	$(0.05, 50)$	$(0.04, 30)$	$(0.05, 30)$	—	—
Running time	18 s	29 s	35 s	773 s	8 h

The EST sequences are obtained from NCBI [23]. We randomly select 40.000 sequences from *H. sapiens* (25 Mbp) and 5.600 sequences from *M. musculus* (2 Mbp). The poly-X tails (X = {A, C, G, T}) typically found in EST sequences due to sequencing errors are trivially masked to Ns. The q-gram length is set to 11 in both BLAST and SWIFT. All programs use match/mismatch scores ± 1, and gap open and extension penalties are set to 5 and 1, respectively. The local alignment score threshold is 16. Searches with SSEARCH are conducted on a cluster with 50 UltraSparcIIe/500 MHz nodes, whereas the searches with BLAST and SWIFT run on a 2 GHz AMD Athlon-XP Linux PC. Table 2 lists the running times for the different programs and Fig. 3 compares the sensitivity of BLAST and SWIFT.

Using parameters for typical EST clustering criteria, $\epsilon = 0.05$ and $n_0 = 50$, the sensitivity is a bit lower for SWIFT than for BLAST. However, this is expected as SWIFT requires and guarantees the presence of an ϵ-match before an alignment is recorded. In other words, the extra alignments found by BLAST do not conform

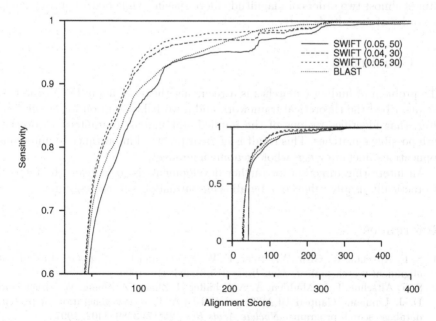

Fig. 3. Sensitivity of BLAST and SWIFT. The horizontal line in 1.0 corresponds to the sensitivity of SSEARCH. For SWIFT, the (ϵ, n_0) parameters are shown in parentheses. The inset shows the complete sensitivity range

Table 3. Filtration ratios and times for EST all-against-all comparison

| (ϵ, n_0) | SWIFT | | QUASAR | | | |
| | Filtration | | Filtration, best ratio | | Filtration, best time | |
	Ratio	Time (s)	Ratio	Time (s)	Ratio	Time (s)
$(0.05, 50)$	$6.5 \cdot 10^{-6}$	6.0	$4.5 \cdot 10^{-4}$	36.1	$2.1 \cdot 10^{-3}$	4.2
$(0.04, 30)$	$4.5 \cdot 10^{-6}$	5.0	$4.0 \cdot 10^{-4}$	69.0	$3.1 \cdot 10^{-3}$	4.4
$(0.05, 30)$	$5.4 \cdot 10^{-6}$	6.1	$4.3 \cdot 10^{-4}$	68.5	$3.5 \cdot 10^{-3}$	4.4

to the query criteria. Additionally, note that by using more agressive (ϵ, n_0) parameters, SWIFT can in general attain sensitivity levels comparable to or better than BLAST, while still being more than 25 times faster.

The filtration efficiency is the primary gauge of the expected running time for the application-specific post-processing of the filter output. We therefore also measure the filtration ratio, which we define as the total area of the unfiltered regions, i.e. the reported parallelograms, divided by the total size of the implied edit matrix. Table 3 shows the resulting filtration ratios and times for SWIFT. For comparison, we include the times and ratios for the closest precursor to our filter, QUASAR. We repeated each run with the QUASAR block size giving the best possible filtration ratio (128 in first; all others 64) and with the smallest block size giving filtration time equal or superior to that of SWIFT (1024 for all runs).

As stressed by Table 3, our filter is very efficient. Compared to QUASAR, our filter is almost two orders of magnitude more specific while being approximately one order of magnitude faster.

6   Conclusion

The problem of finding ϵ-matches is a recurring theme in many DNA searches. We described the theoretical framework and a workable solution for an efficient filter, that identifies regions of the implied edit matrix guaranteed to overlap with possible ϵ-matches. This result is of great practical importance to numerous applications, including e.g. whole-genome alignment.

An interesting direction for further developments is to increase the interval of practically usable values of ϵ by allowing mismatches in the q-grams.

References

1. S. F. Altschul, W. Gish, W. Miller, E. W. Myers, and D. J. Lipman. Basic local alignment search tool. *J. Mol. Biol.*, 215(3):403–410, 1990.
2. S. F. Altschul, T. L. Madden, A. A. Schäffer, J. Zhang, Z. Zhang, W. Miller, and D. J. Lipman. Gapped BLAST and PSI-BLAST: a new generation of protein database search programs. *Nucleic Acids Res.*, 25(17):3389–3402, 1997.
3. S. Burkhardt, A. Crauser, P. Ferragina, H.-P. Lenhof, E. Rivals, and M. Vingron. q-gram based database searching using a suffix array. In *Proc. of the 3rd Annu. Int. Conf. on Computational Molecular Biology (RECOMB'99)*, pages 77–83, 1999.

4. S. Burkhardt and J. Kärkkäinen. Better filtering with gapped q-grams. In *Proc. of CPM'01*, volume 2089 of *LNCS*, pages 73–85, 2001.
5. A. Califano and I. Rigoutsos. FLASH: a fast look-up algorithm for string homology. In *Proc. of the 1st Int. Conf. on Intelligent Systems for Molecular Biology (ISMB'93)*, pages 56–64, 1993.
6. W. I. Chang and E. L. Lawler. Sublinear expected time approximate string matching and biological applications. *Algorithmica*, 12(4/5):327–344, 1994.
7. D. Eppstein, Z. Galil, R. Giancarlo, and G. F. Italiano. Sparse dynamic programming I: linear cost functions. *J. ACM*, 39(3):519–545, 1992.
8. D. Gusfield. *Algorithms on Strings, Trees and Sequences: Computer Science and Computational Biology*. Cambridge University Press, 1997.
9. P. Jokinen and E. Ukkonen. Two algorithms for approximate string matching in static texts. In *Proc. of MFCS'91*, volume 520 of *LNCS*, pages 240–248, 1991.
10. W. J. Kent. BLAT – the BLAST-like alignment tool. *Genome Res.*, 12(4):656–664, 2002.
11. M. Li, B. Ma, D. Kisman, and J. Tromp. PatternHunter II: Highly Sensitive and Fast Homology Search. In *Proc. of the 14th Annu. Int. Conf. on Genome Informatics (GIW'03)*, pages 164–175, 2003.
12. B. Ma, J. Tromp, and M. Li. PatternHunter – faster and more sensitive homology search. *Bioinformatics*, 18:440–445, 2002.
13. E. Myers. A sublinear algorithm for approximate keyword searching. *Algorithmica*, 12(4/5):345–374, 1994.
14. E. Myers. A fast bit-vector algorithm for approximate string matching based on dynamic programming. *J. ACM*, 46(3):539–553, 1999.
15. E. Myers and R. Durbin. A table-driven, full-sensitivity similarity search algorithm. *J. Comp. Bio.*, 10(2):103–118, 2003.
16. Z. Ning, A. J. Cox, and J. C. Mullikin. SSAHA : A fast search method for large DNA databases. *Genome Res.*, 11(10):1725–1729, 2001.
17. W. R. Pearson and D. J. Lipman. Improved tools for biological sequence comparison. *Proc. Natl. Acad. Sci. USA*, 85:2444–2448, 1988.
18. W. R. Pearson. Searching protein sequence libraries: Comparison of the sensitivity and selectivity of the Smith-Waterman and FASTA algorithms. *Genomics*, 11:635–650, 1991.
19. T. F. Smith and M. S. Waterman. Identification of common molecular subsequences. *J. Mol. Biol.*, 147(1):195–197, 1981.
20. E. Ukkonen. Approximate string-matching with q-grams and maximal matches. *Theor. Comput. Sci.*, 92(1):191–211, 1992.
21. Z. Zhang, P. Berman, and W. Miller. Alignments without low-scoring regions. In *Proc. of the 2nd Annu. Int. Conf. on Computational Molecular Biology (RECOMB'98)*, pages 294–301, 1998.
22. ftp://ftp.ncbi.nih.gov/toolbox/ncbi_tools/ncbi.tar.gz.
23. ftp://ftp.ncbi.nlm.nih.gov/blast/db/FASTA/est_{human,mouse}.gz.
24. ftp://ftp.virginia.edu/pub/fasta.
25. http://bibiserv.techfak.uni-bielefeld.de/swift.

A Polynomial Time Solvable Formulation of Multiple Sequence Alignment

Sing-Hoi Sze[1,2], Yue Lu[2], and Qingwu Yang[1]

[1] Department of Computer Science
[2] Department of Biochemistry and Biophysics,
Texas A&M University, College Station, TX 77843, USA

Abstract. Since traditional multiple alignment formulations are NP-hard, heuristics are commonly employed to find acceptable alignments with no guaranteed performance bound. This causes a substantial difficulty in understanding what the resulting alignment means and in assessing the quality of these alignments. We propose an alternative formulation of multiple alignment based on the idea of finding a multiple alignment of k sequences which preserves $k - 1$ pairwise alignments as specified by edges of a given tree. Although it is well known that such a preserving alignment always exists, it did not become a mainstream method for multiple alignment since it seems that a lot of information is lost from ignoring pairwise similarities outside the tree. In contrast, by using pairwise alignments that incorporate consistency information from other sequences, we show that it is possible to obtain very good accuracy with the preserving alignment formulation. We show that a reasonable objective function to use is to find the shortest preserving alignment, and, by a reduction to a graph-theoretic problem, that the problem of finding the shortest preserving multiple alignment can be solved in polynomial time. We demonstrate the success of this approach on three sets of benchmark multiple alignments by using consistency-based pairwise alignments from the first stage of two of the best performing progressive alignment algorithms TCoffee and PROBCONS, and replace the second heuristic progressive step of these algorithms by the exact preserving alignment step (we ignore the iterative refinement step in this study). We apply this strategy to TCoffee and show that the new approach outperforms TCoffee on two of the three test sets. We apply the strategy to a variant of PROBCONS with no iterative refinements and show that the new approach achieves a similar accuracy except on one test set. The most important advantage of the preserving alignment formulation is that we are certain that we can solve the problem in polynomial time without using a heuristic. A software program implementing this approach (PSAlign) is available at http://faculty.cs.tamu.edu/shsze.

1 Introduction

The goal of the multiple alignment problem is to bring similar regions from different sequences as closely together as possible, with applications in diverse types

S. Miyano et al. (Eds.): RECOMB 2005, LNBI 3500, pp. 204–216, 2005.

of biosequence analysis (Taylor [24]; Carillo and Lipman [2]; Thompson et al. [26]; Gotoh [8]; Morgenstern et al. [15]; Stoye [23]; Notredame et al. [18]; Lee et al. [14]; Do et al. [4]; Edgar [5]; Van Walle et al. [28]). Since traditional multiple alignment formulations are NP-hard (Just [11]), it is unreasonable to expect that one will ever be able to find an efficient approach that always returns an optimal alignment. The best known exact algorithm employs dynamic programming techniques with time complexity $O(n^k)$ (Carillo and Lipman [2]), where n is the maximum sequence length and k is the number of sequences, and thus is useful only when k is small. Stoye [23] proposed a divide-and-conquer heuristic to limit the search space by subdividing the input sequences into shorter segments, but it is not efficient enough for large-scale applications.

The inherent difficulty of the multiple alignment problem leads naturally to the development of heuristic approaches. Among the most successful of these are progressive approaches, which combine the given sequences in some order to obtain a multiple alignment (Feng and Doolittle [6]; Thompson et al. [26]; Notredame et al. [18]; Do et al. [4]; Edgar [5]). These heuristics are often coupled with iterative refinement of the initial multiple alignment to obtain improved performance (Gotoh [8]; Do et al. [4]; Edgar [5]). Alternatively, other non-progressive approaches assemble a final multiple alignment from short alignments of local similarities (Morgenstern et al. [15]; Van Walle et al. [28]). Although in most cases, a scoring scheme and an accompanying objective function can be defined for these heuristics, it is often unclear how close the final alignment is to the optimal. Some efforts have been spent to develop approximation algorithms for multiple alignment with guaranteed performance bound, but the theoretical bound is usually too weak to reflect the actual performance (Gusfield [9]).

We propose an alternative formulation of multiple alignment that is solvable in polynomial time. Such a formulation is very important as it makes it possible to know what the alignment means and also ensures that the optimal solution can be found. Instead of employing objective functions that are very difficult to optimize, the formulation is based on the idea of finding a multiple alignment of k sequences which preserves $k-1$ pairwise alignments as specified by edges of a given tree. In particular, one can use the optimum spanning tree that includes the best $k-1$ pairwise alignments covering all the sequences. Although it is well known that such a preserving alignment always exists (Feng and Doolittle [6]; Gusfield [9]; Pevzner [21]), it did not become a mainstream method for multiple alignment since it seems that a lot of information is lost from ignoring pairwise similarities outside the tree.

The preserving alignment approach can be seen as a restricted version of a broader class of consistency-based approaches, which aim to maximize the consistency between the resulting multiple alignment and a given set of pairwise alignments on aligned residue pairs (Kececioglu [12]; Notredame et al. [19]). A distinct advantage of these consistency-based approaches is that once the pairwise alignments are fixed, the multiple alignment follows logically without the need to define a multiple alignment score (Notredame et al. [19]). Since the pairwise alignments are not restricted to be on a tree in this more general formulation, they can be conflicting and thus the objective function is likely to be more accurate, but it is also likely to be intractable to optimize. In another direction, by incorporating consistency information from other sequences when computing individual pairwise alignments, these consistency-based pairwise alignments have been successfully used in the pairwise stage of progressive approaches to give some of the best performing multiple alignment approaches to date (Notredame et al. [18]; Do et al. [4]; Edgar [5]). By employing these consistency-based pairwise alignments, we will show that it is possible to obtain very good accuracy even with the more restrictive preserving alignment formulation. Other studies that made use of the notion of consistency include Gotoh [7] and Vingron and Argo [29].

A complication with the preserving alignment formulation is that there may be many multiple alignments which preserve the given $k-1$ pairwise alignments and previous studies did not suggest how to choose among them. Ideally, one would like to maximize the similarity level over all columns. However, many of the objective functions that attempt to exploit these similarities are likely to be intractable to optimize. One natural way that allows us to develop a tractable approach is to find the shortest preserving multiple alignment with the smallest number of columns, corresponding to adding as few gaps as possible while preserving the pairwise alignments along the tree edges. Without being able to control the similarity level in each individual column, this formulation discourages gaps (similar to other traditional formulations) while making sure that the resulting multiple alignment resembles the given pairwise alignments. One important advantage of the preserving alignment formulation is that once the tree and the pairwise alignments are fixed, no additional parameters or a scoring scheme for multiple alignment are needed. This makes it possible to use any tree or pairwise alignments directly from other approaches, including ones that have made use of structural information. Similar ideas of utilizing structural pairwise alignments have been proposed by a few studies (O'Sullivan et al. [20]; Van Walle et al. [28]).

We will show, by a reduction to a graph-theoretic problem, that the problem of finding the shortest preserving multiple alignment can be solved in polynomial time, and, by using consistency-based pairwise alignments, that the accuracy of this exact approach is comparable to the best heuristic multiple alignment approaches on three sets of benchmark multiple alignments from Thompson et al. [27], Edgar [5] and Van Walle et al. [28], thus justifying the use of the proposed polynomial time formulation over other NP-hard formulations. In particular, we reduce the multiple alignment problem to finding a topological partial ordering in a directed acyclic graph where each vertex represents a partially aligned column and unordered vertices are allowed to share the same label. The label assigned to each vertex from the ordering specifies its position in the multiple alignment and the ordering itself represents a preserving multiple alignment.

2   Problem Formulation

Let $S = \{s_1, \ldots, s_k\}$ be a given set of sequences. Assume that we are given a tree T with k vertices where each vertex of T is labeled by a distinct sequence and each edge (i, j) of T represents a pair of sequences s_i and s_j, and we are also given a pairwise alignment P_{ij} between sequences s_i and s_j for each edge (i, j) of T. A multiple alignment M of S is said to preserve all the $k - 1$ pairwise alignments on T if for each edge (i, j) of T, the induced pairwise alignment on sequences s_i and s_j is the same as P_{ij} when the columns containing only gap characters are removed. Since the tree T specifies pairwise alignments that can be simultaneously preserved, it is obvious that such a preserving multiple alignment M always exists (Feng and Doolittle [6]; Gusfield [9]; Pevzner [21]). Likewise, a multiple alignment M of S is said to preserve all matches and mismatches (or just the matches) in the $k - 1$ pairwise alignments on T if for each edge (i, j) of T, each column containing a match or a mismatch (or just a match) has to stay in the same column in M. We formulate the multiple alignment problem as follows: given a tree T and a pairwise alignment P_{ij} for each edge (i, j) of T, find a preserving multiple alignment M with the smallest number of columns.

The simplest way to obtain pairwise alignments is by applying standard techniques, including global (Needleman and Wunsch [17]) and local (Smith and Waterman [22]) approaches, or a combination of these approaches. However, we found that it is often better to use other kinds of pairwise alignments such as those that have incorporated consistency information from other sequences. These consistency-based pairwise alignments can be obtained from the pairwise stage of a few progressive approaches (Notredame et al. [18]; Do et al. [4]). From these pairwise alignments, one reasonable tree T to use is an optimum spanning tree on the complete graph C_k where each vertex is labeled by a distinct sequence and each edge (i, j) is labeled by the score of P_{ij}, which can either be the pairwise alignment score of P_{ij} or other scores given by the approach generating the pairwise alignments. To compute the optimum spanning tree from C_k, Prim's algorithm can be used which has time complexity $O(k^2 \log k)$ (Cormen et al. [3]). Alternatively, one can use a star tree with a central vertex and $k - 1$ leaves (Gusfield [9]; Pevzner [21]). Although computational results show that using a star tree works well when all the given sequences are closely related, it does not give very good performance when none of the given sequences can act as the center, such as when there is more than one cluster of closely related sequences. In contrast, by using a general tree, it is possible to utilize the best non-conflicting pairwise alignments.

Note that the tree used here represents each sequence as a vertex, which is different from the phylogenetic tree (Morrison [16]) or the guide tree (Thompson et al. [27]) used in traditional progressive approaches in which sequences are represented only at the leaves. As a result, only alignments between sequences are needed in the preserving alignment formulation, which is similar to the comparisons made between sequences during the greedy extension step of the multiple alignment algorithm of Taylor [24] and Taylor [25], while progressive approaches need to consider alignments between alignments (Altschul and Lipman [1]). However, it is possible to make use of a phylogenetic tree when one is given. Instead of using pairwise alignment scores, we can use the distances between sequences

s_i and s_j on the phylogenetic tree to compute an optimum spanning tree. In this case, we only need to compute $k-1$ pairwise alignments along the spanning tree.

3   Exact Algorithm

For simplicity of analysis, assume that each of the input sequences is of the same length n. Given a tree T and a pairwise alignment P_{ij} for each edge (i,j) of T, we give an algorithm to solve the shortest preserving multiple alignment problem in linear time by two successive graph reductions. Gusfield [9] gave an algorithm to solve the problem in the important special case when the given tree is a star. We first consider preserving only the matches and mismatches instead of entire pairwise alignments. We will show that this strategy also preserves the indel columns under normal situations. Let s_{ij} be the letter at the jth position of sequence s_i. The first reduction constructs an undirected graph $G = (V, E)$ as follows (see Fig. 1(a)–(d)):

Let $V = \{v_{ij}\}$, where v_{ij} represents the jth position of sequence s_i, and $E = \{\{v_{ip}, v_{jq}\} \mid (i,j) \in T$ and (s_{ip}, s_{jq}) is a match or a mismatch column in $P_{ij}\}$.

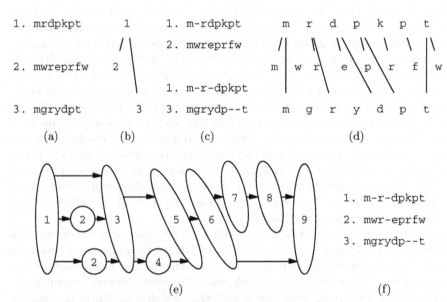

Fig. 1. Illustration of the exact algorithm. (a) Set of sequences S. (b) Tree T. (c) Pairwise alignments P_{12} and P_{13}. (d) Undirected graph G constructed from S, T, P_{12} and P_{13}. (e) Directed graph G' (transitively reduced) constructed from G by taking connected components in G as vertices. Labels of vertices in G' are assigned by the topological partial ordering algorithm. (f) Shortest preserving alignment by interpreting labels as columns

Intuitively, E contains all the match and mismatch columns within the pairwise alignments along the edges of T, and thus specifies exactly all the preservation constraints. The observation below follows directly from T being a tree and P_{ij} being pairwise alignments.

Proposition 1. *Each connected component C in G is a tree and contains at most one vertex from each sequence s_i.*

To obtain a preserving multiple alignment, letters within each connected component in G must be put into the same column. On the other hand, we are free to put two different connected components in the same column as long as they do not contain vertices from the same sequence. Also, when assigning components to different columns to obtain a multiple alignment, the order of the letters within each sequence must be maintained. To represent these constraints precisely, the second reduction constructs a directed graph $G' = (V', E')$ from G as follows (see Fig. 1(e) for a transitively reduced version of G'):

Let V' be the set of all connected components C in G and let $s(C)$ be the set of sequences that the vertices in C reside. Connect a component C_1 to another component C_2 by a directed edge in E' if $s(C_1) \cap s(C_2) \neq \phi$ and for every sequence $s_i \in s(C_1) \cap s(C_2)$ shared by C_1 and C_2, the vertex v_{ip} in C_1 appears before the vertex v_{iq} in C_2 (i.e., $p < q$).   (1)

Note that two connected components that contain vertices from the same sequence are strictly ordered and thus will be connected by an edge (in one of the directions), since if there are two vertices v_{ip} and v_{jq} in C_1 and two vertices v_{ir} and v_{js} in C_2 with $p < r$, then we must have $q < s$. The reasoning is as follows. Let $v_{ip} = u_1, \ldots, u_t = v_{jq}$ be the unique path between v_{ip} and v_{jq} in C_1 and $v_{ir} = w_1, \ldots, w_t = v_{js}$ be the unique path between v_{ir} and v_{js} in C_2. For $1 \leq l < t$, since u_l and w_l are on the same sequence, the adjacent pairs (u_l, u_{l+1}) and (w_l, w_{l+1}) represent two match or mismatch columns within one single pairwise alignment. The fact that $p < r$ and $q \geq s$ contradicts with these being columns in the pairwise alignments. A more elaborate argument gives the following.

Proposition 2. *G' is a directed acyclic graph.*

Proof. Let $C_1, \ldots, C_t$ be a cycle in G'. Then there exists sequences $s_{i_1}, \ldots, s_{i_t}$ such that $v_{i_1 p_1}$ is in C_1 and $v_{i_1 q_1}$ is in C_2 with $p_1 < q_1$, $v_{i_2 p_2}$ is in C_2 and $v_{i_2 q_2}$ is in C_3 with $p_2 < q_2$, and so on, until finally, $v_{i_t p_t}$ is in C_t and $v_{i_t q_t}$ is in C_1 with $p_t < q_t$. Between these vertices, there is a unique path $v_{i_1 q_1}, \ldots, v_{i_2 p_2}$ on C_2, $v_{i_2 q_2}, \ldots, v_{i_3 p_3}$ on C_3, and so on, until finally, $v_{i_t q_t}, \ldots, v_{i_1 p_1}$ on C_1. Thus the cycle can be represented by the path $v_{i_1 p_1} \xrightarrow{j} v_{i_1 q_1} \xrightarrow{w} v_{i_2 p_2} \xrightarrow{j} v_{i_2 q_2} \xrightarrow{w} \cdots \xrightarrow{w} v_{i_t p_t} \xrightarrow{j} v_{i_t q_t} \xrightarrow{w} v_{i_1 p_1}$, where $\xrightarrow{j}$ denotes a jump to a later connected component on the same sequence, and $\xrightarrow{w}$ denotes walking along the tree edges in T within a connected component which visits each sequence at most once. On this path, whenever a sequence s is visited again, the position of visit on s must increase, since a jump increases the position of visit on the same sequence from p_l to q_l, at least one jump has to occur before s is visited again, and whenever a walk moves from s to another sequence t along a tree edge in T, the only way to return to s is through t along the same edge in T. Walking along T this way with no increase in the position of visit on s when s is revisited contradicts the given pairwise alignments. In particular, this is true for sequence s_{i_1}, a contradiction to the above cycle which keeps the position of visit on s_{i_1} at p_1.   □

Since the primary purpose of G' is to specify ordering constraints, a transitively reduced version of G' suffices (see Fig. 1(e)). Instead of performing the transitive reduction step, such a graph G' can be obtained directly from G by requiring further that there exists a sequence s_i such that $p + 1 = q$ in (1). This reduces the number of edges in G' substantially and we will be using this definition in what follows. The above results suggest that a multiple alignment can be obtained by finding a topological partial ordering in G'.

Definition 1. *A topological partial ordering of a directed acyclic graph $G' = (V', E')$ is an assignment of an integer label $l(v)$ to each vertex $v \in V'$ such that for each edge $(u, v) \in E'$, $l(u) < l(v)$.*

Since it is possible that two vertices are assigned the same label, the result is not necessarily a total order (a total order corresponds to a topological sorting (Knuth [13])). Without loss of generality, assume that the labels are consecutive integers from 1 to m. From a given graph G', there are many ways to realize such an ordering. For each fixed ordering, a multiple alignment can be obtained by putting each letter within a connected component C in G (C is a vertex in G') in column $l(C)$ and filling other unassigned spaces by gap characters (see Fig. 1(f)). The set of all possible ways to do this represents the solution space of all preserving alignments.

Proposition 3. *Each topological partial ordering of G' specifies a multiple alignment preserving all matches and mismatches in the $k - 1$ pairwise alignments on T.*

By making additional assumptions, it is possible to ensure that the resulting multiple alignment preserves the given pairwise alignments entirely, which includes the indel columns in addition to the match and the mismatch columns, without requiring an algorithm change. The observation below follows directly from the constraints imposed on the placement of gap characters between two match or mismatch columns that must be preserved and are separated only by indel columns.

Proposition 4. *If for each pairwise alignment P_{ij} on T, there does not exist two adjacent indel columns (without match or mismatch columns in between) such that the gap character is on sequence s_i in column l and on sequence s_j in column $l-1$ or $l+1$, then each topological partial ordering G' specifies a multiple alignment preserving the $k - 1$ pairwise alignments on T.*

It is very rare to be given pairwise alignments that violate the condition given in Proposition 4, and thus in most cases the resulting multiple alignment also preserves the given pairwise alignments entirely. In the other direction, one can relax the constraints by requiring only the match columns (or the columns with a positive score with respect to a given substitution matrix) to be preserved, which can be achieved by allowing only edges representing these columns to be added to G. Computational results show that simply finding the shortest preserving multiple alignment in this case does not give very good performance since the flexibility in the placements of the resulting smaller connected components in G becomes excessive. The following observation completes the reduction.

Proposition 5. *A topological partial ordering of G' that uses the smallest number of labels specifies a shortest preserving multiple alignment.*

Since edges in G correspond to match or mismatch columns in the given $k-1$ pairwise alignments along the given tree T and each pairwise alignment is of length $O(n)$, there are $O(kn)$ vertices and edges in G. Thus there are $O(kn)$ connected components in G which make the vertices in G'. These connected components can be obtained in $O(kn)$ time by a depth-first search on G. In the simplest case, each connected component in G is of size one (which represents one position on a single sequence), and edges in G' are constructed between components that represent adjacent positions within the same sequence, resulting in a total of $O(kn)$ edges in G'. The graph G' in any other case with larger connected components can be obtained from this simplest case by merging the corresponding vertices and collapsing each resulting multi-edge into a single edge, and thus the number of edges in G' is $O(kn)$ in all cases. To find a topological partial ordering that uses the smallest number of labels in G', an algorithm very similar to the standard topological sorting algorithm (Knuth [13]) can be used: initially, all vertices are unmarked. Repeatedly find an unmarked vertex v with all its incoming vertices marked. Set the label of v to be one plus the maximum label among all its incoming vertices and mark v (see Fig. 1(f)). If a count is kept in each vertex representing the number of remaining unmarked incoming vertices, the algorithm can be implemented in time linear in the number of edges in G'. Thus, with appropriate data structures, the overall time complexity of the entire procedure is $O(kn)$, which is linear in the input size. If it is not important to obtain a totally ordered multiple alignment, it is possible to return the graph G' directly as a partially ordered multiple alignment, which is similar in concept but different in structure to the notion of partial order multiple alignment proposed in Lee et al. [14]. In this case, there is no need to define any objective function.

4   Performance

We evaluate the accuracy of the preserving alignment algorithm (PSAlign) on three sets of benchmark multiple alignments: BAliBASE from Thompson et al. [27], PREFAB from Edgar [5], and SABmark from Van Walle et al. [28]. We compare our performance to TCoffee (Notredame et al. [18]) and PROBCONS

(Do et al. [4]), which are currently considered to be among the best multiple alignment algorithms. To make fair comparisons, we compare our performance (PSAlign[TCoffee]) with TCoffee when pairwise alignments from TCoffee are used and we compare our performance (PSAlign[PROBCONS]) with PROB-CONS when pairwise alignments from PROBCONS are used. For TCoffee, we use pairwise alignments computed from the extended library that have incorporated consistency information from other sequences (Notredame et al. [18]). For PROBCONS, we use pairwise alignments computed after consistency transformation (Do et al. [4]). All these pairwise alignments incorporate consistency information from other sequences and it has been shown that this significantly improves the quality of the pairwise alignments with respect to the overall consistency. We then compute an optimum spanning tree from these consistency-based pairwise alignments using pairwise scores given by the two algorithms (normalized by the length of each pairwise alignment) and apply the preserving alignment step. Since our goal is to show that the heuristic progressive step of these approaches can be replaced by the exact preserving alignment step, we compare with a variant of PROBCONS with no iterative refinements.

Following Thompson et al. [27], two score measures are used to evaluate the accuracy of each algorithm in finding the core blocks in BAliBASE (which are annotated regions that can be reliably aligned): the sum-of-pairs score (SPS) measures how well an algorithm can align pairs of residues within the same column correctly, while the column score (CS) measures how well an algorithm can align entire columns correctly. For PREFAB, we follow Edgar [5] and use the Q score, which has the same meaning as SPS used in BAliBASE. For SABmark, we define the Q score for each test case as the average Q score over all pairs of reference sequences. For both PREFAB and SABmark, the reference alignments are based on pairwise comparisons and thus the CS score is not applicable. For each test set, we compare average accuracy over meaningful subsets and use the Wilcoxon matched-pairs signed-ranks test (Wilcoxon [30]) to check if there are significant performance differences with $p = 0.05$ as cutoff. Note that in the preserving alignment computation, the shortest solution is not necessarily unique and we simply report an arbitrary one.

Tables 1, 2 and 3 show performance comparisons of the various algorithms on BAliBASE, PREFAB and SABmark respectively. A general trend was that PROBCONS($ir = 0$) tends to perform better than TCoffee, whether PSAlign is used or not. When we consider the ability of PSAlign[TCoffee] to replace TCof-fee and the ability of PSAlign[PROBCONS] to replace PROBCONS($ir = 0$), PSAlign was a much better replacement when used in conjunction with TCoffee than with PROBCONS and the only case where PSAlign[TCoffee] is worse than TCoffee was in reference 4 of BAliBASE. Also, PSAlign was a better replace-ment when used on SABmark than on BAliBASE, but PSAlign[PROBCONS] was not an adequate replacement when used on PREFAB.

On BAliBASE, the use of PSAlign improved accuracy on reference 1V1 in all cases but this was offset by worse accuracy on references 3 and 4. Al-though there was no noticeable difference in overall accuracy, the Wilcoxon

Table 1. Average SPS and CS scores (in %) on BAliBASE. Reference 1 is further subdivided into three subsets: V1 ($< 25\%$ identity), V2 (20%–40% identity) and V3 ($> 35\%$ identity). Comparisons are made between TCoffee and PSAlign utilizing TCoffee pairwise alignments (PSAlign[TCoffee]) and between PROBCONS and PSAlign utilizing PROBCONS pairwise alignments (PSAlign[PROBCONS]). No iterative refinements are performed for PROBCONS ($ir = 0$). Default parameters are used otherwise

SPS	1V1 1V2 1V3	1 (Overall)	2	3	4	5	Overall
TCoffee	63.4 95.0 98.5	87.3	88.5	77.4	91.9	95.8	87.8
PSAlign[TCoffee]	66.3 94.9 98.4	88.1	89.6	78.6	87.3	97.2	88.2
PROBCONS($ir = 0$)	68.8 96.1 98.6	89.3	91.6	84.2	88.0	98.1	89.9
PSAlign[PROBCONS]	70.6 96.5 98.6	90.0	91.1	81.7	88.0	97.3	89.9
CS	1V1 1V2 1V3	1 (Overall)	2	3	4	5	Overall
TCoffee	41.9 90.5 96.9	79.0	43.0	51.3	74.9	90.3	71.4
PSAlign[TCoffee]	46.8 90.5 96.6	80.3	44.9	54.6	64.5	89.5	71.8
PROBCONS($ir = 0$)	50.1 92.4 97.0	82.1	53.3	62.5	66.9	91.9	75.3
PSAlign[PROBCONS]	54.1 93.4 97.2	83.7	50.5	54.7	66.2	92.1	75.0

Table 2. Average Q scores (in %) on PREFAB. Each subset includes all structure pairs with identity within the specified range

	0%–20%	20%–40%	40%–70%	70%–100%	Overall
TCoffee	50.0	85.3	97.2	98.2	63.8
PSAlign[TCoffee]	51.3	86.4	98.0	99.1	65.0
PROBCONS($ir = 0$)	53.7	87.9	97.8	98.7	66.9
PSAlign[PROBCONS]	50.9	86.2	96.1	97.1	64.4

Table 3. Average Q scores (in %) on SABmark. The FP variant of the two subsets includes false positive sequences

	Superfamily	Superfamily-FP	Twilight	Twilight-FP	Overall
TCoffee	52.9	45.6	23.7	17.0	39.8
PSAlign[TCoffee]	54.8	54.1	25.8	25.4	45.0
PROBCONS($ir = 0$)	56.9	53.6	28.4	23.8	45.7
PSAlign[PROBCONS]	56.1	53.8	28.0	26.2	45.8

matched-pairs test revealed that PROBCONS($ir = 0$) performed better than PSAlign[PROBCONS] in the SPS score with $p = 0.02$. The differences in all the other cases on the overall accuracy were insignificant with TCoffee or the CS score. We did not apply the Wilcoxon test to any of the subsets of BAliBASE due to their small sizes. On PREFAB, PSAlign[TCoffee] had a better accuracy than TCoffee in all five categories and these improvements were significant for the subsets with 0% to 20% identity ($p < 0.002$), with 70% to 100% identity ($p < 0.0001$), and for the entire set ($p < 0.0001$). On the other hand, PROBCONS($ir = 0$) performed significantly better than PSAlign[PROBCONS] in all five categories ($p < 0.001$ for the subset with 40% to 70% identity and $p < 0.0001$ in all other categories). Surprisingly, PSAlign[TCoffee] had a better accuracy than PSAlign[PROBCONS] in all five categories. On SABmark, PSAlign[TCoffee] showed highly significant improvements over TCoffee with $p < 0.0001$ in all five categories. However, on the Superfamily and Twilight subsets, PROBCONS($ir = 0$) performed significantly better than PSAlign[PROBCONS] ($p < 0.0001$ for the Superfamily subset and $p < 0.01$ for the Twilight subset). On the other hand, the situation was reversed on the Twilight-FP subset when PSAlign[PROBCONS] performed significantly better than PROBCONS($ir = 0$) with $p < 0.0001$. Other differences with PROBCONS($ir = 0$), including the overall accuracy, were insignificant. One advantage of PSAlign is that it had a much smaller accuracy decrease when either the Superfamily or Twilight subset is replaced by its FP variant with false positive sequences.

Overall, PSAlign[TCoffee] performed at least as well as TCoffee on BAliBASE and was much better than TCoffee on PREFAB and SABmark. When compared with PROBCONS($ir = 0$), PSAlign[PROBCONS] achieved a similar accuracy on BAliBASE and SABmark, but did not perform as well on PREFAB. These results did not lead to a conclusive statement that shows that using PSAlign has a definite advantage or disadvantage, and thus it is hard to predict whether there will be significant accuracy differences if we replace the heuristic progressive step of some given multiple alignment algorithm by the exact preserving alignment step. Nevertheless, the most important advantage of the preserving alignment formulation is that we are certain that we can solve the problem in polynomial time without using a heuristic. Since the time complexity is dominated by the computations of the pairwise alignments, the preserving alignment step does not add much to the running time. What we actually observe was at most a two times slowdown when PSAlign was used to replace TCoffee or PROBCONS($ir = 0$), due to the need to compute all consistency-based pairwise alignments to obtain the optimum spanning tree.

5   Discussion

The proposed multiple alignment formulation divides the multiple alignment problem into two subproblems. The first subproblem requires the computation of pairwise alignments and a tree, which can be defined systematically so that optimal solutions can be computed in polynomial time. For example, one can use a technique similar to that used in Notredame et al. [18] to compute consistency-based pairwise alignments based on comparisons of three sequences so that each of them can be computed in $O(kn^2)$ time. It is especially important to obtain high quality pairwise alignments in this stage, since we found that good accuracy cannot be obtained when simple non-consistency-based pairwise alignments are used. This was confirmed by a much bigger decrease in the performance of PSAlign[PROBCONS] as compared to PROBCONS($ir = 0$) when the consistency transformation in PROBCONS was disabled. The second stage computes a shortest preserving multiple alignment from this information, which can be used to replace the progressive step of any approach in which unified pairwise alignments are available before the progressive step, or as a second step to construct multiple alignments for algorithms that only produce a set of pairwise alignments from the given sequences (Heger et al. [10]; Van Walle et al. [10]), without requiring additional parameters.

The graph-theoretic technique employed allows further extensions to consider more general models of pairwise similarity. In its full generality, all we need from each pairwise comparison is an ordered list of non-intersecting connections (representing matches or mismatches) that reflect significant pairwise similarities. With these inputs, the preserving alignment approach naturally returns either local or incomplete multiple alignments. To further improve accuracy, it is possible to consider formulations other than finding the shortest solution, although many of these objective functions may be intractable to optimize. One possible strategy is to employ various heuristics to find a preserving alignment from G' that tries to assign related connected components in G to the same column as much as possible. Other directions include improving the quality of the pairwise alignments and devising strategies to perform iterative refinements (Gotoh [8]; Do et al. [4]; Edgar [5]) after the preserving multiple alignment is obtained.

Acknowledgments

This work was supported by NSF grants CCR-0311590 and DBI-0421815.

References

Altschul, S.F., Lipman, D.J.: Trees, stars, and multiple biological sequence alignment. SIAM J. Appl. Math. **49** (1989) 197–209

Carillo, H., Lipman, D.: The multiple sequence alignment problem in biology. SIAM J. Appl. Math. **48** (1988) 1073–1082

Cormen, T.H., Leiserson, C.E., Rivest, R.L., Stein, C.: Introduction to Algorithms, Second Edition. The MIT Press (2001)

Do, C., Brudno, M., Batzoglou, S.: PROBCONS: probabilistic consistency-based multiple alignment of amino acid sequences. Proc. 12th Int. Conf. Intelligent Systems Mol. Biol./3rd European Conf. Comp. Biol. (ISMB/ECCB'2004)

Edgar, R.C.: MUSCLE: multiple sequence alignment with high accuracy and high throughput. Nucleic Acids Res. **32** (2004) 1792–1797

Feng, D., Doolittle, R.: Progressive sequence alignment as a prerequisite to correct phylogenetic trees. J. Mol. Evol. **25** (1987) 351–360

Gotoh, O.: Consistency of optimal sequence alignments. Bull. Math. Biol. **52** (1990) 509–525

Gotoh, O.: Significant improvement in accuracy of multiple protein sequence alignments by iterative refinement as assessed by reference to structural alignments. J. Mol. Biol. **264** (1996) 823–838

Gusfield, D.: Efficient methods for multiple sequence alignment with guaranteed error bounds. Bull. Math. Biol. **55** (1993) 141–154

Heger, A., Lappe, M., Holm, L.: Accurate detection of very sparse sequence motifs. Proc. 7th Ann. Int. Conf. Res. Comp. Mol. Biol. (RECOMB'2003) 139–147

Just, W.: Computational complexity of multiple sequence alignment with SP-score. J. Comp. Biol. **8** (2001) 615–623

Kececioglu, J.D.: The maximum weight trace problem in multiple sequence alignment. Lect. Notes Comp. Sci. **684** (1993) 106–119

Knuth, D.E.: The Art of Computer Programming, Volume 1: Fundamental Algorithms, Third Edition. Addison-Wesley (1997)

Lee, C., Grasso, C., Sharlow, M.F.: Multiple sequence alignment using partial order graphs. Bioinformatics **18** (2002) 452–464

Morgenstern, B., Dress, A., Werner, T.: Multiple DNA and protein sequence alignment based on segment-to-segment comparison. Proc. Natl. Acad. Sci. USA **93** (1996) 12098–12103

Morrison, D.A.: Phylogenetic tree-building. Int. J. Parasitology **26** (1996) 589–617

Needleman, S.B., Wunsch, C.D.: A general method applicable to the search for similarities in the amino acid sequence of two proteins. J. Mol. Biol. **48** (1970) 443–453

Notredame, C., Higgins, D.G., Heringa, J.: T-Coffee: a novel method for fast and accurate multiple sequence alignment. J. Mol. Biol. **302** (2000) 205–217

Notredame, C., Holm, L., Higgins, D.G.: COFFEE: an objective function for multiple sequence alignments. Bioinformatics **14** (1998) 407–422

O'Sullivan, O., Suhre, K., Abergel, C., Higgins, D.G., Notredame, C.: 3DCoffee: combining protein sequences and structures within multiple sequence alignments. J. Mol. Biol. **340** (2004) 385–395

Pevzner, P.A.: Computational Molecular Biology: an Algorithmic Approach. The MIT press (2000)

Smith, T.F., Waterman, M.S.: Identification of common molecular subsequences. J. Mol. Biol. **147** (1981) 195–197

Stoye, J.: Multiple sequence alignment with the divide-and-conquer method. Gene **211** (1998) GC45–56

Taylor, W.R.: Multiple sequence alignment by a pairwise algorithm. Comp. Appl. Biosci. **3** (1987) 81–87

Taylor, W.R.: A flexible method to align large numbers of biological sequences. J. Mol. Evol. **28** (1988) 161–169

Thompson, J.D., Higgins, D.G., Gibson, T.J.: CLUSTAL W: improving the sensitivity of progressive multiple sequence alignment through sequence weighting, position specific gap penalties and weight matrix choice. Nucleic Acids Res. **22** (1994) 4673–4680

Thompson, J.D., Plewniak, F., Poch, O.: A comprehensive comparison of multiple sequence alignment programs. Nucleic Acids Res. **27** (1999) 2682–2690

Van Walle, I., Lasters, I., Wyns, L.: Align-m — a new algorithm for multiple alignment of highly divergent sequences. Bioinformatics **20** (2004) 1428–1435

Vingron, M., Argos, P.: Motif recognition and alignment for many sequences by comparison of dot-matrices. J. Mol. Biol. **218** (1991) 33–43

Wilcoxon, F.: (1947) Probability tables for individual comparisons by ranking methods. Biometrics **3** (1947) 119–122

A Fundamental Decomposition Theory for Phylogenetic Networks and Incompatible Characters

Dan Gusfield[1] and Vikas Bansal[2]

[1] Department of Computer Science,
University of California, Davis
gusfield@cs.ucdavis.edu
[2] Department of Computer Science and Engineering,
University of California, San Diego
vibansal@cs.ucsd.edu

Abstract. Phylogenetic networks are models of evolution that go beyond trees, allowing biological operations that are not consistent with tree-like evolution. One of the most important of these biological operations is recombination between two sequences (homologous chromosomes). The algorithmic problem of reconstructing a history of recombinations, or determining the minimum number of recombinations needed, has been studied in a number of papers [10, 11, 12, 23, 24, 25, 16, 13, 14, 6, 9, 8, 18, 19, 15, 1]. In [9, 6, 10, 8, 1] we introduced and used "conflict graphs" and "incompatibility graphs" to compute lower bounds on the minimum number of recombinations needed, and to efficiently solve constrained cases of the minimization problem. In those results, the non-trivial connected components of the graphs were the key features that were used.

In this paper we more fully develop the structural importance of nontrivial connected components of the incompatibility graph, to establish a fundamental decomposition theorem about phylogenetic networks. The result applies to phylogenetic networks where cycles reflect biological phenomena other than recombination, such as recurrent mutation and lateral gene transfer. The proof leads to an efficient $O(nm^2)$ time algorithm to find the underlying maximal tree structure defined by the decomposition, for any set of n sequences of length m each. An implementation of that algorithm is available. We also report on progress towards resolving the major open problem in this area.

Keywords: Molecular Evolution, Phylogenetic Networks, Perfect Phylogeny, Ancestral Recombination Graph, Recombination, Gene-Conversion, SNP.

1 Introduction to Phylogenetic Networks and Problems

With the growth of genomic data, much of which does not fit ideal evolutionary-tree models, and the increasing appreciation of the genomic role of such phenomena as recombination, recurrent and back mutation, horizontal gene transfer,

S. Miyano et al. (Eds.): RECOMB 2005, LNBI 3500, pp. 217–232, 2005.
© Springer-Verlag Berlin Heidelberg 2005

cross-species hybridization, gene conversion, and mobile genetic elements, there is greater need to understand the algorithmics and combinatorics of phylogenetic networks on which extant sequences were derived [20]. Recombination is particularly important in deriving chimeric sequences in a population of individuals of the same species. Recombination in populations is the key element underlying techniques that are widely hoped to locate genes influencing genetic diseases.

Formal Definition of a Phylogenetic Network.

There are four components needed to specify a phylogenetic network that allows multiple-crossover recombination (see Figure 1).

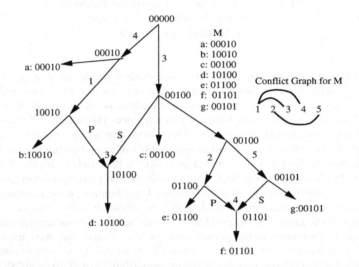

Fig. 1. A phylogenetic network that derives the set of sequences M. The two recombinations shown are single-crossover recombinations, and the crossover point is written above the recombination node. In general the recombinant sequence exiting a recombination node may be on a path that reaches another recombination node, rather than going directly to a leaf. Also, in general, not every sequence labeling a node also labels a leaf

A phylogenetic network N is built on a directed acyclic graph containing exactly one node (the root) with no incoming edges, a set of internal nodes that have both incoming and outgoing edges, and exactly n nodes (the leaves) with no outgoing edges. Each node other than the root has either one or two incoming edges. A node x with two incoming edges is called a *recombination* node.

Each integer (site) from 1 to m is assigned to exactly one edge in N, but for simplicity of exposition, none are assigned to any edge entering a recombination node. There may be additional edges that are assigned no integers. We use the terms "column" and "site" interchangeably.

Each node in N is labeled by an m-length binary sequence, starting with the root node which is labeled with some sequence R, called the "root" or the "ancestral" sequence. Since N is acyclic, the nodes in N can be topologically sorted into a list, where every node occurs in the list only after its parent(s). Using that list, we can constructively define the sequences that label the non-root nodes, in order of their appearance in the list, as follows:

a) For a non-recombination node v, let e be the single edge coming into v. The sequence labeling v is obtained from the sequence labeling v's parent by changing the state (from 0 to 1, or from 1 to 0) of the value at site i, for every integer i on edge e. This corresponds to a mutation at site i occurring on edge e.

b) For the recombination at node x, let Z and Z' denote the two m-length sequences labeling the parents of x. Then the "recombinant sequence" X labeling x can be any m-length sequence provided that at every site i, the character in X is equal to the character at site i in (at least) one of Z or Z'.

The "event" that creates X from Z and Z' is called a "multiple-crossover recombination". To fully specify the event, we must specify for every position i whether the character in X "comes from" Z or Z'. This specification is forced when the characters in Z and Z' at position i are different. When they are the same, a choice must be specified. For a given event, we say that a *crossover* occurs at position i if the characters at positions $i-1$ and i come from different parents. It is easy to determine the minimum number of crossovers needed to create X by a recombination of Z and Z'.

The sequences labeling the leaves of N are the extant sequences, i.e., the sequences that can be observed. We say that an (n, m)-phylogenetic network N *derives (or explains)* a set of n sequences M if and only if each sequence in M labels one of the leaves of N.

With these definitions, the classic "perfect phylogeny" [4] is a phylogenetic network without any recombinations. That is, each site mutates exactly once in the evolutionary history, and these is no recombination between sequences.

There are two restricted forms of recombination that are of particular biological interest. One is where X is formed from a *prefix* of one of its parent sequences (Z or Z') followed by a *suffix* of the other parent sequence. This is called "single-crossover recombination" since it uses exactly one crossover, and it is the definition of recombination used in [9, 8]. The other case is when X is formed from a prefix of one parent sequence, followed by an internal segment of the other parent sequence, followed by a suffix of the first parent sequence. This is a two-crossover recombination, which occurs during "gene-conversion" in meiosis, and during some forms of "lateral gene-transfer". Multiple-crossover recombination allows the modeling of complex biological phenomena, and hence the main result in this paper applies to many causes of incompatibility besides recombination.

What we have defined here as a phylogenetic network with single-crossover recombination is the digraph part of the stochastic process called an "ancestral recombination graph (ARG)" in the population genetics literature.

In the context of meiotic recombination, the assumption that the sequences are binary is motivated today by the importance of SNP data, where each site can take on at most two states (alleles) [2]. In the context of macroevolution, complex evolutionary characters are usually considered to be binary (either present or absent)[3].

Rooted and Root-Unknown Problems. Problems of reconstructing phylogenetic networks, given an input set of binary sequences M, can be addressed either in the rooted case, or the root-unknown case. In the *rooted* phylogenetic network problem, a required root or ancestral sequence R for the network is specified in advance. In the *root-unknown* phylogenetic network problem, no ancestral sequence is specified in advance, and the algorithm must select an ancestral sequence.

2   A Fundamental Decomposition Theory for Phylogenetic Networks and Incompatible Characters

In this section we define and derive the main result of this paper, that for any input M, there always is a phylogenetic network of an important, natural structure. We believe this to be a very fundamental fact about phylogenetic networks that will have many applications. We now begin the needed definitions that lead to the statement of the main result.

In a phylogenetic network N, let w be a node that has two paths out of it that meet at a recombination node x. Those two paths together define a "recombination cycle" Q. Node w is called the "coalescent node" of Q, and x is the recombination node of Q. In Figure 1, the nodes labeled 00000 and 00100 are coalescent nodes of two different recombination cycles.

If a recombination cycle in a phylogenetic network N is not isolated (a "gall" in the terminology of [9]), it shares at least one edge with some other recombination cycle. We can add another cycle to that blob if the new cycle shares an edge with at least one cycle already on the blob. Continuing in this way, we ultimately get a maximal set of recombination cycles in N that form a single connected subgraph of N, and each cycle shares at least one edge with some other cycle in the set. We call such a maximal set of cycles a "blob".

Clearly, because of maximality, the blobs in a phylogenetic network N are well-defined. Moreover, if we contract each blob in N to a single point, the resulting network is a directed tree T'. This follows because if the resulting graph had a cycle (in the underlying undirected graph) that cycle would correspond to a recombination cycle which should have been contracted. We call T' a "tree of blobs" or a "blobbed tree". So every phylogenetic network N can be viewed as a blobbed tree. The edges in T' are called "tree edges" of N.

2.1   The Main Tools

The main tools that we used in [9, 10, 1] and other papers were two graphs representing "incompatibilities" and "conflicts" between sites. We introduce these graphs here.

Given a set of binary sequences M, two columns i and j in M are said to be *incompatible* if and only if there are four rows in M where columns i and j contain all four of the ordered pairs 0,1; 1,0; 1,1; and 0,0. For example, in Figure 1 columns 1 and 3 of M are incompatible because of rows a, b, c, d. The test for the existence of all four pairs is called the "four-gamete test" in the population genetics literature. A site that is not involved in any incompatibility is called a "compatible site".

Given a sequence S, two columns i and j in M are said to *conflict (relative to S)* if and only if columns i and j contain all three of the above four pairs that differ from the i, j pair in S.

The classic Perfect Phylogeny Theorem (in the terminology of this paper) is that there is a root-unknown phylogenetic network without any recombination cycles, that derives a set of binary sequences M, if and only if there is no incompatible pair of columns. Similarly, there is a phylogenetic network with ancestral sequence S, without any recombination cycles, that derives M, if and only if there is no pair of columns that conflict relative to S. For one exposition of this classic result, see [5].

Incompatibility and Conflict Graphs

We define the "incompatibility graph" $G(M)$ for M as a graph containing one node for each column (site) in M, and an edge connecting two nodes i and j if and only if columns i and j are incompatible. Similarly, given a sequence S, we define the "conflict graph" $G_S(M)$ for M (relative to S) as a graph containing one node for each column in M, and an edge connecting two nodes i and j if and only if columns i and j conflict relative to S. Figure 1 shows the conflict graph relative to the all-zero sequence S. This conflict graph is also the incompatibility graph for M.

A "connected component" (or "component" for short), C, of a graph is a maximal subgraph such that for any pair of nodes in C there is at least one path between those nodes in the subgraph. A "trivial" component has only one node, and no edges. The conflict graph in Figure 1 has two components.

2.2   Main Result

Theorem 1. *Let $G(M)$ be the incompatibility graph for M. Then, there is a phylogenetic network N that derives M where every blob contains all and only the sites of a single non-trivial connected component of $G(M)$, and every compatible site is on a tree edge of N.*

Stated another way, for any input M, there is a blobbed-tree that derives M, where the blobs are in one-one correspondence with the non-trivial connected components of $G(M)$, and if B is the blob corresponding to component C, then B contains all and only the sites in C. We call a network "fully-decomposed" if it has the structure specified in Theorem 1.

Theorem 1 is an extension of the stronger theorem proved in [9] about galled-trees. In the case of galled-trees, *every* reduced galled-tree for M *must* be fully-decomposed. A galled-tree is "reduced" if every recombination cycle contains some incompatible sites. When there is a galled-tree for M, there is a reduced galled-tree for M, and the program galledtree.pl will produce one (see Section 3).

There is an analogous theorem to Theorem 1 in the case that the ancestral sequence S is known in advance. In that case, there is a phylogenetic network N that derives M, with ancestral sequence S, where the blobs in N are in one-one correspondence with the non-trivial connected components of $G_S(M)$, and any non-conflicting site is on a tree edge of N.

3   Proof of Theorem 1

Let C and C' be two connected components in the incompatibility graph $G(M)$. Note that either C or C' or both may be a trivial connected component, i.e., consist of only a single node.

For any $i \in C, i' \in C'$ let $(X, \overline{X})$ and $(Y, \overline{Y})$ be the respective bipartitions (of the rows of M), associated with sites i and i'. The two bipartitions cannot be identical, for otherwise sites i and i' would have exactly the same incompatibilities and so be in the same connected component. Each of the four subsets $X, \overline{X}, Y, \overline{Y}$ is called a "class" of the bipartition it is part of. Sites i and i' are not incompatible, so one class of the i bipartition must strictly contain one class of the i' bipartition, and the other class of the i' bipartition must strictly contain the other class of the i bipartition. Without loss of generality, suppose $X \supset Y$ and $\overline{Y} \supset \overline{X}$. We say that X is the "dominant" class of i, and $\overline{X}$ is the "dominated" class, with respect to the pair i, i'. Similarly, $\overline{Y}$ is the dominant class of i', and Y is the dominated class, with respect to the pair i, i'.

Lemma 1. *Let i, i', X, and Y be as above. Let j' be any site in C', and let $(Z, \overline{Z})$ be the bipartition associated with j'. Then, the dominant class of i with respect to the pair i, j' is the dominant class of i with respect to the pair i, i'. That is, either $X \supset Z$ or $X \supset \overline{Z}$.*

Proof. The Lemma is vacuously true if C' is a trivial connected component, so assume C' is non-trivial, and consider a site $k' \in C'$ that is incompatible with i'. Such a site k' must exist since C' is connected. Let $(W, \overline{W})$ be the bipartition defined by site k' If X is not dominant with respect to i, k', then $\overline{X}$ is dominant with respect to i, k', and so either $\overline{X} \supset W$ or $\overline{X} \supset \overline{W}$. Suppose that $\overline{X} \supset \overline{W}$, so $W \supset X$. But then $W \supset Y$ since $X \supset Y$, and so $Y \cap \overline{W} = \emptyset$, and i and k' can't be incompatible, which is a contradiction. Similarly, if $\overline{X} \supset W$, then $\overline{W} \supset X$, so $\overline{W} \supset Y$, and $W \cap \overline{Y} = \emptyset$, a contradiction. So the dominant class, X, with respect to i, i' is the dominant class with respect to i, k', where k' is any site that is incompatible with i'. The Lemma now follows by transitivity, because C' is a connected component, so from i' it is possible to reach any $j' \in C'$ by a series of incompatibility relations.   □

Lemma 1 establishes that for any $i \in C$, one class of i is dominant with respect to *all* sites in C', and symmetrically, for any $i' \in C'$ one class of i' is dominant with respect to *all* sites in C. So, with respect to the (C, C') pair of connected components, each site in $C \cup C'$ has a well-defined dominant class, and a well-defined dominated class.

Now return focus to the sequences in M and the sites in C and C'. For a site $i \in C$, the bipartition $(X, \overline{X})$ is encoded with 0's and 1's, where all the rows in X have one character at site i and all the rows in $\overline{X}$ have the other character at site i. So, with respect to the (C, C') pair of connected components, and a specific set of sequences M, each site in C has a well-defined *dominant character* (either 0 or 1). For example, in Figure 2, the dominant character is 0 in all sites except 3, where the dominant character is 1.

For $i \in C$, let $D(i)$ be the rows in the dominated class with respect to (C, C'), and similarly, for $i' \in C'$, let $D(i')$ be the rows in the dominated class with respect to (C, C'). Let $D[C, C']$ be the union of the rows in any dominated class of C, with respect to (C, C'). Similarly, let $D[C', C]$ be union of the rows in any dominated class of C', with respect to (C, C').

Let $M(C)$ and $M(C')$ be the sequences in M, restricted to the sites in C and C' respectively. Then Lemma 1 implies

Theorem 2. *Every row in $D[C, C']$ has the same sequence in $M(C')$. In particular, in each row of $D[C, C']$, every site $i' \in C'$ has the dominant character with respect to (C, C'). Similarly, every row in $D[C', C]$ has the same sequence in $M(C)$. In particular, in each row of $D[C', C]$, every site $i \in C$ has the dominant character with respect to (C, C').*

Given Theorem 2, we can define the *dominant sequence* in $M(C)$ with respect to (C, C') as the sequence in $M(C)$ where each site has the dominant character with respect to (C, C'). Similarly, we can define the dominant sequence in $M(C')$ with respect to (C, C').

Corollary 1. *Let C and C' be two connected components of $G(M)$. There is no row in M which contains both a non-dominant sequence in $M(C)$ and a non-dominant sequence in $M(C')$ with respect to (C, C').*

Figure 2 illustrates Lemma 1, Theorem 2 and Corollary 1. Note that a row can have the dominant sequence in $M(C)$ and the dominant sequence in $M(C')$. Row c in Figure 2 is an example of this.

Lemma 1, Theorem 2 and Corollary 1 establish a structure that exists in M, imposed by the partition of the columns of M by the connected components of

	1	3	4	2	5
a	0	0	1	0	0
b	1	0	1	0	0
d	1	1	0	0	0
c	0	1	0	0	0
e	0	1	0	1	0
f	0	1	0	1	1
g	0	1	0	0	1

Fig. 2. The sites in the two connected components from Figure 1. We denote the component with sites $\{1, 3, 4\}$ as C, and the component with sites $\{2, 5\}$ as C'. The dominant sequence for C is 010, and the dominant sequence for C' is 00. The rows in $D[C, C']$ are $\{a, b, d\}$, and the rows in $D[C', C]$ are $\{e, f, g\}$. Note that row c is in neither $D[C, C']$ nor $D[C', C]$, since row c has the dominant sequence in both its C and C' sides

$G(M)$, the incompatibility graph of M. We begin now to exploit that structure to prove the main theorem. We will create a new, binary, matrix MB from M and $G(M)$. Let C be a connected component of $G(M)$ and let $M(C)$ be the sequences in M restricted to the sites in C. Create one column in MB for each *distinct* sequence in $M(C)$. Each such new column, associated with a sequence $S \in M(C)$ say, encodes a bipartition of the rows of M, where one side of the bipartition contains all the rows that have sequence S in $M(C)$, and the other side of the bipartition contains the remaining rows. More specifically, and without loss of generality, in the new column we assign value 1 to each row which contains sequence S in $M(C)$, and assign value 0 to each row that does not. The new column defines a binary character derived from M and $G(M)$. Note that if C is a trivial connected-component, so it only contains one site, then MB will have two columns derived from that one site, but those columns define the same bipartition. That will cause no problems, and one can be removed for simplicity.

Matrix MB is defined by the columns described above, over the set of all connected components in $M(G)$. We call a character (column, site) of MB a "super-character". We want to use these super-characters to build a tree that will prove Theorem 1. We start by showing

Lemma 2. *No pair of super-characters are incompatible.*

Proof. Let p and q be super-characters in MB, and let $(P, \overline{P})$ and $(Q, \overline{Q})$ be the bipartitions associated with p and q. If p and q originate from the same connected component C in $G(M)$, then without loss of generality, the rows in P all have the same sequence in $M(C)$, and the rows in Q all have the same sequence in $M(C)$, and those two sequences are different. Hence, $P \cap Q = \emptyset$, and so p and q are not incompatible.

Now suppose p and q originate from two different connected components C and C' in $G(M)$. If p and q both originate from non-dominant sequences of C and C', then Corollary 1 guarantees that there is no row with $1, 1$ in columns p and q, and so p and q cannot be incompatible. Symmetrically, if p and q both originate from dominant sequences in C and C', then there is no row with $0, 0$ in columns p and q. If p originates from the dominant sequence of C and q originates from a non-dominant sequence of C', then there can be no $0, 1$ in columns p and q. The remaining case is symmetric.   □

Hence, by the Perfect Phylogeny Theorem, there is a *unique* perfect phylogeny $\overline{T}$ where each super-character labels an edge in $\overline{T}$, and each edge is labeled by one or more super-characters[1].

We now develop the structure of $\overline{T}$ to both complete the proof of Theorem 1, and to constructively show how to build a network N for M from $\overline{T}$. A "split of edge e" is defined as the bipartition of the leaves resulting from the removal of

---

[1] It is of independent interest to note that we have established that the super-characters defined by the connected components of $G(M)$ generalize the standard (tree) characters and play a role in the theory of phylogenetic networks, that tree characters play in the theory of phylogenetic trees.

edge e from $\overline{T}$. Note that all the splits of the edges in $\overline{T}$ are distinct. The removal of any edge e in $\overline{T}$ creates two connected subtrees, whose leaves correspond to the two classes of the split of edge e. If e is labeled by super-character C, we define the "1-side" of e as the subtree of $\overline{T} - e$ that contains the leaves for rows in MB that have value 1 for super-character C. The other side is called the "0-side" of the split.

Lemma 3. *In $\overline{T}$, there is a node v_C such that all the edges labeled by super-characters that originate from the same connected component C in $G(M)$ are incident with v_C. That is, these edges form a star around a single central node v_C. Further, v_C is on the 0-side of each split defined by every super-character that originates from C.*

Proof. First, the Lemma is trivialy true if C is a trivial component. Note, however, that any non-trivial connected component has at least four distinct super-characters. Consider such a connected component C and any three of its super-chars, and let e_1, e_2, e_3 be the three edges in $\overline{T}$ labeled with those super-chars. Note that every row in MB has value 1 in exactly one column of $MB(C)$, so every leaf of $\overline{T}$ is on the 1-side of exactly one edge labeled by a super-character that derives from C. Hence, no leaf in $\overline{T}$ can be on the 1-side of two of the edges e_1, e_2, e_3.

If e_1 and e_2 are incident with each other, sharing a node v, then there must be another edge incident with node v, and hence there must be a leaf l_v that is reachable from v without going through e_1 or e_2. If this were not true, then e_1 and e_2 would define the same splits in $\overline{T}$, which is not possible. If e_1 and e_2 are not incident with each other, then there is a unique shortest path P from an endpoint of e_1 to an endpoint of e_2. Clearly, path P does not contain edge e_1 or e_2. There must be a node v on P and a leaf l_v that is reachable from v via a path that does not go through e_1 or e_2. If this were not true, then again there would be two adjacent edges that define the same splits in $\overline{T}$.

Now we claim that node l_v must be on the 0-side of both e_1 and e_2. We have already established that it cannot be on the 1-side of both, since no leaf can be on the 1-side of two splits derived from the super-characters of C. However, suppose without loss of generality, that l_v is on the 1-side of e_1 and the 0-side of e_2. Then consider the endpoint u of e_2 that is on the 1-side of e_2, and consider a leaf l_u that is reachable from u without going through e_2. Leaf l_u would be on the 1-side of both e_1 and e_2, which is not possible. Hence the 1-sides of both e_1 and e_2 point "away" from each other. It also follows that path P cannot go through edge e_3. If it did, then some leaf on the 1-side of e_3 would also be on the 1-side of e_1 or e_2.

So edges e_1 and e_2 are either incident with each other, or there is an edge e which is incident with e_1 on path P, where e is not labeled by a super-character from C. We will show that such an edge e cannot exist. All internal edges in $\overline{T}$ are labeled by some super-character, so suppose e exists and is labeled by a super-character that derives from a connected component C'. Let v be the common endpoint of e_1 and e. As above, there must be a leaf l_v that is reachable from v without going through either edge e or e_1, for otherwise e and e_1 define the same split and should not be separate edges in $\overline{T}$. Recall that each super-character and each split in $\overline{T}$ that derives from C or C' corresponds to a sequence in $M(C)$ or $M(C')$, and with respect to the pair (C, C'), there is a dominant sequence S in $M(C)$ and a dominant sequence S' in $M(C')$. Let $e(S)$ be the edge in $\overline{T}$ labeled by the super-character for S, and let $e(S')$ be the edge in $\overline{T}$ labeled by the super-chararacter for S'. Now e_1 is either $e(S)$ or not, and e is either $e(S')$ or not, so we have four cases to consider.

Case 1: Suppose e_1 is $e(S)$ and e is $e(S')$. We know that $l(v)$ is on the 0-side of e_1, so it must be on the 1-side of e to obey Corollary 1. But then, all leaves on the 1-side of e will be on the 0-side of both e and e_1, which contradicts Corollary 1. The other cases are similar. So e cannot exist, and hence e_1 and e_2 are incident with each other. The three other cases are similar and omitted.

Since e_1 and e_2 were arbitrary edges labeled by super-characters derived from connected component C, every pair of edges labeled by super-characters from C must be incident with each other. But in a tree, that is only possible if all those edges share exactly one endpoint, and so form a star around a single center. That endpoint is the claimed node v_C. Also, we established that if there are two distinct edges labeled with super-characters derived from C, then the 1-sides of these edges point away from each other. This holds for any pair of edges labeled with super-characters derived from C, so v_C is on the 0-side of every such edge.   □

To finish the proof of Theorem 1, we first arbitrarily select a leaf in $\overline{T}$ to act as the root node, and we direct every edge away from that leaf. This also defines an ancestral sequence for the phylogenetic network we will construct. Next, we need to inflate each node v_C in $\overline{T}$ that is the central node of the star associated with the super-characters of a *non-trivial* connected component of $M(G)$. We can identify such central-star nodes by the fact that for some non-trivial connected component C, all of the edges labeled by the super-characters that derive from C are incident with a node v, and hence that node must be v_C. Each such edge may also be labeled with the super-character that derives from another connected component or with a compatible character. However, every leaf is on the 1-side of exactly one super-character that derives from C, and v_C is on the 0-side of each such super-character, so there can be no no edge $e = (v_C, v')$ in $\overline{T}$ that is labeled only by a compatible site. If there was such an edge, then a leaf reached from the v' without going through v_C would not be on the 1-side of any super-character that derives from C.

Note that each central-star node v_C has exactly one edge directed into it. We call the sequence in $M(C)$ on that edge the "ancestral sequence" of v_C. Now, any sequence in $M(C)$ can be derived from the ancestral sequence of v_C using at most one mutation per site, if enough recombinations are allowed. So, each central-star node v can be inflated into a blob B_v containing one node labeled by each distinct sequence in $M(C)$ (and other nodes if needed). Then for each distinct sequence in $M(C)$ we connect the node in B_v labeled with that sequence to the edge (incident with v_C) that is labeled by the super-character for that sequence in $M(C)$.

After inflating each central-star node in $\overline{T}$, the end result is a phylogenetic network N where each blob contains all and only the sites from one connected component of $G(M)$. Every compatible site labels a tree edge of N. This completes the proof of Theorem 1.

Uniqueness. We leave the proof to the reader, but it is also true that if N is a fully-decomposed network and T' is created by contracting each blob of N to a single node, then after the directed edges in T' are made undirected, the resulting tree is necessarily $\overline{T}$. So $\overline{T}$ is the *invariant* underlying structure of any fully-decomposed phylogenetic network for M.

Programs. The above proof of the existence of $\overline{T}$ can be converted into an efficient, constructive method[2] for finding $\overline{T}$ from any input M. The program galledtree.pl, available at wwwcsif.cs.ucdavis.edu/~gusfield/ takes in a set of sequences M and tries to build a galled-tree for M. If it succeeds, then it has produced a complete phylogenetic network for M where each blob is a single cycle, and the cycles are node disjoint. Hence, the program produces a fully-decomposed phylogenetic network for M. If the program determines that there is no galled-tree for M, then it outputs the tree $\overline{T}$ for M. The running time for the program is $O(nm^2 + m^3)$, but the time used to build $\overline{T}$ is just $O(nm^2)$.

4   What Is the "Most Tree-Like" Phylogenetic Network?

When a set of sequences M fails the four-gametes test and hence cannot be generated on a perfect phylogeny, one would still like to derive the sequences on a phylogenetic network that is the "most tree-like". There is no accepted definition of "treeness", and under many natural definitions, the problem of finding the most tree-like network would likely be computationally difficult. In this section, we introduce a measure of treeness and relate it to Theorem 1.

Given a phylogenetic network N we first modify N so that no two blobs share a node. The only way that two blobs can share a node v is if v is the "root" of one of the blobs, so we can always add a new edge to separate the two blobs. We can also assume that N has no node with in and out-degrees that are both one. Then if each blob is contracted to a single node, the number of edges in the resulting directed tree measures the "treeness" of N. In other words, the "treeness" of N is measured by the size of the tree in the underlying tree structure of N. For example, if all the sites in M are in a single blob in N, then N is less tree-like than a network where the sites are distributed between several blobs, connected by several edges in a tree structure.

---

[2] It may seem that $\overline{T}$ can be obtained by simply building a perfect phylogeny T using one site from each connected component of $G(M)$. This does not work because the edge structure of T may be very different from that of $\overline{T}$. For example, in the tree T created from sites 1 and 2 in Figure 1, the two edges labeled with those sites are adjacent, while they are not adjacent in $\overline{T}$.

With the above definition of "treeness", we claim that a phylogenetic network N is "the most tree-like" if and only if $\overline{T}$ is the resulting undirected tree, after the blobs of N are contracted, and all the edges are made undirected. This follows from Theorem 1 and that fact that all the sites in a single non-trivial connected component of $G(M)$ *must* be together in a single blob in any phylogenetic network. This second fact is proven in [9].

This definition of "most tree-like" is somewhat crude because it does not consider any details inside of a blob, but it has the advantage of being easy to compute and allowing a clear identification of the most tree-like networks. Further, it seems reasonable that any other natural definition of "most tree-like" would identify a *subset* of the networks identified by the definition considered here.

5 Alternative Proofs of Theorem 1

We believe that Theorem 1 has not previously been stated in any published literature, but Mike Steel has pointed out that Theorem 1 can be proven by using Buneman graphs [21], and the details of this approach have been worked out by Yufeng Wu at UC Davis. However, it takes exponential time in worst case to build a Buneman graph from M, and so this is not an efficient constructive approach. Andreas Dress and Hiroshi Hirai have also pointed out that Theorem 1 can be derived from the framework of block-decompositions in T-theory. Finally, Daniel Huson and Mike Steel have (subsequent to the development of Theorem 1) recently developed a related decomposition theory for splits graphs, where the input to the problem is not a set of sequences, but a set of trees that must be subtrees in a constructed phylogenetic network. Problems of that type have been studied in [19, 15].

6 Theorem 1 Applies in Diverse Biological Contexts

Theorem 1 was proven in the context of multiple crossover recombination when explicit binary sequences are given as input. This is most directly motivated by the evolution of sequences of SNPs (single nucleotide polymorphisms). SNP sequences evolve in a population by site mutation and by (meiotic) recombination of homologous chromosomes (single crossover recombination), and by gene conversion (a specific kind of two crossover recombination). However, multiple crossover recombination can be considered as a mathematical operation on binary sequences, rather than a biological event, and can be used to model biological events that don't explicitly involve recombination. As a consequence, Theorem 1 holds in many biological contexts where diverse biological events cause incompatibility between sites (or more generally, incompatibility between binary evolutionary "characters"). Three such biological events are "back-mutation", "recurrent-mutation", and "lateral gene transfer", and we consider the first two of those below.

6.1   Back and Recurrent Mutation

"Back-mutation" occurs when the state of a site mutates back from its derived state to its ancestral state. "Recurrent-mutation" occurs when the state of a site is permitted to mutate from its ancestral state more than once in an evolutionary history. Because there is no explicit recombination, the underlying graph of a network with back or recurrent mutation is a tree. Generally, when back or recurrent mutation is the cause of incompatibility, we seek an evolutionary tree that derives a given set of sequences using as few back or recurrent mutations as possible. Such a tree is called a "maximum parsimony tree" and it is a solution to the maximum parsimony problem [3, 21].

While biologically unrelated to recombination, each occurrence of back-mutation or recurrent-mutation of a site i in a sequence S can be *modeled* as a two-crossover recombination between S and some appropriate sequence, in the intervals $i - 1, i$ and $i, i + 1$. Modeling back and recurrent mutations in this way explicitly creates recombination cycles and blobs, and shows explicitly how Theorem 1 applies when back-mutation and/or recurrent mutation cause incompatibilities. The consequence is that one can derive M using a separate tree for the sites in each non-trivial connected component C of $G(M)$. The tree for each C derives the sequences in $M(C)$ using recurrent and/or back mutation if needed, and the separate trees can be connected using $\overline{T}$.

Note that when back or recurrent-mutation is modeled in this way, each recombination only changes a single site, so the linear order of the site has no impact on the permitted recombinations, and the ordering of the sites can be arbitrary. This allows Theorem 1 to apply to (binary) "evolutionary characters" which may experience back and/or recurrent mutation, but have no natural order.

7   Open Question and Conjecture

The main open question related to Theorem 1 is the following

> **Decomposition Optimality Conjecture:** For any M, there is always a fully-decomposed phylogenetic network for M that minimizes the number of recombinations used, over all possible phylogenetic networks for M.

Note, that the conjecture does not say that the minimum number of recombinations is equal to $\sum_C cc(C)$, where $cc(C)$ is the minimum number of recombinations needed in a phylogenetic network for $M(C)$. Such a stronger claim has been shown to be false [22]. The difficulty is that the separate solutions may choose ancestral sequences that cannot be combined into a single network.

The Decomposition Optimality Conjecture can be proven when the recombinations model recurrent and back mutations, as discussed earlier (one proof is based on the Buneman graph of M). Because of this result, when incompatibilities are caused by recurrent and/or back mutation, one can solve the parsimony problem separately for each connected component of $M(G)$, and then connect the trees as specified by $\overline{T}$. Since the parsimony problem is itself NP-hard, and the only known methods to solve it take exponential time in worst-case, decomposing the problem into several smaller problems may allow larger problems to be solved in practice.

If the Decomposition Optimality Conjecture is true in general (for any multiple-crossover recombinations), we could follow a similar approach to finding phylogenetic networks that minimize the number of recombinations. It is easiest to exploit the conjecture (if proved) in the case that an ancestral sequence A is given. In that case, we know the root of $\overline{T}$ and hence the ancestral sequence for each of the blobs in the network for M. Hence we could solve a single (rooted) problem for each component of $G_A(M)$. When no ancestral sequence is known in advance, this approach would need to be repeated for each choice of root position in $\overline{T}$. If the conjecture is true, it would also follow that we could compute lower bounds on the number of needed recombinations by computing bounds separately for the sites on each connected component of $M(G)$, and then add these bounds together for a correct overall bound. This would be correct no matter what lower bound method is used. This approach has been proven correct for two specific lower bounds [1], strengthening the belief that the above conjecture is true.

Progress on Proving the Conjecture. We have recently proven [7] a weaker version of the Decomposition Optimality Conjecture. We say that a node v in a phylogenetic network N for M is "visible" if the sequence labeling node v in N is a sequence in M.

Theorem 3. *If every node v in N is visible, then there is a fully-decomposed network for M which uses the same number of recombinations, or fewer, than does N.*

The theorem can be proven with somewhat weaker conditions than the visibility of all nodes in N. Also, a sufficient (but not necessary) condition for the visibility of all nodes is that the "haplotype lower bound" [16] on the minimum number of recombinations equals the true minimum. Simon Myers has shown [17] that under the neutral coalescent model with recombination, the expected difference between the haplotype bound and the true minimum is bounded by a *constant* as the number of sequences goes to infinity. Thus, there may be optimal phylogenetic networks where all nodes are visible, more often than might at first be assumed.

Acknowledgements

We would like to thank S. Eddhu for writing the program to construct $\overline{T}$. We thank Dean Hickerson, Chuck Langley, Vineet Bafna, Yun Song, Yufeng Wu, Zhihong Ding and Gabriel Valiente for helpful discussions and comments. Research of the first author is partly supported by NSF grant EIA-0220154.

References

1. V. Bafna and V. Bansal. The number of recombination events in a sample history: conflict graph and lower bounds. *IEEE/ACM Transactions on Computational Biology and Bioinformatics*, 1:78–90, 2004.
2. A. Chakravarti. It's raining SNP's, hallelujah? *Nature Genetics*, 19:216–217, 1998.
3. J. Felsenstein. *Inferring Phylogenies*. Sinauer, Sunderland, MA., 2004.
4. D. Gusfield. Efficient algorithms for inferring evolutionary history. *Networks*, 21:19–28, 1991.
5. D. Gusfield. *Algorithms on Strings, Trees and Sequences: Computer Science and Computational Biology*. Cambridge University Press, Cambridge, UK, 1997.
6. D. Gusfield. Optimal, efficient reconstruction of Root-Unknown phylogenetic networks with constrained recombination. Technical report, Department of Computer Science, University of California, Davis, CA, 2004.
7. D. Gusfield. On the decomposition optimality conjecture for phylogenetic networks. Technical report, UC Davis, Department of Computer Science, 2005.
8. D. Gusfield, S. Eddhu, and C. Langley. The fine structure of galls in phylogenetic networks. *INFORMS J. on Computing, special issue on Computational Biology*, 16:459–469, 2004.
9. D. Gusfield, S. Eddhu, and C. Langley. Optimal, efficient reconstruction of phylogenetic networks with constrained recombination. *J. Bioinformatics and Computational Biology*, 2(1):173–213, 2004.
10. D. Gusfield and D. Hickerson. A new lower bound on the number of needed recombination nodes in both unrooted and rooted phylogenetic networks. Report UCD-ECS-06. Technical report, University of California, Davis, 2004.
11. J. Hein. Reconstructing evolution of sequences subject to recombination using parsimony. *Math. Biosci*, 98:185–200, 1990.
12. J. Hein. A heuristic method to reconstruct the history of sequences subject to recombination. *J. Mol. Evol.*, 36:396–405, 1993.
13. R. Hudson and N. Kaplan. Statistical properties of the number of recombination events in the history of a sample of DNA sequences. *Genetics*, 111:147–164, 1985.
14. J. D. Kececioglu and D. Gusfield. Reconstructing a history of recombinations from a set of sequences. *Discrete Applied Math.*, 88:239–260, 1998.
15. B. Moret, L. Nakhleh, T. Warnow, C.R. Linder, A. Tholse, A. Padolina, J. Sun, and R. Timme. Phylogenetic networks: Modeling, reconstructibility, and accuracy. *IEEE/ACM Transactions on Computatational Biology and Bioinformatics*, pages 13–23, 2004.
16. S. R. Myers and R. C. Griffiths. Bounds on the minimum number of recombination events in a sample history. *Genetics*, 163:375–394, 2003.
17. Simon Myers. *The detection of recombination events using DNA sequence data*. PhD thesis, University of Oxford, Oxford England, Department of Statistics, 2003.
18. L. Nakhleh, J. Sun, T. Warnow, C.R. Linder, B.M.E. Moret, and A. Tholse. Towards the development of computational tools for evaluating phylogenetic network reconstruction methods. In *Proc. of 8'th Pacific Symposium on Biocomputing (PSB 03), pages 315-326*, 2003.
19. L. Nakhleh, T. Warnow, and C.R. Linder. Reconstructing reticulate evolution in species - theory and practice. In *Proc. of 8'th Annual International Conference on Computational Molecular Biology*, pages 337–346, 2004.
20. D. Posada and K. Crandall. Intraspecific gene genealogies: trees grafting into networks. *Trends in Ecology and Evolution*, 16:37–45, 2001.

21. C. Semple and M. Steel. *Phylogenetics*. Oxford University Press, UK, 2003.
22. Y. Song. Personal Communication.
23. Y. Song and J. Hein. Parsimonious reconstruction of sequence evolution and haplotype blocks: Finding the minmimum number of recombination events. In *Proc. of 2003 Workshop on Algorithms in Bioinformatics*, Berlin, Germany, 2003. Springer-Verlag LNCS.
24. Y. Song and J. Hein. On the minimum number of recombination events in the evolutionary history of DNA sequences. *Journal of Mathematical Biology*, 48:160–186, 2004.
25. L. Wang, K. Zhang, and L. Zhang. Perfect phylogenetic networks with recombination. *Journal of Computational Biology*, 8:69–78, 2001.

Reconstruction of Reticulate Networks from Gene Trees

Daniel H. Huson[1], Tobias Klöpper[1], Pete J. Lockhart[2], and Mike A. Steel[3]

[1] Center for Bioinformatics (ZBIT),
Tübingen University, Sand 14, 72076 Tübingen, Germany
[2] Institute of Molecular BioSciences,
Massey University, Palmerston North, New Zealand
[3] Biomathematics Research Centre,
University of Canterbury, Christchurch, New Zealand

Abstract. One of the simplest evolutionary models has molecular sequences evolving from a common ancestor down a bifurcating phylogenetic tree, experiencing point-mutations along the way. However, empirical analyses of different genes indicate that the evolution of genomes is often more complex than can be represented by such a model. Thus, the following problem is of significant interest in molecular evolution: Given a set of molecular sequences, compute a reticulate network that explains the data using a minimal number of reticulations. This paper makes four contributions toward solving this problem. First, it shows that there exists a one-to-one correspondence between the tangles in a reticulate network, the connected components of the associated incompatibility graph and the netted components of the associated splits graph. Second, it provides an algorithm that computes a most parsimonious reticulate network in polynomial time, if the reticulations contained in any tangle have a certain overlapping property, and if the number of reticulations contained in any given tangle is bounded by a constant. Third, an algorithm for drawing reticulate networks is described and a robust and flexible implementation of the algorithms is provided. Fourth, the paper presents a statistical test for distinguishing between reticulations due to hybridization, and ones due to other events such as lineage sorting or tree-estimation error.

1 Introduction

One of the most powerful approaches for developing an understanding of the evolution of genomes is phylogenetic analysis of genes at independent loci. However, optimal phylogenetic reconstructions for such genes are not always concordant [1]. One possible reason for this is that at the level of organisms, hybridization between diverging evolutionary lineages is a fundamental process important in the evolution of organisms [2]. Unfortunately, at the level of individual gene analyses, the interpretation of hybrid genomes is complicated by phylogenetic error, gene conversion (events of non-reciprocal recombination) and lineage sorting. Despite

S. Miyano et al. (Eds.): RECOMB 2005, LNBI 3500, pp. 233–249, 2005.

this complexity, the importance of the issue has motivated the following problem: Given a set of phylogenetic data that have evolved under a reticulate model of evolution, what is the most efficient way to reconstruct the underlying reticulate network. There has been much interest in this topic, see e.g. [3, 4, 5, 6, 7, 8]. Computationally, the goal is to construct a reticulate network using a minimum number of reticulations that accounts for the given data. Alternatively, one may attempt to give lower or upper bounds for the number of reticulations required to explain the data [9, 10, 11, 7].

In this paper we describe a general frame-work for studying reticulate networks, which is based on the theory of splits and splits graphs [12, 13]. In this setting, a *reticulate network* N is a generalization of a phylogenetic tree in which we additionally allow certain *reticulation* nodes and edges. The set of *splits associated with* N is defined as $\Sigma = \bigcup_{T \in \mathcal{T}(N)} \Sigma(T)$, where $\mathcal{T}(N)$ is the set of all trees that are induced by N and $\Sigma(T)$ is the split encoding of T. We present four new results.

Our first main result is a Decomposition Theorem that implies the existence of a one-to-one correspondence between the tangles of a reticulate network N, the non-trivial connected components of the incompatibility graph $IG(\Sigma(N))$ and the netted components of the splits graph $SG(\Sigma(N))$. This is related to similar results that have been previously described in the context of recombination of binary sequences under the infinite sites model [14, 11], for which we give a new formulation, interpretation and proof.

We say that a reticulate network N has the *overlapping* property, if every set of tangled reticulations in N has the property that all reticulation cycles intersect "nicely" along a common "backbone". Our second main result is an algorithm that computes the most parsimonious reticulate network this type for a given input set in polynomial time, if we limit the number of reticulations contained in any given tangle to k.

Our third main result is an algorithm for drawing reticulate networks. We have developed a robust and flexible implementation of our approach, which is freely available as a plug-in for the program SplitsTree [15]. It takes as input either a set of trees, partial trees or splits and produces as output a (rooted or unrooted) phylogenetic network indicating both the splits contained in the input and also the possible ways to resolve each netted component of the splits graph into a collection of reticulations.

Our fourth main result is a new statistical test for distinguishing between reticulations due to hybridization, on the one hand, and ones due to other events such as lineage sorting or tree estimation error, on the other.

To illustrate our algorithms, we apply them to two different gene trees for New Zealand alpine *Ranunculus* (buttercups) species based on the nuclear ITS gene and the chloroplast J_{SA} region [16]. In an Appendix we provide a second application, namely to haplotype data for the alcohol dehydrogenase locus of *Drosophila melanogaster* [11, 17].

We would like to thank Dan Gusfield for a number of very useful discussions.

2   Phylogenetic Trees and Reticulate Networks

Let X denote a set of taxa. A *phylogenetic tree for* X, or X*-tree*, consists of a tree $T = (V, E)$ in which every node v is either a *leaf* of degree 1 or an *internal* node of degree ≥ 3, together with a node labeling $\nu : X \to V$ such that every leaf of T obtains a label [18]. Additionally, we may designate one of the taxa $o \in X$ to be an *outgroup* and then consider the tree to be "rooted" at the midpoint ρ of the pendant edge leading to $\nu(o)$, in the usual sense. We choose to define the root in this indirect way because our approach is based on the concept of splits, i.e. bipartitionings of the taxon set X, for which it is awkward to specify a root node explicitly.

Let X denote a set of taxa. A *reticulate network* $N = (V, E, \nu)$ consists of a graph (V, E) with node set V and edge set E and a labeling of the nodes by taxa $\nu : X \to V$. The node set $V = V_R \cup V_T$ is partitioned into a set of *reticulation nodes* V_R and *tree nodes* V_T, and the edge set $E = E_R \cup E_T$ is partitioned into a set of *reticulation edges* E_R and *tree edges* E_T. The labeling ν only assigns labels to nodes in V_T and every leaf of N obtains a label. Additionally, we require the following five properties:

(R1)   All nodes have degree $\neq 2$.

(R2)   Every reticulation node $v \in V_R$ is incident to precisely two reticulation edges, denoted by $p(v)$ and $q(v)$, respectively.

(R3)   Every reticulation edge $e \in E_R$ is incident to exactly one reticulation node.

(R4)   Every subgraph of N obtainable by deleting precisely one reticulation edge $p(v)$ or $q(v)$ for every reticulation node $v \in V_R$, is an X-tree. We will use $\mathcal{T}(N)$ to denote the set of all such trees *induced* by N.

(R5)   We will always assume that an outgroup $o \in X$ has been specified and will require for all trees $T \in \mathcal{T}(N)$ that every reticulation node $v \in V_R$ is separated from $\nu(o)$ by either $p(v)$ or $q(v)$.

As in the case of trees, we will usually consider N to be rooted at the center of the pendant edge leading to the node $\nu(o)$ labeled by the outgroup o. In Figure 1 we show an example of a reticulate network N with three reticulations. In Figure 2(a–b) we show two different trees induced by N. We explicitly allow the graph to contain *unresolved* nodes of degree > 3, labeled internal nodes and nodes with multiple labels.

It follows from these definitions that each reticulation node (or *reticulation*, for short) $v \in V_R$ is contained in one or more cycles of the form $C = (v, p(v), w_1, e_1, \ldots, e_{k-1}, w_k, q(v), v)$, $C = (v, p(v), w_1, e_1, \ldots, e_{k-1}, w_k, v)$ or $C = (v, q(v), w_1, e_1, \ldots, e_{k-1}, v)$, with $w_i \in V$ and $e_i \in E \setminus \{p(v), q(v)\}$ for all i. Any such cycle C is called a *reticulation cycle* and we define its *backbone* as $B(C) = (w_1, e_1, \ldots, e_{k-1}, w_k)$. Note that a reticulation v possesses at most one reticulation cycle C whose backbone B contains only tree edges and in this case we call C a *tree cycle*.

We say that two different reticulations $v \in V_R$ and $v' \in V_R$ are *dependent*, if they are contained in reticulation cycles C and C', respectively, such that C and C' share at least one edge. Otherwise, they are called *independent*. (Previous

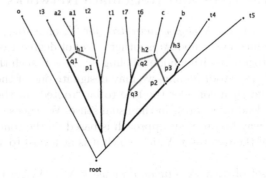

Fig. 1. A reticulate network N displaying three reticulations at nodes h_1, h_2 and h_3, each involving edges p_i and q_i, with $i = 1, 2, 3$, respectively. The reticulation at node h_1 is independent of the reticulations at nodes h_2 and h_3, whereas the latter two are tangled. The backbone edges of each reticulation are highlighted by heavier lines

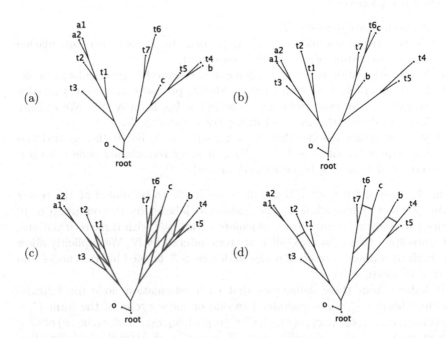

Fig. 2. Exactly eight different "gene trees" are induced by the reticulate network N in Figure 1, and we depict two such trees, T_1 in (a) and T_2 in (b). In (c) we show a splits graph SG representing the union of the splits $\Sigma(T_1) \cup \Sigma(T_2)$ of the two trees T_1 and T_2. This graph has two netted components, highlighted by heavy lines, and these correspond precisely to the two sets of tangled reticulations contained in N. In (d) we show the reticulate network N' reconstructed from SG using the algorithm described in the text, with reticulation edges highlighted by heavy lines

definitions of independence were more restrictive by requiring node-disjointness [5, 8].) Additionally, we call v and v' *tangled*, if there exists a chain of reticulation cycles $C = C_1, C_2, \ldots, C_k = C'$ such that C_i and C_{i+1} are dependent for each $i = 1, \ldots, k - 1$. A reticulation that is independent of all other reticulations is also called a *gall* [6] and a reticulate network N that contains only galls is called a *galled tree*.

A reticulate network N gives rise to a *reticulate model* of evolution as follows: Starting at the root, molecular sequences evolve along tree edges in the usual fashion, experiencing point-mutations along the way [19]. However, the sequence that arises at a reticulation node v is obtained as a mixture of sequences along the two reticulation edges $p(v)$ and $q(v)$. The three main biological mechanisms that may operate here are *hybridization, horizontal gene transfer (HGT)* and *recombination*. In hybridization or HGT, we think of a sequence as consisting of an unordered set of genes and the net result of an hybridization or HGT event is a mixture of complementary genes or sites. In the case of recombination within a population, we consider individual sites ordered along the sequences and a resulting recombinant sequence is usually obtained by *cross-over* events, in which a prefix and a suffix of two ancestor sequences are combined together.

Our definitions capture the essence of the graphs used to describe hybridization and HGT scenarios [8], and the underlying graph employed to describe recombination scenarios [3, 5, 6]. However, the latter possess additional structure, namely a labeling of the tree edges by mutation sites and a labeling of the reticulation nodes by cross-over positions, and thus our approach requires further development before it will be able to fully address the reconstruction of recombination scenarios.

Throughout this paper we will use the term *gene* to mean a segment of sequence that is atomic with respect to the mechanism of reticulate evolution in operation. Hence, the evolutionary history of any given gene will be a single phylogenetic tree $T \in \mathcal{T}(N)$ [20].

3   Splits, Incompatibility and Splits Graphs

Suppose we are given a set of taxa X. A *split* (or, more precisely, X-*split*) is a bipartitioning of X into two non-empty sets A and B, denoted by $S = \frac{A}{B} (= \frac{B}{A})$.

For a given X-tree T, deletion of any single edge e will produce a graph with exactly two connected components and this defines a split $\sigma_T(e) = \frac{A}{B}$, given by the two sets of taxa labeling the two components [12]. The set of all splits obtainable in this way is called the *splits encoding* $\Sigma(T)$ of T. For a given set H of X-trees, we define $\Sigma(H) = \bigcup_{T \in H} \Sigma(T)$.

Two X-splits $S = \frac{A}{B}$ and $S' = \frac{A'}{B'}$ are called *compatible*, if one of the four possible intersections $A \cap A', A \cap B', B \cap A'$ and $B \cap B'$ is empty. A set of X-splits Σ is called *compatible*, if all pairs of splits in Σ are compatible. The *incompatibility graph* $IG(\Sigma) = (V, E)$ has node set $V = \Sigma$ and edge set $E \subseteq \binom{V}{2}$, in which any two nodes S and S' are connected, if and only if they are incompatible.

It is a well-known result that a set of X-splits Σ is pairwise compatible, if and only if there exists a unique X-tree T with $\Sigma = \Sigma(T)$. In this case we say that T *represents* Σ. Moreover, an arbitrary set of splits Σ, not necessarily compatible, can also be represented by a graph. Such a *splits graph* $SG(\Sigma)$ consists of a connected graph (V, E) together with a node labeling $\nu : X \to V$ and an edge coloring $\sigma : \Sigma \to E$, whose essential property is that deleting all edges colored by a given split $S = \frac{A}{B} \in \Sigma$ will produce precisely two connected components, labeled by A and B, respectively, see [13] for details. (This splits graph is not uniquely defined, however we will refer to it as *the* splits graph $SG(\Sigma)$ representing Σ, as the differences are inconsequential.) For example, the splits graph depicted in Figure 2 (c) represents the union of the split encodings of the two trees shown in Figure 2 (a–b).

Suppose we are given a set of splits Σ. A *netted component* Z of $SG(\Sigma)$ is a maximum set of nodes such that any two nodes $v, w \in Z$ are connected by two different node-disjoint paths in $SG(\Sigma)$ (in graph-theoretic terminology, a 2-connected component). The splits graph depicted in Figure 2 (c) has two netted components, each highlighted by heavy lines. Any node v that is contained in some netted component Z and is labeled, or is incident to an edge that is not contained in Z, is called a *gate node*. For example, in Figure 2(c), the left-hand netted component has precisely five gate nodes and the right-hand one has seven.

It is a simple observation that any two splits $S, S' \in \Sigma$ are incompatible, if any only if the edges representing S and S' are contained in the same netted component [12]. More precisely:

Lemma 1. *Suppose we are given a set of X-splits Σ. There exists a one-to-one correspondence between the netted components of the splits graph $SG(\Sigma)$ and the non-trivial connected components of the incompatibility graph $IG(\Sigma)$.*

4   Parsimonious Reconstruction Problem

An X-tree T' is called a *refinement* of an X-tree T, if $\Sigma(T) \subseteq \Sigma(T')$, that is, if $T' = T$ or if T can be obtained by contracting some edges of T'. Given a (usually unknown) reticulate network N. We say that a set of trees H was *sampled* from N, or that N *supports* H, if each tree $T \in H$ possesses a refinement $T' \in \mathcal{T}(N)$. Similarly, we say that a set of splits Σ' was *sampled* from N, or that N *supports* Σ', if $\Sigma' \subseteq \bigcup_{T \in \mathcal{T}(N)} \Sigma(T)$.

In molecular evolution we are interested in the following problem: Given a collection of gene trees that have evolved from a common ancestor under a reticulate model of evolution, reconstruct the underlying reticulate network. We address this as follows:

Problem 1. Parsimonious Reticulate Network from Gene Trees Problem: *Given a set of X-trees H. Construct a reticulate network N that supports H and contains a minimal number of reticulations.*

This is a purely combinatorial problem. It is easy to see that a solution always exists [21]. In practice a key issue is how to obtain a collection of sufficiently accurate gene trees that contain all edges necessary to be able to detect the reticulations in the underlying network. A second key issue is that, even if we can construct such a network N, it is not certain that a reticulation node v was indeed caused by a hybridization event, as we discuss in Section 8.

Problem 1 in its most general form is known to be computationally intractable [5]. The following simpler version of the problem is tractable and different solutions have been proposed [20, 8, 6, 7]:

Problem 2. Parsimonious Independent Reticulate Network from Gene Trees Problem: *Given a set of X-trees H. Construct a reticulate network N containing only independent reticulations that supports H and contains a minimal number of reticulations, if one exists.*

This problem is a special case of the following more general problem. Let N be a reticulate network. We say that two reticulation nodes v and v' *overlap*, if they possess tree cycles C and C', respectively, that intersect "nicely", that is, whose backbones B and B' overlap either in prefixes or suffixes of each other, or for which one backbone is contained in the other.

Problem 3. Parsimonious Overlapping Reticulate Network from Gene Trees Problem: *Given a set of X-trees H. Construct a reticulate network N containing only independent or overlapping reticulations that supports H and contains a minimal number of reticulations, if one exists.*

One of our main results is that this problem is computationally tractable, if we limit the maximum number of reticulations contained in any given tangle, and we present an algorithm to solve it in Section 6.

5   The Decomposition Theorem

Suppose that $N = (V, E, \nu)$ is a reticulate network. We define $\Sigma(N) := \bigcup_{T \in \mathcal{T}(N)} \Sigma(T)$, and for any edge $e \in E$, we take $\Sigma(e) := \{\sigma_T(e) \mid e \text{ is edge of } T \in \mathcal{T}(N)\} \subseteq \Sigma(N)$ to be the set of all splits generated by e in trees induced by N.

There is a close relationship between reticulate networks and splits graphs. More precisely, there exists a one-to-one relationship between the tangles of a reticulate network N, the connected components of the incompatibility graph $IG(\Sigma(N))$ and the netted components of the splits graph $SG(\Sigma(N))$. This is implied by the following result:

Theorem 1. *[Decomposition Theorem] Suppose N is a reticulate network. Two tree edges $e, f \in E_T$ are contained in a cycle in N, if and only if there exist two splits $S \in \Sigma(e)$ and $S' \in \Sigma(f)$ that are contained in the same connected component of the incompatibility graph $IG(\Sigma(N))$.*

Similar results, using different definitions, interpretations and proofs, are reported in $[11, 14]^1$.

To formulate our proof, we first introduce some additional definitions and results. Suppose that N is a reticulate network and C is a path or cycle in N. We say that C *fully contains* a reticulation $v \in V_R$, if C contains both $p(v)$ and $q(v)$ (consecutively, of course), and we use $s(C)$ to denote the number of reticulations fully contained in C.

Lemma 2. *For every cycle C in a reticulate network N we have $s(C) \geq 1$.*

Proof: Direct all edges away from the root of N. The edges of C cannot all be oriented in the same direction and thus there must exist two consecutive edges e_1 and e_2 in C that are oriented toward their common vertex v. By definition of N, we must have $v \in V_R$ and $\{p(v), q(v)\} = \{e_1, e_2\}$, and thus $s(C) \geq 1$.   □

Lemma 3. *Consider a reticulate network N. If $e, f \in E_T$ are two different tree edges contained in a common cycle C with $s(C) = 1$, then there exist two splits $S \in \Sigma(e)$ and $S' \in \Sigma(f)$ that are incompatible.*

Proof: Consider two edges $e, f \in E_T$ and assume they are both part of a cycle C that contains precisely one reticulation $v \in V_R$ for which both $p(v)$ and $q(v)$ are edges in C. Then there exist two trees $T_p \in \mathcal{T}(N)$ and $T_q \in \mathcal{T}(N)$ that contain all edges of C except for $q(v)$ and $p(v)$, respectively, and differ only by these two edges. Assume that $C = (e_0 = p(v), w_1 = v, e_1 = q(v), w_2, \ldots, w_\alpha, e_\alpha = e, w_{\alpha+1}, \ldots, w_\beta, e_\beta = f, w_{\beta+1}, \ldots, w_k)$ for appropriate α, β. For any node u in C, let V_u denote the set of all nodes that can be reached in T_p or T_q from a

---

[1] In [14], the input to the problem is a set M of binary sequences of length n that are assumed to have been generated under the infinite sites model. If we discard all constant sites, then the set of remaining columns of M is equivalent (up to a choice of ancestral states) to a set of splits Σ and the definition of an *incompatibility graph* in [14] is equivalent to our definition. A *phylogenetic network*, as defined in [14] (and perhaps better termed a *recombination network*), is based on a directed acyclic graph with a specified root, and certain coalescence and recombination nodes. Such a network N is considered to explain an input set M, if there exists a labeling of the leaves by M and a labeling of the internal nodes by additional sequences of length n, together with a labeling of the edges by columns of M (that is, splits) and a labeling of the recombination nodes by certain recombination events, such that each split occurs precisely once and the implied mutations and recombinations give rise to the specified labeling of the nodes by sequences. Because there is some choice in the placement of individual splits within such a network, the Decomposition Theorem in [14] holds only in one direction, namely *if two splits are contained in the same connected component of the incompatibility graph, then there exists a phylogenetic network such that the corresponding edges are contained in a cycle*, but not vice-versa. In our definition of a reticulate network, we do not explicitly label edges by splits. Rather, each edge implicitly corresponds to a set of splits that is defined via the set of trees that can be sampled from the network. This lack of choice explains why our version of the Decomposition Theoren holds in *both* directions.

node u without using any edge in C. Note that all four sets V_v, V_{w_α}, V_{w_β} and $V_{w_{\beta+1}}$ must be disjoint, as both T_p and T_q are cycle-free. Let X_u denote the set of taxa that occur as labels in V_u. The split $S = \sigma_{T_p}(e)$ induced by e in T_p separates $X_v \cup X_{w_\alpha}$ from $X_{w_\beta} \cup X_{w_{\beta+1}}$. Similarly, the split $S' = \sigma_{T_q}(f)$ induced by f in T_q separates $X_{w_\alpha} \cup X_{w_\beta}$ from $X_{w_{\beta+1}} \cup X_v$. Thus, S and S' are incompatible. □

Now we prove Theorem 1:

Proof: "⇒": Suppose we are given a reticulate network N and two different tree edges $e, f \in E_T$ that are contained in a cycle C. Orient all edges away from the root and use g^- and g^+ to indicate the implied start and end of an edge $g \in E$. We have two cases, which we both prove by induction:

Case 1: The edges e and f have opposite orientations in C. By Lemma 2 we have $s(C) > 0$. If $s(C) = 1$, then Lemma 3 implies the result. So assume that $s(C) = n > 1$ and $C = (e^-, e, e^+, \ldots, p(v_1), v_1, q(v_1), w, \ldots, f^+, f, f^-, \ldots)$, where v_1 is the first encountered reticulation that is fully contained in C. By properties (R4–R5) of N, for any node u there exists a path P_u from the root ρ to u with $s(P_u) = 0$. Let $-P_u$ denote the reversal of P_u. Construct a cycle C' by concatenating P_{e^+}, the section of C that links e^+ to w, and $-P_{w_k}$. It has $s(C') = 1$. Construct a second cycle C'' by concatenating P_{w_k}, the section of C that links w to f^+, and $-P_{f^+}$. It has $s(C'') < n$. Let f' be the edge contained in P_w that is adjacent to w. This must be a tree edge, because $w_k \in V_T$. By Lemma 3, there exist two incompatible splits $S \in \Sigma(e)$ and $S' \in \Sigma(f')$. By induction, there exists a chain of pairwise incompatible splits from S' to some split $S'' \in \Sigma(f)$. Hence, the claim follows.

Case 2: This is dealt with in the same way, but using slightly different paths. "⇐": If e and f are two tree edges not contained in a cycle, then there exists a cut-edge h (or at least a cut-vertex, which can be refined to provide a cut-edge h) that separates e and f. The edge h induces the same split $S = \frac{A}{B}$ in every tree $T \in \mathcal{T}(N)$. Thus, every split $S' \in \Sigma(N)$ subdivides either A or B, but not both sets. This implies the claim. □

6   Algorithms

Suppose we are given a set of gene trees H sampled from a reticulate network N and would like to solve Problem 3. By Theorem 1 and Lemma 1, we can assume that N consists of precisely one tangle. We can reduce the problem size further by assuming that X is Σ-*separated*, that is, that for every pair of distinct taxa $x, y \in X$ there exists a split $S = \frac{A}{B} \in \Sigma(N)$ with $|A \cap \{x, y\}| = 1$.

Lemma 4. *To solve the posed computational problems, it suffices to consider the reduced case of a collection of X-trees H such that X is Σ-separated and the incompatibility graph $IG(\Sigma)$ consists of precisely one connected component, or, equivalently, the splits graph $SG(\Sigma)$ consists of exactly one netted component.*

In the following we restrict our attention to the reduced case by virtue of Lemma 4. Given a taxon set X, we define a *reticulation scenario* (R, B, I) to consist of a set of *reticulate taxa* $R \subset X$ and an ordered list of *backbone taxa* $B = (b_1, b_2, \ldots, b_k)$, so that X equals the disjoint union $R \cup \{b_1, \ldots, b_k\}$, together with a mapping I from R to distinct intervals of backbone taxa in $(b_2, \ldots, b_{k-1})$. We require that the outgroup taxon $o \in X$ is contained in B, if specified. Moreover, we set

$$\Sigma(R, B, I) = \left\{ \tfrac{R^-(i) \cup \{b_1, \ldots, b_i\} \cup R_1(i)}{R_2(i) \cup \{b_{i+1}, \ldots, b_k\} \cup R^+(i)} \mid i = 1, \ldots, k-1 \right\}$$
$$\cup \left\{ \tfrac{R^-(i) \cup \{b_1, \ldots, b_i\} \cup R_2(i)}{R_1(i) \cup \{b_{i+1}, \ldots, b_k\} \cup R^+(i)} \mid i = 2, \ldots, k \right\},$$

with

$$R_1(i) \cup R_2(i) \text{ is any partitioning of } R(i) := \{r \in R \mid b_i \in I(r)\} \neq \emptyset,$$
$$R^-(i) := \{r \in R \mid I(r) \subseteq \{b_1, \ldots, b_{i-1}\}\},$$
$$\text{and } R^+(i) := \{r \in R \mid I(r) \subseteq \{b_{i+1}, \ldots, b_k\}\}.$$

We interpret $B = (b_1, \ldots, b_k)$ as the joint "super-backbone" along which all overlapping reticulations are arranged. Every taxon $r \in R$ corresponds to a reticulation node and the associated interval $I(r) \subseteq B$ defines precisely which part of B will form the backbone of the tree cycle associated with r.

Given a reticulation scenario (R, B, I), we can construct a corresponding reticulate network $N(R, B, I)$ in polynomial time, as follows:

Algorithm 1. *Assign a reticulate node $v(r)$ to each reticulate taxon $r \in R$ and a tree node $v(b)$ to each backbone taxon $b \in B$. Additionally, for each consecutive pair of taxa b_i, b_{i+1} in B, define a tree node $v(b_i, b_{i+1})$, and then connect $v(b_i)$ to $v(b_i, b_{i+1})$, and $v(b_i, b_{i+1})$ to $v(b_{i+1})$, using tree edges $(i = 1, \ldots, k-1)$. Finally, connect each reticulate node $v(r)$ to the two nodes $v(b_i, b_{i+1})$ and $v(b_j, b_{j+1})$ using reticulate edges, where i, j are chosen such that $I(r) = \{b_{i+1}, b_{i+2}, \ldots, b_{j-1}\}$ holds.*

By construction we have:

Lemma 5. *A reticulation scenario (R, B, I) corresponds to a solution $N(R, B, I)$ of Problem 3, if and only if H can be sampled from $N(R, B, I)$ and $|R|$ is minimal.*

The following useful observation follows directly from the definition of $\Sigma(R, B, I)$:

Lemma 6. *Let Σ denote the set of splits in the reduced case. If (R, B, I) corresponds to a solution of Problem 3, then every split $S \in \Sigma$ must separate b_1 and b_k, that is, we must have $|\{b_1, b_k\} \cap A| = 1$ for all $S = \tfrac{A}{A'} \in \Sigma$.*

The following polynomial-time algorithm takes as input a set of trees H in the reduced case and produces as output a minimal reticulation scenario, if it exists:

Algorithm 2. *Consider all possible subsets R of X of size $\leq k$ in order of increasing cardinality, and set $B = X \setminus R$. There are $O(|X|^k)$ such subsets. Define $\Sigma|_B := \left\{ \frac{A \cap B}{A' \cap B} \mid \frac{A}{A'} \in \Sigma \right\} \setminus \{\frac{\emptyset}{B}, \frac{B}{\emptyset}\}$. To obtain an ordering of B, first determine b_1 using Lemma 6. Let $\Sigma|_B(b_1) = \{A \in S \in \Sigma|_B : b_1 \in A\}$ denote the set of split parts in $\Sigma|_B$ containing b_1. Then, these must be ordered by inclusion $\{b_1\}, \{b_1, b_2\}, \{b_1, b_2, b_3\}, \ldots$ and we thus obtain an ordering $b_1, b_2, \ldots$ of B. To compute the set $I(r) \subseteq B$ for a given reticulate taxon $r \in R$, determine the set of all $b \in B$ with the property that $\frac{A \cup \{b, r\}}{A'} \in \Sigma|_{B \cup \{r\}} \Leftrightarrow \frac{A \cup \{b\}}{A' \cup \{r\}} \in \Sigma|_{B \cup \{r\}}$. Finally, check whether $\Sigma(H) \subseteq \Sigma(R, B, I)$ holds.*

From this algorithm one can derive the following consequence:

Theorem 2. *If the number of tangled reticulations is limited to k, then Problem 3 is solvable in polynomial time. In particular, Problem 2 is solvable in polynomial time.*

A visualization of a reticulate network N can be obtained using the following algorithm:

Algorithm 3. *Given a set of splits Σ, compute the splits graph $IG(\Sigma)$ using [13]. For each component Z of the incompatibility graph $IG(\Sigma)$, use Algorithm 2 to compute a reticulation scenario for Z. Then, use Algorithm 1 to compute the topology of the reticulate network $N(Z)$ associated with Z. Let Z' denote the netted component in $SG(\Sigma)$ associated with Z. Replace Z' by $N(Z)$ by first removing all non-gate nodes and any edge contained in Z and then mapping the appropriate nodes of $N(Z)$ onto the gate nodes of Z'.*

Although the computation of $SG(\Sigma)$ can take exponential time for an arbitrary split set Σ, for our graph-layout purposes it suffices to compute the splits graph for a subset of splits chosen in such a way that the associated splits graph contains no cliques of size greater than 4, say, which can be done in polynomial time (unpublished result).

7   Implementation and Application

We have implemented the algorithms described in Section 6, and our implementation is freely available as a plug-in for the program SplitsTree [15]. We thus provide a robust and flexible tool for biologists to explore real datasets for evidence of reticulate evolution. Given a set of taxa X, the input can be either a set of X-trees, a set of *partial* X-trees (that is, a set of X'-trees for different subsets of taxa $X' \subseteq X$) or a set of X-splits. In the context of investigating gene trees, it is seldom the case that all trees are full X-trees and so the capability to process partial trees is of particular importance. We achieve this in practice by first applying the Z-closure method [22] to a given set of partial trees to obtain a set of full X-splits.

An important aspect of our implementation is that we provide a visualization of the computed reticulate network. It is based on an algorithm for constructing splits graphs [13] and provides a useful visualization of the complete input

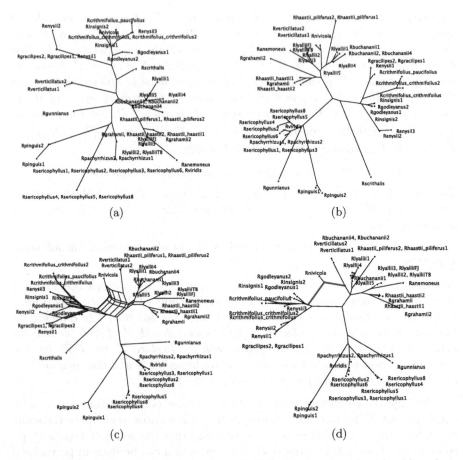

Fig. 3. In (a) and (b) we show two gene trees for 46 buttercups, based on the chloroplast J_{SA} region and nuclear ITS gene [16]. In (c) we show the splits graph of all splits of both trees. Removal of one taxon (*R.scrithalis*) and five interfering splits leads to a configuration that is recognized by our algorithm as a reticulation that gives rise to *R.nivicola*, as shown in (d)

data. Netted regions of the graph that can be explained by a set of overlapping reticulations are drawn as such and the others remain netted, see Figure 2(c–d).

In Figure 3(a–b) we depict two different gene trees for New Zealand alpine *Ranunculus* (buttercup) species based on the nuclear ITS gene and the chloroplast J_{SA} region [16]. The splits graph in Figure 3(c) suggests that *R.nivicola* may be a hybrid between the evolutionary lineages on the left- and right-hand side of the splits graph. However, our algorithm fails to explain the netted region by a collection of overlapping reticulations. This failure is due to additional incompatible splits in both the left- and right-hand side of the graph that extend into the netted component containing *R.nivicola*. Additionally, the placement of *R.scrithalis* is problematic. Interactive deletion of one split on the right-hand side and four splits on the left-hand side, and removal of *R.scrithalis*, leads to a simplified netted component that had a detectable reticulation that may correspond to a hybridization event, Figure 3(d). For this specific case, studies of morphological variation and chromosome numbers had earlier suggested that *R.nivicola* was an allopolyploid (hybrid) formed between *R.insignis* and *R.verticillatus* [16].

8   A Statistical Test for Reticulate Evolution

The mere incompatibility of gene trees does not necessarily constitute evidence for reticulate evolution, as gene phylogenies may conflict by (at least) three other processes: firstly, the gene trees may not be historically accurate due to (i) model misspecification or inappropriate methodology, or due to (ii) sampling effects (insufficient sites to compensate for site saturation or short interior edges); alternatively the gene phylogenies may be historically correct but differ from the species tree due to (iii) the population-genetic effect known as *lineage sorting* [20, 23, 24]. One scenario that could easily be mistaken for reticulation under process (i) is the following. Suppose the evolutionary history of the taxa is accurately described by a tree T (i.e. no reticulate evolution occurred) and this tree describes the history of gene 1 and gene 2 (so there is no lineage sorting effect for these two genes). Suppose further that two taxa x and y that are widely separated in T have independently acquired a strong compositional bias (such as increased GC richness). Then most tree reconstruction methods will reconstruct a phylogeny for gene 2 that is different to T (grouping taxa x and y as sister taxa) but which together with the (correct) tree for gene 1 would be explained by a single reticulation event - namely that taxon x (or y) was a hybrid. In this case one can test the null hypothesis that the two trees differ simply due to compositional variation, against the alternative of genuine reticulation, by adapting the statistical test described by [25].

Lineage sorting (process (iii)) can also be distinguished from reticulation, either by a parametric approach based on divergence time estimates [24], or by a non-parametric approach when the number of gene trees is large. This non-parametric procedure can also allow sampling effects (process (ii)) as well as, or in place of, lineage sorting - provided the substitution process follows a molecular clock. To illustrate this approach – for a sequence of gene trees $(g_1, g_2, \ldots, g_k)$– suppose we have just three taxa a, b, c and a hypothesized rooted species tree $(ab)c$. If n_1 of these gene trees support $(ac)b$ and n_2 of them support $(bc)a$, let $m = n_1 + n_2$, and consider the test statistic

$$\Delta := |n_1 - n_2|.$$

We describe how one can use Δ to test the null hypothesis H_o that $(ab)c$ is the species tree and that the underlying process that resulted in the m other trees is independent occurrences of lineage sorting (or sampling effects subject to a molecular clock) against the alternative hypothesis H_1 that there has also been a reticulation event involving the transfer of some genes from one lineages in the past across to another. Let $I := \{i : g_i \text{ supports } (ac)b \text{ or } (bc)a\}$ so we can write $\Delta = |\sum_{i \in I} \delta_i|$ where $\delta_i = 1$ (resp. -1) if gene g_i supports tree $(ac)b$ (resp. tree $(bc)a$). We will regard the δ_i values as realizations of a sequence of m independent outcomes $D_1, \ldots, D_m$ that take values in $\{1, -1\}$ (i.e. lineage sorting or sampling effects that change the tree structure are assume to be independent from gene to gene). Then under H_o, $\mathbb{E}[D_i] = 0$ since when a lineage sorting event occurs that results in a gene tree at variance with the species tree, then that gene tree is equally likely to be either of two alternative trees (this follows from population-genetic considerations [26, 23]), while for sampling effects subject to a molecular clock $\mathbb{E}[D_i] = 0$ also holds by symmetry. In contrast, under H_1 we would expect $\mathbb{E}[D_i]$ to be systematically negative or positive depending (respectively) on whether $(ac)b$ or $(bc)a$ is the other dominant gene tree involved in the reticulation. Furthermore, $|\Delta - \Delta'| \leq 2$ if Δ and Δ' differ on just value of $i \in I$. Thus, regarding Δ as a function of m independent random variables $(D_1, \ldots, D_m)$ we can apply the Azuma-Hoeffding inequality [27] to deduce that under H_o,

$$\mathbb{P}[\Delta \geq a] \leq 2 \exp(-a_2/8m).$$

This allows us to use Δ as a test statistic to test H_o against H_1. As an example, suppose we find that $n_1 = 60, n_2 = 10$. Then $\Delta(g) = 60 - 10 = 50$, and $m = 70$, so $\mathbb{P}[\Delta \geq 50] \leq 2 \exp(-50^2/8 \cdot 70) = 0.023$, and consequently we could reject H_o at the 5% significance level.

References

1. Holland, B., Huber, K., Moulton, V., Lockhart, P.J.: Using consensus networks to visualize contradictory evidence for species phylogeny. Molecular Biology and Evolution **21** (2004) 1459–1461
2. Rieseberg, L.H., Raymond, O., Rosenthal, D.M., Lai, Z., Livingstone, K., Nakazato, T., Durphy, J.L., Schwarzbach, A.E., Donovan, L.A., Lexer, C.: Major ecological transitions in annual sunflowers facilitated by hybridization. Science **301** (2003) 1211–1216
3. Hein, J.: Reconstructing evolution of sequences subject to recombination using parsimony. Math. Biosci. (1990) 185–200
4. Hein, J.: A heuristic method to reconstruct the history of sequences subject to recombination. J. Mol. Evol. **36** (1993) 396–405
5. Wang, L., Zhang, K., Zhang, L.: Perfect phylogenetic networks with recombination. Journal of Computational Biology **8** (2001) 69–78
6. Gusfield, D., Eddhu, S., Langley, C.: Efficient reconstruction of phylgenetic networks with constrained recombination. In: Proceedings of the 2003 IEEE CSB Bioinformatics Conference. (2003)
7. D. Gusfield, S.E., Langley, C.: The fine structure of galls in phylogenetic networks. to appear in: INFORMS J. of Computing Special Issue on Computational Biology (2004)
8. Nakhleh, L., Warnow, T., Linder, C.R.: Reconstructing reticulate evolution in species - theory and practice. RECOMB'04 (2004) 337–346
9. Hudson, R.R., Kaplan, N.L.: Statistical properties of the number of recombination events in the history of a sample of DNA sequences. Genetics **111** (1985) 147–164

10. Myers, S.R., Griffiths, R.C.: Bounds on the minimal number of recombination events in a sample history. Genetics **163** (2003) 375–394
11. Bafna, V., Bansal, V.: The number of recombination events in a sample history: conflict graph and lower bounds. IEEE/ACM Transactions in Computational Biology and Bioinformatics **1** (2004) 78–90
12. Bandelt, H.J., Dress, A.W.M.: A canonical decomposition theory for metrics on a finite set. Advances in Mathematics **92** (1992) 47–105
13. Dress, A.W.M., Huson, D.H.: Constructing splits graphs. IEEE/ACM Transactions in Computational Biology and Bioinformatics **1** (2004) 109–115
14. Gusfield, D., Bansal, V.: A fundamental decomposition theory for phylogenetic networks and incompatible characters. To appear in: Proceedings of RECOMB'2005 (2004)
15. Huson, D.H., Bryant, D.: Estimating phylogenetic trees and networks using SplitsTree 4. Manuscript in preparation, software available from `www-ab.informatik.uni-tuebingen.de/software` (2004)
16. Lockhart, P.J., McLenachan, P.A., Havell, D., Glenny, D., Huson, D.H., Jensen, U.: Phylogeny, dispersal and radiation of New Zealand alpine buttercups: molecular evidence under split decomposition. Ann Missouri Bot Gard **88** (2001) 458–477
17. Kreitman, M.: Nucleotide polymorphism at the alcohol dehydrogenase locus of Drosophila melanogaster. Genetics **11** (1985) 147–164
18. Semple, C., Steel, M.A.: Phylogenetics. Oxford University Press (2003)
19. Jukes, T.H., Cantor, C.R.: Evolution of protein molecules. In Munro, H.N., ed.: Mammalian Protein Metabolism. Academic Press (1969) 21–132
20. Maddison, W.P.: Gene trees in species trees. Syst. Biol. **46** (1997) 523–536
21. Baroni, M., Semple, C., Steel, M.A.: A framework for representing reticulate evolution. Annals of Combinatorics (In press)
22. Huson, D.H., Dezulian, T., Kloepper, T., Steel, M.A.: Phylogenetic super-networks from partial trees. IEEE/ACM Transactions in Computational Biology and Bioinformatics, in press (2004)
23. Rosenberg, N.A.: The probability of topological concordance of gene trees and species trees. Theor. Pop. Biol. **61** (2002) 225–247
24. Sang, T., Zhong, Y.: Testing hybrization hypotheses based on incongruent gene trees. System. Biol. **49** (2000) 422–424
25. Steel, M.A., Lockhart, P., Penny, D.: Confidence in evolutionary trees from biological sequence data. Nature **364** (1993) 440–442
26. Tajima, F.: Evolutionary relationships of DNA sequences in finite populations. Genetics **105** (1983) 437–460
27. Alon, N., Spencer, J.H.: The Probabilistic Method. 2nd edn. John Wiley (2000)

Appendix

The examples shown in this paper are based on gene trees, and Figures 2–3 were generated using our software. As stated in Section 7, our implementation can also process other types of input, for example binary sequences representing haplotype data. We illustrate this using the data presented in [11], which was reportedly taken from the alcohol dehydrogenase locus from 11 chromosomes of *Drosophila melanogaster* [17]. This data consists of a reduced set of 9 haplotypes typed at 16 sites:

	1	2	3	4	5	6	7	8	9	10	11	12	13	14	15	16
a	0	0	0	1	0	0	0	1	0	0	0	0	0	0	0	0
b	0	1	0	0	0	0	0	1	0	0	0	0	0	0	0	0
c	0	0	0	0	0	0	0	0	0	0	0	0	0	0	1	0
d	0	0	0	0	0	0	1	0	0	0	0	0	0	0	1	0
e	0	0	1	1	1	1	1	0	0	0	0	0	0	0	0	1
f	0	1	0	0	0	1	0	0	0	1	0	1	0	1	1	1
g	0	1	0	0	0	1	0	0	1	1	1	1	1	1	0	1
h	1	1	1	1	1	1	0	0	1	1	1	1	1	1	0	1
i	1	1	1	1	0	1	0	0	1	1	1	1	1	1	0	1

Each column of this matrix defines a split of the taxon set $X = \{a, b, \ldots, i\}$ and let Σ denote the set of all such splits. This data can be entered directly into the SplitsTree program. The resulting splits graph $SG(\Sigma)$ is shown in Figure 4(a) (with all trivial splits added for clarity). This figure indicates that the configuration of splits is quite complex and, as a consequence, our algorithms fail to detect a reticulate network for this data.

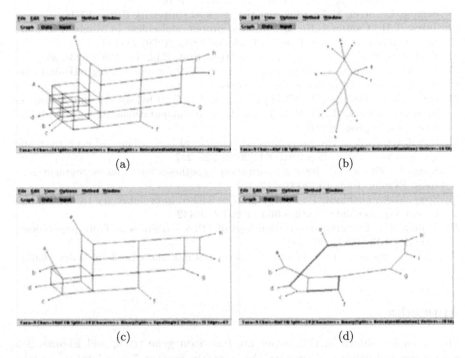

(a)

(b)

(c)

(d)

Fig. 4. In (a) we show the splits graph associated with the full sixteen columns of haplotype data taken from [11]. In (b), we show the splits graph for the four columns $\{1, 4, 5, 6\}$. It consists of two cycles with one reticulation per cycle. In (c), we show the splits graph for 14 columns of the data, with columns 2 and 4 removed. In (d), we show the reticulate network computed from this reduced data set, involving two reticulations, with reticulation edges highlighted by heavy lines

Our implementation allows one to easily add or remove sites from the analysis. In Figure 4(b), we show the splits graph for the subset of sites $\{1, 4, 5, 6\}$. It contains two netted components. In this case, the graph topology alone does not determine which nodes are to be interpreted as reticulation nodes. Declaring one of the taxa to be an outgroup will reduce the number of possible choices of reticulate nodes, but even then there will still be more than one choice.

Let Σ' denote the 14 splits that remain after removing sites 2 and 4. We show the resulting splits graph $SG(\Sigma')$ in Figure 4(c). Application of Algorithm 3 to this reduced set of splits Σ' results in the detection of a solution involving precisely two overlapping reticulations, as shown in Figure 4(d).

Inspection of the sequences reveals that the reticulation displayed at sequence e can be interpreted as a recombination, as e can be obtained from sequences d and h via combination of the first seven (or five, not counting sites 2 and 4) positions of h and the last nine positions of d, with only one mutation in either area. However, it is not possible to obtain sequence f from c and g via a single cross-over and a small number of mutations.

A Hybrid Micro-Macroevolutionary Approach to Gene Tree Reconstruction

Dannie Durand[1], Bjarni V. Halldórsson[2], and Benjamin Vernot[3]

[1] Departments of Biological Sciences and Computer Science,
Carnegie Mellon University
durand@cmu.edu
[2] Department of Mathematical Sciences, Carnegie Mellon University,
Current address: deCode Genetics, Sturlugata 8, 101 Reykjavik, Iceland
bjarni.halldorsson@decode.is
[3] Department of Biological Sciences, Carnegie Mellon University
bvernot@cs.cmu.edu

Abstract. Gene family evolution is determined by microevolutionary processes (e.g., point mutations) and macroevolutionary processes (e.g., gene duplication and loss), yet macroevolutionary considerations are rarely incorporated into gene phylogeny reconstruction methods. We present a dynamic program to find the most parsimonious gene family tree with respect to a macroevolutionary optimization criterion, the weighted sum of the number of gene duplications and losses. The existence of a polynomial time algorithm for duplication/loss phylogeny reconstruction stands in contrast to most formulations of phylogeny reconstruction, which are NP-complete.

We next extend this result to obtain a two-phase method for gene tree reconstruction that takes both micro- and macroevolution into account. In the first phase, a gene tree is constructed from sequence data, using any of the previously known algorithms for gene phylogeny construction. In the second phase, the tree is refined by rearranging regions of the tree that do not have strong support in the sequence data to minimize the duplication/lost cost. Components of the tree with strong support are left intact. This hybrid approach incorporates both micro- and macroevolutionary considerations, yet its computational requirements are modest in practice because the two phase approach constrains the search space.

We have implemented these algorithms in a software tool, NOTUNG 2.0, that can be used as a unified framework for gene tree reconstruction or as an exploratory analysis tool that can be applied *post hoc* to any rooted tree with bootstrap values. NOTUNG 2.0 also has a new graphical user interface and can be used to visualize alternate duplication/loss histories, root trees according to duplication and loss parsimony, manipulate and annotate gene trees and estimate gene duplication times.

1 Introduction

The evolutionary history of a gene family is determined by a combination of microevolutionary events (sequence evolution) and macroevolutionary events. The

S. Miyano et al. (Eds.): RECOMB 2005, LNBI 3500, pp. 250–264, 2005.

macroevolutionary events considered here are speciation, gene duplication and gene loss. Gene tree reconstruction should be based on a model that incorporates both micro- and macroevolutionary events [12], yet few phylogeny reconstruction tools based on such a unified model are available. Furthermore, reconciliation of a gene tree and a species tree can be used to investigate a variety of questions related to macroevolution, such as inferring gene duplications and losses, estimating upper and lower bounds on times these events occurred and determining whether a given pair of homologs is orthologous or paralogous. Algorithmic and software support for both these tasks is needed.

In the current work, we present a dynamic programming algorithm to find all most parsimonious phylogenies with respect to a macroevolutionary model of gene duplication and loss. Given the number of gene family members found in each species as input, our algorithm will construct a tree with the fewest duplications and losses required to explain the data. We show that this problem can be solved in polynomial time, in contrast to most phylogeny reconstruction problems, which are NP-complete [4, 6, 7].

Using this result, we develop a two-phase approach to gene tree reconstruction that incorporates sequence evolution, gene duplication and gene loss in the evaluation of alternate phylogenies. In phase one, a tree is constructed based on a microevolutionary model only. In phase two, regions of the tree that are not strongly supported by the sequence data are refined with respect to a macroevolutionary parsimony model, while regions with strong support are left intact. By reserving consideration of macroevolutionary events until phase two and focusing only on those areas where the sequence data cannot resolve the topology, this hybrid approach reduces the search space, leading to a method that incorporates both types of events, yet has modest computational requirements.

We have implemented these algorithms in a software tool called NOTUNG 2.0, which can be used as a unified framework for gene tree reconstruction or as an exploratory analysis tool that can be applied *post hoc* to any rooted tree with bootstrap values. This extension of an earlier program [3] has the following new features:

- a polynomial time algorithm for finding *all* most parsimonious trees with respect to a weighted gene duplication and loss cost.
- a new graphical interface for tree manipulation and visualization, with features especially designed for analysis of gene duplications, including identification of duplicated nodes, lost leaves and/or subtrees, annotation of subfamilies and color highlighting of rootable edges.

In addition, NOTUNG 2.0 retains all features of the original program including high throughput identification of gene duplications with time estimates and rooting unrooted trees based on duplication/loss score. NOTUNG 2.0, described on the supplementary web site (http://www.cs.cmu.edu/ durand/Lab/Notung/), is available for free, public distribution.

In Section 2, we discuss previous work and review reconciliation of a gene tree with a species tree. A taxonomy of macro- and microevolutionary models is presented in Section 3. We state our gene tree reconstruction problems formally in Section 4 and present algorithms to solve these problems in Section 5. In Section 6, we summarize the capabilities of our software tool, NOTUNG 2.0, and discuss the application of NOTUNG 2.0 to two large data sets to investigate the role of gene duplication in genetic adaptation to environmental change.

2   Previous Work on Reconciliation

The problem of disagreement between gene trees and species trees was first raised in the context of inferring a species tree from a gene tree that may contain paralogies [12]. These concepts were further developed and formalized in [13,14, 16,17,20,21,28,29]. Formally, given a set of rooted gene trees, the problem is to find the species tree that optimizes an evaluation criterion. Several optimality criteria have been proposed (see [9,10] for a comparative survey), all of which attempt to capture the notion that gene duplication and subsequent loss are rare events. The problem of finding an optimal species tree is NP-hard [16] for the optimality criteria considered so far.

A related body of work deals with algorithms and software tools to analyze the history of duplications and losses in the evolution of a gene family. Page and Charleston [19,22] have developed two software packages, COMPONENT and GENETREE, that will compute and display duplication histories for rooted gene trees, as well as inferring species trees. Zmasek and Eddy's Forester package performs some reconciliation functions and provides a tree visualization tool [30,31].

All of these approaches are based on reconciliation of a gene tree with a species tree, a procedure that can be used to infer the number of duplications and losses and estimate the times at which they occurred. We review reconciliation here. Let T_G be a gene tree and let T_S be a species tree of taxa from which the gene sequences were sampled. The identification of duplication nodes requires constructing a mapping from every node in T_G to a target node in T_S. Each leaf node in T_G is mapped to the node in T_S representing the species from which the sequence was obtained. (Leaf nodes in T_G represent sequences, whereas leaf nodes in T_S represent species.) Each internal node in T_G is mapped to the least common ancestor (*lca*) of the target nodes of its children. In Fig. 1(a), for example, the leaf nodes in the rightmost subtree are mapped to MOUSE and HUMAN. The root of this subtree is mapped to *eut* (eutherian), since the *lca* of MOUSE and HUMAN in the species tree is *eut*.

The number of duplications can be determined by observing that under the mapping, a node in T_G is a *speciation* node if its children are mapped to independent lineages in T_S. If the children of v are mapped to the same lineage (either they are mapped to the same species or one's mapping is an ancestor of the other's), then v is a *duplication* node. In the gene tree in Fig. 1(a), both nodes labeled *eut* are speciation nodes since rodents and primates are separate lineages in T_S. The root of the tree is a duplication node because it has the same label as both of its children. The number of losses can also be determined from the mapping. Details of both the duplication and loss count computations are given in [3].

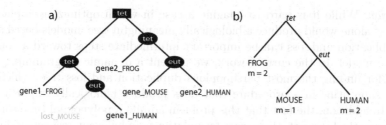

Fig. 1. (a) A gene tree with one duplication and one loss. Internal nodes represent duplication (boxes) or speciation (ovals). Lost genes are shown in gray. (b) A species tree for frog, human and mouse. Leaves of the species tree are labeled with the number of gene family members found in each species. Internal nodes in the species tree are labeled *eut* (eutherian) and *tet* (tetrapod)

The cost of duplications and losses can be expressed by the following optimization function:

Definition 1. *The **D/L Score** of a gene tree is $c_\lambda L + c_\delta D$, the weighted sum of the number of duplications, D, and the number of losses, L, in the tree.*

3   A Taxonomy of Models

The goal of phylogeny reconstruction is to determine the hypothesis, expressed as an evolutionary tree, that best explains the data with respect to a model of evolutionary change. Given a set of sequences from a gene family, which may include paralogous sequences from the same species and orthologous sequences from different species, a gene tree can be reconstructed according to various evolutionary models.

Microevolutionary model: *Find the tree that best explains the data with respect to a model of sequence evolution only.* In practice, most gene trees are constructed based on a model of sequence evolution alone, as this approach requires a less complex model, is less computationally intensive, and may be achieved with one of the many tools currently available for sequence-based phylogeny reconstruction. However, information about macroevolutionary events is not incorporated in this approach. In many cases, workers subsequently infer the macroevolutionary history implied by the sequence-based tree, either by inspection or by using a software tool for exploratory analysis (e.g., [3,23,30,31]), thereby treating the gene tree as though it were data rather than a hypothesis. If bootstrap values [8] indicate weak support for some edges in the tree, some practitioners discuss alternate hypotheses motivated by *post hoc* macroevolutionary considerations, partially mitigating this problem.

Macroevolutionary model: *Find the tree that best explains the data with respect to a model of duplications and losses with no consideration of sequence* *evolution.* While it is hard to imagine a case in which optimizing duplication and loss alone would produce a biologically meaningful tree, models based solely on duplication and loss can be important intermediate steps toward a comprehensive model. In the current work, we present a dynamic programming algorithm for finding the most parsimonious duplication and loss tree, which is a useful subroutine in a procedure that does produce biologically important results. In addition, the fact that this problem admits a polynomial time solution is of theoretical interest, since most formulations of phylogeny reconstruction are NP-complete [4, 6, 7]. In a Bayesian context, such models have also been used to investigate related questions such as estimating rates of duplication and loss, determining the posterior probability of a tree with respect to a reconciliation and probabilistic ortholog identification [1, 11].

Unified model: *Find the tree that best explains the data with respect to a model that takes both sequence evolution and duplication and loss into account.* In a 1979 landmark paper, Goodman and colleagues [12] pointed out the importance of using a model of both micro- and macroevolutionary events for constructing gene trees and introduced the term "reconciliation" to describe fitting a gene tree to a species tree. They implemented a heuristic search procedure for obtaining parsimony trees that optimize a cost function based on nucleotide replacement and gene duplication and loss. However, it is difficult to determine how to incorporate events occurring on very different spatial and temporal scales in a single model, and most gene tree reconstruction in the intervening 25 years has been based on sequence evolution alone. Very recently, some Bayesian approaches to a unified model have appeared. Arvestad *et al.* [2] have presented a maximum likelihood method that evaluates a set of sequences with respect to a model that includes the gene tree and the reconciliation, parameterized by birth, death and substitution rates.

The Bayesian framework permits a unified model of macro- and microevolution facilitating an approach that uses both types of information in the reconstruction process. It also has the advantage that it allows us to test evolutionary models and infer parameters of those models. On the down side, Bayesian approaches are notoriously computationally intensive and require sufficient data to obtain reasonable estimates of the parameters. Furthermore, a unified, Bayesian model is a strength when both sequence evolution and gene duplication and loss can be modeled by a neutral, stochastic process, but less natural for data sets under strong selective pressure. Evolutionary change in this latter regime is not stochastic and parsimony may be a more appropriate model. With these considerations in mind, we propose the following approach:

Hybrid model: *Obtain an initial tree using a microevolutionary model and refine it with respect to macroevolutionary considerations.* More specifically, we construct an initial tree based on sequence data alone and assess the support for each edge using bootstrapping [8] (or some other method of significance estimation). Subsequently, we rearrange the resulting tree around edges with weak support in the sequence data (e.g., bootstrap values below some threshold) to optimize the number of duplications and losses needed to explain the tree, while preserving the structure of the tree at edges with strong support.

This hybrid approach incorporates both micro- and macroevolutionary considerations in phylogeny reconstruction. Although in pathological cases an exponential number of trees must be considered, in practice its computational and data requirements are modest because the two-phase approach constrains the search space. Furthermore, it makes it possible to decouple the micro- and macroevolutionary models. Since duplication and loss occur rarely relative to sequence mutation, parsimony may be a more appropriate macroevolutionary model than maximum likelihood for many data sets. In this paper, we present a particular implementation of this approach where the initial sequence-based tree is constructed using any standard phylogeny reconstruction method and, hence, any microevolutionary model. The refinement step is based on a parsimony model of duplication and loss.

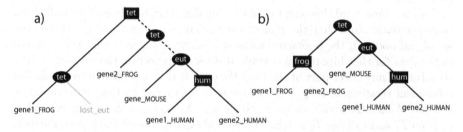

Fig. 2. (a) A gene tree with two duplication nodes and one lost gene. Weak edges are shown as dashed lines. The edge between the *eut* and *human* nodes is robust. The edges adjacent to leaf nodes are neither weak nor robust, since these edges cannot be evaluated using bootstrapping. (b) An alternate topology for the same gene family, with two duplications and no losses

This hybrid approach is illustrated by the hypothetical gene tree in Fig. 2(a), which shows a gene family with two members in frog, two in human and one in mouse. The topology of the tree indicates that two duplications occurred in this gene family. The first occurred in the common ancestor of all three species. One copy was retained in all three species, while the second was retained in frog and lost in mouse and human. We represent this as a single loss in their common ancestor (*eut*), since this is the most parsimonious explanation for the two missing genes. A second duplication occurred within the human lineage. The total score for this tree is two duplications and one loss. However, the edge grouping *gene2_FROG* with the mouse and human genes is weak, suggesting that the sequence evidence does not, in fact, strongly support this topology. Rearranging the tree around the weak edge to place *gene2_FROG* in the left subtree with *gene1_FROG* (Fig. 2(b)) results in a tree that requires two duplications and no losses to explain.

4   Formal Macroevolutionary Reconstruction Problems

The problem of finding all most parsimonious trees with respect to a duplication/loss cost function alone, ignoring sequence information, can be formally stated as follows:

Macroevolutionary Phylogeny

> **Input:** A rooted species tree, T_S with s leaves; a list of multiplicities $m_1 \ldots m_s$, where m_l is the number of gene family members found in species l; weights c_λ and c_δ.
> **Output:** The set of all rooted gene trees $\{T_G\}$ with $\sum_{l=1}^{s} m_l$ leaves such that the **D/L Score** of T_G is minimal.

We provide a polynomial time algorithm to solve this problem, showing that, unlike most formulations of phylogeny reconstruction, which are NP-complete [4, 6,7], an optimal solution to the **Macrophylogeny** problem can be obtained in polynomial time.

Second, we extend this result to obtain an algorithm to refine a tree built from sequence data. Note that the removal of any edge, e, in a tree bipartitions the set of leaf nodes. If the bootstrap value of e is low, it suggests that the evidence in the data for that bipartition is weak. It does not reflect on the certainty of the structure of any other part of the tree. We exploit this observation to obtain our refinement strategy. Let $T_1 = (V_1, E_1)$ and $T_2 = (V_2, E_2)$ be trees with the same leaf set. We say T_1 and T_2 agree at edges $e_1 \in E_1$ and $e_2 \in E_2$ if the removal of e_1 from T_1 and e_2 from T_2 results in identical bipartitions of the leaves. Further, T agrees with T_1 at e_1, if there is some edge in T that agrees with e_1.

Definition 2. *Let $T_G = (V, E)$ be a rooted tree and let $R \subseteq E$ be a set of robust edges (e.g., edges with bootstrap values above some threshold, θ). We define $T_{G,R}$ to be the set of all rooted trees that agree with T_G at every edge in R and call $T \in T_{G,R}$ a rearrangement tree of T_G. $T_{G,R}^* \subset T_{G,R}$ denotes the subset of trees with minimum **D/L Score**.*

We now state the reconstruction problem for the more general, hybrid parsimony model. Note that in the case where $R = \emptyset$, $T_{G,R}$ is simply the set of all trees with $|L(T_G)|$ leaves, where $L(T_G)$ is the leaf set of T_G, and the **Hybrid Micro-Macrophylogeny** problem reduces to the **Macrophylogeny** problem.

Hybrid Micro-Macrophylogeny

> **Input:** A rooted gene tree, T_G with robust edges $R \subseteq E$; a rooted species tree, T_S; weights c_λ and c_δ.
> **Output:** $T_{G,R}^*$

Due to several types of degeneracy there are potentially a large number of equally parsimonious trees in $T_{G,R}^*$. We define a duplication/loss history to be a set of event, edge pairs, where an event is a duplication or a loss and the associated edge in T_S specifies when the event occurred. There may be more

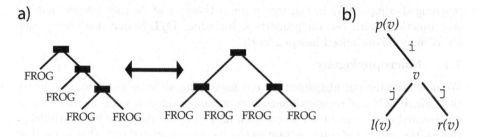

Fig. 3. (a) Two equally parsimonious arrangements of four FROG genes, each with three duplications. (b) Schematic showing the meaning of variables in the algorithm in Fig. 4

than one history with the same **D/L Score**. For example, if $c_\lambda = c_\delta = 1$, then the gene trees in Fig. 1(a) and Fig. 2(b) both have a **D/L Score** of two, although they have different histories (one duplication and one loss versus two duplications).

Additional degeneracy arises because the same history may correspond to more than one tree. This occurs when labels in the gene tree can be permuted without changing the score. For example, in Fig. 1(a), exchanging *gene1*_HUMAN and *gene2*_HUMAN results in a different tree for the same history. Multiple trees for the same history can also occur when subtrees with different topologies within a single species have the same number of duplication nodes. This is illustrated in Fig. 3(a).

Our algorithm generates one tree for each distinct history. We present all other trees for each history to the user through NOTUNG 2.0's graphical user interface. Nodes that may be swapped without changing the **D/L Score** are highlighted, allowing the user to generate alternate minimum cost permutations using a point and click interface.

5   Reconstruction Algorithms

In this section, we first present a dynamic program for reconstructing a gene tree based on macroevolutionary considerations only. Since sequence evolution is not taken into account, the only information about the gene family required is the number of family members observed in each species in the species tree. Given the species tree, these multiplicities and positive weights c_λ and c_δ as inputs, our algorithm determines the minimum **D/L Score** and generates all most parsimonious histories.

In Section 5.2, we discuss how to incorporate these results in an algorithm for optimizing the reconciliation of a gene tree with weak edges. Each connected component of weak edges (CCW) in this tree is a binary, rooted tree whose leaves are strong subtrees in the original tree. An extension of the dynamic program presented in 5.1 is used to obtain all minimum cost rearrangements of the embedded rooted tree associated with each CCW. We then reinsert the rearranged components in the tree using a theorem of [3] that proves that for cost functions with certain properties, including **D/L Score** used here, each *CCW* may be optimized independently.

5.1 Macrophylogeny

We now describe our algorithm for reconstructing all most parsimonious histories, where only macroevolutionary events are considered. Each history can be represented as a species tree, where each node is annotated with its multiplicity; that is, the number of copies extant in the species associated with that node. The multiplicity of the root must be one, while the multiplicities of the leaves, denoted $m_1 \ldots m_s$, are specified in the input. An example of this is shown in Fig. 1(b).

Our dynamic program considers all possible multiplicities for each internal node. The variable i is the number of gene copies a node inherits from its parent, $p(v)$, and j is the number of gene copies a node sends to its children, $l(v)$ and $r(v)$, as shown in Fig. 3(b). The values of i and j range from one to $\hat{m} \leftarrow \max_{\forall l \in L(T_S)} \{m_l\}$, where $L(T_S)$ is the set of leaves in T_S[1]. For each node, v, the dynamic program calculates the minimum **D/L Score** of the subtree rooted at v for all possible values of i and j. These values are stored in the minimum cost table, $cost_v^{min}[i] \leftarrow \min_{\forall j} \{cost_v[i,j]\}$. The array $cost_v^{min}[i]$ is kept to enable quick lookup of the minimum cost given that v inherits i genes from its parent.

Pseudocode for this algorithm is shown in Fig. 4.

RECONSTRUCT, the main loop, calls the procedure ASCEND, which annotates the minimum cost tables for all nodes of T_S. To enumerate all alternate histories from these tables, RECONSTRUCT repeatedly calls DESCEND followed by CONSTRUCT. In each iteration, DESCEND extracts a new annotation that corresponds to a distinct optimal history. CONSTRUCT then builds a gene tree to represent the history marked by DESCEND. This gene tree is output and DESCEND is called again to find the next history. The procedure terminates when DESCEND returns *true* to indicate that all histories have been enumerated.

The enumeration scheme requires that each node keep track of those lowest cost entries in its cost table that have already been selected for alternate histories by DESCEND. Each node maintains persistent state variables including *v.dups*, the optimal number of duplicated genes in this species; *v.losses*, the optimal number of lost genes in this species; and *v.out*, the optimal number of genes for v to pass to its children. DESCEND also uses the variable *v.done*, which is *true* if v has no more histories to return for its current value of i. This value is reinitialized by calling DESCEND with a new i and the status variable *reset* set to *true*. Additional bookkeeping information is stored on each node for use by the NOTUNG user interface, to enable the user to generate alternate minimum cost permutations, if they exist, for each history displayed.

---

[1] To see this, note that if the number of gene copies in any vertex, v, of T_S is larger than $\hat{m}$ there will have to be losses to reduce the gene count below v in T_S. As long as c_δ and c_λ are positive this will increase the **D/L Score** of the resulting gene tree. Therefore, it is never necessary to consider more than $\hat{m}$ gene copies in any vertex of T_S.

```
RECONSTRUCT[T_S, {m_1 ... m_s}]
    ASCEND[root(T_S)];  root(T_S).done ← true;
    repeat {
        root(T_S).done ← DESCEND[root(T_S), root(T_S).done, 1];
        ** reset construction counters **
        if( !root(T_S).done ) T_G ← CONSTRUCT[root(T_S)]; output(T_G);
    } until( root(T_S).done )
```

```
ASCEND[v]
if v is not a leaf: ASCEND[l(v)]; ASCEND[r(v)];
```
$\forall i, j \ s.t. \ 1 \le i \le \hat{m}, \ 0 \le j \le \hat{m}$
```
    if v is a leaf:
```
$cost_v^{min}[i] \leftarrow c_\delta * \max(m_v - i, 0) + c_\lambda * \max(i - m_v, 0);$
```
    if v is not a leaf:
```
$cost_v[i, j] \leftarrow c_\delta * \max(j - i, 0) + c_\lambda * \max(i - j, 0) + cost_{l(v)}^{min}[j] + cost_{r(v)}^{min}[j];$
$\forall i \ cost_v^{min}[i] \leftarrow \min_{\forall j} \{cost_v[i, j]\};$

```
DESCEND[v, reset, i]
if v is a leaf:
```
$v.losses \leftarrow \max((i - m_v), 0); \ v.dups \leftarrow \max((m_v - i), 0);$
```
    v.out ← 0; return false;
if (!reset)
    if ( v.done ) return true ** no more histories. **
    if ( !(l(v).done ← DESCEND[l(v), false, v.out]) ) return false;
    if ( !(r(v).done ← DESCEND[r(v), false, v.out]) ) return false;
    ** go on to next optimal out degree **
    repeat { v.out ++ } until ( cost_v[i, v.out] == cost_v^min[i] OR v.out > m̂ );
    if( v.out > m̂ ) return v.done ← true;
else ** reset == true **
    v.out ← 0; repeat { v.out ++; } until ( cost_v[i, v.out] == cost_v^min[i] );
    l(v).done ← DESCEND[l(v), true, v.out]; r(v).done ← DESCEND[l(v), true, v.out];
```
$v.losses \leftarrow \max((i - v.out), 0); \ v.dups \leftarrow \max((v.out - i), 0);$
```
return false;
```

```
CONSTRUCT[s]
g ← new gene node; g.species ← s
if (s.currDup < s.dups)
    s.currDup ++; l(g) ← CONSTRUCT[s]; r(g) ← CONSTRUCT[s];
else if (s.currLoss < s.losses)
    s.currLoss ++;
else if (s.currSpec < s.out)
    s.currSpec ++;
    if s is not a leaf: l(g) ← CONSTRUCT[l(s)]; r(g) ← CONSTRUCT[r(s)];
return g;
```

Fig. 4. The Macrophylogeny Reconstruction Algorithm

Lemma 1. *The time required for* RECONSTRUCT *to find a single optimal history is* $O(n\hat{m}^2)$, *where* n *is the number of nodes in the species tree and* $\hat{m}$ *is the maximum number of genes drawn from any species. The time complexity for reporting* k *optimal histories is* $O(n\hat{m}(k + \hat{m}))$.

Proof. RECONSTRUCT calls ASCEND once. The leaves of the species tree can be annotated with multiplicities in $O(n)$ time. The main computational cost in ASCEND is calculating the cost matrix in the internal species nodes, and the minimum cost vector in the leaf nodes. Each cost matrix is of size $\hat{m}(\hat{m}+1)$, and each minimum cost vector is of size $\hat{m}$. Thus, the time complexity of ASCEND is $O(n\hat{m}^2)$.

DESCEND and CONSTRUCT are called once per optimal history. The time required for DESCEND to find the value of j that minimizes the cost at given a node for a given value of i takes time $O(\hat{m})$. Each node in the species tree will be visited at most two times, so the total complexity for DESCEND is $O(n\hat{m})$. CONSTRUCT inserts duplication and loss nodes in the new tree, which can number in total no more than $\hat{m}$ per node in T_S. Hence, the total complexity for CONSTRUCT is $O(n\hat{m})$. □

Lemma 2. RECONSTRUCT *finds all histories with minimum* **D/L Score**.

Proof. We first show that ASCEND correctly calculates $cost_v^{min}[i]$, the minimum cost, for all nodes v and all possible number of inherited genes, i. We prove this by induction on the root of every subtree of T_S. Note that the algorithm is trivially correct for all trees containing only one node, independent of i. In a species tree with more than one node, either a duplication or a loss can occur at the same node in the species tree, but never both, as such a history will never be optimal. For a given i and j, the duplication/loss cost for a species node, v, is the optimal cost of the left and right subtrees, given that they inherit j gene copies from v, plus the cost of losses $(c_\lambda \cdot \max(i - j, 0))$ or duplications $(c_\delta \cdot \max(j - i, 0))$, whichever event occurs at this node. The minimum cost is then the lowest value, taken over all j from 1 to $\hat{m}$. Then, by the induction hypothesis, ASCEND calculates the minimum **D/L Score** for each of subtree, for all i.

We next show by induction that a call to DESCEND will then always result in an optimal history, given that ASCEND has correctly annotated the species tree. For a single node, DESCEND will correctly select each optimal history calculated by ASCEND. For each internal node, v, in a non-trivial tree, for a given i, and for every j which minimizes the cost of the subtree, DESCEND requests optimal histories from each child of v until they both report done. Thus, DESCEND considers and reports every combination of minimum cost histories from every node in a tree. □

5.2   Hybrid Micro-Macrophylogeny

In order to incorporate the algorithm in Fig. 4 in a framework for optimizing a gene tree T_G with both strong and weak edges, we must make a few

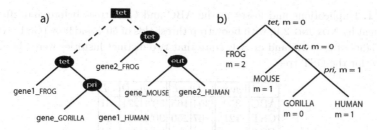

Fig. 5. (a) A gene tree with a weakly connected component, shown here by dotted edges. (b) Annotated tree for the species in (a). Note that the internal species node *pri* (primate) has a multiplicity, while GORILLA does not

modifications. A preprocessing step is needed to extract the rooted tree corresponding to each CCW and record its multiplicities in the species tree. Note that the leaves of a CCW may be strong subtrees in T_G. Thus, unlike the algorithm in Fig. 4, both the internal nodes and leaves of the species tree are annotated with multiplicities. An example of a gene tree with a strong subtree and the corresponding annotated species tree can be seen in Fig. 5. Additional bookkeeping is required in order to reconnect the reconstructed CCWs to strong parts of the tree. The complete code for this algorithm is available at (http://www.cs.cmu.edu/~durand/Lab/Notung/).

6   Experimental Results

The algorithms described in the previous sections have been implemented in a Java program called NOTUNG 2.0. Given a rooted gene family tree, a rooted species tree, a bootstrap threshold, θ, and costs c_δ and c_λ, NOTUNG 2.0 computes all optimal rearrangement histories and presents one tree for each optimal history. The user can generate all other trees for the same history using a point and click interface. Input trees are represented in Newick format [18] and must be binary. The graphical user interface was constructed using the tree visualization library provided by ATV (version 1.92) [30].

In a previous study [3], we developed a test set of thirteen trees discussed in three recent articles on large scale duplication [15,24,25]. For each rooted tree in the test set, we compared the results automatically generated by NOTUNG 2.0 with those of the original authors. The duplication histories generated were consistent with the analyses of the authors of the original papers for all trees considered.

In addition, we used NOTUNG 2.0 to analyze two large data sets, summarized in Table 1. The first data set [5] includes ATP-binding cassette (ABC) transporter sequences from *D. discoideum*, *A. thaliana* and *S. cerevisiae* and a number of *Plasmodia* species, provided by Roxana Cintron, Dr. Adelfa Serrano and colleagues (University of Puerto Rico). The second, provided by Dr. Hugh Nicholas (Pittsburgh Supercomputing Center) is derived from the Glutathione-

Table 1. Duplications and losses in the ABC and GST trees before and after rearrangement by NOTUNG 2.0, with bootstrap thresholds of 50% and 90% (GST tree only) and weights of $c_\delta = 1.5$ and $c_\delta = 1$. Note that two distinct histories were obtained for $\theta = 90\%$ for the GST tree

	leaves	D	L	θ	D	L
ABC	350	304	165	50%	271	61
GST	121	67	250	90%	51	58
GST				90%	49	61
GST				50%	55	164

S Transferases [27], a superfamily of detoxification enzymes. Initial sequence trees for both data sets were constructed using Neighbor Joining [26]. Bootstrap replicates were obtained using the SEQBOOT program from Felsenstein's Phylip package (v. 3.6.1) available at the Pittsburgh Supercomputing Center. The divergence in both families was substantial resulting in a large number of edges with low bootstrap values. NOTUNG 2.0 was used to count the number of duplications and losses in the resulting sequence trees and then to rearrange the trees. The results given in Table 1 show that macroevolutionary rearrangement substantially reduces the number of duplications and losses needed to explain the data.

The trees analyzed here provide a concrete example of Notung's utility in functional as well as evolutionary applications. Gene duplication followed by functional differentiation is a mode of adaptation to environmental change. Identification of recent duplications in specific lineages indicates genes that are sites of rapid change and, when combined with ecological and biochemical data, the environmental forces to which they are responding. This information can be used to plan additional experimental studies, suggest strategies for circumventing drug and pesticide resistance in parasites, identify potential detoxification enzymes for use in bioremediation and design breeding programs to enhance pest resistance in cash crops.

Drug and toxin resistance in the ATP-binding Cassette (ABC) transporter family in *Arabidopsis thaliana* provides a concrete example of this process. Analysis and visualization of ABC tree [5] with NOTUNG 2.0 revealed a large number of recently duplicated genes in the Multi-Drug Resistance subfamily, suggesting a pattern of recent, lineage-specific adaptation, possibly in response to new pesticides. These genes are potential industrial targets for bioremediation.

Acknowledgments

We thank R. Cintron, M. Farach-Colton, R. Hoberman, D. Miranker, H. B. Nicholas, Jr. and R. Ravi for helpful discussions, R. Ravi for his contributions to algorithm development and R. Cintron, H. B. Nicholas, Jr. and A. E. Serrano for providing the ABC transporter and GST data sets. D.D. and B.V were supported by NIH grant 1 K22 HG 02451-01 and a David and Lucille Packard

Foundation fellowship. B.V.H. was supported by a Merck Computational Biology and Chemistry Program Graduate Fellowship from the Merck Company Foundation.

References

1. Lars Arvestad, Ann-Charlotte Berglund, Jens Lagergren, and Bengt Sennblad. Bayesian gene/species tree reconciliation and orthology analysis using MCMC. *Bioinformatics*, 19 Suppl 1:i7–15, 2003.
2. Lars Arvestad, Ann-Charlotte Berglund, Jens Lagergren, and Bengt Sennblad. Gene tree reconstruction and orthology analysis based on an integrated model for duplications and sequence evolution. In *Proceedings of the Eighth Annual International Conference on Computational Molecular Biology*, pages 326–335. ACM Press, 2004.
3. K. Chen, D. Durand, and M. Farach-Colton. Notung: A program for dating gene duplications and optimizing gene family trees. *Journal of Computational Biology*, 7(3/4):429–447, 2000.
4. B. Chor and T. Tuller. Maximum likelihood of evolutionary trees is hard. In Satoru Miyano, editor, *Proceedings of the Ninth Annual International Conference on Computational Molecular Biology (RECOMB 2005)*, Lecture Notes in Bioinformatics. Springer Verlag, 2005.
5. R. Cintron, H. B. Nicholas Jr., I. Ferrer, R. Gonzalez, B. Vernot, D. Durand, and A. E. Serrano. The ABC superfamily: Evolutionary implications for drug resistance in extitPlasmodia. In *Functional Genomics and Bioinformatics Approaches to Infectious Disease Research*. American Society for Microbiology, 2004. Abstract book.
6. W. H. Day. Computational complexity of inferring phylogenies from dissimilarity matrices. *Bull Math Biol*, 49(4):461–7, 1987.
7. W. H. E. Day, D. S. Johnson, and D. Sankoff. The computational complexity of inferring rooted phylogenies by parsimony. *Math Biosci*, 81(33):33–42, 1986.
8. Bradley Efron and Gail Gong. A leisurely look at the bootstrap, jackknife, and cross-validation. *The American Statistician*, 37(1):36–48, 1983.
9. O. Eulenstein, B. Mirkin, and M. Vingron. Comparison of a annotating duplication, tree mapping, and copying as methods to compare gene trees with species trees. *Mathematical Hierarchies and Biology, DIMACS Series in Discrete Mathematics and Theoretical Computer Science*, 37:71–93, 1996.
10. O. Eulenstein, B. Mirkin, and M. Vingron. Duplication-based measures of difference between gene and species trees. *Journal of Computational Biology*, 5:135–148, 1998.
11. Joseph Felsenstein. *Inferring Phylogenies*, chapter 29, page 514. Sinauer, 2003.
12. M. Goodman, J. Czelusniak, G.W. Moore, A.E. Romero-Herrera, and G Matsuda. Fitting the gene lineage into its species lineage, a parsimony strategy illustrated by cladograms constructed from globin sequences. *Syst Zool*, 28:132–163, 1979.
13. R. Guigo, I. Muchnik, and T.F. Smith. Reconstruction of ancient phylogenies. *Molecular Phylogenetics and Evolution*, 6:189–213, 1996.
14. M. T. Hallett and J. Lagergren. New algorithms for the duplication-loss model. In *RECOMB2000, Fourth Annual International Conference on Computational Molecular Biology*, 2000.

15. A. L. Hughes. Phylogenetic tests of the hypothesis of block duplication of homologous genes on human chromosomes 6, 9, and 1. *MBE*, 15(7):854–70, 1998.

16. B. Ma, M. Li, and L. Zhang. From gene trees to species trees. *SIAM J. on Comput.*, 2000.

17. B. Mirkin, I. Muchnik, and T.F. Smith. A biologically consistent model for comparing molecular phylogenies. *Journal of Computational Biology*, 2:493–507, 1995.

18. The Newick tree format, 1986. http://evolution.genetics.washington.edu/phylip/newicktree.html

19. R. D. Page. GeneTree: comparing gene and species phylogenies using reconciled trees. *Bioinformatics*, 14(9):819–20, 1998.

20. R. D. M. Page. Maps between trees and cladistic analysis of historical associations among genes, organisms and areas. *Syst Zool*, 43:58–77, 1994.

21. R. D. M. Page and M. A. Charleston. Reconciled trees and incongruent gene and species trees. *Mathematical Heirarchies and Biology, DIMACS Series in Discrete Mathematics and Theoretical Computer Science*, 37:57–70, 1996.

22. R. D. M. Page and M. A. Charleston. From gene to organismal phylogeny: Reconciled trees and the gene tree/species tree problem. *Molecular Phylogenetics and Evolution*, 7:231–240, 1997.

23. R. D. M. Page and M. A. Charleston. Trees within trees: phylogeny and historical associations. *Trends in Ecology and Evolution*, 13(9):356–359, 1998.

24. M.-J. Pebusque, F. Coulier, D. Birnbaum, and P. Pontarotti. Ancient large-scale genome duplications: phylogenetic and linkage analyses shed light on chordate genome evolution. *MBE*, 15(9):1145–59, 1998.

25. I. Ruvinsky and L. M. Silver. Newly indentified paralogous groups on mouse chromosomes 5 and 11 reveal the age of a t-box cluster duplication. *Genomics*, 40:262–266, 1997.

26. N. Saitou and M. Nei. The neighbor-joining method: A new method for reconstructing phylogenetic trees. *Molecular Biology and Evolution*, 4:406–425, 1987.

27. D. Sheehan, G. Meade, V. M. Foley, and C. A. Dowd. Structure, function and evolution of glutathione transferases: implications for classification of non-mammalian members of an ancient enzyme superfamily. *Biochem J*, 360(Pt 1):1–16, Nov 2001.

28. U. Stege. Gene trees and species trees: The gene-duplication problem is fixed-parameter tractable. In *Proceedings of the 6th International Workshop on Algorithms and Data Structures (WADS'99)*, 1999.

29. L. Zhang. On a Mirkin–Muchnik–Smith conjecture for comparing molecular phylogenies. *Journal of Computational Biology*, 4:177–188, 1997.

30. C. M. Zmasek and S. R. Eddy. ATV: display and manipulation of annotated phylogenetic trees. *Bioinformatics*, 17(4):383–4, Apr 2001.

31. C. M. Zmasek and S. R. Eddy. A simple algorithm to infer gene duplication and speciation events on a gene tree. *Bioinformatics*, 17(9):821–8, Sep 2001.

Constructing a Smallest Refining Galled Phylogenetic Network

Trinh N.D. Huynh, Jesper Jansson, Nguyen Bao Nguyen, and Wing-Kin Sung

School of Computing, National University of Singapore,
3 Science Drive 2, Singapore 117543
{huynhngo, jansson, nguyenba, ksung}@comp.nus.edu.sg

Abstract. Reticulation events occur frequently in many types of species. Therefore, to develop accurate methods for reconstructing phylogenetic networks in order to describe evolutionary history in the presence of reticulation events is important. Previous work has suggested that constructing phylogenetic networks by merging gene trees is a biologically meaningful approach. This paper presents two new efficient algorithms for inferring a phylogenetic network from a set $\mathcal{T}$ of gene trees of arbitrary degrees. The first algorithm solves the open problem of constructing a refining galled network for $\mathcal{T}$ (if one exists) with no restriction on the number of hybrid nodes; in fact, it outputs the smallest possible solution. In comparison, the previously best method (SpNet) can only construct networks having a single hybrid node. For cases where there exists no refining galled network for $\mathcal{T}$, our second algorithm identifies a minimum subset of the species set to be removed so that the resulting trees can be combined into a galled network. Based on our two algorithms, we propose two general methods named RGNet and RGNet+. Through simulations, we show that our methods outperform the other existing methods neighbor-joining, NeighborNet, and SpNet.

1 Introduction

A phylogenetic network is a generalization of a phylogenetic tree that allows internal nodes to have more than one parent. Phylogenetic networks are used to describe the evolutionary history of species when traditional tree-based models are known to be insufficient due to the occurrence of reticulation events such as hybrid speciation or horizontal gene transfer [2, 6, 8, 9, 11] that tend to occur frequently in certain types of organisms [6, 9]. Phylogenetic networks are also used in order to visualize several conflicting phylogenetic trees at the same time to represent ambiguity and to make it easier to identify parts of the trees that agree [1, 3, 4], which is helpful because different trees constructed from different datasets often contain parts that contradict each other and because many tree construction methods (e.g., bootstrapping) produce collections of trees rather than a single tree. Hence, development of reliable and efficient methods for constructing phylogenetic networks is crucial in the study of phylogenetics.

S. Miyano et al. (Eds.): RECOMB 2005, LNBI 3500, pp. 265–280, 2005.
© Springer-Verlag Berlin Heidelberg 2005

Several phylogenetic network reconstruction methods have been proposed recently. The different methods use different types of input data. Bryant and Moulton [1] proposed a method called NeighborNet to construct a phylogenetic network from a given distance matrix for the species. Gusfield et al. [2] and Wang et al. [11] showed how to construct a *galled* phylogenetic network (defined below) given character-based data. Jansson, Nguyen and Sung [5] considered how to infer a phylogenetic network which is consistent with a given set of rooted triplets. Huson et al. [4] and Nakhleh et al. [9] proposed to infer phylogenetic networks and galled phylogenetic networks, respectively, by combining a given set of gene trees, obtained, e.g., via applying maximum likelihood to sequence data. This paper follows the approach of Huson et al. and Nakhleh et al. since combining gene trees into a phylogenetic network is a promising direction. In addition, this approach is biologically sound, as stated below.

Maddison [7] observed that if a phylogenetic network for a set of species contains a single hybrid node (i.e., a node with indegree greater than one) then each gene present in all of the species must evolve according to one of the two trees embedded in the network. (Maddison also described how to construct such a phylogenetic network from two given gene trees and hypothesized that this method can be extended to construct phylogenetic networks containing more than just one hybrid node.) Based on this observation, Nakhleh, Warnow, and Linder [9] considered the following general approach to reconstructing a phylogenetic network for a set L of species from two given gene datasets for L:

> For each of the two gene datasets, infer a gene tree; next, if the two trees are identical then return that tree, else find a phylogenetic network with as few hybrid nodes as possible that contains both trees.

In particular, Nakhleh et al. proposed two efficient algorithms for reconstructing a structurally restricted phylogenetic network from two given gene trees, corresponding to the last step above.

The first algorithm of Nakhleh et al. [9] constructs a galled phylogenetic network, if one exists, having the minimum number m of hybrid nodes that induces the two given binary phylogenetic trees on a leaf set L. It runs in $O(mn)$ time, where $n = |L|$. The algorithm does not work when there exists no galled network which induces both trees in the input, for example due to errors in the estimated gene trees. The second algorithm of Nakhleh et al. [9] is designed to handle this issue. It assumes that the input is two (not necessarily binary) phylogenetic trees t_1 and t_2, obtained by first inferring a set of "good" trees for each of the two gene datasets (e.g., by using maximum parsimony or maximum likelihood) and then taking the strict consensus of each set. The algorithm outputs in $O(n)$ time a galled phylogenetic network N with a single hybrid node (if one exists) that *refines* t_1 and t_2, meaning that N contains as induced trees two binary phylogenetic trees T_1 and T_2 such that t_1 and t_2 are contractions of T_1 and T_2.

The method SpNet (short for "Species Network") proposed in [9] is a method for reconstructing phylogenetic networks from sequence datasets that uses maximum likelihood to infer the gene trees and then applies the algorithm above. The simulations studies in [9] indicate that SpNet performs very well compared to other existing methods (e.g., neighbor-joining [10] and NeighborNet [1]). However, their algorithm which SpNet is based on can only construct a phylogenetic network with *one* hybrid node (if such a network exists), which is a severe restriction. The case that is more likely to occur in practice, i.e., to construct a galled phylogenetic network having more than one hybrid node, was left as an important open problem.

We present a simple and efficient algorithm that solves the main open problem of Nakhleh *et al.* [9]. [1] It takes as input two phylogenetic trees of arbitrary degree with the same leaf set, and outputs a galled phylogenetic network (if one exists) which refines both of them. In fact, whenever such a network exists, our algorithm returns a refining galled phylogenetic network having the minimum possible number of hybrid nodes. Moreover, our algorithm can easily be extended to more than two input trees, whereas both algorithms of Nakhleh *et al.* will only work for exactly two input trees. We term the corresponding computational problem *the smallest refining galled network (SRGN) problem.* We also give an algorithm for situations where there exists no refining galled phylogenetic network for the given set of trees and we instead need to identify a largest possible subset L' of the leaf set such that if the input trees are topologically restricted to leaves in L', then they can be combined into a refining galled network (for example by using our algorithm for the SRGN problem mentioned above). We call this new problem *the maximum galled network-compatibility (MGNC) problem.* Our algorithm for the MGNC problem runs in polynomial time as long as the number of input trees is bounded by a constant and their maximum degree is at most logarithmic in the number of leaves. Finally, we apply our two proposed main algorithms for the SRGN problem and the MGNC problem to obtain two general methods for inferring galled networks from gene sequence datasets which we name RGNet and RGNet+. We show by practical simulation studies that RGNet outperforms the other existing methods neighbor-joining, NeighborNet, and SpNet, while RGNet+ can be used even when the data does not fit into a galled network.

1.1   Problem Definitions

A *phylogenetic tree* is a rooted, unordered tree whose leaves are distinctly labeled. A *phylogenetic network* is a generalization of a binary phylogenetic tree formally defined as a rooted, connected, directed acyclic graph in which: (1) exactly one node has indegree 0 (the *root*), and all other nodes have indegree 1 or 2; (2) all nodes with indegree 2 (referred to as *hybrid nodes*) have outdegree 1, and

---

[1] An alternative approach to solving this problem may be by constructing the set $\mathcal{R}$ of all rooted triplets that are consistent with at least one of the two given trees T_1 and T_2 and then applying the algorithm of [5]. However, note that the algorithm in [5] requires at least one rooted triplet to be specified for each $\{a, b, c\} \subseteq L$, so this method might not work when T_1 and T_2 are non-binary. Furthermore, the running time would be $O(kn^3)$ which is impractical for large n. Also note that an arbitrary galled network N which is consistent with $\mathcal{R}$ does not always induce T_1 and T_2, so extra care would need to be taken to ensure the correctness of this approach.

all other nodes have outdegree 0 or 2; and (3) all nodes with outdegree 0 (the *leaves*) are distinctly labeled. For any phylogenetic network N, let $rn(N)$ be the number of hybrid nodes in N. Next, let $\mathcal{U}(N)$ be the undirected graph obtained from N by replacing each directed edge by an undirected edge. N is said to be a *galled phylogenetic network* (or *galled network*, for short) if all cycles in $\mathcal{U}(N)$ are node-disjoint. Galled networks are an important type of phylogenetic networks suitable for describing evolutionary history when the frequency of reticulation events is moderate (see [2] for a discussion), and are also known in the literature as *topologies with independent recombination events* [11], *galled-trees* [2], *gt-networks* [9], and *level-1 phylogenetic networks* [5].

Let N be a phylogenetic network and let T be a phylogenetic tree. T is said to be an *induced tree of* N (symmetrically, N is said to *induce* T) if T can be obtained from N by deleting a set of edges and then, for every node with outdegree 1 and indegree less than 2, contracting its outgoing edge.

Let T and t be two (not necessarily binary) phylogenetic trees. t is called a *contraction* of T if t can be obtained from T by performing a series of edge contractions. In this case, T is also said to *refine* t. A phylogenetic network N *refines* a phylogenetic tree t if N induces a binary phylogenetic tree T such that T refines t. Two phylogenetic trees t_1 and t_2 are called *tree-compatible* (or just *compatible*) if there exists a phylogenetic tree which refines both t_1 and t_2, or *galled network-compatible* if there exists a galled network which refines t_1 and t_2.

For any phylogenetic network N with a leaf set L and a subset $L' \subseteq L$, *the topological restriction* of N to L', denoted by $N \mid L'$, is defined as the phylogenetic network obtained by first deleting all nodes which are not on any directed path from the root to a leaf in L' along with their incident edges, and then, for every node with outdegree 1 and indegree less than 2, contracting its outgoing edge (any resulting set of multiple edges between two nodes is replaced by a single edge). Given a set $\mathcal{N}$ of phylogenetic networks with a leaf set L and a subset $L' \subseteq L$, we let $\mathcal{N} \mid L'$ denote the set $\{N \mid L' : N \in \mathcal{N}\}$.

We define *the smallest refining galled network (SRGN) problem* as follows: Given a set $\mathcal{T} = \{t_1, t_2, \ldots, t_k\}$ of phylogenetic trees of arbitrary degree having a leaf set L, construct a galled phylogenetic network N with leaf set L (if one exists) that refines every $t_i \in \mathcal{T}$ and minimizes $rn(N)$. *The maximum galled network-compatibility (MGNC) problem* is: Given a set $\mathcal{T} = \{t_1, t_2, \ldots, t_k\}$ of phylogenetic trees of arbitrary degree having a leaf set L, compute a maximum subset L' of the leaf set L such that $\mathcal{T} \mid L'$ has a refining galled network. In the rest of the paper, we let n and k denote the cardinality of L and $\mathcal{T}$, respectively, in the problem definitions above and let d denote the maximum degree (i.e., the maximum number of children of any node) of all trees in $\mathcal{T}$.

1.2   Our Contributions

In this paper, we first present a polynomial-time algorithm for the SRGN problem restricted to two input trees. Thus, we are able to solve the open problem posed in [9]. The running time of our new algorithm is $O(n^2)$. Next, we show how to extend our algorithm from 2 to k input trees to run in $O(k^2 n^2)$ time.

When a set of phylogenetic trees cannot be combined into a galled phylogenetic network which refines each of them, it is useful to remove as few leaves as
possible from the leaf set of the trees so that the resulting trees admit a solution. Therefore, we also consider the optimization problem MGNC. We give an
algorithm that solves the MGNC problem in $O(2^{3kd}n^{2k})$ time.

Based on our algorithm for the SRGN problem, we propose a new method for
inferring a galled phylogenetic network from gene sequence datasets which we
name RGNet. We combine RGNet with our algorithm for the MGNC problem
to obtain an even more general method named RGNet+, and demonstrate the
usefulness of our methods by evaluating and comparing their accuracy to those
of several existing methods through extensive simulation studies.

2   Terminology and Notation

Let N be a phylogenetic network with a hybrid node h. Every ancestor s of h such
that h can be reached using two disjoint directed paths starting at the children
of s is called a *split node of h*. If s is a split node of h then any path starting at s
and ending at h is called a *merge path of h*. From the above, it follows that in
a galled network, each split node corresponds to exactly one hybrid node, and
each hybrid node has exactly one split node.

Given a galled network N, we write $\Lambda(N)$ to denote the set of leaf labels
in N and $s(N)$ to denote the set of children of the root of N. Given a node u in
N, we write $child(u)$ to denote the set of children of u and $N[u]$ to denote the
subnetwork of N rooted at u, i.e., the minimal subgraph of N which includes
all nodes and directed edges of N reachable from u. $N[u]$ is said to be *attached*
to a merge path P in N if u does not belong to P but u is a child of a node
belonging to P. If N' is a subnetwork of N then $N \setminus N'$ is the network obtained
by removing N' (and all incident edges) from N, and then, for every node with
outdegree 1 and indegree less than 2, contracting its outgoing edge.

Given a tree T and a node v in T, for any nonempty subset $A \subsetneq child(v)$,
we write $T[v, A]$ to denote the subtree obtained from $T[v]$ by removing all leaves
that are not reachable from any node in A and all incident edges. We call $T[v, A]$
a *restricted subtree rooted at v in T*. Finally, $T[A]$ is short for $T[v, A]$ if v is the
root of N.

3   Solving the SRGN Problem

This section solves the SRGN problem in $O(k^2n^2)$ time. For explanation of the
algorithm, we solve the problem when $k = 2$. Some technical lemmas which are
needed to prove our main result can be found in Section 4 and Section 5.

Definition 1. *Given two trees T_1 and T_2 with leaf set L, T_1 and T_2 admit a leaf
set bipartition (X, Y) of L if, for $i = 1, 2$, there is a partition (A_i, B_i) of $s(T_i)$
such that $X = \Lambda(T_i[A_i])$ and $Y = \Lambda(T_i[B_i])$. (See Figure 3(b) for an example.)*

Algorithm *BuildGalledNetwork*

Input: Two trees T_1 and T_2 leaf-labeled by L.

Output: A SRGN N for T_1 and T_2, if exists.

1 **if** T_1 and T_2 admits a leaf set bipartition (X, Y) **then**

1.1 Let $N_X = $ BuildGalledNetwork$(T_1|X, T_2|X)$ and $N_Y = $ BuildGalledNetwork$(T_1|Y, T_2|Y)$.

1.2 If any of N_X and N_Y is null, return null.

1.3 Return N obtained by attaching both N_X and N_Y to a common root.

  elseif T_1 and T_2 admits a leaf set tripartition (X, Y, Z) **then**

1.4 For $i = 1, 2$, let u_i be the root of $T_i|Y$ in T_i.

1.5 Let $N_X = $ BuildSide$(T_1|X, T_2|X, u_2)$, $N_Z = $ BuildSide$(T_2|Z, T_1|Z, u_1)$, $N_Y = $ BuildGalledNetwork$(T_1|Y, T_2|Y)$.

1.6 If any of N_X, N_Y, N_Z is null, return null.

1.7 Let N' be the network formed by attaching N_X and N_Z to a common root.

1.8 Let w_X and w_Z be the nodes in N', where $\Lambda(N'[w_X]) = \Lambda(T_2[u_2]) \cap X$ and $\Lambda(N'[w_Z]) = \Lambda(T_1[u_1]) \cap Z$, and let v_X and v_Z be the parent of w_X and w_Z, respectively.

1.9 For $i = X, Z$, create a new node q_i by subdividing the edge (v_i, w_i).

1.10 Create a new node h, make h a child of both q_X and q_Z, and add an edge from h to the root of N_Y.

1.11 Let N be the resulting network and return N.

  endif

End *BuildGalledNetwork*

Fig. 1. Framework for constructing phylogenetic network

Algorithm *BuildSide*

Input: Two trees T_1 and T_2 and a node u in T_2, where T_1 is side compatible with (T_2, u).

Output: A SRGN N for T_1 and T_2, where there is no split node on the path from the root of N to a node v, excluding v, where $\Lambda(N[v]) = \Lambda(T_2[u])$.

1 If u is the root of T_2, return BuildGalledNetwork(T_1, T_2).

2 Find leaf set bipartition (X, Y) admitted by T_1 and T_2 where $T_2|X$ contains u.

3 If there is no such partition, return null.

4 Let $N_1 = $ BuildSide$(T_1|X, T_2|X, u)$ and $N_2 = $ BuildGalledNetwork$(T_1|Y, T_2|Y)$.

5 If any of N_1 and N_2 is null, return null.

6 Return N obtained by attaching both N_1 and N_2 to a common root.

End *BuildSide*

Fig. 2. Algorithm BuildSide

Definition 2. *Given two trees T_1 and T_2 with leaf set L and u be a node in T_2. We say that T_1 is* side compatible *with (T_2, u) if either (1) u is the root of T_2; or (2) there is a restricted subtree t_1 rooted at the root of T_1 and a restricted subtree*

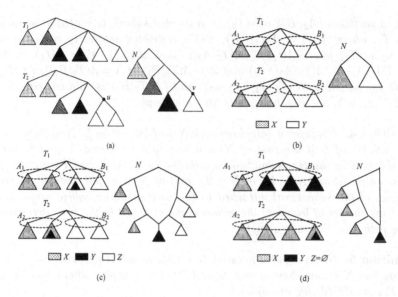

Fig. 3. (a) The left figure shows an example of T_1 and T_2 such that T_1 is side compatible with (T_2, u). Sets of leaves with the same pattern in both trees are identical. Based on Lemma 2, if T_1 and T_2 are galled network-compatible, then there exists a SRGN N for T_1 and T_2 having no split node on the path from the root to the parent of v, where $\Lambda(N[v]) = \Lambda(T_2[u])$. (b) (X, Y) is a leaf set bipartition admitted by T_1 and T_2. By Lemma 4, there exists a SRGN N for T_1 and T_2 that satisfies (X, Y). (c) (X, Y, Z) is a leaf set tripartition admitted by T_1 and T_2. Note that Y is the set of leaves of some restricted subtrees in both T_1 and T_2. By Lemma 5, there exists a SRGN N for T_1 and T_2 that satisfies (X, Y, Z). (d) This figure is similar to (c) except that $Z = \emptyset$

t_2 *rooted at the root of* T_2, *where* t_2 *does not contain* u, *such that* $\Lambda(t_1) = \Lambda(t_2)$ *and* $T_1 \setminus t_1$ *is side compatible with* $(T_2 \setminus t_2, u)$. *(see Figure 3(a).)*

Given that T_1 is side compatible with (T_2, u), we have the following properties.

Lemma 1. *Given that* T_1 *is side compatible with* (T_2, u). *Then* T_1 *and* T_2 *admit a leaf set bipartition. Furthermore, for any leaf set bipartition* (X, Y) *admitted by* T_1 *and* T_2, *where* $T_2|X$ *contains* u, $T_1|X$ *is side compatible with* $(T_2|X, u)$.

Lemma 2. *Given that* T_1 *is side compatible with* (T_2, u). *Suppose* T_1 *and* T_2 *are galled network-compatible, then there exists a SRGN* N *for* T_1 *and* T_2, *where there is a node* v *in* N *such that* $\Lambda(N[v]) = \Lambda(T_2[u])$ *and there is no split node on the path from the root of* N *to* v, *excluding* v.

Proof. An algorithm for building such an N is shown in Figure 2. That is $N = $BuildSide$(T_1, T_2, u)$. Its correctness comes from Lemma 1 and Lemma 7. □

Definition 3. *Given two trees* T_1 *and* T_2 *with leaf set* L. T_1 *and* T_2 *admit a leaf set tripartition of* L *into* (X, Y, Z) *if there exist a partition* (A_1, B_1) *of*

$s(T_1)$, a partition (A_2, B_2) of $s(T_2)$, a restricted subtree t_1 rooted at some node u_1 in T_1, where $\Lambda(t_1) \subseteq \Lambda(T_1[B_1])$, and a restricted subtree t_2 rooted at some node u_2 in T_2, where $\Lambda(t_2) \subseteq \Lambda(T_2[A_2])$, such that (1) $Y = \Lambda(t_1) = \Lambda(t_2)$, $X = \Lambda(T_1[A_1]) = \Lambda(T_2[A_2] \setminus t_2)$ and $Z = \Lambda(T_1[B_1] \setminus t_1) = \Lambda(T_2[B_2])$; (2) $T_1[A_1]$ is side compatible with $(T_2[A_2] \setminus t_2, u_2)$; and (3) $T_2[B_2]$ is side compatible with $(T_1[B_1] \setminus t_1, u_1)$. See Figure 3(c)(d) for an example.

Definition 4. Consider a galled network N leaf-labeled by L. N satisfies a partition (X, Y) of L if the root of N is a non-split node and X and Y are the sets of leaves in the two subnetworks attached to the root (see Figure 3(b)). N satisfies a partition (X, Y, Z) of L if the root of N is a split node, Y is the set of leaves of the subnetwork attached to the corresponding hybrid node, and X and Z are the set of leaves of the subnetworks attached to the two corresponding merge path (see Figure 3(c)(d)).

Definition 5. Given a galled network N satisfying a leaf set partition (X, Y, Z). We say that N is non-skew if both X and Z are nonempty, otherwise it is skew. (See Figure 3(c)(d) for examples.)

Below three lemmas describe the properties of both leaf set bipartition and leaf set tripartition.

Lemma 3. If T_1 and T_2 admit neither a leaf set bipartition nor a leaf set tripartition, then there is no refining galled network for T_1 and T_2.

Lemma 4. Consider two trees T_1 and T_2 leaf-labeled by L. A leaf set bipartition (X, Y) admitted by T_1 and T_2, if exists, can be computed in $O(n)$ time. Furthermore, if T_1 and T_2 are galled network-compatible, then there is a SRGN for T_1 and T_2 that satisfies (X, Y).

Proof. Follows from Lemmas 7 and 8.   □

Lemma 5. Consider two trees T_1 and T_2 which do not admit any leaf set bipartition. A leaf set tripartition (X, Y, Z) admitted by T_1 and T_2, if exists, can be computed in $O(n)$ time. Furthermore, if T_1 and T_2 are galled network-compatible, there is a SRGN for T_1 and T_2 that satisfies (X, Y, Z).

Proof. Follows from Lemmas 10, 13.   □

Based on the above lemmas, a SRGN for T_1 and T_2 can be computed by Algorithm BuildGalledNetwork as shown in Figure 1. If T_1 and T_2 admit a leaf set bipartition (X, Y), the algorithm first builds recursively the SRGN N_X and N_Y for $(T_1|X, T_2|X)$ and $(T_1|Y, T_2|Y)$ respectively. Then, based on Lemma 4, we return the network formed by attaching N_X and N_Y by a common root.

Otherwise, the algorithm checks if T_1 and T_2 admit a leaf set tripartition (X, Y, Z). If yes, it builds a SRGN N for T_1 and T_2 that satisfies (X, Y, Z) by following Step 1.4-1.11. It does this by recursively building a SRGN N_X for $(T_1|X, T_2|X)$, a SRGN N_Z for $(T_1|Z, T_2|Z)$, and a SRGN N_Y for $(T_1|Y, T_2|Y)$.

N_X, N_Y, and N_Z are combined into N such that each of N_X and N_Z constitutes a merge path from the root of N to the corresponding hybrid node and N_Y is the subnetwork rooted at the hybrid node. Since N is a galled network, N_X and N_Z must be constructed such that there is no split node on both merge paths. Thus, by Lemma 2, the algorithm calls BuildSide to construct N_X and N_Z.

Otherwise, by Lemma 3, there is no galled network refining T_1 and T_2.

Note that in Step 1.4-1.11, the algorithm assumes both X and Z are nonempty. But a slight technical change can be made to remove this assumption.

In summary, we have the following theorem.

Theorem 1. *Given two trees T_1 and T_2, a SRGN for T_1 and T_2 can be computed by Algorithm BuildGalledNetwork in $O(n^2)$ time.*

The algorithm can be extended to more than two trees.

Theorem 2. *The SRGN problem can be solved in $O(k^2 n^2)$ time.*

4   Computing the Leaf Set Bipartition of T_1 and T_2

This section is devoted to prove Lemma 4.

4.1   Relationship Between Leaf Set Bipartition and SRGN

Suppose T_1 and T_2 are galled network-compatible. This section shows that if there exists a leaf set bipartition (X, Y) admitted by T_1 and T_2, then there is a SRGN for T_1 and T_2 that satisfies (X, Y).

Lemma 6. *Let N be a SRGN for T_1 and T_2. For any hybrid node h of N, we have $\Lambda(N[h]) \subseteq \Lambda(T_1[a])$ for some $a \in s(T_1)$ or $\Lambda(N[h]) \subseteq \Lambda(T_2[b])$ for some $b \in s(T_2)$.*

Lemma 7. *Suppose T_1 and T_2 are galled network-compatible. Assume T_1 and T_2 admit a leaf set bipartition (X, Y). Then, we can construct a SRGN N for T_1 and T_2 that satisfies (X, Y).*

Proof. Let N_X and N_Y be SRGN for $(T_1|X, T_2|X)$ and $(T_1|Y, T_2|Y)$, respectively. Let N be the network formed by attaching N_X and N_Y to a common root, then N satisfies (X, Y). By contrary, assume there is a SRGN N^* for T_1 and T_2 and $rn(N^*) < rn(N)$. For every hybrid node $h \in N^*$, by Lemma 6, we conclude that either $\Lambda(N^*[h]) \subseteq X$ or $\Lambda(N^*[h]) \subseteq Y$. Hence, the set of hybrid nodes in $N^*|X$ and $N^*|Y$ should be disjoint. In other word, $rn(N^*|X) + rn(N^*|Y) \leq rn(N^*)$. As $rn(N_X) \leq rn(N^*|X)$ and $rn(N_Y) \leq rn(N^*|Y)$, we have $rn(N) = rn(N_X) + rn(N_Y) \leq rn(N^*)$. Thus, we arrived at contradiction and the lemma follows.   □

4.2   Linear Time Algorithm for Computing a Leaf Set Bipartition

Next, we describe how to compute a leaf-set bipartition admitted by T_1 and T_2 in $O(n)$ time. Our computation is based on the bipartite graph $G(T_1, T_2)$ where

$s(T_1)$ and $s(T_2)$ are the vertex set on the left and on the right, respectively. In addition, (u, v) is an edge in $G(T_1, T_2)$ if and only if $u \in s(T_1)$, $v \in s(T_2)$ and $\Lambda(T_1[u]) \cap \Lambda(T_2[v]) \neq \emptyset$. Note that $G(T_1, T_2)$ can be constructed in $O(n)$ time. A partition (X, Y) is a leaf set bipartition admitted by T_1 and T_2 if and only if $G(T_1, T_2)$ can be divided into two disjoint subgraphs G_1 and G_2 such that $X = \cup_{(u,v) \in E(G_1)} \Lambda(T_1[u])$ and $Y = L \setminus X$. Thus we get the following lemma.

Lemma 8. *A leaf set bipartition admitted by T_1 and T_2, if exists, can be computed in $O(n)$ time.*

5   Computing the Leaf Set Tripartition of T_1 and T_2

This section is devoted to prove Lemma 5.

5.1   Relationship Between Leaf Set Tripartition and SRGN

Given T_1 and T_2. For any leaf set L', let $rn^*(L')$ denote the number of hybrid nodes in a SRGN for $T_1|L'$ and $T_2|L'$.

Lemma 9. *Suppose T_1 and T_2 with leaf set L do not admit any leaf set bipartition and are galled network-compatible. If T_1 and T_2 admit a leaf set tripartition (X, Y, Z), then $rn^*(X) + rn^*(Y) + rn^*(Z) + 1 \leq rn^*(L)$.*

Lemma 10. *Suppose T_1 and T_2 with leaf set L do not admit any leaf set bipartition and are galled network-compatible. If T_1 and T_2 admit a leaf set tripartition (X, Y, Z), then there is a SRGN N for T_1 and T_2 that satisfies (X, Y, Z).*

Proof. We construct N by following Steps 1.4-1.11 shown in Figure 1. Then N satisfies (X, Y, Z) and is a galled network. Let N' be the network obtained from N by removing the edge (q_X, h) and let N'' be the network obtained from N by removing the edge (q_Z, h). Then N' refines T_1 and N'' refines T_2, which implies N refines both T_1 and T_2. By the construction of N, we have $rn(N) = rn^*(X) + rn^*(Y) + rn^*(Z) + 1$. From Lemma 9, we conclude $rn(N) = rn^*(L)$.   □

5.2   Linear Time Algorithm for Computing a Leaf Set Tripartition

To find leaf set tripartition for T_1 and T_2, we examine the graph $G(T_1, T_2)$ again.

Lemma 11. *Suppose T_1 and T_2 did not admit any leaf set bipartition while there exists a non-skew network refining T_1 and T_2 that satisfies (X, Y, Z). Then, $G(T_1, T_2)$ contains an edge (u_1, u_2) such that $Y = \Lambda(T_1[u_1]) \cap \Lambda(T_2[u_2])$. Also, $G(T_1, T_2) - \{(u_1, u_2)\}$ consists of two disjoint nontrivial star graphs G' and G'' (where a star graph is a connected graph with at most one node whose degree is larger than 1) such that $X = \cup_{(u,v) \in G'} \Lambda(T_1[u])$ and $Z = \cup_{(u,v) \in G''} \Lambda(T_1[u])$.*

Lemma 12. *Given three disjoint sets X, Y, Z such that $L = X \cup Y \cup Z$. We can check if (X, Y, Z) is a leaf set tripartition admitted by T_1 and T_2 in $O(n)$ time.*

Lemma 13. *Suppose T_1 and T_2 did not admit any leaf set bipartition. We can compute a leaf set tripartition admitted by T_1 and T_2, if one exists, in $O(n)$ time.*

Proof. When there exist a non-skew network refining T_1 and T_2, by Lemma 11, there exists an edge $(u, v) \in G(T_1, T_2)$ such that its deletion divides $G(T_1, T_2)$ into two star graphs. (u, v) corresponds to a set $Y = \Lambda(T_1(u)) \cap \Lambda(T_2(v))$. As its deletion divides G into two star graphs, we know that $T_1|L - Y$ and $T_2|L - Y$ admit a leaf set bipartition (X, Z). By Lemma 12, we can determine if (X, Y, Z) is a leaf set tripartition in $O(n)$ time. Note that every $G(T_1, T_2)$ contains at most two edges (u, v) such that $G(T_1, T_2) - \{(u, v)\}$ consists of two disjoint nontrivial star graphs. Hence, a tripartition for T_1 and T_2 can be computed in $O(n)$ time.

By using a similar idea, we can compute a leaf set tripartition in $O(n)$ time when there exists a skew network refining T_1 and T_2.   □

6 RGNet: A New Technique for Inferring Galled Networks

In this section, we describe a method we call RGNet (short for Refining Galled Network) for inferring galled networks from sequence datasets and compare its performance to other methods. RGNet is based on the approach proposed by Nakhleh *et al.* [9] called SpNet, but unlike that approach, RGNet is capable of inferring networks with more than one hybrid node. Given two gene datasets of a set of taxa, RGNet tries to infer a smallest galled phylogenetic network for the set of taxa by utilizing our algorithm for the SRGN problem as follows:

- **Step 1:** For each gene dataset, use maximum likelihood to construct the best two trees for the dataset,
- **Step 2:** For each dataset, compute the strict consensus of the two best trees, thus producing the trees t_1 and t_2, and
- **Step 3:** If t_1 and t_2 are compatible, combine the datasets and analyze the combined dataset using neighbor-joining (NJ) and return a tree. Else, apply our algorithm for the SRGN problem to t_1 and t_2. If possible, return a galled network with minimum reticulations that refines t_1 and t_2; if no such network exists, we try to *reroot* t_1 and t_2 (described below) to get t_1' and t_2', respectively, and apply our algorithm again. If we still cannot get any SRGN for t_1' and t_2', we apply NJ to the concatenated dataset and return a tree.

Rerooting the strict consensus trees. Our algorithm assumes rooted trees and networks. So we use an outgroup to obtain rooted estimates of gene trees inferred by maximum likelihood method (in Step 1). However, the estimates are sometimes rooted incorrectly. This makes the inference of phylogenetic networks impossible. To overcome this problem, whenever there do not exist any SRGN for t_1 and t_2 (in Step 3), we try to reroot t_1 and t_2. To reroot both trees, we find an edge (u_1, v_1) in t_1 and (u_2, v_2) in t_2, where u_i is the parent of v_i for $i = 1, 2$, such that $\Lambda(t_1[v_1]) = \Lambda(t_2) \setminus \Lambda(t_2[v_2])$ and the absolute difference between $|\Lambda(t_1[v_1])|$ and $|\Lambda(t_1) \setminus \Lambda(t_1[v_1])|$ is as small as possible. Then, if they exist, for $i = 1, 2$, we create a new node p_i in t_i by subdividing the edge (u_i, v_i) and reroot t_i at p_i.

6.1   Experimental Evaluation

To evaluate and compare the performance of the four methods neighbor-joining, NeighborNet, SpNet, and RGNet for phylogenetic reconstruction, we have carried out extensive simulations. For the simulations, we have used the same methodology as Nakhleh *et al.* [9], as discussed below.

Experimental Settings. We used the model from [8] to generate random networks. Within each generated galled network, we produced two induced gene trees and simulated sequence evolution on these trees under the GTR+ Γ+I (gamma distributed rates, with invariable sites) model of evolution, using the settings of [12]. The two separate sequence datasets then were used to run SpNet and RGNet, and the combined sequence dataset was given to NeighborNet and neighbor-joining. As in [9], we used split-based false positive and false negative rates to measure the error rates of the methods.

    Topological accuracy Given a phylogenetic tree T leaf-label by a set L of taxa. Each edge e in T induces a *split* $\{A(e), B(e)\}$ on L, where $A(e)$ is the set of taxa which are descendants of e, and $B(e)$ is the set containing the rest of the taxa. We define the set of splits of T, denoted by $C(T)$, as the set of all splits induced by edges in T. Generalizing the above, let N be a phylogenetic network whose set of induced trees is referred to as $T(N)$. The set of splits of N, denoted by $C(N)$, is defined as $C(N) = \bigcup_{T \in T(N)} C(T)$. Given a model network N_1 and an inferred network N_2, the *false positive rate* and *false negative rate* are defined as: $FP(N_1, N_2) = \frac{|C(N_2) - C(N_1)|}{|C(N_1)|}$ and $FN(N_1, N_2) = \frac{|C(N_1) - C(N_2)|}{|C(N_1)|}$.

Experimental Results. We have done extensive experiments to evaluate the performances of the four methods. We focus here on some of the experimental results on 40-taxon galled model networks, shown in Figure 4. The performance

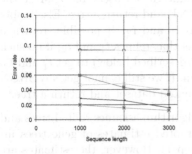
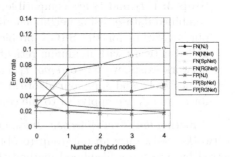

Fig. 4. FN and FP error rates of neighbor-joining (NJ), NeighborNet(NNet), SpNet, and RGNet on 40-taxon galled model networks. The left graph shows the error rates as a function of concatenated sequence length (c.s.l) on model networks with expected diameter (e.d) 0.5 and 3 hybrid nodes. The right graph shows the error rates as a function of the number of hybrid nodes in model networks with e.d=0.2 and c.s.l=2000. For clarity, the results is shown without the FP rates of NeighborNet

of SpNet and neighbor-joining (NJ) are essentially identical, which is expected since SpNet uses NJ whenever it cannot infer a one-hybrid network [9].

As indicated by the figures, the FN error rates of RGNet are comparable to those of NeighborNet and are better than those of NJ and SpNet. Furthermore, the FN rates of RGNet are stable as the number of hybrid nodes increases, while the FN rates of NJ and SpNet increase significantly as the number of hybrid nodes increases. In all cases, NeighborNet shows very poor false positive rates which are always more than 70%, while NJ, SpNet and RGNet show very good false positive rates (almost always less than 3%). The FP rates of RGNet are marginally higher than those of NJ and SpNet. To summarize, RGNet outperforms the other three methods in the combined view of both FN and FP rates.

7 Solving the MGNC Problem

In this section, we solve the MGNC problem when $k = 2$. Given an instance T, we call the optimal solution network N for the problem a *maximum compatible galled network* (MCGN) for T, and $\Lambda(N)$ a *maximum compatible set* (MCS). We present here an algorithm to compute the MCGN. Extension to $k > 2$ is straightforward.

7.1 The Algorithm

Let T_1 and T_2 be two input trees. For any restricted subtrees $T_1[u_1, A_1]$ and $T_2[u_2, A_2]$ of T_1 and T_2, respectively, we define the function $MCS(T_1[u_1, A_1], T_2[u_2, A_2])$ to be the cardinality of the MCS of $T_1[u_1, A_1]$ and $T_2[u_2, A_2]$. Furthermore, suppose $T_1[v_1, V_1]$ is a restricted subtree of $T_1[u_1, A_1]$, let N be a MCGN of $T_1[u_1, A_1] \setminus T_1[v_1, V_1]$ and $T_2[u_2, A_2]$ such that there is a node v in N, where $\Lambda(N[v]) \subseteq \Lambda(T_1[v_1] \setminus T_1[v_1, V_1])$, and any node in N, which is on the path from the root to v and excluding v, should be a non-split node. We define $MCS^*(T_1[u_1, A_1], T_1[v_1, V_1], T_2[u_2, A_2])$ to be the cardinality of $\Lambda(N)$.

Below lemmas show the recursive equations for computing $MCS(T_1[u_1, A_1], T_2[u_2, A_2])$ and $MCS^*(T_1[u_1, A_1], T_1[v_1, V_1], T_2[u_2, A_2])$.

Lemma 14. *For any restricted subtrees $T_1[u_1, A_1]$ and $T_2[u_2, A_2]$ of T_1 and T_2, respectively, where u and v are non-leaf nodes, $MCS(T_1[u_1, A_1], T_2[u_2, A_2]) =$*

$$\max \begin{cases} \max\{MCS(T_1[u_1, A_1], T_2[v_2, child(v_2)]) : v_2 \in A_2\} \\ \max\{MCS(T_1[v_1, child(v_1)], T_2[u_2, A_2]) : v_1 \in A_1\} \\ MCS_1(T_1[u_1, A_1], T_2[u_2, A_2]) \\ MCS_2(T_1[u_1, A_1], T_2[u_2, A_2]) \end{cases}$$

where

- $MCS_1(T_1[u_1, A_1], T_2[u_2, A_2]) = \max\{MCS(T_1[u_1, B_1], T_2[u_2, B_2]) + MCS$
  $(T_1[u_1, A_1 - B_1], T_2[u_2, A_2 - B_2]) : B_i$ *is some nonempty proper subset of A_i, for $i = 1, 2\}$, and*

- $MCS_2(T_1[u_1, A_1], T_2[u_2, A_2]) = \max\{MCS^*(T_1[u_1, B_1], T_1[v_1, V_1], T_2[u_2, A_2$
$-B_2]) + MCS^*(T_2[u_2, B_2], T_2[v_2, V_2], T_1[u_1, A_1 - B_1]) + MCS(T_1[v_1, V_1],$
$T_2[v_2, V_2]) :$ B_i is a subset of A_i, v_i is some node in T_i, V_i is a nonempty
proper subset of $child(v_i)$ such that $T_i[v_i, V_i]$ is a subtree in $T_i[u_i, B_i]$, for
$i = 1, 2\}$.

For the base cases, in which u_1 or u_2 is a leaf, $MCS(T_1[u_1, A_1], T_2[u_2, A_2])$ equals
$|\Lambda(T_1[u_1, A_1]) \cap \Lambda(T_2[u_2, A_2])|$.

Lemma 15. Consider any restricted subtrees $T_1[u_1, A_1]$ and $T_2[u_2, A_2]$ of T_1
and T_2, respectively. Suppose $T_1[v_1, V_1]$ is a restricted subtree of $T_1[u_1, A_1]$. Then,
$MCS^*(T_1[u_1, A_1], T_1[v_1, V_1], T_2[u_2, A_2])$ equals the maximum of the following three
terms.

- $max\{MCS^*(T_1[u_1, A_1], T_1[v_1, V_1], T_2[w_2, child(w_2)]) : w_2 \in A_2\}$,
- $MCS^*(T_1[w_1, child(w_1)], T_1[v_1, V_1], T_2[u_2, A_2])$ where $w_1 \in A_1$ and $T_1[v_1, V_1]$
is a subtree in $T_1[w_1]$,
- $max\{MCS^*(T_1[u_1, B_1], T_1[v_1, V_1], T_2[u_2, B_2]) + MCS(T_1[u_1, A_1 - B_1], T_2[u_2,$
$A_2 - B_2]) :$ B_1 and B_2 are nonempty proper subset of A_1 and A_2, respectively,
and $T_1[v_1, V_1]$ is a subtree in $T_1[u_1, B_1]$ $\}$.

For base cases, in which $v_1 = u_1$, then $MCS^*(T_1[u_1, A_1], T_1[v_1, V_1], T_2[u_2, A_2]) =$
$MCS(T_1[u_1, A_1 - V_1], T_2[u_2, A_2])$.

By applying dynamic programming on the above two recursive equations and
simple backtracking, we get the following.

Theorem 3. The MCGN of T_1 and T_2 can be computed in $O(2^{6d}n^4)$ time.

In general, the above algorithm can be extended to get the following result.

Theorem 4. The MGNC problem can be solved in $O(2^{3kd}n^{2k})$ time.

8   RGNet+: Combining RGNet and Our Algorithm for the MGNC Problem

The simulations in Section 6 demonstrate that RGNet has superior performance
over the other existing methods when the model network is a galled network.
However, when the model network is not galled, there is a high chance that
RGNet cannot construct any galled phylogenetic network from the gene tree
estimates. In these cases, RGNet would return a phylogenetic tree by calling
neighbor-joining. However, trees are not sufficient to accurately describe the
relationships among species that are represented in model networks, leading to
poor results in these cases.

To overcome this problem, we can employ our algorithm for the MGNC
problem. When RGNet cannot infer a galled phylogenetic network, we do not
call neighbor-joining. Instead, we apply our algorithm for the MGNC problem

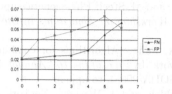

rn(N*)	|(N)|	|(N)|/|(N*)|
0	28.919	96.40%
1	27.1455	90.49%
2	26.9526	89.84%
3	25.4353	84.78%
4	24.7273	82.42%
5	23.3529	77.84%
6	22.6	75.33%

Fig. 5. Illustrating the performance of RGNet+. Experiments done on model networks with 30 taxa, 0 to 6 hybrid nodes, concatenated sequence length=2000, and expected diameter=0.5. The table shows the average number of leaves in inferred galled networks as a function of the number of hybrid nodes in model networks. The graph shows the average false positive rate and average false negative rate of inferred galled networks as a function of the number of hybrid nodes in model networks

to remove as few species as possible from the input so that the resulting trees can be merged to a galled network, and then proceed as before. In Section 6, we have shown that RGNet has very good performance with regards to both the false positive error rate and the false negative error rate. Therefore, the network returned by our approach which combines our algorithms for the MGNC problem and the SRGN problem is highly likely to show the relationships among a subset of species that are represented in the true network. We name this combined approach RGNet+.

We have done extensive experiments to evaluate RGNet+. We used the same experimental setting as in Section 6.1. We simulated on general (i.e, not restricted to galled) model networks. For each model network N^*, we inferred a galled network N with as many taxa as possible. We used $FN(N^*|\Lambda(N), N)$ and $FP(N^*|\Lambda(N), N)$ to measure the topological accuracy of N.

Figure 5 shows our experimental results. The FN and FP error rates are less than 7%, which are good. Hence, the inferred networks estimated with high accuracy the relationships among subsets of the species. We also see that on average, the inferred galled networks kept a majority of the species (over 80% when $rn(N^*) \leq 4$). This implies that even when the true networks are not restricted to be galled, the evolutionary relationships among a majority of the species can be represented by a galled network.

References

1. D. Bryant and V. Moulton. Neighbor-Net: An agglomerative method for the construction of phylogenetic networks. *Molecular Biology and Evolution*, 21(2):255–265, 2004.
2. D. Gusfield, S. Eddhu, and C. Langley. Efficient reconstruction of phylogenetic networks with constrained recombination. In *Proc. of Computational Systems Bioinformatics* (CSB2003), pages 363–374, 2003.
3. B. Holland and V. Moulton. Consensus networks: A method for visualising incompatibilities in collections of trees. In *Proc. of the 3rd Workshop on Algorithms in Bioinformatics* (WABI 2003), pages 165–176, 2003.

4. D. H. Huson, T. Dezulian, T. Klöpper, and M. Steel. Phylogenetic super-networks from partial trees. In *Proc. of the 4ᵗʰ Workshop on Algorithms in Bioinformatics* (WABI 2004), pages 388–399, 2004.

5. J. Jansson, N. B. Nguyen, and W.-K. Sung. Algorithms for combining rooted triplets into a galled phylogenetic network. In *Proc. of the 16ᵗʰ Annual ACM-SIAM Symposium on Discrete Algorithms* (SODA 2005), to appear.

6. C. R. Linder, B. M. E. Moret, L. Nakhleh, and T. Warnow. Network (reticulate) evolution: Biology, models, and algorithms. Tutorial presented at *the 9ᵗʰ Pacific Symposium on Biocomputing* (PSB 2004), 2004.

7. W. P. Maddison. Gene trees in species trees. *Systematic Biology*, 46(3):523–536, 1997.

8. L. Nakhleh, J. Sun, T. Warnow, C. R. Linder, B. M. E. Moret, and A. Tholse. Towards the development of computational tools for evaluating phylogenetic reconstruction methods. In *Proc. of the 8ᵗʰ Pacific Symposium on Biocomputing* (PSB 2003), pages 315–326, 2003.

9. L. Nakhleh, T. Warnow, and C. R. Linder. Reconstructing reticulate evolution in species – theory and practice. In *Proc. of the 8ᵗʰ Annual International Conference on Research in Computational Molecular Biology* (RECOMB 2004), pages 337–346, 2004.

10. N. Saitou and M. Nei. The neighbor-joining method: a new method for reconstructing phylogenetic trees. *Molecular Biology and Evolution*, 4(4):406–425, 1987.

11. L. Wang, K. Zhang, and L. Zhang. Perfect phylogenetic networks with recombination. *Journal of Computational Biology*, 8(1):69–78, 2001.

12. D. Zwickl and D. Hillis. Increased taxon sampling greatly reduces phylogenetic error. *Systematic Biology*, 51(4):588–598, 2002.

Mapping Molecular Landscapes Inside Cells

Wolfgang Baumeister

Max-Planck-Institute of Biochemistry,
Am Klopferspitz 18, D-82152 Martinsried, Germany

Electron Tomography (ET) is uniquely suited to obtain three-dimensional (3-D) images of large pleiomorphic structures, such as supramolecular assemblies, organelles, or even whole cells. While the principles of ET have been known for decades, its use has gathered momentum only in recent years. Technological advances (namely computer controlled transmission electron microscopes and large area CCD cameras) have made it possible to develop automated data acquisition procedures. This, in turn, allowed to reduce the total electron dose to levels low enough for studying radiation sensitive biological materials embedded in vitreous ice. As a result, we are now poised to combine the power of high-resolution 3-D imaging with the best possible preservation of the specimen.

In the past, ET has mainly been used to examine thin sections of plastic-embedded materials. This approach has provided valuable insights into cellular architecture, but it falls short of revealing the macromolecular organization inside cells. Chemical fixation, staining with contrast enhancing heavy atom compounds, dehydration and plastic embedding affect the specimen significantly and make the interpretation of such tomograms at the molecular level very problematic, if not impossible. The use of high-pressure freezing instead of chemical fixation improves specimen preservation significantly but it does not eliminate the problems arising from the intricate interactions between heavy atom stains and molecular structures. Obviously, cryo-sectioning of vitrified material is the method of choice for large ($> 1\mu$m) objects (cells, tissues), but it remains a challenging task despite recent progress.

It is often possible to isolate macromolecular complexes, virus particles or organelles using procedures that maintain their structural integrity. Such nanoscale specimens are suitable for direct analysis by ET. The resolution obtained allows the docking of high resolution component structures obtained by X-ray crystallography or NMR. Hybrid approaches of this kind can be used to generate pseudoatomic maps of assemblies that are too large or variable for direct high-resolution structural studies. ET of frozen-hydrated whole prokaryotic cells or thin eukaryotic cells grown directly on EM grids provides 3-D images of macromolecular structures unperturbed and in their functional environment. Currently resolution is limited to 4-6 nm but with instrumental advances, such as liquid He cooling and CCD cameras optimized for intermediate voltage TEMs, we are now entering the realm of molecular resolution (2-4 nm).

High resolution tomograms of organelles or cells contain vast amounts of information; essentially they are 3-D images of the cell's entire proteome and they should ultimately enable us to map the spatial relationships of macromolecules

S. Miyano et al. (Eds.): RECOMB 2005, LNBI 3500, pp. 281–282, 2005.

in a cellular context, the 'interactome'. However, it is no trivial task to retrieve this information because of the poor signal-to-noise ratio of such tomograms and the crowded nature of the cytoplasm and many organelles. Denoising procedures can help to combat noise and to facilitate visualization, but advanced pattern recognition methods are needed for detecting and identifying with high fidelity specific macromolecules based on their structural signature (size and shape, for example). Provided that high- or medium-resolution structures of the molecules of interest are available, they can be used as templates for a systematic interrogation of the tomograms.

Experiments with phantom cells, i.e. lipid vesicles encapsulating a known set of proteins have shown that such a template-matching approach is feasible. Once the challenges of obtaining sufficiently good resolution and of creating efficient data-mining algorithms are met, and comprehensive libraries of template structures become available, we will be able to map the supramolecular landscape of cells systematically and thereby provide a new perspective for the function of cellular systems.

Information Theoretic Approaches to Whole Genome Phylogenies*

(Extended Abstract)

David Burstein, Igor Ulitsky, Tamir Tuller**, and Benny Chor

School of Computer Science, Tel Aviv University
{davidbur, ulitskyi, tamirtul, bchor}@post.tau.ac.il

Abstract. We describe a novel method for efficient reconstruction of phylogenetic trees, based on sequences of whole genomes or proteomes. The core of our method is a new measure of pairwise distances between sequences, whose lengths may greatly vary. This measure is based on information theoretic tools (Kullback-Leibler relative entropy). We present an algorithm for efficiently computing these distances. The algorithm uses suffix arrays to compute the distance of two ℓ long sequences in $O(\ell)$ time. It is fast enough to enable the construction of the phylogenomic tree for hundreds of species, and the phylogenomic forest for almost two thousand viruses. An initial analysis of the results exhibits a remarkable agreement with "acceptable phylogenetic truth". To assess our approach, it was implemented together with a number of alternative approaches, including two that were previously published in the literature. Comparing their outcome to ours, using a "traditional" tree and a standard tree comparison method, our algorithm improved upon the "competition" by a substantial margin.

Keywords: Phylogenomics, whole genome and proteom phylogenetic, tree reconstruction, divergence, Kullback-Leibler relative entropy, distance matrix.

1 Introduction

The elucidation of the evolutionary history of extinct and extant species is a major scientific quest, dating back to Darwin [7] and before. Early approaches were based on morphological and palaeontological data, but with the advent of molecular biology, the emphasis has shifted to molecular (amino acid and nucleotide) sequence data. Rapid sequencing technologies have produced the complete genome sequence of over 200 organisms, and many more projects are underway [13]. Full genomes contain huge amounts of sequence data that should undoubtedly be useful in constructing phylogenetic trees. However, this insight is not yet reflected in the practice of phylogenetic trees reconstruction. The

---

* Research supported by ISF grant 418/00.
** Corresponding author.

S. Miyano et al. (Eds.): RECOMB 2005, LNBI 3500, pp. 283–295, 2005.
© Springer-Verlag Berlin Heidelberg 2005

vast majority of published works are based on a single gene or protein. Most others are based on combining a few gene trees to a species tree [21], on quartet methods [27, 1], or on super-tree methods [2].

There are many compelling reasons to consider whole genomes or proteomes as a basis for phylogenetic reconstruction, and in some contexts, such methods are essential. One example are the viruses, where different families often have very few or hardly any genes in common, and it is undesirable to base the whole phylogenetic reconstruction on one or very few genes. Traditional methods (such as maximum parsimony or maximum likelihood) are thus inapplicable for viruses, and whole genome methods are naturally called for.

An alternative method that encorporates global, genome wide information, is building trees based on gene order. Here, the goal is to find the shortest rearrangement distance between two genomes, and to use it for tree reconstruction. This approach has been used for quite some time, initially employing heuristics [10]. In 1995, Hannanelly and Pevzner [16] made a breakthrough, finding a polynomial time algorithm for the problem of computing the inversion distance. Following numerous improvements and refinements, it is now possible to compute the inversion distance "metric" for a large number of species. A number of trees based on inversion distances have been constructed, e.g. [22, 4, 10], but their biological significance is still under investigation. Furthermore, before the inversion distance methods can be applied, an identification of the "genetic units" (usually genes) under study is required in each genome. A mapping of each unit to its counterpart in the other genome should follow. By way of contrast, no such preprocessing (which might require manual intervention) is needed in the sequence based approaches we use.

In this work we apply information theoretic methods to construct phylogenies that are based on complete genomes or proteomes. These methods are essentially distance methods: the first step is to compute all the pairwise "distances" between species. We implicitly assume the two strings were generated by *unknown* Markov processes, and use the divergence, or Kullback-Leibler relative entropy [6], of the two distributions. The divergence is a measure of the distance between two distributions induced by two Markov processes. Given two strings, our method estimates this divergence, without knowing the parameters of any of the Markov processes. Since DNA and proteins sequences can be modelled as a Markovian random process [11], the use of this measure is natural. Furthermore, it should be emphasized that the algorithm employs string operations that can be applied to any set of sequences, regardless of its origin. This situation is similar to the Lempel–Ziv compression algorithm [18, 19], whose properties were proved under the assumption of an underlying finite state Markovian source, but is then applicable to sequences of any source.

Our algorithm takes $O(\ell)$ steps to compute the distance between two $O(\ell)$ long genomes. This runtime is fast enough to compute the $\binom{n}{2}$ pairwise distances between n species for large n. We then apply a common distance-based pyhlogenetic reconstruction method to build a tree from the $n \times n$ distance matrix. The efficiency of our algorithm enables us to generate trees for all $n = 191$ organisms whose complete proteome appears in the NCBI database. These include archea, bacteria, and eukaryotes. A forest of phylogentic trees for $n = 1,865$ viruses has also been constructed.

Prior to our work, only about six works for constructing trees from complete genomes/proteomes sequences were published. Stuart et al. [32] used singular value decomposition (SVD) of large sparse data matrices. Each proteome is represented as a vector of tetrapeptide frequencies. The distance between two species is determined by the cosines of the angle between the species' vectors. A similar idea was used by Qi et al. [26]. In their method the frequencies of amino acid K-strings in the complete proteomes of two species determines the distance between them. The main drawback of these two methods is their inflexibility: the analysis is based on *fixed length* K-mers (usually $3 \leq K \leq 8$). Otu et al. [25] used Lempel-Ziv complexity as a basis of a strings' distance. This method is closer in nature to the one we use.

In a series of two papers, Chen et al. [5] and Li et al. [20] develop tools that tools that are inspired (even though there are no direct relations) by Kolmogorov complexity to compress biosequences, and then to compute pairwise distance based on the compression outcome. Since Kolmogorov complexity is incomputable, their GenCompress algorithm uses a generalizion of the Lempel-Ziv algorithm [18, 19]. This compression algorithm reportedly outperforms other DNA compression methods. It has been applied to construct whole mitochondrial genome phylogeny, as well as to sequences of non biological source.

We compared our approach to others, published in the literature [25, 26], by implementing and applying the different approaches to the a dataset of proteomes and genomes of 75 organisms whose whole genomes and proteomes were published [13]. The performance of the various methods has been compared using a standard measure of phylogenetic trees comparison (the Robinson-Foulds method) with respect to a "reference" tree (based on small ribosomal sybunit rRNA) . Our algorithm outperformed all other ones. Compared to the best alternative method, it improvement was 2% on genome sequences, and as much as 17% on proteome sequences. We then ran our algorithm to produce a tree of all 191 available proteome sequences, and a forest of 1,865 viral genomes. The results in general were good, exhibiting high agreement with the accepted taxonomies. We examined some portions of these large trees (*e.g.* the retroviral tree), and observed that in some cases where the exact placement disagrees with the accepted taxonomy, there is support in the literature to this alternative placement.

The remaining of this work is organized as follows: In section 2 we give a brief mathematical intuition for our method, and describe the algorithm and its properties. In section 3 we describe the results of running our algorithm on real data sets: in subsection 3.1 we compare it to other methods on 75 species, in subsection 3.2 we present the results of our algorithm on all 191 known proteomes, and in subsection 3.3 the results on a large set of $1,865$ viruses. Finally, Section 4 contains concluding remarks and directions for further research.

2   Mathematical Background

A natural way to measure the distance between two string of characters is the amount of bits we need for describing one sequence, given the other sequence. It is well known that if a sting was generated by a finite state Markovian process, asymptotically the minimum number of bits that is needed for describing it is proportional to its entropy [6]. If a string was generated by a Markovian process, there are compression algorithms, like Lempel-Ziv [30], which asymptotically achieve the optimal compression ratio. Using the optimal dictionary of one string to compress the other string asymptotically achieves compression ratio proportional to the so called divergence, or Kullback-Leibler (KL) relative entropy of these two strings [6]. In this work we develop phylogeny reconstructing algorithms that use the divergence between the empirical distribution of amino acid, or nucleotides, in two proteomes or genomes, respectively, as a distance measure between them. The divergence (KL) between two i.i.d probability distributions, $p(x)$ and $q(x)$, is defined as:

$$D(p||q) = \sum_{x \in X} p(x) \log \frac{p(x)}{q(x)} = E_p(\log \frac{p(X)}{q(X)}) \tag{1}$$

In general D is *not* a metric. For example $D(p||q) \neq D(q||p)$, and the triangle inequality may not hold.

If our variable, x, ranges over single letters from two sequences, then very different genomes may be at distance 0 or close to 0, provided their statistics of single letters are close. To overcome this, one may try K-mer distributions. These are more informative, but not sensitive enough due to the fixed length of the K-mers. We are looking for an operator that will captor the Markovian dependencies in the genomes/proteomes. Specifically, we want to compute $E_p \left(\log \frac{p(X)}{q(X)} \right)$, where p and q are Markovian distributions.

An additional complication is that we do not know the two underlying probability distributions or their parameters. What we do have is the two sequences that were generated by these distribution. We want to estimate this operator, using these sequences. Before we describe our main method, we provide the mathematical background for our algorithm, based on [36, 14, 35].

Let X_1^n be a finite state Markov chain with probability law p, and let Y_1^n be a Markov chain with probability law q. Let $H(q) = -E_q(\log p(X))$ denote the entropy of the Markovian probability distribution p. The entropy of a random process measures the number of bits that are needed for describing it. If there is less "order" in the process more bits are needed. The waiting time for the first L letters in Y, Y_1^L, to appears in X_1^n, is defined as $W_L = \inf\{m : X_{m+1}^{m+L} = Y_1^L\}$. The following theorem characterize for the waiting time:

Theorem 1. $\lim_{L \longrightarrow \infty} \frac{\log(W_L)}{L} = H(q) + D(q||p)$

By Theorem 1, two distributions are more similar if on average we wait a shorter time for observing in the second subsequences that were generated by the first. In many cases it is easy to consider instead of the waiting time the "longest match". By using the duality between the longest match and waiting times, we get the following relations: Let $D_k = X^0_{-k}$ denote a database containing a chain of length k that was emitted by the process p. Let $L^x = \inf\{\ell : X_1^\ell \subsetneq D_k\}$, and let $L^y = \inf\{\ell : Y_1^\ell \subsetneq D_k\}$. By Theorem 2, process have more "order" if longer strings tend to repeat in sequences it generate. By Theorem 3, two Markovian distributions are closer if longer strings that are generated by one distribution tend to appear in strings that are generated by the other.

Theorem 2. *As* $k \longrightarrow \infty$, $|E[L^x] - \frac{\log k}{H(p)}| = O(1)$.

Theorem 3. *As* $k \longrightarrow \infty$, $|E[L^y] - \frac{\log k}{H(q)+D(q||p)}| = O(1)$.

Our method is based on the above theorems and use these operators to estimate the divergence. To find a distance measure between two genomes, g_1 and g_2, we use the divergence $D(g_1||g_2)$ and $D(g_2||g_1)$. Since $D(g_1||g_2) \neq D(g_2||g_1)$ in general, we use the estimation $D(g_1||g_2) + D(g_2||g_1)$ as a distance measure.

Theorem 4. *The measure* $D_s(g_1, g_2) \equiv D(g_1||g_2) + D(g_2||g_1)$ *satisfies the first two of the following conditions but, in general, not the third one:*

1. $D_s(g_1, g_2) \geq 0$, *and* $D_s(g_1, g_2) = 0$ *if and only if* $g_1 = g_j$.
2. $D_s(g_1, g_2) = D_s(g_2, g_1)$.
3. $D_s(g_1, g_2) \leq D_s(g_1, g_3) + D_s(g_3, g_2)$

To estimate the divergence, we use the following steps: Let $(g_1)_{i+1}^{i+k+1}$ denote the string that starts in position $i+1$ and ends in position $i+k+1$ in g_1. Let

$$L_i(g_2, g_1) = \min_k((g_1)_{i+1}^{i+k+1} \subsetneq g_2).$$

We search the shortest subsequence that start in position i in g_1 and do not appear in g_2. Let $\bar{L}(g_2, g_1) = \frac{1}{|g_2|} \sum_i L_i(g_2, g_1)$ be the average of $L_i(g_2, g_1)$ on i. According to theorems 2 and 3, we can estimate the divergence, $D(g_1||g_2)$, for long enough genomes, by

$$\log|g_2|/\bar{L}(g_2, g_1) - \log|g_1|/\bar{L}(g_1, g_1) .$$

Our algorithm compute this estimation. We call our method the *average common string*, or for brevity ACS.

Even though there is a divergence based metric for i.i.d distributions [12], such a metric does not exist when the distributions are Markovian. In general the triangle inequality does not hold. However, it turned out empirically that for out estimators the triangle inequality does hold for "our" genomic and proteomic datasets. We used our method (ACS) to generate distance matrix for dataset of 75 full genomes and 75 proteomes of including archea, bacteria, and eukaroyte. We checked the triangle inequality for each triplet in the dataset and it holds in all the cases (compared to about 25% violation for a random "distances", where each entry $D(i, j)$ uniform iid).

Given a set of DNA or amino acid sequences, our algorithm computes the pairwise distances for this set according to our metric. In order to efficiently perform the subsequence search, suffix arrays have been chosen as the data structure. The suffix array is a lexicographically sorted array of the suffixes of a string. We used the lightweight suffix array implementation [31]. Creating the suffix array for each sequence requires $O(\ell)$ time for a sequence of length ℓ. It is created once for each sequence, stored, and then used for all comparisons. As the suffix array provides a sorted array of the suffixes of the sequence, it allows to search for each subsequence in amortized $O(1)$ time. Thus searching all the subsequences of a sequence of length ℓ_1 in a sequence of length ℓ_2, takes $O(\ell_1 + \ell_2)$ time. All in all, comparing m sequences of length up to ℓ to each other, takes $O(m^2 \cdot \ell)$ time.

In terms of space complexity, each suffix array requires $O(\ell)$ space, and an additional $O(\ell/\sqrt{\log \ell})$ is required in the construction stage and then reclaimed. The suffix arrays are stored in secondary memory (disk) and loaded to primary memory only when needed for a pairwise comparison session.

We also developed and implemented a different, new method. In this method we use the divergence between two probability mass functions of K-mers (fixed K) in the two genomes. It is faster and simpler than our ACS method, but proved inferior in experiments. We compared our method to methods that were suggested in other works. The first method is a LZ-based [6, 23, 18, 19], where the distance between two strings is inferred by the compressibility of one string given the other's dictionary, using the LZ algorithm. This method is due to Otu et al. [25]. The second method is by Qi et al. [26]. They first calculate the vector of K-mer frequencies in each genome, and then use scalar products of these vectors to generate a distance measures. We implemented these two methods, and used these implementations in our comparisons.

The methods have been applied to sets of genomic and proteomic sequences. For species with multiple chromosomes the genomic sequence is a concatenation of all the chromosomes with delimiters recognized as end points by the algorithms. The proteomic sequences are a concatenation of all the known amino-acid sequences for an organism, also with delimiters. All the sequences have been obtained from the NCBI Genome database [13] in FASTA format (.fna and .faa files).

3   Experimental Results

In this section we describe the results of running our algorithm on three sets of species. The first dataset has 75 species, and we used both genome and the proteome sequences. We have also implemented other reconstruction methods, including two of the leading methods from the literature. Our algorithm outperformed all the alternative ones. We also ran it on two larger datasets: The first contains all known proteomes (191 species), the second includes the genomes of 1, 865 viruses.

3.1 Comparison to Other Methods

We compared the tree that was constructed by our and other methods to a "traditional tree", generated from the sequences of the small ribosomal database project (RDP) [17]. The "traditional tree" , as appearing in the site [17], is the maximum likelihood tree for the aligned set of small ribosomal subunit rRNA.

We obtained a dataset of 75 full genome and proteome sequences. This dataset contains archea, bacteria, and eukaryote for which both genomic and proteomic sequences are available, and that appear in the Ribosomal Database Project. The different "competing methods" were applied, generating different distance matrices. Phylogenetic trees have been constructed from the distance matrices using the Neighbor Joining algorithm (NJ) [29]. We used the Robinson-Foulds measure to compare each tree to the reference, ribosomal tree. Each edge in a tree partition the leaves, or species, to two disjoint sets. The Robinson-Fouldes (RF) method counts the number of partitions that are *not* common to both trees. For two trees on n leave, the RF "distance" is in the range $[0, 2 \cdot n - 6]$ and is always even.

We implemented the Qi-Wang-Hao method *et al.* [26] and the method of Otu *et al.* [25] (an LZ - based method). We also implemented a K-mer based divergence method (FLS method), and a method which produces a matrix with random distances (entries are independently, uniformly and identically distributed), for comparison purposes. Furthermore, a method of comparison based solely on the relative length of sequences has been implemented, in order to test possible correlations between certain methods and the sequence length. The results of all the methods are summarized in Table 1. We chose K-mers of size 5 for the FLS method, this parameter gave the best results for this method.

Table 1. Comparing the ACS to other methods

Method	RF distance from the reference tree	
	Genomes	Proteomes
Random	140	144
Length	142	142
LZ	126	114
FLS	120	96
Qi-Wang-Hao	110	92
Our method (ACS)	108	76

It should be observed that our results improve upon all other methods for both genomes and proteomes. But while the genome improvement compared to [26] is only about 2%, the improvement for the proteomes is about 17% in the RF measure (all with respect to the reference tree).

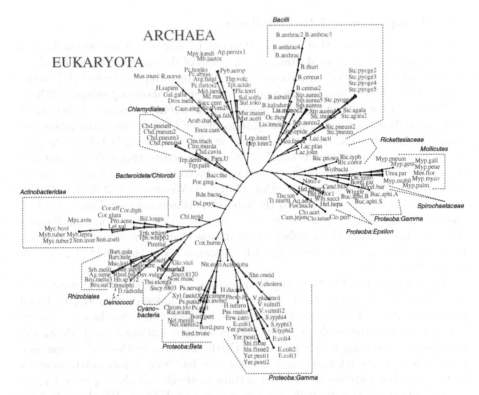

Fig. 1. Tree of 200 proteomes generated by the ACS method. The tree has been drawn using DRAWTREE program of the PHYLIP package [15]

3.2   A Tree Based on All Existing Proteomes

We collected all 191 available proteome sequences from the NCBI databank [13]. This dataset includes 19 proteomes of archea, 161 proteomes of bacteria, and 11 proteomes of eukaroytes. Our tree is presented in Fig. 1, for the sake of clarity all the branches have the same length.

The ACS method has correctly partitioned the species into the 3 main domains: *Eukaryota*, *Archaea* and *Prokayota*, with the exception of 2 archaeal species which will be discussed further. Within the 11 eukaroyte species in the dataset, the *Fungi*, *Eumetazoa*, plants and rodents are all correctly separated by the algorithm. This isn't surprising, given the major differences between the representative eukaryote genomes that have been completely sequenced to date. Thus, more challenging and interesting is the correspondence between the ACS algorithm results and the known taxonomic division within the bacterial and archeal domains. This correspondence has been examined using the taxonomic information found in *NCBI Taxonomy Database* [8]. At large, the tree correctly distinguishes between most of the taxonomic groups in the dataset, making the disagreements between the trees a fertile ground for further comparison between the known taxonomic and the phylogeny revealed using genome comparison.

In the *Archaea* domain , we found a clear separation of genera represented by several species such as the *Pyrococcus* and *Methanosarcina*. The organization of the genera into classes, orders and families is less evident, possibly due to the relatively small number of specimen examined, as discussed below. Two archean species seem to be "misplaced" in the tree - *Nanoarchaeum equitans* (Nano.eq) and *Halobacterium NRC-1* (Hb.spY12) are both mapped on the tree within a mixture of prokaryote species. For *Nanoarchaeum* the reason may be the fact that it is one of a few known archeal parasites, lacking genes for lipid, cofactor, amino acid, or nucleotide biosynthesis [34], making it more problematic in light of our complete genome comparisons. The *Halobacterium sp.NRC-1* is the only archeal species present from the entire *Halobacteria* class, which might explain the difficulty of its classification. Its is found in the tree close to other stress-resistant species such as *D.radiodurans* and *T.Thermophilus* belonging to the *Deinococci* class.

In the *Prokaryota* domain, the *Actinobacteria* class (high G+C Gram-positive bacteria) is clustered together with a correct separation of the majorly represented *Mycobacterium* (5 species) and *Corynebacterium* (3 species) genera. The same holds for the *Chlamydiae* class and *Cyanobacteria* phylum (including correct separation of *Prochlorales* and *Chroococcales*. The largely represented *Firmicutes* phylum (Gram-positive bacteria) is clustered almost entirely on a single branch of the tree, with all the species within the *Clostridia* and the *Bacilli* classes clustered together including divisions into orders, families and genera which is largely at agreement with the taxonomic knowledge. Within the *Proteobacteria* (purple non-sulfur bacteria),the *Beta* and *Epsilon* classes are accurately separated, in the *Alpha* class the represented Rhizobiales class and Rickettsiaceae are both monophyletic groups in the tree (but not clustered together), the small *Delta* class is split, as is the large *Gamma* class, which is divided into two major branches on the tree.

Overall, the algorithm's ability to provide a phylogeny in good agreement with the taxonomic knowledge (which is largely based on the 16S rRNA sequences) is good at the lower levels of genera, families and classes. The method accuracy is decreasing for higher taxonomic groups, a common problem to the whole-genomic approach to phylogenetic inference, as has been reported in [26]. It is expected that the performance on these taxonomy groups will improve as more genomic sequences will become available, as we have experienced with the gradual increase in the number of species that were used in this study. It is natural that the tree construction Neighbor Joining algorithm will perform better provided with more specimen from each group.

3.3   Viruses Forests

Viruses are known to be partitioned to a few superfamilies according to the their nuclear acid description: DNA or RNA, double strand or single strand, positive or negative. Each of these superfamily is believed to have a different evolutionary origin [24]. We used our method for generating a forest for a large viruses' dataset. We collected 1,865 viral genomes, where for 1,837 of the viruses

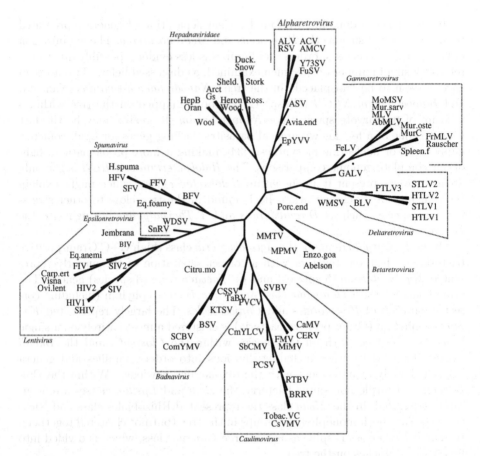

Fig. 2. A tree of the retroid family generated using the ACS method. The common shortcut name is used where available

we had prior knowledge about their superfamily. We partitioned the viruses with known superfamily to one of following families: dsDNA (double stranded DNA), dsRNA (double stranded RNA), retroid (reverse transcriptase viruses), ssDNA (single stranded DNA), ssRNAplus (single stranded RNA plus), ssRNAminus (single stranded RNA minus), and satellite nucleic acids. We attributed each virus with unknown family to the family that is closest to it, according the average ACS distance between the members of a family and the unknown virus. We use the reasonable assumption of larger distance between different trees in the forest than distances in a tree. Then, we applied the ACS method and generated a tree for each of these families. For lack of space, only one of the resulting trees is presented (Fig. 2), for the sake of clarity all the branches have the same length. The other trees will appear in the full version of this work.

In order to evaluate the consistency of the viral trees with the current taxonomic knowledge, the phylogenetic tree constructed using the 83 viral genomes classified as *Retroid viruses* by NCBI Taxonomy. The tree has been compared against the taxonomy appearing in the NCBI Taxonomy and ICTVdb [9]. The partition of the viruses to the 3 main families of reverse transcriptases : Hepadnaviridae, Caulimoviridae (Circular dsDNA reverse transcriptases) and Retroviridae (ssRNA reverse transcriptase), has been fully supported by our tree.

Within the *Hepadnaviridae* family (Hepatitis B viruses) the algorithm distinguished between the *Orthohepadnavirus* (mammalian) and the *Avihepadnavirus* (avian) genera. This included the *Ross Goose Hepatitis B virus* which is currently not classified, with evidence of belonging to the Avihepadnavirus genus [33] and the *Arctic Squirrel Hepatitis virus* (classified with the *Orthohepadnaviruses*).

Among the *Caulimoviridae* family, the two main genera - the *Badnaviruses* (bacilliform DNA viruses) and *Caulimoviruses* are separated in full accordance with the taxonomic data. The *Cestrum Yellow Leaf Curling virus*, considered a tentative member of the genus is clustered together with the rest of the *Caulimoviruses*. Another genus of the *Caulimoviridae* - the *Petunia Vein Clearing virus*, is clustered close to the Caulimoviruses, as is the *Soybean Chlorotic Mottle virus*, *Peanut Chlorotic Streak virus*, *Cassava Vein Mosaic virus*, and the *Rice Tungro Bacilli-form-like viruses*. The location of these families suggests them, according to the genomic data as possible members of the *Caulimovirus* genus.

The *Retroviridae* Family is correctly separated according to the *Orthoretrovirinae* and *Spumaretrovirinae* subfamilies. The sub-division of the Orthorerovirinae to the Alpha, Beta, Gamma, Delta, and Epsilon genera also fits the taxonomic data, except for some spreading of the *Gammaretroviruses*. The widely studied *Procine Endogenous retrovirus*, which is currently classified as a *Mammalian Type-C virus*, is classified among the Gammaretroviruses, fitting existing evidence of its protease resembling the protease of *MLV* within the gamma genus [3]. The yet unclassified *Avian Endogenous Retrovirus EAV-HP* clusters close to the Alpharetrovirus family, following a sequence identity previously reported in [28]. The *Lentivirus* genus members are clustered together, with a clear separation of the Primate (containing the *HIV*), Avian and Bovine species of the viruses. In the *Spumaretrovirinae* family the *Spumavirus* genus (foamy viruses) is clustered together.

The observations above suggest our method is a valid tool deserves further study of in the context of viral taxonomy, overcoming the shortcoming of various traditional methods.

4   Concluding Remarks and Further Research

We believe that our ACS approach is promising, and that its outcomes are interesting, so that overall this is an important step in the of direction constructing whole genome or proteome phylogenies. The experimental results indicate that the proposed method should be explored further. However, this work is certainly not the last algorithmic word in this direction, and many improvements remain to be discovered and developed. For example, our distance matrices were generated using either proteomic or genomic data. We believe that combining those two sources of information can improve the quality of the reconstruction. However, theoretical based approaches for combining such two different sources of information are still missing.

Two similar genomes may share many reversed subsequences (subsequences that have direction that is reversed in one genome compared to the other) and not only subsequences with the same orientation. Another interesting direction is to generalize our algorithm to deal with this observation.

Acknowledgements. Thanks to Eran Bacharach, Tal Pupko, and Jacob Ziv for helpful discussions.

References

1. A. Ben-Dor, B. Chor, D. Graur, R. Ophir, and D. Pelleg. Constructing phylogenies from quartets: elucidation of eutherian superordinal relationships. *Journal of computational Biology*, 5:377–390, 1998.
2. O. Bininda-Emonds. *Phylogenetic Supertrees: Combining Information to Reveal the Tree of Life*. Kluwer series in Computational Biology, 2004.
3. J. H. Blusch, S. Seelmeir, and K. V. Helm. Molecular and enzymatic characterization of the porcine endogenous retrovirus protease. *Virol*, 76(15):7913–17, August 2002.
4. G. Bourque and P. A. Pevzner. Genome-scale evolution: Reconstructing gene order in ancestral species. *Genome Research*, 12:26–36, 2002.
5. X. Chen, S. Kwong, and M. Li. A compression algorithm for dna sequences and its applications in genome comparison. *RECOMB*, pages 107–117, 2000.
6. T. M. Cover and J. A. Thomas. *Elements of Information Theory*. J.Wiley and sons, New York, 1991.
7. C. Darwin. *On the origin of species*. First edition edition, Nov 24 1859.
8. NCBI Taxonomy Database. http://www.ncbi.nlm.nih.gov/entrez/linkout/tutorial/taxtour.html.
9. The Universal Virus Database. http://www.ncbi.nlm.nih.gov/ictvdb/ictvdb/.
10. S. Downie and J. Palmer. Use of chloroplast dna rearangements in reconstructing plant phylogeny. *In: P. Soltis and D. Soltis and J. Doyle (Eds.), Plant Molecular Systematics, Chapman and Hall*, pages 14–35, 1992.
11. R. Durbin, S. R. Eddy, A. Krogh, and G. Mitchison. *Biological sequence analysis: probabilistic models of proteins and nucleic acids*. Cambridge University Press, 1998.
12. D. M. Endres and J. E. Schindelin. A new metric for probability distribution. *IEEE Tran. Inf. Theory.*, 49(7), 2003.
13. NCBI Genome Entrez. http://www.ncbi.nlm.nih.gov/entrez/query.fcgi?db=genome.
14. M. Farach, M. Noordewier, S. Savari, L. Shepp, A. Wyner, and J. Ziv. On the entropy of dna: Algorithms and measurements based on memory and rapid. *Symposium on Discrete Algorithms*, 1994.
15. J. Felsenstein. Phylip (phylogeny inference package) version 3.5c. *Distributed by the author. Department of Genetics, University of Washington, Seattle*, 1993.
16. S. Hannenhalli and P. Pevzner. Transforming cabbage into turnip (polynomial algorithm for sorting signed permutations by reversals). *In Proc. 27th Annual ACM Symposium on the Theory of Computing*, pages 178–189, 1995.

17. Ribosomal Database Project II. http://rdp.cme.msu.edu/html/.
18. A. Lempel and J. Ziv. On the complexity of finite sequences. *IEEE Trans. Inf. Theory*, 22:75–88, 1976.
19. A. Lempel and J. Ziv. A universal algorithm for sequential data compression. *IEEE Trans. Inf. Theory.*, 1977.
20. M. Li, J. Badger, X. Chen, S. Kwong, P. Kearney, and H. Zhang. An information-based sequence distance and its application to whole mitochondrial genome phylogeny. *Bioinformatics*, 17(2):149–154, 2001.
21. B. Ma, M. Li, and L. Zhang. From gene trees to species trees. *SIAM*, 1998.
22. B. M. E. Moret, L. S. Wang, T. Warnow, and S. K. Wyman. New approaches for reconstructing phylogenies from gene order data. *bioinformatics*, 17:165–173, 2001.
23. M. Nelson. *LZW Data Compression*. 1989.
24. Origins of viruses. http://www.mcb.uct.ac.za/tutorial/virorig.html.
25. H. H. Otu and K. Sayood. A new sequence distance measure for phylogenetic tree construction. *Bioinformatics*, 19(16), 2003.
26. J. Qi, B. Wang, and B. Hao. Whole proteome prokaryote phylogeny without sequence alignment: a k-string composition approach. *J. Mol. Evol.*, 58(1):1–11, 2004.
27. P. T. Raul, B. Gordon, and E. Oliver. *in Bininda-Emonds, Olaf R.P. (ed), Phylogenetic Supertrees: Combining Information to Reveal the Tree of Life*, chapter Quartet Supertrees, pages 173–191. Kluwer Academic (In Press), Dordrecht, the Netherlands, 2004.
28. M. A. Sacco, D. M. J. Flannery, K. Howes, and K. Venugopal. Avian endogenous retrovirus eav-hp shares regions of identity with avian leukosis virus subgroup j and the avian retrotransposon art-ch. *J Virol*, 74(3):1296–1306, February 2000.
29. N. Saitou and M. Nei. The neighbor-joining method: a new method for reconstructing phylogenetic trees. *Mol. Biol. Evol.*, 4:406–425, 1987.
30. K. Sayood. *Introduction to data compression.* second edition.
31. B. Stefan and J. Kärkkäinen. Fast lightweight suffix array construction and checking. pages 55–69, 2003.
32. G. W. Stuart and M. W. Berry. A comprehensive whole genome bacterial phylogeny using correlated peptide motive defined in a high dimensional vector space. *Journal of Bioinformatics and Computational Biology*, 1(3):475–493, 2003.
33. M. Triyatnib, P. L. Ey, T. Tran, M. L. Mire, M. Qiao, C. J. Burrell, and A. R. Jilbert. Sequence comparison of an australian duck hepatitis b virus strain with other avian hepadnaviruses. *Journal of General Virology*, 82:373–378, 2001.
34. E. Waters, M. J. Hohn, I. Ahel, D. E. Graham, M. D. Adams, M. Barnstead, K. Y. Beeson, L. Bibbs, R. Bolanos, M. Keller, K. Kretz, X. Lin, E. Mathur, J. Ni, M. Podar, T. Richardson, G. G. Sutton, M. Simon, D. Söll, K. O. Stetter, J. M. Short, and M. Noordewier. The genome of nanoarchaeum equitans: Insights into early archaeal evolution and derived parasitism. *Proc Natl Acad Sci U S A*, 100(22):12984–12988, October 2003.
35. A. D. Wyner and A. J. Wyner. An improved version of lempel-ziv algorithm. *IEEE Tran. Inf. Theory.*, 1995.
36. A. J. Wyner. *String matching theorems and applications to data compression and statistics.* Ph.d., Stanford, 1993.

Maximum Likelihood of Evolutionary Trees Is Hard[*]

(Extended Abstract)

Benny Chor and Tamir Tuller[**]

School of Computer Science, Tel Aviv University
{tamirtul, bchor}@post.tau.ac.il

Abstract. Maximum likelihood (ML) is an increasingly popular optimality criterion for selecting evolutionary trees (Felsenstein, 1981). Finding optimal ML trees appears to be a very hard computational task, but for tractable cases, ML is the method of choice. In particular, algorithms and heuristics for ML take longer to run than algorithms and heuristics for the second major character based criterion, maximum parsimony (MP). However, while MP has been known to be NP-complete for over 20 years (Day, Johnson and Sankoff [5], reduction from vertex cover), such a hardness result for ML has so far eluded researchers in the field.

An important work by Tuffley and Steel (1997) proves quantitative relations between parsimony values and the corresponding log likelihood values. However, a direct application of it would only give an *exponential time* reduction from MP to ML. Another step in this direction has recently been made by Addario-Berry *et al.* (2004), who proved that *ancestral maximum likelihood* (AML) is NP-complete. AML "lies in between" the two problems, having some properties of MP and some properties of ML.

We resolve the question, showing that "regular" ML on phylogenetic trees is indeed intractable. Our reduction follows those for MP and AML, but starts from an approximation version of vertex cover, known as GAP VC. The crux of our work is not the reduction, but its correctness proof. The proof goes through a series of tree modifications, while controlling the likelihood losses at each step, using the bounds of Tuffley and Steel. The proof can be viewed as correlating the value of any ML solution to an arbitrarily close approximation to vertex cover.

1 Background

Molecular data, and even complete genomes, are being sequenced at an increasing pace. This newly accumulated information should make it possible to resolve long standing questions in evolution, such as reconstructing the phylogenetic tree of placental mammals and estimating the times of species divergence. The

---

[*] Research supported by ISF grant 418/00.
[**] Corresponding author.

analysis of this data flood requires sophisticated mathematical tools and algorithmic techniques. Two character-based methods are widely used in practice: MP (*maximum parsimony*, Fitch, 1971 [9]) and ML (*maximum likelihood*, Felsenstein, 1981 [7]). It is known that ML is *consistent*, namely the correct tree is the tree maximizing the likelihood. Consistency does not hold for MP, and in fact for certain families of trees (the so called *Felsenstein zone* [8]) MP will reconstruct the *wrong* trees, even for arbitrarily long input sequences. The two methods are known to be computationally intensive, and exact algorithms are limited to just about $n = 20$ sequences. This forces practitioners to resort to heuristics. For both exact algorithms and heuristics, ML seems a harder problem than MP.

In the absence of concrete lower bound techniques, the major tool for demonstrating computational intractability remains NP hardness proofs. Both MP and ML have well-defined objective functions, and the related decision problems (or at least discretized versions of them) are in the complexity class NP. It has been known for over 20 years that MP is NP-complete [5, 10, 4, 6], using a reduction from vertex cover (VC). However, no such result has been found for ML to date. This is particularly frustrating in light of the intuition among practitioners that ML is harder than MP.

Tuffley and Steel have investigated the relations between MP and ML [16]. In particular, they showed that if the n sequences are padded with sufficiently many zeroes, the ML and MP trees coincide. Since parsimony is invariant under padding by zeroes, this approach could in principle lead to a reduction from MP to ML. Unfortunately, the upper bound provided in [16] on the padding length is *exponential* in n. A step in a different direction was taken by Addario-Berry *et al.* [1]. They studied the complexity of AML (ANCESTRAL MAXIMUM LIKELIHOOD) [12, 17]. This variant of ML is "between" MP and ML in that it is a likelihood method (like ML) but it reconstructs sequences for internal vertices (like MP). They showed that AML is NP-complete, using a reduction from (exact) VERTEX COVER.

Our NP hardness proof of ML uses ingredients from both [16] and [1], as well as new insights on the behavior of the likelihood function on trees. The reduction itself is essentially identical to that given for MP by Day, Johnson, and Sankoff [5], and also used in the AML paper. However, our starting point is not exact VC but the *gap* version of it [2]. The proof of correctness for this reduction relative to ML is different, and substantially more involved. We define a family of *canonical trees*. Every such tree is associated with a unique cover in the original graph. We show that if L is the likelihood of the canonical tree, n is the number of vertices in the original graph, m is the number of edges in the original graph, and c is the size of the associated cover, then as $n \to \infty$,

$$\frac{\log(L)}{-(m+c)\log(n)} \to 1 .$$

In particular, this gives an inverse relation between likelihood and cover size: Larger L implies smaller c, and vice versa.

In the proof, we want to show two directions: ($\Rightarrow$) If there is a small cover than there is a tree with high likelihood, and ($\Leftarrow$) that the existence of a tree with high likelihood implies a small cover. The first direction is easy, using the canonical tree related to the small vertex cover. It is the other direction that is hard, because there is no obvious relation between the log likelihood of a *non-canonical* tree and the size of any cover. What we do, starting from any ML tree, is to apply a sequence of modifications that leads it to a *canonical tree*. The whole series of modifications may actually *decrease* the likelihood of the resulting, canonical tree vs. the original, ML one. We use the techniques of [16] to infer likelihood properties from parsimony ones, and show that in every step, the log likelihood decreases by at most $O(\log n)$ bits. Here we rely on the fact that at each step we only modify a small size sub-forest ($2 \log \log n$ leaves at most). Finally, we show that the total number of modifications is not too large – at most $n/\log \log n$. This implies that the ratio of the log likelihood of the last, canonical tree, and the log likelihood of the ML tree, approaches 1 as $n \to \infty$. This proves that log ML is tightly related to an approximate vertex cover, establishing NP hardness of ML.

Remark: Due to space limitations, we had to omit all figures and many proofs. Full version can be found on the authors' web sites.

2 Proof's Overview

In this section we give a high level description of the hardness proof. The reduction is from the GAP VERTEX COVER problem on graphs whose degree is at most three, a problem proved NP-hard in 1999 by Berman and Karpinski [2].

Given a graph $G = (V, E)$ of max degree 3 with $n = |V|$ nodes and $m = |E| \leq 1.5n$ edges, we construct an ML instance, consisting of $m + 1$ binary strings of length n. The goal is to find a tree with the $m + 1$ sequences at its leaves, and an assignment of substitution probabilities to the edges of that tree (edges' length), such that the likelihood of generating the given sequences is maximized. The proof relates the approximate max log likelihood value to the size of an VC in G. This approximation is tight enough to enable solving the original gap problem.

Our reduction follows the one for maximum parsimony given by Day, Johnson and Sankoff [5] and for ancestral ML, given by Addario-Berry *et al.* [1]. Both reductions were from the (exact) VERTEX COVER problem. In this reduction we generate one string with only 0s, and m "edge strings" that contain exactly two 1s each. Consider all unrooted weighted trees with $m + 1$ leaves that have the given sequences at their leaves. We say that such tree is in *canonical form* if the following properties hold

Definition 1.

1. *There is an internal node (called the "root" for clarity, even though the trees are unrooted) that has the all zero leaf as a son, and the length of the edge going to this leaf is 0.*
2. *All leaves are at distance 1 or 2 from the root.*

3. *If a leaf is at distance* 2 *from the root, then the subtree that contains that leaf has one, two, or three leaves. In the latter cases, all two or three sequences at the leaves share a "1" in the same position.*

The reason we force the root to be connected to the all zero leaf with an edge of weight 0 is that this way the root itself is "effectively forced" to the all zero label (with probability 1). This enables us to express the likelihood of a canonical form tree as a product of the likelihoods of its subtrees. In particular, there is no influence, or dependency between different subtrees. Canonical trees uniquely define a vertex cover, where each subtree corresponds to one, two, or three edges that are covered by one node. Consequently, given a tree in canonical form, we can quantify the size of the corresponding vertex cover of the original graph.

The major part of the proof is showing that given any ML tree, T_{ML}, with the "reduction sequences" at its leaves, there is a series of local modifications on trees with the given sequences at their leaves, such that in each modification the log likelihood of the resulting tree is decreased by at most $O(\log n)$ per step, and the final tree, T_{Ca}, is in canonical form. The number of modifications is small enough to establish a tight ratio $1 - o(1)$ between the max log likelihood and the log likelihood of the final, canonical tree. In each step, we transform one tree to another, we identify a small subforest, containing between $\log \log n$ and $2 \log \log n$ leaves. Such a subforest is a union of subtrees with a common internal node. We show that the parsimony score of this subforest when its root is labeled by the whole zero string can be worse by at most a constant $B < 8^4$ than the score with any other root labeling. Using the results of Tuffley and Steel [16], and employing the small size of the subtree, it is possible to unroot this subforest, rearrange it, and connect it directly to the root in a "canonical way", such that the overall log likelihood of the whole tree decreases by at most $B \log n + o(\log n)$. Over the series of $n/ \log \log n$ modifications, the overall decrease is at most $Bn \log n/ \log \log n + o(n \log n/ \log \log n)$. We show that the log likelihood of the final canonical tree, T_{Ca}, is sufficiently large that despite such decrease,

$$\frac{\log L(S|T_{Ca})}{\log L(S|T_{ML})} = 1 - o(1) .$$

Every tree in canonical form naturally corresponds to a vertex cover in the original graph. The tight relation between $L(S|T_{ML})$ and $L(S|T_{Ca})$ implies a tight relationship between the size of an approximate vertex cover in the original graph and the maximum likelihood tree on the given sequences we produced, and establishes the NP hardness of maximum likelihood on phylogenetic trees.

3   Model, Definitions and Notations

In this section we describe the model and basic definitions that we will use later.

Definition 2. *Phylogenetic trees, characters [16]*
A phylogenetic tree is a tree $T = (V(T), E(T))$ having no vertices of degree two and such that each leaf (degree one vertex) is given a unique label from $1, ..., n$

(where n is the number of leaves of T, we denote [n] = 1, ..., n). For convenience, we identify each leaf with its label. A non leaf vertex is called an internal vertex. A function $\lambda : [n] \to \{0,1\}$ is called a state function for T; if $\hat{\lambda}$ is such that $\hat{\lambda}$ agrees with λ on the leaves of T then $\hat{\lambda}$ is called an extension of λ (on T). In a similar way we define the function $\lambda^k : [n] \longmapsto \{0,1\}^k$ and $\hat{\lambda}^k : V(T) \longmapsto \{0,1\}^k$. This function is called a labelling function for T. If $\lambda^k(i) = s$ we say that the s is the labelling of the i-th leaf.

Given a labelling $\hat{\lambda}^k$, let $d_e(\hat{\lambda}^k)$ be the number of differences between two labellings in the endpoints of an edge $e \in E(T)$.

Definition 3. *(Maximum parsimony tree score)*
Given a set S of length k binary strings, and a tree T, let $\hat{\lambda}^k_{pars} : V(T) \longmapsto \{0,1\}^k$ be a labelling function for T that minimizes the expression $\sum_{e \in E(T)} d_e$. The parsimony score for T, $pars(S,T)$, is the value of this sum. A maximum parsimony tree(or trees) for a set S of binary strings is a tree (or trees) with the minimal sum above.

In the likelihood setting, we use the Neyman two states model [13]. For a tree T, let $\mathbf{p} = [p_e]_{e \in E(T)}$ be the edge probabilities. According to this model:

- The probability of a net change of state (from '1' to '0' or vice versa) occurring across an edge e (a "mutation event") is given by p_e (the "length" of edge e).
- Mutation events on different edges are independent.
- The "edge probability" p_e satisfies $0 \leq p_e \leq \frac{1}{2}$.
- Different sites are independent.

Let $S = [s(1), s(2), s(3), \ldots, s(n)] \in \{0,1\}^{n \times k}$ be the observed (given) sequences of length k over n taxa (on n leaves). The likelihood of observing such S, given the tree T with $r \leq n - 2$ internal nodes and the edge probabilities $\mathbf{p}$, $L(S|T, \mathbf{p})$, is defined as

$$L(S|T, \mathbf{p}) = \prod_{i=1}^{k} \sum_{\mathbf{a} \in \{0,1\}^r} \prod_{e \in E(T)} m(p_e, S_i, a_i) \,, \tag{1}$$

where $\mathbf{a}$ ranges over all combinations of assigning characters states (0 or 1) to the r internal nodes of T. This notion of ML is termed maximum *average* likelihood in Steel and Penny [15]. Each term $m(p_e, S_i, a_i)$ is either p_e or $(1 - p_e)$, depending on whether in the i-th site of S and $\mathbf{a}$, the two endpoints of e are assigned different characters states (and then $m(p_e, S_i, a_i) = p_e$) or the same characters states (and then $m(p_e, S_i, a_i) = 1 - p_e$). The ML solution(or solutions) for a specific tree T is the point (or points) in the edge space $\mathbf{p} = [p_e]_{e \in E(T)}$ (where $0 \leq p_e \leq 1/2$) that maximizes the expression $L(S|T, \mathbf{p})$. The global ML solution(or solutions) is the pair (or pairs) $(T, \mathbf{p})$, maximizing the likelihood over all trees T of n leaves and all edge probabilities $\mathbf{p}$, see [7], Steel

[14], and Tuffley and Steel [16] for more details. It is easy to see that by site independence, an equivalent way to define the likelihood of observing S in the tree T is:

$$L(S|T, \mathbf{p}) = \sum_{\lambda \in \{0,1\}^{k \times r}} \prod_{e \in E(T)} p_e^{d_e(\lambda)} \cdot (1 - p_e(\lambda))^{k - d_e(\lambda)} \qquad (2)$$

In the rest of the paper we use this definition for likelihood.

4   Properties of Maximum Likelihood Trees

In this section we prove some useful properties of ML trees. We start with properties of general trees and continue with canonical ones.

General Properties of ML Trees. In our NP-hardness proof we want to show that the ML tree for a set of strings, which will be described on the next section, have log likelihood arbitrarily close to a tree of the canonical form. We achieve this by continuously pruning sub-forests that satisfy certain conditions, and rearranging them in a canonical way around a certain internal node. Theorem 1 describes bound on the decrease in the log likelihood by such rearrangements.

The following Lemma is used several times in the rest of this paper.

Lemma 1. *Let T be a phylogenetic tree with edge probabilities $\mathbf{p}$, let S (set of binary string of length k) denote the labelling for the leaves of the tree. Suppose F_1 and F_2 are two disjoint subforests that have x as their common root, and their union is all of T. Let S_1 and S_2 be the leaf labelling of F_1 and F_2, respectively. Let ℓ_x denote labelling of x. Then the likelihood of observing S given T and p is:*

$$L(S|T, p) = \sum_{s \in \{0,1\}^k} L(S_1, \ell_x = s | F_1, p) \cdot L(S_2, \ell_x = s | F_2, p).$$

For "standard" phylogenetic trees, the internal nodes do not have any specified labelling, or state, while leaves are labelled by a k long sequence. In the course of our modifications we could have a leaf with no labelling

The next Lemma states that such "unlabelled" leaves can be pruned without effecting the likelihood.

Lemma 2. *Let T be a phylogenetic tree with an unlabelled leaf. By pruning this leaf (and the edge connecting it to its ancestor) we will get a tree, T', with equal likelihood.*

Lemma 3. *Let T be a phylogenetic tree with an internal node, h, of degree two, and let x, y be its neighbors. Then h can be eliminated to create an (x, y) edge without changing the likelihood.*

Theorem 1. *Let T^+ be the tree consisting of T_{new}, hung off a node at distance 0 from the all zero leaf. Let T^- be a tree where $T_1, .., T_j$ hang off a common root h that is labelled by the length k sequence z Suppose there is W such that for every labelling s of h, W times the likelihood of T^- is less than the likelihood of T^+. Let $T_{original}$ be the original tree, and $T_{arranged}$ the tree resulting from uprooting $T_1, .., T_j$ arranging them to T_{new} (the subtrees $T_1, .., T_j$ and the tree T_{new} are over the same taxa) and hanging them off a node at distance zero from an all zero leaf. Then the likelihood of $T_{arranged}$ is at least as large as W times the likelihood of $T_{original}$.*

For any tree T on n leaves and any observed sequences S, we denote by $\mathbf{p}^* = \mathbf{p}^*(S,T)$ the edge probability that maximize $L(S|T,p)$. The following Theorem is a restatement of Theorem 7 from the work of Tuffley and Steel [16].

Theorem 2. *Let S be a set of binary strings of length $k = k_c + k_{nc}$, where k_c is the number of constant characters in S (i.e. positions that have the same value for all the strings). Then for large enough k_c, the maximum likelihood and the maximum parsimony tree for S are identical. Furthermore, for every tree T:*

$$2^{-\log(k_c) \cdot pars(S,T) - C^d_{T,pars(S,T)}} \leq Pr(S|p^*,T) \leq 2^{-\log(k_c) \cdot pars(S,T) - C^u_{T,pars(S,T)}}$$

and

$$\lim_{k_c \to \infty} \frac{-\log(Pr(S|p^*,T))}{\log(k_c)} = pars(S,T)$$

where $C^u_{T,pars(S,T)}$ and $C^d_{T,pars(S,T)}$ are subquadratic functions of the size of $|V(T)|$ (the number of vertices in the tree) and of $pars(S,T)$ (the parsimony score for the tree given S). They do not depend on the number of constant sites, k_c (notice that by increasing k_c we do not change $pars(S,T)$).

Corollary 1. *Let T_a and T_b be two tree topologies for a string set S. S Contain n sequences of length k. Suppose that the strings in S have a large enough number of constant characters, k_c, i. e. k_c is doubly exponential in n, $pars(S,T_a)$, and $pars(S,T_b)$. If $pars(S,T_a) < pars(S,T_b)$, then $Pr(S|p^*,T_a) > Pr(S|p^*,T_b)$ (Notice that changing k_c does not effect the parsimony score).*

We remark that in general equality in the parsimony score does not imply equality in the likelihood. The next corollary generalizes the previous one to general trees with one an internal node that is labelled.

Corollary 2. *Let S be a string set with length k strings. Suppose that S has k_c constant positions. The likelihood, $Pr(S_1, l_h = z|p^*, F)$, of a subforest F with r subtrees $T_1, .., T_r$ and with a label $l_h = z$ at the root of the subforest*

$$Pr(S, l_h = z|p^*, F) = \prod_{i=1}^{r} Pr(S_1^i, l_h = z|p^*, T^i)$$

thus:

$$Pr(S, l_h = z|p^*, F) \geq \prod_{i=1}^{r} 2^{-\log(k_c) \cdot pars(S^i \cup z, T_i) - C_{T_i, pars(S^i \cup z, T_i)}^{d}},$$

and

$$Pr(S, l_h = z|p^*, F) \leq \prod_{i=1}^{r} 2^{-\log(k_c) \cdot pars(S^i \cup z, T_i) - C_{T_i, pars(S^i \cup z, T_i)}^{u}},$$

where $C_{T_i, pars(S^i \cup z, T_i)}^{u}$ and $C_{T_i, pars(S^i \cup z, T_i)}^{d}$ are functions for the sub tree T_i with the strings set $S^i \cup z$, as was defined in Theorem 2. Let $pars(S \cup z, F) = \sum_i pars(S^i \cup z, T_i)$, and let $C_{S \cup z, F}^{u} = \sum_i C_{T_i, pars(S^i \cup z, T_i)}^{u}$ and let $C_{S \cup z, F}^{d} = \sum_i C_{T_i, pars(S^i \cup z, T_i)}^{d}$. Then

$$2^{-\log(k_c) \cdot pars(S \cup z, F) - C_{S \cup z, F}^{d}} \leq Pr(S, l_h = z|p^*, F) \leq 2^{-\log(k_c) \cdot pars(S \cup z, F) - C_{S \cup z, F}^{u}}.$$

Corollary 3. *Let T_1 and T_2 be two subforests with the same number of leaves, each with its common ancestor as described in Corollary 2. Let $Pr(S_1, l_{h1} = z_1|p_1^*, T_1)$ and $Pr(S_2, l_{h2} = z_2|p_2^*, T_2)$ be the maximum likelihood scores of S_1, S_2, for T_1 and T_2, respectively. We fix two labels at the roots of T_1, T_2 (not necessarily the same label). If under this setting the parsimony scores of the two subforests are equal, then the likelihood ratio is "sandwiched" between two functions C_1, C_2:*

$$C_1 \leq \lim_{k_c \to \infty} \frac{Pr(S_2, l_{h2} = z_2|p_2^*, T_2)}{Pr(S_1, l_{h1} = z_1|p_1^*, T_1)} \leq C_2$$

where C_1, C_2 are sub quadratic functions of $|V(T_1)|$, $|V(T_2)|$, $pars(S_1, T_1)$, and $pars(S_2, T_2)$.

Properties Related to Canonical ML Trees. We now study properties related to canonical ML trees (definition 1), which play an important role in our reduction.

Definition 4. *Let T_{C_i} ($i = 1, 2$, or 3) be a phylogenetic tree with $i + 1$ leaves, and one internal node (i.e. T_{C_i} has the star topology). Suppose one of the strings in the leaves is the all zero string (of length $k = n$). The other i strings are all of weight 2 (two 1s), and for $i > 1$ they all share one "1" position. Let $ML_i(n)$ be the log ML score of T_{C_i}. Let S_{C_i} denote the strings in the leaves of tree T_{C_i}*

It is easy to see that $ML_i(n)$ does not depend on the specific choice of strings in T_{C_i}

Lemma 4. *Let $C_1^u, C_2^u, C_3^u, C_1^d, C_2^d, C_3^d$ denote constants, then for n large enough the following properties hold:*

1. $-2 \cdot \log(n) + C_1^d \leq ML_1(n) \leq -2 \cdot \log(n) + C_1^u$
2. $-3 \cdot \log(n) + C_2^d \leq ML_2(n) \leq -3 \cdot \log(n) + C_2^u$
3. $-4 \cdot \log(n) + C_3^d \leq ML_3(n) \leq -4 \cdot \log(n) + C_3^u$

Theorem 3. *Let T_a and T_b be two canonical trees with $m + 1$ leaves labelled by S, where S contains strings of length k. Let d_a and d_b denote the degree of the root of these trees, respectively. Let p_a^* and p_b^* be optimal edges weights for these trees, respectively. Then*

$$\log(P(S|p_b^*, T_b)) = -(d_b + m) \cdot \log n + d_b \cdot C_b$$

and

$$\log(P(S|p_a^*, T_a)) = -(d_a + m) \cdot \log n + d_a \cdot C_a \ ,$$

where $C_b = \theta(n)$, and $C_a = \theta(n)$.
So for large enough n:

$$lim_{n \to \infty} \frac{\log(P(S|p_a^*, T_a))}{\log(P(S|p_b^*, T_b))} = \frac{d_a + m}{d_b + m}.$$

In particular if $d_a = d_b$ then:

$$lim_{n \to \infty} \frac{\log(P(S|p_a^*, T_a))}{\log(P(S|p_b^*, T_b))} = 1.$$

5   NP-Hardness of Maximum Likelihood

The decision version of maximum likelihood is the following:

Problem 1. Maximum likelihood, (ML)
Input: S, A set of binary strings, all of length k, and a negative number L.
Question: Is there a tree, T, such that $\log(Pr(S|p^*(S,T), T)) > L$?

A gap vertex cover problem is the following:

Definition 5. *Gap problem for vertex cover, $gap - VC[C_1, C_2]$*
Input: *A graph, $G = (V, E)$, two positive numbers, C_1 and C_2.*
Task: *Does G have a vertex cover smaller than C_1? Or is the size of each vertex cover is larger than C_2 ? (If the minimum vertex cover is in the intermediate range, there is no requirement.)*

Our proof uses a reduction from the gap version of vertex cover, restricted to degree 3 graphs, to maximum likelihood. We use the following hardness result of Karpinski and Berman [2].

Theorem 4. *[2] The following problem, $gap - VC3[\frac{144}{284} \cdot n, \frac{145}{284} \cdot n]$, is NP-hard: Given a degree 3 graph, G, on n nodes, is the minimum VC of G smaller than $\frac{144}{284} \cdot n$? Or is it larger than $\frac{145}{284} \cdot n$?*

We reduce the version of $gap - VC3$ above to ML.

5.1   Reduction and Proof Outline

Given an instance $< G = (V, E) >$ of $gap-VC_3$, denote $|V| = n$, $|E| = m$, $m_1 = \frac{144}{284} \cdot n$ and $m_2 = \frac{145}{284} \cdot n$. We construct an instance $< S, L >$ of ML such that S is a set of $m+1$ strings, each string of length $k = n$, and $L = -(m + \frac{m_1 + m_2}{2}) \cdot \log n$.

The first string in S consists of all zeros (the all zeros string), i.e.,

$$\underbrace{00...0...00}_{k}$$

and for every edge $e = (i, j) \in E$ there is a string, $S(e)$,

$$\underbrace{\overbrace{00..00}^{i-1} 1 \overbrace{00..00}^{j-i-1} 1 \overbrace{00..00}^{k-j}}_{k}$$

where only the i-th and the j-th positions are set to 1. These m strings are called "edge strings". From now on, the trees we refer to have leaves with labels generated by this construction.

We use asymptotic properties of likelihood of trees, so most claims will hold when the input graph is large enough (*i.e.* $n = |V|$ is large enough). In our proof, we deal with small size subtrees or subforests, containing at most $2 \cdot \log \log n$ leaves. We will need the following relation for the expressions in the likelihood of the subforests to hold (see Corollary 2): $\lim_{n \to \infty} [C^d_{S \cup z, T}/(\log(k_c) \cdot pars(S \cup z, T))] = 0$ and $\lim_{n \to \infty} [C^u_{S \cup z, T}/(\log(k_c) \cdot pars(S \cup z, T))] = 0$. According to our reduction the parsimony score (and k_{nc}, the number of non-constant sites) of such subtrees and subforests is no more than $4 \cdot \log \log n$. So according to Theorem 2 and Corollaries 1, 2, and 3, it is enough that $k_c = k - k_{nc}$ will be doubly exponential in these parameters (the size and the parsimony of these subforests) to get these relations.

The proof strongly relies on quantitative relations between parsimony and likelihood as proved in [16].

5.2   Likelihood of Canonical Trees

In this section we show that for every $\varepsilon > 0$ there is an $n_0 > 0$ such that for $n > n_0$, the ratio between the log likelihood and the maximum log likelihood of some canonical tree is upper bounded by $(1 + \varepsilon)$.

Given an ML tree, T, if it is in canonical form, we are done. Otherwise we locate subtrees of T, $T_1, T_2, ..., T_\ell$ with a common root, such that the number of leaves in $\bigcup_{i=1}^{l} T_i$ is in the interval $[\log \log n, 2 \cdot \log \log n]$. Notice that this is a subforest as there may be other subtrees rooted at the same node. It is easy show that such a subforest always exists (Lemma 5). On the next step we show that the ratio of the log-likelihood of such subforest when, the all zero labelling is placed in its root, and the log-likelihood of the same subforest with any other labellings in its root, is small.

Lemma 5. *Suppose T is a rooted tree and v is an internal node such that the number of leaves below v is at least q. Then v has a descendent, u, such that* *u has a forest consisting of ℓ subtrees $T_1, T_2, ..., T_\ell$ ($\ell \geq 1$) rooted at u, and the number of leaves in the forest $\bigcup_{i=1}^{\ell} T_i$ is in the interval $[q, 2 \cdot q]$.*

Lemma 6. *Let h be the root of a subforest $F \subseteq T$. Let u be an internal node in F ($u \neq h$), whose degree is $r \geq 9$, and let $s \in \{0,1\}^k$. Consider an assignment of labels to internal nodes of F, where h is assigned s. Among such assignments, those that optimize the parsimony score label u with 0^k.*

The proof of the following lemma is similar to that of Lemma 6.

Lemma 7. *Let h be the root of a subforest $F \subseteq T$. Suppose the degree of h is $r \geq 9$. Consider the parsimony score of F when h is assigned $s \in \{0,1\}^k$, and when h is assigned 0^k. The latter score is better (smaller).*

Lemma 8. *Let h be the root of a subforest F (h has at least two children in F). Suppose that in each position, the leaves labelled with "1" are at distance at least 4 from h. Then the max parsimony score on F is achieved with the all zero labelling in h.*

Corollary 4. *Let h be a root of a subforest F. Suppose all leaves having "1" in the position are either at distance ≥ 4 from the root, or have an internal node of degree ≥ 9 in the path to the root. Then the parsimony score of F when labelling the root h with 0 at this position is at least as good as when labelling the root with 1.*

Theorem 5. *Let T be a tree whose leaves are labelled by a subset of the edge strings. Let F be a subforest of T, rooted at h, and let $s \in \{0,1\}^k$ be a label of h. The parsimony score of F with the 0^k label at the root is worse by less than 8^4 than the parsimony score of F with label s at its root.*

The following Lemma was proved by Day et al. [5,1].

Lemma 9. *Let $S' \subseteq S$ be a subset of the "reduction strings", which contains the all zero string. The structure of the best parsimony tree for S' is canonical.*

Theorem 6. *For every $\varepsilon > 0$ there is an n_0 such that for all $n \geq n_0$, if $S \subseteq 0, 1^n$ is a set of $m+1$ reduction strings on a graph with n nodes, the following hold: Let T_{ML} denote an ML tree for S, and let p^*_{ML} and be an optimal edges length for this ML tree. Then there is a canonical tree for S, T_{Ca}, with optimal edges length p^*_{Ca}, such that:*

$$\frac{\log(P(S|p^*_{ML}, T_{ML}))}{\log(P(S|p^*_{Ca}, T_{Ca}))} > (1 - \varepsilon) .$$

Proof. We start from any ML tree, T_{ML}, and show how to transform it to a canonical tree, T_{Ca}, with "close enough" log likelihood, in a sequence of up to $n/\log\log(n)$ steps. Each step involves a small, local change to the current tree:

We identifying a subforest with a common root and number of leaves in the interval $[\log\log(n), 2 \cdot \log\log(n)]$. By Lemma 5, if the root of the whole tree has a subtree with more than $\log\log(n)$ leaves, we can find such a subforest. In such case, we first uproot this subforest and move it to the root. By Theorem 5 the parsimony score of such subforest with the 0^k label at the root is worse by less than $B \equiv 8^4$ than the parsimony score of F with any $s \in \{0,1\}^k$ label at its root. Since the number of leaves in F is at most $2\log\log(n)$, the number of *constant sites* k_c, is at least $n - 4\log\log(n)$, so $\log(k_c) = \log(n) - o(\log n)$. Applying Corollary 2 to $l_h = 0^k$ and $l_h = s \in \{0,1\}^k$:

$$2^{-\log(k_c) \cdot pars(S \cup 0^k, F) - C^d_{S \cup 0^k, F}} \leq Pr(S, l_h = 0^k | p^*, F) \leq 2^{-\log(k_c) \cdot pars(S \cup 0, F) - C^u_{S \cup 0^k, F}}.$$

$$2^{-\log(k_c) \cdot pars(S \cup s, F) - C^d_{S \cup s, F}} \leq Pr(S, l_h = s | p^*, F) \leq 2^{-\log(k_c) \cdot pars(S \cup s, F) - C^u_{S \cup s, F}}.$$

The parsimony score on such F with $l_h = 0^k$ at its root is no more than the number of "1" entries, which is bounded by $4\log\log(n)$. The size of F (the number of vertices) is at most $4\log\log(n)$. The function $C^u_{S \cup z, F}$ is a positive, sub-quadratic function of F's size and the parsimony score. Thus, are $C^u_{S \cup 0^k, F} = o((\log\log(n))^2)$.

$$\log Pr(S, l_h = s | p^*, F) - \log Pr(S, l_h = 0^k | p^*, F)$$
$$\leq -\log(k_c) \cdot pars(S \cup s, F) - C^u_{S \cup s, F} - (-\log(k_c) \cdot pars(S \cup 0^k, F) - C^d_{S \cup 0^k, F})$$
$$= -\log(k_c) \cdot (pars(S \cup s, F) - pars(S \cup 0^k, F)) + C^u_{S \cup 0^k, F} - C^d_{S \cup s, F}$$
$$\leq B\log(k_c) + C^u_{S \cup 0^k, F} - C^d_{S \cup s, F}$$
$$\leq B\log(k_c) + O(\log^2\log(n))$$
$$= B\log(n) + o(\log n)$$

Therefore, according to Theorem 1, when moving this forest to the root, the total log likelihood of the tree decreases by less then $B\log(n) + o(\log n)$. According to Lemma 9 we can rearrange such a subforest with the all zero root in a canonical form such that its parsimony score will not become worse, let F_c denote such such canonical rearrangement. Thus by Corollary 2 and Corollary 3:

$$\log Pr(S, l_h = 0^k | p^*, F) - \log Pr(S, l_h = 0^k | p^*, F_c)$$
$$\leq -\log(k_c) \cdot pars(S \cup 0^k, F) - C^u_{S \cup 0^k, F} - (-\log(k_c) \cdot pars(S \cup 0^k, F_c) - C^d_{S \cup 0^k, F_c})$$
$$\leq C^u_{S \cup 0^k, F} - C^d_{S \cup s, F_c}$$
$$\leq O(\log^2\log(n))$$
$$= o(\log n)$$

Therefore, such rearrangement can decrease the log likelihood of the tree by less $O((\log\log(n))^2) = o(\log n)$. If we reached a situation where all subtrees are smaller than $\log\log n$, we rearrange each subforest of size in the range $[\log\log(n), 2 \cdot \log\log(n)]$ separately. According to Theorem 3, the log-likelihood of all canonical trees is larger than $-n\log(n)$. We just showed the existence of a canonical tree whose log likelihood differs from the log likelihood of any ML tree by less than $Bn\log(n)/\log\log(n)$ (for large enough n). Thus there must be a constant $K > 0$ such that the log-likelihood of any ML tree is at most $-K \cdot n\log(n)$, and consequently there is a canonical tree such that the ratio between the log likelihood of the ML tree and this tree is

$$\frac{-K \cdot n\log(n)}{-K \cdot n\log(n) - O(n \cdot \log n / \log\log(n))} = 1 + O(\frac{1}{\log\log n}) < 1 + \varepsilon.$$

5.3   Correctness of the Reduction

In this section we complete our proof by showing that indeed we have a reduction from $GAP - VC_3$ to ML. The basic idea is to show that if G has a small enough cover, then the likelihood of the corresponding canonical tree is high (this is the easy direction), and if the likelihood is high, then there is a small cover (the harder direction). The translation of sizes, from covers to log likelihood, and vice versa, is not sharp but introduces some slack. This is why a hard approximate version of vertex cover is required as our starting point.

The next Lemma establishes a connection between MP and VC, and was used in the NP-hardness proof of MP.

Lemma 10 ([5, 1]). $G = (V, E)$ has a vertex cover of size c if and only if there is a canonical tree with parsimony score $c + m$, where c is the degree of the root.

Theorem 7. For every $0 < \varepsilon$ there is an n_0 such that for $n \geq n_0$, given a degree 3 graph $G = (V, E)$ on n nodes and m edges, with a cover of size at most c, the following holds: There is a tree T such that the log-likelihood of the tree satisfies

$$\log(Pr(S|p^*(S, T), T)) \geq -(1 + \varepsilon) \cdot (m + c) \cdot \log n.$$

On the other hand, if the the size of every cover is $\geq c$ then the log likelihood of each tree, T, satisfies

$$\log(Pr(S|p^*(S, T), T)) \leq -(1 - \varepsilon) \cdot (m + c) \cdot \log n.$$

Proof. Suppose G has a vertex cover of size $\leq c$. Since G's is of bounded degree 3: $m/3 \leq c \leq m$ and $n \leq m \leq 1.5 \cdot n$. According to Lemma 10, there is a canonical tree, T, with parsimony score $c + m$, such that the degree of its root is c. According to Theorem 3, this tree has log likelihood $-(c + m) \cdot \log(n) + \theta(n)$. Since $m, c = \theta(n)$, we have $n = o((c + m) \cdot \log(n))$, so $\log(Pr(S|p^*(S, T), T)) = -(c + m) \cdot \log(n) + \theta(n)$ implies tat for every $\varepsilon > 0$ and large enough n: $\log(Pr(S|p^*(S, T), T)) > -(m + c) \cdot \log(n) \cdot (1 + \varepsilon)$.

For the other direction, suppose the size of every cover of G is $\geq c$. According to Lemma 10 the parsimony score of each canonical tree is at least $c + m$. Thus the likelihood of each canonical tree is at most $-(c + m) \cdot \log(n) + C$ where $C \leq m \cdot \max\{C_1^u, C_2^u, C_3^u\} = \theta(n)$. Since $m, c = \theta(n)$ we get that the likelihood of each canonical tree is at most $-(c+m)\cdot\log(n)+\theta(n) \leq -(c+m)\cdot\log(n)\cdot(1-\varepsilon_1)$

according to Theorem 6 the likelihood of the best tree is $\leq -(c+m) \cdot \log(n) \cdot (1 - \varepsilon_1)(1 - \varepsilon_2)$ where $\varepsilon_1, \varepsilon_2$ are arbitrarily small, thus for every ε there is n_0 such that for $n > n_0$ the likelihood of the best tree is $\leq -(c+m) \cdot \log(n) \cdot (1-\varepsilon)$.

Theorem 8. *ML is NP-hard.*

Proof. Let $1 + \varepsilon_c = 1.0069 = \frac{m_2}{m_1}$, let c be the size of the minimal cover in G. Suppose $c < m_1$, then according to Theorem 7 there is a tree whose likelihood is at least $-(1 + \varepsilon) \cdot (m + c) \cdot \log n$, where ε is arbitrarily small. Since $\frac{m_1 + m_2}{2} = m_1 \cdot (1 + \frac{\varepsilon_c}{2})$, and $\frac{m}{3} \leq m_1 \leq m$ (the degree of the graph is at most 3). For small enough ε we get $L = -(\frac{m_1+m_2}{2} + m) \cdot \log n = -(m_1 \cdot (1 + \frac{\varepsilon_c}{2}) + m) \cdot \log n \leq -(m + m_1) \cdot \log n \cdot (1 + \varepsilon) \leq -(m + c) \cdot \log n \cdot (1 + \varepsilon)$. Thus $(S, L) \in ML$.

Suppose every cover of G is larger than $c > m_2$, according to Theorem 7 the log likelihood of each tree is less than $-(1 - \varepsilon) \cdot (m + c) \cdot \log n$, where ε is arbitrarily small. Since $\frac{m_1 + m_2}{2} = m_2 \cdot (1 + \frac{1}{\varepsilon_c})$, $c > m_2$ and $\frac{m}{3} \leq m_2, m_1 \leq m$. For small enough ε we get

$$-(1 - \varepsilon) \cdot (c + m) \cdot \log n \leq -(1 - \varepsilon) \cdot (m_2 + m) \cdot \log n \leq$$

$$-(1 - \varepsilon) \cdot ((\frac{m_1 + m_2}{2} + m) \cdot \frac{1 + \frac{1}{\varepsilon_c}}{4}) \cdot \log(n) \leq -(m + \frac{m_1 + m_2}{2}) \log(n) = L.$$

Thus $(S, L) \notin ML$.

6   Conclusions and Further Research

In this work, we proved that ML reconstruction of phylogenetic trees is computationally intractable. We used the simplest model of substitution - Neyman two states model [13]. Furthermore, we generalised our NP-hardness proof to the Jukes-Cantor model [11] (details omitted due to space limitation) . This model is a special case of Kimura and other models of DNA substitution.

While resolving a 20 year old problem, this work raises a number of additional ones. Our proof techniques can be extended to show that there is a constant $\delta > 0$ such that finding a tree whose *log likelihood* is within $(1 + \delta)$ of the optimum is hard. Can such inapproximability be proved with respect to the *likelihood* itself? Vertex cover, which is the starting point for the reduction, has a simple 2-approximation result. What about approximation algorithms for log likelihood? What can be said about the complexity of ML when restricted to trees under a *molecular clock*?

And finally, it would be nice to identify regions where ML is *tractable*. However, it is not even known what is the complexity of *small ML*, where the sequences and the unweighted tree are given, and the goal is to find optimal edge lengths. In practice, local search techniques such as EM or hill climbing seem to perform well, but no proof of performance is known, and multiple maxima [14,3] shed doubts even on the (worst case) correctness of this approach.

Acknowledgements

We wish to thank Isaac Elias for helpful discussions, and Sagi Snir for reading early drafts of the manuscript.

References

1. L. Addario-Berry, B. Chor, M. Hallett, J. Lagergren, A. Panconesi, and T. Wareham. Ancestral maximum likelihood of evolutionary trees is hard. *Jour. of Bioinformatics and Comp. Biology*, 2(2):257–271, 2004.
2. P. Berman and M. Karpinski. On some tighter inapproximability results. *Proc. 26th ICALP*, 1999.
3. B. Chor, M. D. Hendy, B. R. Holland, and D. Penny. Multiple maxima of likelihood in phylogenetic trees: An analytic approach. *Mol. Biol. Evol.*, 17(10):1529–1541, 2000.
4. W. Day. The computational complexity of inferring phylogenies from dissimilarity matrix. *Bulletin of Mathematical Biology*, 49(4):461–467, 1987.
5. W. Day, D. Johnson, and D. Sankoff. The computational complexity of inferring rooted phylogenies by parsimony. *Mathematical Biosciences*, 81:33–42, 1986.
6. W. Day and D. Sankoff. The computational complexity of inferring phylogenies by compatibility. *Systematic Zoology*, 35(2):224–229, 1986.
7. J. Felsenstein. Evolutionary trees from DNA sequences: A maximum likelihood approach. *J. Mol. Evol.*, 17:368–376, 1981.
8. J. Felsenstein. Inferring phylogenies from protein sequences by parsimony, distance, and likelihood methods. *Meth. in. Enzym.*, 266:419–427, 1996.
9. W. M. Fitch. Toward defining the course of evolution: minimum change for specified tree topology. *Systematic Zoology*, 20:406–416, 1971.
10. L. Foulds and R. Graham. The steiner problem in phylogeny is np-complete. *Advances in Applied Mathematics*, 3:43–49, 1982.
11. T. H. Jukes and C. R. Cantor. Evolution of protein molecules. *In H. N. Munro, editor, Mammalian protein metabolism*, pages 21–132, 1969.
12. M. Koshi and R. Goldstein. Probabilistic reconstruction of ancestral nucleotide and amino acid sequences. *Journal of Molecular Evolution*, 42:313–320, 1996.
13. J. Neyman. Molecular studies of evolution: A source of novel statistical problems. *In S. Gupta and Y. Jackel, editors, Statistical Decision Theory and Related Topics*, pages 1–27., 1971. Academic Press, New York.
14. M. Steel. The maximum likelihood point for a phlogenetic tree is not unique. *Syst. Biol.*, 43:560–564, 1994.
15. M. Steel and D. Penny. Parsimony, likelihood and the role of models in molecular phylogenetics. *Mol. Biol. Evol.*, 17:839–850, 2000.
16. C. Tuffley and M. Steel. Link between maximum likelihood and maximum parsimony under a simple model of site substitution. *Bulletin of Mathematical Biology*, 59(3):581–607, 1997.
17. Z. Yang, S. Kumar, and M. Nei. A new method of inferring of ancestral nucleotide and amino acid sequences. *Genetics*, 141:1641–1650, 1995.

Graph Theoretical Insights into Evolution of Multidomain Proteins

Teresa Przytycka[1], George Davis[2], Nan Song[3], and Dannie Durand[3,4]

[1] National Center for Biotechnology Information,
US National Library of Medicine, National Institutes of Health,
Bethesda, MD 20894
przytyck@mail.nih.gov
[2] Program in Computation, Organizations and Society,
Carnegie Mellon University
[3] Department of Biological Sciences, Carnegie Mellon University
[4] Computer Science Department, Carnegie Mellon University
durand@cmu.edu

Abstract. We study properties of multidomain proteins from a graph theoretical perspective. In particular, we demonstrate connections between properties of the domain overlap graph and certain variants of Dollo parsimony models. We apply our graph theoretical results to address several interrelated questions: do proteins acquire new domains infrequently, or often enough that the same combinations of domains will be created repeatedly through independent events? Once domain architectures are created, do they persist? In other words, is the existence of ancestral proteins with domain compositions not observed in contemporary proteins unlikely? Our experimental results indicate that independent merges of domain pairs are not uncommon in large superfamilies.

1 Introduction

Protein domains are elementary units of protein structure and evolution. About two thirds of proteins in prokaryotes and eighty percent in eukaryotes are multidomain proteins [1]. On average, a protein has two to three domains, but there are proteins for which the domain count exceeds one hundred [15,31].

There is no agreement on a precise definition of protein domain. The definition adopted in this work assumes that domains are conserved evolutionary units that are (1) assumed to fold independently, (2) observed in different proteins in the context of different neighboring domains, and are (3) minimal units satisfying (1) and (2).

Multidomain proteins pose a challenge in the analysis of protein families. Traditional approaches for studying the evolution of sequences were not designed with multidomain proteins in mind. For example, gene family evolution is typically modeled as a tree built from multiple sequence alignment. However, it is not clear how to construct such an alignment for a family with heterogeneous

S. Miyano et al. (Eds.): RECOMB 2005, LNBI 3500, pp. 311–325, 2005.

domain composition. Another challenge arises in graph theoretical approaches to protein family classification [22, 19, 34]. This approach typically models the protein universe as a similarity graph, $G = (V, E)$, where V is the set of all amino acid sequences and two vertices are connected by an edge if the associated sequences have significant similarity. The idea is first to identify all pairs of homologous proteins and then apply a clustering technique to construct protein families. In an ideal world, protein families would appear as cliques in such a graph, where every member of the family is related to all other members and to no other protein. However, relationships in this graph are not always transitive. First, it may be impossible to detect sequence homology between related but highly diverged sequences. In addition, lack of transitivity can result from domain chaining in multidomain proteins. A protein containing domain A is a neighbor of a protein containing domains A and B, which in turn is connected to a protein containing only domain B, but there would be no direct relationship between the proteins containing only A and only B, respectively. Consequently, in the presence of multidomain proteins, protein families identified by graph clustering methods may contain completely unrelated proteins. More methods that deal explicitly with multidomain proteins are needed.

In order to focus on the properties of multidomain proteins and the relationships between them, we introduce the *protein overlap graph* and its dual, the *domain overlap graph*. In the protein overlap graph, the vertices are proteins represented by their domain architectures, where domains are represented by probabilistic models of multiple sequence alignments, such as PSSMs [14] or HMMs [5, 24]. Two vertices are connected by an edge if the corresponding proteins share a domain. In the domain overlap graph, the vertices are protein domains and two domains are connected by an edge if there is a protein that contains both domains. These abstractions allow us to focus on domain architectures.

In the current work, we study the structure of domain overlap graphs to gain insight into evolution of multidomain architectures. Multidomain proteins can be formed by gene fusion [20, 23, 32], domain shuffling [1, 4, 25, 27] and retrotransposition of exons [26]. We abstract these biological mechanisms into two operations: domain merge and domain deletion. We use the term domain merge to refer to any process that unites two or more previously separate domains in a single protein. Domain deletion refers to any process in which a protein loses one or more domains. We represent a domain architecture by the set of its domains. Obviously, this abstraction neglects the fact that multidomain proteins are also subject to domain rearrangement, tandem duplication, and sequence divergence. However in the case of domain pairs it has been observed that only about 2% of such pairs occur in both possible orders [4]. Nevertheless, we must keep in mind our simplifying assumptions while interpreting the results.

We apply the graph theoretic tools developed in this paper to genomic data to consider two questions: First, is domain merging a rare event or is it common for the same pair of domains to arise repeatedly through independent events? Second, once domain architectures are created do they persist? In other words, do the majority of ancestral architectures occur as subsets of some contemporary protein architectures? It has been argued that the vertex degree for domain overlaps graphs can be reasonably approximated by power law [33,2]. The most popular method of modeling such distribution is using the preferential attachment model [3]. Can this model be applied to multidomain proteins? We investigate these questions using the following approach:

1. We define two parsimony models for multidomain family evolution based on the concept of Dollo parsimony, which we call conservative and static Dollo parsimony. The existence of a conservative Dollo parsimony for a protein family is consistent with a history in which every instance of a domain pair observed in contemporary members of the family arose from a single merge event. The existence of a static Dollo parsimony is consistent with a history in which no ancestor contains a domain combination not seen in a contemporary taxon.
2. We establish a relationship between these parsimony models and particular structures in the domain overlap graph, namely chordality and the Helly property. (Rigorous definitions of these concepts are given in the body of the paper.)
3. We adapt fast algorithms for testing chordality and the Helly property previously developed by other authors to obtain fast existence tests for conservative and static Dollo parsimony and reconstruction of corresponding trees.
4. Using a result from random graph theory, we design a method for selecting a statistically informative test set. We also test the agreement of preferential attachment model with the data.
5. We apply these tests to genomic data and determine the percentage of protein families that can be explained by static or conservative Dollo parsimony.

The paper is organized as follows. First, we review the relevant phylogenetic models and introduce our restrictions on the Dollo parsimony in Section 2. In Section 3, we introduce the graph theoretical concepts used in the paper and show how they apply to the domain overlap graph. We also provide an elegant link between these concepts and parsimony models introduced in Section 2. The application of the theoretical results to genomic data is presented in Section 4. Finally, we provide conclusions and directions for future research.

2   Tree Models

Gene family evolution is traditionally modeled by phylogenetic trees, where leaves are sequences and internal nodes are either speciation or duplication events. Gene trees are traditionally built from multiple sequence alignments (MSAs). However, it is not clear how to construct an MSA for a family with heterogeneous domain composition. One approach is to use the MSA of one domain only (see for example [17, 30]). There is no guarantee, however, that the resulting tree will capture large scale changes in domain composition. Therefore, in this work we will consider parsimony models, where the primary evolutionary events are domain insertion and deletions.

In general, parsimony methods assume that each taxon is characterized by a set of characters or attributes. Each character can assume a finite set of possible states and can change state in the course of evolution. The *maximum parsimony tree* is a tree with leaves labeled with character states associated with the input taxa, and internal nodes labeled with the inferred character states of ancestral taxa, such that the total number of character changes along its branches is minimized. Additional restrictions on the type, number and direction of changes lead to a variety of specific parsimony models [12]. In this work, we focus on binary characters, characters that take only the values zero or one, usually interpreted as the presence or absence of the attribute in the taxa.

The most restrictive parsimony assumption is *perfect phylogeny*: a tree in which each character state change occurs at most once [12]. One method of testing the existence of perfect phylogeny tree is based on the *compatibility criterion*. For a given set of taxa, two characters A and B are *compatible* if and only if there do not exist four different taxa respectively representing all four possible combination of character states for A and B (that is, $(0,0), (0,1), (1,0), (1,1)$). The appearance of all four combinations indicates that one of the two characters must have changed state twice. A set of taxa admits a perfect phylogeny if and only if every pair of taxon is compatible [12].

In *Dollo parsimony*, a character may change state from zero to one only once, but from one to zero multiple times [21]. This model is appropriate for complex characters such as restriction sites and introns which are hard to gain but relatively easy to lose [12].

We model multidomain protein evolution in terms of domain insertion and loss. In our model, the taxa are domain architectures and each domain defines a single binary character, where state one corresponds to the presence, and zero to the absence, of the domain in a given architecture. Thus, a state change from zero to one corresponds to an insertion and from one to zero to a deletion. This model focuses on the evolution of domain architecture, ignoring sequence evolution and thus obviating the problem of constructing an appropriate MSA for tree reconstruction. Figure 1 (a) shows a domain architecture phylogeny for the protein tyrosine kinases, based on a tree constructed from an MSA of the kinase domain [30]. Note that the tree is not optimal with respect to a parsimony criterion minimizing the total number of insertions and deletions. For example, if architectures INSR and EGFR were siblings (the only two architectures containing the Furin-like cysteine rich and Receptor lingand binding domains) the number of insertions and deletions would be smaller.

The general maximum parsimony and the Dollo parsimony problems are optimization problems: an optimal tree satisfying the given parsimony criterion is sought. In contrast, the perfect phylogeny problem asks whether a perfect phylogeny exists. If such a tree does exist, it is guaranteed to be optimal. Finding the most parsimonious tree (both in the general setting as well with the Dollo restriction) is NP-complete [10]. The existence of a perfect phylogeny can be

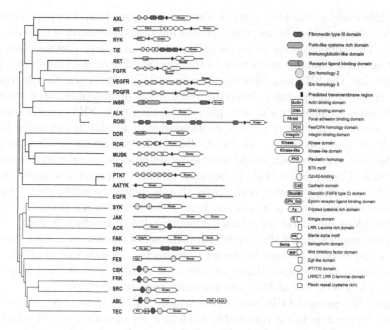

Fig. 1. Phylogenetic tree of family protein tyrosine kinase family, adopted from the tree presented in [30] constructed from an MSA of the kinase domain

solved in $O(nm)$ time, where m is the number of characters (i.e., domains) and n is the number of taxa (i.e., architectures) [11].

In contrast to perfect phylogeny, it is always possible to construct a Dollo phylogeny by positing an ancestral taxon where the state of every character is one. Since there is no restriction on the number of transitions from one to zero, any combination of character states found in the leaf taxa can be generated. Such a tree makes no sense in the context of multidomain evolution, since it implies the existence of an ancient protein containing all domains seen in any contemporary protein in the family. Can we put a restriction on the Dollo phylogeny tree so that the existence of such restricted Dollo parsimony is both informative and computationally tractable? In this paper we propose two such restrictions:

Static Dollo Parsimony is a Dollo parsimony with the following restriction: for any ancestral taxon the set of characters in state one in this taxon is a subset of the set of characters in state one in some leaf taxon (hence, the term "static"). We assume here that more than one character can change in one step.

Conservative Dollo Parsimony is a Dollo parsimony with the following restriction: for any ancestral taxon and any pair of characters that appear in state one in this taxon, there exists a leaf taxon where these two characters are also in state one.

Clearly every static phylogeny is also conservative. From the perspective of multidomain proteins, the intuition motivating the conservative restriction is as follows. The simultaneous presence of two domains in one protein often suggests that these domains contribute to the functionality of the protein as a pair. For instance, the SH2 and SH3 domains frequently appear together in various signal transduction proteins involving recognition of phosphorylated tyrosine. SH2 domains localize tyrosine phosphorylated sites while SH3 domain binds to target proteins through sequences containing proline and hydrophobic amino acids. If the domains acting in concert offer a selective advantage, it is unlikely that the pair, once formed, would later separate in all contemporary protein architectures. Conservative Dollo parsimony provides a correct parsimony model when this possibility is excluded. (Note that it may be possible that a pair of domains does not form a functional unit without additional domains but we do not explore such intricate relationships here due to insufficient data). Static Dollo parsimony additionally requires that the set of characters in state "one" in an ancestral taxa is a subset of the set of characters in state one in at least one contemporary taxa. Consequently, an ancestral architecture (defined as a set of domains) is a subset of at least one contemporary architecture.

Unlike the general Dollo parsimony which can be always inferred (even in the case where it is not a reasonable model), a set of taxa may not admit the conservative Dollo parsimony. Such a failure can be interpreted in two ways: the single insertion assumption is not reasonable, or conservative assumption is too strong. Thus non-existence of conservative Dollo parsimony provides a proof that at least one of these two assumptions is incorrect. On the other hand, existence of conservative Dollo tree does not provide a proof of correctness of the model but only evidence that the assumptions are consistent with the data.

In this paper, we show that there is an elegant link between existence of static and conservative Dollo phylogenies and some graph theoretical properties of the domain overlap graph. This leads to fast algorithms for testing existence of such restricted phylogenies and, in the case when the respective phylogenetic tree exists, constructing that tree.

3   Graph Theoretical Properties of Domain Overlap Graphs and Their Relation to Restricted Dollo Parsimony

In this section we present our theoretical results. We start with the analysis of the domain overlap graph. Stated formally, the *domain overlap graph* for a given family of multidomain proteins is the graph $G = (V, E)$ such that V is the set of all domains represented in the data base and $(u, v) \in E$ if there exists a protein p in the set such that both u and v appear in p. Below, we state the definition of chordality and discuss its importance in the context of the domain overlap graph. Subsequently, we review the Helly property and its relation to chordal graphs. Finally, we show how these concepts can be exploited to answer the questions stated in the introduction and discuss related statistical issues.

3.1   Chordal Graphs and Their Properties

Chordal graphs constitute an important and well studied graph family [16]. A *chord* in a graph is any edge that connects two non-consecutive vertices of a cycle. A *chordal graph* is a graph which does not contain chordless cycles of length greater than three. Intuitively, any cycle of length greater than three in a chordal graph is triangulated, that it is partitioned (not necessary uniquely) into triangles using edges of the graph. This motivates another term for chordal graphs, namely *triangulated graphs*. Figure 2 (a) shows the domain overlap graph for a set of domains in protein kinase family shown in Figure 1. For simplicity, only domains that occur in more than one architecture are used. Note that this graph is chordal.

The important property that is explored directly in this paper is the relation between chordal graphs and trees. To elucidate this relation we need to introduce the concept of *intersection graph*.

Let $\mathcal{F}$ be a family of objects, such as intervals on a coordinate line or rectangles in space. A graph $G = (V, E)$ is called an *intersection graph* of $\mathcal{F}$ if each vertex, $v \in V$, corresponds to an object in $\mathcal{F}$ and $(u, v) \in E$ if and only if the objects in $\mathcal{F}$ corresponding to u and v intersect.

We will consider a special family of intersection graphs where the objects are subtrees of some (usually unrooted) tree. We will refer to the tree as the *guide tree*. Here, by a subtree of a tree we understand any connected subgraph of a tree. Furthermore, our family typically does not contain all possible subtrees of a tree.

Theorem(Gavrill [13]). A graph G is chordal if and only if there exists a tree T and a family of subtrees of this tree such that G is the intersection graph of this family.

Our key observation is stated in the following theorem:

Theorem 1. There exists a conservative Dollo parsimony tree for a given set of multidomain architectures, if and only if the domain overlap graph for this set is chordal.

Proof (sketch): We argue that if a given set of architectures admits conservative Dollo parsimony then the corresponding domain overlap graph is chordal. The argument in the opposite direction is left to the full version of the paper.

Assume that conservative Dollo parsimony tree exists and take this tree as the guide tree for an intersection graph construction. For any domain consider all nodes (leaves or ancestral) that contain this domain. Since the guide tree is a Dollo tree, these nodes form a connected subtrees. Consider the family of such subtrees for all considered domains. We argue that the intersection graph of this family of subtrees is exactly the domain overlap graph. By the definition, the nodes of this intersection graph correspond to protein domains and there is an edge between two such domains if and only if there exists a node in the Dollo tree containing both domains. Thus if two domains belong to the same protein they are connected by an edge in the intersection graph. We need to show that

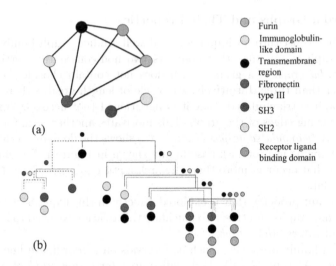

Fig. 2. a) The domain overlap graph for a selected domains from the human tyrosine kinase family. Chosen domains belong to more than one architecture (under assumption that architectures containing the same set of domains are considered to be the same). The kinase domain is omitted since it is present in all these architectures. b) Representation of the domain overlap graph as an intersection graph of subtrees of a tree. The correspondence between subtrees and domains is indicated by corresponding colors. The label of a node indicates which subtrees intersect in this node

if two domains do not occur together in at least one protein architecture, then there is no edge between them in the intersection graph. Assume towards a contradiction that there exists an edge between two domains that do not belong to the same architecture. This means that the corresponding subtrees of the guide tree intersect in an internal node but they don't intersect in a leaf. This contradicts the assumption that the tree is conservative. Thus the intersection graph is exactly equal to the domain overlap graph. By Gavril's theorem, the domain overlap graph is chordal.   **QED**

Figure 2 shows a domain overlap graph and a corresponding Dollo parsimony tree. Note that the tree does not need to be unique. For example the order of internal no nodes in the SH2/SH3 subtree can be switched without inducing a change in the overlap graph.

3.2   The Helly Property

As mentioned before, it is not always possible to construct a Dollo phylogeny tree without introducing ancestral nodes with domain compositions not observed in the leaves. For example, if we have three domains A, B, and C and three proteins AB, BC, CA then we cannot construct a Dollo parsimony tree without introducing an ancestral protein ABC. This property is equiva-

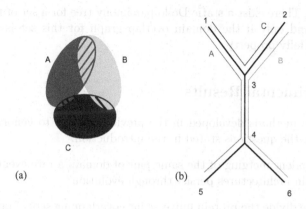

Fig. 3. (a) Three ovals that do not satisfy the Helly property; (b) and three subtrees of a tree which satisfy the Helly property

lent to the Helly property, named after the Austrian mathematician Eduard Helly [9]:

A family $\{T_i \mid i \in I\}$ of subsets of a set T is said to satisfy the **Helly property** if, for any collection of sets from this family, $\{T_i \mid j \in J \subseteq I\}$, $\cap_{j \in J} T_j \neq \emptyset$, whenever $T_j \cap T_k \neq \emptyset$, $\forall j, k \in J$.

In Figure 3 (a) shows an example of a family of three sets that do not satisfy the Helly property: Each pair intersects but there is no intersection point common to all three. In contrast, Figure 3 (b) shows an example where the three subtrees A, B, C (respectively with vertices $\{1,3,4,5\}$, $\{2,3,4,6\}$, and $\{1,2,3\}$) of tree T. The subtrees pairwise intersect and also have a common intersection point in vertex 3. Thus they satisfy the Helly property. The last fact is true for any set of subtrees of a tree: there is no way to have such subtrees pairwise intersect but not intersect in a common point.

Consistent with the above definition of the Helly property, we introduce the Helly property for a domain overlap as follows:

Defintion (Helly property for domain overlap graph). A domain overlap graph satisfies the Helly property if and only if for every clique in this graph there exists a protein architecture that contains all domains of this clique.

To see why this definition is consistent with the set theoretical definition for each domain i consider set T_i of architectures that contain this domain. Then an edge between i and j corresponds to $T_i \cap T_j \neq \emptyset$. Subsequently a clique J has property that all $i, j \in J$ $T_i \cap T_j \neq \emptyset$. The existence of an architecture that contains all domains in the clique J ensures that $\cap_{j \in J} T_j \neq \emptyset$ since this architecture belongs to all sets T_j in this clique.

The relation of the Helly property to the static Dollo parsimony is provided by the following theorem (the proof will be given in the full version of the paper):

Theorem 2. There exist a static Dollo parsimony tree for a set of multidomain proteins, if and only if the domain overlap graph for this set is chordal and satisfies the Helly property.

4 Experimental Results

We apply the methods developed in the previous section to genomic data sets to investigate the questions stated in the introduction:

- Is independent merging of the same pair of domain a rare event?
- Do domain architectures persist through evolution?

To do this, we divide the protein universe into overlapping sets of proteins called superfamilies. Each domain defines one superfamily, namely the set of all proteins that contain the given domain. For example, all proteins containing the kinase domain form one superfamily, proteins containing the SH2 domain form another superfamily and these two superfamilies intersect. It is important for our argument that each superfamily have a common reference point - here the common domain. This reference point allows us to interpret each merge as an insertion with respect to this domain. In particular, multiple independent insertions correspond to multiple independent merges of the inserted domain and the reference domain. For each superfamily in our data set, we determine whether it satisfies the perfect phylogeny and conservative and static Dollo criteria. To estimate the significance of our results, we also investigate the probability of observing conservative Dollo parsimony in two null models, uniform random graphs (Erdos-Renyi model) and random scale free graphs generated using preferential attachment random model.

Null Models. The existence of a conservative Dollo parsimony tree for a given domain superfamily is a necessary but not a sufficient condition for concluding that no repeated, independent merges occurred in the history of the family. We therefore estimate the probability that a superfamily admits a conservative Dollo phylogeny by chance under two different null models. Note, that this is equivalent to determining the probability that a graph of with a given number of vertices is chordal under our null hypotheses.

All graphs with less than four vertices are chordal, as are all acyclic graphs (i.e., graphs which are collections of trees). Since a random, sufficiently sparse graph will be acyclic with high probability, such a graph is also likely to be chordal. In fact, a random graph with edge probability $p < \frac{c}{n}$, where n is number of vertices, is almost certainly acyclic when $c < 1$, while almost all vertices of such a graph belong to a cycle when $c > 1$ and the phase transition occurs at $p = \frac{1}{n}$ [7]. Consequently, since we are interested in graphs that are unlikely to be chordal by chance, we consider only graphs with at least four vertices that have at least as many edges as vertices. We define a *complex superfamily* to be a superfamily whose domains overlap graph satisfies these criteria and restrict our analysis to complex superfamilies in our data sets. To determine the probability of observing conservative Dollo parsimony in complex superfamilies by chance, we collected statistics to estimate the value of c for domain overlap graphs in our data set. We then used simulation (1000 runs) to estimate the probability that a random graph with uniform edge probability $p = \frac{c}{n}$ is chordal.

Several papers have suggested that the domain overlap graphs have scale free properties [2,33]. We therefore also considered a null model based on preferential attachment, a classical random model for scale free graphs [3]. Under this model, a random graph is constructed iteratively. At each step, a new vertex is connected to an existing vertex with probability proportional to the degree of that vertex. We simulated the preferential attachment model taking care that the parameters are chosen in such a way that the edge density of the resulting random graphs is approximately the same as that in domain overlap graphs of the same size.

Data. We use two different data sets derived from SwissProt version 44 released in 09/2004 [6] (http://us.expasy.org/sprot/). The first contains all mouse proteins, thus all homologous proteins in this set are paralogs. In contrast, the second test set consists of all non redundant (nr90) proteins in SwissProt, and thus contains both paralogs and orthologs. The architectures of each protein in both sets were identified using CDART [14] based on PSSM domain models. The domains identified by CDART as similar have been clustered using single linkage clustering and subsequently considered as one *superdomain*. The proteins that contained no recognizable domain were removed, leaving 256,937 proteins with 5,349 distinct domains in the nr90 data set and 6,681 proteins with 1951 distinct domains in the mouse data set. Of these, 2,896 nr90 and 983 mouse superfamilies have at least one partner domain. We let Mouse.c and nr90.c denote the set of complex superfamilies in mouse and nr90, respectively. To determine the effect of superfamily size on the results, we defined Mouse.c.x-y and nr90.c.x-y to be sets of superfamily in Mouse.c and nr90.c, respectively, containing at least x and at most y domains.

There is always a danger of inaccuracy when working with large, automatically annotated, data sets. Since errors in domain architecture identification could result in incorrect conclusions concerning domain insertion and loss, we also tested our approach on a hand curated data set, namely the kinase superfamily, which has been heavily studied and for which it is possible to obtain highly reliable domain annotations. We compared the set of complete human protein sequences, obtained from SwissProt along with their symbols and Pfam codes, with a list of designated kinase gene symbols and Pfam codes (PF00069, PF001163 and PF01633) derived from three recent, genomic analyses of the kinase superfamily [30,18,8]. A protein was judged to be a kinase if it was annotated with a known kinase gene symbol or Pfam code. This procedure resulted in a set of 378 human kinase sequences. The domain architectures of these kinases were then obtained from CDART [14]. From this curated set, we analyzed the kinase superfamily, and all superfamilies that overlapped with it.

Analysis. To test the consistency of the data with the perfect phylogeny, static Dollo parsimony, and conservative Dollo parsimony models, we implemented the

Table 1. The percentage of superfamilies that are consistent with the perfect phylogeny (PP), static Dollo parsimony (SDP) and conservative Dollo parsimony (CDP) criteria. Abbreviations: PA - preferential attachment; NE - not estimated

set	# super-families	% PP	% SDP	% CDP	% random uniform	% random PA
Mouse	983	95	99	99.7	NE	NE
Mouse.c.4-5	88	99	100	100	80	98
Mouse.c.6-8	37	84	100	100	31	66
Mouse.c.9-10	11	66	100	100	17	25
Mouse.c.11-20	23	31	96	96	1.7	1.0
Mouse.c.21-30	9	0	66	100	0	0
Mouse.c.31- *	8	0	50	75	0	0
Nr90	2896	80	98	99.9	NE	NE
Nr90.c.4-5	143	57	99	99.5	80	98
Nr90.c.6-8	130	37	99	100	31	66
Nr90.c.9-10	40	28	100	100	17	25
Nr90.c.11-20	104	13	87	99	1.7	1.0
Nr90.c.21-30	34	6	53	88	0	0
Nr90.c.30- *	28	0	15	50	0	0
Human Kin	101	11	100	100	NE	NE

algorithms discussed in the previous sections using the LEDA platform [29]. The agreement with perfect phylogeny criterion was tested using compatibility criterion [12]. To test conservative Dollo parsimony, we implemented a chordality test and for static Dollo parsimony we additionally tested if the Helly property is satisfied. Using these tools, we test our data for these criteria and asked under what circumstances could at least 90% of superfamilies be explained by a given evolutionary model. The results are summarized in Table 1.

Not surprisingly, with the exception of very small (in terms of number of different domains or equivalently the size of domain overlap graph) superfamilies in mouse perfect phylogeny does not meet this standard suggesting that it is not a suitable model for multidomain protein evolution. In contrast, 95% or more of complex superfamilies up to size 20 in mouse and size 10 in nr90 could be explained by static Dollo parsimony. All but the largest complex superfamilies (greater than 30 in mouse and greater than 20 in nr90) were consistent with conservative Dollo parsimony. In contrast, the probability of observing conservative Dollo parsimony by chance was much lower in both null models. Furthermore, our results show that domain overlap graphs of real multidomain superfamilies do not have the same the topological structure as random scale free graphs of the same size and edge density constructed according to preferential attachment random model.

While the vast majority of small and medium size superfamilies admit conservative and static Dollo parsimony, a significant percentage large superfamilies do not. A less restrictive evolutionary model that allows multiple insertions is needed to explain the data. Furthermore, our simplifying assumptions may result in underestimation of the number of independent merges since only merges that violate chordality are detected. For the mouse data set, the superfamilies that do not satisfy conservative Dollo parsimony are FER2, Trypsin, and EGF. For nr90, this set contains 34 superfamilies including TRK, IG, PH, EGF, SH3, C2, and a large superdomain containing several ATPases (the largest superfamily in the nr90 set). Several of these are known to be "promiscuous" domains, which also supports the hypothesis of repeated independent merges in large families [28]. While the quality of domain recognition and incompleteness of the data may be affecting our results, the results for the curated kinases family are consistent with the results for non-curated data (the sizes of all but one domain overlap graphs for this set, are less than 20).

5   Conclusions and Future Research

In this paper, we formulated two new parsimony models and showed their connection to properties of domain overlap graphs. Previous analysis of these graphs focused on counting vertex degrees and statistical analysis of connectivity [2,33]. We demonstrated that these graphs frequently have interesting topological properties, and in fact the topology of domain overlap graphs can provide information about evolution of a multidomain protein family. We applied our new graph theoretical tools to test whether independent merging of the same pair of domains is a rare event and whether domain architectures persist through evolution? In the case of small and medium sizes superfamilies, the data is consistent with this hypothesis. However, our results do not support the hypothesis in the case of large families. We also demonstrate that the topological properties of domain overlap graphs of multidomain superfamilies are very different from those of random scale free graphs of the same size and density. Based on these results, we reject preferential attachment as a mechanism for multidomain protein evolution. This also prompts the question: what evolutionary model for multidomain proteins will explain the observed behavior?

We show that the independent domain mergers can be detected by testing if the corresponding domain overlap graph is chordal. An intriguing question is whether the minimal set of domains which must be removed to obtain a chordal domain overlap graph is related to the set of does this minimal set tend to be promiscuous domains.

Although the focus of this study is evolution of protein architectures, applicability of the methods developed in this paper goes beyond the analysis of multidomain protein superfamilies. They can be applied to analysis of any set of taxa with binary character states.

Another interesting direction of future research is to study of properties of protein overlap graphs. While the domain overlap graph is dual to the protein overlap graph, this duality is not symmetric. Given a protein overlap graph, we can construct the corresponding domain overlap graph, but given a domain overlap graph we cannot reconstruct the initial protein overlap graph. The domain overlap graph thus contains less information than the protein overlap graph.

Therefore, direct analysis of protein overlap graphs may bring new insights in analyzing evolution of multidomain proteins.

Acknowledgments

We thank L. Y. Geer and S. H. Bryant (NCBI) for providing the complete set of CDART domain architectures, D. Ullman, R. Jothi, and E. Zotenko for valuable discussions. T.P. was supported by the intramural research program of the National Institutes of Health. D.D., G.D. and N.S. were supported by NIH grant 1 K22 HG 02451-01 and a David and Lucille Packard Foundation fellowship.

References

1. G. Apic, J. Gough, and S.A. Teichmann. Domain combinations in archaeal, eubacterial and eukaryotic proteomes. *J Mol Biol*, 310:311–325, 2001.
2. G. Apic, W. Huber, and S.A. Teichmann. Multi-domain protein families and domain pairs: Comparison with known structures and a random model of domain recombination. *J. Struc. Func. Genomics*, 4:67–78, 2003.
3. A.-L. Barabasi and R. Albert. Emergence of scaling in random networks. *Science*, 286:509–512, 1999.
4. M. Bashton and C. Chothia. The geometry of domain combination in proteins. *J Mol Biol*, 315:927–939, 2002.
5. A. Bateman, E. Birney, R. Durbin, S.R. Eddy, K.L. Howe, and E.L. Sonnhammer. The Pfam protein families database. *Nucleic Acids Res.*, 28(1):263–266, 2000.
6. B. Boeckmann, A. Bairoch, R. Apweiler, M.-C. Blatter, A. Estreicher, E. Gasteiger, M.J. Martin, K. Michoud, C. O'Donovan, I. Phan, S. Pilbout, and M. Schneider. The SWISS-PROT protein knowledgebase and its supplement TrEMBL in 2003. *Nucleic Acids Res.*, 31:365–370, 2003.
7. B. Bollobas. *Random Graph Theory*. Cambridge University Press, 2001.
8. S. Cheek, H. Zhang, and N. V. Grishin. Sequence and structure classification of kinases. *J Mol Biol*, 320(4):855–881, Jul 2002.
9. L. Danzer, B. Grunbaum, and V. Klee. Helly's theorem and its relatives. *Convexity, AMS*, 7:101–180, 1963.
10. W.H.E. Day, D. Johnson, and D. Sankoff. The computational complexity of inferring rooted phylogenies by parsimony. *Mathematical Biosciences*, 81:33–42, 1986.
11. D.Gusfield. Efficient methods for inferring evolutionary history. *Networks*, 21:19–28, 1991.
12. J. Felsenstein. *Inferring Phylogenies*. Sinauer Associates, 2004.
13. F. Gavril. The intersection graphs of subtrees in trees are exactly the chordal graphs. *J. Comb. Theory (B)*, 16:47–56, 1974.
14. L.Y. Geer, M. Domrachev, D.J. Lipman, and S.H. Bryant. CDART: protein homology by domain architecture. *Genome Res.*, 12(10):1619–23, 2002.
15. M. Gerstein. How representative are the known structures of the proteins in a complete genome? A comprehensive structural census. *Fold des.*, 3:497–512, 1998.
16. M. Golumbic. *Algorithmic Graph Theory and Perfect Graphs*. Academic Press, New York, 1980.

17. J. Gu and X. Gu. Natural history and functional divergence of protein tyrosine kinases. *Gene*, 317:49–57, 2003.

18. S.K. Hanks. Genomic analysis of the eukaryotic protein kinase superfamily: a perspective. *Genome Biol*, 4(5):111, 2003.

19. A. Heger and L. Holm. Exhaustive enumeration of protein domain families. *J. Mol Biol*, 328:749–767, 2003.

20. I.Yanai, Y.I. Wolf, and E.V. Koonin. Evolution of gene fusions: horizontal transfer versus independent events. *Genome Biol*, 3, 2002. research:0024.

21. J.S.Farris. Phylogenetic analysis under Dollo's law. *Systematic Zoology*, 26(1):77–88, 1977.

22. A. Krause, J. Stoye, and M. Vingron. The SYSTERS protein sequence cluster set. *Nucleic Acids Res.*, 28(1):270–272, 2000.

23. S. Kummerfeld, C. Vogel, M. Madera, and S. Teichmann. Evolution of multi-domain proteins by gene fusion and fission. *ISMB 2004*, 2004.

24. I. Letunic, L. Goodstadt, N.J. Dickens, T. Doerks, J. Schultz, R. Mott, F. Ciccarelli, R.R. Copley, C.P. Ponting, and P. Bork P. Recent improvements to the SMART domain-based sequence annotation resource. *Nucleic Acids Res.*, 31(1):242–244, 2002.

25. Y. Liu, M. Gerstein M, and D.M. Engelman. Evolutionary use of domain recombi-nation: a distinction between membrane and soluble proteins. *Proc Natl Acad Sci USA*, pages 3495 – 3497, 2004.

26. M. Long. Evolution of novel genes. *Curr Opin Genet Dev*, 11(6):673–680, 2001.

27. L.Patthy. Genome evolution and the evolution of exon-shuffling–a review. *Gene*, 238:103–114, 1999.

28. E.M. Marcotte, M. Pellegrini, H.L. Ng, D.W. Rice, T.O. Yeates, and D. Eisen-berg. Detecting protein function and protein-protein interactions from genome sequences. *Science*, 285:751–53, 1999.

29. K. Mehlhorn and S. Naher. *The LEDA Platform of Combinatorial and Geometric Computing*. Cambridge University Press, 1999.

30. D.R. Robinson, Y.M. Wu, and S.F. Lin. The protein tyrosine kinase family of the human genome. *Oncogene*, 19(49):5548–5558, 2000.

31. S.A.Teichmann, J. Park, and C. Chothia. Structural assignments to the my-coplasma genitalium proteins show extensive gene duplications and domain re-arrangements, 1998.

32. B. Snel, P. Bork, and M. Huynen. Genome evolution gene fusion versus gene fission. *Trends Genet*, 16:9–11, 2002.

33. S. Wuchty. Scale-free behavior in protein domain networks. *Mol. Biol. Evol.*, 18:1694–1702, 2001.

34. G. Yona, N. Linial, and Linial M. Protomap: Automatic classification of protein sequences, a hierarchy of protein families, and local maps of the protein space. *Proteins: Structure, Function and Genetics*, 37:360–378, 1999.

Peptide Sequence Tags for Fast Database Search in Mass-Spectrometry

Ari Frank[1], Stephen Tanner[2], and Pavel Pevzner[1]

[1] Department of Computer Science and Engineering,
University of California, San Diego. 9500 Gilman Drive,
La Jolla, CA 92093-0114, USA
{arf, ppevzner}@cs.ucsd.edu
[2] Department of Bioinformatics, University of California,
San Diego, 9500 Gilman Drive,
La Jolla, CA 92093-0419, USA
stanner@ucsd.edu

Our ability to generate data now far outstrips our ability to analyze it ...

- Scott Patterson. *Data Analysis - The Achilles Heel of Proteomics.*
Nat. Biotech., 21 (2003), 221-222.

Abstract. Filtration techniques, in the form of rapid elimination of candidate sequences while retaining the true one, are key ingredients of database searches in genomics. Although SEQUEST and Mascot are sometimes referred to as "BLAST for mass-spectrometry", the key algorithmic idea of BLAST (filtration) was never implemented in these tools. As a result MS/MS protein identification tools are becoming too time-consuming for many applications including search for post-translationally modified peptides. Moreover, matching millions of spectra against all known proteins will soon make these tools too slow in the same way that "genome vs. genome" comparisons instantly made BLAST too slow. We describe the development of filters for MS/MS database searches that dramatically reduce the running time and effectively remove the bottlenecks in searching the huge space of protein modifications. Our approach, based on a probability model for determining the accuracy of sequence tags, achieves superior results compared to GutenTag, a popular tag generation algorithm.

1 Introduction

Computational mass-spectrometry is traditionally divided into two areas: peptide identification via database search (that is responsible for the lion's share of applications) and de novo peptide sequencing (see [1, 24] for recent reviews). However, this separation is an artificial one since ultimately de novo sequencing is simply a search in a database of all possible peptides. In the last 2 years the boundary between these two techniques has started to blur. We envision that these two directions will soon become highly interconnected and even merge.

S. Miyano et al. (Eds.): RECOMB 2005, LNBI 3500, pp. 326–341, 2005.

Although MS/MS database search algorithms such as SEQUEST [12] and Mascot [27] offer high-throughput peptide identification, the current database search techniques do not give a complete solution to this problem, particularly in the case of spectra from *Post-Translationally Modified* (PTM) peptides. The requirement to consider every modification over every possible peptide, not only makes the process too slow, but also makes it harder to distinguish the true peptide from false hits. As a result, the mass-spectrometrists currently face a challenging computational problem: given a large collection of spectra, find out which modifications are present in each protein in the sample.

In a sense, the protein identification problem is similar to one faced by the genomics community in their search for sequence similarities. The solution is to use *filters* that quickly eliminate much of the database, while retaining the true hits. Most sequences are rejected by the filters and *filtration efficiency* is measured by the fraction of retained sequences in the filtered database. The 20-year history of database search in genomics is essentially the history of designing more and more efficient and statistically sound filters. From this limited perspective, the history of MS/MS database search is at the very beginning. Good filters for MS/MS database search are not trivial, and until recently the studies of peptide identification have concentrated primarily on scoring [42, 43, 3, 32, 38, 31, 27, 8, 19], or the reliability of the peptide assignments [23, 17, 26, 30], with little focus on filters.[1]

We argue that a study of filtration is central to PTM identification. At first glance, this is counter-intuitive since there is no apparent connection between reducing the number of candidates and identifying modified peptides. Note, however, that aggressive (but accurate) filtration allows us to apply more sophisticated and computationally intensive algorithms and scoring to the few remaining candidates. Indeed, the current approaches to PTM analysis are based on generating huge "virtual databases" of all PTM variants. As Yates and colleagues remarked in [43], extending this approach to a larger set of modifications remains an open problem. However, if the database is reduced to a few peptides, one can afford to consider all possible PTMs for every peptide in the filtered database.

Mass-spectrometrists routinely use *Peptide Sequence Tags* (PSTs) for spectra interpretation.The idea of using Peptide Sequence Tags as filters is not novel (see Mann and Wilm, 1994 [25]) and one may wonder whether we are trying to reinvent the wheel, particularly since in the last year Tabb et al., 2003 [37], Hernandez et al., 2003 [16], Sunyaev et al., 2003 [36], Searle et al., 2004 [33], and Day et al., 2004 [10] studied PSTs. In particular, John Yates' group recently released their GutenTag algorithm for PST generation and raised the possibility that searching with PSTs as filters results in additional identifications [37]. However, while these new tools greatly improve on early heuristics for PST generation, they do not explicitly measure the filtration efficiency versus accuracy trade-off, and the final analysis produces fewer hits than SEQUEST, albeit with

---

[1] At best, the use of parent peptide mass and trypsin end-point specificity could be thought of as simple filters in existing database search engines.

fewer false positives. We emphasize that not every set of PSTs considered in [37, 36, 33] forms a filter since it does not necessarily satisfy the *covering* property that ensures with high probability that the correct peptide is not filtered out. As a result these tools cannot eliminate a need to run a time-consuming database search (like SEQUEST) but rather provide *additional candidates* that SEQUEST may miss. Our goal is to completely substitute SEQUEST with a filtration algorithm that is faster by a few orders of magnitude. Below we describe the first steps toward this goal. In particular, our tag generation algorithms significantly improve on GutenTag [37] and lead to efficient database filtration.

Our new MS/MS filtration tool combines the following components (i) de novo peptide sequencing, (ii) accurate PST generation, (iii) trie-based database search with PSTs, (iv) extension of tags to generate peptide candidates, including modified ones, and (v) scoring of the PTM candidates in the filtered database. Our approach is based on a probabilistic model that can accurately assign probabilities to each amino acid in de novo predictions. These probabilities are further used to determine the reliability of the generated peptide sequence tags.

The paper is organized as follows. In section 2 we briefly review our PepNovo sequencing algorithm [13] that serves as a tag generation engine. Section 3 describes the probabilistic model for assessing reliability of individual amino acids in the de novo interpretations. In section 4 we capitalize on the performance of PepNovo and use it for reliable tag generation. Section 5 briefly sketches our filtering approach. In Section 6 we present experimental results and benchmark tag generation algorithms.

2   De Novo Peptide Sequencing

The process of MS/MS spectrum generation can be viewed as sampling the spectrum S from the space of all possible spectra for peptide P, according to a distribution $Prob(S|P)$. The goal of de novo algorithms is to find a peptide P that maximizes $Prob(S|P)$ amongst all possible peptides. Since the distribution $Prob(S|P)$ is too complex to model, sequencing algorithms resort to using approximations in the form of *scoring functions*. Until recently mass-spectrometrists used *ad-hoc* scoring functions but a more rigorous scoring is based on learning the parameters of the hidden distribution $Prob(S|P)$.

The first such approach in MS/MS studies was proposed by Dancik et al., 1999 [9] who automatically learned probabilities of ions and developed the likelihood-based approach to scoring. The Dancik scoring was further generalized by Bafna and Edwards [3], Havilio et al., [15], and Colinge et al. [6] who demonstrated that probabilistic scoring improves upon SEQUEST's performance. Elias et al. [11] proposed yet another machine learning model (decision trees) and demonstrated a significant improvement compared to SEQUEST scoring. However, all these new developments are designed specifically for database search, and it is unclear how to adapt them for de novo algorithms and PST generation.

With de novo sequencing [40, 9, 41, 5, 21, 4, 22, 34, 20], a reconstruction of the peptide sequence is done without a protein database. In addition, the introduction of modified amino acids into the reconstruction is usually less prohibitive than in database searches. Amongst the de novo algorithms being used today are Lutefisk [40, 41] (a publicly available program), Sherenga [9] (available from Agilent), and Peaks [22] (available from Bioinformatics Solutions, Inc.).

PepNovo [13] utilizes a scoring based on a graphical model whose structure reflects the rules governing peptide fragmentation. The scoring function is designed in such a way that it can take advantage of the antisymmetric path algorithm [5, 18] in the *spectrum graph* [9].At the heart of our scoring is a likelihood ratio test that compares two competing hypotheses concerning a mass m and spectrum S. The first is the Collision Induced Dissociation hypothesis (CID) which states that a cleavage occurred at mass m in the peptide that created the spectrum S. We use a probabilistic network (Fig. 1) that models the automatically learned fragmentation rules to determine the probability $P_{CID}(\boldsymbol{I}|m, S)$ of detecting an observed vector of fragment intensities $\boldsymbol{I}$, given that mass m is a cleavage site in the peptide that created S ($\boldsymbol{I}$ represents intensities of all possible fragment ions produced by the cleavage at mass m). The competing hypothesis is the Random Peaks hypothesis ($RAND$), which assumes that the peaks in the spectrum are random. Our score is the logarithm of the likelihood ratio of these two hypotheses,

$$Score(m, S) = \log \frac{P_{CID}(\boldsymbol{I}|m, S)}{P_{RAND}(\boldsymbol{I}|m, S)} \ . \tag{1}$$

Our probabilistic network (Fig. 1) includes three factors that were not adequately represented in Dancik et al., 1999 scoring: (i) correlations between types of fragment ions and their intensities, (ii) the positional influence of the cleavage site, and (iii) the influence of the flanking amino acids.

A benchmark of PepNovo against Sherenga [9], Lutefisk [41], and Peaks [22] is described in [13]. PepNovo performed better than the other algorithms both in terms of prediction accuracy (the ratio of the number of correctly predicted amino acids compared with the total number of predicted amino acids), and in terms of the success in predicting subsequences of the true peptides. An important feature of PepNovo is that it assigns reliability probabilities to each predicted amino acid, something that is missing in other de novo approaches besides the Peaks software.[2] This feature is crucial for reliable tag generation.

While PepNovo improves on the existing de novo algorithms, using de novo predictions as text-based filters is hardly possible since they often contain errors caused by limitations of scoring and other reasons. A single sequencing error can render the de novo results useless in the text-based filtration mode for a database searches. However, it is not necessary to reconstruct the entire peptide sequence that created a spectrum. The filtration can be done equally well with a partial sequence if it is known to be correct with high probability. Below we

---

[2] Peaks is a commercial software and no information is available on how Peaks reliability scores are obtained.

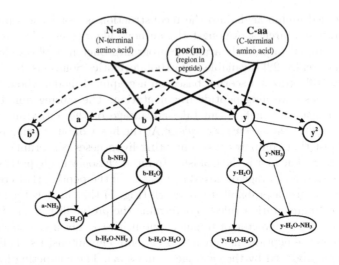

Fig. 1. The probabilistic network for the CID fragmentation model (doubly charged tryptic peptides sub-network) reflecting three different types of relations: (1) correlations between fragment ions (regular arrows); (2) relative position of the cleavage site (dashed arrows); (3) influence of flanking amino acids (bold arrows). N-aa and C-aa vertices represent flanking amino acids and $pos(m)$ vertex represents the relative region in the peptide in which the cleavage occurs (the intensity of peaks is correlated with the region in the peptide in which the peaks appear)

formulate the problem of constructing such high probability PSTs that can serve as filters.

The paths in the properly constructed spectrum graph represent all possible peptides. While de novo algorithms find the best-scoring path amongst all these paths (de novo sequencing), the path corresponding to the true peptide is not necessarily the highest scoring one. It is not clear how to find the best-scoring path among paths corresponding to the database peptides (peptide identification) without testing each such "database path" in a case-by-case fashion. The tag generation offers an alternative that alleviates this time consuming step.

If the score of the best-scoring "database path" is δ, let $\mathcal{P} = \mathcal{P}(\delta)$ be a set of all paths in the spectrum graph whose score is larger than or equal to δ. We call this set δ-*suboptimal* and emphasize that while δ is not known in advance, one can compute a lower bound for δ. The question then arises whether there exists a simple characterization of the set $\mathcal{P}$. For example, if all paths in $\mathcal{P}$ contain a tri-peptide SEQ then one can safely filter the database retaining only the peptides with SEQ at a certain prefix mass and score these few remaining peptides. In reality, one PST may not be sufficient and we are interested in a *covering set* of PSTs X such that each path from $\mathcal{P}$ contains at least one PST from X. The problem of finding PSTs can now be formulated as follows.

Tag Generation Problem. Let G be a spectrum graph, l be a parameter defining the length of PSTs, and δ be a score threshold. For a set of δ-suboptimal paths $\mathcal{P}(\delta)$, find a covering set of PSTs (each of length l) of minimal size.

For $l = 1$, the solution of the Tag Generation Problem can be approximated by constructing the minimum cut [7] in the "winnowed" spectrum graph.[3] However, the tags of length 1 (complemented by prefix masses) do not provide a significant improvement in filtration efficiency compared to filtering according to the prefix mass alone. The extension of this combinatorial approach for $l > 1$ remains an open problem. Moreover, it may be impractical since it may generate a rather large and *unranked* set of tags. In this paper we consider a probabilistic version of the Tag Generation Problem and approximate its solution by ranking PSTs and simply selecting top-ranked PSTs from the ranked list. The question arises on how to identify the most reliable amino acids in the de novo reconstruction and use them for tag generation.

3 Reliability of Amino Acids in De Novo Predictions

Every predicted amino acid corresponds to an edge in the spectrum graph. One way to determine if an edge is correct is by assessing how important it is for obtaining a high scoring path in the spectrum graph. If we remove a correct edge from the spectrum graph, the true peptide's path no longer exists. A subsequent run of the de novo algorithm on the modified graph should yield a de novo reconstruction with a significantly lower score. However, if the edge is incorrect, its removal should not cause a large score reduction since there should exist a relatively good alternative (the correct path). It turns out that the ratio of the reduction between these two scores (called **Score Reduction due to Edge Removal**) correlates well with the reliability of predicted amino acids. The Score Reduction due to Edge Removal is not the only *feature* correlated with reliability of amino acids and below we describe some other features. The transformation of these features into a probabilistic estimate is not a trivial problem. We use the logistic regression method [14] to transform the combination of feature values into probabilities.

In order to determine the reliability of amino acid assignments, we view this task as a classification problem. The training samples in this scenario are pairs of variables of the form (x, y), where x is an amino acid from an input space $\mathcal{X}$, and y is a class variable from $\mathcal{Y} = \{0, 1\}$, which states if the sample is correct ($y = 1$) or incorrect ($y = 0$). The reliability assessment task is reduced to creating a probabilistic model that determines for an unknown sample $x \in \mathcal{X}$ the probability $p(y = 1|x)$, which is the probability that x is correct amino acid. To use the logistic regression, we map each sample x into a point $\bar{x} \in \mathbb{R}^n$, using n *feature functions* of the form $f : \mathcal{X} \to \mathbb{R}$. The probability function derived from the regression model is

---

[3] Winnowing of the spectrum graph amounts to removing all edges that do not belong to δ-optimal paths (such edges can be found by dynamic programming).

$$p_\lambda(y = 1|x) = \frac{e^{\lambda_0 + \sum_{i=1}^{n} \lambda_i \bar{x}_i}}{1 + e^{\lambda_0 + \sum_{i=1}^{n} \lambda_i \bar{x}_i}} \qquad (2)$$

where the parameters λ are fit according to the training data using a nonlinear Conjugate Gradient method [35], and $\bar{x}_i$ are the feature values given to x by the n feature functions. Logistic regression models maximize the training data's log-likelihood, given by $\sum_{(x,y)} \log p_\lambda(y|x)$ where the sum is over all training samples. The success of these models in assigning points to their correct classes depends on the features' ability to capture the nuances that distinguish between correct and incorrect predictions. Following is a description of features we use in our model.

Score Reduction due to Edge Removal. (described in the first paragraph of this section).

Cleavage Site Scores. We observed that edges connecting two high scoring vertices in the spectrum graph are likely to be correct, while erroneous edges often connect high and low scoring vertices. If the lower score amongst the two is still relatively high, this is a good indicator that we have a correct amino acid (the two features we create are *high* and *low* PepNovo vertex scores).

Consecutive Fragment Ions. Correct interpretations of spectra often contain runs of b and y-ions. Incorrect interpretations have fewer such runs due to spurious cleavage sites. Therefore the detection of b-ions or y-ions at both ends of the edge is an indication of an accurate assignment. The bb feature is an indicator function equal to 1 iff the b-ions on both ends of the edge are detected. The yy feature is defined similarly.

Peak offsets. Since the peaks' positions in MS/MS spectra are imprecise, we use a tolerance of ± 0.5 Daltons in the peak location. However, the series of b or y-ions tend to have accurate offsets from each other (i.e., the mass difference between the consecutive ions is close to the mass of the amino acid). These offsets are usually larger for incorrect amino acid assignments. We define two features, the first is the squared offset of the b-ions, and the second is the squared offset of the y-ions.

4   Generating Tags

A sequence tag is a short amino acid sequence with a prefix mass value that designates its starting position. Even a short correct tag of length 3 achieves very high filtration efficiency when combined with a prefix mass offset (see Section 6).

In de novo sequencing, the goal is to select a single path, as long and accurate as possible. For filtering, we are interested in a small covering set of shorter *local* paths (tags). Generation of small covering sets of tags is a tricky problem and the recent approaches to tag generation [37, 36, 33] did not explicitly address the covering condition.[4] We argue that PST generation may greatly benefit from algorithms developed specifically for de novo sequencing.

Starting from Mann and Wilm, 1994 [25], all approaches to PST generation search for *local* tags without checking whether the tags are part of a *global* de novo interpretation. Our results below indicate that this approach may have inherent limitations since the tag generation algorithms based on global rather than local approach seem to perform much better. An optimal local path may not be extensible into an optimal global path, or indeed into any global path at all. We call such a misleading local path a *garden path* (referring to 18th century English maze gardens with many dead-end paths.)

Similarly to estimating the reliability of amino acids, we want to estimate reliabilities of tags. The approach that simply multiplies the probabilities of the individual amino acids in the tag does not produce good results, since it assumes that the amino acids are independent. In many instances this is not the case (e.g., in a tag SEQ, the amino acid E shares a cleavage site both with S and with Q). We take a different approach based on the principle *"A chain is only as strong as its weakest link"* because all it takes to render a tag incorrect is for one of its amino acids to be wrong.

The features we use in the tag generation model are as follows: (i) The lowest probability amongst the amino acids in the tag (the weakest link); (ii) The probability of the neighbor of the weakest link (if it has two neighbors we choose the neighbor with the lowest probability amongst the two); (iii) The geometric mean of the probabilities of the remaining amino acids in the tag. Using these features we train a logistic regression model to evaluate the reliability of a tag.

We explored three different approaches for tag generation. The first one (called PepNovoTag) exploits the fact that PepNovo is quite accurate in its predictions (in our test data, 72.7% of the amino acids are correct and 53.9% of all substrings of length 3 are correct). Therefore, it is likely that tags derived from PepNovoTag will also be correct. PepNovoTag extracts all substrings of the desired length from the PepNovo reconstruction and assigns probabilities to these tags using the logistic regression model. Because its PSTs are taken from a de novo reconstruction, PepNovoTag is not misled by garden paths.

In the second method (called LocalTag), the vertices of the spectrum graph are scored according to PepNovo's scoring. The spectrum graph is then searched, all sub-paths of the desired length are extracted as tags, and probabilities are assigned to the tags using the regression model. This tag generating method requires changes to the previously described probability models we use to asses the reliability of amino acids and tags. The "Score Reduction due to Edge Removal" feature in the amino acids model cannot be used since it requires the amino acid in question to be part of a complete high scoring path. Dropping this feature reduces to some degree the capability of the model to make distinguishing probability assignments. For instance, the log likelihood of 2884 amino

---

[4] Searle et al., 2004 [33] use Lutefisk that has rather low accuracy or Peaks that is rather accurate but is not designed for generating covering sets of tags.

acids from the test set falls to -1286.1 from -1187.5 when the feature is removed. Another change that we made is to add a feature to the tag probability models. In PepNovoTag, because the tags are derived from a de novo path, they almost always have the correct orientation. However, when we extract tags from the spectrum graph in LocalTag, it is likely that both a correct tag and its mirror[5] have a high score and we cannot really tell which one is correct. Usually the tag with the correct orientation has a slightly higher score (it is typical to detect stronger y-ions than b-ions, and PepNovo's scoring accounts for this). Therefore, we added to the LocalTag method a feature which measures the ratio of scores between the tag and its mirror (if it exists). Adding this feature increases the log likelihood of 2307 tags of length 3 from the test set from -1270.4 to -1151.3.

The third tag generating method, LocalTag+, merges PepNovoTag's and LocalTag's results into a combined list sorted according to the tags' probabilities.

5   Database Filtering

In combinatorial pattern matching, matching a thousand patterns against a database takes roughly the same time as matching a single pattern. This speedup can be achieved using Aho-Corasick [2] algorithm that preprocesses the set of patterns to construct a *trie*. We construct a trie of all PSTs in multiple spectra and use it to search the protein database for all the spectra's tags simultaneously. While scan time does not increase with a larger number of tags, the number of peptide candidates increases, which in turn increases the scoring time. Therefore, we also employ a *tag extension* step, analogous to *seed extension* in sequence similarity search. The sequence tag has the prefix mass, and a scan can tell us if the prefix substrings have the right mass. This is trickier in the presence of PTMs [28], and we use dynamic programming to scan efficiently. Further details regarding the database scanning, tag extension and candidate scoring will be given in a forthcoming publication (Tanner et al. [39]).

6   Results

6.1   Data Set

Our dataset is composed from doubly charged tryptic peptides from ISB dataset [17] and the Open Proteomics Database [29] (low energy ion trap LC/MS/MS runs). In total we obtained 1252 spectra of peptides with unique sequences which were identified by SEQUEST with $X_{corr} > 2.5$. From this set 280 spectra were set aside as the test set (the peptide assignments to these spectra were verified using an independent run of SEQUEST against a non-redundant 20Mb database of protein sequences using non-specific digestion).

---

[5] The mirror tag is the tag obtained when the roles of the b and y-ions are reversed.

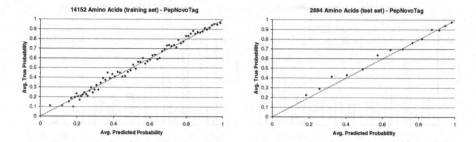

Fig. 2. Comparison of the average predicted accuracy and the true predicted accuracy of amino acids. Results are reported for the training set (*left*) and the test set (*right*)

6.2   Reliability of Individual Amino Acids

We conducted the following experiment to asses the quality of our amino acid probability assignments. For each spectrum in the data set, we obtained a de novo prediction using PepNovo. The amino acids in the training (test) set were sorted according to decreasing predicted accuracy and divided into bins containing 200 amino acids each. Each point in Fig. 2 represents a bin, its x coordinate is the average predicted probability that the samples in the bin are correct amino acids (calculated using the regression models), and the y coordinate is the true proportion of samples in the bin that are correct amino acids. The diagonal dash line represents the region where the predicted probabilities equal the true probabilities.

In an ideal figure, obtained using an oracle for probability assignments, we would find two dense clusters in the graphs. The first, located near (0,0), would contain the incorrect amino acids, and the other, located near (1,1), would contain the correct amino acid assignments. However, in many cases it is difficult for our model to be that discriminating, and when confronted with questionable amino acids it resorts to assigning them probabilities throughout the [0,1] range. However, the fact that the points in the figure are located in the vicinity of the diagonal line shows that our model is not biased in these probability assignments. In addition, the fact that this happens both in the training and test set, can indicate that our models are not over-fitted.

6.3   Reliability of Tags

Figure 3 compares PepNovoTag with LocalTag (for tags of length 3). Two separate sets of tags were generated as follows. For each of the 280 spectra in the test set, PepNovo-generated tags were placed in the PepNovoTag tag list. In addition, an equal number of highest probability tags was extracted from the spectrum graph, and placed in the LocalTag tag list. Note that the composition of tags is not the same in both sets. Only 32.8% of the tags predicted by LocalTag are correct, compared to 53.9% correct tags predicted by PepNovoTag. In addition, the PepNovoTag probability model is much more robust than the LocalTag model.

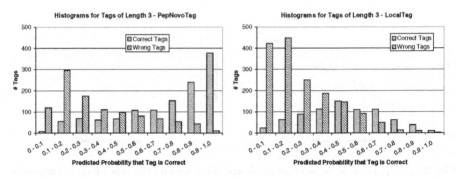

Fig. 3. Histograms of predicted tag probabilities. Both histograms show the probabilities for 2307 tags of length 3 derived from the 280 spectra in the training set. On the left is the histogram of PepNovoTag tags (53.9% of these tags are correct), and on the right is the histogram of the LocalTag tags (32.8% of these tags are correct)

The mean probability assigned to a correct tag by PepNovoTag's model is 0.722 (with 30.2% of the correct tags given probability greater than 0.9), whereas the mean probability assigned to the correct tags in the LocalTag is 0.473 (with only 1% of the correct tags being assigned probabilities above 0.9). It is apparent that the PepNovoTag has an advantage over the LocalTag, since tags derived from a de novo prediction have an *a priori* higher probability of being correct.

6.4   Benchmarking of Tag Generation Algorithms

Table 1 compares the performance of PepNovoTag, LocalTag, and LocalTag+ with GutenTag [37]. PepNovoTag outperforms other methods when the number of tags that are generated is small. The highest scoring tag of length 3 generated by PepNovoTag is correct in 80.4% of the cases, compared to only 49.3% with GutenTag's highest scoring tag. Using 5 tags of length 3 generated by PepNovoTag, we obtain in 93.9% of the cases at least one correct tag compared to only 81.1% of the cases for GutenTag. Table 1 suggests that if the desired number of tags is small (typically less than 5), PepNovoTag should be used. However, for larger number of tags, LocalTag+ performs better.

Since interpreting mass spectra is a high-throughput process, it is worthwhile to discuss the running time required to generate the tags. Typically it takes PepNovoTag or LocalTag+ less than 0.1 seconds to generate a set of tags. LocalTag+ running time scales well with increasing tag lengths, where generating tags of length 6 takes less than 0.2 seconds. GutenTag on the other hand, doesn't scale well with increasing tag length. While it can generate tags of length 3 at a decent pace, it takes an average of 1 minute to generate tags of length 5, and in some cases more than 30 minutes to generate tags of length 6 for a single spectrum.

Table 1. Comparison of tag generating methods (280 spectra in the test sample). For each tag length, algorithm and number of generated tags, the table displays the proportion of test spectra with least one correct tag. Since the number of tags that can be generated by PepNovoTag is limited by the length of the predicted sequence, usually no more than 10 tags were predicted. GutenTag was run with the default settings for ion trap tryptic peptides. Due to the long time required to generate tags of length 6, this data was not collected for GutenTag

Tag Length	Algorithm	Number of Generated Tags						
		1	3	5	10	25	50	100
3	LocalTag	0.529	0.764	0.929	0.957	0.971	0.975	0.979
	PepNovoTag	**0.804**	**0.925**	0.932	0.946	-	-	-
	LocalTag+	0.725	0.855	**0.939**	**0.961**	**0.979**	**0.979**	**0.982**
	GutenTag	0.493	0.732	0.811	0.893	0.914	0.936	0.950
4	LocalTag	0.464	0.714	0.771	0.850	0.932	0.943	0.954
	PepNovoTag	**0.732**	**0.850**	0.864	0.871	-	-	-
	LocalTag+	0.700	0.811	**0.871**	**0.900**	**0.946**	**0.954**	**0.964**
	GutenTag	0.418	0.614	0.711	0.782	0.832	0.861	0.879
5	LocalTag	0.410	0.593	0.678	0.786	0.836	0.854	0.879
	PepNovoTag	**0.664**	**0.764**	**0.775**	0.800	-	-	-
	LocalTag+	0.571	0.696	0.736	**0.803**	**0.846**	**0.864**	**0.893**
	GutenTag	0.318	0.464	0.539	0.643	0.736	0.761	0.775
6	LocalTag	0.332	0.489	0.593	0.661	0.739	0.771	**0.804**
	PepNovoTag	**0.579**	**0.632**	**0.639**	0.654	-	-	-
	LocalTag+	0.527	0.546	0.593	**0.671**	**0.743**	**0.779**	**0.804**
	GutenTag	-	-	-	-	-	-	-

6.5 Database Search Results

With no prefix mass values (in PTM detection mode), a tag of length 3 had on average 30000 hits to the 54Mb SWISS-PROT database. Because we consider peptides of various lengths including non-tryptic peptides, the effective database size is roughly 550 million entries. Such a large number of entries requires efficient filtration in order to obtain results in reasonable time. Table 2 gives examples of the efficiency of our tag filtration. For instance, using a single tag of length 3 as a filter yields on average 129 candidate peptides having both a correct parent mass and a correct prefix mass for the tag. Of course, a single tag often does not satisfy the covering condition, particularly for low-quality spectra. Increasing the number of generated tags to 10 ensures with high probability that the resulting set satisfies the covering condition and still provides high filtration efficiency of 1.8×10^{-6}. This is almost two thousandfold more efficient than using only the parent mass as a filter (which has 0.003 filtration efficiency).

Considering post-translational modifications does not impact the number of initial matches, but affects the chances of a successful extension (and hence the scoring time). As annotated spectra of modified peptides are not readily available, we report statistics from a simulated dataset with phosphorylations

Table 2. Filtration efficiency of tag-based filtering. The spectra were searched against the SWISS-PROT database (54 Mb) using a standard desktop PC (3GHz CPU). The search permitted one or both endpoints to be non-tryptic, and allowed missed cleavages; requiring tryptic endpoints would further improve filtration efficiency

PTMs	Tag Length	# Tags	# Candidates	Filtration Efficiency	Runtime
None	3	1	129	2.3×10^{-7}	1.4s
	3	10	1011	1.8×10^{-6}	2.5s
	4	1	6.5	1.2×10^{-8}	1.3s
	4	10	50	9.1×10^{-8}	1.8s
Phosphorylation	3	1	199	9.0×10^{-8}	1.4s
	3	10	1521	6.9×10^{-7}	2.9s
	3	25	3298	1.5×10^{-6}	4.2s

introduced to the ISB spectra in a realistic probabilistic setting. Tag-based filters provide far greater efficiency in the presence of PTMs. For a case of up to two phosphorylations, 10 PSTs of length 3 are 1500 times as efficient as basic parent mass filtering. Each possible modification enriches the spectrum graph with more edges. For instance, phosphorylation adds three new masses to our "alphabet" of possible edge masses (we considered phosphorylations of Serine, Threonine, and Tyrosine). Therefore, some increase in number of tags generated is necessary in order to maintain the same high sensitivity for medium-quality spectra. Twenty-five tags on the phosphorylated data set produce accuracy equivalent to ten tags on the unmodified data set (data not shown).

Although this test is run on simulated spectra, to the best of our knowledge, it is the first systematic benchmarking for speed and sensitivity of PTM identifications. Previous studies report identification of PTMs, but not how many PTMs are missed in the analysis. Thus, the sensitivity of these algorithms remains unknown. We plan to supplement our study through analysis of real PTMs as soon as large data sets of spectra of modified peptides become publicly available.

Acknowledgments

We are indebted to Vineet Bafna who is a key contributor to this project. We would like to thank Andrew Keller for supplying the protein mixture data and to Edward Marcotte for making the Open Proteomics Database publicly available. We would also like to thank Richard Johnson for supplying the Lutefisk software, Bin Ma for running the Peaks benchmarks, and Karl Clauser for running the Sherenga benchmarks. We benefitted a lot from our discussions with Sanjoy Dasgupta, Nuno Bandeira, Qian Peng, Tim Chen and Haixu Tang. This project was supported by NIH grant NIGMS 1-R01-RR16522.

References

1. R. Aebersold and M. Mann. Mass spectrometry-based proteomics. *Nature*, 422:198–207, 2003.
2. A.V. Aho and M.J. Corasick. Efficient string matching: an aid to bibliographic search. *Communications of the ACM*, 18:333–340, 1975.
3. V. Bafna and N. Edwards. SCOPE: a probabilistic model for scoring tandem mass spectra against a peptide database. *Bioinformatics*, 17 Suppl 1:13–21, 2001.
4. V. Bafna and N. Edwards. On de-novo interpretation of tandem mass spectra for peptide identification. *Proceedings of the Seventh Annual International Conference on Computational Molecular Biology*, pages 9–18, 2003.
5. T. Chen, M.Y. Kao, M. Tepel, J. Rush, and G.M. Church. A dynamic programming approach to de novo peptide sequencing via tandem mass spectrometry. *J Comput Biol*, 8:325–337, 2001.
6. J. Colinge, A. Masselot, M. Giron, T. Dessingy, and J. Magnin. OLAV: towards high-throughput tandem mass spectrometry data identification. *Proteomics*, 3:1454–1463, 2003.
7. T.H. Cormen, C.H. Leiserson, R.L. Rivest, and C. Stein. *Introduction to Algorithms*. The MIT Press, 2nd edition, 2001.
8. D.M. Creasy and J.S. Cottrell. Error tolerant searching of uninterpreted tandem mass spectrometry data. *Proteomics*, 2:1426–1434, 2002.
9. V. Dancík, T.A. Addona, K.R. Clauser, J.E. Vath, and P.A. Pevzner. De novo peptide sequencing via tandem mass spectrometry. *J Comput Biol*, 6:327–342, 1999.
10. R.M. Day, A.Borziak, and A. Gorin. Ppm-chain de novo peptide identification program comparable in performance to sequest. In *Proceedings of 2004 IEEE Computational Systems in Bioinformatics (CSB 2004)*, pages 505–508, 2004.
11. J.E. Elias, F.D. Gibbons, O.D. King, F.P. Roth, and S.P. Gygi. Intensity-based protein identification by machine learning from a library of tandem mass spectra. *Nat Biotechnol*, 22:214–219, 2004.
12. J.K. Eng, A.L. McCormack, and J.R. Yates. An Approach to Correlate Tandem Mass-Spectral Data of Peptides with Amino Acid Sequences in a Protein Database. *Journal Of The American Society For Mass Spectrometry*, 5:976–989, 1994.
13. A. Frank and P. Pevzner. Pepnovo: De novo peptide sequencing via probabilistic network modeling. *Anal Chem.*, 77:964–973, 2005.
14. T. Hastie, R. Tibshirani, and J. Friedman. *The Elements of Statistical Learning*. Springer, 2001.
15. M. Havilio, Y. Haddad, and Z. Smilansky. Intensity-based statistical scorer for tandem mass spectrometry. *Anal Chem*, 75:435–444, 2003.
16. P. Hernandez, R. Gras, J. Frey, and R.D. Appel. Popitam: towards new heuristic strategies to improve protein identification from tandem mass spectrometry data. *Proteomics*, 3:870–8, 2003.
17. A. Keller, S. Purvine, A.I. Nesvizhskii, S. Stolyar, D.R. Goodlett, and E. Kolker. Experimental protein mixture for validating tandem mass spectral analysis. *OMICS*, 6:207–212, 2002.
18. B. Lu and T. Chen. A suboptimal algorithm for de novo peptide sequencing via tandem mass spectrometry. *J Comput Biol*, 10:1–12, 2003.
19. B. Lu and T. Chen. A suffix tree approach to the interpretation of tandem mass spectra: applications to peptides of non-specific digestion and post-translational modifications. *Bioinformatics*, 19 Suppl 2:113–113, 2003.

20. B. Lu and T. Chen. Algorithms for de novo peptide sequencing via tandem mass spectrometry. *Drug Discovery Today: BioSilico*, 2:85–90, 2004.
21. O. Lubeck, C. Sewell, S. Gu, X. Chen, and D. Cai. New computational approaches for de novo peptide sequencing from MS/MS experiments. *IEEE Proc. on Challenges in Biomedical Informatics*, 90:1868–1874, 2002.
22. B. Ma, K. Zhang, C. Hendrie, C. Liang, M. Li, A. Doherty-Kirby, and G. Lajoie. PEAKS: powerful software for peptide de novo sequencing by tandem mass spectrometry. *Rapid Commun Mass Spectrom*, 17:2337–2342, 2003.
23. M.J. MacCoss, C.C. Wu, and J.R. Yates. Probability-based validation of protein identifications using a modified SEQUEST algorithm. *Anal Chem*, 74:5593–5599, 2002.
24. M. Mann and O.N. Jensen. Proteomic analysis of post-translational modifications. *Nat Biotechnol*, 21:255–261, 2003.
25. M. Mann and M. Wilm. Error-tolerant identification of peptides in sequence databases by peptide sequence tags. *Analytical Chemistry*, 66:4390–4399, 1994.
26. A.I. Nesvizhskii, A. Keller, E. Kolker, and R. Aebersold. A statistical model for identifying proteins by tandem mass spectrometry. *Anal Chem*, 75:4646–4658, 2003.
27. D.N. Perkins, D.J. Pappin, D.M. Creasy, and J.S. Cottrell. Probability-based protein identification by searching sequence databases using mass spectrometry data. *Electrophoresis*, 20:3551–3567, 1999.
28. P.A. Pevzner, Z. Mulyukov, V. Dancik, and C.L. Tang. Efficiency of database search for identification of mutated and modified proteins via mass spectrometry. *Genome Res.*, 11:290–9, 2001.
29. J. T. Prince, M. W. Carlson, R. Wang, P. Lu, and E. M. Marcotte. The need for a public proteomics repository (commentary). *Nature Biotechnology*, April 2004.
30. J. Razumovskaya, V. Olman, D. Xu, E. Uberbacher, N.C. VerBerkmoes, R.L. Hettich, and Y. Xu. A computational method for assessing peptide-identification reliability in tandem mass spectrometry analysis with sequest. *Proteomics*, 4:961–969, 2004.
31. R.G. Sadygov and J.R. Yates. A hypergeometric probability model for protein identification and validation using tandem mass spectral data and protein sequence databases. *Anal Chem*, 75:3792–3798, 2003.
32. F. Schutz, E.A. Kapp, R.J. Simpson, and T.P. Speed. Deriving statistical models for predicting peptide tandem ms product ion intensities. *Biochem Soc Trans*, 31:1479–83, 2003.
33. B.C. Searle, S. Dasari, M. Turner, A.P. Reddy, D. Choi, P.A. Wilmarth, A.L. McCormack, L.L. David, and S.R. Nagalla. High-throughput identification of proteins and unanticipated sequence modifications using a mass-based alignment algorithm for MS/MS de novo sequencing results. *Anal Chem*, 76:2220–2230, 2004.
34. A. Shevchenko, S. Sunyaev, A. Liska, P. Bork, and A. Shevchenko. Nanoelectrospray tandem mass spectrometry and sequence similarity searching for identification of proteins from organisms with unknown genomes. *Methods Mol Biol*, 211:221–234, 2003.
35. J.R. Shewchuk. An introduction to the conjugate gradient method without the agonizing pain. http://www-2.cs.cmu.edu/~jrs/jrspapers.html, 1994.
36. S. Sunyaev, A.J. Liska, A. Golod, A. Shevchenko, and A. Shevchenko. MultiTag: multiple error-tolerant sequence tag search for the sequence-similarity identification of proteins by mass spectrometry. *Anal Chem.*, 75:1307–1315, 2003.
37. D.L. Tabb, A. Saraf, and J.R. Yates. GutenTag: High-throughput sequence tagging via an empirically derived fragmentation model. *Anal Chem*, 75:6415–6421, 2003.

38. D.L. Tabb, L.L. Smith, L.A. Breci, V.H. Wysocki, D. Lin, and J.R. Yates. Statistical characterization of ion trap tandem mass spectra from doubly charged tryptic peptides. *Anal Chem*, 75:1155–1163, 2003.

39. S. Tanner, H. Shu, A. Frank, M. Mumby, P. Pevzner, and V. Bafna. Inspect: Fast and accurate identification of post-translationally modified peptides from tandem mass spectra. submitted, 2005.

40. J.A. Taylor and R.S. Johnson. Sequence database searches via de novo peptide sequencing by tandem mass spectrometry. *Rapid Commun Mass Spectrom*, 11:1067–1075, 1997.

41. J.A. Taylor and R.S. Johnson. Implementation and uses of automated de novo peptide sequencing by tandem mass spectrometry. *Anal Chem*, 73:2594–2604, 2001.

42. J.R. Yates, J.K. Eng, and A.L. McCormack. Mining genomes: correlating tandem mass spectra of modified and unmodified peptides to sequences in nucleotide databases. *Anal Chem*, 67:3202–3210, 1995.

43. J.R. Yates, J.K. Eng, A.L. McCormack, and D. Schieltz. Method to correlate tandem mass spectra of modified peptides to amino acid sequences in the protein database. *Anal Chem*, 67:1426–1436, 1995.

A Hidden Markov Model Based Scoring Function for Mass Spectrometry Database Search

Yunhu Wan[1] and Ting Chen[2],*

[1] Department of Mathematics, University of Southern California,
Los Angeles, CA 90089-1113, USA
ywan@usc.edu
[2] Department of Biological Sciences, University of Southern California,
University of Southern California, Los Angeles, CA 90089-1113, USA
tingchen@usc.edu

Abstract. An accurate scoring function for database search is crucial for peptide identification using tandem mass spectrometry. Although many mathematical models have been proposed to score peptides against tandem mass spectra, we design a unique method (called HMMscore) that combines information on consecutive ions, ie, b_i and b_{i+1}, and complementary ions, ie, b_i and y_{n-i} for a peptide of n amino acids, plus information on machine accuracy and mass peak intensity, into a hidden Markov model (HMM). In addition, we develop a way to calculate statistical significance of the HMM scores. We implement the method and test them on experimental data. The results show that the accuracy of our method is high compared to MASCOT, and that the false positive rate of HMMscore is low.

1 Introduction

Mass spectrometry, especially tandem mass spectrometry, has become the most widely used method for high-throughput identification of peptides and proteins. In an experiment, proteins of interest are first extracted. Then, the proteins are digested into peptides by using an enzyme such as trypsin, and run through High Performance Liquid Chromatography(HPLC) for separation. The complete mixtures from HPLC are analyzed by a mass spectrometer, and a mass peak (called a precursor ion) for some unknown peptide above the background level is selected for further analysis. In order to identify this peptide, another procedure is then applied: fragmentation of the peptide. The peptide is broken into smaller molecules, the masses of which are measured again using the mass spectrometer. The most abundant N-terminal ions are b ions and the most abundant C-terminal ions are y ions. In this procedure (called tandem mass spectrometry), a tandem mass (MS/MS or MS2) spectrum is produced for the peptide.

---

* To whom the correspondence should be addressed.

S. Miyano et al. (Eds.): RECOMB 2005, LNBI 3500, pp. 342–356, 2005.
© Springer-Verlag Berlin Heidelberg 2005

Multiple MS/MS spectra for peptides of a protein sequence constitute a specific data set that can be used to accurately identify this protein.

Generally, there are two methods for identification of peptide sequences from tandem mass spectra: *de novo* peptide sequencing and database search. De novo peptide sequencing, which derives sequences of peptides directly from MS/MS spectra, has been developed by several groups [3, 5, 10, 11]. Due to limited information content of data, poor fragmentation, and inaccuracy of mass measurements, *de novo* sequencing is not broadly used. The more widely used method is the database search. In a database search framework, candidate peptides from a protein database are generated using specific enzyme digestion. A scoring scheme is used to rate the quality of matching between an experimental mass spectrum and a hypothetical spectrum that is generated from a candidate peptide sequence directly. If the protein database is a complete annotation of all coding sequences in a genome, a good scoring function is able to identify the right peptide sequence with the best score.

Database search programs that have been developed differ in their methods of computing the correlation score between a spectrum and a peptide sequence. The first program, SEQUEST, was developed by Eng et al. (1993) [18]. It first generated a hypothetical spectrum for each peptide in a database and then compared it against the experimental spectrum by using a cross-correlation scoring function. Perkins et al. (1999) [12] later developed a program called MASCOT, which introduced a p-value based probabilistic scheme to access the significance of peptide identification. Keller et al. (2002)[8] derived a probabilistic model called PeptideProphet to improve the SEQUEST scores. Other available programs include SCOPE[2], SHERENGA[5], Profound[20], ProbID[19], and GuntenTag[17].

The ultimate interest of mass spectrometry is actually the identification and characterization of proteins. The foundation of them is the identification of peptides. Ideally, we expect that the sequence with the highest score found by a database scoring function would be the correct identification. However, developing such an accurate scoring function remains a challenging task because of the following facts.

- *Charges of Precursor Ions.* Knowing charges of precursor ions is important for database search because the charges will affect the calculation of the masses of the target peptide sequences that are used in the database search. However, Detecting charges accurately is difficult, especially charge +2 and charge +3.
- *Consecutive and Composite Ions.* It has been observed that for the same ion type, consecutive (or neighboring) ions, ie, b_i and b_{i+1} ions appear together frequently. It has also been observed that for each fragmentation, both the N-terminal ion, ie, b_i, and the C-terminal ion, ie, y_{n-i} for a peptide of n amino acids appear together frequently. This means that if a peptide sequence matches a spectrum with more consecutive ions and composite ions it is more likely to be the target peptide.
- *Peak Intensity.* It has been observed that in mass spectra high abundance peaks are more likely to be real peaks, while low abundance peaks are more

likely to be noise. Thus, a scoring function should take into account the peak intensity effect.

- *Machine Accuracy (or Match Tolerance).* The accuracies of mass measurement by mass spectrometers vary from one to another. The difference between a mass peak and the actual mass of the corresponding ion can be as high as 1 m/z. We call this difference "match tolerance." However, the match tolerance of most mass spectrometry machines is larger than the mass difference between some pairs of amino acids such as (I, L), (K, Q), (I, N), (L, N), (D, N), and (E, Q). Match tolerance is an important criteria for determining whether there is a match or not.
- *Isotopic Ions.* Every ion has multiple isotopic forms, mainly contributed by ^{12}C and ^{13}C. Isotopes shift an ion mass by 1 Da, which generally is distorted by the machine accuracy.
- *Others.* Other challenges are post-translational modifications (PTM) of target peptides such as phosphorylation, incomplete and inaccurate sequences in protein databases that were constructed from coding sequences found by applying gene-finding programs for sequenced genomes, and limited knowledge about the peptide fragmentation process.

In this paper, we will develop a hidden Markov model(HMM) based scoring function (called HMMscore) that calculates the probability mass function that a spectrum s is generated by a peptide p, $\Pr(s|p)$. This scoring function combines information on consecutive and composite ions and on peak intensity and match tolerance. The model automatically detects whether there is a match between a mass peak and a hypothetical ion resulting from the fragmentation of the peptide. The detection is based on the local information on the intensity of the matched mass peak and the match tolerance, and also on the global information on all matches between the spectrum and the peptide. Because $\Pr(s|p)$ varies in accordance with the density of s, the distribution of the peak intensities, and the mass of the precursor ion, we convert $\Pr(s|p)$ into a Z-score Z that measures the ranking of the score of this peptide among all possible peptides that have the same mass. For a given database, we can easily calculate the E-score E, the expected number of peptides that have a score better than Z.

We implement this scoring function, train it, and test it on several data sets from the Institute of Systems Biology (ISB). Compared with MASCOT, our program has a lower error rate (3.6% versus 8.5% for MASCOT). We also access the false positive rate of our prediction.

2   Material and Methods

2.1   Datasets

We obtained a mass spectra data set from ISB [9]. Two mixtures, A and B, were obtained by mixing together 18 purified proteins of different physicochemical properties with different relative molar amounts and modifications. Twenty-two runs of LC/MS/MS were performed on the data sets, of which 14 runs were

Runs Peptides	18 runs	4 runs
144	Training	
36		Testing

Fig. 1. Partitioning the data into training and testing

performed on mixture A and 8 on mixture B. The data sets were analyzed by SEQUEST and other in-house software tools, with 18,496 peptides assigned to spectra of $[M+2H]^{2+}$, 18,044 to spectra of $[M+3H]^{3+}$, and 504 to spectra of $[M+H]^+$. The peptide assignments were then manually scrutinized to determine whether they were correct. The final data set contains 1,649 curated $[M+2H]^{2+}$ spectra and 1,010 curated $[M+3H]^{3+}$ spectra. In this study, we only consider $[M+2H]^{2+}$ spectra.

We apply the following rules to partition the $[M+2H]^{2+}$ data set into a training set and a testing set in order to minimize bias, as shown in Figure 1: (1) if a spectrum from a run is used in the training set, all spectra from this run are excluded from the testing set, and vice versa; (2) if a spectrum of a peptide is used in the training set, all spectra of this peptide from this protein are excluded from the testing set, and vice versa.

2.2   Analysis of Peak Intensity and Match Tolerance

The distributions of peak intensity and match tolerance play a crucial role in the scoring function. These distributions determine the quality of a match that is whether a mass peak matches a hypothetical ion of a peptide.

To obtain information on peak intensity, we use different formats to plot the peak intensity: relative intensity, absolute intensity, and relative ranking. The relative ranking is obtained by dividing the ranking of the intensity in descending order by the total number of peaks. The best characterization of the intensity information is to use the relative ranking. Figure 2 shows the distribution of B ion intensities using the relative ranking (Y ions show the same trend). Clearly, the relative rank of a matched mass peak conforms to an exponential distribution, as shown in Figure 2(a). Figure 2(b) shows a uniform distribution for noise, obtained by excluding the matched mass peaks from the training data set.

The distribution of the match tolerance of B and Y ions is shown in Figure 3 (Y ions shows the same trend). Figure 3 shows that this distribution agrees with a normal distribution except that the right-hand side has a small peak at around +1 mass/charge, at which isotopic peaks appear. For simplicity, we use

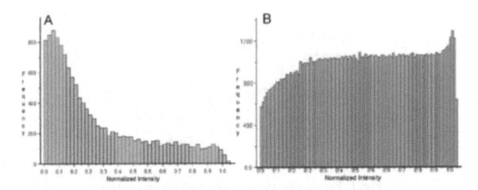

Fig. 2. Distribution of peak intensity in the training set. (a) B ion intensity. (b) Noise intensity

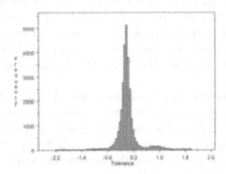

Fig. 3. Distribution of match tolerance of B and Y ions in the training set

the normal distribution to model the match tolerance for all ions and a uniform distribution for noise.

2.3   Framework of Database Search

For the database search, we use a non-redundant protein sequence database called MSDB, which is maintained by the Imperial College, London. We downloaded the release (20042301) that has 1,454,651 protein sequences from multiple organisms. We pre-process the database as follows: for a specific enzyme such as trypsin, we digest *in silico* every protein sequence into small peptide sequences, and then we index these peptide sequences by their masses. The procedure of the database search follows a standard framework shown in the following paragraphs.

1. *Extracting Peptides.* For a given spectrum, we identify candidate peptide sequences the masses of which are within 2 Da of the precursor ion mass, m. The indexing of peptide masses can greatly speeded up this process.
2. *Generating Hypothetical Spectra.* For each candidate peptide, p, we generate a hypothetical spectrum h without weights (or intensities). In fact, the weights

are embedded in the HMM framework. We consider the following seven ions: b ion, y ion, b-H_2O, y-H_2O, a-ion, b^{2+} and y^{2+}.

3. *Computing the HMM Score.* We compare ions in the hypothetical spectrum with mass peaks in the experimental spectrum. The comparison results in three mutually exclusive groups: *match*; *missing*, where an ion does not match to any peak in the experimental spectrum; and *noise*, where a mass peak does not match any ion in the hypothetical spectrum. Initially, we use a simple match tolerance threshold $(+/- 2\ m/z)$ to classify the comparison into these three groups. Then we apply the initial classification as input for HMMscore. The HMM automatically determines whether they are actual matches, missings, or noise, and it returns a score $\Pr(s|p)$. The details of HMMscore are described in Section 2.4.

4. *Computing the Z-score.* We simulate 1,000 random peptides the masses of which are within $[m - 2, m + 2]$, and we calculate HMM scores for these peptides using the above procedure. This simulation is done once for this spectrum. We adjust the HMM scores by the length of the peptides, and we calculate the mean μ and the standard deviation σ. Based on μ and σ, we compute a Z-score Z for peptide p.

5. *Computing the E-value.* Given the size of the database, we calculate the expected number of peptides for which the Z-scores are better than Z.

2.4   HMMscore

HMM Structure. We model the information of consecutive and composite ions into an HMM framework as shown in Figure 4. For each fragmentation (or position), there are four possible assignments corresponding to four hidden states: (1) both the b ion and the y ion are observed, (2) the b ion is observed but the y ion is missing, (3) the y ion is observed but the b ion is missing,

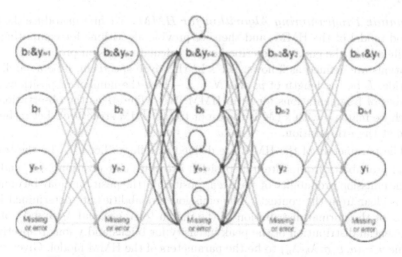

Fig. 4. The hidden Markov model for the scoring function

(4) neither of the two ions is observed. The information on consecutive ions is modelled into the transition probabilities between states, and the information on the composite ions is modelled into hidden states. Due to the limited size of the training set, we only consider b and y ions in the HMM. Other kinds of ions can also be added into this framework, and we will discuss this later. In order to deal with different lengths of peptides in the same fashion, we only include five positions of fragmentations here, the first two peptide bonds, the middle peptide bonds, and the last two peptide bonds. Analysis of the training data shows that the middle peptide bonds have similar properties: percentages of observed b ions and y ions, percentages of consecutive ions, and percentages of composite ions.

The input to the HMM are: sets of matches, missings, and noise. Each match is associated with an observation (T, I), where T is the match tolerance and I is the peak intensity. We model the information of (T, I) into the emission of each state. In our case, an observation state is the observed (T, I), and a hidden state is the true assignment of this observation. Since (T, I) can be an error observation, the fourth state emits an error observation. On the other hand, if there is no observation, the fourth state emits a missing observation. For each pair of a spectrum and a peptide (s, p), a dynamic programming algorithm can calculate the probability that s is generated by p.

The HMM method has two main advantages. First, the model emphasizes the global assignments of matches. True assignments of observations (the optimal path in HMM) are automatically selected through a dynamic programming algorithm along with the learned parameters. Second, we do not use a hard threshold for match tolerance or peak intensity. Instead, we model them into probability mass functions, of which parameters can be trained through an expectation-maximization (EM) algorithm.

Other types of ions are also considered in our model. Due to the limited size of the training data set, we assume that the appearance of other types of ions is independent.

Dynamic Programming Algorithm for HMM. We first introduce the likelihood model of the HMM, and then we provide algorithms for computing the likelihood. At the end, we describe how to calculate emission probabilities. The notations are defined as follows. Let s be the given spectrum, p be a candidate peptide, L be the length of p, and $N = L - 1$ be the number of positions (the number of fragmentations for p) in HMM. We define M as the observations in which we match p with s (M is the input to the HMM structure). Let π_i be the state at the i-th position.

The parameters of the HMM are defined as follows. Let $a_{u,v}^{(i)}$, be the transitional probability from state u at position i to state v at position $i + 1$, and $e_u^{(i)}$ be the emission probability of state u at position i (the position i may be omitted if it is clear from the context). The emission probabilities are determined by μ and σ of the normal distribution of the match tolerance and λ_b and λ_y of the exponential distribution of the peak intensity for b ions and y ions respectively. Define $\theta = (a, \mu, \sigma, \lambda_b, \lambda_y)$ to be the parameters of the HMM model. Given θ, we can find the highest-score peptide from a database D for s:

$$\max_{p \in D} \Pr(s|p, \theta).$$

Let a path at HMM be $\pi = \pi_1 \pi_2 \cdots \pi_N$ (or a configuration of states). We add π_0 at the beginning of π to indicate the starting state. The probability of this path,

$$\Pr(\pi, s|p, \theta) = \Pr(\pi, M|\theta)$$
$$= \prod_{i=0}^{N-1} (a_{\pi_i, \pi_{i+1}} e_{\pi_{i+1}}) * \eta^{\#noise},$$

where a_{π_0, π_1} is the prior probability of the first state π_1, $a_{\pi_i, \pi_{i+1}}$ is the transitional probability from state π_i to state π_{i+1}, e_{π_i} is the emission probability of state π_i, and η is the probability of observing a noise peak. Considering all possible paths,

$$\Pr(s|p, \theta) = \Pr(M|\theta) = \sum_{\pi} \Pr(\pi, M|\theta).$$

Because there are an exponential number of possible paths, we can use either the forward algorithm or the backward algorithm to calculate this probability. These two algorithms are also used in estimating the parameters θ.

Forward Algorithm. Define $f_v^{(i)}$ as the probability of all possible paths from the first position to state v at position i, or

$$f_v^{(i)} = \sum_{\pi_1, \dots, \pi_{i-1}} \Pr(\pi_1, \dots, \pi_{i-1}, \pi_i = v, M|\theta).$$

$f_v^{(i)}$ can be calculated by the following recursion,

$$f_v^{(i)} = e_v^{(i)} \sum_{u} (f_u^{(i-1)} a_{u,v}^{(i-1)}),$$

with an initial value $f_v^{(1)} = a_{0,v} e_v^{(1)}$. Using a dynamic programming algorithm, we can compute the final probability

$$\Pr(M|\theta) = \sum_{v} f_v^{(N)}.$$

Backward Algorithm. Define $b_u^{(i)}$ as the probability of all possible paths from state u at position i to any state at the final position, or

$$b_u^{(i)} = \Pr(\pi_{i+1}, \cdots, \pi_N, \pi_i = u, M|\theta).$$

Similarly, $b_u^{(i)}$ can be calculated by the following recursion,

$$b_u^{(i)} = \sum_{v} (a_{u,v}^{(i)} e_v^{(i+1)} b_v^{(i+1)}),$$

with an initial value $b_u^{(N)} = 1$. Using a dynamic programming algorithm, we can compute the final probability

$$\Pr(M|\theta) = \sum_u (a_{0,u} e_u^{(1)} b_u^{(1)}).$$

Calculating Emission Probability. The probability of observing (T_b, I_b) and (T_y, I_y) at state 1 (observation of both b and y ions) of position i is

$$\begin{aligned}
e_1^{(i)} &= \Pr((T_b, I_b), (T_y, I_y)|\theta) \\
&= \Pr((T_b, I_b)|\theta) \Pr((T_y, I_y)|\theta) \\
&= \Pr(T_b|\theta) \Pr(I_b|\theta) \Pr(T_y|\theta) \Pr(I_y|\theta),
\end{aligned}$$

where $\Pr(T_b|\theta)$ and $\Pr(T_y|\theta)$ can be calculated by the normal distribution $N(\mu, \sigma)$, and $\Pr(I_b|\theta)$ and $\Pr(I_y|\theta)$ can be calculated by the exponential distributions with means λ_b and λ_y respectively. The emission probabilities of other states can be calculated similarly.

EM Algorithm. Assume the training set includes a total of K pairs of peptides and mass spectra, $(p_1, s_1), \cdots, (p_K, s_K)$. We obtain K observations $M = M_1, \cdots, M_K$ by matching peptides with spectra. Our goal is to find θ that maximize the likelihood $l(\theta|M)$. Define $\{\gamma, \xi\}$ as missing data, where $\gamma_v^{(i)}(k)$ is the probability of state v at position i for the k-th spectra given the observation M_k and θ, and $\xi_{u,v}^{(i)}(k)$ is the probability of state u at position i and state v at position $i + 1$ for the k-th spectra given the observation M_k and θ. The EM algorithm estimates the expectation of the missing data as follows (E-step):

$$\begin{aligned}
\gamma_v^{(i)}(k) &= \Pr(\pi_i = v|M_k, \theta) \\
&= \frac{f_v^{(i)}(k) \, b_v^{(i)}(k)}{\Pr(M_k|\theta)},
\end{aligned}$$

$$\begin{aligned}
\xi_{u,v}^{(i)}(k) &= \Pr(\pi_i = u, \pi_{i+1} = v|M_k, \theta) \\
&= \frac{f_u^{(i)}(k) \, a_{u,v}^{(i)}(k) \, e_v^{(i+1)}(k) \, b_v^{(i+1)}(k)}{\Pr(M_k|\theta)},
\end{aligned}$$

Using the missing data we compute θ to maximize the likelihood (M-step). The transitional probability,

$$\hat{a}_{uv}^{(i)} = \frac{\sum_{k=1}^{K} \xi_{u,v}^{(i)}(k)}{\sum_{k=1}^{K} \gamma_u^{(i)}(k)}.$$

Let $(T_b^{(i)}(k), I_b^{(i)}(k))$ and $(T_y^{(i)}(k), I_y^{(i)}(k))$ be the observations of the b ion and y ion matches at the i-th fragmentation (or position at HMM) for the k-th peptide. The parameters for the normal distribution of the match tolerance,

$$\mu = \frac{\sum_{k=1}^{K} \sum_{i=1}^{N} (\gamma_1^{(i)}(k) + \gamma_2^{(i)}(k)) T_b^{(i)}(k) + (\gamma_1^{(i)}(k) + \gamma_3^{(i)}(k)) T_y^{(i)}(k)}{\sum_{k=1}^{K} \sum_{i=1}^{N} (2\gamma_1^{(i)}(k) + \gamma_2^{(i)}(k) + \gamma_3^{(i)}(k))}$$

$$\sigma^2 = \frac{\sum_{k=1}^{K} \sum_{i=1}^{N} (\gamma_1^{(i)}(k) + \gamma_2^{(i)}(k))(T_b^{(i)}(k) - \mu)^2 + (\gamma_1^{(i)}(k) + \gamma_3^{(i)}(k))(T_y^{(i)}(k) - \mu)^2}{\sum_{k=1}^{K} \sum_{i=1}^{N} (2\gamma_1^{(i)}(k) + \gamma_2^{(i)}(k) + \gamma_3^{(i)}(k))}.$$

The parameters for the exponential distribution of the b ion intensities and the y ion intensities,

$$\lambda_b = \frac{\sum_{k=1}^{K} \sum_{i=1}^{N} (\gamma_1^{(i)}(k) + \gamma_2^{(i)}(k))}{\sum_{k=1}^{K} \sum_{i=1}^{N} I_b^{(i)}(k)(\gamma_1^{(i)}(k) + \gamma_2^{(i)}(k))},$$

$$\lambda_y = \frac{\sum_{k=1}^{K} \sum_{i=1}^{N} (\gamma_1^{(i)}(k) + \gamma_3^{(i)}(k))}{\sum_{k=1}^{K} \sum_{i=1}^{N} I_y^{(i)}(k)(\gamma_1^{(i)}(k) + \gamma_3^{(i)}(k))}.$$

The EM algorithm iterates between the E-step and the M-step until it converges.

2.5   Significance of HMM Scores

The HMM scores vary in accordance with the length of peptides, the densities of spectra, the distributions of peak intensities, and so on. Here, we propose a general way to compute the significance of an HMM score. This method can be applied to any other scoring function. The central idea is to compute the ranking of a score among all possible scores. Given a spectrum s with precursor ion mass m and machine accuracy δ, we consider all peptides with masses within the range of $[m - \delta, m + \delta]$ as a complete set Q. If we can score every peptide in Q against s using our HMMscore, we can obtain a complete set of HMM scores, and easily compute the ranking of a score. However, in general there are an exponential number of peptides in Q, so just listing every peptide in Q is already unrealistic. For simplicity, we assume that the size of Q is infinite and that the HMM scores (logarithm) of peptides in Q follow a normal distribution. In the following, we describe how to compute the mean and standard deviation for the normal distribution, with which we can calculate the significance of a score.

1. *Building a Mass Array.* Without loss of generality, we assume that all masses are integers and that every amino acid is independently and identically distributed. Let A be a mass array, where $A[i]$ equals the number of peptides with mass exactly i. We compute A in linear time using the following recursion:

$$A[i] = \sum_{aa} A[i - mass(aa)], \quad A[0] = 1,$$

where aa is one of the 20 amino acids and $mass(aa)$ returns the mass of aa. The size of A depends upon the accuracy and the measurement range of mass spectrometry machines. In our study, we build an array with an accuracy of 0.01 Da and a range of up to 3,000 Da. The size of A is 300,000. We build this array once for all applications. We can easily adapt our method to the case that difference amino acids have different frequencies.

2. *Sampling Random Peptides.* We describe how to generate a random peptide. First, we randomly select a peptide mass $m' \in [m - \delta, m + \delta]$ using the following probability:

$$A[m']/ \sum_{i=m-\delta}^{m+\delta} A[i].$$

With m', we generate amino acids from the last one to the first one. The last amino acid aa is selected using the following probability:

$$A[m' - mass(aa)]/A[m'].$$

We repeat this process to generate a random peptide with mass m'. We sample 1,000 random peptides, calculate the HMM scores for them, and compute the mean and standard deviation of the normal distribution. This step is done once for a spectrum.

3. *Calculating the Z-score.* We use the above normal distribution to calculate the Z-score for each HMM score. The Z-score is a measure of the distance in standard deviations of a sample from the mean.

This approach to the significance of a score is unique in that it assumes a database of random sequences, and computes the ranking of a score as its significance. Given a specific database, we can calculate an E-score, the expected number of peptides with scores better than the Z-score.

3   Results

Training of Parameters. Using 1,310 training spectra, the EM algorithm converges after 40 iterations. The parameters for the normal distribution of match tolerance are $\mu = -0.0385$ and $\sigma = 0.119$. The parameters for the exponential distributions of peak intensities are $\lambda = 4.223$ for b ions, and $\lambda = 6.421$ for y ions.

Preliminary Test. The purpose of the preliminary test is to study whether we should consider ion types other than b and y ions and how we incorporate them into the model. We start with the simplest model (Model 1) which includes only b and y ions. Then add other ion types, b-H_2O, y-H_2O, a, b^{2+} and y^{2+}. A mass peak that matches to one of these ion types is given a probability of 1 (Model 2). In this way, we basically exclude these peaks from the HMM scoring function. The last model (Model 3) considers each of these ions independently and computes the probability of observing a mass peak (T, I) given an ion. We implement these three models and train the corresponding parameters for the HMMs. We use 1,310 spectra from ISB for training, and 157 spectra for testing. Then we search each of these spectra in MSDB using the three scoring functions based on Models 1, 2 and 3. We consider the prediction of a spectrum an error if the best peptide found in MSDB is not the real peptide. The number of errors made by Models 1, 2 and 3 are 18, 8, and 7, respectively. The results prove that

Table 1. Comparison of HMMscore and MASCOT

Groups	# of Testing Spectra	Errors by HMMscore	Errors by MASCOT
1	170	2	16
2	173	11	17
3	171	4	12
4	176	8	14
5	171	4	14
6	181	7	18
7	182	6	14
8	171	7	14
9	166	8	15
10	177	5	14
Sum	1,738	62 (3.6%)	148(8.5%)

including the probability of observing other ion types will improve the accuracy of the scoring function.

Comparison with MASCOT. MASCOT [4] is generally considered to be the best available program for mass spectrometry database search. We compare the accuracy of our program against that of MASCOT. The testing sets are constructed as follows. We randomly partition the 1,649 curated charge +2 mass spectra from ISB into a training set and a test set using the rules described in Section 2.1 to minimize bias. We repeat this partition 10 times on the original data set to obtain 10 groups of training and testing sets. For each group, we train the HMM and use the trained HMM for prediction. The parameters trained by the EM algorithm are very similar across all the training sets. We also run MASCOT from its website on these testing spectra. Both programs use MSDB for searches. Table 1 lists the number of testing spectra and the number of errors by HMMscore and MASCOT for each of the ten groups. It can be seen that HMMscore outperforms MASCOT in each group. The average error rate for HMMscore (3.6%) is less than half that of MASCOT (8.5%).

Accessing False Positives. It is also important to estimate the false positive rate of HMMscore for unknown mass spectra. To calculate the false positive rate, we need to construct a *positive* set with annotated mass spectra and a database, as well as a *negative* set in which spectra and the database do not match. We choose the above ISB data using 18 purified proteins and the human protein database plus the 18 purified proteins as a positive set. At the same time, we choose a set of published mass spectra of human proteins from ISB using ICAT experiments [16] and the reversed human protein database as the negative set. This human ICAT data set contains 21,592 charge +2 spectra from 41 runs. The reverse human database contains the reversed protein sequences of human protein sequences. Given a threshold, any match found in the negative set is incorrect.

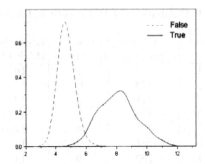

Fig. 5. HMM Z-score distribution for the negative set and the positive set

Table 2. False positive rates and true positive rates for different Z-score thresholds

Z-Score	False positive rate	True positive rate
4	88.21%	100%
4.5	61.07%	100%
5	26.74%	99.88%
5.5	7.63%	99.26%
6	1.66%	97.06%
6.5	0.3%	93.50%

The histogram of the Z-score distribution of the positive set and the negative set is shown in Figure 5. The false positive rates and the true positive rates are shown in Table 2. Table 2 reveals that even with a high threshold HMMscore still has a high true positive rate, while the false positive rate becomes very small.

4 Discussion

We develop an HMM-based scoring function, HMMscore, for mass spectra database search. We show that this scoring function is very accurate with a low false positive rate. The comparison with MASCOT was done in a small test set and thus the result is not sufficient to support the claim that HMMscore always outperforms MASCOT. More test data for comparison are needed in order to draw a conclusion. However, the comparison does suggest that a properly designed scoring function can outperform MASCOT under a right setting. Given sufficient training data, we can develop scoring functions for charge +1 and charge +3 spectra. The HMM structure is flexible in such a way that other ion types can be included. Currently, we do not separate charge +2 peptides into mobile, half-mobile and non-mobile due to the limited size of the training data. We do not use the sequence information that is useful for predicting mass peak intensities because we do not have training data for this purpose. How we can incorporate these data into our model remains an open question. Another challenge is to score a mass spectrum with post-translational modifications. The next step of this study is to predict proteins rather than peptides.

Acknowledgements

We thank Andrew Keller and Alexey Nesvizhskii from the Institute of Systems Biology for providing us data sets. This research is partially supported by NIH NIGMS 1-R01-RR16522-01, "New MS/MS approach to protein sequencing and identification", and NSF ITR EIA - 0112934, "Algorithmic study in computational proteomics".

References

1. Aebersold, R., Mann, M.:Mass spectrometry-based proteomics, Nature, Vol 422, 6928 (2003), 198-207.
2. Bafna, V., Edwards, N.:SCOPE: a probabilistic model for scoring tandem mass spectra against a peptide database,Bioinformatics,Vol 17, Suppl 1 (2001),S13-21.
3. Chen, T., et al.:A dynamic programming approach to de novo peptide sequencing via tandem mass spectrometry,J Comput Biol,Vol 8,3 (2001),325-37.
4. Creasy, D. M.,Cottrell, J. S.:Error tolerant searching of uninterpreted tandem mass spectrometry data,Proteomics,Vol 2,10 (2002),1426-34.
5. Dancik, V., et al.:De novo peptide sequencing via tandem mass spectrometry,J Comput Biol,Vol 6,3-4 (1999),327-42.
6. Eng, J. K., et al.: An approach to correlate tandem mass spectral data of peptides with amino acid sequences in a protein database,Journal of the American Society for Mass Spectrometry,Vol 5,11 (1994),976-989.
7. Havilio, M., Haddad, Y.,Smilansky, Z.:Intensity-based statistical scorer for tandem mass spectrometry,Anal Chem,Vol 75,3 (2003),435-44.
8. Keller,A., et al.:Empirical statistical model to estimate the accuracy of peptide identifications made by MS/MS and database search,Anal Chem,Vol 74,20 (2002),5383-92.
9. Keller,A., et al.:Experimental protein mixture for validating tandem mass spectral analysis,Omics,Vol 6,2 (2002),207-12.
10. Lu, B., Chen, T.,A suboptimal algorithm for de novo peptide sequencing via tandem mass spectrometry,J Comput Biol,Vol 10,1 (2003),1-12.
11. Ma, B.,Doherty-Kirby,A., Lajoie, G.,:PEAKS: powerful software for peptide de novo sequencing by tandem mass spectrometry,Rapid Commun Mass Spectrom,Vol 17,20 (2003),2337-42.
12. Perkins, D. N., et al.: Probability-based protein identification by searching sequence databases using mass spectrometry data,Electrophoresis,Vol 20,18 (1999),3551-67.
13. Pevzner, P. A., et al.:Mutation-tolerant protein identification by mass spectrometry,J Comput Biol,Vol 7,6 (2002),777-87.
14. Pevzner, P. A., et al.:Efficiency of database search for identification of mutated and modified proteins via mass spectrometry,Genome Res,Vol 11,2 (2001),290-9.
15. Rabiner, L. R.:A Tutorial on Hidden Markov-Models and Selected Applications in Speech Recognition,Proceedings of the IEEE,Vol 77,2 (1989),257-286.

16. Von Haller, P. D., et al.:The Application of New Software Tools to Quantitative Protein Profiling Via Isotope-coded Affinity Tag (ICAT) and Tandem Mass Spectrometry: II. Evaluation of Tandem Mass Spectrometry Methodologies for Large-Scale Protein Analysis, and the Application of Statistical Tools for Data Analysis and Interpretation,Mol Cell Proteomics,Vol 2,7 (2003),428-42.
17. Tabb, D.L., Saraf, A., Yates, J. R. 3rd. :GutenTag: high-throughput sequence tagging via an empirically derived fragmentation model. Anal Chem. Vol 75, 23 (2003), 6415-21.
18. Yates, J. R. 3rd et al.:Method to correlate tandem mass spectra of modified peptides to amino acid sequences in the protein database,Anal Chem,1995,67,8,1426-36.
19. Zhang, N., et al.:ProbID: a probabilistic algorithm to identify peptides through sequence database searching using tandem mass spectral data,Proteomics,Vol 2,10 (2002),1406-12.
20. Zhang, W.,Chait, B. T.:ProFound: an expert system for protein identification using mass spectrometric peptide mapping information,Anal Chem,Vol 72,11 (2000),2482-9.

EigenMS: De Novo Analysis of Peptide Tandem Mass Spectra by Spectral Graph Partitioning

Marshall Bern and David Goldberg

Palo Alto Research Center,
3333 Coyote Hill Rd., Palo Alto, CA 94304, USA
{bern, goldberg}@parc.com

Abstract. We report on a new *de novo* peptide sequencing algorithm that uses spectral graph partitioning. In this approach, relationships between m/z peaks are represented by attractive and repulsive springs, and the vibrational modes of the spring system are used to infer information about the peaks (such as "likely b-ion" or "likely y-ion"). We demonstrate the effectiveness of this approach by comparison with other *de novo* sequencers on test sets of ion-trap and QTOF spectra, including spectra of mixtures of peptides. On all data sets we outperform the other sequencers. Along with spectral graph theory techniques, EigenMS incorporates another improvement of independent interest: robust statistical methods for recalibration of time-of-flight mass measurements. Robust recalibration greatly outperforms simple least-squares recalibration, achieving about three times the accuracy for one QTOF data set.

1   Introduction

Mass spectrometry (MS) has emerged as the dominant technique for analysis of protein samples. The basic, single-MS, technique digests the protein into peptides, ionizes the peptides, and then measures the mass-to-charge ratio (m/z) of the ions. In *tandem* mass spectrometry (MS/MS), the initial round of m/z measurement is followed by fragmentation of peptide ions of a selected small m/z range and then by m/z measurement of the fragment ions. Protein identification can be accomplished in several ways [22, 24, 38].

- **Mass Fingerprinting** matches single-MS peaks with peptide masses from a database of known proteins.
- **Database Search** extracts candidate peptides of about the right mass from a protein database. It then scores these candidates by matching observed MS/MS peaks with predicted fragments.
- **De Novo Peptide Sequencing** finds the amino acid sequence among all possible sequences that best explains observed MS/MS peaks. The result can then be used as a probe to search for homologous proteins in a database.

S. Miyano et al. (Eds.): RECOMB 2005, LNBI 3500, pp. 357–372, 2005.

- **Sequence Tagging (or Partial De Novo)** computes a short subsequence (a "tag") using *de novo* methods. It extracts candidates containing this subsequence from a database, and scores the candidates using a polymorphism- and modification-tolerant matching algorithm.

The dominant method is database search, as embodied in programs such as SEQUEST [15] and Mascot [30], because it is more sensitive than mass fingerprinting, capable of identifying a protein from only one or two peptides, and can use much lower quality spectra than *de novo* sequencing.

Database search, however, fails in certain situations, and *de novo* analysis offers a viable alternative. First, if databases are of low quality, as in the case of unsequenced organisms [36], then *de novo* analysis can outperform database search. Second, sometimes the unknown peptide—rather than the parent protein—is itself the object of interest, for example, in screening for pharmaceutical peptides. Finally and most importantly, if a peptide includes a polymorphism or post-translational modification, then many of its spectral peaks will be shifted from their usual positions, and straightforward database search will fail.

It is possible to adapt database search to find polymorphisms and modifications, either by expanding all [28, 45] or part [10] of the database to include anticipated modifications, or by using a fast mutation-tolerant matching algorithm [31, 32]. These methods, however, have limitations in the modifications they can find, in speed, or in sensitivity and specificity. *De novo* analysis has the potential to overcome these limitations. In this approach, the results of a *de novo* analysis—either a sequence tag [21, 29, 35, 40] or a complete but error-ful sequence [19, 34]—are used to extract candidates from the database, which are then scored with a sensitive mutation-tolerant matching algorithm. Searle *et al.* [34] recently reported excellent identification rates for a combination of the PEAKS *de novo* sequencer [26, 27] and a specialized search tool for aligning sequences based on masses. (BLAST is a relatively weak search tool for this purpose, because it is not aware of common *de novo* sequencing errors, such as transposition and close-mass substitutions.)

In this paper we apply two mathematical techniques popular in other fields to *de novo* analysis. The first technique is spectral graph partitioning (Section 3), which we use to disambiguate b- and y-ion peaks before generating candidate amino acid residue sequences. The second technique is robust regression (Section 4), which we use to recalibrate time-of-flight mass measurements based on tentative peak identifications. The resulting sequencer, called EigenMS, outperforms PEAKS [26, 27] and greatly outperforms Lutefisk [42] on our test data. As with other *de novo* sequencers, performance depends upon the type of mass spectrometer, with much better results on quadrupole time-of-flight (QTOF) spectra, which have on the order of ten times better mass accuracy than ion-trap spectra.

2   Background and Our Contributions

Almost all *de novo* sequencers factor the problem into two phases: *candidate generation* and *scoring*.

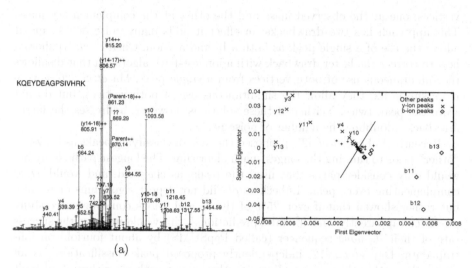

(a)

Fig. 1. (a) An ion-trap spectrum of the peptide KQEYDEAGPSIVHRK. This high-quality spectrum has few noise peaks and many b-ions (prefixes) and y-ions (suffixes). The peak marked y_{10} is the ion EAGPSIVHRK and b_{11} is KQEYDEAGPSI. Peaks for y_6, b_9, and b_{10} are missing due to poor fragmentation around proline. Peaks marked **??** are chemical noise peaks, not identifiable as any of the standard ion types (a-, b-, y-, neutral losses, internal fragments, etc.). *De novo* analysis infers residue identities from the mass differences between peaks. For example, the difference between b_{11} at 1218.45 and b_{12} at 1317.55 is 99.10, which matches the 12th residue V (true mass 99.07). Note that V could also be deduced from the complementary ions y_4 and y_3.   (b) This plot shows spectral graph partitioning for separating the b- and y-ion peaks. Splitting the peaks by the line shown (or simply by first eigenvector) gives a perfect partition into two sets, one containing all 10 y-ions and the other containing all 8 b-ions. This removes the need to complement each peak in the longest-path algorithm

As shown in Figure 1(a), residue identities can be inferred from the mass differences between peaks. The informative peaks are those corresponding to *b-ions*, prefixes, and *y-ions*, suffixes of the residue sequence. In Figure 1, y_3 is HRK and b_{12} is KQYEDEAGPSIV. These two ions are *complementary*, meaning that together they sum to the mass of the *parent ion*, the entire peptide.

An ideal spectrum would contain a complete "ladder" of b-ions or y-ions from which we could simply read off the residue sequence or its reversal. Hence candidate generation is usually formulated as a longest (or best) path problem [4, 5, 7, 11, 26, 42] in a *peak graph*, which has a vertex for each peak in the spectrum and an edge connecting two peaks if they differ by the mass of a residue. Due to missed cleavages, peak graphs might use vertices corresponding to small mass ranges rather than peaks [27], and/or include edges between vertices differing by the mass of a pair of residues. It is very rare, however, for a spectrum to have a complete ladder of b- or y-ions, and hence most sequencers attempt to form a ladder from a mixture of the two types. Lutefisk [42] turns each peak into two vertices, one at the observed mass and the other at the complementary mass. This approach has two drawbacks: in effect it adds many noise peaks, and it allows the use of a single peak as both a b- and a y-ion. Chen *et al.* [7] showed how to correct the latter drawback with a longest-path algorithm that disallows the simultaneous use of both vertices from a single peak. Ma *et al.* [26] gave a more subtle fix; they allow the simultaneous use of both vertices, but do not score the peak twice. Neither of these solutions, however, addresses the larger drawback—doubling the number of noise peaks.

Recently, Lubeck *et al.* [25] proposed the idea of classifying peaks as b-, y-, or "other" prior to running the longest-path algorithm. The longest-path algorithm would only consider paths that used the peaks as classified, and would avoid complementing every peak. Lubeck *et al.* did not report algorithms or results, but rather showed that if even 25% of the b- and y-ions could be successfully predicted (apparently with 0% false predictions), this would improve the success rate of their *de novo* sequencer (called PepSUMS) by about fourfold on ion-trap data. Day *et al.* [12] independently proposed peak classification (as an alternative rather than as an adjunct to the longest-path algorithm), and built a three-class classifier keying on b/y pairs and the "neutral loss neighborhood" (about 30 Daltons) around each peak; they applied this classifier to sequence tagging, but did not report error rates. Yan *et al.* [44] also came up with the same idea, and formulated the classification problem as a graph tripartition problem. They gave an exponential-time algorithm for an exact solution to the problem. They report about 10% error rate in classifying about 40 peaks per spectrum on extremely high accuracy (1–3 ppm) Fourier transform mass spectrometry (FTMS) data.

Actually classifying peaks as b-ions, y-ions and "others", however, is unnecessary to improve *de novo* analysis. When using classification in conjunction with the longest-path algorithm, it suffices to solve a much easier problem: divide the peaks into two groups, such that one group contains all the b-ions and the other contains all the y-ions. Such a bipartition allows one group to be complemented (subtracted from the parent ion mass) and thereby removes the *b/y ambiguity* of peaks without doubling the number of noise peaks. We apply spectral graph partitioning—a polynomial-time algorithm—to this bipartition problem. We obtain much lower error rates than Yan *et al.* [44], for example, 0.6% error rate on our high-quality QTOF data set, which has about 10–40 ppm mass accuracy at the time of the peak partitioning. (This comparison is unfair in the sense that we are comparing our two-class results with their three-class results.) After preprocessing using spectral graph partitioning, we generate only 296 candidates, including the correct sequence, for the example in Figure 1. Complementing each peak, however, the same path algorithm does not generate the correct sequence among the top 10,000, usually going wrong at the missing y_6/b_9 cleavage.

Our approach to the scoring phase of *de novo* sequencing is fairly conventional, scoring the number, intensity, and mass accuracy of identified peaks. We use fragmentation and intensity statistics as in [14, 20, 39, 40]. Our most important contribution to scoring is extremely accurate recalibration of time-of-flight mass measurements, achieving errors with standard deviation 0.0034 Daltons on our high-quality QTOF data set, for which straightforward least-squares recalibration achieves 0.0106, three times worse.

3   Candidate Generation Using Spectral Graph Theory

Our approach uses two graphs, a *spring system* for the partitioning problem and a *peak graph* for the longest path problem. The basic steps of the candidate generation phase are the following:

1. In a preprocessing step, estimate the parent ion mass using complementary pairs, and—in the case of time-of-flight spectra—perform a coarse recalibration of measured masses.
2. Select the peaks of the spectrum that will be vertices in the spring system.
3. Place positive and negative weighted edges between vertices in adjacency matrix $\mathbf{A} = \{a_{ij}\}$.
4. Compute the first two eigenvectors *E1*, *E2* (smallest eigenvalues) of the generalized eigenvalue problem: $(\mathbf{D} - \mathbf{A})x = \lambda \mathbf{D}x$, where $\mathbf{D}$ is the diagonal matrix with $d_{ii} = \sum_j |a_{ij}|$.
5. Project each vertex v_i of the spring system into $\mathbb{R}^2$ by mapping v_i to $(E1_i, E2_i)$. Partition the vertices into Y and B with a line through the origin.
6. Complement the masses of the vertices in Y, obtaining $\overline{Y}$. Build a peak graph on the vertices in B and $\overline{Y}$, with edges between two peaks that differ by the mass of one or two amino acid residues. Larger masses are allowed for the initial and final gaps.
7. Compute the longest path in the peak graph, and generate a list of candidate amino acid sequences consistent with the longest path.

We now give more details on each of the steps. In Step 1, for coarse recalibration, we compute corrected masses as simply proportional to the measured masses, using pairs of peaks that differ by residue masses to determine the multiplier.

In Step 2, we do not choose all peaks, but only those peaks with high intensity (relative to those of similar m/z) that do not appear to be isotopes, doubly charged, or water losses (determined by their mass and intensity relationships to nearby peaks). Generally we consider only the top $8n$ peaks by intensity, where n is the length of the residue sequence, estimated simply by $ParentMass/100$. Spectral graph partitioning is fast enough to handle thousands of peaks, but accuracy falls off if the number of true b- and y-ion peaks drops below about 20% of all peaks in the spring system.

In Step 3, we use several different types of edges. We place a positive edge between two vertices if they differ by the mass of an amino acid residue. The weight of the edge is $Strong \cdot MassAcc$, where $Strong$ is a constant to be discussed later, and $MassAcc$ is 1.0 for an exact match to the residue mass, falling linearly down to 0.0 at a user-supplied tolerance (0.1 for QTOF and 0.3 for ion-trap).

Similarly we place a positive edge of weight $Weak \cdot MassAcc$ between two vertices if they differ by the mass of a pair of residues. We place a negative edge of weight $Strong$ if two peaks differ by an "impossible" mass, that is, a mass that is not the sum of residue masses (up to the tolerance). We place a negative edge of weight $Strong \cdot MassAcc$ between two complementary peaks.

We also create artificial vertices, called Art-B and Art-Y, connected by a doubly $Strong$ negative edge. If a peak has the mass of a pair of residues (plus a proton), it is attached to Art-B; if it has the complementary mass, it is attached to Art-Y; weights are $Weak \cdot MassAcc$. We connect a vertex to Art-B if it has mass 27.995 greater than another high-intensity peak; this mass (that of carbon monoxide) is the difference between a b-ion and its associated a-ion. The weight is $Strong \cdot MassAcc$ if the peak has the mass of a pair of residues (indicative of the frequently observed a_2 / b_2 pair), and $Weak \cdot MassAcc$ otherwise. Similarly we place $Strong$ negative edges to Art-B or Art-Y if a low-mass or low-complementary-mass peak cannot be a b- or y-ion. Finally a doubly charged peak such as y_{14}^{++} in Figure 1(a) causes its singly-charged counterpart to be connected to Art-Y with a $Weak$ edge, as doubly charged y-ions are more common than doubly charged b-ions. In high-resolution spectra, doubly charged ions can be recognized by the spacing of their isotope series.

In Step 4, we used a symmetric eigenvalue routine based on the LASO software and iterative Lanczos algorithm [13, 23]. This algorithm is overkill; our matrices are smaller than 200×200 and not very sparse (about 30% non-zeros).

Spectral graph partitioning used in Step 5 has a long history [8, 13, 16, 37], and several papers give mathematical analyses to explain its empirical success. The first eigenvector $E1$ solves a least-squares problem: find vector x (an embedding of the spring system in $\mathbb{R}$) to minimize

$$\frac{1}{x^t \mathbf{D} x} \left(\sum_{i,j} (x_i - x_j)^2 a_{ij} \right) = \frac{x^t (\mathbf{A} - \mathbf{D}) x}{x^t \mathbf{D} x}.$$

This follows from (8.7.4) in [18]. By Hooke's law, the expression to be minimized is the potential energy of a set of attractive (spring constants $a_{ij} > 0$) and repulsive ($a_{ij} < 0$) springs connecting vertices with coordinates $\{x_i\}$. For our partitioning, however, we use both $E1$ and $E2$ to determine an embedding of the spring system in $\mathbb{R}^2$. We first fit a line ℓ to the points in $\mathbb{R}^2$ using ordinary least squares, and we then use the line orthogonal to ℓ and passing through the origin as the partitioning line, as shown in Figure 1(b). This partitioning method outperforms the first eigenvector alone; the first eigenvector gives 2.5% error rate for QTOF1 rather than the 0.6% reported in Table 1. Higher eigenvectors correspond to vibrational modes [8, 13]. Researchers in other fields [1] also report better partitioning results using more than one eigenvector.

In Step 6, not only do we complement one group, we reweight the vertices to reflect information gained from the spring system. Specifically, a peak v_i is given the weight $1/(2 \cdot YRank(v_i) + Rank(v_i))$, where $Rank(v_i)$ is its intensity rank (1 for most intense), and $YRank(v_i)$ is its position projected along line ℓ, for example, 1 for y_3, and 2 for y_{12} in Figure 1(b). The vertices v_i in group B all have $YRank(v_i)$ equal to $|Y|$. This choice seems strange—why not count back down after crossing the origin?—but it gave superior results, and we shall explain the reason below.

In Step 7, we compute the best path using Bellman-Ford dynamic programming [7], assuming that each peak has been correctly complemented. We compute the single best path 6 times, varying which half of the partition is B and which is Y (occasionally the two halves are switched), and whether the sequence ends in R, K, or an unspecified letter. (R and K are the common C-terminus residues in tryptic digests.) Other *de novo* sequencers compute the k best paths for fairly large values of k; for example, PEAKS [26, 27] computes 10,000 paths. (We used $k = 1$ in order to demonstrate the great effectiveness of spectral graph partitioning; the released version of EigenMS will use larger k.) We fill in the path of masses with residues to create candidate sequences. A single path can create a number of candidates, for example, a mass gap of 262.1 Daltons could correspond to DF, FD, YV, VY, or MM. We generate candidate sequences for each longest path, typically 1–1000 candidates. Long initial and final gaps can cause the number of candidate sequences to explode; we cut off the number at 10,000 by choosing the longest subpath of masses with at most 10,000 candidates and leaving mass gaps at the beginning and/or end of the mass path.

Spectral graph partitioning is remarkably successful at solving the problem of b/y ambiguity, as shown in Table 1. We count an error if a true b-ion peak ends up in Y or a true y-ion peak ends up in B. Since the path is computed under both assumptions, we switch the names of B and Y if this gives a lower number of errors. Switching is uncommon; for example, only 3 of 74 QTOF1 spectra had the partition backwards. One source of errors is unlucky coincidences, such as the b_5 ion having the same mass as y_8 (within the resolution of the instrument and peak-picking algorithm).

Resolving the b/y ambiguity is a tremendous aid in sequencing. On 26 of the 74 QTOF1 spectra the correct sequence is the unique sequence for the (correctly oriented) best path. As a point of comparison, if we leave out steps 3–6 and

Table 1. Spectral graph partitioning results. The four data sets are described further in Section 5. "Mass Accuracy" gives approximate accuracy (maximum error) after the initial recalibration in Step 1. For MIX we only scored the dominant peptide, envisioning a system that would identify one peptide at a time. An error is a b-ion in Y or a y-ion in B. The column labeled "% Errors" gives the combined error rate for b- and y-ion peaks over all spectra. "# Spectra Correct" gives the number of spectra without any errors. Errors tend to be concentrated in the lower-quality spectra, which are hard to identify anyway

	Mass Accuracy	b-ions	y-ions	% Errors	# Spectra Correct
QTOF1	0.03 Dalton	3 / 268	2 / 546	0.6%	69 / 74
QTOF2	0.1 Dalton	30 / 283	22 / 593	5.9%	53 / 81
LTQ	0.3 Dalton	70 / 572	50 / 504	11.2%	40 / 96
MIX	0.2 Dalton	2 / 42	6 / 61	7.8%	3 / 10

build a peak graph containing the complements of all the peaks selected in Step 2, the single best path is complete and correct on none of the QTOF1 spectra.

Now we return to an issue deferred: how do we set *Weak* and *Strong*? We simply set *Strong* to be 6 times *Weak*, because (at 0.2 Dalton resolution) there are about 6 times as many distinct residue-pair masses as distinct residue masses. Surely, this naive approach could be improved. For example, Yan *et al.* trained positive weights based on *in silico* peptides from a given genome. We found, however, that graph partitioning is very robust under different weights. For example, setting *Strong* = *Weak* = 1 gives 36 b-ion errors and 27 y-ion errors on QTOF2 instead of the 30 and 22 reported in Table 1. And setting *Weak* = 0 gives 34 and 32 errors respectively.

Finally, why is the negative weight between two peaks separated by an impossible mass only *Strong* rather than much stronger or even infinite? If we use extremely strong impossible edges, then a chemical noise peak with, say, 3 impossible edges to y-ion peaks and 5 impossible edges to b-ion peaks will be pushed far to the Y side (where it will then push away some real y-ions). This reasoning originally led us to attempt a tripartition, b, y, and "other", as in [44], in which "other" peaks would neither attract nor repel each other. (This work was actually independent of [44].) None of our attempts, however, gave error rates as good as those in Table 1. So far the power of spectral graph partitioning compensates for its lack of flexibility in classification logic.

We now believe that bipartitioning is the most robust formulation of the peak classification problem. Many spectra have only two significant classes: y-ions and "others", or (less commonly) b-ions and "others". (This observation explains why it is best to leave $YRank(v_i) = |Y|$ for all of B.) Some spectra have more than three classes; this occurs if there is more than one peptide, as in the MIX data set, or if there are many internal fragments (contiguous subsequences from the middle of the residue sequence). Such spectra are difficult for the pure longest-path approach (see PEAKS's performance on MIX), which can easily "jump tracks" from one peptide to another. With spectral graph partitioning, however, the strongest class (largest number of intense peaks) repels all the others, and the attraction between noise peaks, such as internal fragments, that are separated by residue masses, is an aid to classification.

4   Scoring Using Robust Recalibration

A *scorer* is a real-valued function taking two arguments: a spectrum and a candidate sequence. Like other scorers [3, 14, 20, 26, 27, 40, 42], EigenMS's scorer sums benefits for observed m/z peaks predicted by the sequence, along with penalties for predicted ions not observed. As above, we use a mass tolerance to decide if an observed peak matches a predicted ion. For time-of-flight mass measurement, the tolerance used in scoring can be much tighter than the one used in candidate generation, because recalibration of observed masses based on tentatively identified peaks can correct most of the measurement error [17, 27, 42]. A key idea, used in both Lutefisk and PEAKS, is to recalibrate the spectrum with respect to each candidate sequence. The correct sequence recalibrates the spectrum well, so that a large number of peaks correspond to predicted ions, whereas an incorrect sequence recalibrates the spectrum less well, and a smaller number of peaks correspond to ions. Reasonable tolerances after recalibration are 0.015 for QTOF1, 0.080 for QTOF2, and 0.35 for ion-trap.

In our scorer, the benefit for a predicted peak is a product of three factors: *MassAcc*, *IntensityFactor*, and *APriori*. As above, *MassAcc* is 1.0 for peaks with zero error and then falls off linearly, reaching 0 at the mass tolerance. Thus for QTOF1 spectra, this triangular window has roughly the shape of the "Weak Peaks" curve in Figure 3. *IntensityFactor* for a peak p is max$\{0, 32 - (2/n) \cdot Rank(p)\}$, where n is the length of the candidate sequence. Thus for a 10-residue sequence the top 160 peaks are scored. The constants 32 and 2 were chosen so that *IntensityFactor* reflects the log likelihood that a peak will be a b- or y-ion [6]. Finally, *APriori* reflects the probability of observing the ion, based upon its position within the peptide and the residues on either side of the cleavage. We used the statistics from [39] to set *APriori* for b- and y-ions. Generally we simplified the statistics; for example the only residue that changes *APriori* is proline. We have not seen statistics for a-ions, neutral losses, doubly charged ions, and so forth, so we set *APriori* for these to be the same small value. The penalty for not observing a predicted ion is simply twice *APriori*.

Of course, all of the above is rather *ad hoc*. Our primary contribution to scoring, however, is very well founded. We use Least Median of Squares (LMS) regression for recalibration. This method [33] finds a regression line that minimizes the median squared residual (rather than the mean, as in ordinary least squares) for a set of tentatively identified peaks. In Figure 2, the tentatively identified peaks are all peaks within 180 ppm of a predicted a-, b-, y-, or water loss ion. This first regression line is used to classify points as *outliers* and *non-outliers*. For statistical efficiency [33, 46] ordinary least-squares is then used to fit a second regression model (not necessarily linear) to the non-outliers. We tried both linear and cubic models for the second regression, but chose to use linear as it gave slightly better scoring results. Observing that intense peaks are intrinsically more accurate than weak peaks (due to better counting statistics), we also use weighted least squares in the second regression.

In Figure 3 we show the distribution of mass errors of all identified a-, b-, and y-ion peaks, and water losses of these ions, over all 74 recalibrated QTOF1 spectra. Here a peak p is "identified" if it is within .015 Daltons of the mass of a predicted ion; it is "intense" if $Rank(p) \leq 20$ and "weak" if $Rank(p) > 20$. Almost all intense peaks are within ±0.005 Daltons, exceeding the "theoretical" mass accuracy of 0.010 for QTOF instruments [38].

For comparison, ordinary (unweighted) least-squares line fitting, as used by Lutefisk and PEAKS, gives significantly worse results. For the 74 QTOF1 spectra, one round of least-squares—that is, fitting a line to all peaks within 180 ppm of predicted peaks—gives mass errors of 0.0106 compared to 0.0034 for the LMS procedure. These numbers are the medians (over spectra) of the standard deviation (over peaks within a spectrum) of the residuals after recalibration. Two

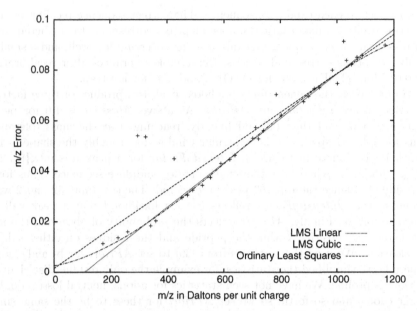

Fig. 2. This plot shows the mass errors of all peaks in a QTOF1 tandem mass spectrum of KVPQVSTPTLVEVSR within 180 ppm of the true mass of an a-, b-, or y-ion, or a water loss from such an ion. There are outliers—likely misidentifications—at masses 297, 325, and 425. Both linear and cubic LMS recalibration give almost all errors below .004. The outliers tilt the ordinary least-squares regression line, giving a large number of errors in the range .005–.010 Daltons

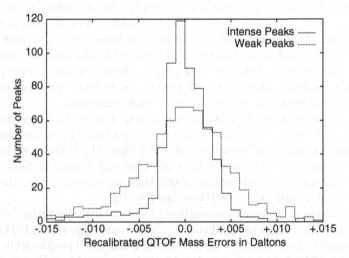

Fig. 3. This histogram shows the mass errors of peaks in all 74 QTOF1 spectra, after recalibration. Intense peaks (those with *Rank* at most 20) are more accurate than weak peaks (*Rank* > 20)

rounds of ordinary least-squares, using thresholds of 180 ppm and .025 Daltons, does much better than one round, achieving 0.0039. But even two rounds of ordinary least squares is significantly worse than LMS on the worst QTOF1 spectrum, 0.0076 instead of 0.0058.

Scoring time-of-flight spectra is surprisingly sensitive to recalibration. We scored each of the 74 QTOF1 spectra against 200-1000 likely decoys, sequences with transposed letters and close-mass substitutions such as K (128.095 Daltons) for Q (128.059) and DY (278.090) for MF (278.109). No recalibration at all gives 22.5 correct answers, meaning that the highest scoring sequence is indeed the true sequence. (If there is two-way tie for the top score, we count one-half.) Ordinary least-squares gives 39, two-round least-squares gives 54, and our LMS procedure gives 65 correct answers.

5   Experimental Results

We measured the performance of EigenMS on the four data sets shown in Table 2. The characteristic of the data sets most pertinent to performance is the m/z accuracy. Spectra in QTOF1 are initially accurate to about 150 ppm, but as we saw above can be recalibrated to below 0.010. QTOF2 spectra are initially accurate to about 300 ppm, but can be recalibrated to about 0.060. LTQ ion-trap spectra are accurate to about 0.3 and MIX to about 0.2, with or without recalibration. More specifically, the standard deviation of mass errors for well-identified peaks in LTQ runs from 0.047 to 0.153 from the best to worst spectrum with a median of 0.107. For MIX the same numbers are 0.046 to 0.099 with a median of 0.066. These data sets, however, are from exactly the same set-up and MS instrument. Why the difference? The spectra in MIX were taken in "profile" mode, that is, we received numbers for all m/z bins (spaced at intervals of .09), rather than just the peaks picked by the instrument software. We then ran our own peak picker, which models the shapes of peaks in order to improve accuracy and resolution. The 40% better accuracy in MIX suggests that the instrument manufacturer's software leaves room for improvement.

Table 2. Four test sets. "Final Acc" gives the approximate mass accuracy (maximum error) after recalibration with reference to the correct sequence. QTOF2 is from a tissue sample (mouse *zona pellucida*); the others are all from mixtures of reference proteins. (This may explain the big difference in mass accuracies between QTOF1 and QTOF2.) MIX contains two peptides per spectrum with close parent-ion m/z's. In all cases, identifications were made by database search against a large decoy database

Name	Mass Spec Instrument	Digest	Final Acc	#Spectra	#Residues
QTOF1	Micromass Q-TOF II	Tryptic	0.01	74	7–15
QTOF2	ABI QSTAR Pulsar i	Tryptic	0.06	81	6–23
LTQ	ThermoFinnigan LTQ linear ion-trap	Various	0.3	96	8–29
MIX	ThermoFinnigan LTQ linear ion-trap	Various	0.2	10	10–21

Table 3. Results on the four test sets. We score only the dominant peptides in MIX; even so, MIX spectra are difficult because the second peptide's peaks act as noise. "Corr" means that the *de novo* sequencer's top answer is completely correct. "Tag" means that the top answer includes a correct contiguous subsequence of length 5 or greater. "Miss" is everything else (including many decent results such as correct 4-residue subsequences and reversed 5-residue subsequences). In case of a two-way tie for the top answer, we always chose the better answer; we never observed three-way ties

	EigenMS			PEAKS			Lutefisk		
	Corr	Tag	Miss	Corr	Tag	Miss	Corr	Tag	Miss
QTOF1	63	5	6	42	26	6	19	31	24
QTOF2	20	38	23	13	31	37	7	16	58
LTQ	5	30	61	1	33	62	3	3	90
MIX	0	4	6	0	0	10	0	1	9

Table 3 gives results for EigenMS and two well-known *de novo* sequencers, Lutefisk1900 (v1.3.2, 2001) and PEAKS (QTOF1 run in April, 2004, and the other data sets run at Bioinformatics Solutions in October, 2004). We believe PEAKS to be the best *de novo* sequencer currently available. Both Lutefisk and EigenMS—but not PEAKS—sometimes emit answers of the form [548.25]DEAG-PSIVH[284.19], where [548.25] means an unidentified set of residues with total mass 548.25 Daltons. EigenMS uses these "mass gaps" only at the beginnings and ends of sequences, and only when filling in the gap with residues would generate more than 10,000 candidates. The three sequencers have comparable running times, about 1–5 seconds per spectrum.

We consider a sequence partially correct if it contains a correct contiguous subsequence of 5 or more residues. (L for I is correct, but K for Q is not.) This definition is conservative, as even 3- or 4-residue tags can be useful; for example, GutenTag [40] uses multiple (errorful) 3-residue tags. With specialized search tools [19, 34], even lower-quality results often suffice to make an identification, such as two 2-residue tags separated by a transposed pair. Other definitions of partially correct results correlate well with the one we used, so Table 3 gives a good comparison of the three *de novo* sequencers, but a less precise estimate of the number of peptide identifications possible.

As can be seen in Table 3, EigenMS gives the best performance on all four data sets. For QTOF1 the better performance of EigenMS over PEAKS is primarily due to robust recalibration, as PEAKS often generates the correct sequence but does not give it the highest score. For MIX our advantage is spectral graph partitioning. EigenMS has a smaller edge over PEAKS in identification of short peptides (say 5–10 residues), as PEAKS's use of 10,000 candidates seems to give good coverage of the space of possibilities. The combination of EigenMS and PEAKS is more powerful than either one alone. PEAKS finds tags for two of EigenMS's six QTOF1 misses (both on the short peptide AEFVEVTK), and EigenMS gives correct sequences for two of PEAKS's six misses. At least one of PEAKS and EigenMS finds a tag for 44 of the LTQ spectra.

One cannot expect many completely correct answers on LTQ and MIX (ion-trap spectra), because of the low mass accuracy and the lack of data at low m/z. Ion-trap spectra often have no peaks below 300; for example, the first b- or y-ion in Figure 1(a) appears at 404.4. In this spectrum, there are y-ions (y_{12} and y_{13} at 1371.66 and 1500.63) that fill in for the missing b-ions, and vice versa. In other cases, however, we are not so lucky, and no algorithm can successfully bridge such a large gap, as 404.4 matches hundreds of subsequences.

6   Discussion

De novo analysis has traditionally been a niche method, competitive with database search only in rare situations. With the advent of high-accuracy machines (QTOF and FTMS) and advanced software (such as PEAKS), this situation is changing. With better peak picking as mentioned above, better *de novo* analysis as presented here, and error-sensitive search tools [19, 34], partial *de novo*—with the machine deciding just how "partial"—could become the preferred method, even for ion-trap machines. Better fragmentation methods, such as controlled hydrolysis using microwave irradiation [47], could lead to complete *de novo* sequencing of proteins. Spectral graph partitioning could play a key role in *de novo* analysis of long-peptide spectra, as "deconvolution" of b- and y-ion peaks [47], and of mixtures of peptides, appears to be the central algorithmic problem.

Another important direction for future proteomics tools is the integrated analysis of complete data sets, rather than individual spectra. For example, low-intensity peaks can be enhanced relative to noise by combining successive spectra in a liquid chromatography run [2]. Peptide identification can be improved by correlating similar spectra [41], perhaps even archived spectra from previous protein samples. Finally, a practical approach to finding post-translational modifications [10, 31] is to look for modifications of peptides only from proteins already identified by unmodified peptides.

We plan to distribute EigenMS, or at least provide it as a Web service. Check the home page of the first author.

Acknowledgments

We would like to thank Tony Day (Genencor), Anne Dell (Imperial College), and John R. Yates III (Scripps) for providing us with mass spectra; Ravi Kolluri (UC-Berkeley) for spectral graph partitioning advice and software; and Iain Rogers (Bioinformatics Solutions) for running our spectra through PEAKS.

References

1. C. Alpert, A. Kahng, and S. Yao. Spectral partitioning: the more eigenvectors, the better. *Discrete Applied Math.* 90 (1999), 3–26.
2. V.P. Andreev, *et al.* A universal denoising and peak picking algorithm for LC-MS based on matched filtration in the chromatographic time domain. *Anal. Chem.* 75 (2003) 6314–6326.

3. V. Bafna and N. Edwards. SCOPE: a probabilistic model for scoring tandem mass spectra against a peptide database. *Bioinformatics* 17 (2001), S13–S21.

4. V. Bafna and N. Edwards. On *de novo* interpretation of tandem mass spectra for peptide identification. *RECOMB 2003*, 9–18.

5. C. Bartels. Fast algorithm for peptide sequencing by mass spectrometry. *Biomedical and Environmental Mass Spectrometry* 19 (1990), 363–368.

6. M. Bern and D. Goldberg. Automatic quality assessment of peptide tandem mass spectra. *Bioinformatics*, ISMB special issue, 2004.

7. T. Chen, M.-Y. Kao, M. Tepel, J. Rush, and G.M. Church. A dynamic programming approach to de novo peptide sequencing by mass spectrometry. *J. Computational Biology* 8 (2001), 325–337.

8. F.R.K. Chung. *Spectral Graph Theory*, CBMS Series #92, American Mathematical Society, 1997.

9. K.R. Clauser, P.R. Baker, and A.L. Burlingame. The role of accurate mass measurement (+/- 10 ppm) in protein identification strategies employing MS or MS/MS and database searching. *Anal. Chem.* 71 (1999), 2871–2882.

10. D.M. Creasy and J.S. Cottrell. Error tolerant searching of uninterpreted tandem mass spectrometry data. *Proteomics* 2 (2002), 1426–1434.

11. V. Dančik, T.A. Addona, K.R. Clauser, J.E. Vath, and P.A. Pevzner. De novo peptide sequencing via tandem mass spectrometry. *J. Computational Biology* 6 (1999), 327–342.

12. R.M. Day, A. Borziak, A. Gorin. PPM-Chain – de novo peptide identification program comparable in performance to Sequest. *Proc. IEEE Computational Systems Bioinformatics*, 2004, 505–508.

13. J. Demmel. Lecture notes on graph partitioning. http://www.cs.berkeley.edu/~demmel/cs267/lecture20/lecture20.html

14. J.E. Elias, F.D. Gibbons, O.D. King, F.P. Roth, and S.P. Gygi. Intensity-based protein identification by machine learning from a library of tandem mass spectra. *Nature Biotechnology* 22 (2004), 214–219.

15. J.K. Eng, A.L. McCormack, and J.R. Yates, III. An approach to correlate tandem mass spectral data of peptides with amino acid sequences in a protein database. *J. Am. Soc. Mass Spectrom.* 5 (1994), 976–989.

16. M. Fiedler. A property of eigenvectors of nonnegative symmetric matrices and its applications to graph theory. *Czech. Math. J.* 25 (1975), 619–633.

17. J. Gobom, M. Mueller, V. Egelhofer, D. Theiss, H. Lehrach, and E. Nordhoff. A calibration method that simplifies and improves accurate determination of peptide molecular masses by MALDI-TOF MS. *Anal. Chem.* 74 (2002), 3915–3923.

18. Gene H. Golub and Charles F. Van Loan. *Matrix Computations*, 3rd Edition. The Johns Hopkins University Press, 1996.

19. Y. Han, B. Ma, and K. Zhang. SPIDER: software for protein identification from sequence tags with de novo sequencing error. *Proc. IEEE Computational Systems Bioinformatics*, 2004, 206–215.

20. M. Havilio, Y. Haddad, and Z. Smilansky. Intensity-based statistical scorer for tandem mass spectrometry. *Anal. Chem.* 75 (2003), 435–444.

21. M. Havilio. Automatic peptide identification using de novo sequencing and efficient indexing. Poster presentation, *Fifth International Symp. Mass Spectrometry in the Health and Life Sciences*, San Francisco, 2001.

22. M. Kinter and N.E. Sherman. *Protein Sequencing and Identification Using Tandem Mass Spectrometry*. John Wiley & Sons, 2000.

23. C. Lanczos. An iteration method for the solution of the eigenvalue problem of linear differential and integral operators. *J. Res. Nat. Bur. Stand.* 45 (1950), 255–282. http://www.netlib.org/laso/

24. D.C. Liebler. *Introduction to Proteomics: Tools for the New Biology.* Humana Press, 2002.

25. O. Lubeck, C. Sewell, S. Gu, X. Chen, and D.M. Cai. New computational approaches for de novo peptide sequencing from MS/MS experiments. *Proc. IEEE* 90 (2002), 1868–1874.

26. B. Ma, K. Zhang, and C. Liang. An effective algorithm for the peptide de novo sequencing from MS/MS spectrum. *Symp. Comb. Pattern Matching* 2003, 266–278.

27. B. Ma, K. Zhang, C. Hendrie, C. Liang, M. Li, A. Doherty-Kirby, and G. Lajoie. PEAKS: powerful software for peptide de novo sequencing by tandem mass spectrometry. *Rapid Comm. in Mass Spectrometry* 17 (2003), 2337–2342. http://www.bioinformaticssolutions.com

28. M.J. MacCoss *et al.* Shotgun identification of protein modifications from protein complexes and lens tissue. *Proc. Natl. Acad. Sciences* 99 (2002), 7900-7905.

29. M. Mann and M. Wilm. Error-tolerant identification of peptides in sequence databases by peptide sequence tags. *Anal. Chem.* 66 (1994), 4390–4399.

30. D.N. Perkins, D.J.C. Pappin, D.M. Creasy, and J.S. Cottrell. Probability-based protein identification by searching sequence databases using mass spectrometry data. *Electrophoresis* 20 (1999), 3551-3567.

31. P.A. Pevzner, V. Dančik, and C.L. Tang. Mutation-tolerant protein identification by mass spectrometry. *J. Comput. Bio.* 7 (2000), 777–787.

32. P.A. Pevzner, Z. Mulyukov, V. Dančik, and C.L. Tang. Efficiency of database search for identification of mutated and modified proteins via mass spectrometry. *Genome Research* 11 (2001), 290–299.

33. P.J. Rousseeuw and A.M. Leroy. *Robust Regression and Outlier Detection.* John Wiley & Sons, 1987.

34. B.C. Searle *et al.* High-throughput identification of proteins and unanticipated sequence modifications using a mass-based alignment algorithm for MS/MS de novo sequencing results. *Anal. Chem.* 76 (2004), 2220–2230.

35. A. Shevchenko, M. Wilm, and M. Mann. Peptide mass spectrometry for homology searches and cloning of genes. *J. Protein Chem.* 5 (1997), 481–490.

36. A. Shevchenko *et al.* Charting the proteomes of organisms with unsequenced genomes by MALDI-quadrupole time-of-flight mass spectrometry and BLAST homology searching. *Anal. Chem.* 73 (2001), 1917–1926.

37. J. Shi and J. Malik. Normalized cuts and image segmentation. *IEEE Trans. Pattern Anal. Machine Intell.* 22 (2000), 888–905.

38. G. Siuzdak. *The Expanding Role of Mass Spectrometry in Biotechnology.* MCC Press, 2003.

39. D.L. Tabb, L.L. Smith, L.A. Breci, V.H. Wysocki, D. Lin, and J.R. Yates, III. Statistical characterization of ion trap tandem mass spectra from doubly charged tryptic digests. *Anal. Chem.* 75 (2003), 1155–1163.

40. D.L. Tabb, A. Saraf, and J.R. Yates, III. GutenTag: high-throughput sequence tagging via an empirically derived fragmentation model. *Anal. Chem.* 75 (2003), 6415–6421.

41. D.L. Tabb, M.J. MacCoss, C.C. Wu, S.D. Anderson, and J.R. Yates, III. Similarity among tandem mass spectra from proteomic experiments: detection, significance, and utility. *Anal. Chem.* 75 (2003), 2470–2477.

42. J. Taylor and R. Johnson. Implementation and uses of automated de novo peptide sequencing by tandem mass spectrometry. *Anal. Chem.* 73 (2001), 2594–2604.

43. S. Uttenweiler-Joseph, G. Neubauer, S. Christoforidis, M. Zerial, and M. Wilm. Automated de novo sequencing of proteins using the differential scanning technique. *Proteomics* 1 (2001), 668–682.
44. B. Yan, C. Pan, V.N. Olman, R.L. Hettich, Y. Xu. Separation of ion types in tandem mass spectrometry data interpretation – a graph-theoretic approach. *Proc. IEEE Computational Systems Bioinformatics*, 2004, 236–244.
45. J.R. Yates, III, J. Eng, A. McCormack, and D. Schietz. Method to correlate tandem mass spectra of modified peptides to amino acid sequences in a protein database. *Anal. Chem.* 67 (1995), 1426–1436.
46. Z. Zhang. Least median of squares. Web-site tutorial. http://www-sop.inria.fr/robotvis/personnel/zzhang/Publis/Tutorial-Estim/node25.html
47. H. Zhong, Y. Zhang, Z. Wen, and L. Li. Protein sequencing by mass analysis of polypeptide ladders after controlled protein proteolysis. *Nature Biotechnology* 22 (2004), 1291–1296.

Biology as Information

Eric Lander

Director, Broad Institute of MIT and Harvard
320 Charles Street, Cambridge, MA 02141, USA

Biology is being rapidly transformed by the ability to collect comprehensive information about cellular and organismal processes, as well as to systematically manipulate pathways. This keynote talk will survey the biological implications of such research programs as: genome-scale sequence from many related organisms; large-scale human genotyping to study the basis of disease; systematic recognition of signatures of cellular responses; comprehensive analysis of mutations in cancer; and systematic inhibition of genes through RNAi interference. It will address the laboratory and computational issues in fulfilling the opportunities presented by this revolution.

S. Miyano et al. (Eds.): RECOMB 2005, LNBI 3500, p. 373, 2005.
© Springer-Verlag Berlin Heidelberg 2005

Using Multiple Alignments to Improve Gene Prediction

Samuel S. Gross[1,2] and Michael R. Brent[1]

[1] Department of Computer Science and Engineering,
Washington University, St. Louis, MO 63130, USA
`brent@cse.wustl.edu`
[2] Current address: Computer Science Department,
Stanford University, Stanford, CA 94305, USA
`ssgross@cs.stanford.edu`

Abstract. The multiple species de novo gene prediction problem can be stated as follows: given an alignment of genomic sequences from two or more organisms, predict the location and structure of all protein-coding genes in one or more of the sequences. Here, we present a new system, N-SCAN (a.k.a. TWINSCAN 3.0), for addressing this problem. N-SCAN has the ability to model dependencies between the aligned sequences, context-dependent substitution rates, and insertions and deletions in the sequences. An implementation of N-SCAN was created and used to generate predictions for the entire human genome. An analysis of the predictions reveals that N-SCAN's predictive accuracy in human exceeds that of all previously published whole-genome de novo gene predictors. In addition, predictions were generated for the genome of the fruit fly *Drosophila melanogaster* to demonstrate the applicability of N-SCAN to invertebrate gene prediction.

1 Introduction

Two recent developments have increased interest in de novo gene prediction. First, the availability of assemblies of several non-human vertebrate genomes has created the possibility for further significant improvements in human gene prediction through the use of comparative genomics techniques. Second, traditional experimental methods for identifying genes based on 5' EST sampling and cDNA clone sequencing are now reaching the point of diminishing returns far short of the full gene set [1]. As a result, efforts to identify new genes by RT-PCR from predicted gene structures are taking on greater importance.

A major advantage of de novo gene predictors is that they do not require cDNA or EST evidence or similarity to known transcripts when making predictions. This allows them to predict novel genes not clearly homologous to any previously known gene, as well as genes that are expressed at very low levels or in only a few specific tissue types, which are unlikely to be found by random sequencing of cDNA libraries. De novo gene predictors are therefore well-suited to

S. Miyano et al. (Eds.): RECOMB 2005, LNBI 3500, pp. 374–388, 2005.

the task of identifying new targets for RT-PCR experiments aimed at expanding the set of known genes.

One of the first de novo systems to perform well on typical genomic sequences containing multiple genes in both orientations was GENSCAN [3]. GENSCAN uses a generalized hidden Markov model (GHMM) to predict genes in a given target sequence, using only that sequence as input. GENSCAN remained one of the most accurate and widely used systems prior to the advent of dual-genome de novo gene predictors. The initial sequencing of the mouse genome made it possible for the first time to incorporate whole-genome comparison into human gene prediction [2]. This led to the creation of a new generation of gene predictors, such as SLAM [4], SGP2 [5], and TWINSCAN [6, 7, 8], which were able to improve on the performance of GENSCAN by using patterns of conservation between the human and mouse genomes to help discriminate between coding and noncoding regions. These programs are the best-performing de novo gene predictors for mammalian genomes currently available.

Recently, there has been an effort to create systems capable of using information from several aligned genomes to further increase predictive accuracy beyond what is possible with two-genome alignments. Programs such as EXONIPHY [9], SHADOWER [10], and the EHMMs of Pedersen and Hein [11] fall into this category. While many important advances have been made in this area, no system of this type has yet managed to robustly outperform two-sequence systems on a genomic scale.

The gene prediction system presented here, N-SCAN (or TWINSCAN 3.0), extends the TWINSCAN model to allow for an arbitrary number of informant sequences as well as richer models of sequence evolution. N-SCAN is descended from TWINSCAN 2.0, which is in turn descended from the GENSCAN GHMM framework. However, instead of emitting a single DNA sequence like GENSCAN or a target DNA sequence and a conservation sequence like TWINSCAN, each state in the N-SCAN GHMM emits one or more columns of a multiple alignment. N-SCAN uses output distributions for the target sequence that are similar to those used by TWINSCAN 2.0 and GENSCAN. It augments these with Bayesian networks which capture the evolutionary relationships between organisms in the multiple alignment. The state diagram is also extended to allow for explicit modeling of 5' UTR structure as well as other conserved noncoding sequence.

2   Methods

2.1   Overview

Whereas TWINSCAN's GHMM outputs a target genomic sequence and a conservation sequence, N-SCAN's GHMM outputs a multiple alignment $\{\mathbf{T}, \mathbf{I}^1, ..., \mathbf{I}^N\}$ of the target sequence, $\mathbf{T}$, and the N informant sequences, $\mathbf{I}^1$ through $\mathbf{I}^N$. The target sequence consists of the four DNA bases, while the informant sequences can also contain the character "_", representing gaps in the alignment and ".", representing positions in the target sequence to which the informant sequence does not align. The states in the N-SCAN GHMM correspond to functional categories in the target sequence only. Therefore, N-SCAN annotates only one sequence at a time. If annotations are desired for more than one of the sequences in the alignment, the system can be run multiple times with different sequences designated as the target.

One component of the model defines, for each GHMM state, the probability

$$P(T_i | T_{i-1}, ..., T_{i-o}) \tag{1}$$

of outputting a particular base in the target genome at the current position in the sequence, given the previous o bases. Here $T_1, ..., T_L$ is the full target sequence $\mathbf{T}$ and o is the model order. This probability is implicitly dependent on the GHMM state at base i. States represent sequence features, such as start and stop codons, splice sites, and coding sequence. N-SCAN uses these target genome models in combination with a set of Bayesian networks to define, for each state, the probability

$$P(T_i, I_i^1, ..., I_i^N | T_{i-1}, I_{i-1}^1, ..., I_{i-1}^N, ..., T_{i-o}, I_{i-o}^1, ..., I_{i-o}^N) \tag{2}$$

of outputting a column in the alignment given the previous o columns. This is accomplished by multiplying the probability from the target genome model by

$$P(I_i^1, ..., I_i^N | T_i, T_{i-1}, I_{i-1}^1, ..., I_{i-1}^N, ..., T_{i-o}, I_{i-o}^1, ..., I_{i-o}^N) \tag{3}$$

This quantity can be computed from the Bayesian network associated with the state.

We assume that the probability of outputting a base in the target sequence is independent of the values of the previous o positions in all of the informants, given the values of the previous o positions in the target. That is,

$$P(T_i | T_{i-1}, ..., T_{i-o}) = P(T_i | T_{i-1}, I_{i-1}^1, ..., I_{i-1}^N, ..., T_{i-o}, I_{i-o}^1, ..., I_{i-o}^N) \tag{4}$$

Given (4), we can multiply (1) by (3) to obtain (2).

2.2   Phylogenetic Bayesian Networks

The Bayesian network representation used in N-SCAN is similar to the phylogenetic models described in [12], with a few important differences. First, the N-SCAN model uses a six-character alphabet consisting of the four DNA bases plus characters representing gaps and unaligned positions. In addition, the substitutions between nodes in the model need not take place via a continuous time Markov process. Finally, the two models use slightly different underlying graphs. For now, we will only discuss Bayesian networks that define a distribution involving single columns of a multiple alignment; context-dependence will be introduced later.

Consider a phylogenetic tree such as the one shown in Fig. 1, left. Leaf nodes represent present-day species, while non-leaf nodes represent ancestral species

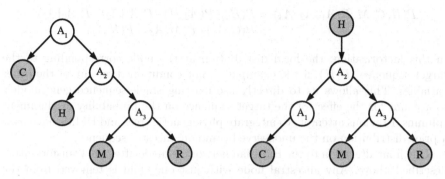

Fig. 1. A phylogenetic tree relating chicken (C), human (H), mouse (M), and rat (R). The graph can also be interpreted as a Bayesian network (left). The result of transforming the Bayesian network (right)

which no longer exist. The same graph can also be interpreted as a Bayesian network describing a probability distribution over columns in a multiple alignment. In that case, the nodes represent random variables corresponding to characters at specific rows in a multiple alignment column and the edges encode conditional independence relations among the variables. The independencies represented by the phylogenetic tree are quite natural – once we know the value of the ancestral base at a particular site in the alignment column, the probabilities of the bases in one descendant lineage are independent of the bases in other descendant lineages. These independence relations allow us to factor the joint distribution as follows:

$$P(H, C, M, R, A_1, A_2, A_3) = P(A_1) \cdot P(C|A_1) \cdot P(A_2|A_1) \cdot P(H|A_2) \cdot$$
$$P(A_3|A_2) \cdot P(M|A_3) \cdot P(R|A_3)$$

By taking advantage of the conditional independence relations present in the seven-variable joint distribution, we can express it as a product of six local conditional probability distributions (CPDs) that have two variables each and a marginal distribution on one variable. In general, factoring according to the independencies represented by a phylogenetic tree leads to an exponential reduction in the number of parameters required to specify the joint distribution. Of course, a real multiple alignment will only consist of sequences from currently existing species. Therefore, we treat the ancestral variables as missing data for the purposes of training and inference (see below). Rather than using a Bayesian network with the same structure as the phylogenetic tree, however, we apply a transformation to the phylogenetic tree graph to create the Bayesian network structure used in N-SCAN.

To transform the graph, we reverse the direction of all the edges along the path from the root of the graph to the target node. This results in a new Bayesian network with a tree structure rooted at the target node (see Fig. 1, right). The new Bayesian network encodes the same conditional independence relations as the original, but it suggests a new factorization of the joint distribution. For the example network shown in Fig. 1, this factorization is:

$$P(H, C, M, R, A_1, A_2, A_3) = P(H) \cdot P(A_2|H) \cdot P(A_1|A_2) \cdot P(A_3|A_2) \cdot$$
$$P(C|A_1) \cdot P(M|A_3) \cdot P(R|A_3)$$

In this factorization, the local distribution at the node corresponding to the target sequence ($P(H)$ in the example) is not conditioned on any of the other variables. This allows us to directly use existing single-sequence gene models to account for the effect of the target sequence on the probability of alignment columns. Previous attempts to integrate phylogenetic trees and HMMs have used a prior distribution on the unobserved common ancestor sequence.

One final alteration to the Bayesian network is made after the transformation described above. Any ancestral node with just one child is removed from the network along with its impinging edges. For each removed node, a new edge is added from the removed node's parent to its child. In the example, we remove A_1 and add an edge from A_2 to C. Again, it is not difficult to show that this transformation does not affect the expressive power of the network. We can write the local CPD at the removed node's child as a sum over the all possible values of the removed node. In the example,

$$P(C|A_2) = \sum_{A_1} P(C|A_1)P(A_1|A_2)$$

In effect, we have implicitly summed out some of the unobserved variables in the distribution. In general, we are only interested in computing the probability of an assignment to the observed variables. When making such a computation, we explicitly sum out all the unobserved variables in the distribution. The transformation described above makes this computation more efficient by reducing the number of explicit summations required.

2.3   Context-Dependent Models

Following [12], we can extend the models presented above to incorporate context dependence by redefining the meaning of the variables in the network. For a model of order o, we interpret the random variables in the network to represent the value of $o+1$ adjacent positions in a row in the alignment. The entire network then defines a joint distribution over sets of $o + 1$ adjacent columns, which can be used to determine the probability of a single column given the previous o columns.

Inference in the network can be accomplished using a modified version of Felsenstein's algorithm [13]. First, consider the problem of calculating the probability of an assignment to all the informant nodes in the network, given the value of the target node. For a given assignment, we define $L_u(a)$ to be the joint probability of all the observed variables that descend from node u, given that node u has value a. If $C(u)$ is the set of children of u and $V(u)$ is the set of possible values of u, we can calculate $L_u(a)$ according to the following recursive formula:

$$L_u(a) = \begin{cases} M(u, a) & \text{if } u \text{ is a leaf} \\ \prod_{c \in C(u)} \left(\sum_{b \in V(c)} Pr(c = b | u = a) L_c(b) \right) & \text{otherwise} \end{cases}$$

Here, M is called the match function, and is defined as follows:

$$M(u, a) = \begin{cases} 1 & \text{if node } u \text{ has value } a \\ 0 & \text{otherwise} \end{cases}$$

If T is the target node, and t is its observed assignment, then $L_T(t)$ is the probability of the informant assignments given the target. To calculate all the L_u's for each node in the network, we can visit the nodes in postorder and calculate all the L_u's for a particular node at once.

We can use essentially the same algorithm for a conditional probability query. We define the partial match function, M', for a model of order o as follows:

$$M'(u, a) = \begin{cases} 1 & \text{if the first } o \text{ characters of the value of node } u \\ & \quad \text{match the first } o \text{ characters of } a \\ 0 & \text{otherwise} \end{cases}$$

We define the quantity $L'_u(a)$ exactly as we did $L_u(a)$, except we substitute M' for M in the recursive definition. $L'_T(t)$ is then the probability of the first o characters of the informant assignments. Once we know the values of $L_T(t)$ and $L'_T(t)$, expression (3) is just

$$\frac{L_T(t)}{L'_T(t)}$$

Each call to the inference algorithm visits each node in the network exactly once, and requires $O(6^{2(o+1)})$ operations per internal node. Thus, the overall time complexity of inference is $O(N \cdot 6^{2(o+1)})$. Adding additional informants only results in a linear increase in the complexity of inference, but we pay an exponential cost for increasing the model order.

2.4   Training

The Bayesian networks for all of N-SCAN's GHMM states share a single topology determined by the phylogenetic tree relating the target and the informant genomes, which we assume to be known. However, the local CPDs for each node in a particular network will depend on the GHMM state with which the network is associated. The CPDs are not known in advance, and must be estimated from training data.

Suppose we had a multiple alignment of all the genomes represented in the phylogenetic tree, with each column labeled to indicate which GHMM state produced it. For a particular Bayesian network of order o, we could treat each set of $o + 1$ adjacent columns ending with a column labeled by the GHMM state associated with the network as an instantiation of the network variables. Once we extract a list of all the instantiations that occur in the multiple alignment, along with the number of times each instantiation occurs, it is a simple matter to produce a maximum likelihood estimate for all the CPDs in the network.

Since the GHMM states correspond to gene features, we can construct a labeled multiple alignment by combining the output of a whole-genome multiple aligner with a set of annotations of known genes. However, the alignment will contain only the genomes that correspond to the root and leaves of the Bayesian network graph. The ancestral genomes are no longer available for sequencing and so must be treated as missing data.

We can still estimate the CPDs despite the missing data by using the EM algorithm. For each network, we begin with an initial guess for the CPDs. We then calculate, for each CPD, the expected number of times each possible assignment to its variables occurs in the multiple alignment. This can be done efficiently using a variation of the inside-outside algorithm essentially the same as the one presented in [12]. Next, the initial guess is replaced with a maximum likelihood estimate of the CPDs based on the expected occurrences. This process is repeated until the maximum likelihood estimate converges. At convergence, the maximum likelihood estimate is guaranteed to be a stationary point of the likelihood function of the multiple alignment.

2.5   CPD Parameterizations

We have not yet described a method for obtaining a maximum likelihood estimate of the CPDs from a set of observations (or expected observations). If no restrictions are placed on the form taken by the CPDs, there exist simple closed-form expressions for the value of each entry in each CPD. A Bayesian network with completely general CPDs can represent any joint distribution in which the conditional independence relations it encodes hold. However, a complete description of such a network requires a relatively large number of parameters. Each N-SCAN Bayesian network of order o has $(2N-1)(6^{o+1})(6^{o+1}-1)$ free parameters if its CPDs are unrestricted. If the amount of training data (i.e., columns in the multiple alignment with the appropriate label) available is small, this may be too many parameters to fit accurately.

It is possible to reduce the number of parameters to fit by specifying the CPDs using fewer than $(6^{o+1})(6^{o+1}-1)$ parameters each. Only a subset of all possible CPDs will be expressible by any given non-general parameterization, but we hope the real CPDs, or ones close to them, will be expressible by the parameterization we choose. Depending on the parameterization chosen, we may be able to derive analytical expressions with which to estimate the values of the parameters. Otherwise, we can use numerical optimization techniques to obtain an estimate.

In the experiments below, we use a parameterization with a form similar to the general reversible rate matrices used in traditional phylogenetic models. The zero-order version of this parameterization, which we call a *partially reversible* model, is shown below. A cell (i, j) in the matrix represents $P(j|i)$, the probability of a particular child node having value j from the alphabet $\{A, C, G, T, \_, .\}$, given that its parent has value i.

$$\begin{pmatrix} - & a\pi_C & b\pi_G & c\pi_T & g & h \\ a\pi_A & - & d\pi_G & e\pi_T & g & h \\ b\pi_A & d\pi_C & - & f\pi_T & g & h \\ c\pi_A & e\pi_C & f\pi_G & - & g & h \\ i\pi_A & i\pi_C & i\pi_G & i\pi_T & - & j \\ k\pi_A & k\pi_C & k\pi_G & k\pi_T & l & - \end{pmatrix}$$

Here, the π_i's are the background frequency of the bases; they are estimated directly from the multiple alignment and are not considered to be free parameters. The model has 12 free parameters, as opposed to 30 in a general parameterization. Note that the 4x4 upper-left submatrix is identical to the general reversible rate matrix used in continuous time Markov process models of sequence evolution [14]. The probability of a deletion is the same for each base, as is the probability of a base becoming unaligned. The probability of a base being inserted or becoming realigned is proportional to the background frequency of the base.

To generalize the partially reversible parameterization to higher orders, we make use of the concept of a gap pattern. We define the gap pattern of an $(o + 1)$-mer to be the string that results from replacing all the bases in the $(o+1)$-mer with the character "X". For example, the trimers "GA_", "GC_", and "AT_" all have the gap pattern "XX_". For substitution probabilities involving an $(o+1)$-mer that contain gaps or unaligned characters, the partially reversible model considers only the gap pattern of the $(o + 1)$-mer. Let $\mathcal{D}$ be the set of all possible $(o + 1)$-mers that contain only the four DNA bases, and $\mathcal{G}$ be the set of all possible gap patterns of length $o + 1$ that contain at least one gap or unaligned character. The substitution probabilities $P(j|i)$ have the following properties:

1. If $j \in \mathcal{D}$ and $i \in \mathcal{D}$, then $P(j|i)\pi_i = P(i|j)\pi_j$.
2. If $j \in \mathcal{D}$ and $i \in \mathcal{G}$, then $P(j|i) = \alpha_i \pi_j$.
3. If $j \in \mathcal{G}$ and $i \in \mathcal{D}$, then $P(j|i) = \beta_j$.

It can be shown that a sequence evolving according to a substitution process that has these three properties will have constant expected values for the relative frequencies of the $(o + 1)$-mers in $\mathcal{D}$. The first-order partially reversible model, which can be described by a 36x36 matrix, has 170 free parameters, far fewer than the 1260 in the general first-order model.

Partially reversible models are able to capture significantly more information about patterns of selection than the conservation sequence approach used in TWINSCAN 2.0, which considers only patterns of matches, mismatches, and unaligned positions. For example, a first-order partially reversible model can model insertions and deletions separately from base substitutions, and can take into account the difference between the rates of transitions and transversions as well as the increased rate of mutation of CpG dinucleotides [15]. Furthermore, unlike TWINSCAN 2.0, N-SCAN uses a separate conservation model for each codon position in coding sequence, allowing it to model differences in substitution rates between the three positions.

2.6   Conservation Score Coefficient

Like TWINSCAN, N-SCAN uses log-likelihood scores rather than probabilities internally. The score of a particular column i in the multiple alignment given a state S can be written as

$$\log\left(\frac{T_S(i)}{T_{Null}(i)}\right) + k \cdot \log\left(\frac{C_S(i)}{C_{Null}(i)}\right)$$

Here, T_S and T_{Null} are the target sequence probabilities of the form shown in expression (1) for state S and the null model, respectively. Likewise, C_S and C_{Null} are the conservation model probabilities, as in expression (3). k is an arbitrary constant called the conservation score coefficient which can be used to increase or decrease the impact of the informant sequences on N-SCAN's predictions. Empirical results show that a value of k between 0.3 and 0.6 leads to the best predictive performance. This may be due to the potential of conserved noncoding regions to contribute to a large number of false positive predictions (see below).

2.7   State Diagram

Figure 3 shows the N-SCAN state diagram. The 5' UTR and CNS states allow N-SCAN to avoid false positives that would occur if these sequence features were not modeled explicitly. Without these states, conserved noncoding regions would tend to be annotated as coding exons due to their high conservation scores. Instead, N-SCAN tends to annotate conserved regions with a low coding score as CNS. Furthermore, the 5' UTR states allow N-SCAN to predict exon/intron structure in 5' UTRs. Simultaneous 5' UTR and coding region prediction by N-SCAN will be discussed in more detail in a forthcoming paper devoted to the subject [16].

Since we lacked a reliable set of annotations of conserved noncoding regions, we used the null target sequence model for the CNS state, which effectively assigns the neutral score of zero to all target sequences under the CNS model. Thus, the score of a putative CNS region is determined entirely by the CNS conservation model, which was estimated from 5' UTRs. While this resulted in an acceptable model for some types of CNS, more highly-conserved CNS was probably not modeled accurately using this method.

2.8   Experimental Design

The human gene prediction experiments presented below were performed on the May 2004 build of the human genome (hg17), and the January 2003 build of the

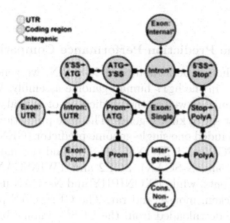

Fig. 2. The N-SCAN state diagram. Intron and exon states with asterisks represent six states each, which are used tracking reading frame and partial stop codons. Only forward strand states are shown; on the reverse strand all non-intergenic states are duplicated, and initial (ATG → 3′ SS), not terminal (5′ SS → Stop), states track phase and partial stop codons

Drosophila melanogaster genome. Both were obtained from the UCSC genome browser [17]. Each group of experiments used a set of annotations consisting of known genes, which was constructed as follows. The annotation set initially contained the mappings of RefSeqs to the genome in question provided by the UCSC genome browser. This set was then filtered to exclude annotations believed likely to have errors. All genes with non-standard start or stop codons, in-frame stop codons, total coding region length not a multiple of three, non-standard donor sites lacking a GT, GC, or AT consensus, or non-standard acceptor sites lacking an AG or AC consensus were discarded. After filtering, the human set contained 16,259 genes and 20,837 transcripts, while the *D. melanogaster* set contained 13,091 genes and 18,591 transcripts.

For the human experiments, N-SCAN used an eight-way whole-genome alignment of human (hg17), blowfish (fr1), chicken (galGal2), chimp (panTro1), dog (canFam1), mouse (mm5), rat (rn3), and zebrafish (danRer1) created by MULTIZ [18]. A four-way MULTIZ alignment of *Drosophila melanogaster*, *Drosophila yakuba*, *Drosophila pseudoobscura*, and *Anopheles gambiae* was used for the *D. melanogaster* experiments. The alignments were downloaded from the UCSC genome browser. For the human experiments, we used only the rows of the alignment corresponding to human, chicken, mouse, and rat, and discarded the other four rows. Columns in either alignment with gaps in the target sequence were also discarded.

All predictions made by N-SCAN were four-fold cross validated. The first-order partially reversible parameterization was used for all of N-SCAN's Bayesian network CPDs, and N-SCAN's conservation score coefficient was set to 0.4.

3   Results

3.1   Human Gene Prediction Performance Comparison

To evaluate the predictive performance of N-SCAN, we generated predictions for every chromosome in the hg17 human genome assembly. We then compared the N-SCAN predictions, as well as the predictions of several other de novo gene predictors, to our test set of known genes. The gene prediction systems involved in this experiment included one single-genome predictor (GENSCAN), two dual-genome predictors (SGP2 and TWINSCAN 2.0), and two multiple-genome predictors (EXONIPHY and N-SCAN). SGP2 and TWINSCAN 2.0 made use of human-mouse alignments, while EXONIPHY and N-SCAN used multiple alignments of human, chicken, mouse, and rat. The GENSCAN predictions used in this experiment were downloaded from the UCSC genome browser. The SGP2 predictions were downloaded from the SGP2 web site [19]. The EXONIPHY predictions were obtained from one of EXONIPHY's creators (A.C. Siepel, personal communication). EXONIPHY does not link exons into gene structures, so its performance at the gene level was not evaluated.

We evaluated both sensitivity and specificity at the gene, transcript, exon, and nucleotide levels. Since none of the gene predictors involved in the experiment had the ability to predict alternative transcripts, a gene prediction was counted as correct at the gene level if it exactly matched any of the transcripts in the test set. The results of the experiment are shown in Table 1. For each performance metric, the result from the best-performing predictor is shown in bold. Note that the specificity numbers are underestimates, since all predicted genes not in the test set were counted as incorrect. N-SCAN achieved substantially better performance on both the gene and exon levels than the other four predictors involved in the experiment. On the nucleotide level, N-SCAN had the highest sensitivity, but a lower specificity than EXONIPHY.

We also evaluated the ability of the systems to predict long introns, a feat notoriously difficult for de novo gene predictors. The results in Table 2 show that N-SCAN has the greatest sensitivity for each length range we tested. Furthermore, N-SCAN's performance drops off much more slowly with length than

Table 1. Whole-genome gene prediction performance in human

	GENSCAN	TWINSCAN 2.0	N-SCAN	SGP2	EXONIPHY
Gene Sn	0.10	0.25	**0.35**	0.16	-
Gene Sp	0.04	0.15	**0.21**	0.08	-
Transcript Sn	0.08	0.21	**0.29**	0.14	-
Transcript Sp	0.04	0.15	**0.21**	0.08	-
Exon Sn	0.69	0.71	**0.84**	0.73	0.57
Exon Sp	0.33	0.61	**0.63**	0.52	0.50
Nucleotide Sn	0.86	0.84	**0.90**	0.86	0.76
Nucleotide Sp	0.40	0.64	0.65	0.62	**0.68**

Table 2. Intron sensitivity by length

Length (Kb)	Count	GENSCAN	TWINSCAN 2.0	N-SCAN	SGP2
0 - 10	157757	0.68	0.77	**0.86**	0.74
10 - 20	9519	0.50	0.46	**0.77**	0.69
20 - 30	3317	0.41	0.22	**0.71**	0.68
30 - 40	1742	0.30	0.08	**0.64**	0.60
40 - 50	992	0.28	0.02	**0.64**	0
50 - 60	652	0.20	0.01	**0.53**	0
60 - 70	447	0.16	0	**0.52**	0
70 - 80	314	0.12	0	**0.50**	0
80 - 90	268	0.10	0	**0.39**	0
90 - 100	211	0.11	0	**0.48**	0

Table 3. Performance with different combinations of informants. The human informants are chicken (C), mouse (M), and rat (R). The *D. melanogaster* informants are *A. gambiae* (G), *D. Pseudoobscura* (P), and *D. Yakuba* (Y)

	Human				D. melanogaster			
	C	M	R	C, M, R	G	P	Y	G, P, Y
Gene Sn	0.21	0.34	0.32	0.34	0.39	0.53	0.49	0.55
Gene Sp	0.16	0.23	0.23	0.22	0.39	0.52	0.48	0.53
Transcript Sn	0.18	0.29	0.27	0.29	0.32	0.43	0.40	0.45
Transcript Sp	0.16	0.23	0.23	0.22	0.39	0.52	0.48	0.53
Exon Sn	0.75	0.81	0.80	0.82	0.67	0.74	0.73	0.77
Exon Sp	0.57	0.61	0.62	0.61	0.67	0.76	0.73	0.75
Nucleotide Sn	0.87	0.87	0.88	0.88	0.92	0.93	0.92	0.93
Nucleotide Sp	0.61	0.63	0.64	0.64	0.92	0.94	0.94	0.94

that of the other gene predictors. In fact, N-SCAN is able to correctly predict approximately half of the introns in the test set with lengths between 50Kb and 100Kb.

3.2 Informant Effectiveness

To test the effect of multiple informants on N-SCAN's predictive accuracy, we generated four sets of predictions each for human and *D. melanogaster*. The first three sets for each organism use a single informant, while the final set uses all three informants simultaneously. Predictions were generated for human chromosomes 1, 15, 19, 20, 21, and 22, and for the entire *D. melanogaster* genome. The results of this experiment are shown in Table 3. In *D. melanogaster*, N-SCAN achieved a small but significant boost in performance by using all three informants together. However, in human, using the three informants at once appears to be no better than using mouse alone.

4   Discussion

We have presented a system, N-SCAN, for de novo gene prediction that builds on an existing system by incorporating several new features, such as richer substitution models, states for 5' UTR structure prediction, a conserved noncoding sequence state, and the ability to use information from multiple informant sequences. N-SCAN achieved significantly better performance than several other de novo gene predictors in a test of whole-genome gene prediction in human. In addition, N-SCAN was successfully applied to gene prediction in *D. melanogaster* without the need for any special modifications.

N-SCAN incorporates information from multiple informant sequences in a novel way which we believe has several potential advantages. First, N-SCAN builds on existing single-sequence models of a target genome. These single-sequence models can be quite sophisticated. For example, the donor splice site model used in GENSCAN and TWINSCAN 2.0 is able to take into account the effect of non-adjacent positions in the splice site signal through the use of a maximal dependence decomposition model [3]. In addition, single-sequence models of a given order generally require fewer parameters than multiple-sequence models of the same order. Therefore, it is possible to use high-order single-sequence models in combination with conservation models of a lower order while maintaining a good fit. In the experiments presented above, some of the N-SCAN target genome models had orders as high as five, while the conservation models were all of order one. Furthermore, because the target sequence is observed, it is possible to obtain a globally optimal estimate of the distributions in the target genome models. The EM estimates for the conservation models are only guaranteed to be locally optimal, and could in principle be far from a global optimum.

Also important is N-SCAN's treatment of gaps and unaligned characters. Instead of treating these characters as missing data, or modeling gap patterns using additional states in the GHMM [9], N-SCAN deals with them directly in its conservation models. This allows the very significant information they contribute to be taken into account in a natural and efficient way. The price for this ability is that the continuous time Markov process model of substitution must be abandoned. Continuous time Markov processes are a good model for base mutations between aligned positions in DNA sequences, but do not accurately model the nonlinear process of positions becoming unaligned over time. For the sake of illustration, consider an ancestor sequence and a descendant sequence that differ by only a single point mutation. It is not possible for the sequences to have unalignable bases. The alignment will have a gap if the single mutation is an insertion or deletion, but the surrounding regions will provide enough information to align the gap with the right base in the other species. Thus, the instantaneous rates of substitutions leading to unaligned characters are all zero. Yet as divergence increases, a point will be reached where even small changes to the sequence can lead to a whole region becoming unalignable. Therefore, rather than assuming substitutions occur as a result of a continuous time Markov process, N-SCAN uses the more general framework of Bayesian networks for its conservation models. This results in a substantial increase in the required number of parameters, but with appropriate parameterizations, this number is still manageable for context-dependent models of the type presented here.

Relaxing the assumption that substitutions occur via a continuous time Markov process also allows N-SCAN to accurately model alignment columns in which the aligned positions do not share a single functional state. In such a case, patterns of substitution across different branches of the phylogenetic tree are likely to vary significantly, reflecting different evolutionary constraints. This situation cannot be represented by a continuous time Markov process model, which uses the same substitution rate matrix for each branch in the tree. In practice, positions in an alignment column may have different functions as a result of a function-changing mutation, alignment error, or sequencing error. The latter two causes are of particular concern when the alignment contains highly diverged or draft-quality sequences.

We are currently pursuing a number of approaches for improving N-SCAN. First, the use of higher-order conservation models has the potential to increase N-SCAN's predictive accuracy. Second-order models for coding sequence, for example, could perfectly distinguish between silent and missense mutations. Although both inference and training are far more expensive in second-order models than in first-order models, the use of second-order models for gene prediction in human and *D. melanogaster* appears feasible. Second, better models of conserved noncoding sequence should lead to better performance and perhaps remove the need for a conservation score coefficient. Finally, the role of multiple informants merits further investigation. Although the use of multiple informants in *D. melanogaster* gene prediction improved performance beyond what was achieved with any single informant, the same effect was not observed in human. This may be due to the specific characteristics of the informant sequences that were used for each organism, differences in the properties of the target genomes, or some other factor. Future experiments on a variety of target genomes using different combinations of informants should shed some light on this issue.

Acknowledgments

We thank Mikhail Velikanov for providing the set of filtered RefSeq genes used for training and evaluation. We also thank Adam Siepel for converting the EX-ONIPHY predictions on hg16 to hg17 coordinates. Finally, thanks to all the members of the Brent lab who developed and maintained the TWINSCAN code base, from which N-SCAN was developed. This work was supported by grant HG02278 from the NIH to M.R.B.

References

1. The MGC Project Team. 2004. The status, quality, and expansion of the NIH full-length cDNA project: The Mammalian Gene Collection (MGC). *Genome Res.*, 14:2121-2127.

2. Waterston et al. 2002. Initial sequencing and comparative analysis of the mouse genome. *Nature,* 420:520-562.
3. C. Burge and S. Karlin. 1997. Prediction of complete gene structures in human genomic DNA. *J. Mol. Biol.,* 268:78-94.
4. M. Alexandersson, S. Cawley, and L. Pachter. 2003. SLAM: Cross-species gene finding and alignment with a generalized pair hidden Markov model. *Genome Res.,* 13:496-502.
5. G. Parra, P. Agarwal, J.F. Abril, T. Wiehe, J.W. Fickett, and R. Guigo. 2003. Comparative gene prediction in human and mouse. *Genome Res.,* 13:108-117.
6. I. Korf, P. Flicek, D. Duan, and M.R. Brent. 2001. Integrating genomic homology into gene structure prediction. *Bioinformatics,* 17 Suppl. 1:S140-148.
7. P. Flicek, E. Keibler, P. Hu, I. Korf, and M.R. Brent. 2003. Leveraging the mouse genome for gene prediction in human: From whole-genome shotgun reads to a global synteny map. *Genome Res.,* 13:46-54.
8. A.E. Tenney, R.H. Brown, C. Vaske, J.K. Lodge, T.L. Doering, and M.R. Brent. 2004. Gene prediction and verification in a compact genome with numerous small introns. *Genome Res.,* in press.
9. A.C. Siepel and D. Haussler. 2004. Computational identification of evolutionary conserved exons. In *RECOMB 2004.*
10. J.D. McAuliffe, L. Pachter, and M.I. Jordan. 2003. Multiple-sequence functional annotation and the generalized hidden Markov phylogeny. Technical Report 647, Department of Statistics, University of California, Berkeley.
11. J.S. Pedersen and J. Hein. 2003. Gene finding with a hidden Markov model of genome structure and evolution. *Bioinformatics,* 19:219-227.
12. A. Siepel and D. Haussler. 2004. Phylogenetic estimation of context-dependent substitution rates by maximum likelihood. *Mol. Biol. Evol.,* 21:468-448.
13. Felsenstein, J. 1981. Evolutionary trees from DNA sequences. *J. Mol. Evol.,* 17:368-376.
14. P. Lió and N. Goldman. 1998. Models of molecular evolution and phylogeny. *Genome Res.,* 8:1233-1244.
15. M. Bulmer. 1986. Neighboring base effects on substitution rates in pseudogenes. *Mol. Biol. Evol.,* 3:322-329.
16. R.H. Brown, S.S. Gross, and M.R. Brent. 2005. Begin at the beginning: predicting genes with 5' UTRs. Submitted.
17. W.J. Kent, C.W. Sugnet, T.S. Furey, K.M. Roskin, T.H. Pringle, A.M. Zahler, and D. Haussler. 2003. The human genome browser at UCSC. *Genome Res.,* 12:996-1006.
18. M. Blanchette, W.J. Kent, C. Riemer, L. Elnitski, A.F.A. Smith, K.M. Roskin, R. Baertsch, K. Rosenbloom, H. Clawson, E.D. Green, D. Haussler, and W. Miller. 2004. Aligning multiple genomic sequences with the threaded blockset aligner. *Genome Res.,* 14:708-715.
19. SGP2 home page. (http://genome.imim.es/software/sgp2).

Learning Interpretable SVMs for Biological Sequence Classification

S. Sonnenburg[1], G. Rätsch[2], and C. Schäfer[1]

[1] Fraunhofer Institute FIRST,
Kekuléstr. 7, 12489 Berlin, Germany
[2] Friedrich Miescher Lab, Max Planck Society,
Spemannstr. 39, Tübingen, Germany

Abstract. We propose novel algorithms for solving the so-called Support Vector Multiple Kernel Learning problem and show how they can be used to understand the resulting support vector decision function. While classical kernel-based algorithms (such as SVMs) are based on a single kernel, in Multiple Kernel Learning a quadratically-constraint quadratic program is solved in order to find a sparse convex combination of a set of support vector kernels. We show how this problem can be cast into a semi-infinite linear optimization problem which can in turn be solved efficiently using a boosting-like iterative method in combination with standard SVM optimization algorithms. The proposed method is able to deal with thousands of examples while combining hundreds of kernels within reasonable time.

In the second part we show how this technique can be used to understand the obtained decision function in order to extract biologically relevant knowledge about the sequence analysis problem at hand. We consider the problem of splice site identification and combine string kernels at different sequence positions and with various substring (oligomer) lengths. The proposed algorithm computes a *sparse* weighting over the length and the substring, highlighting which substrings are important for discrimination. Finally, we propose a bootstrap scheme in order to reliably identify a few statistically significant positions, which can then be used for further analysis such as consensus finding.

Keywords: Support Vector Machine, Multiple Kernel Learning, String Kernel, Weighted Degree Kernel, Interpretation of SVM results, Splice Site Prediction.

1   Introduction

Kernel based methods such as Support Vector Machines (SVMs) have been proven powerful for sequence analysis problems frequently appearing in computational biology (e.g. [27, 11, 26, 15]). They employ a so-called kernel function $k(\mathbf{s}_i, \mathbf{s}_j)$ which intuitively computes the similarity between two sequences $\mathbf{s}_i$ and

S. Miyano et al. (Eds.): RECOMB 2005, LNBI 3500, pp. 389–407, 2005.

$\mathbf{s}_j$. The result of SVM learning is a $\boldsymbol{\alpha}$-weighted linear combination of N kernel elements and the bias b:

$$f(\mathbf{s}) = \text{sign}\left(\sum_{i=1}^{N} \alpha_i y_i k(\mathbf{s}_i, \mathbf{s}) + b\right).$$

One of the problems with kernel methods compared to probabilistic methods (such as position weight matrices or interpolated Markov models [6]) is that the resulting decision function (1) is hard to interpret and, hence, difficult to use in order to extract relevant biological knowledge from it (see also [14, 26]). We approach this problem by considering the use of convex combinations of M kernels, i.e.

$$k(\mathbf{s}_i, \mathbf{s}_j) = \sum_{j=1}^{M} \beta_j k_j(\mathbf{s}_i, \mathbf{s}_j)$$

with $\beta_j \geq 0$ and $\sum_j \beta_j = 1$, where each kernel k_j uses only a distinct set of features of the sequence. For appropriately designed sub-kernels, the optimized combination coefficients can then be used to understand which features of the sequence are of importance for discrimination. This is an important property missing in current kernel based algorithms.

Sequence analysis problems usually come with large number of examples and potentially many kernels to combine. Unfortunately, algorithms proposed for Multiple Kernel Learning (MKL) so far are not capable of solving the optimization problem for realistic problem sizes (e.g. $\geq 10,000$ examples) within reasonable time. Even recently proposed SMO-like algorithms for this problem, such as the one proposed in [1], are not efficient enough since they suffer from the inability to keep all kernel matrices ($K_j \in \mathbb{R}^{N \times N}, j = 1, \ldots, M$) in memory.[1] We consider the reformulation of the MKL problem into a semi-infinite linear problem (SILP), which can be iteratively approximated quite efficiently. In each iteration one only needs to solve the classical SVM problem (with one of the efficient and publicly available SVM implementations) and then performs an adequate update of the kernel convex combination weights $\boldsymbol{\beta}$. Separating the SVM optimization from the optimization of the kernel coefficients can thus lead to significant improvements for large scale problems with general kernels. We will, however, show how one can take advantage of the special structure of string kernels (in particular the one below).

We illustrate the usefulness of the proposed algorithm in combination with a recently proposed string kernel on DNA sequences — the so-called *weighted degree* (WD) kernel [23]. Its main idea is to count the (exact) co-occurrence of k-mers at position l in the sequence between two compared DNA sequences. The kernel can be written as a linear combination of d parts with coefficients β_k ($k = 1, \ldots, d$):

---

[1] Note that also kernel caching becomes insufficient if the number of combined kernels is large.

$$k(\mathbf{s}_i, \mathbf{s}_j) = \sum_{k=1}^{d} \beta_k \sum_{l=1}^{L-k} \mathbf{I}(\boldsymbol{u}_{k,l}(\mathbf{s}_i) = \boldsymbol{u}_{k,l}(\mathbf{s}_j)),$$

where L is the length of the $\mathbf{s}$'s, d is the maximal oligomer order considered and $\boldsymbol{u}_{k,l}(\mathbf{s})$ is the oligomer of length k at position l of sequence $\mathbf{s}$. One question is how the weights β_k for the various k-mers should be chosen. So far, only heuristic settings in combination with expensive cross-validation have been used. The MKL approach offers a clean and efficient way to find the optimal weights β. One would define d kernels

$$k_k(\mathbf{s}_i, \mathbf{s}_j) = \sum_{l=1}^{L-k} \mathbf{I}(\boldsymbol{u}_{k,l}(\mathbf{s}_i) = \boldsymbol{u}_{k,l}(\mathbf{s}_j)),$$

and then optimize the convex combination of these kernels by the newly proposed algorithm. The optimal weights β indicate which oligomer lengths are important for the classification problem at hand. Moreover, one would expect a slight performance gain for optimized weights and since the β's are sparse, also an increased prediction speed.

Additionally, it is interesting to introduce an importance weighting over the position of the subsequence. Hence, we define a separate kernel for each position and each oligomer order, i.e.

$$k_{k,l}(\mathbf{s}_i, \mathbf{s}_j) = \mathbf{I}(\boldsymbol{u}_{k,l}(\mathbf{s}_i) = \boldsymbol{u}_{k,l}(\mathbf{s}_j)),$$

and optimize the weightings of the combined kernel, which may be written as

$$k(\mathbf{s}_i, \mathbf{s}_j) = \sum_{k=1}^{d} \sum_{l=1}^{L-k} \beta_{k,l} \mathbf{I}(\boldsymbol{u}_{k,l}(\mathbf{s}_i) = \boldsymbol{u}_{k,l}(\mathbf{s}_j)) = \sum_{k,l} \beta_{k,l} k_{k,l}(\mathbf{s}_i, \mathbf{s}_j).$$

Obviously, if one would be able to obtain an accurate classification by a sparse weighting $\beta_{k,i}$, then one could easily interpret the resulting decision function. For instance for signal detection problems (such as splice site detection), one would expect a few important positions with long oligomers near the site and some additional positions only capturing nucleotide compositions (short nucleotides).

By bootstrapping and employing a combinatorial argument, we derive a statistical test that discovers the most important kernel weights. On simulated pseudo-DNA sequences with two hidden 7-mers we elucidate which k-mers in the sequence were used for the SVM decision. Finally we apply the method to splice sites classification of *C. elegans* and show that the optimized convex kernel combination may help extracting biological knowledge from the data.

2   Methods

2.1   Support Vector Machines

We use Support Vector Machines[5] which are extensively studied in literature (e.g. [19]). Their classification function can be written as in (1) The α_i's are Lagrange multipliers and b is the usual bias which are the results of SVM training. The kernel k is the *key ingredient* for learning with SVMs. It implicitly defines the feature space and the mapping Φ via

$$k(\mathbf{s}, \mathbf{s}') = (\Phi(\mathbf{s}) \cdot \Phi(\mathbf{s}')).$$

In case of the afore mentioned WD kernel, Φ maps into a *feature space* $\mathbb{R}^D$ of all possible k-mers of length up to d for each sequence position ($D \approx 4^{d+1}L$). For a given sequence $\mathbf{s}$, a dimension of $\phi(\mathbf{s})$ is 1, if it contains a certain substring at a certain position. The dot-product between two mapped examples then counts the co-occurrences of substrings at all positions.

For a given set of training examples $(\mathbf{s}_i, y_i)$ $(i = 1, \ldots, N)$, the SVM solution is obtained by solving the following optimization problem that maximizes the soft margin between both classes:

$$\min \ \frac{1}{2}\|\mathbf{w}\|^2 + C \sum_{i=1}^{N} \xi_i \tag{1}$$

$$\text{w.r.t. } \mathbf{w} \in \mathbb{R}^D, b \in \mathbb{R}, \boldsymbol{\xi} \in \mathbb{R}_+^N$$

$$\text{s.t. } y_i((\mathbf{w} \cdot \Phi(\mathbf{s}_i)) + b) \geq 1 - \xi_i, \quad i = 1, \ldots, N,$$

where the parameter C determines the trade-off between the size of the margin and the margin errors ξ_i. The dual optimization problem is as follows:

$$\max \ \sum_{i=1}^{N} \alpha_i - \frac{1}{2} \sum_{i,j=1}^{N} \alpha_i \alpha_j y_i y_j k(\mathbf{s}_i, \mathbf{s}_j), \tag{2}$$

$$\text{w.r.t. } \boldsymbol{\alpha} \in \mathbb{R}_+^N \text{ with } \boldsymbol{\alpha} \leq C \text{ and } \sum_{i=1}^{N} \alpha_i y_i = 0.$$

Note that there exist a large variety of different software packages that can efficiently solve the above optimization problem even for more than hundred thousands of examples.

2.2   The Multiple Kernel Learning Optimization Problem

Idea In the Multiple Kernel Learning (MKL) problem one is given N data points $(\tilde{\mathbf{s}}_i, y_i)$ $(y_i \in \{\pm 1\})$, where $\tilde{\mathbf{s}}_i$ is subdivided into M components $\tilde{\mathbf{s}}_i = (\mathbf{s}_{i,1}, \ldots, \mathbf{s}_{i,M})$ with $\mathbf{s}_{i,j} \in \mathbb{R}^{k_j}$ and k_j is the dimensionality of the j-th component. Then one solves the following convex optimization problem [1], which is equivalent to the linear SVM for $M = 1$:

$$\min \frac{1}{2} \left(\sum_{j=1}^{M} d_j \beta_j \|\mathbf{w}_j\|_2 \right)^2 + C \sum_{i=1}^{N} \xi_i \tag{3}$$

w.r.t. $\mathbf{w} = (\mathbf{w}_1, \ldots, \mathbf{w}_M), \mathbf{w}_j \in \mathbb{R}^{k_j}, \boldsymbol{\xi} \in \mathbb{R}_+^N, \boldsymbol{\beta} \in \mathbb{R}_+^M, b \in \mathbb{R}$

$$\text{s.t. } y_i \left(\sum_{j=1}^{M} \beta_j \mathbf{w}_j^{\top} \mathbf{s}_{i,j} + b \right) \geq 1 - \xi_i, \forall i = 1, \ldots, N$$

$$\sum_{j=1}^{M} \beta_j = 1,$$

where d_j is a prior weighting of the kernels (in [1], $d_j = 1/\sum_i (\mathbf{s}_{i,j} \cdot \mathbf{s}_{i,j})$ has been chosen such that the combined kernel has trace one). For simplicity, we assume that $d_j = 1$ for the rest of the paper and that the normalization is done within the mapping ϕ (if necessary). Note that the ℓ_1-norm of $\boldsymbol{\beta}$ is constrained to one, while one is penalizing the ℓ_2-norm of $\mathbf{w}_j$ in each block j separately. The idea is that ℓ_1-norm constrained or penalized variables tend to have sparse optimal solutions, while ℓ_2-norm penalized variables do not [21]. Thus the above optimization problem offers the possibility to find sparse solutions on the block level with non-sparse solutions within the blocks.

Reformulation as a Semi-infinite Linear Program. The above optimization problem can also be formulated in terms of support vector kernels [1]. Then each block j corresponds to a separate kernel $(K_j)_{r,s} = k_j(\mathbf{s}_{r,j}, \mathbf{s}_{s,j})$ computing the dot-product in feature space of the j-th component. In [1] it has been shown that the following optimization problem is equivalent to (3):

$$\min \frac{1}{2} \gamma^2 - \sum_i \alpha_i \tag{4}$$

$$\text{w.r.t. } \gamma \in \mathbb{R}, \boldsymbol{\alpha} \in \mathbb{R}^N$$

$$\text{s.t. } 0 \leq \boldsymbol{\alpha} \leq C, \sum_i \alpha_i y_i = 0$$

$$\underbrace{\sum_{r,s} \alpha_r \alpha_s y_r y_s (K_j)_{r,s}}_{=:S_j(\boldsymbol{\alpha})} \leq \gamma^2$$

In order to solve (4), one may solve the following saddle point problem (Lagrangian):

$$L := \frac{1}{2} \gamma^2 - \sum_i \alpha_i + \sum_{j=1}^{M} \beta_j (S_j(\boldsymbol{\alpha}) - \gamma^2) \tag{5}$$

minimized w.r.t. $\boldsymbol{\alpha} \in \mathbb{R}_+^N, \gamma \in \mathbb{R}$ (subject to $\boldsymbol{\alpha} \leq C$ and $\sum_i \alpha_i y_i = 0$) and maximized w.r.t. $\boldsymbol{\beta} \in \mathbb{R}_+^M$. Setting the derivative w.r.t. to γ to zero, one obtains the constraint $\sum_j \beta_j = \frac{1}{2}$ and (5) simplifies to:

$$L := \underbrace{\frac{1}{2} \sum_{j=1}^{M} \beta_j S_j(\boldsymbol{\alpha})}_{=:S(\boldsymbol{\alpha})} - \sum_i \alpha_i \qquad (6)$$

Assume $\boldsymbol{\alpha}^*$ would be the optimal solution, then $\theta^* := S(\boldsymbol{\alpha}^*) - \sum_i \alpha_i$ is minimal and, hence, $S(\boldsymbol{\alpha}) - \sum_i \alpha_i \geq \theta^*$ for all $\boldsymbol{\alpha}$ (subject to the above constraints). Hence, finding a saddle-point of (6) is equivalent to solving the following semi-infinite linear program:

$$\max \ \theta \qquad (7)$$

$$\text{w.r.t. } \theta \in \mathbb{R}, \boldsymbol{\beta} \in \mathbb{R}_+^M \text{ with } \sum_j \beta_j = 1$$

$$\text{s.t. } \sum_{j=1}^{M} \beta_j \left(\frac{1}{2} S_j(\boldsymbol{\alpha}) - \sum_i \alpha_i \right) \geq \theta$$

$$\text{for all } \boldsymbol{\alpha} \text{ with } 0 \leq \boldsymbol{\alpha} \leq C \text{ and } \sum_i y_i \alpha_i = 0$$

Note that there are infinitely many constraints (one for every vector $\boldsymbol{\alpha}$). Typically algorithms for solving semi-infinite problems work by iteratively finding violated constraints, i.e. $\boldsymbol{\alpha}$ vectors, for intermediate solutions $(\boldsymbol{\beta}, \theta)$. Then one adds the new constraint (corresponding to the new $\boldsymbol{\alpha}$) and resolves for $\boldsymbol{\beta}$ and θ [10] (see next section for details).

Fortunately, finding the constraint that is most violated corresponds to solving the SVM optimization problem for a fixed weighting of the kernels:

$$\sum_{j=1}^{M} \beta_j \left(\frac{1}{2} S_j(\boldsymbol{\alpha}) - \sum_i \alpha_i \right) = \sum_{r,s} \alpha_r \alpha_s y_r y_s K_{r,s} - \sum_i \alpha_i,$$

where $K = \sum_j \beta_j K_j$. Due to the number of efficient SVM Optimizers, the problem of finding the most violated constraint can be solved efficiently, too.

Finally, one needs some convergence criterion. Note that the problem is solved when all constraints are satisfied while the β's and θ are optimal. Hence, it is a quite natural choice to use the normalized maximal constraint violation as a convergence criterion. In our case this would be:

$$\epsilon := \left| 1 - \frac{\sum_{j=1}^{M} \beta_j^k \left(\frac{1}{2} S_j(\boldsymbol{\alpha}^{k+1}) - \sum_i \alpha_i^{k+1} \right)}{\theta^k} \right|,$$

where $(\boldsymbol{\beta}^k, \theta^k)$ is the optimal solution at iteration k and $\boldsymbol{\alpha}^{k+1}$ corresponds to the newly found maximally violating constraint of the next iteration (i.e. the SVM solution for weighting $\boldsymbol{\beta}$). We usually only try to approximate the optimal solution and set ϵ to either 10^{-4} or 10^{-2} in our experiments.

Column Generation and Boosting. There are several ways for solving (7). As outlined above, one may iteratively extend and solve a reduced problem, only

Algorithm 1 The column generation algorithm (left) employs a linear programming solver and the boosting-like algorithm (right) uses exponential updates to iteratively solve the semi-infinite linear optimization problem (7). The accuracy parameter ϵ is assumed to be given to the algorithm

---

$D^0 = 1, \theta^1 = 0, \beta_k^1 = \frac{1}{M}$ for $k = 1, \ldots, M$

for $t = 1, 2, \ldots$ **do**
  obtain SVM's $\boldsymbol{\alpha}^k$ with kernel

$$k^t(\mathbf{s}_i, \mathbf{s}_j) := \sum_{k=1}^{M} \beta_k^t k_k(\mathbf{s}_i, \mathbf{s}_j)$$

for $k = 1, \ldots, M$ **do**
  $D_k^t = \frac{1}{2} \sum_{r,s} \alpha_r^t \alpha_s^t y_r y_s k_k(\mathbf{s}_r, \mathbf{s}_s) - \sum_r \alpha_r^t$
end for
$D^t = \sum_{k=1}^{M} \beta_k^t D_k^t$

$(\boldsymbol{\beta}^{t+1}, \theta^{t+1}) = \arg\max \theta$
  w.r.t. $\boldsymbol{\beta} \in \mathbb{R}_+^M, \theta \in \mathbb{R}$ with $\sum_k \beta_k = 1$
    s.t. $\sum_{k=1}^{M} \beta_k D_k^r \geq \theta$ for $r = 1, \ldots, t$
if $|1 - \frac{D^t}{\theta^t}| \leq \epsilon$ **then break**
end for

$D^0 = 1, \rho^1 = \tau_k^1 = 0, \beta_k^1 = \frac{1}{M}$ for $k = 1, \ldots, M$

for $t = 1, 2, \ldots$ **do**
  obtain SVM's $\boldsymbol{\alpha}^k$ with kernel

$$k^t(\mathbf{s}_i, \mathbf{s}_j) := \sum_{k=1}^{M} \beta_k^t k_k(\mathbf{s}_i, \mathbf{s}_j)$$

for $k = 1, \ldots, M$ **do**
  $D_k^t = \frac{1}{2} \sum_{r,s} \alpha_r^t \alpha_s^t y_r y_s k_k(\mathbf{s}_r, \mathbf{s}_s) - \sum_r \alpha_r^t$
end for
$D^t = \sum_{k=1}^{M} \beta_k^t D_k^t$

$\gamma_t = \arg\min_{\gamma \in [0,1]} \sum_{k=1}^{M} \beta_k^t \exp\{\gamma(D_k - \rho^t)\}$
for $k = 1, \ldots, M$ **do**
  $\tau_k^{t+1} = \tau_k^t + \gamma_t D_k^t$
  $\beta_k^{t+1} = \beta_k^t \exp(\gamma_t D_k^t) / \left(\sum_{k'} \beta_{k'}^t \exp(\gamma_t D_{k'}^t) \right)$
end for
$\rho^{t+1} = \max_k \tau_k^{t+1} / \sum_{r=1}^{t} \gamma_r$
if $|1 - \frac{\rho^{t+1}}{D^t}| \leq \epsilon$ **then break**
end for

---

using t constraints corresponding to $\boldsymbol{\alpha}_t$ ($t = 1, 2, \ldots$). By repeatedly adding the most violating constraints one typically converges fast to the optimal solution [2]. Note, however, that there are no known convergence rates for such algorithms [10], but it often converges to the *optimal* solution in a small number of iterations [2, 22].

We would like to consider the use of a boosting-like technique [8] which has been used to solve semi-infinite problems [24]. It has been shown that in order to solve a semi-infinite problem like (7), one needs at most $T = \mathcal{O}(\log(M)/\hat{\epsilon}^2)$ iterations (i.e. SVM optimization), where $\hat{\epsilon}$ is the unnormalized constraint violation and the constants may depend on the kernels and the number of examples N. At least for not too small values of $\hat{\epsilon}$ this technique produces reasonably fast good approximate solutions. We cannot go into detail of the derivation of the algorithm, but only state the boosting-like algorithm that has the above property and is similar to the Arc-GV algorithm [4] and to AdaBoost* [21].

The pseudo codes for both algorithms are given for completeness Algorithm 1.

An SMO-Like Algorithm for Simultaneous Optimization of α and β
Usually it is infeasible to use standard optimization tools (e.g. MINOS, CPLEX,

LOQO) for solving the SVM training problems on datasets containing more than a few thousand examples. So-called decomposition techniques overcome this limitation by exploiting the special structure of the SVM problem. The key idea of decomposition is to freeze all but a small number of optimization variables (*working set*) and to solve a sequence of constant-size problems (subproblems of (2)).

The general idea of Sequential Minimal Optimization (SMO) algorithm has been proposed by [20] and is implemented in many SVM software packages. Here we would like to propose an extension of the SMO algorithm to optimize the kernel weights β and the example weights α at the same time. The algorithm is motivated from an insufficiency of the column-generation algorithm described in the previous section: If the β's are not optimal yet, then the optimization of the α's until optimality is not necessary and therefore inefficient. It would be considerably faster if for any newly obtained α in the SMO iterations, we could efficiently recompute the optimal β and then continue optimizing the α's using the new kernel weighting.

Recomputing β involves solving a linear program and the problem grows with each additional α-induced constraint. Hence, after many iterations solving the LP may become infeasible. Fortunately, there are two facts making it still possible: (1) only a small number of the added constraints are active and one may for each newly added constraint remove an old inactive one — this prevents the LP to grow arbitrarily and (2) for Simplex-based LP optimizers such as CPLEX there exists the so-called *hot-start feature* which allows one efficiently recompute the new solution, if one, for instance, only adds an additional constraint.

The SVM optimizer internally needs the output $\hat{f}_j = \sum_i \alpha_i y_i k(\mathbf{s}_i, \mathbf{s}_j)$ for all training examples in order to select the next variables for optimization [12]. However, if one changes the kernel weights, then the stored $\hat{f}_j$ values become invalid and need to be recomputed. In order to avoid the full re-computation one has to additionally store a $M \times N$ matrix $f_{k,j} = \sum_i \alpha_i y_i k_k(\mathbf{s}_i, \mathbf{s}_j)$, i.e. the outputs for each kernel separately. If the β's change, then $\hat{f}_j$ can be quite efficiently recomputed by $\hat{f}_j = \sum_k \beta_k f_{k,j}$.

Finally, in each iteration the SMO optimizer may change a subset of the α's. In order to update $\hat{f}_j$ and $f_{j,k}$ one needs to compute full rows j of each kernel for every changed α_j. Usually one uses kernel-caching to reduce the computational effort of this operation, which is, however, in our case not efficient enough. Fortunately, for the afore mentioned WD kernel there is a way to avoid this problem by using so-called suffix trees (as similarly proposed in [17]). Due to a lack of space we cannot go into more detail here. We provide, however, the pseudo-code of the algorithm which takes the above discussion into account and provides a few more details in Algorithm 2.

Empirically we noticed that the proposed SMO-like algorithm is often 3-5 times faster than the column-generation algorithm proposed in the last section, while achieving the same accuracy. In the experiments in Section 3 we therefore only used the SMO-like algorithm.

Algorithm 2 Outline of the SMO algorithm that optimizes α and the kernel weighting β simultaneously. The accuracy parameter ϵ and the subproblem size Q are assumed to be given to the algorithm. For simplicity we omit the removal of inactive constraints. Also note that from one iteration to the next the LP only differs by one additional constraint. This can usually be exploited to save computing time for solving the LP

$f_{k,i} = 0$, $\hat{f}_i = 0$, $\alpha_i = 0$ for $k = 1, \ldots, M$ and $i = 1, \ldots, N$
for $t = 1, 2, \ldots$ **do**
    Check optimality conditions and stop if optimal
    select Q suboptimal variables $i_1, \ldots, i_Q$ based on $\hat{\mathbf{f}}$ and α
    $\alpha^{old} = \alpha$
    solve (2) with respect to the selected variables and update α
    create suffix trees to prepare efficient computation of $g_k(\mathbf{s}) = \sum_{q=1}^{Q} (\alpha_{i_q} -$
    $\alpha_{i_q}^{old}) y_{i_q} k_k(\mathbf{s}_{i_q}, \mathbf{s})$
    $f_{k,i} = f_{k,i} + g_k(\mathbf{s}_i)$ for all $k = 1, \ldots, M$ and $i = 1, \ldots, N$
    for $k = 1, \ldots, M$ **do**
        $D_k^t = \frac{1}{2} \sum_r f_{k,r} \alpha_r^t y_r - \sum_r \alpha_r^t$
    end for
    $D^t = \sum_{k=1}^{M} \beta_k^t D_k^t$

    if $|1 - \frac{D^t}{\theta^t}| \geq \epsilon$
        $(\beta^{t+1}, \theta^{t+1}) = \arg\max \ \theta$
            w.r.t. $\beta \in \mathbb{R}_+^M, \theta \in \mathbb{R}$ with $\sum_k \beta_k = 1$
            s.t. $\sum_{k=1}^{M} \beta_k D_k^r \geq \theta$ for $r = 1, \ldots, t$
    else
        $\theta^{t+1} = \theta^t$
    end if
    $\hat{f}_i = \sum_k \beta_k^{t+1} f_{k,i}$ for all $i = 1, \ldots, N$
end for

2.3 Estimating the Reliability of a Weighting

Finally we want to assess the reliability of the learned weights scheme β. For this purpose we generate T bootstrap samples and rerun the whole procedure resulting in T weightings β^t.

To test the importance of a weight $\beta_{k,i}$ (and therefore the corresponding kernels for position and oligomer length) we apply the following method: define a Bernoulli variable $X_{k,i}^t \in \{0, 1\}$, $k = 1, \ldots, d, i = 1, \ldots, L, t = 1, \ldots, T$ by

$$X_{k,i}^t = \begin{cases} 1, & \beta_{k,i}^t > \tau := \mathbf{E}_{k,i,t} X_{k,i}^t \\ 0, & \text{else} \end{cases}.$$

The sum $Z_{k,i} = \sum_{t=1}^{T} X_{k,i}^t$ has a binomial distribution $\text{Bin}(T, p_0)$, p_0 unknown. We estimate p_0 with $\hat{p}_0 = \#(\beta_{k,i}^t > \tau)/T \cdot M$, i.e. the empirical probability to observe $P(X_{k,i}^t = 1)$, $\forall k, i, t$. We test whether $Z_{k,i}$ is as large as could be expected under $\text{Bin}(T, \hat{p}_0)$ or larger: $\mathcal{H}_0 : p \leq c^*$ vs $\mathcal{H}_1 : p > c^*$. Here c^* is defined as $\hat{p}_0 + 2\text{Std}_{k,i,t} X_{k,i}^t$ and can be interpreted as an upper bound of the confidence interval for p_0. This choice is taken to be adaptive to the noise level of the data and hence the (non)-sparsity of the weightings β^t. The hypotheses are tested with a Maximum-Likelihood test on an α-level of $\alpha = 0.05$; that is c^{**} is the minimal value for that the following inequality hold:

$$0.05 = \alpha \geq P_{\mathcal{H}_0}(\text{reject } \mathcal{H}_0) = P_{\mathcal{H}_0}(Z_{k,i} > c^{**}) = \sum_{j=c^{**}}^{T} \binom{T}{j} \hat{p}_0 (1 - \hat{p}_0).$$

For further details on the test see [18] or [16]. This test is carried out for every $\beta_{k,i}^t$. (We assume independence between the weights in one single β, and hence assume that the test problem is the same for every $\beta_{k,i}$). If $\mathcal{H}_0$ can be rejected, the kernel learned at position i on the k-mer is important for the detection and thus (should) contain biologically interesting knowledge about the problem at hand.

3   Results and Discussion

The main goal of this work is to provide an explanation of the SVM decision rule, for instance by identifying sequence positions that are important for discrimination. In the first part on our evaluation, however, we show that the proposed algorithm optimizing kernel weights does perform also slightly better than the standard kernel and leads to SVM classification functions that are computationally more efficient. A detailed comparison of the WD kernel approach with other state-of-the-art methods is provided in [23] and not considered in this work. In the remaining part we show how the weights can be used to obtain a deeper understanding of how the SVM classifies sequences and match it with knowledge about the underlying biological process.

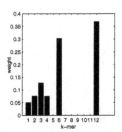

Fig. 1. Optimized WD kernel weights

3.1   Comparison with the Original WD Kernel

To compare classification performance and running time with the original WD kernel, we trained SVMs on the *C. elegans* acceptor splice dataset using $100,000$ sequences in training, $100,000$ examples for validation and $60,000$ examples to test the classifiers performance (cf. Appendix A.2). In this dataset each sequence is a window centered around the true splice site containing 141 nucleotides. Using this setup we perform cross-validation over the parameters $M \in \{10, 12, 15, 17, 20\}$ and $C \in \{0.5, 2, 5, 10\}$. Accuracy was set to $\epsilon = 0.01$.

We find on the validation set that for the original WD kernel $M = 20$ and $C = 0.5$ gives best classification performance (ROC Score 99.66), while for the SVM using the WD kernel that also learns the weighting give best results ($M = 12$

and $C = 12$; ROC Score also 99.66).[2] The figure on the left shows the weights the proposed WD kernel has learned, suggesting that 12-mers and 6-mers seem to be of high importance. On the test dataset the original WD kernel performs as good as on the validation dataset (ROC Score 99.66% again), while the new WD kernel achieves a 99.67% ROC Score. Astonishingly training the new WD kernel SVM (i.e. with weight optimization) was 1.5 times faster than training the the original SVM, which might be due to using a suffix tree requiring no kernel cache. Also note that the resulting classifier provided by the new algorithm is considerably faster than the one obtained by the classical SVM since many weights are zero [7].

3.2   Relation to Positional Weight Matrices

An interesting relation of the learned weightings to the relative entropy between Positional Weight Matrices can be shown with the following experiment: We train an SVM with a WD kernel that consists of 60 first-order sub-kernels on acceptor splice sites from *C. elegans* (100,000 sequences for training, 160,000 sequences for validation). For the SVMs, $C = 1$ and for the WD Kernel accuracy $\epsilon = 0.01$ were chosen. The AG consensus is at position $31 - 32$ and a window of length ± 29 around the consensus was chosen. The learned weights β_k are shown in figure 2 (left). For comparison we computed the Positional Weight Matrices

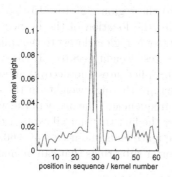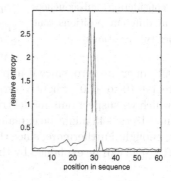

Fig. 2. (left) Value of the learned weightings of an SVM with a WD kernel of 60 first-order sub-kernels, (right) relative entropy obtained between the Positional Weight Matrices for the positive and the negative class, both trained for acceptor splice site detection

for the positive and the negative class separately and computed the relative entropy Δ_i between the two probability estimates $p_{i,j}^+$ and $p_{i,j}^-$ at each position j by $\Delta_i = \sum_{i=1}^4 p_{i,j}^+ \log(p_{i,j}^+/p_{i,j}^-)$, which is shown in Figure 2 (right). The shape

---

[2] While the model selection results are at the border of the parameter range checked, the obtained ROC scores are always $\geq$ 99.65%.

of both plots is very similar, i.e. both methods consider upstream information, as well as a position directly after the splice site to be highly important. As a major difference the WD-weights in the exons remain on a high level. Note that both methods use only first order information. Nevertheless the classification accuracy is extremely high. On the separate validation set the SVM already achieves a ROC score of 99.07% and the Positional Weight Matrices a ROC score of 98.83%.

In another experiment we considered the larger region from -50 to +60nt around the splice site and used a kernel of degree 15. We defined a kernel for every position that only accounts for substrings that start at the corresponding position (up to length 15). To get a smoother weighting and to reduce the computing time we only used $\lceil 111/2 \rceil = 56$ weights (combining every two positions to one weight).

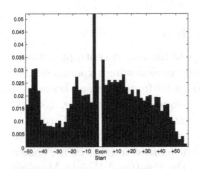

Fig. 3. Optimized WD kernel weights considering subsequences starting at different positions (one weight per two positions)

The average computed weighting on ten bootstrap runs trained on around $65,000$ examples is shown in Figure 3. Several regions of interested can be identified: a) The region -50 to -40, which corresponds to the donor splice site of the previous exon (many introns in *C. elegans* are very short, often only 50nt), b) the region -25 to -15 that corresponds to the location of the branch point, c) the intronic region closest to the splice site with greatest weight (-8 to -1; the weights for the AG dimer are zero, since it appears in splice sites and decoys) and d) the exonic region (0 to $+50$). Slightly surprising are the high weights in the exonic region, which we suspect only model triplet frequencies. The decay of the weights seen from $+15$ to $+45$ might be explained by the fact that not all exons are actually long enough. Furthermore, since the sequence ends in our case at $+60$nt, the decay after $+45$ can be explained by the shorter substrings that can be matched.

3.3   Detecting Motifs in a Toy Dataset

As a prove of concept, we test the our method on a toy datasets at four different noise levels (for a detailed description of the data see Appendix A.1). For every noise level, we train on 100 bootstrap replicates ($C = 2$, $\epsilon = 0.001$) and learn $M = 350$ WD kernel parameters in each run. On the resulting 100 weightings we performed the reliability test (cf. Section 2.3). The results are shown in Figure 4 (columns correspond to different noise levels — increasing from left to right). Each figure shows a kernel weighting β, where columns correspond to weights used at a certain sequence position and rows to the k-mer length used at that position. The plots in the first row show the weights that are detected to be important at a significance level of $\alpha = 0.05$ in bright (yellow) color. The likelihood for every weight to be detected by the test and thus to reject $\mathcal{H}_0$ is illustrated in the plots in the second row. Here bright color means it is more likely to reject $\mathcal{H}_0$.

As long as the noise level does not exceed 2/7, higher order matches of length 3 and 4 seem sufficient to distinguish sequences containing motifs from the rest. However, only the 3-mer is detected with the test procedure. When more nucleotides in the motifs are replaced with noise, more weights are determined to be of importance. This becomes especially obvious in column 3 were 4 out of 7 nucleotides within each motif were randomly replaced, but still an average ROC score of 99.6% is achieved. In the last column the ROC score drops down to 83%, but only weights in the correct range 10 . . . 16 and 30 . . . 36 are found to be significant.

3.4   Detecting Motifs in Splice Data

As the results on the toy dataset are promising, we now apply our method to a more realistic problem. We consider the classification of acceptor splice sites against non-acceptor splice sites (with AG dimer) from the *C. elegans* (cf. Appendix A.2 for details on the generation of the data sets).

We trained our Multiple Kernel Learning algorithm ($C = 2$, $\epsilon = 10^{-4}$) on 5,000 randomly chosen sequences of length 111 with a maximal oligomer length of $d = 10$. This leads to $M = 1110$ kernels in the convex combination. Figure 5 shows the results obtained for this experiment (similarly organized as Figure 4). We can observe (cf. Figure 5 b & c) that the optimized kernel coefficients are biologically plausible: longer significant oligomers were found close to the splice site position, oligomers of length 3 and 4 are mainly used in the exonic region (modeling triplet usage) and short oligomers near the branch site. Note, however,

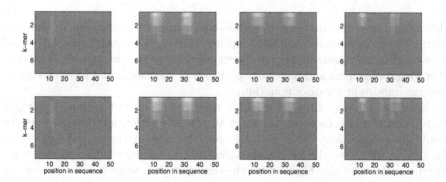

Fig. 4. In this "figure matrix", columns correspond to the noise level, i.e. different numbers of nucleotides randomly substituted in the motif of the toy dataset cf. Appendix A.1. Each sub-plot shows the kernel weighting β, where columns correspond to weights used at a certain sequence position and rows to the oligomer length used at that position. The first row shows the kernel weights that are significant, while the second row describes the likelihood of every weight to be rejected under $\mathcal{H}_0$

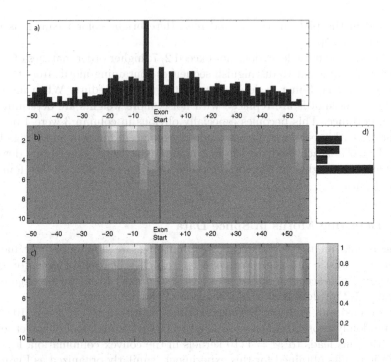

Fig. 5. Figure a) shows the average weight (over 10 runs) of the weights per position (fixed degree weighting; one weight for two positions) and d) the averaged weights per degree (uniform position weighting). Figures b) displays the position/degree combinations that were found to be significantly used (40 bootstrap runs). Figure c) shows the likelihood for rejecting $\mathcal{H}_0$. All runs only used 5,000 training examples

that one should use more of the available examples for training in order to extract more meaningful results (adapting 1110 kernel weights may have lead to overfitting). In some preliminary tests using more training data we observed that longer oligomers and also more positions in the exonic and intronic regions become important for discrimination.

Note that the weight matrix would be the outer product of the position weight vector and the oligomer-length weight vector, if position and oligomer length would be independent. This is clearly not the case: it seems very important (according to the weight for oligomer-length 5) to consider longer oligomers for discrimination (see also Figure 1) in the central region, while it is only necessary and useful to consider monomers and dimers in other parts of the sequence.

4 Summary

In this work we have developed a novel Multiple Kernel Learning algorithm for large-scale sequence analysis problems. The reformulation as a semi-infinite linear programming problem allowed us to reuse efficient SVM optimization implementations for finding the optimal convex combination of kernels. We proposed a simple, easy-to-implement but yet effective boosting-like technique to solve the resulting optimization problem, which comes with good convergence guarantees. The suggested column-generation technique, however, is in practice often faster but requires an efficient LP solver (such as CPLEX). Additionally, we showed that for kernels like the WD kernel an efficient SMO-like algorithm can be derived which in turn is even more efficient than the proposed column-generation algorithm, as it exploits special properties of string kernels.

In experiments on toy and splice-site detection problems we illustrated the usefulness of the Multiple Kernel Learning approach. The optimized kernel convex combination gives valuable hints at which positions discriminative oligomers of which length are hidden in the sequences. This solves to a certain extend one of the major problems with Support Vector Machines: now the decisions become interpretable. On the toy data set we re-discovered hidden sequence motifs even in presence of a large amount of noise. In first preliminary experiments on the acceptor splice site detection problem we discovered patterns in the optimized weightings which are biologically plausible. It is future work to perform a more extensive computational evaluation on splice sites and other signal detection problems.

Acknowledgments. The authors gratefully acknowledge partial support from the PASCAL Network of Excellence (EU #506778), DFG grants JA 379 / 13-2 and MU 987/2-1. We thank Alexander Zien and K.-R. Müller for great discussions and proof reading the manuscript.

N.B. The appendix contains details regarding the data generation and estimates of the computational complexity of the proposed SMO-like optimization technique. Additional information about this work can be found at `http://ida.first.fraunhofer.de/~sonne/mkl_splice`.

References

1. Francis R. Bach, Gert R. G. Lanckriet, and Michael I. Jordan. Multiple kernel learning, conic duality, and the SMO algorithm. In *Twenty-first international conference on Machine learning*. ACM Press, 2004.
2. K.P. Bennett, A. Demiriz, and J. Shawe-Taylor. A column generation algorithm for boosting. In P. Langley, editor, *Proceedings, 17th ICML*, pages 65–72, San Francisco, 2000. Morgan Kaufmann.
3. M.S. Boguski and T.M. Lowe C.M. Tolstoshev. dbEST–database for "expressed sequence tags". *Nat Genet.*, 4(4):332–3, 1993.
4. L. Breiman. Prediction games and arcing algorithms. Technical Report 504, Statistics Department, University of California, December 1997.
5. C. Cortes and V.N. Vapnik. Support vector networks. *Machine Learning*, 20:273–297, 1995.
6. A.L. Delcher, D. Harmon, S. Kasif, O. White, and S.L. Salzberg. Improved microbial gene identification with GLIMMER. *Nucleic Acids Research*, 27(23):4636–4641, 1999.

7. Y. Engel, S. Mannor, and R. Meir. Sparse online greedy support vector regression. In *ECML*, pages 84–96, 2002.

8. Y. Freund and R.E. Schapire. A decision-theoretic generalization of on-line learning and an application to boosting. In *EuroCOLT: European Conference on Computational Learning Theory*. LNCS, 1994.

9. Harris, T.W. et al. Wormbase: a multi-species resource for nematode biology and genomics. *Nucl. Acids Res.*, 32, 2004. Database issue:D411-7.

10. R. Hettich and K.O. Kortanek. Semi-infinite programming: Theory, methods and applications. *SIAM Review*, 3:380–429, September 1993.

11. T. Jaakkola, M. Diekhans, and D. Haussler. A discriminative framework for detecting remote protein homologies. *J Comput Biol.*, 7(1-2):95–114, February 2000.

12. T. Joachims. Making large–scale SVM learning practical. In B. Schölkopf, C.J.C. Burges, and A.J. Smola, editors, *Advances in Kernel Methods — Support Vector Learning*, pages 169–184, Cambridge, MA, 1999. MIT Press.

13. W.J. Kent. Blat–the blast-like alignment tool. *Genome Res.*, 12(4):656–64, 2002.

14. R. Kuang, E. Ie, K. Wang, K. Wang, M. Siddiqi, Y. Freund, and C. Leslie. Profile-based string kernels for remote homology detection and motif extraction. In *Computational Systems Bioinformatics Conference 2004*, pages 146–154, 2004.

15. G.R.G. Lanckriet, T. De Bie, N. Cristianini, M.I. Jordan, and W.S. Noble. A statistical framework for genomic data fusion. *Bioinformatics*, 2004.

16. E.L. Lehmann. *Testing Statistical Hypotheses*. Springer, New York, second edition edition, 1997.

17. C. Leslie, E. Eskin, and W.S. Noble. The spectrum kernel: A string kernel for SVM protein classification. In *Proceedings of the Pacific Symposium on Biocomputing*, Kaua'i, Hawaii, 2002.

18. A.M. Mood, F.A. Graybill, and D.C. Boes. *Introduction to the Theory of Statistics*. McGraw-Hill, third edition edition, 1974.

19. K.-R. Müller, S. Mika, G. Rätsch, K. Tsuda, and B. Schölkopf. An introduction to kernel-based learning algorithms. *IEEE Transactions on Neural Networks*, 12(2):181–201, 2001.

20. J. Platt. Fast training of support vector machines using sequential minimal optimization. In B. Schölkopf, C.J.C. Burges, and A.J. Smola, editors, *Advances in Kernel Methods — Support Vector Learning*, pages 185–208, Cambridge, MA, 1999. MIT Press.

21. G. Rätsch. *Robust Boosting via Convex Optimization*. PhD thesis, University of Potsdam, Computer Science Dept., August-Bebel-Str. 89, 14482 Potsdam, Germany, 2001.

22. G. Rätsch, A. Demiriz, and K. Bennett. Sparse regression ensembles in infinite and finite hypothesis spaces. *Machine Learning*, 48(1-3):193–221, 2002. Special Issue on New Methods for Model Selection and Model Combination. Also NeuroCOLT2 Technical Report NC-TR-2000-085.

23. G. Rätsch and S. Sonnenburg. *Accurate Splice Site Prediction for Caenorhabditis Elegans*, pages 277–298. MIT Press series on Computational Molecular Biology. MIT Press, 2003.

24. G. Rätsch and M.K. Warmuth. Marginal boosting. NeuroCOLT2 Technical Report 97, Royal Holloway College, London, July 2001.

25. Wheeler, D.L. et al. Database resources of the national center for biotechnology. *Nucl. Acids Res*, 31:38–33, 2003.

26. X.H. Zhang, K.A. Heller, I. Hefter, C.S. Leslie, and L.A. Chasin. Sequence information for the splicing of human pre-mrna identified by support vector machine classification. *Genome Res*, 13(12):637–50, 2003.

27. A. Zien, G. Rätsch, S. Mika, B. Schölkopf, T. Lengauer, and K.-R. Müller. Engineering Support Vector Machine Kernels That Recognize Translation Initiation Sites. *BioInformatics*, 16(9):799–807, September 2000.

A   Data Generation

A.1   Toy Data

We generated $11,000$ sequences of length 50, where the symbols of the alphabet $\{A, C, G, T\}$ follow a uniform distribution. We chose $1,000$ of these sequences to be positive examples and hid two motifs of length seven: at position 10 and 30 the motifs GATTACA and AGTAGTG, respectively. The remaining $10,000$ examples were used as negatives. Thus the ratio between examples of class $+1$ and class -1 is $\approx 9\%$. In the positive examples, we then randomly replaced $s \in \{0, 2, 4, 5\}$ symbols in each motif. Leading to four different data sets which where randomly permuted and split such that the first $1,000$ examples became training and the remaining $10,000$ validation examples.

A.2   Splice Site Sequences

EST and cDNA Sequences. We collected all known *C. elegans* ESTs from Wormbase [9] (release WS118; 236,868 sequences), dbEST [3] (as of February 22, 2004; 231,096 sequences) and UniGene [25] (as of October 15, 2003; 91,480 sequences). Using *blat* [13] we aligned them against the genomic DNA (release WS118). The alignment was used to confirm exons and introns. We refined the alignment by correcting typical sequencing errors, for instance by removing minor insertions and deletions. If an intron did not exhibit the consensus GT/AG or GC/AG at the 5' and 3' ends, then we tried to achieve this by shifting the boundaries up to 2 nucleotides (nt). If this still did not lead to the consensus, then we splitted the sequence into two parts and considered each subsequence separately. For each sequence we determined the longest open reading frame (ORF) and only used the part of each sequence within the ORF. In a next step we merged alignments, if they did not disagree and shared at least one complete exon. This lead to a set of 135,239 unique EST-based sequences.

We repeated the above procedure with all known cDNAs from Wormbase (release WS118; 4,848 sequences) and UniGene (as of October 15, 2003; 1,231 sequences), which lead to 4,979 unique sequences. We removed all EST matches fully contained in the cDNA matches, leaving 109,693 EST-base sequences.

Clustering. We clustered the sequences in order to obtain independent training, validation and test sets. In the beginning each of the above EST and cDNA sequences were in a separate cluster. We iteratively joined clusters, if any two sequences from distinct clusters (a) match to the genome at most 100nt apart (this includes many forms of alternative splicing) or (b) have more than 20% sequence overlap (at 90% identity, determined by using *blat*). We obtained 17,763 clusters with a total of 114,672 sequences. There are 3,857 clusters that contain at least one cDNA. Finally, we removed all clusters that showed alternative splicing.

Splitting into Training, Validation and Test Sets. Since the resulting data set is still pretty large, we only used sequences from randomly chosen 20% of clusters with cDNA and 30% of clusters without cDNA to generate true acceptor splice site sequences (15,507 of them). Each sequence is 398nt long and has the AG dimer at 200. Negative examples were generated from any occurring AG within the ORF of the sequence (246,914 of them were found).

For experiments on *C. elegans* we used a random subset of 60,000 examples for testing, 100,000 examples for parameter tuning and up to 100,000 examples for training (unless stated otherwise).

B   Estimating the Computational Complexity

The computational complexity of the proposed algorithm is determined by the complexity of training a SVM and by the number of iterations. For boosting type algorithms the number of iterations in order to achieve an accuracy of ϵ with M variables/kernels can be bounded by $\log(M)/\epsilon^2$ [24]. Moreover, the running time of SVM optimization is often between N^2 and N^3 for state-of-the-art SVM optimizers (such as *SVM-light* [12]). Hence, for the boosting-like algorithm one would expect a computational complexity of $\mathcal{O}(N^2 M \log(M)/\epsilon^2)$ (one needs an additional factor of M to prepare the combined kernels). In practice, however, the column-generation algorithm is considerably faster than the boosting-like algorithm [2]. Furthermore, for the algorithm, which performs the optimization w.r.t. the α's and β's simultaneously, one would expect an additional speedup.

In order to obtain a more realistic estimate of the computational complexity, we ran the latter algorithm (simultaneous optimization) over a wide range of $N \in [10^2, 2 \cdot 10^5]$, $M \in [10, 398]$ and $\epsilon \in [3 \cdot 10^{-7}, 1]$ (cf. Figure 6). From these running times we coordinate-wisely estimated the powers of N, M and $\log(1/\epsilon)$,

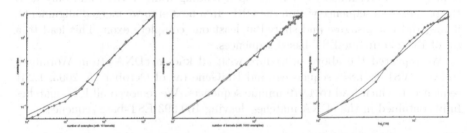

Fig. 6. Running times in seconds for various numbers of examples N (number of kernels $M = 10$, $\epsilon = 10^{-2}$), different numbers of kernels (number of examples $N = 10,00$, $\epsilon = 10^{-2}$) and some values of ϵ (the relative gap; $N = 10,000$ and $M = 100$)

and obtained the following running time estimate for standard PC hardware (*Athlon 2200+*):

$$t(N, M, \epsilon) = cM^{2.22}N^{1.68}\log_2(1/\epsilon)^{2.52} + \mathcal{O}(1),$$

where $c \approx 5 \cdot 10^{-11}$.

One can observe that the measured computational complexity is considerably lower than expected theoretically (at least w.r.t. N and ϵ). In order to make our algorithm's running times comparable to the ones given in [1], we give two examples: For $N = 6212$, $M = 4$ and $\epsilon = 10^{-4}$, we need approximately 2 seconds, while [1] needs 670 seconds. Furthermore, for $N = 1605$, $M = 48$ and $\epsilon = 10^{-4}$, we need about 44 seconds, while [1] needs 618 seconds.

Segmentation Conditional Random Fields (SCRFs): A New Approach for Protein Fold Recognition

Yan Liu[1], Jaime Carbonell[1], Peter Weigele[2], and Vanathi Gopalakrishnan[3]

[1] School of Computer Science, Carnegie Mellon University, Pittsburgh, PA, USA
{yanliu, jgc}@cs.cmu.edu
[2] Biology Department, Massachusetts Institute of Technology, Cambridge, MA, USA
pweigele@mit.edu
[3] Center for Biomedical Informatics, University of Pittsburgh, PA, USA
vanathi@cbmi.pitt.edu

Abstract. Protein fold recognition is an important step towards understanding protein three-dimensional structures and their functions. A conditional graphical model, i.e. segmentation conditional random fields (SCRFs), is proposed to solve the problem. In contrast to traditional graphical models such as hidden markov model (HMM), SCRFs follow a discriminative approach. It has the flexibility to include overlapping or long-range interaction features over the whole sequence, as well as global optimally solutions for the parameters. On the other hand, the segmentation setting in SCRFs makes its graphical structures intuitively similar to the protein 3-D structures and more importantly, provides a framework to model the long-range interactions directly.

Our model is applied to predict the parallel β-helix fold, an important fold in bacterial infection of plants and binding of antigens. The cross-family validation shows that SCRFs not only can score all known β-helices higher than non β-helices in Protein Data Bank, but also demonstrate more success in locating each rung in the known β-helix proteins than BetaWrap, a state-of-the-art algorithm for predicting β-helix fold, and HMMER, a general motif detection algorithm based on HMM. Applying our prediction model to Uniprot database, we hypothesize previously unknown β-helices.

1 Introduction

It is believed that protein structures reveal important information about the protein functions. One key step towards modeling a tertiary structure is to identify how secondary structures as building blocks arrange themselves in space, i.e. the supersecondary structures or protein folds. There has been significant work on predicting some well-defined types of structural motifs or functional units, such as $\alpha\alpha$- and $\beta\beta$-hairpins [1, 2, 3, 4]. The task of protein fold recognition is the following: given a protein sequence and a particular fold or super-secondary structure, predict whether the protein contains the structural fold and if so, locate its exact positions in the sequence.

S. Miyano et al. (Eds.): RECOMB 2005, LNBI 3500, pp. 408–422, 2005.

The traditional approach for protein fold prediction is to search the database using PSI-BLAST [5] or match against an HMM profile built from sequences with the same fold by HMMER [4] or SAM [3]. These approaches work well for short motifs with strong sequence similarities. However, there exist many important motifs or folds without clear sequence similarity and involving the long-range interactions, such as folds in β class [6]. These cases necessitate a more powerful model, which can capture the structural characteristics of the protein fold. Interestingly, the protein fold recognition task parallels an emerging trend in machine learning community, i.e the *structure* prediction problem, which predict the labels of each node in a graph given the observation with particular structures, for example webpage classification using the hyperlink graph or object recognition using grids of image pixels. The *conditional* graphical models prove to be one of the most effective tools for this kind of problem [7, 8].

In fact, several graphical models have been applied to protein structure prediction. One of the early approaches is to apply simple hidden markov models (HMMs) to protein secondary structure prediction and protein motif detection [3, 4, 9]; Delcher et al. introduced probabilistic causal networks for protein secondary structure modeling [10]. Recently, Liu et al. applied conditional random fields (CRFs), a discriminative graphical model based on undirected graph, for protein secondary structure prediction [11]; Chu et al. extended segmental semi-Markov model (SSMM) under the Baysian framework for protein secondary structures [12].

The bottleneck for protein fold prediction is the long-range interactions, which could be either two β-strands with hydrogen bonds in a parallel β-sheet or helix pairs in coupled helical motifs. Generative models, such as HMM or SSMM, assume a particular generating process, which makes it difficult to consider overlapping features and long-range interactions. Discriminative graphical models, such as CRFs, assume a single residue as an observation. Thus they fail to capture the features over a whole secondary structure element or the interactions between adjacent elements in 3-D, which may be distant in the primary sequence. To solve the problem, we propose segmentation conditional random fields (SCRFs), which retain all the advantages of original CRFs and at the same time can handle observations of variable length.

2   Conditional Random Fields (CRFs)

Simple graphical chain models, such as hidden markov models (HMMs), have been applied to various problems. As a "generative" model, HMMs assume that the data are generated by a particular model and compute the joint distribution of the observation sequence $\mathbf{x}$ and state sequence $\mathbf{y}$, i.e. $P(\mathbf{x}, \mathbf{y})$. However, generative models might perform poorly with inappropriate assumptions. In contrast, discriminative models, such as neural networks and support vector machines (SVMs), estimate the decision boundary directly without computing the underlying data distribution and thus often achieve better performance.

Recently, several discriminative graphical models have been proposed by the machine learning community, such as Maximum Entropy Markov Models (MEMMs) [13] and Conditional Random fields (CRFs) [14]. Among these models, CRFs proposed by Lafferty et al., are very effective in many applications, including information extraction, image processing and so on [8, 7].

CRFs are "undirected" graphical models (also known as *random fields*, as opposed to directed graphical models such as HMMs) to compute the conditional likelihood $P(\mathbf{y}|\mathbf{x})$ directly. By the Hammersely-Clifford theorem [15], the conditional probability $P(\mathbf{y}|\mathbf{x})$ is proportional to the product of the potential functions over all the cliques in the graph,

$$P(\mathbf{y}|\mathbf{x}) = \frac{1}{Z_0} \prod_{c \in C(\mathbf{y},\mathbf{x})} \Phi_c(\mathbf{y}_c, \mathbf{x}_c),$$

where $\Phi_c(\mathbf{y}_c, \mathbf{x}_c)$ is the potential function over the clique c, and Z_0 is the normalization factor over all possible assignments of $\mathbf{y}$ (see [16] for more detail). For a chain structure, CRFs define the conditional probability as

$$P(\mathbf{y}|\mathbf{x}) = \frac{1}{Z_0} \exp(\sum_{i=1}^{N} \sum_{k=1}^{K} \lambda_k f_k(y_{i-1}, y_i, \mathbf{x}, i)), \tag{1}$$

where f_k is an arbitrary feature function over $\mathbf{x}$, N is the number of observations and K is the number of features. The model parameters λ_k are learned via maximizing the conditional likelihood of the training data.

CRFs define the clique potential as an exponential function, which results in a series of nice properties. First, the conditional likelihood function is convex so that finding the global optimum is guaranteed [14]. Second, the feature function can be arbitrary, including overlapping features and long-range interactions. Finally, CRFs still have efficient algorithms, such as forward-backward or Viterbi, as long as the graph structures are sequences or trees.

Similar to HMMs, we can define the forward-backward probability for CRFs. For a chain structure, the "forward value" $\alpha_i(y)$ is defined as the probability of being in state y at time i *given* the observation up to i. The recursive step is:

$$\alpha_{i+1}(y) = \sum_{y'} \alpha_i(y') \exp(\sum_{k} \lambda_k f_k(y', y, \mathbf{x}, i+1)).$$

Similarly, $\beta_i(y)$ is the probability of starting from state y at time i *given* the observation sequence after time i. The recursive step is:

$$\beta_i(y') = \sum_{y} \exp(\sum_{k} \lambda_k f_k(y', y, \mathbf{x}, i+1)) \beta_{i+1}(y).$$

The forward-backward and Viterbi algorithms can be derived accordingly [17].

3   Segmentation Conditional Random Fields (SCRFs)

Protein folds are frequent arrangement pattern of several secondary structure elements: some elements are quite conserved or prefer a specific length, while

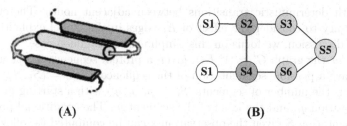

Fig. 1. Graph structure of β-α-β motif (A) 3-D structure (B) Protein structure graph: node: Green=β-strand, yellow=α-helix, cyan=coil, white=non-β-α-β (I-node); edge: $E_1 = \{$black edges$\}$ and $E_2 = \{$red edges$\}$

others might form hydrogen-bonds with each other, such as two β-strands in a parallel β-sheet. To model the protein fold better, it would be natural to think of each secondary structure element as one observation (or node) and the edges between elements indicating their interactions in 3-D. Then, given a protein sequence, we can search for the best segmentation defined by the graph and determine if the protein has the fold.

3.1   Protein Structural Graph

Before covering the algorithm in detail, we first introduce a special kind of graph, called protein structural graph. Given a protein fold, a structural graph is defined as $G = \; <V, E_1, E_2>$, where $V = U \bigcup \{I\}$, U is the set of nodes corresponding to the secondary structure elements within the fold and I is the node to represent the elements outside the fold. E_1 is the set of edges between neighboring elements in primary sequences, and E_2 is the set of edges indicating the potential long-range interactions between elements in tertiary structures. Figure 1 shows an example of the structural graph for β-α-β motif. Notice that there is a clear distinction between edges in E_1 and those in E_2 in terms of probabilistic semantics: similar to HMMs, the E_1 edges indicate transitions of states between adjacent nodes. On the other hand, the E_2 edges are used to model the long-range interactions, which is unique to the structural graph.

In practice, one protein fold might correspond to several reasonable structural graphs given different semantics for one node. There is always a tradeoff between the graph complexity, fidelity of model and the real computational costs. Therefore a good graph is the most expressive one that captures the properties of the protein folds while retaining as much simplicity as possible. There are several ways to simplify the graph, for example we can combine multiple nodes with similar properties into one, or remove those E_2 edges that are less important or less interesting to us. We give a concrete example of β-helix fold in Section 4.

3.2   Segmentation Conditional Random Fields

Since a protein fold is regular arrangement of its secondary structure elements, the general topology is often known apriori and we can easily define a structural graph with deterministic transitions between adjacent nodes. Therefore it is not necessary to consider the effect of E_1 edges in the model explicitly. In the following discussion, we focus on this simplified but common case.

Consider the graph $G' = \langle V, E_2 \rangle$, given a protein sequence $\mathbf{x} = x_1 x_2 \ldots x_N$, we can have a possible segmentation of the sequence, i.e. $S = (S_1, S_2, \ldots, S_M)$, where M is the number of segments, $S_i = \langle p_i, q_i, y_i \rangle$ with a starting position p_i, an end position q_i, and the label of the segment y_i. The conditional probability of a segmentation S given the observation x can be computed as follows:

$$P(S|\mathbf{x}) = \frac{1}{Z_0} \prod_{c \in G'(S, \mathbf{x})} \exp(\sum_k \lambda_k f_k(\mathbf{x}_c, S_c)),$$

where Z_0 is a normalization factor. If each subgraph of G' is a chain or a tree (an isolated node can also be seen as a chain), then we have

$$P(S|\mathbf{x}) = \frac{1}{Z_0} \exp(\sum_{i=1}^{M} \sum_{k=1}^{K} \lambda_k f_k(\mathbf{x}, S_i, S'_{i-1})), \tag{2}$$

where S'_{i-1} is the direct forward neighbor of S_i in graph G'.

We estimate the parameters λ_k by maximizing the conditional log-likelihood of the training data:

$$L_\Lambda = \sum_{i=1}^{M} \sum_{k=1}^{K} \lambda_k f_k(\mathbf{x}, S_i, S'_{i-1}) - \log Z_0 + \frac{\lambda_k^2}{2\sigma^2},$$

where the last term is a Gaussian prior over the parameters as a smoothing term to deal with sparsity problem in the training data. To perform the optimization, we need to seek the zero of the first derivative, i.e.

$$\frac{\partial L_\Lambda}{\partial \lambda_k} = \sum_{i=1}^{M} (f_k(\mathbf{x}, S_i, S'_{i-1}) - E_{P(s|x)}[f_k(\mathbf{x}, S_i, S'_{i-1})]) + \frac{\lambda_k}{\sigma^2}, \tag{3}$$

where $E_{P(s|\mathbf{x})}[f_k(x, S_i, S'_{i-1})]$ is the expectation of feature $f_k(\mathbf{x}, S_i, S'_{i-1})$ over all possible segmentations of x. The convexity property guarantees that the root corresponds to the optimal solution. However, since there is no closed-form solution to (3), it is not straightforward to find the optimal. Recent work on iterative searching algorithms for CRFs suggests that L-BFGS converges much faster than other commonly used methods, such as iterative scaling or conjugate gradient [17], which is also confirmed in our experiments.

Similar to CRFs, we still have an efficient inference algorithm as long as each subgraph of G' is a chain. We redefine the forward probability $\alpha_{<l,y_l>}(r, y_r)$ as the conditional probability that a segment of state y_r ends at position r given the observation $x_{l+1} \ldots x_r$ and a segment of state y_l ends at position l. The recursive step can be written as:

$$\alpha_{<l,y_l>}(r, y_r) = \sum_{p, p', q'} \alpha_{<l,y_l>}(q', y') \alpha_{<q',y>}(p - 1, \overline{y_r}) \exp(\sum_k \lambda_k f_k(\mathbf{x}, S, S')),$$

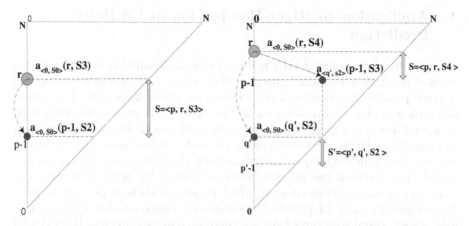

Fig. 2. An example of forward algorithm for the graph defined in Figure-1(B). x/y-axis: index of starting/end residue position; green circle: target value; red circle: intermediate value. (Left) calculation for $\alpha_{<0,S_0>}(r, S3)$ for segment S_3 with no direct forward neighbor; (right) calculation for $\alpha_{<0,S_0>}(r, S4)$ for segment S_4 with direct forward neighbor S_2

where S' is the direct forward neighbor of S in graph G' (if any), $S = \langle p, r, y_r \rangle$, $S' = \langle p', q', y' \rangle$, "$\rightarrow$" is the operator for next state and "$\leftarrow$" for previous state (the value is known since the state transition is deterministic). The range over the summation is $\sum_{p=r-\ell_1+1}^{r-\ell_2+1} \sum_{q'=l+\ell'_1-1}^{p-1} \sum_{p'=l}^{q'-\ell'_1+1}$, where $\ell_1 = \max \text{length}(y)$, $\ell_2 = \min \text{length}(y)$. Then the normalizer $Z_0 = \alpha_{<0,y_{\text{start}}>}(N, y_{\text{end}})$. Figure 2 shows a toy example of how to calculate the forward probability in detail.

Similarly, we can define the backward probability $\beta_{<r,y_r>}(l, y_l)$ as the probability of $x_{l+1} \ldots x_r$ given a segment of state y_l ends at l and a segment of state y_r ends at r. Then we have

$$\beta_{<r,y_r>}(l, y_l) = \sum_{q', p, q} \beta_{<r,y_r>}(p-1, \overleftarrow{y}) \beta_{<p'-1, \overleftarrow{y}>}(q', \overrightarrow{y_l}) \exp(\sum_k \lambda_k f_k(\mathbf{x}, S, S')),$$

where $S = \langle p, q, y \rangle$, $S' = \langle l+1, q', \overrightarrow{y_l} \rangle$. Given the backward and forward algorithm, we can compute the expectation of each feature f_k in (3) accordingly.

For a test sequence, we search for the segmentation that maximizes the conditional likelihood $P(S|x)$. Similar to CRFs, we define:

$$\delta_{<l,y_l>}(r, y_r) = \sum_{p, p', q'} \delta_{<l,y_l>}(q', y') \delta_{<q',y>}(p-1, \overleftarrow{y_r}) \exp(\sum_k \lambda_k f_k(\mathbf{x}, S, S')).$$

The best segmentation is the path traced back by $\max \delta_{<0,y_{\text{start}}>}(N, y_{\text{end}})$, where N is the number of residues in the sequence.

In general, the computational cost of SCRFs for the forward-backward probability and Viterbi algorithm will be polynomial to the length of the sequence N. However, in most real applications of protein fold prediction, the number of possible residues in each node is much smaller than N or fixed. Therefore the final complexity will be approximately $O(N^2)$.

4   Application to Right-Handed Parallel β-Helix Prediction

The right-handed parallel β-helix fold is an elongated helix-like structure with a series of progressive stranded coilings (called *rungs*), each of which is composed of three parallel β-strands to form a triangular prism shape [18]. The typical 3-D structure of a β-helix is shown in Fig. 3(A-B). As we can see, each basic structural unit, i.e. a rung, has three β-strands of various lengths, ranging from 3 to 5 residues. The strands are connected to each other by loops with distinctive features. One loop is a unique two-residue turn which forms an angle of approximately 120° between two parallel β-strands (called *T-2 turn*). The other two loops vary in size and conformation, which might contain helix or even β-sheets. There currently exist 14 protein sequences with three-stranded right-hand β-helix whose crystal structures have been deposited in Protein Data Bank (PDB) (See Table 1). The β-helix structures are significant in that they include pectate lyases, which are secreted by pathogens and initiate bacterial infection of plants; the phage P22 tailspike adhesin that binds the O-antigen of Salmonella typhimurium; and the P.69 pertactin toxin from Bordetella pertussis, the cause of Whooping Cough. Therefore it would be very interesting if we can accurately predict other unknown β-helix structure proteins.

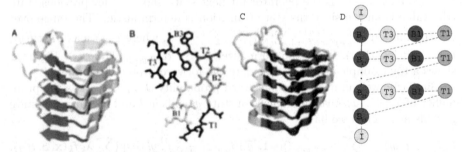

Fig. 3. 3-D structures and side-chain patterns of β-helices; (A) Side view (B) top view of one rung (C) Segmentation of 3-D structures (D) protein structural graph. E1 = {black edge} and E2 = {red edge}

Traditional methods for protein family classification, such as threading, PSI-BLAST and HMMs, fail to solve the β-helix recognition problem across different families [19]. Recently, a computational method called BetaWrap, has been proposed to predict the β-helix specifically [19]. The algorithm "wraps" the unknown sequences in all plausible ways and check the scores to see if any wrap makes sense. The cross-validation results in the protein data bank (PDB) seem promising. However, the BetaWrap algorithm suffers from hand-coding many biological heuristic rules. Hence it is prone to over-fit the known β-helix proteins and hard to generalize for other structural prediction tasks.

4.1   Protein Structural Graph for β-Helix

From previous literature on β-helix, there are two facts important for accurate prediction: 1) the β-strands of each rung have patterns of pleating and hydrogen bonding that are well conserved across the superfamily; 2) the interaction of the strand side-chains in the buried core are critical determinants of the fold [20, 21]. Therefore we define the protein structural graph of β-helix as in Fig.3-(D).

There are 5 states in the graph altogether, i.e. s-B23, s-T3, s-B1, s-T1 and s-I. The state s-B23 is a union of B2, T2 and B3 because these three segments are all highly conserved in pleating patterns and a combination of conserved evidence is generally much easier to detect. We fix the length of S-B23 and S-B1 as 8 and 3 respectively for two reasons: first, these are the number of residues shared by all known β-helices; second, it helps limit the search space and reduce the computational costs. The states s-T3 and s-T1 are used to connect s-B23 and s-B1. It is known that the β-helix structures will break if the insertion is too long. Therefore we set the length of s-T3 and s-T1 so that it varies from 1 to 80. s-I is the non-β-helix state, which refers to all those regions outside the β-helix structures. The red edge between s-B23 is used to model the long-range interaction between adjacent β-strand pairs. For a protein without any β-helix structures, we define the protein structural graph as a single node of state s-I.

4.2   SCRFs for β-Helix Fold Prediction

In Section 3.2, we made two assumptions in the SCRFs model: a) the state transition is deterministic; b) each subgraph of $G' = < V, E_2 >$ is a chain or a tree. For β-helix, we cannot directly define a structural graph with deterministic state transitions, since the number of rungs in a protein is unknown beforehand. In Fig.3, it seems that the previous state of s-B23 can be either s-I or s-T1. However, notice that s-I can appear only at the beginning or the end of a sequence, therefore s-I can be the previous state of s-B23 iff the previous segment starts at the first residue in the sequence. Similarly, s-I can be the next state of s-B23 iff the next segment ends at the last residue. Therefore *the state transition is deterministic given the constraint we have for s-I.* As for assumption b), it is straightforward that graph G' consists of a chain and a set of isolated nodes. Therefore the algorithm discussed in Section 3.2 can be applied accordingly.

To determine whether a protein sequence has the β-helix fold, we define the score ρ as the log ratio of the probability of the best segmentation to the probability of the whole sequence as one state s-I, i.e. $\rho = \log \frac{\max_s P(S|x)}{P(<1,N,s-I>|x)}$. The higher the score ρ, the more likely that the sequence has a β-helix fold. We did not consider the long-range interactions between B1 strands explicitly since the effect is relatively weak given only 3 residues in s-B1 segments. However, we use the B1 interactions as a filter in Viterbi algorithm: specifically, $\delta_t(y)$ will be the highest value whose corresponding segmentation also have alignment scores for B1 higher than some threshold set using cross-validation.

4.3   Feature Extraction

SCRFs provide an expressive framework to handle long-range interactions for protein fold prediction. However, the choice of feature function f_k plays a key role in accurate predictions. We define two types of features for β-helix prediction, i.e. *node features* and *inter-node features*.

Node features cover the properties of an individual segment, including:

a) **Regular expression template:** Based on the side-chain alternating patterns in B23 region, BetaWrap generates a regular expression template to detect β-helices, i.e. $\Phi X \Phi X X \Psi X \Phi X$, where Φ matches any of the hydrophobic residues as {A, F, I, L, M, V, W, Y}, Ψ matches any amino acids except ionisable residues as {D, E, R, K} and X matches any amino acid [19]. Following similar idea, we define the feature function $f_{RST}(x, S)$ equal to 1 if the segment S matches the template, and 0 otherwise.

b) **Probabilistic HMM profiles:** The regular expression template as above is straightforward and easy to implement. However, sometimes it is hard to make a clear distinction between a true motif and a false alarm. Therefore we built a probabilistic motif profile using HMMER [4] for the s-B23 and s-B1 segments respectively. We define the feature function $f_{HMM1}(x, S)$ and $f_{HMM2}(x, s)$ as the alignment scores of S against the s-B23 and s-B1 profiles.

c) **Secondary structure prediction scores:** Secondary structures reveal significant information on how a protein folds in three dimension. The state-of-art prediction method can achieve an average accuracy of 76 - 78% on soluble proteins. We can get fairly good prediction on alpha-helix and coils, which can help us locate the s-T1 and s-T3 segments. Therefore we define the feature function $f_{ssH}(x, S)$, $f_{ssE}(x, S)$ and $f_{ssC}(x, S)$ as the average of the predicted scores over all residues in segment S, for helix, sheet and coil respectively by PSIPRED [22].

d) **Segment length:** It is interesting to notice that the β-helix structure has strong preferences for insertions within certain length ranges. To consider this preference in the model, we did parametric density estimation. Several common functions are explored, including Poisson distribution, negative-binomial distribution and asymmetric exponential distribution, which consists for two exponential functions meeting at one point. We use the latter one since it provides a better estimator than the other two. Then we define the feature function $f_{L1}(x, S)$ and $f_{L3}(x, S)$ as the estimated probability of the length of segment S as s-T1 and s-T3 respectively.

Inter-node features capture long-range interactions between adjacent β-strand pairs, including:

a) **Side chain alignment scores:** BetaWrap calculates the alignment scores of residue pairs depending on whether the side chains are buried or exposed. In this method, the conditional probability that a residue of type X will align with residue Y, given their orientation relative to the core, is estimated from a β-structure database developed from the whole PDB [19]. Following similar idea, we define the feature function $f_{SAS}(x, S, S')$ as the weighted sum of the side chain alignment scores for S given S' if both are s-B23 segments, where a weight of 1 is given to inward pairs and 0.5 to the outward pairs.

b) Parallel β-sheet alignment scores: In addition to the side chain position, another aspect is to study the different preferences for parallel and anti-parallel β-sheets. Steward & Thornton [23] derived the "pairwise information values" (V) for a residue of type X given the residue Y on the pairing parallel (or anti-parallel) strand and the offsets of Y from the paired residue Y' of X. The alignment score for two segments $x = X_1 \ldots X_m$ and $y = Y_1 \ldots Y_m$ is defined as

$$score(x, y) = \sum_i \sum_j (V(X_i|Y_j, i - j) + V(Y_i|X_j, i - j)).$$

Compared with the side chain alignment scores, this score also takes into account the effect of neighboring residues on the paired strand. We define the feature function $f_{PAS}(x, S, S') = score(S, S')$ if both S and S' are s-B23 and 0 otherwise.

c) Distance between adjacent s-B23 segments There are also different preferences for the distance between adjacent s-B23 segments. It is difficult to get an good estimation of this distribution since the range is too large. Therefore we simply define the feature function as the normalized length, i.e. $f_{DIS}(x, S, S') = \frac{dis(S,S')-\mu}{\sigma}$, where μ is the mean and σ^2 is the variance.

It is interesting to notice that some features defined above are quite general, not limited to predicting β-helices only. For example, an important aspect to discriminate a specific protein fold with others is to build HMM profiles or identify regular expression templates for conserved regions if they exist; the secondary structure assignments are essential in locating the elements within a protein fold; if some segments have strong preferences for certain length range, then length are also informative. For internode features, the β-sheet alignment scores are useful for folds in β-family while hydrophobicity is important for α- or $\alpha\beta$-family.

5   Experiments

In our experiments, we followed the setup described in [19]. A PDB-minus dataset was constructed from the PDB protein sequences (July 2004 version) [24] with less than 25% similarity to each other and no less than 40 residues in length. Then the β-helix proteins are removed from the dataset, resulting in 2094 sequences in total. The proteins in PDB-minus dataset will serve as negative examples in the cross-family validation and discovery of new β-helix proteins. Since negative data dominate the training set, we subsample 15 negative sequences that are most similar to the positive examples in sequence identity so that SCRFs can learn a better decision boundary than randomly sampling.

5.1   Cross-Family Validation

A leave-family-out cross-validation was performed on the nine β-helix families of closely related proteins in the SCOP database [1]. For each cross, proteins in the one β-helix family are placed in the test set while the remainder are placed in the training set as positive examples. Similarly, the PDB-minus was also randomly partitioned into nine subsets, one of which are placed in the test set while the rest serve as the negative training examples. We compare our results with BetaWrap, a state-of-art algorithm for predicting β-helices, and HMMER, a general motif detection algorithm based on a simple graphical model, i.e. HMMs. The input to HMMER is a multiple sequence alignment. The best multiple alignments are typically generated using 3-D structural information, although this is not strictly sequence-based method. Therefore we generated two kinds of alignments for comparison: one is the multiple structural alignments using EC-MC [25], the other is purely sequence-based alignments by CLUSTALW[26].

Table 1 shows the output scores by different methods and the relative rank for the β-helix proteins in the cross-family validation. From the results, we can see that the SCRFs model can successfully score all known β-helices higher than non β-helices in PDB. On the other hand, there are two proteins (i.e. 1ktw and 1ea0) in our validation sets that are crystallized recently and thus are not included in the BetaWrap system. We test these two sequences on BetaWrap and get a score of -23.4 for 1ktw and -24.87 for 1ea0. These values are significantly

Table 1. Scores and rank for the known right-handed β-helices by HMMER, BetaWrap and SCRFs. 1: the scores and rank from BetaWrap are taken from [3] except 1ktw and 1ea0; 2: the bit scores in HMMER are not directly comparable

SCOP family	PDB-id	Struct-based HMMs		Seq-based HMMs		BetaWrap[1]		SCRFs	
		Bit score[2]	Rank	Bit score[2]	Rank	Score	Rank	ρ-score	Rank
P.69 pertactin	1dab	-73.6	3	-163.4	75	-17.84	1	10.17	1
Chondroitinase B	1dbg	-64.6	5	- 171.0	55	-19.55	1	13.15	1
Glutamate synthase	1ea0	-85.7	65	-109.1	72	-24.87	N/A	6.21	1
Pectin methylesterase	1qjv	-72.8	11	-123.3	146	-20.74	1	6.12	1
P22 tailspike	1tyu	-78.8	30	-154.7	15	-20.46	1	6.71	1
Iota-carrageenase	1ktw	-81.9	17	- 173.3	121	-23.4	N/A	8.07	1
Pectate lyase	1air	-37.1	2	-133.6	35	-16.02	1	16.64	1
	1bn8	180.3	1	-133.7	37	-18.42	3	13.28	2
	1ee6	-170.8	852	-219.4	880	-16.44	2	10.84	3
Pectin lyase	1idj	-78.1	14	-178.1	257	-17.99	2	15.01	2
	1qcx	-83.5	28	-181.2	263	-17.09	1	16.43	1
Galacturonase	1bhe	-91.5	18	-183.4	108	-18.80	1	20.11	3
	1czf	-98.4	43	-188.1	130	-19.32	2	40.37	1
	1rmg	-78.3	3	-212.2	270	-20.12	3	23.93	2

Table 2. (Left) Histograms of protein scores of known β-helix proteins against PDB-minus dataset. Blue bar: PDB-minus dataset; green bar: known β-helix proteins. 2076 out of 2098 protein sequences in PDB-minus have a log ratio score ρ of 0, which means that the best segmentation is a single segment in non-β-helix state; (Right) Examples of proteins predicted to form β-helix in UniProt

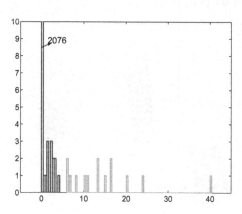

UniProt ID	Description	Score
Q8YK40	All8078 protein	119.7
Q8PRX0	Conserved protein	93.8
Q8WTU9	Hypothetical protein	81.3
Q8DK34	Tlr1036 protein	81.1
Q8RD81	Hypothetical protein	55.1
O26812	Cell surface glycoprotein	54.2
Q6LZ14	Hypothetical protein	43.8
P35338	Exopolygalacturonase precursor	42.2
Q6ZGA1	Putative polygalacturonase	41.6
Q9K1Z6	Hypothetical protein	40.8

lower than the scores of other β-helices and some of the non β-helix proteins, which indicates that the BetaWrap might be overtrained. As expected, HMMER did worse than SCRFs and BetaWrap even using the structural alignments.

Table 2 plots the score histogram for known β-helix sequences against the PDB-minus dataset. Compared with the histograms in similar experiment by BetaWrap [19], our log ratio score ρ indicates a clearer separation of β-helix proteins v.s. non β-helix proteins. Only 18 out of 2094 proteins has a score higher than 0. Among these 18 proteins, 13 proteins belong to the beta class and 5 proteins belong to the alpha-beta class in CATH database [2]. In Table 3 we also cluster the proteins into three different groups according to the segmentation results and show examples of the predicted segmentation in each group.

5.2   Discovery of Potential β-Helix Proteins

New potential β-helix proteins were identified from the UniProt reference databases (UniRef) (a combination of Swiss-Prot Release 44.2 of 30-Jul-2004 and TrEMBL 27.2 of 30-Jul-2004) [27]. We choose the UniRef50 (50% identity) with 490,713 sequences as the discovering set. 93 sequences were returned with scores above a cutoff of 5, which are identified as potential beta-helices. The sequences come from organisms in all domains of life. Of 44 eukaryotic sequences, 25 are from plants. It is interesting to note that none of the known β-helices are from plants. The remaining eukaryotic sequences come from mammals, fungi, nematodes and pathogens from the genus Plasmodium: 4 sequences were viral, including 3 from bacteriophages; 9 sequences are archeal, 7 of which are from methanogens of the genus Methanosarcina. Of the 93 high scoring se-

Table 3. Groups of segmentation results for the known right-handed β-helix

Group	Perfect match	Good match	OK match
Missing rungs	0	1-2	3 or more
PDB-ID	**1czf**	1air, 1bhe, 1bn8, 1dbg, 1ee6(right), 1idj, **1ktw**(left), 1qcx, 1qjv, 1rmg	**1dab**(left), 1ea0, **1tyu**(right)

quences, 48 are likely homologous (BLAST E-val < 0.001) with proteins currently known to contain parallel beta-helix domains. For the rest, most sequences are not homologous to any of the sequences in PDB. The protein sequences with maximal log ratio scores is shown in Table 2 (the full list can be accessed at http://www.cs.cmu.edu/ yanliu/SCRF.html.

Our method also identifies gp14 of Shigella bacteriophage Sf6 as having a parallel beta-helix domain, giving it a score of 15.63. This protein was not included in the UniRef50 dataset because it was incorrectly grouped with the P22 tailspike protein (1tyu), which was used in the training dataset. These two proteins share homologous capsid binding domains at their N-termini which are not parallel beta-helices while their C-terminal domains do not have any sequence identity. A Sf6 gp14 crystal structure has recently been solved and shown to be a trimer of parallel β-helices (R. Seckler, personal communication). Therefore SCRFs not only can identify homologous sequences to the known proteins, but also succeed in discovering proteins with less sequence similarity.

6   Discussion and Conclusion

In [19], BetaWrap was compared with other alternative methods, such as PSI-BLAST and Threader. We repeated their experiments and got similar results confirming that these methods fail to detect β-helix proteins accurately. Now it would be interesting to ask: why is β-helix prediction difficult for these commonly used methods? why can SCRFs model perform better?

We think the β-helix motif is hard to predict because there are long-range interactions in the β-helix fold. In addition, the structural properties unique to β-helix are not reflected clearly in the sequences. For example, the conserved templates for s-B23 segment also appear many times in non β-helix proteins; the side chain alignment propensities in β-sheets are also shared by β-sheets in other structures, such as the β-sandwich. Therefore the commonly used methods based on sequence similarity, such as PSI-BLAST and HMMER, cannot perform well in this kind of task. However, a combination of both sequence and structure characteristics might help to identify a β-helix, which is one of the major reasons why BetaWrap and SCRFs work well. The difference between these two methods is: BetaWrap searches the combination space by defining a series of heuristic rules while SCRFs search automatically by maximizing the conditional likelihood of the training data under a unified graphical model, which guarantees the solution to be global optimally. Therefore the SCRFs model is more general and robust.

There are several directions to improve the SCRFs model, which are interesting both computationally and empirically. One is to extend the SCRFs model for predicting other protein folds, such as the leucine rich repeats (LLR) or triple β-spirals. On the other hand, the 2-D protein structural graph has limited power to capture the dynamic constraints for 3-D protein structures. Therefore it would be interesting to extend the SCRFs model to include protein dynamics. The latter, however, will be a major undertaking.

Acknowledgement

This material is based upon work supported by the National Science Foundation under Grant No. 0225656. We thank Jonathan King for his input and biological insights and anonymous reviewers for their comments.

References

1. Murzin, A., Brenner, S., Hubbard, T., Chothia, C.: SCOP: a structural classification of proteins database for the investigation of sequences and structures. J Mol Biol. **247** (1995) 536–40
2. Orengo, C., Michie, A., Jones, S., Jones, D., Swindells, M., Thornton, J.: CATH–a hierarchic classification of protein domain structures. Structure. **5** (1997) 1093–108
3. Karplus, K., Barrett, C., Hughey, R.: Hidden markov models for detecting remote protein homologies. Bioinformatics **14** (1998) 846–56
4. Durbin, R., Eddy, S., Krogh, A., Mitchison, G.: Biological sequence analysis: probabilistic models of proteins and nucleic acids. Cambridge University Press (1998)
5. Altschul, S., Madden, T., Schaffer, A., Zhang, J., Zhang, Z., Miller, W., Lipman, D.: Gapped BLAST and PSI-blast: a new generation of protein database search programs. Nucleic Acids Res. **25** (1997) 3389–402
6. Menke, M., Scanlon, E., King, J., Berger, B., Cowen, L.: Wrap-and-pack: a new paradigm for beta structural motif recognition with application to recognizing beta trefoils. In: Proceedings of the 8th ACM RECOMB conference. (2004) 298–307

7. Kumar, S., Hebert, M.: Discriminative random fields: A discriminative framework for contextual interaction in classification. In: Proc. IEEE International Conference on Computer Vision (ICCV). (2003) 1150–1159

8. Pinto, D., McCallum, A., Wei, X., Croft, W.B.: Table extraction using conditional random fields. In: Proceedings of the 26th ACM SIGIR conference. (2003) 235–242

9. Bystroff, C., Thorsson, V., Baker, D.: HMMSTR: a hidden markov model for local sequence-structure correlations in proteins. J Mol Biol. **301** (2000) 173–90

10. Delcher, A., Kasif, S., Goldberg, H., Xsu, W.: Protein secondary-structure modeling with probabilistic networks. In: International Conference on Intelligent Systems and Molecular Biology (ISMB'93). (1993) 109–117

11. Liu, Y., Carbonell, J., Klein-Seetharaman, J., Gopalakrishnan, V.: Comparison of probabilistic combination methods for protein secondary structure prediction. Bioinformatics. **20** (2004) 3099–107

12. W. Chu, Z.G., Wild, D.L.: A graphical model for protein secondary structure prediction. In: Proc.of International Conference on Machine Learning (ICML-04). (2004) 161–168

13. McCallum, A., Freitag, D., Pereira, F.C.N.: Maximum entropy markov models for information extraction and segmentation. In: Proc.of International Conference on Machine Learning (ICML-00). (2000) 591–598

14. Lafferty, J., McCallum, A., Pereira, F.: Conditional random fields: Probabilistic models for segmenting and labeling sequence data. In: Proc. 18th International Conf. on Machine Learning, Morgan Kaufmann, San Francisco, CA (2001) 282–289

15. Hammersley, J., Clifford, P.: Markov fields on finite graphs and lattices. Unpublished manuscript (1971)

16. Jordan, M.I.: Learning in Graphical Models. The MIT press (1998)

17. Sha, F., Pereira, F.: Shallow parsing with conditional random fields. In: Proceedings of Human Language Technology, NAACL 2003. (2003)

18. Yoder, M., Keen, N., Jurnak, F.: New domain motif: the structure of pectate lyase c, a secreted plant virulence factor. Science **260** (1993) 1503–7

19. Bradley, P., Cowen, L., Menke, M., King, J., Berger, B.: Predicting the beta-helix fold from protein sequence data. In: Proceedings of 5th Annual ACM RECOMB conference. (2001) 59–67

20. Yoder, M., Jurnak, F.: Protein motifs. 3. the parallel beta helix and other coiled folds. FASEB J. **9** (1995) 335–42

21. Kreisberg, J., Betts, S., King, J.: Beta-helix core packing within the triple-stranded oligomerization domain of the p22 tailspike. Protein Sci. **9** (2000) 2338–43

22. Jones, D.: Protein secondary structure prediction based on position-specific scoring matrices. J. Mol. Biol. **292** (1999) 195–202

23. Steward, R., Thornton, J.: Prediction of strand pairing in antiparallel and parallel beta-sheets using information theory. Proteins. **48** (2002) 178–91

24. Berman, H., Westbrook, J., Feng, Z., Gilliland, G., Bhat, T., Weissig, H., Shindyalov, I., Bourne, P.: The protein data bank. Nucleic Acids Research **28** (2000) 235–42

25. Guda, C., Lu, S., Sheeff, E., Bourne, P., Shindyalov, I.: CE-MC: A multiple protein structure alignment server. Nucleic Acids Res. **In press** (2004)

26. Thompson, J., Higgins, D., Gibson, T.: CLUSTAL W: improving the sensitivity of progressive multiple sequence alignment through sequence weighting, positions-specific gap penalties and weight matrix choice. Nucleic Acids Research **22** (1994) 4673–80

27. Leinonen, R., Diez, F., Binns, D., Fleischmann, W., Lopez, R., Apweiler, R.: Uniprot archive. Bioinformatics. **20** (2004) 3236–7

Rapid Protein Side-Chain Packing via Tree Decomposition

Jinbo Xu

School of Computer Science, University of Waterloo,
Waterloo, Ontario N2L 3G1, Canada
Department of Mathematics, MIT, Cambridge, MA 02139.
j3xu@theory.csail.mit.edu

Abstract. This paper proposes a novel tree decomposition based side-chain assignment algorithm, which can obtain the globally optimal solution of the side-chain packing problem very efficiently. Theoretically, the computational complexity of this algorithm is $O((N + M)n_{rot}^{tw+1})$ where N is the number of residues in the protein, M the number of interacting residue pairs, n_{rot} the average number of rotamers for each residue and $tw(= O(N^{\frac{2}{3}} \log N))$ the tree width of the residue interaction graph. Based on this algorithm, we have developed a side-chain prediction program SCATD (Side Chain Assignment via Tree Decomposition). Experimental results show that after the Goldstein DEE is conducted, n_{rot} is around 3.5, tw is only 3 or 4 for most of the test proteins in the SCWRL benchmark and less than 10 for all the test proteins. SCATD runs up to 90 times faster than SCWRL 3.0 on some large proteins in the SCWRL benchmark and achieves an average of five times faster speed on all the test proteins. If only the post-DEE stage is taken into consideration, then our tree-decomposition based energy minimization algorithm is more than 200 times faster than that in SCWRL 3.0 on some large proteins. SCATD is freely available for academic research upon request.

1   Introduction

The structure of a protein plays an instrumental role in determining its functions. Experimental methods such as X-ray crystallography and NMR techniques cannot generate protein structures in a high throughput way. Protein structure prediction tools have been frequently used by structural biologists and pharmaceutical companies to analyze the structure features and function characteristics of a protein. Typically, in order to overcome the computational challenge, protein structure prediction is decomposed into two major steps. One is the prediction of the backbone atom coordinates and the other is the prediction of side-chain atom coordinates. The former step is usually done using protein threading programs [1, 2, 3, 4, 5, 6, 7, 8, 9, 10] or sequence-based homology search tools such as PDB-BLAST [11]. The task of side-chain prediction is to determine the position of all the side-chain atoms given that the backbone coordinates of a protein are already known. Along with the advancement of protein backbone prediction

S. Miyano et al. (Eds.): RECOMB 2005, LNBI 3500, pp. 423–439, 2005.

techniques, side-chain prediction is becoming more important since the backbone coordinates of a protein can be predicted accurately.

Many side-chain prediction methods have been proposed and implemented in the past two decades [12, 13, 14, 15, 16, 17, 18, 19, 20, 21, 22, 23, 24, 25, 26, 27]. Almost all of them use a rotamer library, which is a set of side-chain conformation candidates. In order to overcome computational difficulty, the side-chain conformation of a residue is discretized into a finite number of states (rotamers). Each rotamer is a representative of a set of similar side-chain conformations. A backbone-independent rotamer only depends on the type of the residue, while a backbone-dependent rotamer also depends on two dihedral angles (ϕ and ψ) associated with the residue. For example, Dunbrack et $al.$ first developed a backbone-independent rotamer library and then a backbone-dependent rotamer library [28].

Given a rotamer library, the side-chain prediction problem can be formulated as a combinatorial search problem. The quality of a side-chain packing can be measured by an energy function, which usually consists of singleton scores and pairwise scores. Singleton score describes the preference of one rotamer for a particular residue and its environment. Singleton score can also describe the interaction between the rotamer atoms and the backbone atoms. Since the position of the backbone atoms are already fixed, this kind of interaction only depends on one movable side-chain atom. Pairwise score measures the interaction between two side-chain atoms. Pairwise score usually is used to avoid inter-atom clashes. A good side-chain packing should avoid as many clashes as possible. In addition, other atom-atom interactions can also be incorporated into the energy function. It is the side-chain atom interaction that makes the side-chain packing problem computationally challenging. The underlying reason is that in order to minimize the conflict, we have to fix the positions of all the interacting side-chain atoms simultaneously.

The side-chain prediction problem has been proved to be NP-hard [29, 30], which justifies the development of many heuristic algorithms such as SCAP [19] and MODELLER [31], and some approximate algorithms [23, 24]. These algorithms are usually computationally efficient, but cannot guarantee to find the side-chain assignment with the lowest system energy. The exact algorithms such as DEE/A^* [14, 32] and integer programming approach [25, 27] can find a globally minimum energy for a given rotamer library and an energy function. However, not many algorithms belong to this category due to the expensive computational time. SCWRL 3.0 [17] guarantees to find a globally optimal solution and also runs very fast for many proteins. SCWRL 3.0 first decomposes the residue interaction graph into some small biconnected components. Any two biconnected components share at most one articulation residue. If the side-chain assignment to the articulation residue is fixed, then the side-chain positioning in one component is independent of the other. As such, SCWRL 3.0 can optimize the side-chain assignment of these components one by one. A big optimization problem is decomposed into some small subproblems. According to their report, most of the biconnected components are quite small, containing no more than

21 residues. However, it still takes a long time to optimize the side-chain assignment to a component of 21 residues even if each residue has only 3.5 possible rotamers, which is the average number in our experiments.

In this paper, we present a novel tree decomposition based approach to the globally optimal side-chain packing problem. The biconnected decomposition of a graph used in SCWRL 3.0 can be considered as only a special case of the tree decomposition. The key point is that we optimize the tree decomposition of a graph such that the resultant components are as small as possible. Theoretically, we can have a polynomial-time algorithm to decompose the residue interaction graph into some components with size $O(N^{\frac{2}{3}} \log N)$ where N is the number of residues, which leads to a globally exact side-chain assignment algorithm with a computational complexity $O(Nn_{rot}^{cN^{\frac{2}{3}} \log N})$ where c is a constant. As far as we know, this is the first globally optimal side-chain packing algorithm with a non-trivial time complexity. In contrast, the biconnected decomposition of a graph can easily result in a large component with size $O(N)$ even for some sparse graphs such as a cycle. Experimental results show that a typical residue interaction graph can be decomposed into components of 4 or 5 residues using our decomposition method. Since each residue has no more than 4 candidate rotamers after the Goldstein criterion DEE [33] is conducted, the optimal side-chain assignment of each component can be done very quickly.

The remainder of this paper is organized as follows. In Section 2, we describe the side-chain assignment problem further and formulate it as a combinatorial optimization problem. Section 3 introduces the concepts of tree decomposition and presents the tree decomposition based side-chain assignment algorithm. In this section, we also describe a low-degree polynomial-time algorithm that can decompose a residue interaction graph into some components of size $O(N^{2/3} \log N)$, and a simple heuristic tree decomposition algorithm, which works very well for our purpose. Section 4 describes a graph reduction technique that can be used to prune those residues interacting with only one or two other residues. This pruning strategy can improve the computational efficiency a little bit. In Section 5, we present the experimental results of our algorithm in detail and compare our side-chain prediction program with SCWRL 3.0 in terms of computational efficiency and accuracy. Finally, Section 6 points out that there are some polynomial-time approximation schemes for the side-chain packing problem, and discusses further development of our side-chain packing system.

2   Problem Description

The side-chain prediction problem can be formulated as follows. We use a *residue interaction graph* $G = (V, E)$ to represent the residues in a protein and their relationship. Each residue is represented by a vertex in the set V. For a residue i, we use $D[i]$ denote the set of possible rotamers for this residue. There is an *interaction edge* $(i, j) \in E$ between two residues i and j if and only if there are two rotamers $l \in D[i]$ and $k \in D[j]$ such that at least one atom in rotamer l conflicts with at least one atom in rotamer k. Two atoms conflict with each other if and only if their distance is less than the sum of their radii. We say that residues i and j interact with each other if there is one edge between i and j in G. For each rotamer $l \in D[i]$, there is an associated singleton score, denoted by $S_i(l)$. In our energy function, $S_i(l)$ is the interaction energy between rotamer l and the backbone of the protein. $S_i(l)$ also includes the preference of assigning one rotamer to a specific residue. For any two rotamers $l \in D[i]$ and $k \in D[j]$ $(i \neq j)$, there is also an associated pairwise score, denoted by $P_{i,j}(l,k)$, if residue i interacts with residue j. Let $E(a,b)$ denote the interaction score between two atoms a and b. We use the method in the SCWRL 3.0 paper [17] to calculate $E(a,b)$ as follows .

$$
\begin{aligned}
E(a,b) &= 0 & r \geq R_{a,b} \\
&= 10 & r \leq 0.8254 R_{a,b} \\
&= 57.273(1 - \tfrac{r}{R_{a,b}}) & \text{otherwise}
\end{aligned}
$$

where r is the distance between atoms a and b and $R_{a,b}$ is the sum of their radii. Let $SC(i)$ and $BB(i)$ denote the set of side-chain atoms and the set of backbone atoms of one residue i, respectively, and $Pr_i(l|\phi, \psi)$ denote the probability of rotamer l given the residue i and two angles ϕ and ψ. Then we calculate $S_i(l)$ and $P_{i,j}(l,k)$ as follows [17].

$$
S_i(l) = -K \log\left(\frac{Pr_i(l|\phi,\psi)}{\max_{l \in D[i]} Pr_i(l|\phi,\psi)}\right) + \sum_{|i-j|>1} \sum_{s \in SC(i)} \sum_{b \in BB(j)} E(s,b) \tag{1}
$$

$$
P_{i,j}(l,k) = \sum_{a \in SC(i)} \sum_{b \in SC(j)} E(a,b) \tag{2}
$$

In Eq. 1, K is optimized to 6 to yield the best prediction accuracy. Please notice that in the above two equations, the position of one side-chain atom depends on its associated rotamer.

Given a side-chain assignment $A(i) \in D[i]$ to residue i $(i \in V)$, the quality of one side-chain packing is measured by the following energy function.

$$
E(G) = \sum_{i \in V} S_i(A(i)) + \sum_{i \neq j, (i,j) \in E} P_{i,j}(A(i), A(j)) \tag{3}
$$

The smaller the system energy $E(G)$ is, the better the side-chain assignment. So our goal is to develop an efficient algorithm to search for the side-chain assignment such that the energy $E(G)$ in Eq. 3 is minimized.

3   Side-Chain Prediction Algorithm

In this section, we will first introduce the concept of tree decomposition of a graph, and then describe how to search for the optimal side-chain assignment based on the decomposition. We will also show that there is a low-degree polynomial-time algorithm to decompose the residue interaction graph into some components of size $O(|V|^{\frac{2}{3}} \log |V|)$. Finally, we describe an efficient heuristic algorithm to find a good tree decomposition of a graph.

3.1   Tree Decomposition Concepts

The notions of tree width and tree decomposition are introduced by Robertson and Seymour [34] in their work on graph minors. The tree decomposition of a sparse graph has been applied to many NP-hard problems such as frequency assignment problem [35] and Bayesian inference [36].

Definition 1. *Let $G = (V, E)$ be a graph. A tree decomposition of G is a pair (T, X) satisfying the following conditions:*

1. *$T = (I, F)$ is a tree with a node set I and an edge set F,*
2. *$X = \{X_i | i \in I, X_i \in V\}$ and $\bigcup_{i \in I} X_i = V$. That is, each node in the tree T represents a subset of V and the union of all the subsets is V,*
3. *for every edge $e = \{v, w\} \in E$, there is at least one $i \in I$ such that both v and w are in X_i, and*
4. *for all $i, j, k \in I$, if j is a node on the path from i to k in T, then $X_i \cap X_k \subseteq X_j$.*

The width of a tree decomposition is $\max_{i \in I}(|X_i| - 1)$. The tree width of a graph G, denoted by $tw(G)$, is the minimum width over all the tree decompositions of G.

According to the above definition, the decomposition of a graph into biconnected components is also a tree decomposition. Each biconnected component corresponds to a node in T and any two biconnected components share at most one articulation vertex in G. However, the width of a biconnected-component decomposition could be $O(|V|)$, which is much bigger than the tree width of a graph G if G is sparse. For example, when a graph is a cycle, this graph has only one biconnected component—itself. In contrast, the tree width of a cycle is only 2. Figure 1, 2 and 3.2 show an example of an interaction graph, its biconnected component decomposition with width 5 and a tree decomposition with width 2. The width of a tree decomposition is a key factor determining the computational complexity of all the tree decomposition based algorithms. The smaller the width of a tree decomposition is, the more efficient the algorithm. Therefore, we need to optimize the tree decomposition of the residue interaction graph such that we can have a very small tree width. In the next subsection, we will describe a tree decomposition based side-chain assignment algorithm and analyze its computational complexity.

3.2   Tree Decomposition Based Side Chain Assignment Algorithm

In this subsection, we describe an algorithm to search for the optimal side-chain assignment based on a tree decomposition (T, X) of a residue interaction graph G.

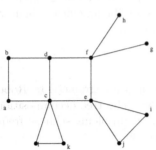

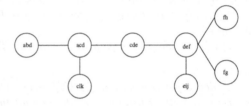

Fig. 1. Example of a residue interaction graph

Fig. 2. Example of the biconnected-component decomposition of a graph. The width of this decomposition is 5

Fig. 3. Example of a tree decomposition of a graph with width 2

For simplicity purpose, we assume that tree T has a root r and that each node is associated with a height. The height of a node is equal to the maximum height of its child nodes plus one. Our tree decomposition based side-chain assignment algorithm consists of two steps. One is the calculation of the optimal energy in a bottom-to-top way and the other is the extraction of the optimal assignment in a top-to-bottom way.

Bottom-to-Top. Suppose we start from a leaf node i in the tree T and node j is the parent of i. Let $X_{i,j}$ denote the intersection between X_i and X_j. Given a side-chain assignment $A(X_{i,j})$ to the residues in $X_{i,j}$, we enumerate all the possible side-chain assignments to the residues in $X_i - X_{i,j}$ and then find the best side-chain assignment such that the energy of the subgraph induced by X_i is minimized. We record this optimal energy as the multi-body score of $X_{i,j}$, which only depends on $A(X_{i,j})$. All the residues in $X_{i,j}$ form a hyper edge, which is added into the subgraph induced by X_j. When the energy of the subgraph induced by X_j is calculated, the multi-body score corresponding to this hyper edge should be included. In addition, we also save the optimal assignment to all the residues in $X_i - X_{i,j}$ for each $A(X_{i,j})$ since in the top-to-bottom step we need it for traceback. For example, in Figure 3.2, if we assume the node acd is the root, then node def is an internal node with parent cde. For each side-chain assignment to residues d and e we can find the best side-chain assignment to residue f such that the energy of the subgraph induced by d, e, and f is minimized. Then we add one hyper edge (d, e) to node cde. In this bottom-to-top process, a tree node can be calculated only after all of its child nodes are calculated. When calculating the root node of T, we enumerate the side-chain assignments to all the residues in it and obtain the optimal assignment such that the energy is minimized. This minimized energy is also the optimal menergy of the whole system.

Top-to-Bottom. After calculating the root node of tree T, we have the optimal assignment to all the residues in the root. Now we trace back from the parent node to its child nodes to extract out the optimal assignment to all the residues in a child node. Assume that the optimal assignment to all the residues in node j are already known and node i is a child node of j. We can easily extract out the optimal assignment to all the residues in $X_i - X_{i,j}$ based on the assignment to the residues in $X_{i,j}$ since we have already saved this assignment in the bottom-to-top step. Recursively, we can track down to the leaf nodes of T to extract out the optimal assignment to all the residues in G. If we want to save the memory consumption, then we do not need to save the optimal assignments in the bottom-to-top step. Instead, in this step we can enumerate all the assignments to $X_i - X_{i,j}$ to obtain the optimal assignment to all the residues in $X_i - X_{i,j}$. The computational effort for this enumeration is much cheaper than that in the bottom-to-top step since there is only one side-chain assignment to $X_{i,j}$.

In addition, based on the definition of tree decomposition, one residue might occur in several tree nodes. To avoid incorporating the singleton score of this residue into the overall system energy more than once, we include the singleton score of this residue into the system only when we are calculating the tree node with the maximal height among all the nodes containing this residue. We can prove that there is one and only one such a tree node. Similarly, an edge in graph G might also occur in several tree nodes. We can use the same method to avoid redundant addition of its pairwise score.

Based upon the above description, we have the following lemma.

Lemma 1. *The tree decomposition based side-chain assignment algorithm has a computational complexity of $O((|V|+|E|)n_{rot}^{1+tw})$ where V is the set of residues in the system, E the set of interaction edges, n_{rot} the average number of rotamers for each residue, and tw the width of the tree decomposition. The space complexity of this algorithm is $O(|V|n_{rot}^{tw})$.*

According to the above lemma, the computational complexity of tree decomposition based side-chain assignment algorithm is exponential to the width of the tree decomposition. Therefore, we need an algorithm to decompose the interaction graph into very small components.

3.3   Construction of Tree Decomposition

The optimal tree decomposition of a general graph has been proved to be *NP*-hard [37], which means it is unlikely to find the optimal decomposition of a graph within polynomial time. However, since the residue interaction graph is a geometric graph, we can very quickly obtain a tree decomposition with width $O(|V|^{\frac{2}{3}} \log |V|)$, based on the following sphere separator theorem [38].

Theorem 1. *Given a residue interaction graph $G = (V, E)$, there is a separator subset U with size $O(|V|^{\frac{2}{3}})$ of V such that removal of U from the graph can partition V into two subsets V_1 and V_2, and the following conditions are satisfied: (1) there is no interaction edge between V_1 and V_2; (2) $|V_i| \leq \frac{4}{5}|V|$ for $i = 1, 2$. In addition, such a subset U can be computed by a deterministic algorithm in random linear time.*

Before presenting the proof of the above theorem, we first introduce the definition of *k-ply neighborhood system* and a *sphere separator* theorem [38].

Definition 2 (k-ply neighborhood system). *A k-ply neighborhood system in $\Re^3$ is a set $\{B_1, B_2, ..., B_n\}$ of closed balls in $\Re^3$ such that no point in $\Re^3$ is strictly interior to more than k of the balls.*

Theorem 2 (Sphere Separator Theorem). *For every k-ply neighborhood system $\{B_1, B_2, ...B_n\}$ in $\Re^3$, there is a sphere separator S such that:(1)$|N_E| \leq \frac{4}{5}n$ where N_E contains all the balls in the exterior of S; (2) $|N_I| \leq \frac{4}{5}n$ where N_I contains all the balls in the interior of S; (3) $|N_o| = O(k^{\frac{1}{3}}n^{\frac{2}{3}})$ where N_o contains all the balls that intersect S. In addition, such an S can be computed by a deterministic algorithm in random linear time.*

Based on the above theorem, we can prove Theorem 1 as follows.

Proof of Theorem 1. Since the distance between any side-chain atom and its associated C_α atom is bounded above by a constant, according to the definition of interaction edge in Section 2, there is a constant $d_u > 0$ such that if the distance between two residues is more than d_u, then there will be no interaction edge between these two residues no matter which rotamer is assigned to them. In this paper, we use the position of C_α atoms to calculate the distance between two residues. For each residue, we construct a ball with radius $d_u/2$ centered at its C_α atom. In a normal protein, the distance between any two residues should be no less than a constant d_l. Therefore, there is a constant k ($\leq (1 + \frac{d_u}{d_l})^3$) such that no point in $\Re^3$ is strictly interior to more than k balls. Based on Theorem 2, we can have a sphere S such that S intersects with only $O(|V|^{2/3})$ balls and all the other balls are located inside or outside S in a balanced way. Let U denote the set of all the residues with its ball intersecting with S. Then we have $|U| = O(|V|^{2/3})$ and U will partition graph G into two subgraphs with balanced size.

Based on Theorem 1, we can prove the following theorem.

Theorem 3. *There is a low-degree polynomial-time algorithm that can find a tree decomposition of the residue interaction graph with a tree width of $O(|V|^{\frac{2}{3}} \log |V|)$.*

Proof. Based on Theorem 1, we can partition G into two subgraphs G_1 and G_2 by removing a separator subset of $O(|V|^{\frac{2}{3}})$ residues such that there is no interaction edge between G_1 and G_2 and $1/4 \leq |V(G_1)|/|V(G_2)| \leq 4$. Recursively, we can also partition G_1 and G_2 into smaller subgraphs until the size of the subgraph is $O(|V|^{2/3})$. Finally, we can have a binary partition tree in which each subtree corresponds to a subgraph, and the root of the subtree is the separator subset of the subgraph. Based on this binary partition tree, we can construct a tree decomposition of G as follows. For each partition tree node, we construct a decomposition component by assembling together all the residues along the path from the partition tree root to this node. We can easily verify that all the components form a tree decomposition of the graph. Since the height of the binary partition tree is $O(\log |V|)$, the tree width of this tree decomposition is $O(|V|^{\frac{2}{3}} \log |V|)$. Each partition step can be finished within linear time, so we can construct such a tree decomposition within low-degree polynomial-time.

Combining Theorem 3 and Lemma 1, we can calculate the computational complexity of the tree-decomposition based algorithm.

Theorem 4. *The tree decomposition based side-chain assignment algorithm has a computational complexity of* $O\left((|V| + |E|)\, n_{rot}^{O\left(|V|^{2/3} \log |V|\right)} \right)$ *where V is the set of residues in the system, E the set of interaction edges, n_{rot} the average number of rotamers for each residue.*

Although we give a low-degree polynomial-time algorithm to find a tree decomposition of G with a theoretically sound tree width, in practice we can use a simple heuristic algorithm to decompose the graph. Later in this paper we will show that the tree decompositions found by the heuristic algorithm are good enough. Many of them have a tree width of only 3 or 4, which leads to a very efficient algorithm for side-chain prediction. In our program, we use "minimum degree" heuristic [39] to recursively partition the graph. Specifically, we choose the vertex with the smallest number of neighbors. Then we add edges to the graph such that any two neighbors of the selected vertex is connected by an edge. Finally, we remove the selected vertex and its adjacent edges from the graph and recursively choose the next vertex. The selected vertex with its neighbors form a partition component of the graph. This algorithm runs very efficiently and also effectively for our purpose.

4   Graph Reduction Technique

After DEE is conducted, many residues interact with only one or two other residues since many rotamers are removed from the candidate lists. That is, the interaction graph contains many vertices with degree one or two, which results in many small components in the tree decomposition of this graph. Although it is extremely fast to compute the optimal energy of these small components, it will incur certain amount of overheads since we need to do many bookkeepings for the calculation of one component. We can use a graph reduction technique to remove these low-degree vertices before applying the tree decomposition to the interaction graph. The reduced system will have the same energy as the original system and can be easily recovered to the original system. For a residue i with degree one, assume i is only adjacent to residue j. Then, we can remove i by modifying the singleton score of j as follows.

$$S_j(k) \leftarrow S_j(k) + \min_{l \in D[i]} \{S_i(l) + P_{i,j}(l,k)\} \tag{4}$$

Since residue i only interacts with residue j, for each rotamer k of residue j, we can find the best rotamer for residue i and then remove residue i from the system. The optimal system energy will not change, and once the rotamer assignment to residue j is fixed, we can also easily obtain the optimal rotamer assignment to residue i.

For a residue with degree two, we can prune it by modifying the pairwise interaction scores of its two adjacent residues j_1, j_2 as follows.

$$P_{j_1,j_2}(k_1,k_2) \leftarrow P_{j_1,j_2}(k_1,k_2) + \min_{l \in D[i]} \{S_i(l) + P_{i,j_1}(l,k_1) + P_{i,j_2}(l,k_2)\} \tag{5}$$

Since residue i only interacts with two residues j_1 and j_2, for each combination of rotamer assignment to j_1 and j_2, we can find the best rotamer for i and then remove i from the system. The optimal system energy will not change, and once the rotamer assignment to residue j_1 and j_2 is fixed, we can also easily obtain the optimal rotamer assignment to residue i.

After applying the above-mentioned graph reduction technique to the interaction graph, the resultant new interaction graph looks more neat. The tree decomposition of new interaction graph will not generate any component of size one and few components of size two. Please notice that the graph reduction technique will not change the tree width of a graph. Therefore, this technique will not improve the theoretical computational complexity of the problem, but will improve the practical computational efficiency a little bit.

5   Experimental Results

We have implemented the idea presented in this paper as a program SCATD. In order to compare SCATD with SCWRL 3.0 [17], we test SCATD on the set of 180 proteins listed in the SCWRL 3.0 paper. The reason that we compare SCATD with SCWRL 3.0 is that both programs use the same rotamer library, similar energy function, same dead-end elimination method and, furthermore, solve the problem to its globally optimal solution.

Just like what is done in SCWRL 3.0 [17], for a particular residue and its associated two angles ψ and ϕ, we rank all the candidate rotamers from the highest probability to the lowest and remove the tail candidate if the probabilities of all the rotamers before it add up to 0.90 or above. Then we apply the Goldstein criterion dead-end elimination technique [33] to remove those rotamers that cannot be a part of the optimal side-chain assignment. Finally, we construct the residue interaction graph according to its definition described in Section 2. We build SCATD using BALL library [40] for some basic objects such as proteins, residues, and atoms, ANN library [41, 42] to determine if two atoms conflict or not, and split package [43] for tree decomposition of a graph.

5.1   Computational Efficiency

The residue interaction graphs are decomposed into small components with size no more than 10, much smaller than that reported in SCWRL 3.0 [17]. Most components have size only 4 or 5. Figure 4 shows the distribution of component sizes after the "minimum-degree" tree decomposition algorithm described in Section 3 is applied to the residue interaction graphs of the 180 proteins. As reported in the SCWRL3.0 paper, the maximum biconnected component size is 21 and there are quite a few of components with size larger than 10. As discussed in Section 3, the computing time of both SCATD and SCWRL 3.0 is exponential to the component size. Therefore, we can expect that our algorithm will be much more efficient than SCWRL 3.0. We also calculated the biconnected decomposition of all the interaction graphs generated by SCATD. Figure 5 illustrates the distribution of biconnected component sizes. As shown in Figure 5, the biconnected components of our interaction graphs have a bigger size than those in SCWRL 3.0 [17]. This indicates that we did not make the problem easier by using a slightly different energy function. Since the average number of rotamers for an active residue is 3.5, the algorithm used in SCWRL 3.0 cannot work very well on our interaction graphs if the biconnected component has size greater than 20.

We also ran SCATD and SCWRL 3.0 on a Debian Linux box with a 1.7GHz Pentium CPU. SCATD can do the side-chain prediction for all the 180 proteins within no more than 5 minutes while SCWRL 3.0 takes approximately 28 minutes. On average, SCATD is more than 5 times faster than SCWRL 3.0. Among these 180 test proteins, the maximum CPU time spent by SCATD on an indi-

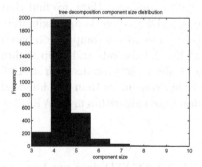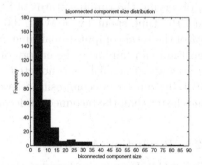

Fig. 4. The component size distribution of "minimum-degree" heuristic decomposition method. The components with size one or two are ignored

Fig. 5. The biconnected component size distribution of residue interaction graph. The components with size one or two are ignored

Table 1. Computational time spent by SCWRL 3.0 and SCATD on the side-chain prediction of several large proteins

protein	# residues	SCWRL 3.0 (s)	SCATD (s)
1gai	472	266.62	2.98
1a8i	812	184.36	8.68
1b0p	2462	300.00	21.03
1xwl	580	26.51	4.64
1bu7	910	56.50	7.62

Table 2. The average CPU time for each step

	library loading	coordinate calculation	interaction score calculation	DEE	tree decomposition based algorithm
time (s)	0.2822	0.2164	0.7924	0.3038	0.0239

vidual protein is 8.68 seconds. We further compare SCATD with SCWRL 3.0 on some large proteins including 1gai, 1xwl, 1a8i, 1bu7 and 1b0p. The results are shown in Table 1. As shown in this table, SCATD is up to 90 times faster than SCWRL 3.0 for large proteins.

SCATD consists of the following major steps: loading rotamer library, converting rotamer angles to atom coordinates, calculating interaction scores between two atoms, dead-end elimination (Goldstein criterion), and energy minimization via tree decomposition. We test the CPU time spent on each step to examine which steps are the CPU bottleneck in SCATD. Table 2 lists the detailed CPU times spent on each step by SCATD. As shown in this table, the tree decomposition based side-chain assignment algorithm is not the bottleneck at all. The average CPU time spent on this step is only 1.5% of the total computational time. Since we do not have the source code of SCWRL 3.0, we have no way to exactly measure the time spent by SCWRL 3.0 on the post-DEE stage. By observing the running status of SCWRL 3.0 on 1a8i and 1b0p, we find that the CPU time spent by SCWRL 3.0 on the post-DEE stage is approximately 70% of the total computational time. Nevertheless, our tree decomposition based algorithm can minimize the energy of 1a8i within 0.4 seconds and 1b0p within 0.7 seconds after DEE is conducted. Therefore, for large proteins such as 1a8i and 1b0p, our tree decomposition based algorithm runs more than two hundred times faster than the biconnected decomposition based algorithm in SCWRL 3.0.

5.2   Prediction Accuracy

While SCATD runs much faster than SCWRL 3.0, SCATD does not lose any accuracy. SCATD uses the same rotamer library as SCWRL 3.0 and a slightly different energy function. Table 3 lists the prediction accuracy of 18 types of amino acids. The prediction accuracy of both programs are very close. The minor difference comes from the fact that the atomic radii in the BALL library is

Table 3. Prediction accuracy of SCATD and SCWRL 3.0 in the 180-protein test set. A prediction is judged as correct if its deviation from the experimental value is no more than 40 degree. For χ_{1+2} to be correct, both χ_1 and χ_2 must be correct

amino acid	SCATD		SCWRL 3.0	
	χ_1 accuracy	χ_{1+2} accuracy	χ_1 accuracy	χ_{1+2} accuracy
ARG	0.7576	0.6135	0.7673	0.6381
ASN	0.7666	0.6479	0.7898	0.6749
ASP	0.7727	0.6668	0.8147	0.7129
CYS	0.7746	–	0.7052	–
GLN	0.7466	0.5086	0.7464	0.5290
GLU	0.7057	0.4922	0.7177	0.5223
HIS	0.8363	0.7711	0.8523	0.7877
ILE	0.9352	0.8376	0.9195	0.8095
LEU	0.9070	0.8279	0.9007	0.8203
LYS	0.7371	0.5607	0.7421	0.5773
MET	0.8183	0.6702	0.8016	0.6657
PHE	0.9306	0.8625	0.9354	0.8728
PRO	0.8511	0.7949	0.8449	0.7875
SER	0.6957	–	0.6730	–
THR	0.8871	–	0.8846	–
TRP	0.8786	0.6482	0.8828	0.6468
TYR	0.9085	0.8460	0.9212	0.8627
VAL	0.9212	–	0.9081	–
overall	0.8256	0.7329	0.8262	0.7374

slightly different from those in SCWRL 3.0. An interesting result is that SCATD does better in predicting CYS and worse in ASN and ASP than SCWRL 3.0.[1]

6   Discussions

Based on the tree-decomposition of protein structures, we not only can give a fast, rigorous and accurate protein side-chain assignment method, but also can develop several polynomial-time approximation schemes (PTAS) to this problem. When an optimization problem admits a PTAS, it means that given an arbitrary error ϵ ($1 > \epsilon > 0$), there is a polynomial-time algorithm to approximate its objective function value within a factor of $(1 \pm \epsilon)$. In contrast, based on a general graph model, Chazelle *et al.* [23] proved that it is *NP*-complete to approximate this problem within a factor of $\Omega(N)$. Due to space limit, we only introduce the following three theorems without giving any proof, which will be presented in the extended version of this paper.

---

[1] We disable the "-u" option of SCWRL 3.0 in order to compare both programs fairly since we have not implemented disulfide bond detection in SCATD. The overall accuracy of SCWRL 3.0 does not improve if "-u" option is enabled.

Theorem 5. *If every energy item in Eq. 3 is negative and the system energy should be minimized, then the side-chain packing problem admits a PTAS.*

Theorem 6. *Assume that all the pairwise energy items in Eq. 3 are positive and the system energy should be minimized. The side-chain packing problem admits a PTAS if the lowest system energy is $\Omega(NP_{max}\sqrt{\frac{\log n_{rot}}{\log N}})$ where P_{max} is the maximum among all $P_{i,j}(A(i), A(j))$.*

Theorem 7. *Assume that all the pairwise scores in Eq. 3 are negative and the system energy should be minimized. The side-chain packing problem admits a PTAS if the lowest system energy is no more than $cNP_{min}\sqrt{\log n_{rot}/\log N}$ where c is a positive constant and P_{min} the minimum among all $P_{i,j}(A(i), A(j))$.*

These theoretical results, especially Theorem 5, will stimulate us to develop a new energy function satisfying the conditions specified in these theorems and also having a good prediction accuracy so that we can apply these polynomial-time approximation algorithms to the problem. With a polynomial-time algorithm, we can deal with a larger rotamer library, which may result in a better prediction accuracy.

In protein structure prediction server RAPTOR [44, 45], we have developed a linear programming (LP) algorithm to obtain the globally optimal solution of the protein threading problem. The LP formulation used by RAPTOR can also be used to formulate the side-chain prediction problem. Mathematically, threading problem and side-chain prediction problem can be formulated in a very similar way. In fact, several research groups have proposed several more or less similar LP formulations for the side-chain packing problem [24, 25, 27]. We have also tested our LP formulation for the side-chain packing problem. In our setting, the tree decomposition based algorithm runs slightly faster than the LP approach. Interestingly, the tree decomposition algorithm proposed in this paper can also be used to the protein threading problem and contact map-based protein structure comparison.

The energy function used by our program is still very simple. In the future, we plan to add the disulfide bond detection into SCATD. We also plan to investigate more involved energy function to see how effective our algorithm is. For example, we can incorporate hydrogen bonds, electrostatics and solvation terms into our energy function. The major contribution of this paper is a novel and very efficient algorithm to the optimal protein side-chain packing problem but not a new energy function.

In this paper, we only test SCATD on the native backbone of the test proteins. The next step is to test SCATD on those backbones predicted by structure prediction programs. After all, a major usage of SCATD is to build the side-chain coordinates for a protein after its backbone coordinates are predicted.

Acknowledgments

The major work presented in this paper was done when the author is affiliated with Ming Li's group at the University of Waterloo. Some experimental analysis and the writing of this paper was done when the author is with Bonnie Berger's group at MIT. The author gratefully acknowledges support from NSERC CRC chair grant and PMMB fellowship.

References

1. Rost, B.: TOPITS: Threading one-dimensional predictions into three-dimensional structures. In Rawlings, C., Clark, D., Altman, R., Hunter, L., Lengauer, T., Wodak, S., eds.: Third International Conference on Intelligent Systems for Molecular Biology, Cambridge, England, AAAI Press (1995) 314–321
2. Xu, Y., Xu, D., Uberbacher, E.: An efficient computational method for globally optimal threadings. Journal of Computational Biology **5** (1998) 597–614
3. Kim, D., Xu, D., Guo, J., Ellrott, K., Xu, Y.: PROSPECT II: Protein structure prediction method for genome-scale applications. Protein Engineering **16** (2003) 641–650
4. Jones, D.: GenTHREADER: An efficient and reliable protein fold recognition method for genomic sequences. Journal of Molecular Biology **287** (1999) 797–815
5. Kelley, L., MacCallum, R., Sternberg, M.: Enhanced genome annotation using structural profiles in the program 3D-PSSM. Journal of Molecular Biology **299** (2000) 499–520
6. Alexandrov, N., Nussinov, R., Zimmer, R.: Fast protein fold recognition via sequence to structure alignment and contact capacity potentials. In: Biocomputing: Proceedings of 1996 Pacific Symposium. (1996)
7. von Ohsen, N., Sommer, I., Zimmer, R., Lengauer, T.: Arby: automatic protein structure prediction using profile-profile alignment and confidence measures. Bioinformatics **20** (2004) 2228–35
8. Shi, J., Tom, L.B., Kenji, M.: FUGUE: Sequence-structure homology recognition using environment-specific substitution tables and structure-dependent gap penalties. Journal of Molecular Biology **310** (2001) 243–257
9. Lathrop, R., Smith, T.: A branch-and-bound algorithm for optimal protein threading with pairwise (contact potential) amino acid interactions. In: Proceedings of the 27th Hawaii International Conference on System Sciences, IEEE Computer Society Press (1994)
10. Akutsu, T., Miyano, S.: On the approximation of protein threading. Theoretical Computer Science **210** (1999) 261–275
11. Li, W., Pio, F., Pawlowski, K., Godzik, A.: Saturated BLAST: detecting distant homology using automated multiple intermediate sequence BLAST search. Bioinformatics **16** (2000) 1105–1110
12. Summers, N., Karplus, M.: Construction of side-chains in homology modelling: Application to the c-terminal lobe of rhizopuspepsin. Journal of Molecular Biology **210** (1989) 785–811
13. Holm, L., Sander, C.: Database algorithm for generating protein backbone and sidechain coordinates from a C_α trace: Application to model building and detection of coordinate errors. Journal of Molecular Biology **218** (1991) 183–194

14. Desmet, J., Maeyer, M.D., Hazes, B., Laster, I.: The dead-end elimination theorem and its use in protein side-chain positioning. Nature **356** (1992) 539–542

15. Desmet, J., Spriet, J., Laster, I.: Fast and accurate side-chain topology and energy refinement (faster) as a new method for protein structure optimization. Protein: Structure, Function and Genetics **48** (2002) 31–43

16. Jr., R.D.: Comparative modeling of CASP3 targets using PSI-BLAST and SCWRL. Protein: Structure, Function and Genetics **3** (1999) 81–87

17. Canutescu, A., Shelenkov, A., Jr., R.D.: A graph-theory algorithm for rapid protein side-chain prediction. Protein Science **12** (2003) 2001–2014

18. Samudrala, R., Moult, J.: Determinants of side chain conformational preferences in protein structures. Protein Engineering **11** (1998) 991–997

19. Xiang, Z., Honig, B.: Extending the accuracy limits of prediction for side-chain conformations. Journal of Molecular Biology **311** (2001) 421–430

20. Bower, M., Cohen, F., Jr., R.D.: Prediction of protein side-chain rotamers from a backbone-dependent rotamer library: A new homology modeling tool. Journal of Molecular Biology **267** (1997) 1268–1282

21. Liang, S., Grishin, N.: side-chain modelling with an optimized scoring function. Protein Science **11** (2002) 322–331

22. Hong, E., Lozano-Perez, T.: Protein side-chain placement: probabilistic inference and integer programming methods. Technical report, Computer Science and Artificial Intelligence Laboratory, Massachusetts Institute of Technology (2004)

23. Chazelle, B., Kingsford, C., Singh, M.: A semidefinite programming approach to side-chain positioning with new rounding strategies. Informs Journal on Computing, Special Issue in Computational Molecular Biology/Bioinformatics (2004) 86–94

24. Kingsford, C.L., Chazelle, B., Singh, M.: Solving and analyzing side-chain positioning problems using linear and integer programming. Bioinformatics (2004)

25. Eriksson, O., Zhou, Y., Elofsson, A.: Side chain-positioning as an integer programming problem. In: Proceedings of the First International Workshop on Algorithms in Bioinformatics, Springer-Verlag (2001) 128–141

26. Dukka, K., Tomita, E., Suzuki, J., Akutsu, T.: Protein side-chain packing problem: a maximum common edge-weight clique algorithmic approach. In: The Second Asia Pacific Bioinformatics Conference. (2004)

27. Althaus, E., Kohlbacher, O., Lenhof, H., Müller, P.: A branch and cut algorithm for the optimal solution of the side-chain placement problem. Technical Report MPI-I-2000-1-001, Max-Planck-Institute für Informatik (2000)

28. Jr., R.D., Karplus, M.: Backbone-dependent rotamer library for proteins: Application to side-chain prediction. Journal of Molecular Biology **230** (1993) 543–574

29. Pierce, N., Winfree, E.: Protein design is NP-hard. Protein Engineering **15** (2002) 779–782

30. Akutsu, T.: NP-hardness results for protein side-chain packing. In Miyano, S., Takagi, T., eds.: Genome Informatics 8. (1997) 180–186

31. Sali, A., Blundell, T.: Comparative protein modelling by satisfaction of spatial restraints. Journal of Molecular Biology (1993) 779–815

32. Leach, A., Lemon, A.: Exploring the conformational space of protein side chains using dead-end elimination and the A^* algorithm. Protein: Structure, Function and Genetics **33** (1998) 227–239

33. Goldstein, R.: Efficient rotamer elimination applied to protein side-chains and related spin glasses. Biophysical Journal **66** (1994) 1335–1340

34. Robertson, N., Seymour, P.: Graph minors. II. algorithmic aspects of tree-width. Journal of Algorithms **7** (1986) 309–322

35. Koster, A., van Hoesel, S., Kolen, A.: Solving frequency assignment problems via tree-decomposition. Research Memoranda 036, Maastricht : METEOR, Maastricht Research School of Economics of Technology and Organization (1999) available at http://ideas.repec.org/p/dgr/umamet/1999036.html.
36. Bach, F., Jordan, M.: Thin junction trees. In Dietterich, T., Becker, S., Ghahramani, Z., eds.: Advances in Neural Information Processing Systems (NIPS). Volume 14. (2002)
37. Arnborg, S., Corneil, D., Proskurowski, A.: Complexity of finding embedding in a k-tree. SIAM Journal on Algebraic and Discrete Methods **8** (1987) 277–284
38. Miller, G.L., Teng, S., Thurston, W., Vavasis, S.A.: Separators for sphere-packings and nearest neighbor graphs. Journal of ACM **44** (1997) 1–29
39. Berry, A., Heggernes, P., Simonet, G.: The minimum degree heuristic and the minimal triangulation process. Volume 2880., Proceedings WG 2003 - 29th Workshop on Graph Theoretic Concepts in Computer Science, Lecture Notes in Computer Science, Springer Verlag (2003) 58–70
40. Kohlbacher, O., Lenhof, H.: BALL - rapid software prototyping in computational molecular biology. Bioinformatics **16** (2000) 815C824
41. Arya, S., Mount, D., Netanyahu, N., Silverman, R., Wu, A.: An optimal algorithm for approximate nearest neighbor searching in fixed dimensions. Journal of ACM **45** (1998) 891–923
42. Mount, D., Arya, S.: ANN: a library for approximate nearest neighbor searching. In: 2nd CGC Workshop on Computational Geometry. (1997)
43. Amir, E.: Efficient approximation for triangulation of minimum treewdith. In: 17th Conference on Uncertainty in Artificial Intelligence (UAI '01). (2001)
44. Xu, J., Li, M., Lin, G., Kim, D., Xu, Y.: Protein threading by linear programming. In: Biocomputing: Proceedings of the 2003 Pacific Symposium, Hawaii, USA (2003) 264–275
45. Xu, J., Li, M., Kim, D., Xu, Y.: RAPTOR: optimal protein threading by linear programming. Journal of Bioinformatics and Computational Biology **1** (2003) 95–117

Recognition of Binding Patterns Common to a Set of Protein Structures

Maxim Shatsky[1,*], Alexandra Shulman-Peleg[1], Ruth Nussinov[2,3], and Haim J. Wolfson[1]

[1] School of Computer Science,
Raymond and Beverly Sackler, Faculty of Exact Sciences,
Tel Aviv University, Tel Aviv 69978, Israel
[2] Sackler Inst. of Molecular Medicine, Sackler Faculty of Medicine,
Tel Aviv University, Tel Aviv 69978, Israel
[3] Basic Research Program, SAIC-Frederick, Inc,
Lab. of Experimental and Computational Biology,
Bldg. 469, Rm. 151, Frederick, MD 21702, USA

Abstract. We present a novel computational method, MultiBind, for recognition of binding patterns common to a set of protein structures. It is the first method which performs a multiple alignment between protein binding sites in the absence of overall sequence, fold or binding partner similarity. MultiBind recognizes common spatial arrangements of physico-chemical properties in the binding sites. These should be important for recognition of function, prediction of binding and drug design. We discuss the theoretical aspects of the computational problem of multiple structure alignment. This problem involves solving a 3D k-partite matching problem, which we show to be NP-Hard. The MultiBind method, applies an efficient Geometric Hashing technique to detect a potential set of multiple alignments of the given binding sites. To overcome the exponential number of possible multiple combinations it applies a very efficient filtering procedure which is heavily based on the selected scoring function. Our method guarantees detection of an approximate solution in terms of pattern proximity as well as cardinality of multiple alignment. We show applications of MultiBind to several biological targets. The method recognizes patterns which are responsible for binding small molecules such as estradiol, ATP/ANP and transition state analogues. The presented computational results agree with the available biological ones.

Availability: http://bioinfo3d.cs.tau.ac.il/MultiBind/.

Keywords: multiple structure alignment of binding sites; consensus binding patterns; pattern matching; pattern discovery, recognition of functional sites; k-partite matching.

---

* To whom correspondence should be addressed, email: maxshats@cs.tau.ac.ils

S. Miyano et al. (Eds.): RECOMB 2005, LNBI 3500, pp. 440–455, 2005.
© Springer-Verlag Berlin Heidelberg 2005

Introduction

Binding sites with similar physico-chemical and geometrical properties may perform similar functions and bind similar binding partners. Such binding sites may be created by evolutionarily unrelated proteins that share no overall sequence or fold similarities. Their recognition has become especially acute with the growing number of protein structures determined by the Structural Genomics project. Multiple alignment of binding sites that are known to have similar binding partners allows recognition of the physico-chemical and geometrical patterns that are responsible for the binding. These patterns may help to understand and predict molecular recognition. Moreover, multiple alignment of binding sites allows analysis of the dissimilarities of the binding sites which are important for the specificity of drug leads.

Sequence patterns have been widely used for comparison and annotation of protein binding sites [1]. Several methods search for patterns of residues that are conserved in their 3D positions and in amino acid identities [2, 3, 4]. However, there are numerous examples of functionally similar binding sites that are neither sequence order dependent nor share common patterns of amino acids [5, 6, 7]. Several methods have been developed for protein multiple structural alignment [8, 9, 10, 11, 12]. To overcome the alignment complexity of large protein structures these methods apply a variety of heuristics as well as some assumptions on properties of protein backbone, e.g. sequentiality of some backbone fragments. However, similar binding site patterns may appear in proteins with different overall folds. In addition, such patterns may be relatively small and can be easily missed when applying heuristic approaches used for protein backbone alignment. Methods for recognition of a pharmacophore common to a set small ligands [13, 14] share some methodological aspects with problems of protein backbone or binding site alignment. However, most developed methods for common pharmacophore detection are optimized for its specific problem definition, e.g. assume a tree-like ligand topology. Consequently, the methods for protein backbone alignment or common ligand pharmacophore detection are generally not suitable for recognition of common patterns of protein binding sites.

Several works have analyzed complexes of different proteins with the same ligand. The superimposition between the binding sites has been obtained by alignment of their common ligands [15, 6]. This approach has several limitations. First, it can analyze only protein structures with exactly the same partner ligands. Second, the same ligand can bind in alternative modes even to the same protein binding site [6]. Therefore, alignment according to ligands may fail to recognize the pattern.

Computational methods have been developed for direct alignment of protein binding sites. From the algorithmic standpoint, this involves solving a problem of spatial labeled/unlabeled pattern detection. Most of the methods apply clique detection algorithms. Some recent examples are the methods described by Kinoshita et al. [16], Schmitt et al. [17] and Shulman-Peleg et al. [7] However, all of these methods perform a comparison of only two molecules. Pairwise alignments may contain large number of features that are not necessarily required for the binding. Multiple alignments of binding sites with the same function may help to recognize the smallest set of features, a *consensus*, that is essential to achieve the desired biological effect. In creation and screening of databases, such *consensus binding patterns* may facilitate the development of efficient ranking schemes and database architectures [7]. Although it is possible to combine the results of various pairwise comparisons, high scoring pairwise solutions do not necessarily lead to a high scoring solution for a set of molecules [18].

Below we review the progress made in Computational Geometry in studying this problem. We start with a description for the case of two structures and continue to multiple structures. Our emphasis is only on subjects related to molecular structures. The problem of pattern detection answers the following question. Given two point sets A and B, find a subset of A that is similar to some subset of B. The optimization problem is to maximize the cardinality of similar subsets. One common way to define the similarity between two point sets is by the *bottleneck* metric [19, 20]. Similar sub-sets are called ϵ-congruent if the maximal distance between the matched points is less than ϵ. The optimization problem, called the *Largest Common Point Set* (LCP) problem, involves finding a transformation, e.g. Euclidean motion, that maximizes the size of two ϵ-congruent sub-sets. For the *bottleneck* metric in 3D it can be solved in $O(n^{32.5})$ time [21], where $n = max(|A|, |B|)$. Obviously, its time complexity is not practical even for small point sets. Therefore, more efficient methods are required.

Approximation techniques can significantly reduce time complexity at the price of solution accuracy. A simple *alignment technique* [22] constructs a finite set of transformations by *aligning* each triplet of points from the first structure with each ϵ-congruent triplet from the second one. For each transformation we can apply the maximal bipartite matching algorithm to compute a bijective mapping of points that are within ϵ distance from each other. Such *alignment technique* guarantees finding LCP under 8ϵ-congruence of cardinality at least of LCP under ϵ-congruence, if the latter exists. The time complexity is $O(n^{8.5})$. This technique was first developed for the Hausdorff distance [23] and later applied for the *bottleneck* distance [19]. Instead of approximating ϵ-congruence the same technique can be applied to approximate the LCP size [24]. Interestingly, an optimal algorithm for solving LCP for a group of transformations limited to rotations only, can improve the approximation factor of the LCP problem for general Euclidean transformations from 8ϵ to 2ϵ, while preserving the complexity of $O(n^{8.5})$ [24]. However, an implementation of such a technique is more complicated and the constant factors of the time complexity become larger.

Extension of the problem to detect a common point set between a set of K structures (from now on the term *point set* and *structure* will be used alternatively) has many important applications for the analysis of protein and drug molecules. However, even in one dimensional space for the case of exact congruence ($\epsilon = 0$) the problem is NP-Hard and it is hard to approximate within the factor $n^{1-\delta}$, for any $\delta > 0$, where n is the size of the smallest structure [18]. The problem is further complicated by the fact that in practice it is impossible to work with *zero-congruence*. Therefore, we face another combinatorial problem. Namely, given a set of superimposed structures, compute the largest common ϵ-congruent sub-set. We will call this problem *K-partite-3D* matching. While for two structures $(K = 2)$ it can be solved by bipartite matching, for $K > 2$ structures it can be solved by *K-partite* matching. However, this problem is known to be NP-Hard even for $k = 3$ in general and hyper graphs [25, 26]. Here, we show that the *K-partite-3D* matching problem is also NP-Hard.

In this paper we present an efficient, practical, method, MultiBind, for identification of common protein binding patterns by solving the multiple structure alignment problem. The problem we aim to solve is NP-Hard, therefore our goal is to find a trade-off between practical efficiency and theoretical bounds of solution accuracy, while, most importantly, validating the biological correctness of the results. We represent the protein binding site as a set of 3D points that are assigned a set of physico-chemical and geometrical properties important for protein-ligand interactions. The implementation of our method includes three major computational steps. The first one is a generation of 3D transformations that *align* the molecular structures. Here we apply the time efficient Geometric Hashing method [27]. The advantage of this method is that it enables to avoid processing of points that can not be matched under any transformation. In other words, its time complexity is proportional to the number of potentially matched points included in the defined set of transformations. The second step is a search for a combination of 3D transformations that gives the highest scoring common 3D core. For this step we provide an algorithm that guarantees to find the optimal solution by applying an efficient filtering procedure which practically overcomes the exponential number of multiple combinations. The final step is a computation of matching between points under multiple transformations, namely *K-partite-3D* matching. Here, we give a fast approximate solution with factor K. The overall scheme guarantees to approximate the ϵ-congruence as well as the cardinality of multiple alignment. We apply MultiBind to some well studied biological examples such as estradiol, ATP/ANP and transition state analogues binding sites. Our computational results agree with the available biological data.

The Largest Common Point Set Problem

We start from the definition of a pure geometric problem and in the next section extend it to the biology related problem. Objects are represented by point sets in 3D Euclidean space. Object S_1 is $\epsilon - congruent$ to S_2 if there exists an Euclidean transformation T and a bijective mapping $m : S_2 \rightarrow S_1$ such that for each point $s \in S_2$, $d(m(s), T(s)) \leq \epsilon$, where $d(.,.)$ is the Euclidean metric. Such similarity measure is called the *bottleneck matching measure*. For simplicity, we will also say that $d(S_1, T(S_2)) \leq \epsilon$ if the two objects are of the same cardinality, $S_1 = \{p_i\}_1^n$ and $S_2 = \{q_i\}_1^n$, and $d(p_i, T(q_i)) \leq \epsilon$, $i = 1, ..., n$, that is the set S_2 is preordered according to some bijective mapping.

Problem 1. Largest Common Point Set (LCP) between 2 Sets. *Given $\epsilon > 0$, and two point sets S_1 and S_2, find a transformation T and equally sized subsets $S_i' \subseteq S_i$ (i=1,2) of maximal cardinality such that $d(S_1', T(S_2')) < \epsilon$.*

Assuming that $|S_1|, |S_2| \leq n$ this problem can be exactly solved in $O(n^{32.5})$ time [21]. Now, we define an approximation version for this problem. Assume that L is the size of the largest common point set of S_1 and S_2 with an error ϵ. An $(\epsilon, \beta, \gamma)$-*approximation* of the LCP problem, $\epsilon \geq 0, \beta \geq 1, \gamma \geq 1$, is to find a common point set of size at least L/γ with an error of at most $\beta\epsilon$. Consider a simple alignment method that works as follows. For each triplet of points from S_1 and for each triplet from S_2 construct a 3D transformation that *aligns* the second triplet with the first one (it is enough to consider only pairs of triangles with maximal triangle side difference $\leq 2\epsilon$). Apply this transformation on S_2 and construct a bipartite graph where the vertices are the points of S_1 and of transformed S_2, and edges are created between the points with distance less than $\beta\epsilon$. Apply a maximal bipartite matching algorithm to compute the largest set of aligned points. The algorithm works in $O(n^3 * n^3 * n^{2.5})$. For this method the approximation ratio depends on the following alignment rule for construction of a 3D transformation based on two triplets of points (p_1, p_2, p_3) and (q_1, q_2, q_3).

Local Reference Frame, $LRF(p_1, p_2, p_3)$: Define a local right hand coordinate system s.t.: $p_1 = (0,0,0)$, $(p_2 - p_1)/|(p_2 - p_1)| = (1,0,0)$, $(p_2 - p_1) \times (p_3 - p_1)/|(p_2 - p_1) \times (p_3 - p_1)| = (0,0,1)$.

Alignment Rule: Define a transformation T' that superimposes $LRF(q_1, q_2, q_3)$ onto $LRF(p_1, p_2, p_3)$, i.e. $p_1 = T'(q_1)$, $(p_2 - p_1)/|p_2 - p_1| = (T'(q_2) - T'(q_1))/|T'(q_2) - T'(q_1)|$ and $sign((p_2 - p_1) \times (p_3 - p_1)) = sign((T'(q_2) - T'(q_1)) \times (T'(q_3) - T'(q_1)))$

This rule gives the approximation ratio $\beta \leq 8$ ($\gamma = 1$) [23, 19]. In this work we extend the LCP problem to multiple sets and we call it the mLCP problem. We also define the multiple LCP problem with respect to a *pivot* structure and we call it the pmLCP problem.

Problem 2. (**mLCP**). Largest Common Point Set between K Sets. *Given $\epsilon > 0$, and K point sets S_i, $i = 1, ..., K$, find transformations $\{T_i\}$, $i = 2, ..., K$, and equal sized sets $\{S_i' \subseteq S_i\}$, $i = 1, ..., K$, of maximal cardinality such that $d(T_i(S_i'), T_j(S_j')) < \epsilon$ ($i \neq j$, $i, j = 1, ..., K$, where T_1 is identity transformation).*

Problem 3. (**pmLCP**). Largest Common Point Set between K Sets with a Pivot Structure. *Given $\epsilon > 0$, a pivot set S_1 and $K - 1$ point sets S_i, $i = 2, ..., K$, find transformations $\{T_i\}$, $i = 2, ..., K$, and equal sized sets $\{S_i' \subseteq S_i\}$, $i = 1, ..., K$, of maximal cardinality such that $d(S_1', T_i(S_i')) < \epsilon$, $i = 2, ..., K$.*

Not surprisingly, both problems are NP-Hard, even in one dimensional space for the case of exact congruence, i.e. $\epsilon = 0$ [18]. In addition, for $\epsilon > 0$ we face another combinatorial problem. Consider a reduced mLCP/pmLCP problem where the transformation search is omitted, i.e. the position of the structures is fixed. Then, for two structures the problem is easily solved by a bipartite matching algorithm. However, for K structures it requires to solve a K-dimensional matching in Euclidean 3D space. In general graphs this problem is NP-Hard even for three sets [25,26]. We show that it is still NP-Hard even for graphs defined on 3D structures, where edges between the nodes from different partitions (structures) are created if and only if the distance between the nodes is less than ϵ.

Definition 1. (K-partite-3D graph). Given $\epsilon > 0$ and K point sets S_i, $i = 1, ..., K$, a K-partite-3D graph $G(S_1, ..., S_K) = (V, E)$ is defined as $V = \{\cup_{i=1}^{K} S_i\}$ and $E = \{(p_i, p_j) : i \neq j, p_i \in S_i, p_j \in S_j, d(p_i, p_j) \leq \epsilon\}$. A matching of a K-partite-3D graph is a set of disjoint K-tuples $\{(p_{t_1}, ..., p_{t_K}) : p_{t_i} \in S_i, p_{t_j} \in S_j, (p_{t_i}, p_{t_j}) \in E\}$.

Definition 2. (K-partite-3D-pivot graph). Given $\epsilon > 0$ and K point sets S_i, $i = 1, ..., K$, of which S_1 is the pivot, a K-partite-3D-pivot graph $G(S_1, ..., S_K) = (V, E)$ is defined as $V = \{\cup_{i=1}^{K} S_i\}$ and $E = \{(p_1, p_j) : j > 1, p_1 \in S_1, p_j \in S_j, d(p_1, p_j) \leq \epsilon\}$. A matching of a K-partite-3D-pivot graph is a set of disjoint K-tuples $\{(p_{t_1}, ..., p_{t_K}) : p_{t_1} \in S_1, p_{t_j} \in S_j, (p_{t_1}, p_{t_j}) \in E\}$.

Theorem 1. *The maximal cardinality matching problem in K-partite-3D and K-partite-3D-pivot graphs is NP-Hard.*

A Sketch of the Proof. First, we briefly present a reduction from *3-SAT* to *3-partite matching* in general graphs (for more details see [25]), and then extend it for the instances of K-partite-3D and K-partite-3D-pivot graphs.

An instance of the 3SAT problem includes a set of variables $U = \{u_1, u_2, ..., u_n\}$ and a set of clauses $C = \{c_1, c_2, ..., c_m\}$. Each clause contains three literals of variables U. The goal of the reduction is to construct three disjoint sets S_1, S_2 and S_3 of equal cardinality, and a set of edges $M \subseteq S_1 \times S_2 \times S_3$ such that M contains a perfect matching if and only if C is satisfiable.

Three classes of edges are created, T - "truth setting and fan-out", C - "satisfaction testing" and G - "garbage collection". The components of T are constructed for each variable u_i. Denote $u_i[j]$ to be a variable u_i in clause j.

$$T_i^t = \{(\bar{u}_i[j], a_i[j], b_i[j]) : 1 \leq j \leq m\}$$
$$T_i^f = \{(u_i[j], a_i[j+1], b_i[j]) : 1 \leq j \leq m\} \cup \{(u_i[m], a_i[1], b_i[m])\}$$
$$\bar{u}_i[j], u_i[j] \in S_1, \quad a_i[j] \in S_2, \quad b_i[j] \in S_3$$

The component T forces a matching to choose between setting u_i true and setting u_i false. Any perfect matching will have to include either all triplets from T_i^t or all triplets from T_i^f, see Figure 1 (a). Next, for each clause c_j a component C_j aims to select a truth setting for one of its three literals: $C_j = \{(u_i[j], s_2[j], s_3[j]) : u_i \in c_j\} \cup \{(\bar{u}_i[j], s_2[j], s_3[j]) : \bar{u}_i \in c_j\}$, where $s_2[j] \in S_2$ and $s_3[j] \in S_3$.

Thus, only one triplet can be contained in any matching assigning the clause c_j to true setting. Finally, the "garbage collection" component aims to compensate the unequal number of nodes created so far in S_1 and in other two partitions S_2 and S_3: $G = \{(u_i[j], g_2[k], g_3[k]), (\bar{u}_i[j], g_2[k], g_3[k]) : 1 \leq k \leq m(n-1), 1 \leq i \leq n, 1 \leq j \leq m, g_2[j] \in S_2, g_3[j] \in S_3\}$.

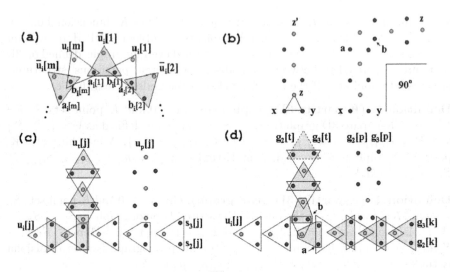

Fig. 1

To summarize, the edges are defined as: $T = \cup_{i=1}^{n}(T_i^t \cup T_i^f)$, $C = \cup_{j=1}^{m}C^j$, $M = T \cup C \cup G$. This completes the reduction from *3-SAT* to *3-partite matching*. Next, we adapt the above reduction for K-partite-3D type graphs.

Notice that the constructed graph M does not belong to the K-partite-3D type of graphs. Only the component T can be drawn in 2D to satisfy this property, i.e. only the point triplets from T can be placed within ϵ distance one from each other (see Figure 1 (a))[1]. The problem is that the nodes of type s_2, s_3 and g_2, g_3 can not be placed in 3D so that their distance from the different nodes of type u_i is less than ϵ. To resolve this problem we introduce *long-distance-edge* gadgets. The basic principle is illustrated in Figure 1 (b). The edge (x, y, z) can be elongated to any distance (z transforming to z') preserving the property for matching. Also, we can *bend* to 90° any *long-distance-edge*. We can bend an edge in two different ways. The first one, as illustrated in Figure 1 (b), continues the edge in the same 2D plane. The second option, bends the edge so that its two parts are in two perpendicular 2D planes (not illustrated). Care should be taken at the bending part, e.g. nodes a and b from Figure 1 (b) should be placed at a distance larger than ϵ, otherwise it introduces ambiguity for matching. The second gadget aims to *split* edges going from nodes of type s_2, s_3 (g_2, g_3). Assume we have three edges $(u_i[j], s_2[j], s_3[j])$, $(u_t[j], s_2[j], s_3[j])$ and $(u_p[j], s_2[j], s_3[j])$. Figure 1 (c) illustrates how these three edges can be constructed. Triangles illustrate possible matching. This gadget guarantees that any perfect matching will select only one node $u[j]$ with combination of nodes from this gadget. However, there are many *long-distance-edges* coming to nodes of type $u[j]$ and there is a

---

[1] In the *3-SAT* to *3-partite matching* reduction, the definition of a hypergraph edge as a triplet of points (a,b,c) is equivalent to three edges (a,b), (a,c) and (b,c) in a regular graph.

need to join them. The *split* gadget is not suitable for this task, therefore we introduce a *join* gadget (see Figure 1 (d)). The *join* gadget guarantees that any perfect matching will connect a node $u[j]$ to only one pair of type s_2, s_3 (g_2, g_3). To complete the construction we need to show how to place in 3D all the *long-distance-edges* and connections between them. The idea is to place the component T in the plane $(x, y, 0)$ (*zero level*), to place the components C_j on *negative levels* $(x, y, -l_j)$ and the components G_k on *positive levels* (x, y, l_k). The *long-distance-edges* are constructed between the levels like water pipes. The whole construction requires polynomial number of components. Due to lack of space we omit the exact details, which will appear elsewhere. Notice, that by selecting the first partition S_1 as a pivot structure and deleting the edges between S_2 and S_3 the same reduction works as well for the instances of K-partite-3D-pivot graphs.

The MultiBind Algorithm

Input Representation: Physico-Chemical Properties. Selection of the proper representation is crucial for the biochemical significance of the recognized patterns. Given the atomic coordinates of a protein structure, we follow Schmitt et al. [17] and for each amino acid we group atoms with similar physico-chemical properties to functional groups. These are localized by 3D points in space, denoted as pseudocenters. Each pseudocenter represents one of the following properties important for protein-ligand interactions: *hydrogen-bond donor, hydrogen-bond acceptor, mixed donor/acceptor, hydrophobic aliphatic and aromatic(pi) contacts*. Since both backbone and side-chain atoms are considered, each amino acid is represented by a set of such pseudocenters. We construct the smooth molecular surface as implemented by Connolly [28] and retain only pseudocenters that represent at least one surface exposed atom. When considering binding sites, we refer only to the surface regions that are within 4Å from the binding partner. In practice, a comparison of the spatial locations of the retained pseudocenters is not sufficient for the accurate prediction of protein-ligand interactions. Thus, we are interested in the maximal number of matching pseudocenters that are most similar in all the physico-chemical and geometrical aspects. For each pair of pseudocenters, p and q, we define a scoring function $PC\text{-}Score(p,q)$ which measures the similarity of the properties important for the specific type of interaction in which they can participate ($PC\text{-}Score(S_1, S_2)$ is defined as a sum of the matched point scores). The exact calculations and default parameters are detailed in Appendix A. Therefore, practically, we look for a solution for a weighted *pmLCP* problem that we define as[2]:

Problem 4. (**Max-Min Weighted *pmLCP* Problem**) *Given $\epsilon > 0$, a scoring function PC-Score, a pivot set S_1 and $K-1$ point sets S_i, $i = 2, ..., K$ find trans-*

---

[2] In our implementation we consider only the pmLCP problem since the K-partite-3D matching of the mLCP problem introduces additional complications even for greedy approaches. We'll address this problem somewhere else.

formations $\{T_i\}$, $i = 2, ..., K$, *and equal sized sets* $\{S_i' \subseteq S_i\}$, $i = 1, ..., K$, *such that* $d(S_1', T_i(S_i')) < \epsilon$, $i = 2, ..., K$, *and* $min_i PC\text{-}Score(S_1', T_i(S_i'))$ *is maximal.*

The Pattern Matching Algorithm. There are three major computational steps: (1) generation of 3D transformations and potential points for matching; (2) combinatorial search for a combination of 3D transformations that gives the highest scoring common 3D core (*Traversal stage*); and (3) computation of *K-partite-3D-pivot* matching.

In our approach we follow the efficient strategy of the Geometric Hashing method [27]. The Geometric Hashing method consists of two stages, *preprocessing* and *recognition*. At the *preprocessing* stage each triplet of pseudocenters, (a, b, c), from each molecule except the pivot is considered as a local reference frame $r = LRF(a, b, c)$. The coordinates of the other points are calculated with respect to the local reference frame r. This information is stored in a *Geometric Hash Table*. The key to the hash table is (x^r, y^r, z^r, p), where (x^r, y^r, z^r) are point coordinates with respect to the local reference frame r, and p is the physico-chemical property of the pseudocenter. Only pseudocenters with the same property can be matched [3]. The data stored in the hash table includes the key itself and the identifiers of the molecule and the reference frame.

In the *recognition* stage the same process as in the *preprocessing* stage is repeated for the pivot molecule. However, instead of storing data in the hash table, all entries close to the key within radius ϵ and with the same physico-chemical property are retrieved. For each reference frame r of the pivot structure a voting table is created. It counts the number of matched points for each reference frame stored in the hash table. For simplicity, we explain the method for the pure geometrical case, i.e. for the pmLCP problem. If a reference frame r' from structure i received v votes that means the following. Define a 3D transformation $T_{r,r'}$ that superimposes the triplets of points r' on r according to the *Alignment Rule*. Applying $T_{r,r'}$ on S_i will result in v point pairs from the pivot and i structure that are within ϵ distance. Thus, the size of a maximal matching between S_{pivot} and $T_{r,r'}(S_i)$ is less than v. Therefore, for the next step it is enough to consider only reference frames that have received a number of votes equal or greater than M^*, the size of the largest multiple solution found so far (initially $M^* = 0$). For each survived transformation T we store the list of matched points, $\{(p, q) : p \in S_{pivot}, q \in S_i, |p - T(q)| < \epsilon\}$.

Traversal stage. For each reference frame of the pivot structure we create a combinatorial bucket that contains transformations that received a high number of votes. Namely, a combinatorial bucket for the reference frame r is defined as $CB_r = \{T^2, T^3, ..., T^K\}$, where $T^i = \{T_{i_j}\}$ is a set of transformations for structure i that received $v > M^*$ votes. A multiple alignment is a combination of $K-1$ transformations, $(T_{i_2}^2, T_{i_3}^3, ..., T_{i_K}^K)$. The number of all possible combinations equals to $|T^2| * |T^3| * ... * |T^K|$, which is exponential with K. However, we have

---

[3] Pseudocenters that can function both as hydrogen bond donors and acceptors are encoded twice, once as donors and once as acceptors.

implemented a branch-and-bound traversal method which in practice is very efficient. First we provide some definitions. Given a transformation vector of the first t structures, $T = (T_{i_2}, ..., T_{i_t})$, create a t-partite-3D-pivot graph, $G(T) = G(S_1, T_{i_2}(S_2), ..., T_{i_t}(S_t))$. Define single sides of the graph $G(T)$, $G(T)[j] = \{p_j : p_j \in S_j, \exists p_1 \in S_1 \ (p_1, p_j) \in G(T) \text{ and } \forall k \leq t \ \exists p_k \in S_k \ (p_1, p_k) \in G(T)\}$. Let $M(G(T))$ be a maximal t-partite-3D-pivot matching of the graph $G(T)$. Obviously, $M(G(T)) \leq M(G(S_{pivot}, T_{i_j}(S_j))) \leq |G(T)[j]|$.

Given a combinatorial bucket $CB = \{T^2, T^3, ..., T^K\}$ we iteratively traverse it in the following manner. Assume that we have created a vector $T = (T_{i_2}, T_{i_3}, ..., T_{i_t})$, $T_{i_j} \in T^j$. We try to extend it with a transformation $T_{i_{t+1}} \in T^{t+1}$, $T^* = (T_{i_2}, T_{i_3}, ..., T_{i_t}, T_{i_{t+1}})$. Clearly, $|G(T^*)[j]| \leq |G(T)[j]|$, $j = 2, ..., t$. Therefore, if for some index j holds $|G(T^*)[j]| \leq M^*$, then we can disregard the vector T^* and start to build another combination of transformations. Essentially, we continue with the vector T and try to add another transformation from T^{t+1}, and so on. The number of traversals may be exponential, however in practice the $M(G(T))$ drops very quickly below M^* as the algorithm advances in iterations in the *recognition* stage [4]. Still, the theoretical bound is $O(n^3 n^{3(K-1)})$.

K-partite-3D-pivot Matching. During the traversal stage, once we reach the last bucket we have a uniquely defined K-partite-3D-pivot graph. The next step is to solve the matching problem. As we have shown above this problem is NP-Hard. We apply a greedy method, which iterates over pivot points and selects K-tuples, from non-selected points. This method gives a K approximation to the largest matching since at each greedy selection of K-tuples it may violate at most K-1 nodes that may belong to the optimal matching[5]. In the context of molecular structures for small ϵ (around 3Å) the maximal node degree is bound by a small constant. Therefore the time complexity of the greedy method is $O(Kn)$.

Theorem 2. *MultiBind algorithm is an* $(\epsilon, 8, K)$*-approximation*[6] *for* **Problem 3** *and has time complexity* $O(n^{3K}nK)$.

In practice, when solving the *Max-Min Weighted pmLCP* we introduce the following modifications. First, we define M^* to be the highest physico-chemical score of the multiple solution found so far. Given equally sized sets $(S_1, ..., S_t)$ the physico-chemical score M is defined as in Problem 4 by $M = min_j PC\text{-}Score(S_1, S_j)$, $j = 2...t$. When traversing the combinatorial buckets, instead of looking at the cardinality of the side j, $|G(T^*)[j]|$, we estimate the upper bound of $PC\text{-}Score(S_1, T_j(S_j))$ as $PC\text{-}Score(G(T^*)[j]) = \sum_{q \in T_j(S_j)} max_p PC\text{-}Score(p, q)$.

---

[4] In the second example from the *Results* section, the total number of combinations for all combinatorial buckets is about $1.3 \cdot 10^{11}$, which shows the exponential nature of the problem. The filtering procedure leaves only 246310 combinations of multiple alignments. Most filtering is done already at the third structure ($t = 3$).

[5] The best known approximation algorithm for hyper-graphs gives K/2 ratio [29]

[6] It is possible to reduce the $\beta = 8$ approximation to any accuracy $c\beta$, $c \leq 1$, by applying a discretization technique of the transformational space [30,31]. However, the payoff is increasing the time complexity factor proportional to $(1/c)^6$.

Therefore, we disregard vectors T^* for which $PC\text{-}Score(G(T^*)[j]) \leq M^*$ for some j. We retain a user defined number of high scoring solutions, which are then evaluated by an additional *Overall Surface Scoring* [7] function which compares the corresponding surfaces of the binding sites.

Biological Results

Below we present examples of application of MultiBind for recognition of patterns required for binding of different ligands. In each of the presented examples, we describe the details of a single solution that received the highest score. An additional example of application of MultiBind to proteins of trypsin and subtilisin folds is presented by Mintz et al [32]. The running times are measured on a standard PC, Intel(R) Pentium(R) IV 2.60GHz CPU with 2GB RAM. The default distance threshold for the ϵ-congruence is 3.0Å .

ATP/ANP Binding Sites of Protein Kinases. To validate the performance of the method on a well studied example we have selected a set of ATP/ANP binding sites extracted from 5 different protein kinases: cAMP-dependent PK (1cdk), Cyclin-dependent PK, CDK2 (1hck), Glycogen phosphorylase kinase (1phk), c-Src tyrosine kinase (2src), Casein kinase-1, CK1 (1csn). We applied MultiBind to perform a multiple alignment of the corresponding ATP/ANP binding sites. These were recognized to share 14 pseudocenters, 4 of which are created by amino acids with the same identity (see Figure 2(a)). The RMSD between the adenine moieties (which are not a part of the input and are used for verification only) under these transformations is less than 1.4Å . The average binding site size is 76 pseudocenters, and the running time is 58 minutes. It must be noted that since these proteins share similar overall folds, the 3D superposition problem of the binding sites can be solved by multiple backbone alignment methods [11,12]. However, these methods do not give solution to the *K-partite-3D* matching problem of physico-chemical features (since these are not-ordered on the protein surface). Below we present two examples for which both the superimposition and the matching problems can not be solved by standard protein backbone alignment methods.

Transition State Analogue Binding Sites. We have selected five binding sites complexes with endo-oxabicyclic transition state analogues (TSA/BAR). The binding sites were extracted from proteins of three different folds: (1) Chorismate mutase II (1ecm, 4csm, 3csm); (2) Bacillus chorismate mutase-like (2cht); (3) Immunoglobulin-like beta-sandwich (1fig). Figure 2(b) presents 8 functional groups that were recognized by MultiBind to be shared by all the binding sites. Two of the compared proteins (1ecm and 4csm) were previously aligned by Schmitt et al [17]. Most of the pseudocenters recognized by Multi-Bind are indeed a subset of those obtained by their pairwise alignment method (except for two donors contributed by 1ecm:Arg28). However, 10 of the functional groups common to a pair of chorismate mutases according to their study, were not recognized to be common to the five structures compared by Multi-Bind. Alignment of multiple structures with different folds helps to identify the minimal set of features required for the binding of endo-oxabicyclic transition state analogues. The average size of a binding site is 29, and the running time is 8 minutes.

Estradiol Binding Sites. Estradiol molecules are known to bind to protein receptors with different overall sequences and folds. The dataset of this study was comprised of the binding sites of 7 proteins from 4 different folds: (1) Nuclear receptor ligand-binding domain (3ert, 1a52, 1err, 1qwr); (2) NAD(P)-binding Rossmann-fold (1fds); (3) Concanavalin A-like lectins/glucanases (1lhu); (4) P-loop containing nucleoside triphosphate hydrolases (1aqu). Two of these structures were crystallized with Raloxifen (1err) and 4-hydroxytamoxifen (3ert), which are different from estradiol. In spite of the conformational changes required to accommodate these ligands, MultiBind has recognized 6 functional groups shared by all the binding sites (see Figure 2(c)). One of them is a conserved Phenylalanine (1lhu:Phe67) with an aromatic property shared by all the binding sites. The mean binding site size is 44 pseudocenters and the running time is 15 minutes.

In order to compare the presented results with those obtained by superimposition of ligand molecules, we performed such an alignment for the above mentioned examples (for the complexes with the same binding partners). In the

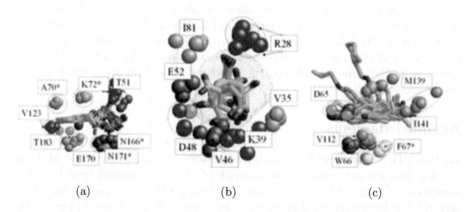

(a)   (b)   (c)

Fig. 2. Multiple alignments done by MultiBind. Matched pseudocenters are represented as balls. Hydrogen bond donors are blue, acceptors - red, donors/acceptors - green, hydrophobic aliphatic - orange and aromatic - white. Matched pseudocenters (backbone or side-chain) from identical amino acids are marked by *. The ligand molecules are presented for verification purpose only and are not a part of the input to MultiBind. **(a)** Multiple alignment of 5 ATP/ANP binding sites, the labeling is according to 1cdk. **(b)** Multiple alignment of five endo-oxabicyclic transition state analogue binding sites. The labeling and the surface (depicted in dots) is according to 1ecm. **(c)** Multiple alignment of eight estradiol binding sites, the labeling is according to 3ert

last two cases alignment by ligands failed to recognize any significant pattern (less than 3 pseudocenters), while MultiBind identified patterns of size 8 and 6.

Conclusions

We have presented a novel computational method, MultiBind, for recognition of physico-chemical binding patterns. The method is practically efficient for multiple alignment of protein binding sites and guarantees to detect an approximate solution for the case of pure geometrical problem. We have shown that the matching problem of K-partite-3D/K-partite-3D-pivot graphs is NP-Hard. We have presented an efficient filtering procedure which in our applications practically overcomes the exponential number of multiple combinations.

We have applied MultiBind to several biological targets, such as the binding sites of estradiol, ATP/ANP and transition state analogues. MultiBind is the first method that performs multiple alignment of binding sites in the absence of overall sequence, fold or binding partner similarity. To the best of our knowledge, the presented results can not be obtained by any other existing computational method. We hope that it will be a useful tool in prediction of molecular recognition and in identification of *consensus binding patterns*. These are important for improvement of architectures of databases of binding sites and development of efficient ranking schemes.

However, from the biological standpoint the method has several limitations. First, there is no explicit treatment of protein flexibility which is introduced only through a set of thresholds to allow variability in locations. Second, due to the hardness of the problem the method is practically limited to point sets of size about 100. Third, scoring functions are known to be one of the major problems in all types of *in silico* predictions. The scoring function of MultiBind suffers from the same limitations [7]. We intend to address these challenges in our future research.

Acknowledgments

We thank D. Schneidman for contribution of software and O. Dror and M. Landau for their critical reading. The research of M.S. is supported by a PhD fellowship in "Complexity Science" from the Yeshaya Horowitz association. This research has been supported in part by the "Center of Excellence in Geometric Computing and its Applications" funded by the Israel Science Foundation. The research of H.J.W. is partially supported by the Hermann Minkowski-Minerva Center for Geometry at TAU. The research of R.N. has been funded in whole or in part with Federal funds from the NCI, NIH, under contract number NO1-CO-12400. The content of this publication does not necessarily reflect the view or policies of the Dep. of Health and Human Services, nor does mention of trade names, commercial products, or organization imply endorsement by the U.S. Government.

References

1. Falquet, L., Pagni, M., Bucher, P., Hulo, N., Sigrist, C.J., Hofmann, K., Bairoch, A.: The PROSITE database, its status in 2002. Nucleic Acids Res. **30** (2002) 235–238
2. Wallace, A.C., Laskowski, R.A., Thornton, J.M.: Derivation of 3D coordinate templates for searching structural databases: application to Ser-His-Asp catalytic triads in the serine proteinases and lipases. Protein Science **5** (1996) 1001–1013
3. Russell, R.: Detection of protein three-dimensional side-chain patterns: new examples of convergent evolution. J. Mol. Biol. **279(5)** (1998) 1211–1227
4. Artymiuk, P.J., Poirrette, A.R., Grindley, H.M., Rice, D.W., Willett, P.: A graph-theoretic approach to the identification of three-dimensional patterns of amino acid side-chains in protein structures. J. Mol. Biol. **243** (1994) 327–344
5. Moodie, S.L., Mitchell, J.B.O., Thornton, J.M.: Protein recognition of adenylate: An example of a fuzzy recognition template. J. Mol. Biol. **263** (1996) 486–500
6. Denessiouk, K.A., Rantanen, V., Johnson, M.: Adenine Recognition: A motif present in ATP-,CoA-,NAD-,NADP-, and FAD-dependent proteins. PROTEINS: Structure, Function and Genetics **44** (2001) 282–291
7. Shulman-Peleg, A., Nussinov, R., Wolfson, H.J.: Recognition of functional sites in protein structures. J. Mol. Biol. **339(3)** (2004) 607–633 http://bioinfo3d.cs.tau.ac.il/SiteEngine/.
8. Russell, R., Barton, G.: Multiple protein sequence alignment from tertiary structure comparison: assignment of global and residue confidence levels. PROTEINS: Structure, Function and Genetics **14** (1992) 309–323
9. Taylor, W.R., Flores, T., Orengo, C.: Multiple protein structure alignment. Protein Science **3** (1994) 1858–1870
10. Leibowitz, N., Nussinov, R., Wolfson, H.: MUSTA-a general, efficient, automated method for multiple structure alignment and detection of common motifs: application to proteins. J Comput Biol. **8** (2001) 93–121
11. Shatsky, M., Nussinov, R., Wolfson, H.: A method for simultaneous alignment of multiple protein structures. Proteins: Structure, Function, and Genetics **56(1)** (2004) 143–156 http://bioinfo3d.cs.tau.ac.il/MultiProt/.
12. Dror, O., Benyamini, H., Nussinov, R., Wolfson, H.J.: MASS: multiple structural alignment by secondary structures. Bioinformatics **19 Suppl. 1** (2003) i95–i104 http://bioinfo3d.cs.tau.ac.il/MASS.
13. Lemmen, C., Lengauer, T.: Computational methods for the structural alignment of molecules. J. of Computer-Aided Mol. Design **14** (2000) 215–232
14. Dror, O., Shulman-Peleg, A., Nussinov, R., Wolfson, H.J.: Predicting molecular interactions in silico: I. A guide to pharmacophore identification and its applications for drug design. Curr. Med. Chem. **11** (2004) 71–90
15. Kuttner, Y.Y., Sobolev, V., Raskind, A., Edelman, M.: A consensus-binding structure for adenine at the atomic level permits searching for the ligand site in a wide spectrum of adenine-containing complexes. PROTEINS: Structure, Function and Genetics **52** (2003) 400–411
16. Kinoshita, K., Nakamura, H.: Identification of protein biochemical functions by similarity search using the molecular surface database ef-site. Protein Science **12** (2003) 1589–1595
17. Schmitt, S., Kuhn, D., Klebe, G.: A new method to detect related function among proteins independent of sequence or fold homology. J. Mol. Biol. **323** (2002) 387–406

18. Akutsu, T., Halldorson, M.M.: On the approximation of largest common subtrees and largest common point sets. Theoretical Computer Science **233** (2000) 33–50
19. Akutsu, T.: Protein structure alignment using dynamic programming and iterative improvement. IEICE Trans. Information and Systems **E79-D** (1996) 1629–1636
20. Efrat, A., Itai, A., Katz, M.J.: Geometry helps in bottleneck matching and related problems. Algorithmica **31** (2001) 1–28
21. Ambuhl, C., Chakraborty, S., Gartner, B.: Computing largest common point sets under approximate congruence. In: Proc. of the 8th Ann. European Symp. on Alg., Springer-Verlag (2000) 52–63
22. Huttenlocher, D., Ullman, S.: Recognizing solid objects by alignment with an image. International Journal of Computer Vision **5(2)** (1990) 195–212
23. Goodrich, M.T., Mitchell, J.S.B., Orletsky, M.W.: Practical methods for approximate geometric pattern matching under rigid motions: (preliminary version). In: Proc. of the 10th Ann. Symp. on Comp. Geom., ACM Press (1994) 103–112
24. Chakraborty, S., Biswas, S.: Approximation algorithms for 3-d commom substructure identification in drug and protein molecules. In: Proc. 6th Int. Workshop on Algorithms and Data Structures, Vancouver, Can., Springer-Verlag (1999) 253–264
25. Garey, M.R., Johnson, D.S.: Computers and Intractability. W. H. Freeman, San Francisco (1979)
26. Hazan, E., Safra, S., Schwartz, O.: On the Complexity of Approximating k-Dimensional Matching. In: Approximation, Randomization, and Combinatorial Optimization. Volume 2764 of LNCS., Springer (2003) 83–97
27. Wolfson, H.J.: Model-Based Object Recognition by Geometric Hashing. In: Proc. of the 1st European Conf. on Comp. Vision (ECCV). LNCS, Springer-Verlang (1990) 526–536
28. Connolly, M.L.: Analytical molecular surface calculation. J. Appl. Cryst. **16** (1983) 548–558
29. Hurkens, C.A.J., Schrijver, A.: On the size of systems of sets every t of which have an sdr, with an application to the worst-case ratio of heuristics for packing problems. SIAM J. Discret. Math. **2** (1989) 68–72
30. Heffernan, P.J., Schirra, S.: Approximate decision algorithms for point set congruence. Comput. Geom. Theory Appl. **4** (1994) 137–156
31. Gavrilov, M., Indyk, P., Motwani, R., Venkatasubramanian, S.: Combinatorial and experimental methods for approximate point pattern matching. Algorithmica **38** (2004) 59–90
32. Mintz, S., Shulman-Peleg, A., Wolfson, H.J., Nussinov, R.: Generation and analysis of a protein-protein interface dataset with similar chemical and spatial patterns of interactions. (submitted) (2004)
33. Connolly, M.L.: Measurement of protein surfaces shape by solid angles. J. Mol. Graph. **4** (1986) 3–6

Appendix A: Physico-Chemical Scoring

Let p and q be the two matched pseudocenters.

- $dist(p, q)$ - the distance between p and q after the superimposition. Default threshold for the maximal distance is $\epsilon = 3.0$Å.
- $chem(p)$ - the physico-chemical property of the point p. There are three types of properties: Hydrogen Bonding (HB), Aliphatic Hydrophobic (ALI) and Aromatic (PII).

- $charge(p)$ - the partial atomic charge of the atom p, which can form hydrogen bonds. $charge(p, q) = |charge(p) - charge(q)|$.
- $shape(p)$ - the average curvature of the surface region created by p. Calculated as an average of the solid angle shape functions [33] with spheres of radius 4,5,6 and 7Å . The sphere centers are located at projection point of p to the surface. $shape(p, q) = |shape(p) - shape(q)|$.
- $n_S(p)$ - normal vector at projection point of p to the surface, $n_S(p, q) = n_S(p) \cdot n_S(q)$.
- $n_{PII}(p)$ - for aromatic pseudocenters denotes the normal to the plane of the aromatic ring. $n_{PII}(p, q) = n_{PII}(p) \cdot n_{PII}(q)$.
- $v_{ALI}(p, q)$ - the overlap of the hydrophobic group spheres of p and q, approximated by the difference between sum of radiuses and the distance between the centers.

Each pair of matched pseudocenters is assigned a score according to the similarity of the properties important for the specific type of interaction:

$$PC\text{-}Score\,(p, q) = \begin{cases} 0, & dist(p, q) > \epsilon \ or \ chem(p) \neq chem(q) \\ 0, & shape(p, q) > 0.2 \ or \ n_S(p, q) > 0.2 \\ dist(p, q)/(1 + charge(p, q)) & chem(p) = HB \\ dist(p, q)/(1 + shape(p, q) + n_{PII}(p, q)) & chem(p) = PII \\ (dist(p, q) + v_{ALI}(p, q))/(2 + 20 * shape(p, q)) & chem(p) = ALI \end{cases}$$

Predicting Protein-Peptide Binding Affinity by Learning Peptide-Peptide Distance Functions

Chen Yanover[1,*] and Tomer Hertz[1,2,*]

[1] School of Computer Science and Engineering
[2] The Center for Neural Computation,
The Hebrew University of Jerusalem, Jerusalem, Israel, 91904
{cheny, tomboy}@cs.huji.ac.il

Abstract. Many important cellular response mechanisms are activated when a peptide binds to an appropriate receptor. In the immune system, the recognition of pathogen peptides begins when they bind to cell membrane Major Histocompatibility Complexes (MHCs). MHC proteins then carry these peptides to the cell surface in order to allow the activation of cytotoxic T-cells. The MHC binding cleft is highly polymorphic and therefore protein-peptide binding is highly specific. Developing computational methods for predicting protein-peptide binding is important for vaccine design and treatment of diseases like cancer.

Previous learning approaches address the binding prediction problem using traditional margin based binary classifiers. In this paper we propose a novel approach for predicting binding affinity. Our approach is based on learning a peptide-peptide distance function. Moreover, we learn a **single** peptide-peptide distance function over an **entire** family of proteins (e.g MHC class I). This distance function can be used to compute the affinity of a novel peptide to any of the proteins in the given family. In order to learn these peptide-peptide distance functions, we formalize the problem as a semi-supervised learning problem with partial information in the form of equivalence constraints. Specifically we propose to use *DistBoost* [1, 2], which is a semi-supervised distance learning algorithm.

We compare our method to various state-of-the-art binding prediction algorithms on MHC class I and MHC class II datasets. In almost all cases, our method outperforms all of its competitors. One of the major advantages of our novel approach is that it can also learn an affinity function over proteins for which only small amounts of labeled peptides exist. In these cases, *DistBoost*'s performance gain, when compared to other computational methods, is even more pronounced.

1 Introduction

Understanding the underlying principles of protein-peptide interactions is a problem of fundamental importance in biology, with application to medicinal

---

* Both authors contributed equally.

S. Miyano et al. (Eds.): RECOMB 2005, LNBI 3500, pp. 456–471, 2005.

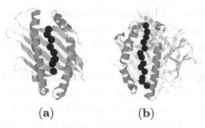

(a)   (b)

Fig. 1. Schematized drawing of a peptide in the binding groove of MHC class I (a) and class II (b) molecules. The peptide backbone is shown as a string of balls, each of which represents a residue

chemistry and drug design. Many cellular responses and activation mechanisms are triggered when a peptide binds to an appropriate receptor which leads to a cascade of downstream events. This communication mechanism is widely used in eukaryotic cells by cytokins, hormones and growth factors. In the immune system, the activation of cytotoxic T-cells is mediated by the binding of foreign antigen peptides to cell membrane Major Histocompatibility Complexes (MHCs). Antigen presenting MHC molecules are usually divided into two major classes: class I and class II (see Fig. 1).   The function of both class I and class II molecules is to carry short peptides to the cell surface for presentation to circulating T-cells. The recognition of these pathogen peptides as non-self by T-cells elicits a cascade of immune responses. MHC binding peptides, therefore, play an important role in diagnosis and treatment of cancer [3].

As a result of two decades of extensive experimental research, there exists empirical evidence on peptides that bind to a specific MHC molecule and peptides that do not bind to it. In some cases, binding peptides are classified as either high-affinity, moderate-affinity or low-affinity binders. Empirical evidence reveals that only 1 out of 100 to 200 peptides actually binds to a particular MHC molecule [4]. Since biochemical assays, which empirically test protein-peptide binding affinity, are not amenable to high throughput analysis, protein-peptide computational prediction methods come into play. Many different computational approaches have been suggested for predicting protein-peptide binding including motif based methods [5, 6], structural methods [7] and machine learning algorithms [8, 9, 10]. While all of these methods obtain promising results, the problem seems far from being solved.

Many machine learning prediction methods [8, 9, 10] are implicitly based on the observation that peptides that bind to a specific protein are similar in some sense. These learning algorithms formalize the problem as a binary (margin based) classification task — binding predictions are provided using a classifier which is trained to separate the binding and non-binding peptides for each protein independently.

In this paper we propose a novel approach for predicting protein-peptide binding affinity. Our approach is based on learning a peptide-peptide distance (or similarity) function[1]. This peptide-peptide distance function can then be used to compute a protein-peptide affinity score. We further propose to pool together information about binding and non-binding peptides from a number of related proteins ("protein family", e.g MHC class I). Our algorithm uses this data to learn a **single** peptide-peptide distance function for an entire protein family. Intuitively, a "good" learnt distance function should assign relatively small values (distances) to pairs of peptides that bind to a specific protein. Thus, given a novel binding peptide, we would expect its average distance to all known binders to be relatively small (as opposed to the average distance of a novel non-binding peptide to the same known binding peptides). We therefore propose the following learning scheme:

1. Compile a dataset of binding and non binding peptides from an entire protein family (e.g MHC class I or MHC class II).
2. Use the *DistBoost* algorithm [1, 2] to learn a single peptide-peptide distance function over this dataset using the information provided about binding and non-binding peptides.
3. Use this learnt distance function to compute the affinity of novel peptides to any of the proteins in the protein family.

We compare our method to various protein-peptide affinity prediction methods on several datasets of proteins from MHC class I and MHC class II. The results show that our method significantly outperforms all other methods. We also show that on proteins for which small amounts of binding peptides are available our improvement in performance is even more pronounced. This demonstrates one of the important advantages of learning a single peptide distance function on an entire protein family.

2   Related Work

Many different computational approaches have been suggested for the protein-peptide binding prediction problem (see [11] for a recent review). These methods can be roughly divided into three categories:

Motif based methods: Binding *motifs* represent important requirements needed for binding, such as the presence and proper spacing of certain amino acids within the peptide sequence. Prediction of protein-peptide binding is usually performed as motif searches [5]. The position specific scoring matrix (PSSM) approach is a statistical extension of the motif search methods, where a matrix represents the frequency of every amino acid in every position along the peptide. Peptide candidates can be assigned scores by summing up the position specific weights. The *RANKPEP* resource [6] uses this approach to predict peptide binding to MHC class I and class II molecules.

---

[1] We do not require that the triangle inequality holds, and thus our distance functions are *not* necessarily metrics.

Structure based methods: These methods predict binding affinity by evaluating the binding energy of the protein-peptide complex [7]. Note that these methods can be applied only when the three-dimensional structure of the protein-peptide complex is known or when reliable molecular models can be obtained.

Machine learning methods: Most of the methods in this category formalize the problem of predicting protein-peptide binding as a binary classification problem — for each specific protein, a classifier is trained to separate binding and non-binding peptides. Many different algorithms have been proposed. Among these are artificial neural networks (NetMHC) [10], hidden Markov models (HMM's) [8] and support vector machines (SVMHC) [9]. These methods require sufficient amounts of training data (i.e peptides which are known to be binders or non-binders) for each of the proteins.

3   From Peptide Distance Functions to Binding Prediction

As mentioned above, we propose to address the protein-peptide binding affinity prediction problem by learning a peptide-peptide distance function. Intuitively, we would expect that peptides that bind to a specific protein would be "similar" (or close) to each other, and "different" (far) from peptides that do not bind to this protein. Following this intuition, our goal is to learn a distance function, which assigns relatively small distances to pairs of peptides that bind to a specific protein and relatively large distances to pairs, consisting of a binder and a non binder (see Fig. 2). We can then predict the binding affinity of a novel peptide to a specific protein, by measuring its average distance to all of the peptides which are known to bind to that protein (see Fig. 2). More formally, we define the affinity between peptide i and protein j to be:

$$\textit{Affinity}(Peptide_i, Protein_j) \equiv e^{\left(-\frac{1}{|Binding_j|}\sum_{k \in Binding_j} \mathcal{D}(Peptide_i, Peptide_k)\right)} \quad (1)$$

where $\mathcal{D}(Peptide_i, Peptide_k)$ is the distance between $Peptide_i$ and $Peptide_k$ and $Binding_j = \{k | Peptide_k \text{ is known to bind to } Protein_j\}$.

For each protein, our training data consists of a list of binding peptides (and, possibly, their binding affinities) and a list of non-binding peptides. This form of partial information can be formally defined as equivalence constraints. Equivalence constraints are relations between pairs of data points, which indicate whether the points in the pair belong to the same category or not. We term a constraint *positive* when the points are known to be from the same class, and *negative* in the opposite case. In our setting, each protein defines a class. Each pair of peptides (data points) which are known to bind to a specific protein (that is, belong to the same class) defines a positive constraint, while each pair of peptides in which one binds to the protein and the other does not — defines a negative constraint.

In this work we propose to use the *DistBoost* algorithm for learning peptide-peptide distance functions. Moreover, we propose to learn a single peptide dis-

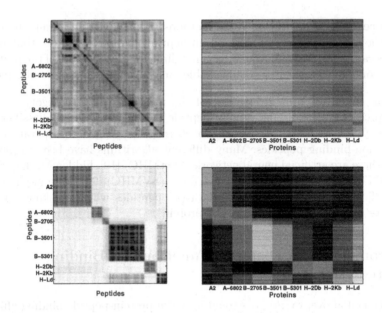

Fig. 2. Left: peptide-peptide distance matrices on MHC class I binding peptides, collected from the MHCBN dataset. Peptides that bind to each of the proteins were grouped together and labeled accordingly. A "good" distance matrix should therefore be block diagonal. Top left: The Euclidean peptide-peptide distance matrix in $\mathbb{R}^{45}$ (see Section 5 for details). Bottom left: The peptide-peptide distance matrix computed using the *DistBoost* algorithm. **Right:** protein-peptide affinity matrices. The affinity between a peptide and a specific protein is computed by measuring the average distance of the peptide to all peptides known to bind to that protein (see eq. 1). Top right: the Euclidean affinity matrix. Bottom right: the *DistBoost* affinity matrix. *DistBoost* was trained on binding peptides from all of the molecules **simultaneously**

tance function using partial information about binding and non binding peptides on several proteins from the same "protein family". When the different classes share common characteristics, learning a single distance function for all classes might benefit from a larger and more diverse training set. Our suggested approach has, therefore, the following potential advantages over the above mentioned computational approaches: (1) it can be used on proteins for which only a small amount of training data is available, (2) it can also be used to predict peptide binding affinity to novel proteins and (3) it can compute the relative binding affinities of a peptide to several proteins from the same protein family.

It should be emphasized that one cannot employ standard multi-class learning techniques in this scenario, since peptides do not have a well defined label. The partial information we have access to cannot be regarded as labeled data for three reasons: (1) if a peptide does not bind to a specific protein, it will not necessarily bind to a different protein from the same protein family, (2) peptides that bind to a specific protein can also bind to other proteins in its family and

(3) most peptide pairs are unlabeled, that is we do not know whether they both bind to some specific protein, or not.

4   The *DistBoost* Algorithm

Let us denote by $\{x_i\}_{i=1}^n$ the set of input data points which belong to some vector space $\mathcal{X}$. The space of all pairs of points in $\mathcal{X}$ is called the "product space" and is denoted by $\mathcal{X} \times \mathcal{X}$. An equivalence constraint is denoted by (x_{i_1}, x_{i_2}, y_i), where $y_i = 1$ if points (x_{i_1}, x_{i_2}) belong to the same class (positive constraint) and $y_i = -1$ if these points belong to different classes (negative constraint). The *DistBoost* algorithm is a semi-supervised algorithm that uses unlabeled data points in $\mathcal{X}$ and equivalence constraints to learn a bounded distance function, $\mathcal{D} : \mathcal{X} \times \mathcal{X} \to [0,1]$, that maps each pair of points to a real number in $[0,1]$.

The algorithm makes use of the observation that equivalence constraints on points in $\mathcal{X}$ are binary labels in the product space, $\mathcal{X} \times \mathcal{X}$. By posing the problem

---

Algorithm 1 The *DistBoost* Algorithm

Input:

    Data points: $(x_1, ..., x_n)$, $x_k \in \mathcal{X}$
    A set of equivalence constraints: (x_{i_1}, x_{i_2}, y_i), where $y_i \in \{-1, 1\}$
    Unlabeled pairs of points: $(x_{i_1}, x_{i_2}, y_i = *)$, implicitly defined by all unconstrained pairs of points

- Initialize $W_{i_1 i_2}^1 = 1/(n^2)$ $i_1, i_2 = 1, \ldots, n$ (weights over pairs of points)
              $w_k = 1/n$ $k = 1, \ldots, n$ (weights over data points)
- For $t = 1, .., T$

    1. Fit a constrained GMM (weak learner) on weighted data points in $\mathcal{X}$ using the equivalence constraints.
    2. Generate a weak hypothesis $\tilde{h}_t : \mathcal{X} \times \mathcal{X} \to [-1, 1]$ and define a weak distance function as $h_t(x_i, x_j) = \frac{1}{2}\left(1 - \tilde{h}_t(x_i, x_j)\right) \in [0, 1]$
    3. Compute $r_t = \sum\limits_{(x_{i_1}, x_{i_2}, y_i = \pm 1)} W_{i_1 i_2}^t y_i h_t(x_{i_1}, x_{i_2})$, only over **labeled** pairs. Accept the current hypothesis only if $r_t > 0$.
    4. Choose the hypothesis weight $\alpha_t = \frac{1}{2}\ln(\frac{1+r_t}{1-r_t})$
    5. Update the weights of **all** points in $\mathcal{X} \times \mathcal{X}$ as follows:

$$W_{i_1 i_2}^{t+1} = \begin{cases} W_{i_1 i_2}^t \exp(-\alpha_t y_i \tilde{h}_t(x_{i_1}, x_{i_2})) & y_i \in \{-1, 1\} \\ W_{i_1 i_2}^t \exp(-\alpha_t) & y_i = * \end{cases}$$

    6. Normalize: $W_{i_1 i_2}^{t+1} = \dfrac{W_{i_1 i_2}^{t+1}}{\sum\limits_{i_1, i_2 = 1}^{n} W_{i_1 i_2}^{t+1}}$
    7. Translate the weights from $\mathcal{X} \times \mathcal{X}$ to $\mathcal{X}$: $w_k^{t+1} = \sum_j W_{kj}^{t+1}$

Output: A final distance function $\mathcal{D}(x_i, x_j) = \sum_{t=1}^{T} \alpha_t h_t(x_i, x_j)$

in product space we obtain a classical binary classification problem: an optimal classifier should assign +1 to all pairs of points that come from the same class, and −1 to all pairs of points that come from different classes. This binary classification problem can be solved using traditional margin based classification techniques. Note, however, that in many real world problems, we are only provided with a sparse set of equivalence constraints and therefore the margin based binary classification problem is semi-supervised.

DistBoost learns a distance function using a well known machine learning technique, called *Boosting* [12]. In Boosting, a set of "weak" learners are iteratively trained and then linearly combined to produce a "strong" learner. Specifically, *DistBoost*'s weak learner is based on the constrained Expectation Maximization (cEM) algorithm [13]. The cEM algorithm is used to generate a "weak" distance function. The final ("strong") distance function is a weighted sum of a set of such "weak" distance functions. The algorithm is illustrated in Fig. 3, and presented in Alg. 1.

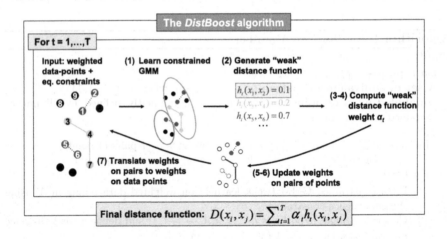

Fig. 3. An illustration of the *DistBoost* algorithm. At each boosting round t the weak learner is trained using weighted input points and some equivalence constraints. In the example above, points $1, 2$ and $5, 6$ are negatively constrained (belong to different classes) and points $3, 4$ and $4, 7$ are positively constrained (belong to the same class). All other pairs of points (e.g $8, 9$ and $1, 4$) are unconstrained. The constrained EM algorithm is used to learn a GMM (step (1)). This GMM is then used to generate a "weak" distance function (step (2)) that assigns a value in $[0, 1]$ to each pair of points. The distance function is assigned a hypothesis weight (steps (3-4)) which corresponds to its success in satisfying the current weighted constraints. The weights of the equivalence constraints are updated (steps (5-6)) – increasing the weights of constraints that were unsatisfied by the current weak learner. Finally, the weights on pairs are translated to weights on data points (step (7)). In the example above, the distance between the negatively constrained points $1, 2$ is small (0.1) and therefore the weight of this constraint will be enhanced

In order to make use of unlabeled data points, *DistBoost*'s weak learner is trained in the original space, $\mathcal{X}$, and is then used to generate a "weak distance function" on the product space. *DistBoost* uses an augmentation of the 'Adaboost with confidence intervals' algorithm [12] to incorporate unlabeled data into the boosting process. Our semi-supervised boosting scheme computes the weighted loss only on **labeled** pairs of points but updates the weights over **all** pairs of points (see steps (3-6)). The weights of the unlabeled pairs decay at least as fast as the weight of any labeled pair. The translation scheme from product space to the original space and vice-versa is presented in steps (2,7) of the algorithm.

DistBoost's weak learner: *DistBoost*'s weak learner is based on the constrained Expectation Maximization (cEM) algorithm [13]. The algorithm uses unlabeled data points and a set of equivalence constraints to find a Gaussian Mixture Model (GMM) that complies with these constraints. A GMM is a parametric statistical model which is given by $p(x|\Theta) = \Sigma_{l=1}^{M} \pi_l p(x|\theta_l)$, where π_l denotes the weight of each Gaussian, θ_l its parameters, and M denotes the number of Gaussian sources in the GMM. Under this model, each data sample originates independently from a weighted sum of several Gaussian sources. Estimating the parameters (Θ) of a GMM is usually done using the well known EM algorithm. The cEM algorithm introduces equivalence constraints by modifying the 'E' (Expectation) step of the algorithm: instead of summing over *all* possible assignments of data points to sources, the expectation is taken only over assignments which comply with the given equivalence constraints.

The algorithm's input is a set of unlabeled points $X = \{x_i\}_{i=1}^n$, and a set of pairwise constraints, Ω, over these points. Denote positive constraints by $\{(p_j^1, p_j^2)\}_{j=1}^{N_p}$ and negative constraints by $\{(n_k^1, n_k^2)\}_{k=1}^{N_n}$. Let $H = \{h_i\}_{i=1}^n$ denote the hidden assignment of each data point x_i to one of the Gaussian sources ($h_i \in \{1, \ldots, M\}$). The constrained EM algorithm assumes the following joint distribution of the observables X and the hiddens H:

$$p(X, H|\Theta, \Omega) = \frac{1}{Z} \prod_{i=1}^{n} \pi_{h_i} p(x_i|\theta_{h_i}) \prod_{j=1}^{N_p} \delta_{h_{p_j^1} h_{p_j^2}} \prod_{k=1}^{N_n} (1 - \delta_{h_{n_k^1} h_{n_k^2}}) \qquad (2)$$

where Z is the normalizing factor and δ_{ij} is Kronecker's delta. The algorithm seeks to maximize the data likelihood, which is the marginal distribution of (2) with respect to H. For a more detailed description of this weak learner see [13].

In order to use the algorithm as a weak learner in our boosting scheme, we modified the algorithm to incorporate weights over the data samples. These weights are provided by the boosting process in each round (see Alg. 1 step 7).

Generating a weak distance function using a GMM: The weak learners' task is to provide a weak distance function $h_t(x_i, x_j)$ over the product space $\mathcal{X} \times \mathcal{X}$. Denote by $p^{MAP}(x_i)$ the probability of the Maximum A-Posteriori (MAP) assignment of point x_i ($p^{MAP}(x_i) = \max_m p(h_i = m|x_i, \Theta)$). We partition the data into M groups using the MAP assignment of the points and define

$$\tilde{h}_t(x_i, x_j) \equiv \begin{cases} +p^{MAP}(x_i) \cdot p^{MAP}(x_j) & \text{if } \text{MAP}(x_i) = \text{MAP}(x_j) \\ -p^{MAP}(x_i) \cdot p^{MAP}(x_j) & \text{if } \text{MAP}(x_i) \neq \text{MAP}(x_j) \end{cases}$$

The weak distance function is given by $h_t(x_i, x_j) = \frac{1}{2}\left(1 - \tilde{h}_t(x_i, x_j)\right) \in [0,1]$. It is easy to see that if the MAP assignment of two points is identical their distance will be in $[0, 0.5]$ and if their MAP assignment is different their distance will be in $[0.5, 1]$.

5   Results

MHC datasets. The first two datasets we compiled (MHCclass1 and MHCclass2) were the same as those described in [6]. Sequences of peptides, that bind to MHC class I or class II molecules, were collected from the MHCPEP dataset [14]. Each entry in the MHCPEP dataset contains the peptide sequence, its MHC specificity and, where available, observed activity and binding affinity. Peptides, that are classified as low binders or contain undetermined residues (denoted by the letter code X), were excluded. We then grouped all 9 amino acid long peptides (9-mers), that bind to MHC class I molecules, to a dataset, called *MHCclass1*. This dataset consists of binding peptides for 25 different MHC class I molecules.

Unlike MHC class I binding peptides, peptides binding to MHC class II molecules display a great variability in length, although only a peptide core of 9 residues fits into the binding groove. Following [6], we first used the MEME program [15] to align the binding peptides for each molecule, based on a single 9 residues motif. We finally filtered out redundant peptides and obtained the *MHCclass2* dataset. This dataset consists of binding peptides for 24 different MHC class II molecules.

Since all peptides in the MHCPEP dataset are binders, we added randomly generated peptides as non-binders to both MHCclass1 and MHCclass2 datasets (amino acid frequencies as in the Swiss-Prot database). The number of nonbinders used in any test set was twice the number of the binding peptides. During the train phase, the number of non-binders was the same as the number of binders.

In order to assess the performance of the prediction algorithms on experimentally determined non-binders, we compiled a third dataset, called *MHCclass1BN*. This dataset consists of binding and non-binding peptides, for 8 different MHC class I molecules, based on the *MHCBN 3.1* website [16] (see Fig. 8 (b)).

Data representation. *DistBoost* requires that the data be represented in some continuous vector feature space. Following [17] each amino acid was encoded using a 5-dimensional property vector (and, thus, each peptide in the MHC datasets is a point in $\mathbb{R}^{45}$). The property vectors for each of the 20 amino acids are based on multidimensional scaling of 237 physical-chemical properties. Venkatarajan and Braun's analysis [17] showed that these 5 properties correlate well with hydrophobicity, size, α-helix preference, number of degenerate triplet codons and the frequency of occurrence of amino acid residues in β-strands. They also showed that the distances between pairs of amino-acids in the 5-dimensional property space are highly correlated with corresponding scores from similarity matrices derived from sequence and 3D structure comparisons.

Evaluation methods. In order to evaluate the algorithms' performance, we measured the affinity of all test peptides to each of the proteins. We present the prediction accuracy (that is how well binders are distinguished from non-binders) of the various algorithms as ROC (Receiver Operating Characteristic) curves. The fraction of the area under the curve (AUC) is indicative of the distinguishing power of the algorithm and is used as its prediction accuracy.

5.1   MHC Binding Prediction on the MHCPEP Dataset

We compared our method to the recently enhanced RANKPEP method [6]. We replicated the exact experimental setup described in [6]: (1) We used the exact same MHC class I and class II datasets. (2) Training was performed using 50% of the known binders for each of the MHC molecules. (3) The remaining binding peptides were used as test data to evaluate the algorithm's performance. These binders were tested against randomly generated peptides as described above.

We trained *DistBoost* in two distinct scenarios: (1) Training using only binding peptides (using only positive constraints). (2) Training using both binding and (randomly generated) non-binding peptides (using both positive and negative constraints). In both scenarios *DistBoost* was trained **simultaneously** on all of the MHC molecules in each class. Fig. 4 presents a comparison of *DistBoost*

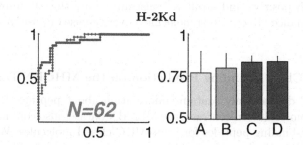

Fig. 4. Comparative results of *DistBoost* and RANKPEP on the H-2Kd MHC class I molecule. The left plot presents ROC curves of the best train score obtained when training on 50% of the entire data (red - using only positive constraints and blue - using both types of constraints). The right plot presents average AUC scores on test data. We compare the two PSSM methods used by RANKPEP (A - PROFILEWEIGHT, B - BLK2PSSM) to *DistBoost* when trained using only positive constraints (C) and when trained using both positive and randomly generated negative constraints (D). The averages were taken over 10 different runs on randomly selected train and test sets. N denotes the total number of binding peptides (of which 50% were used in the training phase and the remaining 50% were used in the test phase)

to both of the PSSM's used in [6] on the H-2Kd MHC class I molecule. Comparative results on the entire MHC class I and class II datasets are presented in Figures 5 and 6. In all these comparisons, the PSSM AUC scores are as reported in [6].

On the MHC class I molecules, our method significantly outperforms both PSSM's used by RANKPEP. On 21 out of the 25 molecules *DistBoost*'s average AUC score, when trained using only positive constraints, is higher than both PSSM methods. The improvement in performance is more pronounced on molecules with relatively small amounts of known binders (e.g. HLA-B27(B*2704) - 10 binders, HLA-A2(A*0205) - 22 binders and HLA-A33(A*3301) - 23 binders). One possible explanation of these results is that the information provided by other proteins within the protein family is used to enhance prediction accuracy, especially in these cases where only small amounts of known binders exist. Additionally, it may be seen that using both positive and negative constraints on this dataset, usually improves the algorithm's performance. Another important advantage of *DistBoost* can be seen when comparing standard deviations (std) of the AUC scores. On 16 out of the 25 molecules the algorithm's std is lower then the std of both PSSM's, implying that our method is more robust.

When tested on the MHC class II molecules, our method obtained similar improvements (see Fig. 6): On 18 out of the 24 molecules *DistBoost*'s average AUC score when trained using only positive constraints is higher then both PSSM methods. In general, it appears that the performance of all of the compared methods is lower than on the MHC class I dataset. It is known that predicting binding affinity on MHC class II is more challenging, partially due to the fact that peptides that bind to class II molecules are extremely variable in length and share very limited sequence similarity [18]. On this dataset, the use of both positive and negative constraints only slightly improved *Dist-Boost*'s performance (13 out of 24 molecules) over the use of positive constraints only.

5.2   MHC Class I Binding Prediction on the MHCBN Dataset

The MHCPEP dataset only contains information about peptides that bind to various MHC molecules. In contrast, the MHCBN dataset also contains information about non-binding peptides for some MHC class I molecules. We used this dataset to evaluate the importance of learning using experimentally determined non-binders (as opposed to randomly generated non binders).

The MHCBN dataset also contains information about binding and non-binding peptides to supertypes which are collections of similar molecules. We compared *DistBoost* to various other computational prediction methods on the HLA-A2 supertype. Specifically we compared the performance of the following methods: (1) The *DistBoost* algorithm. (2) The *SVMHC* web server [9]. (3) The *NetMHC* web server [10]. (4) The RANKPEP resource [6] (5) The Euclidean distance metric in $\mathbb{R}^{45}$. Despite the fact that methods 2-4 are protein specific they also provide predictions on various MHC supertypes including the HLA-A2 supertype.

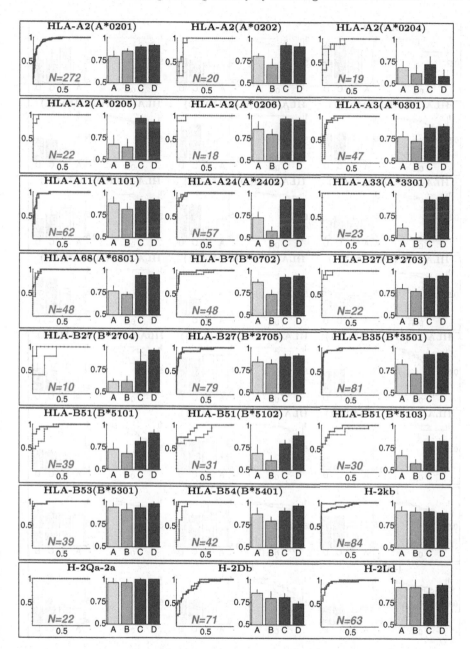

Fig. 5. Comparative results of *DistBoost* (left plot and bars C-D) and RANKPEP (bars A-B) on 24 MHC class I molecules. Plot legends are identical to Fig 4. On 21 out of the 25 molecules (including Fig. 4), *DistBoost* outperforms both PSSM methods. On this data the use of negative constraints also improves performance

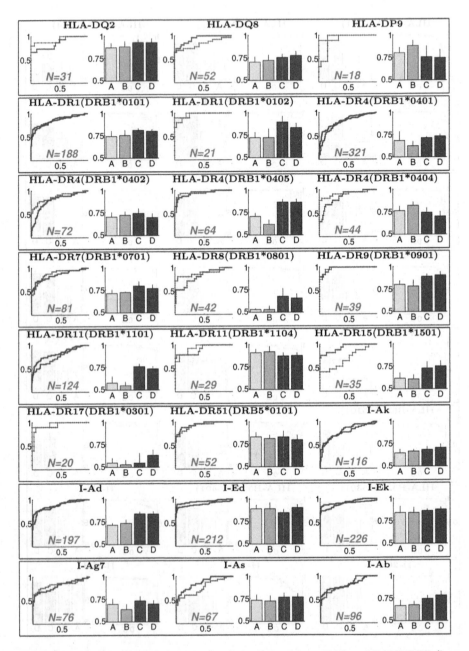

Fig. 6. Comparative results of *DistBoost* (left plot and bars C-D) and RANKPEP (bars A-B) on 24 MHC class II molecules. Plot legends are identical to Fig 4. As may be seen on 18 out of the 24 molecules, *DistBoost* outperforms both PSSM methods. On this dataset the use of negative constraints only slightly improves performance

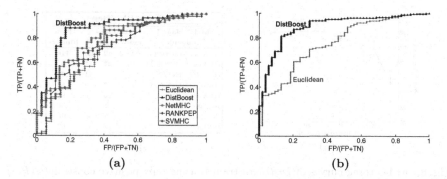

Fig. 7. (a) ROC curves on test data from the HLA-A2 supertype. *DistBoost* is compared to various other prediction algorithms. (b) *DistBoost* and the Euclidean affinity ROC curves on test data from the **entire** *MHCclass1BN* dataset. The rest of the methods are not presented since they were not trained in this multi-protein scenario. In both cases, *DistBoost* was trained on 70% of the data and tested on the remaining 30%. TP (true positives), FP (false positives), TN (true negatives), FN (false negatives). Results are best seen in color

We trained *DistBoost* on 70% of the **entire** *MHCclass1BN* data (including binding and non-binding peptides) and compared its performance to all other methods on the **single** HLA-A2 supertype. The test set, therefore, consists of the remaining 30% of HLA-A2 data. The results are shown in Fig. 7 (a). As may be seen *DistBoost* outperforms all other methods, including SVMHC, NetMHC and RANKPEP, which were trained on this specific supertype. However, it is important to note, that unlike *DistBoost* all of these methods were trained using randomly generated non-binders. The performance of all of these methods when tested against random peptides is much better — AUC scores of 0.947 (SVMHC), 0.93 (NetMHC) and 0.928 (RANKPEP). These results seem to imply that learning using random non-binders does **not** generalize well to experimentally determined non-binders. Interestingly, when we tested *DistBoost* on randomly generated non-binders we obtained an AUC score of 0.923. We can therefore conclude that learning from "real" non-binders generalizes very well to random non-binders.

Our proposed method is trained **simultaneously** on a number of proteins from the same family, unlike methods (2-4). However, our final predictions are protein specific. As the results reveal, we obtain high binding prediction accuracy when tested on a single protein (see Fig. 7 (a)). In order to quantify the overall protein specific binding prediction accuracy, we present ROC curves for *DistBoost* and the Euclidean affinity functions when tested on the **entire** MHC-class1BN dataset (Fig. 7 (b)). The peptide-peptide distance matrices and the protein-peptide affinity matrices of these two methods are presented in Fig. 2. On this dataset *DistBoost* obtained excellent performance.

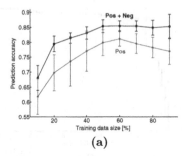

MHCclass1BN dataset			
MHC class I	Binders		Non
molecule	mod. high	?	binders
HLA-A2 supertype	29 42	126	116
HLA-A68 (A*6801)	10 27	17	0
HLA-B27 (B*2705)	7 17	11	7
HLA-B35 (B*3501)	31 47	48	6
HLA-B53 (B*5301)	17 16	22	0
H-2Db	18 11	17	26
H-2Kb	6 7	4	19
H-2Ld	9 16	25	0

(a)   (b)

Fig. 8. (a) Learning curves of *DistBoost* trained using only positive constraints (*Pos*) and using both types of constraints (*Pos+Neg*). Prediction accuracy based on the AUC score, averaged over 20 different randomly selected training sets. (b) The MHCclass1BN dataset

In order to evaluate the stability and learning power of *DistBoost* we ran it on the *MHCclass1BN* dataset, while varying the percentage of training data. Fig. 8 (a), presents the algorithm's learning curves when trained using only positive constraints and when trained using both positive and negative constraints. As may be expected, on average, performance improves as the amount of training data increases. Note that *DistBoost* achieves almost perfect performance with relatively small amounts of training data. Additionally, we can see that on this dataset learning from both types of constraints dramatically improves performance.

6   Discussion

In this paper we showed how the protein-peptide binding affinity problem can be cast as a semi-supervised learning problem in which equivalence constraints can be naturally obtained from empirical data. We then proposed to learn protein peptide affinity by learning a peptide-peptide distance function. Specifically we presented *DistBoost*, an algorithm that can learn distance functions using positive and negative equivalence constraints. Our experiments suggest that binding prediction based on such learned distance functions exhibits excellent performance. We also showed the importance of using negative constraints, which further suggests that information about non-binding peptides should also be published and made publicly available.

Acknowledgements. We thank Daphna Weinshall for many fruitful discussions and comments. C.Y is supported by Yeshaya Horowitz Association through the Center for Complexity Science.

References

1. Hertz, T., Bar-Hillel, A., Weinshall, D.: Learning distance functions for image retrieval. In: CVPR, Washington DC, June 2004. (2004)
2. Hertz, T., Bar-Hillel, A., Weinshall, D.: Boosting margin based distance functions for clustering. In: ICML. (2004)
3. Rammensee., H.G., Friede., T., Stevanoviic, S.: MHC ligands and peptide motifs: first listing. Immunogenetics **41(4)** (1995) 178–228
4. Wang., R.F., Rosenberg, S.A.: Human tumor antigens for cancer vaccine development. Immunol Rev. **170** (1999) 85–100
5. Andersen, M., Tan, L., Sondergaard, I., Zeuthen, J., Elliott, T., Haurum, J.: Poor correspondence between predicted and experimental binding of peptides to class I MHC molecules. Tissue Antigens **55** (2000) 519–531
6. Reche, P.A., Glutting, J., Zhang, H., Reinher, E.: Enhancement to the RANKPEP resource for the prediction of peptide binding to MHC molecules using profiles. Immunogenetics **26** (2004) 405–419
7. Schueler-Furman, O., Altuvia, Y., Sette, A., Margalit, H.: Structure-based prediction of binding peptides to MHC class I molecules: application to a broad range of MHC alleles. Protein Sci **9** (2000) 1838–1846
8. Mamitsuka, H.: Predicting peptides that bind to MHC molecules using supervised learning of hidden Markov models. Proteins **33** (1998) 460–474
9. Donnes, P., Elofsson, A.: Prediction of MHC class I binding. BMC Bioinformatics **3** (2002)
10. Buus, S., Lauemoller, S., Worning, P., Kesmir, C., Frimurer, T., Corbet, S., Fomsgaard, A., Hilden, J., Holm, A., Brunak, S.: Sensitive quantitative predictions of peptide-MHC binding by a 'query by committee' artificial neural network approach. Tissue Antigens **62** (2003) 378–384
11. Flower, D.R.: Towards in silico prediction of immunogenic epitopes. TRENDS in immunology **24** (2003)
12. Schapire, R.E., Singer, Y.: Improved boosting using confidence-rated predictions. Machine Learning **37** (1999) 297–336
13. Shental, N., Bar-Hilel, A., Hertz, T., Weinshall, D.: Computing Gaussian mixture models with EM using equivalence constraints. In: NIPS. (2003)
14. Brusic, V., Rudy, G., Harrison, L.: MHCPEP, a database of MHC-binding peptides: update 1997. Nucl. Acids Res. **26** (1998) 368–371
15. Bailey, T., Elkan, C.: Fitting a mixture model by expectation maximization to discover motifs in biopolymers. In: ISMB. Volume 2. (1994) 28–36
16. Bhasin, M., Singh, H., Raghava, G.P.S.: MHCBN: a comprehensive database of MHC binding and non-binding peptides. Bioinformatics **19** (2003) 665–666 http://www.imtech.res.in/raghava/mhcbn/index.html.
17. Venkatarajan, M.S., Braun, W.: New quantitative descriptors of amino acids based on multidimensional scaling of a large number of physical-chemical properties. Journal of Molecular Modeling **7** (2001) 445–453
18. Madden, D.R.: The three-dimensional structure of peptide-MHC complexes. Annual Review of Immunology **13** (1995) 587–622

Amino Acid Sequence Control of the Folding of the Parallel β-Helix, the Simplest β-Sheet Fold

Jonathan King, Cammie Haase-Pettingell,
Ryan Simkovsky, and Peter Weigele

Department of Biology, MIT,
Cambridge, Massachusetts, USA

Deciphering how amino acid sequences direct polypeptide chains into their native folds remains a major unsolved problem. The most refractory aspect has been understanding the pathways and sequence control of β-sheet folding. This presumably reflects in part that residues that are far apart in the sequence will be intimately interacting in the folded β-structure.

The topologically simplest β-sheet fold is the parallel β-helix, in which the polypeptide chain wraps processively to form successive rungs in an elongated structure. The strands pack orthogonal to the long axis of the fold, forming ribbons of β-sheet, while the buried hydrophobic core is cylindrical rather than globular. This fold provides one of the best models for the packing of β-strands in the amyloid fibrils associated with human disease.

The pathway of the folding and assembly of the trimeric tailspike adhesin of phage P22 has provided an experimental system for probing control of folding. Sequential intermediates have been identified both in vivo and in vitro using purified protein [1]. This allows identification of mutants that act by interfering with chain folding. One class of amino acid substitutions which influence β-helix folding renders the process temperature sensitive - the mutant chains fold at low temperature but not at high. These mutants identify residues - mostly in the turns - which are necessary for successful folding at high temperature. Such conditional information is rarely incorporated in prediction algorithms. A second group of substitutions have more severe defects on β-helix folding. They represent predominantly buried hydrophobic residues that form long stacks within the buried core.

The β-BetaWrap algorithm of Bradley, Berger, Cowen and coworkers [2] very efficiently identifies sequences likely to form parallel β-helices, by weighting these critical core packing interactions, and using a sequential wrapping or scoring process, which proceeds form an initial nucleating set of rungs. This may capture the underlying biological process of formation of a collapsed nucleus, followed by processive folding of the chain onto each newly formed rung.

References

1. Betts, S., King, J.: There's a right way and a wrong way: in vivo and in vitro folding, misfolding and subunit assembly of the P22 tailspike. Structure Fold Des. **7** (1999) R131–139

S. Miyano et al. (Eds.): RECOMB 2005, LNBI 3500, pp. 472–473, 2005.

2. Bradley, P., Cowen, L., Menke, M., King, J., Berger, B.: BETAWRAP: Successful prediction of parallel β-helices from primary sequence reveals an association with many microbial pathogens. Proc. Natl. Acad. Sci. U.S.A. **98** (2001) 14819–14824

A Practical Approach to Significance Assessment in Alignment with Gaps

Nicholas Chia and Ralf Bundschuh

Ohio State University,
Columbus, OH 43210, USA

Abstract. Current numerical methods for assessing the statistical significance of local alignments with gaps are time consuming. Analytical solutions thus far have been limited to specific cases. Here, we present a new line of attack to the problem of statistical significance assessment. We combine this new approach with known properties of the dynamics of the global alignment algorithm and high performance numerical techniques and present a novel method for assessing significance of gaps within practical time scales. The results and performance of these new methods test very well against tried methods with drastically less effort.

Keywords: pairwise sequence alignment, Markov models and/or hidden Markov models, statistics of motifs or strings, statistical significance, Gumbel distribution, extreme value distribution, Kardar-Parisi-Zhang universality class, asymmetric exclusion process.

1 Introduction

Sequence alignment is one of the most commonly used computational tools of molecular biology. Its applications range from identifying the function of newly sequenced genes to the construction of phylogenic trees [43, 18]. Its importance is epitomized by the popularity of the program BLAST [1, 3] which is currently used 300,000 times a day on the NCBI's web site alone.

All alignment algorithms have the drawback that they will find an optimal alignment and an optimal score for *any* pair of sequences — even randomly chosen and thus completely unrelated ones. Thus, it is necessary to assess the significance of a resulting alignment. A popular approach to this problem is to compare the score of the optimal alignment to the scores generated by the optimal alignments of *randomly chosen* sequences. This is quantified by the *p*- or *E*-value. This comparison steadily becomes more important since with the increasing size of the databases the probability for obtaining a relatively large score just by chance increases dramatically.

In order to reliably quote a *p*-value, the *distribution* of optimal alignment scores for alignments of random sequences must be known. In the case of alignment without "gaps", it has been worked out rigorously [24, 25, 26] that this distribution is a Gumbel or extreme value distribution [20]. This distribution is

S. Miyano et al. (Eds.): RECOMB 2005, LNBI 3500, pp. 474–488, 2005.

characterized by two parameters that depend on the scoring system used and on the amino acid frequencies with which the random sequences are generated. For gapless alignment, the dependence of the two Gumbel parameters on the scoring system is completely known.

However, in order to detect weakly homologous sequences, gaps must be allowed in an alignment [35]. Unfortunately, for the case of gapped alignment, there currently exists no theory that describes the distribution of alignment scores for random sequences. However, there remains a lot of numerical evidence as well as a number of heuristic arguments that this distribution is still of the Gumbel form [39, 12, 30, 41, 42, 2]. Nevertheless, even assuming the correctness of the Gumbel form, finding the two Gumbel parameters for a given scoring system turns out to be a very challenging problem. The straightforward method generates a large number of alignment scores by shuffling the two sequences to be compared and taking a histogram of this distribution. But, because of the slow exponential tail of the Gumbel distribution, this method is extremely time consuming. Thus, in practice, the two Gumbel parameters have to be pre-computed for some few fixed scoring systems [2, 3].

Pre-computed Gumbel parameters have the disadvantage that they restrict the user to a few scoring systems (substitution matrices and gap costs) for which the Gumbel parameters have been pre-computed. The necessity of pre-computing the Gumbel parameters definitely becomes problematic if adaptive schemes, e.g., PSI-BLAST [3], are being used. These schemes change their scoring system recursively depending on the sequence data they are confronted with and thus have to be able to find the two Gumbel parameters after each update of the scoring system.

To remedy this problem, a more effective numerical method which estimates the two Gumbel parameters has been proposed [34, 4]. There are also some analytical approximations [31, 37, 32] which are mainly valid for rather large gap costs where the influence of the gaps on the Gumbel parameters is not yet too strong. In addition, an analytical scheme has been used to successfully calculate the Gumbel parameter λ, which describes the tail of the Gumbel distribution, for just one particular scoring system [8, 10]. In this paper, we will present a novel approach that calculates λ for a variety of scoring schemes while drastically reducing the time required to calculate λ and retaining a high degree of precision in the solution. This approach will expand upon and combine the different analytical works devised in [8, 10] and [16, 17], creating a new scheme for calculating λ using the numerical tools of [29]. Once λ is known, it then becomes a simple matter to extract the remaining Gumbel parameter, which characterizes the mean of the score distribution, numerically via e.g., the island method [34, 4] or direct simulation.

In section 2, we will present an abbreviated review of sequence alignment. We then point out that, although λ is intrinsically a quantity of *local* alignments, it may be calculated from solely studying the simpler *global* alignment algorithm. Under some very moderate approximation we then briefly reformulate the problem of finding λ in terms of an eigenvalue equation, as done with more detail in [8, 10]. We then show the feasibility of our novel approach by comparing the results from this new method with established analytical [8] and numerical [4] methods for a variety of scoring systems.

2   Review of Sequence Alignment

In the vast majority of sequence alignment applications, gapped alignment is used as the fundamental alignment technique. Gapped alignment looks for similarities between two sequences $a = a_1 a_2 \ldots a_M$, and $b = b_1 b_2 \ldots b_N$ of length M and N respectively. The letters a_i and b_j are taken from an alphabet of size c. This may be the four letter alphabet $\{A,C,G,T\}$ of DNA or the twenty letter amino-acid alphabet. Here, we consider Smith-Waterman local alignment [38]. In this case, a possible alignment $\mathcal{A}$ consists of two substrings of the two original sequences a and b. These subsequences may have different lengths, since gaps may be inserted in the alignment. For example, the two subsequences GATGC and GCTC may be aligned as GATGC and GCT-C using one gap. Each such alignment $\mathcal{A}$ is assigned a score according to $S[\mathcal{A}] = \sum_{(a,b) \in \mathcal{A}} s_{a,b} - \delta N_g$ where the sum is taken over all pairs of aligned letters, N_g is the total number of gaps in the alignment, δ is an additional scoring parameter, the "gap cost," and $s_{a,b}$ is some given "scoring matrix" measuring the mutual degree of similarity between the different letters of the alphabet. A simple example, the match-mismatch matrix

$$s_{a,b} = \begin{cases} 1 & a = b \\ -\mu & a \neq b \end{cases} \tag{1}$$

is used for DNA sequence comparisons [33]. For protein sequences, normally the 20 x 20 PAM [13] or BLOSUM matrices [21] are used. Practical applications usually use the more complicated affine gap cost. For the purpose of clarity, the following will only consider the case of linear gap cost. However, we want to stress that our approach is applicable to affine gap costs as well as discussed at the end of the manuscript. The computational task is to find the subsequences which give the *highest* total score for a given scoring matrix $s_{a,b}$

$$\Sigma \equiv \max_{\mathcal{A}} S[\mathcal{A}]. \tag{2}$$

The task is to find the alignment $\mathcal{A}$ with the highest score as in Eq. (2). This can be very efficiently done by a dynamic programming method which becomes obvious in the alignment path representation [33]. In this representation, the two sequences to be compared are written on the edges of a square lattice as shown in Fig. 1 where we chose $L \equiv M = N$. Each directed path on this lattice represents one possible alignment. The score of this alignment is the sum over the local scores of the traversed bonds. Diagonal bonds correspond to gaps and carry the score $-\delta$. Horizontal bonds are assigned the similarity scores $s(r,t) \equiv s_{a,b}$ where a and b are the letters of the two sequences belonging to the position (r,t) as shown in Fig. 1.

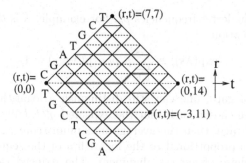

Fig. 1. Local alignment of two sequences. This figure shows the alignment of CGATGCT and TGCTCGA represented as a directed path on the alignment lattice. The highlighted alignment path $r(t)$ corresponds to one possible alignment of two subsequences, GATGC to GCT-C. This path contains one gap. It is also shown how the coordinates r and t are used to identify the nodes of the lattice

If interested in finding the highest scoring *global* alignment of the two sequences a and b, one finds the best scoring path connecting the beginning $(0,0)$ to the end $(0, 2L)$. This path can be found efficiently by defining the auxiliary quantity $h(r, t)$ to be the score of the best path ending in the lattice point (r, t) with initial conditions $h(t, t) = -t\delta = h(-t, t)$. This quantity can be calculated recursively by the Needleman-Wunsch dynamic programming algorithm [33]

$$h(r, t+1) = \max\{h(r, t-1) + s(r, t), \ h(r \pm 1, t) - \delta\}. \tag{3}$$

For *local* alignments, the Smith-Waterman algorithm [38], supplemented by the initial conditions $S(t, t) = 0 = S(-t, t)$, describes the appropriate recursion

$$S(r, t+1) = \max\{S(r, t-1) + s(r, t), S(r \pm 1, t) - \delta, 0\}. \tag{4}$$

The score of the best local alignment is then given by $\Sigma = \max_{r,t} S(r, t)$.

Characterizing the statistical significance of alignments requires the distribution of Σ for the alignment of two *random* sequences whose elements, a_k's and b_k's, are generated independently from the same frequencies p_a as the query sequences, and scored using the scoring matrix $s_{a,b}$. In the gapless limit where $\delta \to \infty$, this distribution of Σ has been worked out rigorously for the regime pertinent to significance assessment — i.e. in the *logarithmic phase* characterized by a negative $\langle s \rangle \equiv \Sigma_{a,b} p_a p_b s_{a,b}$ and $\Sigma \propto \log L$ [7, 25, 26]. For scoring parameters in the logarithmic phase, it is a Gumbel or extreme value distribution given by

$$\Pr\{\Sigma < S\} = \exp(-\kappa e^{-\lambda S}). \tag{5}$$

This distribution is characterized by the two parameters λ and κ with λ giving the tail of the distribution and $\lambda^{-1} \log \kappa$ describing the mean. For gapless alignment, these parameters can be explicitly calculated [25, 26] from the scoring matrix $s_{a,b}$ and the letter frequencies p_a. For example, λ is the unique positive solution of the equation

$$\langle\exp(\lambda s)\rangle \equiv \sum_{a,b} p_a p_b \exp(\lambda s_{a,b}) = 1. \tag{6}$$

In the presence of gaps, one can still distinguish a logarithmic phase [40]. If the parameters are chosen such that the expected *global* alignment score drifts downwards on average, then the average maximum score $\langle\Sigma\rangle$ for gapped *local* alignment remains proportional to the logarithm of the sequence length, as in the logarithmic phase of gapless alignment. The reduced value of $\langle\Sigma\rangle$ in the logarithmic phase makes it the regime of choice for homology detection.

Again, the distribution of Σ must be known for local alignments of random sequences in order to characterize the statistical significance of local alignment. There exists no rigorous theory for this distribution in the presence of gaps. However, a slew of empirical evidence strongly suggests that the distribution of local scores describes the Gumbel distribution [39, 12, 30, 41, 42, 2]. In practice, they have to be determined empirically by time consuming simulations [4]. In the absence of a more efficient means of calculating λ and κ, the use of adaptive schemes such as PSI-BLAST or more finely tuned significance assessment for various letter compositions remains elusive. Below we will present a new method to calculate the parameter λ, as well as an explicit calculation of this parameter for some simple scoring systems, that can resolve this dilemma. Since κ determines the mean and not tail of the distribution, κ can always be determined efficiently by simulation once λ is known. The method outlined here may also be applied directly to more complex scoring schemes, e.g., affine gap costs.

3   Review of Significance Estimation Using Global Alignment as a Dynamic Process

As a first and very crucial step, we will use the fact that, accepting the empirical applicability of the Gumbel distribution to gapped local alignment, the parameter λ, describing the tail of the Gumbel distribution, can be derived solely from studying the much simpler *global* alignment (3). This has been shown in [8, 10]. For our purposes, we will recast the result from [8, 10] in the following form.

Let us define the generating function

$$Z_L(\gamma; \Omega) \equiv \langle\exp[\gamma h(0, L)]\rangle \tag{7}$$

where the brackets $\langle\cdot\rangle$ denote the ensemble average over all choices of random sequences $\boldsymbol{a}$, $\boldsymbol{b}$ and $h(0, L)$ is the *global* alignment score at the end of a lattice of length L as shown in Fig. 2(a), and Ω summarizes the parameters p_a, μ, and δ that contribute to the evaluation of $h(0, L)$. This score can be obtained from the recursion relation (3) with initial condition $h(r, 0) = 0$. Let us now define

$$\Phi(\gamma; \Omega) = \lim_{L\to\infty} \frac{1}{L} \log Z_L(\gamma; \Omega) \tag{8}$$

Then according to [8, 10] the parameter λ of the Gumbel distribution is obtained as the unique positive solution of the equation

$$\Phi(\lambda; \Omega) = 0. \tag{9}$$

Note that this condition reduces simply to Eq. (6) in the case of gapless alignment, since for infinite gap cost δ, we have $\langle\exp[\gamma h(0, L)]\rangle = \langle\exp[\gamma \sum_{k=1}^{L/2} s(0, 2k-1)]\rangle = \langle\exp[\gamma s]\rangle^{\frac{L}{2}}$ and thus $\Phi(\gamma; \Omega) = \frac{1}{2}\log\langle\exp(\gamma S)\rangle$.

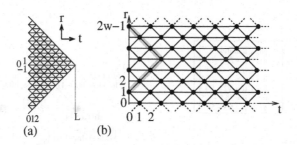

(a)   (b)

Fig. 2. Global alignment lattice used for significance estimation. (a) shows the right half of the lattice from Fig. 1. It can represent all possible paths of length L which end at the point $(r, t) = (0, L)$ and start at $(r, 0)$ for an arbitrary r. (b) shows with the gray lines, how the triangular lattice similar to the one shown in (a) can be embedded into a rectangular alignment lattice of width $2W$ with periodic boundary conditions in the spatial (vertical) direction as long as $L < W$

In order to calculate $\Phi(\gamma; \Omega)$, instead of the triangular alignment lattice shown in Fig. 2(a), we utilize the rectangular lattice of $2W$ lattice points shown in Fig. 2(b). Across the lattice, we apply periodic boundary conditions $h(0, t) = h(2W, t)$ for all t. Defining the generating function of the finite width $Z_{L,W}(\gamma; \Omega)$ by Eq. (7) with $h(r, t)$ calculated on the lattice of width $2W$, we introduce

$$\Phi_W(\gamma; \Omega) = \lim_{L \to \infty} \frac{1}{L} Z_{L,W}(\gamma; \Omega). \tag{10}$$

The function $\Phi(\gamma; \Omega)$ on the original lattice is then given by

$$\Phi(\gamma; \Omega) = \lim_{W \to \infty} \Phi_W(\gamma; \Omega). \tag{11}$$

Thus, our approach will be to first calculate $\Phi_W(\gamma; \Omega)$ for some small W's and then take the limit for large W. Indeed, the major contribution of this work is the procedure for successfully extrapolating the infinite W limit $\Phi(\gamma; \Omega)$ from $\Phi_W(\gamma; \Omega)$ calculated for a few small W's. Further details will be given in section 4. Here, we stress that this methodology may be used in order to calculate $\Phi(\gamma; \Omega)$ from $\Phi_W(\gamma; \Omega)$ regardless of the specific scoring scheme and parameters including affine gap costs. Our method applies equally to *all* means available for calculating $\Phi_W(\gamma; \Omega)$. Ultimately once $\Phi(\gamma; \Omega)$ has been determined, we will use Eq. (9) to infer the value of the parameter λ characterizing local alignment.

In order to illustrate our method as clearly as possible, we will specialize the remaining discussion to the match–mismatch scoring system given by Eq. (1) and even restrict the space of allowable scoring parameters further as discussed below. For this scoring scheme, we can utilize results from [8, 10] to calculate $\Phi_W(\gamma; \Omega)$ for small widths W. Thus, we will next review the appropriate results from [8, 10]. For the reader who is uninterested in the specifics of how $\Phi_W(\gamma; \Omega)$ is calculated here, we suggest skipping forward to the third paragraph of section 4.

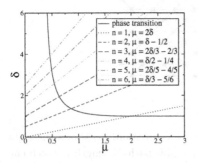 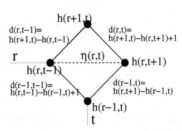

Fig. 3. (a) This figure illustrates the constraint given by Eq. (12). The straight lines plot those μ–δ values that obey the constraint. For any point on these lines, the solution for λ may be obtained using the method given in this presentation. The logarithmic phase is above the solid line denoting the phase transition. The phase transition line was obtained from [7] and has been supplied here for reference. (b) Building blocks of the alignment lattice. By construction r and t are either both even or both odd. This figure shows the relation between the scores at the lattice points and the bond variables $d(r, t)$

In addition to specializing to the match–mismatch scoring system, we constrain μ and δ such that

$$2\delta = n(1 + \mu) - 1 \quad \text{with } n \in \mathbf{N}. \tag{12}$$

This technical condition is necessary in order to utilize the results from [8, 10]. However, it is not a very severe condition since the (μ, δ)-pairs that fulfill this condition can be found all over the μ–δ plane as shown in Fig. 3(a).

The sole approximation neglects the correlations arising between the local scores $s(r, t)$ from the fact that all $M \times N$ local scores are generated by just $M + N$ randomly drawn letters. Instead of taking these correlations into account, we introduce uncorrelated random variables $\eta(r, t) \in \{1, -\mu\}$ replacing the $s(r, t)$ calculated from the letters in the sequences, i.e.,

$$\Pr\{\forall_{r,t}\, \eta(r, t) = \eta_{r,t}\} = \prod_{r,t} \Pr\{\eta(r, t) = \eta_{r,t}\} \tag{13}$$

with match probability $\Pr\{\eta(r,t) = 1\} = \sum_{a=b} p_a^2 \equiv p$ and mismatch probability $\Pr\{\eta(r,t) = -\mu\} = \sum_{a \neq b} p_a p_b = 1 - p$. This approximation, also known as the Bernoulli randomness approximation, is known to change characteristic quantities of sequence alignment only slightly [14, 15, 5, 6, 9, 19, 11]. This general property has been confirmed through numerical studies specifically for the quantity of interest here λ [8]. Numerical evidence for the similarity between the values for λ with and without this approximation [8] is reproduced in Fig. 4.

In [8, 10] it was argued that the calculation of $\Phi_W(\gamma; \Omega)$ can be cast as an eigenvalue problem. Realizing this requires that we introduce *score differences* $d(r,t)$ as defined in Fig. 3(b) and apply them to the finite width picture drawn in Fig. 2(b). Solely from the Needleman-Wunsch recursion relation given by Eq. (3), several important properties of these score differences can be derived [8, 10]: (i) the score differences can only have $n + 1$ different values where n is the natural integer characterizing the choice of μ and δ according to Eq. (12); (ii) the score differences $d(r, t+1)$ can be calculated from the knowledge of the $d(r,t)$ and the random variables $\eta(r,t)$ without reference to the $h(r,t)$; (iii) the score increases $h(r, t+1) - h(r,t)$ can be calculated from the score differences $d(r,t)$ and the random variable $\eta(r,t)$. The first two statements together with the uncorrelated bonds $\eta(r,t)$ assumed in Eq. (13) imply that the dynamics of the score differences $d(r,t)$ can be viewed as a Markov process on the $(n+1)^{2W}$-dimensional state space of the equal time difference vector $(d(0,t), d(1,t), \ldots, d(2W,t))$. This Markov process may be described by a transfer matrix $\hat{T}_W(0; \Omega)$. The entries of this transfer matrix encode the probabilities of the different configurations of the $\eta(r,t)$ in terms of the match probability p and the transitions between the state vectors that these configurations of the $\eta(r,t)$ imply. Finally, property (iii) allows us to modify the transfer matrix in such a way that it keeps track of the changes in the absolute score $h(r,t)$. The curious reader may refer to [10] which provides a detailed explanation of how this p-dependent modified transfer matrix $\hat{T}_W(\gamma; \Omega)$ is obtained. This modified transfer matrix allows us to write

$$Z_{L,W}(\gamma; \Omega) = \boldsymbol{v}^T \hat{T}_W(\gamma; \Omega)^L \boldsymbol{w} \cdot e^{\frac{\gamma L}{2}} \tag{14}$$

with some fixed $(n+1)^{2W}$-dimensional vectors $\boldsymbol{v}$ and $\boldsymbol{w}$. For large L the matrix product is dominated by the largest eigenvalue $\rho_W(\gamma; \Omega)$ which leads to

$$\Phi_W(\gamma; \Omega) = \log \rho_W(\gamma; \Omega) + \frac{\gamma}{2}. \tag{15}$$

For the very simplest scoring system consistent with condition (12), i.e., $n = 1$ where $\mu = 2\delta$, the analytical limit of $\lim_{W \to \infty} \rho_W(\gamma; \Omega)$ can be taken and Eq. (15) yields the closed analytical result [10]

$$\frac{1 + \sqrt{p}\exp[\frac{\lambda}{2}(1+\mu)]}{1 + \sqrt{p}\exp[-\frac{\lambda}{2}(1+\mu)]} \exp[-\frac{\lambda}{2}\mu] = 1. \tag{16}$$

For scoring systems of greater complexity, i.e., larger n, analytic solutions are not readily available for λ. Next, we will present an approach that combines the power of computational numerics and the known analytical properties of the dynamic process described above in order to calculate $\Phi(\gamma; \Omega)$.

4   Numerical Calculation for More Complex Scoring Systems

The main obstacle to obtaining the function $\Phi(\gamma; \Omega)$ (and consequently the Gumbel parameter λ) for more complex scoring systems is the extrapolation (11) of $\Phi(\gamma; \Omega)$ from its finite width counterparts $\Phi_W(\gamma; \Omega)$. In order to get a reliable estimate of the function $\Phi(\gamma; \Omega)$ we need two ingredients: First, we have to be able to calculate $\Phi_W(\gamma; \Omega)$ for as large W as reasonably possible. We will do this using the high performance numerical package ARPACK [29] as described in the next paragraph. Second, we have to extrapolate from as few finite width results as possible toward the infinite width limit $\Phi(\gamma; \Omega)$. The latter is done by using some results from statistical physics and is the main contribution of this manuscript.

The size of the state space, as well as the size of the characteristic matrix $\hat{T}_W$ grows rapidly with the integer n and the width W. Even after exploiting various symmetries, the problem roughly behaves like $(n + 1)^{2W}/nW$. Solving for all eigenvalues in order to discern the greatest quickly becomes exhaustively expensive for $n > 1$. However, two features of this eigenvalue problem succor this otherwise hopeless task for moderate values of n and W. First, the matrix $\hat{T}_W$ is very sparse. The number of non-zero elements grows close to linearly, namely as $O(k \log k)$, where k represents the size of $\hat{T}$. Second, this problem only requires the largest eigenvalue $\rho_W(\gamma; \Omega)$ and not all the eigenvalues. This makes it well suited for the implicitly restarted Arnoldi method (IRAM) [36, 28]. The numerical software package ARPACK [29], which implements IRAM, has been tested as the fastest and most dependable program for finding numerical eigenvalues and eigenvectors [27]. Indeed, in our context ARPACK allows for the quick and specific calculation of only the largest eigenvalue $\rho_W(\gamma; \Omega)$ of the sparse matrix $\hat{T}_W$ for $n < 7$ for at least a few W.

Our accessible numerical solution for $\rho_W(\gamma; \Omega)$, gained via the use of the numerical software package ARPACK, directly gives $\Phi_W(\gamma; \Omega)$ by using Eq. (15). However, the solutions we can obtain for some few small widths W still skirt far from the limit of infinite W in Eq. (11). As such, $\Phi(\gamma; \Omega)$ cannot be straightforwardly approximated from the available $\Phi_W(\gamma; \Omega)$ with any real accuracy. In order to extrapolate from the $\Phi_W(\gamma; \Omega)$ for small finite widths to their infinite limit $\Phi(\gamma; \Omega)$, we make use of two results obtained in the statistical physics community. The first key result is that sequence alignment is a member of the so-called Kardar-Parisi-Zhang (KPZ) universality class [23, 22]. A universality class is a large class of problems that are known to share certain quantitative traits. The second result comes from work by Derrida *et al.*, who were able to calculate an exact solution for what amounts to our $\Phi_W(\gamma; \Omega)$ in a different system of the same universality class. Derrida *et al.* conjecture on general grounds that their exact result for the deviation function $\Phi_W(\gamma; \Omega) - \Phi(\gamma; \Omega)$ from the infinite system is given by a *universal scaling function*, i.e., that it's shape remains the same for *all* members of the KPZ universality class [16, 17]. Together, these two findings imply that our $\Phi_W(\gamma; \Omega) - \Phi(\gamma; \Omega)$ have the same functional form as the Derrida *et al.* deviation function. Expressed in our notation, this means

$$\Phi_W(\gamma; \Omega) = \Phi(\gamma; \Omega) - \frac{a_\Omega G(\gamma W^{1/2} b_\Omega)}{W^{3/2}}. \tag{17}$$

where a_Ω and b_Ω are unknown scaling factors dependent on the particular parameters of the alignment Ω and the scaling function G has been explicitly solved [16,17] (see appendix A). In order to use property (17) to extrapolate $\Phi(\gamma; \Omega)$ from $\Phi_W(\gamma; \Omega)$, a_Ω and b_Ω must be determined. To that end, we take the difference

$$\Phi_W(\gamma; \Omega) - \Phi_{W-1}(\gamma; \Omega) = \frac{a_\Omega G(\gamma W^{1/2} b_\Omega)}{W^{3/2}} - \frac{a_\Omega G(\gamma (W-1)^{1/2} b_\Omega)}{(W-1)^{3/2}} \tag{18}$$

allowing us to eliminate the unknown function $\Phi(\gamma; \Omega)$. We can numerically evaluate the left hand side of this equation as a function of γ. Knowing the exact form of G means that on the right hand side only the scales, controlled by a_Ω and b_Ω remain undetermined. The act of finding these two scaling factors then becomes a matter of fitting the left hand side to the right hand side of Eq. (18). Once a_Ω and b_Ω have been determined, all that remains is to solve for λ using Eqs. (9) and (17)

$$\Phi(\lambda; \Omega) = \Phi_W(\lambda; \Omega) - \frac{a_\Omega G(\lambda W^{1/2} b_\Omega)}{W^{3/2}} = 0. \tag{19}$$

The specifics of the computer algorithm used in determining λ follow. First, we use computer algebra to generate the structure of the transfer matrices $\hat{T}_W$ for different n and W. This process consumes a great deal of time, however, once done for every of the discrete combinations of n and W, the form of the transfer matrices are recorded and can be reused for any choice of mismatch cost μ (which fixes the gap cost δ according to condition (12)) and match probability p. Once μ and p are supplied and the numerical transfer matrix is tabulated, the numerical tool ARPACK obtains the eigenvalues $\rho_W(\gamma; \Omega)$ and $\rho_{W-1}(\gamma; \Omega)$ for $\gamma = 0.8\lambda_{gapless}, 0.9\lambda_{gapless}$ and $\lambda_{gapless}$ where $\lambda_{gapless}$ is the Gumbel parameter λ for the same μ and p in the absence of gaps (as calculated by using Eq. (6)). These initial values, along with the tabulated function G, obtained from KPZ theory, allow for the first approximations of a_Ω and b_Ω to be calculated. The newly found scaling factors are then used along with Eq. (19) in order to choose a new $\gamma \approx \lambda$ by linear extrapolation toward the root. $\rho_W(\gamma; \Omega)$ and $\rho_{W-1}(\gamma; \Omega)$ for this new γ are then evaluated. The whole set of ρ-values then feeds into the reevaluation of a_Ω and b_Ω. This process iterates until γ converges to the solution for λ.

5   Results for More Complex Scoring Systems

Table 1 summarizes the performance of the computer program outlined in section 4. For each value of the integer n, a combination of μ and δ is chosen that leads to a gapped λ of approximately $\lambda \sim 0.8\lambda_{gapless}$. This is considered to

Table 1. This table shows the calculation time and precision with which our algorithm performs. This table was generated using a 2.4 GHz Intel®Xeon™ processor. The error percentages are based on comparisons of results obtained in the range where $\lambda \sim 0.8\lambda_{gapless}$ using the island method [4]. The exception is for $n = 1$, where we have an analytical solution (16), we calculate the error based on the results obtained through the known equation. It should be noted that the percent error inherent in the island method for the simple scoring system described by Eq. (1) is 0.5%

n	μ	δ	W	time(seconds)	error(%)	n	μ	δ	W	time(seconds)	error(%)
1	2.2	1.1	2	0.4	0.1	3	0.9	2.35	2	<0.1	1.6
			3	0.4	0.5				3	0.1	0.3
			4	0.3	0.1				4	0.3	0.2
			5	0.2	<0.1				5	4.3	0.2
			6	0.2	<0.1				6	108.2	0.2
			7	0.2	<0.1	4	0.7	2.9	2	0.1	8.7
			8	0.3	<0.1				3	0.1	<0.1
			9	1.4	<0.1				4	1.0	0.2
			10	4.7	<0.1				5	39.0	0.2
			11	26.4	<0.1	5	0.7	3.75	2	0.1	<0.1
2	1.5	2.0	2	<0.1	0.2				3	0.2	<0.1
			3	0.1	0.4				4	5.3	<0.1
			4	0.3	0.4	6	0.55	4.15	2	0.1	3.2
			5	0.6	0.4				3	1.1	0.3
			6	2.2	0.4				4	39.5	<0.1
			7	26.5	0.4						

be the most relevant region for similarity searches. Most importantly, table 1 shows us that λ converges for $W \geq 4$. (Note that, except for $n = 1$ where the reference value for λ is determined from the exact equation (16), the statistical error on the numerically determined reference values is 0.5% in and of itself.) Our method lands almost all values within the error range of the numerically determined values. This result verifies just how reliably the finite size effects of W are taken into account by the scaling form presented by Derrida *et al.* Secondly, table 1 shows that the evaluation of the Gumbel parameter λ by our new method for all but the largest W (which are unnecessary), finishes in about a second or less. This compares very favorably with the fastest currently available alternatives for obtaining λ, i.e., the island method [4]. The major disadvantage of using the island method for DNA significance assessment lies in the amount of time needed in order to accurately evaluate each data point — the same machine that produced the times in table 1 requires approximately a fortnight in order to obtain an accuracy of 0.5%.

Fig. 4 gives an overview of the dependence of λ on the mismatch cost μ for different n. The lines show the values obtained by our method. In the $n = 1$ case, the solution of Eq. (16) is plotted as well. It is quasi-indistinguishable from the results of our new algorithm. For $n > 1$ the only way to obtain reference values for comparison is by the island method the results of which are shown as the points. Still, our method is within the statistical error of the numerical data of the island method over the whole parameter range.

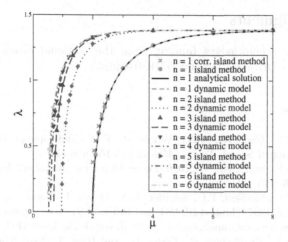

Fig. 4. Values of λ for the DNA alphabet ($p = 0.25$) as a function of the mismatch cost μ. The lines are the results of our new approach; the points are results from stochastic simulation with the island method. The λ-values for $n = 1$ verify well when plotted against solutions of Eq. (16) (also shown as a line barely distinguishable from the line representing the values calculated by our new method) and the island method in [4]. The λ-values for $n = 2, 3, 4, 5$, and 6 displayed here also match well with values obtained from the island method. For $n = 1$, we also include points obtained without use of the uncorrelated approximation Eq. (13). The correlated data obtained via the island method generally compares well with the uncorrelated points and only changes the value of λ slightly. The estimated error of the island method is approximately one quarter of the symbol sizes

6   Conclusions

We have presented a new numerical method to reliably calculate λ with great accuracy and very little computational effort. The efficiency and dependability of this method in characterizing the difficult tail end of the Gumbel distribution removes the major impediment to gapped significance assessment. As previously stated, the remaining Gumbel parameter may be obtained from direct simulation Furthermore, this algorithm grants real time access to the Gumbel parameters and allows for the possibility of updating schemes such as PSI-BLAST to run without resorting to a small set of pre-computed values. The gains in the ability to calculate these parameters not only aids sequence comparison tools but also furthers our ability to discern the most appropriate scoring schemes. We believe that adaptation of these methods is possible for values of μ and δ that do not adhere to the technical condition we imposed for the purpose of our work. This includes the biologically practical and often used affine gap cost schemes. Indeed, future efforts will be directed at using these methods for the more complicated affine gap costs as well as for correlated sequence alignments.

Acknowledgments

RB gratefully acknowledges funding from the National Science Foundation through grants DBI-0317335 and DMR-0404615.

References

1. Altschul, S.F., Gish, W., Miller, W., Myers, E.W., and Lipman, D.J. 1990. Basic Local Alignment Search Tool *J. Mol. Biol.* **215**, 403–410.
2. Altschul, S.F., and Gish, W. 1996. Local Alignment Statistics. *Methods in Enzymology* **266**, 460–480.
3. Altschul, S.F., Madden, T.L., Schäffer, A.A., Zhang, J., Zhang, Z., Miller, W., and Lipman, D.J. 1997. Gapped BLAST and PSI-BLAST: a new generation of protein database search programs. *Nucleic Acids Research* **25**, 3389–3402.
4. Altschul, S.F., Bundschuh, R., Olsen, R., and Hwa, T. 2001. The estimation of statistical parameters for local alignment score distributions. *Nucl. Acids Res.* **29**, 351–361.
5. Boutet de Monvel, J. 1999. Extensive Simulations for Longest Common Subsequences. *Europ. Phys. J. B* **7**, 293–308.
6. Boutet de Monvel, J. 2000. Mean-field Approximations to the Longest Common Subsequence Problem. *Phys. Rev. E* **62**, 204–209.
7. Bundschuh, R., and Hwa, T. 2000. An analytic study of the phase transition line in local sequence alignment with gaps. *Disc. Appl. Math.* **104**, 113–142.
8. Bundschuh, R., 2000. An analytic approach to significance assessment in local sequence alignment with gaps. *Proceedings of the fourth annual international conference on computational molecular biology (RECOMB2000)*, S. Istrail *et al.*, eds., ACM press, (New York, NY), 86–95.
9. Bundschuh, R. 2001. High Precision Simulations of the Longest Common Subsequence Problem. *Europ. Phys. J. B* **22**, 533–541.
10. Bundschuh, R., 2002. Asymmetric exclusion process and extremal statistics of random sequences. *Phys. Rev. E* **65** 031911.
11. Chia, N. and Bundschuh, R. 2004. Finite Width Model Sequence Comparison. *Phys. Rev. E* **70** 021906.
12. Collins, J.F., Coulson, A.F.W., and Lyall, A. 1988. The significance of protein sequence similarities. *CABIOS* **4**, 67–71.
13. Dayhoff, M.O., Schwartz, R.M., and Orcutt, B.C. 1978. A Model of Evolutionary Change in Proteins. In *Atlas of Protein Sequence and Structure*, Dayhoff, M.O., and Eck, R.V., eds., **5** supp. 3, 345–358.
14. Dančík, V., Paterson, M. 1994. Longest Common Subsequences. *Proceedings of 19th International Symposium Mathematical Foundations of Computer Science, Lecture Notes in Computer Science* **841**, 127–142.
15. Dančík, V. 1994. Expected Length of Longest Common Subsequences. *PhD thesis, University of Warwick.*
16. Derrida, B. and Lebowitz, J.L. 1998. Exact Large Deviation Function in the Asymmetric Exclusion Process, *Phys. Rev. Lett.* **80**, 209–213.
17. Derrida, B. and Appert, C. 1999. Universal Large-Deviation Function of the Karder-Parisi-Zhang Equation in One Dimension, *J. Stat. Phys.* **94**, 1–30.
18. Doolittle, R.F. 1996. *Methods in Enzymology* **266**, San Diego, Calif.: Academic Press.

19. Drasdo, D., Hwa, T., and Lassig, M. 2001. Scaling Laws and Similiarity Detection in Sequence Alignment with Gaps. *J. Comp. Biol.* **7**, 115–141.
20. Gumbel, E.J. 1958. *Statistics of Extremes*, Columbia University Press, (New York, NY).
21. Henikoff, S., and Henikoff, J.G. 1992. Amino acid substitution matrices from protein blocks. *Proc. Natl. Acad. Sci. U.S.A.* **89**, 10915–10919.
22. Hwa, T. and Lässig, M. 1996. Similiarity-Detection and Localization. *Phys. Rev. Lett.* **76**, 2591–2594.
23. Kardar, M., Parisi, G., and Zhang, Y.C. 1986. Dynamic Scaling of Growing Surfaces. *Phys. Rev. Lett.* **56**, 889–892.
24. Karlin, S., and Altschul, S.F. 1990. Methods for assessing the statistical significance of molecular sequence features by using general scoring schemes. *Proc. Natl. Acad. Sci. U.S.A.* **87**, 2264–2268.
25. Karlin, S., and Dembo, A. 1992. Limit distributions of the maximal segmental score among Markov-dependent partial sums. *Adv. Appl. Prob.* **24**, 113–140.
26. Karlin, S., and Altschul, S.F. 1993. Applications and statistics for multiple high-scoring segments in molecular sequences. *Proc. Natl. Acad. Sci. U.S.A.* **90**, 5873–5877.
27. Lehoucq, R.B. and Scott, J.A. 1996. An evaluation of software for computing eigenvalues of sparse nonsymmetric matrices. preprint MCS-P547-1195, Argonne National Laboratory, Argonne, IL.
28. Lehoucq, R.B. 1997. Truncated QR algorithms and the numerical solution of large scale eigenvalue problems. preprint MCS-P648-0297, Argonne National Laboratory, Argonne, IL.
29. Lehoucq, R.B., Sorensen, D.C., and Yang, C. 1997. *ARPACK Users' Guide: Solutions of Large Scale Eigenvalue Problems with Implicitly Restarted Arnoldi Methods*, SIAM,(Philadelphia, PA)
30. Mott, R. 1992. Maximum likelihood estimation of the statistical distribution of Smith-Waterman local sequence similarity scores. *Bull. Math. Biol.* **54**, 59–75.
31. Mott, R., and Tribe, R. 1999. Approximate statistics of gapped alignments. *J. Comp. Biol.* **6**, 91–112.
32. Mott, R. 1999. Accurate estimate of *p*-values for gapped local sequence alignment. Private communication.
33. Needleman, S.B., and Wunsch, C.D. 1970. A general method applicable to the search for similarities in the amino acid sequence of two proteins. *J. Mol. Biol.* **48**, 443–453.
34. Olsen, R., Bundschuh, R., and Hwa, T. 1999. Rapid Assessment of Extremal Statistics for Gapped Local Alignment. *Proceedings of the Seventh International Conference on Intelligent Systems for Molecular Biology*, T. Lengauer *et al.*, eds., 211–222, AAAI Press, (Menlo Park, CA).
35. Pearson, W.R. 1991. Searching protein sequence libraries. comparison of the sensitivity and selectivity of the Smith-Waterman and FASTA algorithms. *Genomics* **11**, 635–650.
36. Sorensen, D.C. 1992. Implicit application of polynomial filters in a k-step Arnoldi method. *SIAM J. Matrix Analysis and Applications* **13** 357–385.
37. Siegmund, D., and Yakir, B. 2000. Approximate *p*-values for Sequence Alignments. *Ann. Statist.* **28** 657–680
38. Smith, S.F., and Waterman, M.S., 1981. Comparison of biosequences.*Adv. Appl. Math.* **2**, 482–489.
39. Smith, T.F., Waterman, M.S., and Burks, C. 1985. The statistical distribution of nucleic acid similarities. *Nucleic Acids Research* **13**, 645–656.

40. Waterman, M.S., Gordon, L., and Arratia, R. 1987. Phase transitions in sequence matches and nucleic acid structure, Proc. Natl. Acad. Sci. U.S.A. **84**, 1239–1243.
41. Waterman, M.S., and Vingron, M. 1994. Sequence Comparison Significance and Poisson Approximation. *Stat. Sci.* **9**, 367–381. v
42. Waterman, M.S., and Vingron, M. 1994. Rapid and accurate estimates of statistical significance for sequence database searches. *Proc. Natl. Acad. Sci. U.S.A.* **91**, 4625–4628.
43. Waterman, M.S. 1994. *Introduction to Computational Biology.* London, UK: Chapman & Hall.

A   Deviation Function of the Particle Hopping

The deviation function G_D as solved by Derrida *et al.* is independent of the model parameters and has the following parametric form [16, 17]

$$\beta = \frac{2}{\sqrt{\pi}} \int_0^\infty \epsilon^{1/2} \frac{Ce^{-\epsilon}d\epsilon}{1 + Ce^{-\epsilon}} \tag{20}$$

$$G_D(\beta) = \frac{4}{3\sqrt{\pi}} \int_0^\infty \epsilon^{3/2} \frac{Ce^{-\epsilon}d\epsilon}{1 + Ce^{-\epsilon}}. \tag{21}$$

As C approaches -1 we require a new representation to go beyond $\beta_- = \lim_{C \to -1} \beta$. The analytical continuation of $G_D(\beta)$ is beyond β_- given by the parametric equations [16, 17]

$$\beta = -4\sqrt{\pi} \left[-\ln(-C) \right]^{1/2} - \sum_{q=1}^{\infty} (-C)^q q^{-3/2} \tag{22}$$

$$G_D(\beta) = \frac{8}{3}\sqrt{\pi} \left[-\ln(-C) \right]^{3/2} - \sum_{q=1}^{\infty} (-C)^q q^{-5/2}, \tag{23}$$

as C for these equation varies between 0 and -1, this gives the function $G_D(\beta)$ for all $\beta < \beta_-$.

In the limit as $\beta \to -\infty$ [16, 17],

$$G_D(\beta) \approx -\frac{\beta^3}{24\pi} \tag{24}$$

implying that for large γ,

$$W^{-3/2} G_D(\gamma W^{1/2}) \approx -\frac{\gamma^3}{24\pi}. \tag{25}$$

This term is independent of W, i.e., of finite size effects. In order to appropriately reflect this, we include this W-independent term in Φ. Therefore, the function G used in our methodology relates to the Derrida *et al.* solution for $G_D(\beta)$ via the equation $G(\beta) = G_D(\beta) + \beta^3/(24\pi)$.

Alignment of Optical Maps*

Anton Valouev[1], Lei Li[1], Yu-Chi Liu[2], David C. Schwartz[4], Yi Yang[1],
Yu Zhang[3], and Michael S. Waterman[1]

[1] Department of Mathematics,
University of Southern California, Los Angeles CA 90089, USA
{valouev, lilei, yiy, msw}@usc.edu
[2] Molecular and Computational Biology Program, Department of Biological Sciences,
University of Southern California, Los Angeles CA 90089, USA
ycliu@usc.edu
[3] Department of Statistics, Harvard University, Cambridge MA 02138, USA
yuzhang@bioinfo.stat.harvard.edu
[4] Laboratory for Molecular and Computational Genomics,
Departments of Genetics and Chemistry, University of Wisconsin-Madison,
UW-Biotechnology Center 425 Henry Mall, Madison WI 53706, USA
dcschwartz@facstaff.wisc.ed

Abstract. We introduce a new scoring method for calculation of alignments of optical maps. Missing cuts, false cuts and sizing errors present in optical maps are addressed by our alignment score through calculation of corresponding likelihood ratios. The Sizing error model is derived through the application of CLT and validated by residual plots collected from real data. Missing cuts and false cuts are modeled as Bernoulli and Poisson events respectively. This probabilistic framework is used to derive an alignment score through calculation of likelihood ratio. Consequently, this allows to achieve maximal descriminative power for alignment calculation. The proposed scoring method is naturally embedded within a well known DP framework for finding optimal alignments.

1   Introduction

Optical Mapping is a powerful technology that allows construction of ordered restriction maps. Each optical map represents a single DNA molecule digested by the restriction enzyme and imaged by the optical system. The map is comprised of estimates of fragment sizes in the order they appear on the imaged molecule. Hence, there are two types of information associated with each optical map: sizes of restriction fragments on the molecule and their relative order.

A broad spectrum of problems can be effectively addressed by means of optical mapping. Among the most important are: analysis of genomic variation

---

* Supported by NIH CEGS: Implications of Haplotype Structure in the Human Genome, Grant No. P50 HG002790 and NIH, NHGRI, 2R01 HG00225-10.

S. Miyano et al. (Eds.): RECOMB 2005, LNBI 3500, pp. 489–504, 2005.

(deletions, insertions, rearrangements on scale of thousands of base pairs), construction of complete restriction maps without knowledge of the original DNA sequence, validation of completed sequence contigs, and genomic placement during sequencing projects. Despite the fact that optical mapping lacks single nucleotide resolution like in case sequencing, it is capable of quickly assessing genomic differences as well as genotyping a human in a matter of few hours which is presently not achievable by means of sequencing. The domain of application of optical mapping is somewhat complementary to that of sequencing: quick genome wide analysis with resolution of several hundreds of base pairs is now routinely done for human-sized genomes. Genome wide analysis of genome variations not captured by SNPs is an exciting prospect.

To address these problems by means of optical mapping, tools for revealing homologies between optical maps need to be developed. More generally, such tools should include genomic placement of individual optical maps and potentially be capable of genome wide restriction mapping while remaining computationally feasible. Similar problems have been solved for DNA sequences. Development of sequence alignment dynamic programming (DP) algorithms [11, 16] for accurate comparison of DNA and protein sequences provided tools for genome-wide shotgun sequencing which was successfully implemented and applied to a variety of organisms [6, 15].

It is therefore natural to explore the possibility of applying similar ideas to optical maps and study the feasibility of such methods. A significant amount of work has already been done in this direction. DP algorithms have been successfully used for restriction map alignments in [11, 7, 14]. With the development of the Optical Mapping System [17, 18], much effort has been made to develop methods for genome-wide restriction mapping. Ananthraman et al [4, 8, 9, 10] have designed an extensive Bayesian framework for Optical Mapping with potential for global map assembly. Its use, however, has been limited as laboratories now confront large genomes such as human or mouse.

A key challenge in restriction mapping is to distinguish genomic variations from measurement errors specific to the context of optical mapping system. We present a complete statistical model that enables us to design a likelihood ratio based alignment score with the optimal discriminant power for distinguishing correct alignments from spurious ones. Statistical models corresponding to different error types are incorporated into the likelihoods of observed data in optical maps. Overall, the exact form of the alignment score follows from a probabilistic model associated with the way optical maps are generated. Appropriate conditioning allows a natural decomposition of scores into a sum of two components, first to account for sizing errors, and second to account for presence of false cuts and missing cuts. The designed alignment score is implemented within a standard DP alignment framework similar to that used for sequences [19] and restriction maps [11]. Complexity of finding optimal alignment for two maps with m and n fragments is $O(\delta^2 mn)$ (for all practical purposes $\delta \leq 5$).

2   Models

There are several types of errors associated with optical maps. These include missing cuts and false cuts, missing fragments, sizing errors, and chimeric reads. We will explore each in more detail.

Missing Cuts and False Cuts. The efficiency of DNA digestion by the restriction endonuclease is never perfect. As a result, some restriction sites on DNA remain uncut by endonuclease. After the molecule is imaged, corresponding restriction fragments remain concatenated in the output data, appearing as if they came from a single restriction fragment of the combined size. Digestion rate is monitored after the digestion has taken place. Many copies of a $\lambda-$phage of known size and number of restriction sites are comounted in the solution together with the target DNA, so that digestion rate in the solution can be screened. During the image analysis, if the digestion rate is below 80 percent, the optical maps may not be reported in the output.

False cuts result from random DNA breaks. Under our model assumptions, random breaks show no preference to particular regions of DNA and thus occur equally likely in all regions. False cuts are accounted for by penalizing the regions containing them in calculation of the alignment score.

Missing Fragments. After being extracted from cells, target DNA is deposited into microfluidic channels and attached to the surface by means of capillary action. Digestion results in restriction fragments of various sizes, some of which are too small to firmly hold to the glass surface. These fragments flow away and thus cannot be captured by the imaging system. Most fragments shorter than 0.5 Kb often remain unreported. Also, a large percentage of other small fragments is missing in the data, so that on average about 50 percent of fragments between 0.5 and 1.1 Kb are missing in the data.

Sizing Errors. During the fluorescent marking of DNA, the dye is attached along the span of the molecule in a random fashion. The size of each restriction fragment within the optical map is detected by measuring the fluorescence intensity emitted by the fluors attached to the corresponding piece of DNA. To determine the size of the fragment, its intensity is compared to the measurement standard that corresponds to the amount of intensity associated with the DNA of a known size.

Suppose that the restriction fragment is n times the size of the unit DNA mass. Hence the measured intensity W of this restriction fragment can be written as $W = \sum_{i=1}^{n} W_i$, where W_i is the amount of measured fluorescence per i-th unit DNA mass within the fragment. We model W_i as iid random variables with $EW_i = \mu$ and $Var(W_i) = \gamma^2$. The latter is certainly a strong assumption, for several following reasons. The amount of detected fluorescence may vary across the regions of DNA. Local uptake of fluorescent dye can be affected by local variation of adsorption of DNA chains to the surface. This results in variation of detected fluorescent intensity along the span of DNA chains. Although these factors can have a significant impact on precision of measurements, data confirms that the assumption of iid W_i is valid across an overwhelming proportion of DNA fragments.

By the central limit theorem, W converges in distribution to a normal random variable, namely $W \to N(n\mu, n\gamma^2)$ (recall there are more general conditions for the sums of r.v. being approximately normal). In order to determine sizes of fragments based on the amount of observed fluorescence, the distribution parameter μ needs to be calculated. It is identified as the standard amount of fluorescent intensity per unit DNA mass and is estimated from the lambda DNA comounted along the target DNA. The fragment sizes are calculated by dividing their fluorescent intensity by the standard amount of intensity per unit DNA mass calculated locally relative to the position of mounted target DNA. Therefore, estimated sizes X are identified as $X = W/\mu$. Hence, it is easy to see that as n gets large X converges in distribution to a normal random variable $X \to N(n, n\sigma^2)$, where $\sigma^2 = (\gamma/\mu)^2$. Therefore, if Y is the true underlying size of the restriction fragment, then $X \to N(Y, \sigma^2 Y)$ in distribution.

Further, suppose that ϵ, defined as $\epsilon = X - Y$, is the measurement error. Of course $\epsilon \to N(0, \sigma^2 Y)$ in distribution as n increases. Using alignments, we have collected pairs (X, Y) of single fragments where X is a fragment on optical maps, and Y is its true underlying restriction size. Since according to our model $\epsilon \sim N(0, \sigma^2 Y)$, then for $E = \frac{X-Y}{\sqrt{Y}}$ we have $E \sim N(0, \sigma^2)$. Thus E should be homoscedastic across Y and marginally normal. Figure in the Appendix illustrates that the variance of E is homogenous across Y and E is close to a normal variable.

The described error model agrees well with the data from relatively long fragments ($\geq 4Kb$) observed in the optical maps. However, for small fragments, the normality assumption fails due to lack of convergence to normal density as well as some other effects. To take them into account, the error for short fragments is modelled as $\epsilon \sim N(0, \eta^2)$, where η^2 is significantly larger than σ^2 corresponding to the error variance in the first model.

Chimeric Reads. When two molecules cross, at the resolution of light microscopy, there is no way to confidently disambiguate the strands. As a result, some reads are chimeric, meaning that they correspond to no real maps.

Model Assumptions and Parameters. To summarize, we use the following set of assumptions hereafter in our analysis:

1. Sizes of restriction fragments Y have exponential density with mean $EY = \lambda$, with λ depending on a particular endonuclease. Consequently, the number of restriction sites in s Kb of linear DNA is a Poisson process with the rate s/λ.
2. Observation of restriction site on the optical map is a Bernoulli event with probability p, which we also refer to as digestion rate. Further, we assume that restriction sites are being digested independently, and hence observed restriction sites are due to thinned Poisson process. Digestion rate is usually assumed to be high ($0.5 \leq p \leq 1$) which is generally the case in real mapping projects.

3. The number of false cuts per s Kb of linear DNA is a Poisson process with the parameter ζs.
4. The length discrepancy between the optical fragment of size X and its underlying true size Y, given by $\epsilon = X - Y$ has a normal density $N(0, \sigma^2 Y)$ for $X \geq \Delta$. Further, $\epsilon \sim N(0, \eta^2)$ for $0 < X < \Delta$ with $\eta^2 \gg \sigma^2$. For all practical situations $\Delta = s$ Kb is a reasonable threshold separating the two error models. Also, both X and Y always assume nonnegative values, so negative sizes are never observed.
5. For the purpose of approximations we require $\lambda \gg \sigma^2$ which indeed holds in all practical situations.

3   Probabilistic Framework

Alignments are calculated in order to find matching regions between maps. Calculation of optimal alignments depends profoundly on the alignment score. The utility of the alignment score is in finding the optimal configuration among all possible alignments.

Alignments are based on site matching. A site is the position on the map flanked by two adjacent fragments. According to our model, sites can either result from restriction enzyme cutting DNA at that position or from random DNA breakage.

Consider two maps D_1 and D_2, where at least one of these maps, say D_1 is an optical map. D_2 can either be an optical map, or some other restriction map, perhaps calculated from the DNA sequence. We assume that both maps are obtained using the same restriction enzyme. We define a pair of matching sites $< i, j >$ with the site i located on map D_1 and j located on map D_2. An alignment between D_1 and D_2 is represented by matching regions of D_1 and D_2. Matching regions are defined as regions of two maps flanked by a pair of matching sites. Therefore, matching regions contain no matching sites other than the flanking ones. Naturally, alignment is represented by a set of ordered pairs of matching sites $< i_0, j_0 >$, $< i_1, j_1 >$, ..., $< i_d, j_d >$, where i_t matches j_t and is denoted by $< i_t, j_t >$, where $i_t \in D_1$, $j_t \in D_2$ represent sites on corresponding maps. Here the ordering is taken in the sense that $i_0 < i_1 < ... < i_d$ and $j_0 < j_1 < ... < j_d$. Consequently, regions $I_t = [i_{t-1}, i_t)$ and $J_t = [j_{t-1}, j_t)$ are matching regions if they are flanked by matching sites $< i_{t-1}, j_{t-1} >$ and $< i_t, j_t >$ and contain no internal matching sites. Matching of regions is be denoted by $< I_t, J_t >$.

For alignments, we assume that matching pairs of regions are independent from each other conditioned on underlying match. In other words if $< I_m, J_m >$ and $< I_n, J_n >$, then $Pr(< I_m, J_m >, < I_n, J_n >) = Pr(< I_m, J_m >)Pr(< I_n, J_n >)$. Here $Pr(< I_m, J_m >)$ is understood as the likelihood of observing I_m and J_m within a given matching region. To define the alignment score, consider matching sites $< i_0, j_0 >$, $< i_1, j_1 >$, ..., $< i_d, j_d >$ between maps D_1 and D_2. Matching regions of D_1 and D_2 may result from two different scenarios: true match and spurious (or random) match. True match corresponds to the situation when there is real physical matching between corresponding regions of optical maps. In other words, the matching regions are either derived from the same reference region, or perhaps duplication, or some similarity of interest. Spurious match is a situation when regions of maps are unrelated (have no underlying genomic similarity) and are therefore independent.

Under this setup, we define two competing hypotheses: H_0 and H_a, where H_0 corresponds to a true match and H_a corresponds to the spurious match. We define the alignment score S between regions D_1 and D_2 as $S(D_1, D_2) = -\log(LR(< D_1, D_2 >))$, where LR stands for the likelihood ratio under competing hypotheses H_0 and H_a. Hence,

$$LR(< D_1, D_2 >) = \frac{f(D_1, D_2)_{H_0}}{f(D_1, D_2)_{H_a}} = \prod_{k=1}^{d} \frac{f(I_k, J_k)_{H_0}}{f(I_k, J_k)_{H_a}} = \prod_{k=1}^{d} LR(< I_k, J_k >)$$

with the product taken over d matching regions. It therefore follows that for the total score $S(D_1, D_2) = \sum_{k=1}^{d} S(I_k, J_k) = \sum_{k=0}^{d} \left[-\log \left(\frac{f(I_k, J_k)_{H_0}}{f(I_k, J_k)_{H_a}} \right) \right]$. Furthermore, $S(I_k, J_k)$ for the matching region $< I_k, J_k >$ is calculated based on two pieces of information: total amount of DNA within I_k and J_k and the number of fragments contained in I_k and J_k respectively.

In this paper we consider two different alignment situations. The first one corresponds to the case when optical map is aligned against a reference or other known restriction map. This matching type is therefore designated as reference matching. The second situation refers to the case when an optical map is aligned against another optical map for calculation of overlap. This matching is therefore referred as optical matching. Depending on the type of matching attempted, the calculation of alignment scores is different. We should therefore develop theory for these two cases separately.

Reference Matching. To make things easier to follow, define $M := D_1$ and $R := D_2$ to represent optical and reference maps respectively. Also, we define $M_k := I_k$ for matching regions of optical map and $R_k := J_k$ for matching regions of reference map. Similar $m_k := i_k$ and $r_k := j_k$ will refer to the indices of the matching sites on optical map and reference map respectively.

Denote s_k^M and s_k^R to be the total amount of DNA contained within matching regions M_k and R_k. Also, define f_k^M and f_k^R to be the number of fragments within M_k and R_k respectively.

We can write the likelihood of the observed data as

$$f(M_k, R_k) = f(M_k|R_k) \times f(R_k) = f(s_k^M, f_k^M | s_k^R, f_k^R) \times f(R_k)$$
$$= f(s_k^M | s_k^R, f_k^R, f_k^M) \times f(f_k^M | s_k^R, f_k^R) \times f(R_k),$$

and the likelihood ratio can be rewritten as

$$LR(< M_k, R_k >) = \frac{f(M_k, R_k)_{H_0}}{f(M_k, R_k)_{H_a}} = \frac{[f(M_k|R_k) \times f(R_k)]_{H_0}}{[f(M_k|R_k) \times f(R_k)]_{H_a}}$$
$$= \frac{f(M_k|R_k)_{H_0} \times f(R_k)}{f(M_k|R_k)_{H_a} \times f(R_k)} = \frac{f(M_k|R_k)_{H_0}}{f(M_k|R_k)_{H_a}}.$$

Therefore, we can write

$$
LR(< M_k, R_k >) = \frac{f(s_k^M | s_k^R, f_k^R, f_k^M)_{H_0}}{f(s_k^M | s_k^R, f_k^R, f_k^M)_{H_a}} \times \frac{f(f_k^M | s_k^R, f_k^R)_{H_0}}{f(f_k^M | s_k^R, f_k^R)_{H_a}} =
$$
$$
= LR(s_k^M | s_k^R, f_k^R, f_k^M) \times LR(f_k^M | s_k^R, f_k^R).
$$

Thus, for the total score we have

$$
S(M, R) = \sum_{k=1}^{d} (\alpha_k + \beta_k) = \sum_{k=1}^{d} \left(S(s_k^M | s_k^R, f_k^R, f_k^M) + S(f_k^M | s_k^R, f_k^R) \right),
$$

with the sum taken over d matching regions

Here $\alpha_k = -\log(LR(s_k^M | s_k^R, f_k^R, f_k^M))$ is referred to as the size match score, likewise, $\beta_k = -\log(LR(f_k^M | s_k^R, f_k^R))$ is referred to as the site mismatch score. As can be seen from the decomposition of likelihood ratios, the score for a matching region naturally breaks down into two parts: the score for the total size of the matching region and the score for the total number of non-matching sites per matching region. This makes calculation of the alignment scores more straightforward.

Optical Matching. Optical matching corresponds to the situation when two optical maps D_1 and D_2 are aligned against each other. Recall that I_k and J_k refer to matching regions between of D_1 and D_2. As before, i_k, j_k define matching sites between aligned maps. Let s_k^I and s_k^J be the total amount of DNA contained within matching regions I_k and J_k respectively. Also, define f_k^I and f_k^J to be the number of fragments within I_k and J_k. Both maps D_1 and D_2 define collections of random variables for which we can write the likelihood as follows $f(I_k, J_k) = f(s_k^I, s_k^J, m_k^I, m_k^J) = f(s_k^I, s_k^J | m_k^I, m_k^J) \times f(m_k^I, m_k^J)$, and thus the likelihood ratio can be rewritten as

$$
LR(< I_k, J_k >) = \frac{f(s_k^I, s_k^J | m_k^I, m_k^J)_{H_0}}{f(s_k^I, s_k^J | m_k^I, m^J)_{H_a}} \times \frac{f(m_k^I, m_k^J)_{H_0}}{f(m_k^I, m_k^J)_{H_a}} =
$$
$$
= LR(s_k^I, s_k^J | m_k^I, m_k^J) \times LR(m_k^I, m_k^J).
$$

Hence, we infer that for the score $S(I_k, J_k) = S(s_k^I, s_k^J | m_k^I, m_k^J) + S(m_k^I, m_k^J)$, where $S(s_k^I, s_k^J | m_k^I, m_k^J) = -\log(LR(s_k^I, s_k^J | m_k^I, m_k^J))$ and $S(m_k^I, m_k^J) = -\log(LR$ $(m_k^I, m_k^J))$. Thus the total alignment score can be rewritten as $S(D_1, D_2) = \sum_{k=1}^{d} (\alpha_k + \beta_k) = \sum_{k=1}^{d} \left(S(s_k^I, s_k^J | m_k^I, m_k^J) + S(m_k^I, m_k^J) \right)$. Again, in this paper $\alpha_k = S(s_k^I, s_k^J | m_k^I, m_k^J)$ will be referred as the size match score and $\beta_k = S(m_k^I, m_k^J)$ as the site mismatch score since m_k^I and m_k^J essentially provide information about the number of sites that are not matched within matching regions I_k and J_k.

4   Likelihood Ratios for Various Types of Matchings

The purpose of designing a score for matching regions is to utilize it in the alignment calculation. In standard Smith-Waterman approach to sequence alignment the match/mismatch score corresponds to the score associated with the pair of sequence letters. If the letters match, then the match is rewarded by μ, if the letters do not match, then the mismatch is penalized by $-\delta$. Both match and mismatch scores are associated with the frequency of match/mismatch between compared sequences. Similarly, in case of alignment of optical maps we need to have a score associated with matching sets of DNA fragments. Since different alignment types arise in practical applications, several types of situations need to be addressed depending on what types of maps are being aligned against each other.

In this paper we consider two situations. The first one corresponds to the situation when optical map is being aligned against the reference restriction map. This restriction map can be either derived from DNA sequence using the enzyme of interest or obtained by some other method. The second situation arises when two optical maps are being aligned against each other for the purpose of finding mutual physical overlap between them. Thus two types of matches arise.

Multiple Reference Match. Optical maps are prone to the presence of missing cuts and false cuts. When aligning against the reference, matching several fragments to several is necessary due to mentioned errors. This matching is referred to as multiple reference matching.

Multiple Optical Match. Similar to the situation of multiple reference match, multiple optical matching is needed to address multiple fragments being matched between two optical maps.

These types of matches address all matching configurations in our alignments. The scores will address two types of errors: sizing errors and cut errors. As we have mentioned above, sizing errors are due to errors in measured sizes of regions of DNA, cut errors are due to presence of missing and false cuts.

4.1   Mathematical Facts

In this section we present some preliminary results we use to simplify calculation of the alignment score. Proofs of all results can be found in the Appendix.

Lemma 1 (Distribution of number of cut sites on optical maps). *Under the model assumptions, the number X of cut sites per s Kb of linear DNA on optical map is comprised of restriction cuts and random breaks. It has Poisson distribution with the rate s/τ, where $\tau = \left(\zeta + \frac{p}{\lambda}\right)^{-1}$.*

Lemma 2 (Distribution of fragment sizes on optical maps). *Under the model assumptions, measured sizes X of fragments on optical maps have exponential density with the mean $\theta = \left[\frac{1}{\sigma}\sqrt{\frac{2}{\tau} + \frac{1}{\sigma^2}} - \frac{1}{\sigma^2}\right]^{-1}$ for $X \geq \Delta$.*

Proposition 1. (Size distribution of matching regions between optical and reference maps). *Under the model assumptions, reference sizes of matching regions between optical and reference maps have exponential density with the mean $v = \lambda/p$.*

Proposition 2. (Size distribution of matching regions between optical maps). *Under the model assumptions, reference sizes underlying matching regions between two optical maps have exponential density with the mean $\phi = \lambda/p^2$.*

Proposition 3. (Size distribution for sum of multiple optical fragments). *Consider $X = \sum_{i=1}^{m} X_i$ to be the total size of a region of optical map comprised of m fragments. Then under normality assumptions for error model $X \sim N(Y, \sigma^2 Y)$, where Y is the underlying size of reference region from where $X_1, X_2, \ldots, X_m$ originate.*

4.2   Reference Matching Likelihood Ratios

Consider the situation of matching against the reference map. Suppose that a region of optical map of size X is comprised of m consecutive fragments $X_1, \ldots, X_m$ so that $X = \sum_{i=0}^{m} X_i$. Likewise, suppose Y is the size of the reference region comprised of n consecutive on the reference map ($Y = \sum_{j=1}^{n} Y_j$). In this situation the following test can be constructed.

Theorem 1 (Multiple reference size match LR). *Under the model assumptions the likelihood ratio for the multiple matching of size x of m combined optical fragments against the size y of n combined reference fragments has the form*

$$LR(x|y, n, m) = \frac{\sqrt{2\pi y}\sigma x^{m-1}}{\Gamma(m)\theta^m} exp\left[\frac{(x-y)^2}{2\sigma^2 y} - \frac{x}{\theta}\right].$$

for $x > \Delta$. Furthermore, for $0 < x \leq \Delta$, the likelihood ratio has the form

$$LR(x|y, n, m) = \frac{\sqrt{2\pi}\eta x^{m-1}}{\Gamma(m)\theta^m} exp\left[\frac{(x-y)^2}{2\eta^2} - \frac{x}{\theta}\right].$$

Naturally, alignment score should also account for absence of some restriction sites on optical maps as well as presence of some false cuts due to imperfect digestion and random breakage of DNA. The following theorems explains how to account for such errors.

Theorem 2 (Multiple reference site mismatch LR). *Under the model assumptions the likelihood ratio for matching m combined fragments on the optical map against the reference region of size y comprised of n reference fragments has the form*

$$LR(m|y, n) = \frac{e^{\zeta y}(m-1)! f_M(m)}{(1-p)^{n-1}(\zeta y)^{m-1}},$$

where $f_M(m)$ is the marginal density of m (exact form of $f_M(m)$ is discussed in the proof).

4.3   Optical Matching Likelihood Ratios

The previous section discussed a situation of matching regions of optical map against regions of reference map. Consider now a situation when two optical maps are being aligned. In this situation pairs of optical regions are being matched.

Let X_1 and X_2 be sizes of two regions on optical maps. Suppose these regions are comprised of m_1 and m_2 consecutive optical fragments respectively. We want to design a test for identifying whether X_1 and X_2 are sizes of regions of optical maps with the same underlying reference region.

Theorem 3 (Multiple optical size match LR). *Under the model assumptions the likelihood ratio for the multiple matching of size x_1 of optical map comprised of m_1 fragments against the size x_2 of another optical map comprised of m_2 fragments has the form*

$$LR(x_1, x_2 | m_1, m_2) = \frac{\pi \phi \sigma^2 x_1^{m_1-1} x_2^{m_2-1}}{\Gamma(m_1)\Gamma(m_2)\theta^{m_1+m_2}} \times \frac{exp\left[-\left(\frac{1}{\theta}+\frac{1}{\sigma^2}\right)(x_1+x_2)\right]}{K_0\left(2\sqrt{\frac{1}{\phi}+\frac{1}{\sigma^2}}\sqrt{\frac{x_1^2+x_2^2}{2\sigma^2}}\right)},$$

where $K_\nu(z)$ is a modified Bessel function of the second kind.

Similar to the case of reference matching, we present a theorem that takes into account unmatched sites between regions of two optical maps

Theorem 4 (Multiple optical site mismatch LR). *Consider two regions of optical maps comprised of m_1 and m_2 fragments respectively, then under the model assumptions the likelihood ratio for matching m_1 against m_2 has the form $LR(m_1, m_2) = \frac{A}{B}$, where $A = Pr(m_1, m_2)_{H_0} = f_{M_1, M_2}(m_1, m_2)$, and*

$$B = \frac{1}{\phi}\sum_{n=0}^{\infty}\frac{1}{\lambda^n n!}\sum_{k_1=0}^{m_1-1 \wedge n}\sum_{k_2=0}^{m_2-1 \wedge n-k_1} A(k_1, k_2 | n)$$

$$\times \frac{\zeta^{(m_1+m_2)-(k_1+k_2)-2}}{\left[\frac{1}{\phi}+\frac{1}{\lambda}+2\zeta\right]^{n+(m_1+m_2)-(k_1+k_2)-1}} \times \frac{\Gamma(n+(m_1+m_2)-(k_1+k_2)-1)}{\Gamma(m_1-k_1)\Gamma(m_2-k_2)},$$

where $A(k_1, k_2 | n) = \dfrac{\dbinom{n}{k_1 \; k_2} p^{k_1+k_2}(1-p)^{2n-(k_1+k_2)}}{\displaystyle\sum_{i_1=0}^{n}\sum_{i_2=0}^{n-i_1}\dbinom{n}{i_1 \; i_2} p^{i_1+i_2}(1-p)^{2n-(i_1+i_2)}}.$

Exact form of $f_{M_1, M_2}(m_1, m_2)$ is discussed in the proof.

5   Calculation of the Alignments

Two types of alignments are discussed in this paper: fit and overlap alignment. Other types of alignments are computed similarly with a slight modifications to initialization [3]. Fit alignment to calculate a proper fit of the the optical map into the reference. Indeed, if the whole restriction map of the organism is known, any of its genomic optical maps should properly fit into the reference. Therefore, the fit alignment is used for the purpose of finding the best fit of optical map against the reference to locate region from which that optical map originates.

Overlap alignment allows detection of potential overlaps between optical maps. This corresponds to the situation when two maps can be aligned only partially and one of the tails of each map may remain unaligned.

In this section we outline an alignment algorithm that utilizes an alignment score that we have derived. For the fit alignment, suppose that maps R and M correspond to the reference restriction map with n sites and the optical map with m sites. As before, we count both the start and end positions as separate sites. Additionally, suppose that numbers $0 = q_0 < q_1 < \ldots < q_m$ and $0 = r_0 < r_1 < \ldots < r_n$ mark the positions corresponding to the sites on optical map and reference maps respectively. The fit alignment Π is defined as the ordered set of aligned sites $(i_0, j_0)\ (i_1, j_1) \ldots (i_d, j_d)$, where $0 \leq i_0 < i_1 < \ldots < i_d \leq m$ and $0 = j_0 < j_1 < \ldots < j_d = m$ corresponding to the indices of the restriction sites on the reference and optical maps respectively.

For overlap alignment, consider optical maps O_1 and O_2 with m_1 and m_2 restriction sites respectively. Also, suppose that numbers $0 = q_0 < q_1 < \ldots < q_{m_1}$ and $0 = r_0 < r_1 < \ldots < r_{m_2}$ mark the positions corresponding to the sites on both maps. The overlap Π is defined as the ordered set of aligned sites $(i_0, j_0)\ (i_2, j_2) \ldots (i_d, j_d)$, where either $0 \leq i_0 < i_1 < \ldots < i_d = m_1$ and $0 = j_0 < j_1 < \ldots < j_d \leq m_2$, or $0 = i_0 < i_1 < \ldots < i_d \leq m_1$ and $0 \leq j_0 < j_1 < \ldots < j_d = m_2$ corresponding to the indices of matched sites on maps O_1 and O_2 respectively.

Alignment Algorithm. Let $X(i, j)$ be the score of the largest scoring alignment with the rightmost aligned pair of sites (i, j). The alignment score is calculated according to the Algorithm 1 found in Huang and Waterman [7].

Score. The exact form of the score $S\left(q_i - q_g, r_j - r_h, i - g, j - h\right)$ is computed according to likelihood ratios presented above.

For fit alignment it has the form $S\left(q_i - q_g, r_j - r_h, i - g, j - h\right) = -\log(LR\left(q_i - q_g; r_j - r_h, i - g, j - h\right)) - \log(LR(i - g; r_j - r_h, j - h))$, where the first ratio is given by Theorem 1, and the second is given by Theorem 2. Similarly, for the overlap, score has the form $S\left(q_i - q_g, r_j - r_h, i - g, j - h\right) = -\log(LR(q_i - q_g, r_j - r_h; i - g, j - h)) - \log(LR(i - g, j - h))$, where the first ratio is given by Theorem 3, and the second is given by Theorem 4.

Complexity. Computational complexity of both fit and overlap alignments is $O(\delta^2 mn)$. In practice δ is taken to be small ($\delta \leq 5$) and this both improves the alignments and reduces computation complexity to $O(mn)$, where m and n are numbers of fragments in the maps aligned.

Initialization:
Fit: $X(i,0) \leftarrow -\infty$, $i = 1, \ldots, m$; $X(0,j) \leftarrow 0$, $j = 0, \ldots, n$
Overlap: $X(i,0) \leftarrow 0$, $i = 1, \ldots, m$; $X(0,j) \leftarrow 0$, $j = 0, \ldots, n$

Recursion:
for $i \leftarrow 1$ **to** m **do**
    for $j \leftarrow 1$ **to** n **do**
        $y \leftarrow -\infty$;
        for $g \leftarrow max(0, i - \delta)$ **to** $i - 1$ **do**
            for $h \leftarrow max(0, j - \delta)$ **to** $j - 1$ **do**
                $\mid$ $y \leftarrow max\{y, \ X(g,h) + S(q_i - q_g, r_j - r_h, i - g, j - h)\}$;
            end
        end
        $X(i,j) \leftarrow y$;
    end
end

Algorithm 1: Dynamic programming for calculation of alignments of optical maps

6   Results

We have tested our likelihood score on a variety of organisms and optical data sets. This new scoring is highly capable of producing accurate alignments. Figure 1 illustrates one of such alignments along with the scoring pattern. Overall, spurious alignments have low scores, usually close o less than zero, while

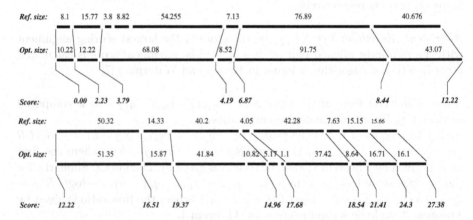

Fig. 1. Representation of alignments. Alignment of part of optical map from Y. pestis against its reference genomic region. Two adjacent pieces of alignment are displayed along with the map fragment sizes and alignment scores. Gaps between fragments represent cut sites, thin lines connect matching sites. Scores are calculated directionally (from left to right), and are displayed at the matched sites. Score for each matching region is obtained by taking the difference of the scores at the two flanking matching sites. Fragment sizes are displayed in Kb

Table 1. Comparative fitting for two scoring schemes (based on 1000 synthetically generated maps from a 40 Mb region of human chromosome 13). Maps are being fitted into the reference. False positives and false negatives rates are calculated for the maps fitted into correct reference location using both methods. False positives represent the percentage of false cuts incorrectly matched to the reference, false negatives represent percentages of unmatched true restriction sites. The corresponding alignment parameters are chosen to be $\mu = 0.2$, $\nu = 1$, $\lambda = 2$ for M.2 and $\sigma^2 = 0.306$ $\lambda = 32$ $p = 0.8$ $\zeta = 0.005$ $\eta^2 = 5$ for M.1

	% of maps fitted into correct reference locations	% of false positive cut sites	% of false negative cut sites
M.1	95.1	7.9	10.1
M.2	92.6	10.5	15.3

correct ones usually produce very high scores, usually in the neighborhood of 100 depending on the map size. Naturally, longer maps are easier to align correctly because they contain more information in the form of fragment sizes.

Next we demonstrate the advantage of our new likelihood based alignment score over the alignment score for restriction maps proposed by Waterman et al [11]. Hereafter Method 1 (M.1) refers to our likelihood based score, and Method 2 (M.2) refers to the heuristic alignment score from [11]. Table 1 demonstrates some comparative results between two alignment types. We have synthetically generated 1000 maps from a region of human chromosome 13. For generation of maps we have used our model assumptions for sizing errors, missing cuts and false cuts. This made it possible to infer exact positions from where maps were generated as well as the exact positions of all true cut sites together with the information of which cuts are spurious. We have then aligned these maps to the reference genomic restriction map of the human chromosome 13. By comparing original and fit positions we were able to infer whether the fits were correct. For correct fits we have collected information about correctly and incorrectly aligned sites and summarized them in Table 1.

As it becomes evident, both alignment methods produce similar results for synthetic maps with our new method (M.1) having a slight advantage. However, the most important feature of the score is its discriminative power which turns out to be significantly different for the two scoring schemes. This is illustrated in Table 2. For each synthetic map that fits into correct reference location (using both M.1 and M.2), we have collected its 20 best alignment scores (10 in each orientation, including the optimal) by declumping dependent alignments. The idea is that if the discriminative power of the score is high, no other independent alignments will have score in the neighborhood of the optimal score (corresponding to the correct fit). To measure this, we counted the number of other alignment scores (from the 19) within k standard deviations of optimal score. Standard deviation of the optimal score was estimated based on optimal scores from 10000 random maps with the same number of fragments as the map of interest. The results for both methods are summarized in Table 2.

Table 2. Optimal score discriminative power for two scoring schemes (based on 1000 artificially generated maps from a 40 Mb region of human chromosome 13). Scoring parameters for both M.1 and M.2 are taken to be the same as in Table 1. Of the 20 top scores of independent alignments (10 in each orientation) the table gives the average number of scores (other than optimal) within k standard deviations ($k = 1, 2, \ldots, 9$) of the optimal score. Hence table entries never exceed 19

stdev num	1 SD	2 SD	3 SD	4 SD	5 SD	6 SD	7 SD	8 SD	9 SD
av. number of scores (M.1)	0.006	0.036	0.17	0.494	1.413	3.206	5.983	8.89	11.82
av. number of scores (M.2)	0.281	2.157	6.275	11.471	15.334	17.546	18.566	18.91	18.99

As it now becomes apparent, likelihood ratio based score (M.1) possesses significantly more discriminative power since even within 3 standard deviations of optimal score on average 0.17 non-optimal scores are observed, while the same number for M.2 is 6.28 ($\approx$ 37 times more).

Not surprisingly, our new alignment score M.1 when compared to M.2 shows even more discriminative power for real optical maps when compared to the same result based on synthetic maps. Hence, our alignment score produces significantly more confident alignments for real optical data compared to the old score (M.2).

We used a Kim strain of *Y. pestis* commonly known as plague (Black Death) to study the performance of the alignment score. The data set was comprised of 251 optical maps. The reference restriction map was inferred from NCBI sequence database. This bacterial genome consists of 4.6 Mb and 267 restriction sites specific to XhoI endonuclease that was used for digestion of its DNA during collection of optical maps.

For real optical maps, we have calculated alignments using both M.1 and M.2. Furthermore, p-value threshold of 0.002 was used to select confident fits. As a result, M.1 allowed 174 maps to pass the p-value threshold, while only 119 maps passed the same threshold using M.2. Notably, our alignment score allowed us

Table 3. Comparative fitting for two scoring schemes (based on optical maps from a 4.6 Mb genome of *Y. pestis*). M.1 refers to likelihood ratio scoring scheme ($\sigma^2 = 0.306$ $\lambda = 17$ $p = 0.8$ $\zeta = 0.005$ $\eta^2 = 5$) and M.2 refers to heuristic score as in Waterman et al (1984) [11] ($\mu = 0.2$ $\lambda = 2$ $\nu = 1$). Optimal score corresponds to the top alignment score for the given map. Optimal p-value corresponds to the p-value of the optimal score. Total of 251 different maps were fitted. For M.1 174 passed 0.002 p-value threshold, for M.2 119 passed 0.002 p-value threshold

map id	0499	1661	0387	0344	0517	0920	0944	1634	1659
fragments	70	34	77	27	85	37	71	23	79
M. 1: optimal	80.56	45.24	62.32	31.79	113.61	51.11	47.54	33.60	90.24
M. 1: optimal p-value	$< 10^{-4}$	$< 10^{-4}$	$< 10^{-4}$	0.0018	$< 10^{-4}$	$< 10^{-4}$	0.00030.00065		$< 10^{-4}$
M. 2: optimal score	-31.76	-13.13	-35.59	-1.53	-39.50	-15.86	-51.22	-1.25	-34.68
M. 2: optimal p-value	0.0005	0.083	0.0020.0074		0.00055	0.181	0.1603	0.0060.00135	

to confidently align 47% more maps than M.2, which is significantly larger than the result for synthetic maps. Selected comparisons are shown in the Table 3.

P-values were computed based on simulation. For an optical map of n fragments, 10000 random maps (each composed of n fragments) are generated by sampling from the fragment size distribution for optical maps in the current mapping project. Then each random map is fitted into the reference and its optimal score is stored. For the optimal alignment score of an optical map, the p-value is based on the number of optimal scores in simulation (out of 10000) exceeding the given alignment score. It took 10000 random maps in order to achieve the p-value accuracy of 10^{-4} in Table 3. Thus, we only need to show results for optical maps with one p-value larger than 10^{-4} and at least one significant score ($p-value < 0.002$). 9 maps satisfying these conditions were selected to illustrate the significance of the results.

References

1. Bateman, H., Erdelyi, A.: Higher Transcendental Functions Mc Graw-Hill Book Company. (1953) 82
2. Abramowitz, M., Stegun, I. A.: Handbook of Mathematical Functions with Formulas, Graphs, and Mathematical Tables, 9th printing. New York: Dover (1972) 374
3. Waterman, M. S.: Introduction to Computational Biology. Chapman & Hall, 1995 201–202
4. Ananthraman, T., Mishra, B., Schwartz, D. C.: Genomics via Optical Mapping II: Ordered Restriction Maps. Journal of Computational Biology **4(2)** (1997) 91–118
5. Grimmett, G., Stirzaker D.: Probability and Random Processes. Oxford University Press (1982)
6. Myers, E.W.: Whole Genome DNA Sequencing Computing in Science and Engineering. (1999) 33–43
7. Huang, X., Waterman, M. S.: Dynamic Programming Algorithms for Restriction Map Comparison. CABIOS **8(5)** (1992) 511–520
8. Antoniotti, M., Ananthraman, T., Paxia, S., Mishra, B.: Genomics via Optical Mapping IV: Sequence Validation via Optical Map Matching. NYU-TR2000-811 (2001)
9. Ananthraman, T., Mishra, B.: A Probabilistic Analysis of False Positives in Optical Map Alignment and Validation. Algorithms in Bioinformatics, First International Workshop, WABI 2001 Proceedings (2001)
10. Ananthraman, T., Schwartz, D. C., Mishra, B.: Genomics via Optical Mapping III: Contiging Genomic DNA and Variations. Proceedings 1th Intl Cnf. on Intelligent Systems for Molecular Biology. (1999)
11. Waterman, M. S., Smith, T. F., Katcher H.: Algorithms for Restriction Map Comparisons. Nucleic Acids Research **12** (1984) 237–242
12. Lim, A., Dimalanta, E. T., Potamousis, K. D., Yen, G., Apodoca, J., Tao, C., Lin, J., Qi, R., Skiadas, J., Ramanathan, A., Perna, N. T., Plunkett, G. 3rd, Burland, V., Mau, B., Hackett, J., Blattner, F. R., Ananthraman, T. S., Mishra, B., Schwartz, D. C.: hotgun optical maps of the whole Escherichia coli O157:H7 genome. Genome Res. **11(9)** (2001) 1584–1593

13. Schwartz, D. C., Li, X., Hernandez, L. I., Ramnarain, S., Huff, E. J., Wang, Y. K.: Ordered restriction maps of Saccharomyces cerevisiae chromosomes constructed by optical mapping. Science **262(5130)** (1993) 110–104
14. Myers, E. W., Huang, X.: An $O(N^2 log N)$ restriction map comparison and search algorithm. Bull. Math. Biol. **54(4)**(1992) 599–618
15. Huang, X., Madan, A.: CAP3: A DNA Sequence Assembly Program. **9(9)** (1999) 868-877
16. Huang, X., Miller, W.: A time-efficient, linear-space local similarity algorithm. **12** (1991) 337–357
17. Dimalanta, E. T., Lim, A., Runnheim, R., Lamers, C., Churas, C., Forrest, D. K., dePablo, J. J., Graham, M. D., Coppersmith, S. N., Schwartz, D. C.: A microfluidic system for large DNA molecule arrays. Anal. Chem. **7** (2004) 5293–5301
18. Zhou, S., Kile, A., Bechner, M., Place, M., Kvikstad, E., Deng, W., Wei, J., Severin, J., Runnheim, R., Churas, C., Forrest, D., Dimalanta, E., Lamers, C., Burland, V., Blattner, F., Schwartz, D.: A single molecule approach to bacterial genomic comparisons via Optical Mapping. J. Bacteriol **in press** (2004)
19. Smith, T. F., Waterman, M. S., Comparison of biosequences. Adv. Appl. Math. **2** (1981) 482–489

Engineering Gene Regulatory Networks: A Reductionist Approach to Systems Biology

James J. Collins

Center for BioDynamics and Department of Biomedical Engineering,
Boston University, 44 Cummington Street, Boston, MA 02215, USA
jcollins@bu.edu
http://www.bu.edu/abl

Many fundamental cellular processes are governed by genetic programs which employ protein-DNA interactions in regulating function. Owing to recent technological advances, it is now possible to design synthetic gene regulatory networks, and the stage is set for the notion of engineered cellular control at the DNA level. Theoretically, the biochemistry of the feedback loops associated with protein-DNA interactions often leads to nonlinear equations, and the tools of nonlinear analysis become invaluable. In this talk, we describe how techniques from nonlinear dynamics and molecular biology can be utilized to model, design and construct synthetic gene regulatory networks. We present examples in which we integrate the development of a theoretical model with the construction of an experimental system. We also discuss the implications of synthetic gene networks for gene therapy, biotechnology, biocomputing and nanotechnology. In particular, we describe how engineered gene networks can be used to reverse-engineer naturally occurring gene regulatory networks. Such methods may prove useful in identifying and validating specific drug targets and in deconvolving the effects of chemical compounds.

S. Miyano et al. (Eds.): RECOMB 2005, LNBI 3500, p. 505, 2005.
© Springer-Verlag Berlin Heidelberg 2005

Modeling the Combinatorial Functions of Multiple Transcription Factors

Chen-Hsiang Yeang and Tommi Jaakkola

Computer Science and Artificial Intelligence Laboratory,
Massachusetts Institute of Technology, Cambridge, MA 02139, USA
chyeang@csail.mit.edu

Abstract. A considerable fraction of yeast gene promoters are bound
by multiple transcription factors. To study the combinatorial interactions
of multiple transcription factors is thus important in understanding gene
regulation. In this paper, we propose a computational method to identify
the co-regulated gene groups and regulatory programs of multiple transcription factors from protein-DNA binding and gene expression data.
The key concept is to characterize a regulatory program in terms of two
properties of individual transcription factors: the function of a regulator
as an activator or a repressor, and its direction of effectiveness as necessary or sufficient. We apply a greedy algorithm to find the regulatory
models which optimally fit the data. Empirical analysis indicates the
inferred regulatory models agree with the known combinatorial interactions between regulators and are robust against the settings of various
free parameters.

1   Introduction

The combinatorial interactions of multiple transcription factors play an essential role in transcriptional regulation. For instance, many genes are regulated by
protein complexes comprised of multiple transcription factors [1]. To model the
combinatorial interactions of transcription factors, it is necessary to relate the
states of activities of transcription factors to the expression levels of regulated
genes. Finding this relation – a regulatory program – between regulators and
regulated genes is a challenging problem since the number of possible regulatory
programs grows rapidly with the number of transcription factors involved. Simplification of possible regulatory programs is therefore important for modeling
the combinatorial interactions of multiple transcription factors.

In this paper, we present a computational method that identifies the regulatory programs of multiple transcription factors and the genes they regulate from
both protein-DNA binding and gene expression data. The results are regulatory
models, each contains a set of transcription factors, genes putatively regulated
by these factors, and the regulatory program specifying the relation between
regulators and regulated gene expressions. We simplify a regulatory program
by characterizing it in terms of the functions and directions of effectiveness of

S. Miyano et al. (Eds.): RECOMB 2005, LNBI 3500, pp. 506–521, 2005.

individual regulators. This characterization gives a simple interpretation of the mechanisms underlying a regulatory program and greatly reduces the model complexity.

Modeling the transcriptional regulation of multiple transcription factors has been addressed in a considerable number of previous works. Most Bayesian network models of gene expression analysis (e.g., [2, 3, 4]) focused only on the structure of a regulatory model and did not directly infer the regulatory program. Some authors considered the effects of single regulators separately and avoided identifying the combinatorial interactions of multiple regulators (e.g., [5]). Some works limited the scope to synergistic or complementary effects of regulator pairs, for example, [6] and [7]. Others attacked the combinatorial functions of multiple regulators with different computational models, such as Boolean networks [8], regression trees [9], and many others. However, since these models targeted only the functional relations of data, the resulting models can be difficult to interpret in terms of the underlying mechanisms. Another approach of modeling the circuitry of multiple regulators is to systematically generate different input states by perturbation and measure the response of regulated genes, for instance, [10]. This approach, though more reliable, is also expensive and time-consuming.

The rest of the paper is organized as follows. We will first introduce the hypotheses and concepts of our gene regulatory model and give it a mathematical definition. Following this introduction we will describe an algorithm to learn the models from binding and gene expression data. We then apply the algorithm to the CHIP-chip binding data and two large-scale gene expression datasets, and demonstrate the modeling results and their validations. Finally we will discuss the pros and cons of the method and directions of future extension.

2   Models of Transcription Regulation

2.1   Modeling Hypotheses and Concepts

We adopt several common hypotheses in the analysis of CHIP-chip and microarray data [2, 3, 11, 9]. First, given that a transcription factor binds to a specific promoter, the activity of the factor is modulated by the factor's mRNA abundance. Second, genes co-regulated by a set of transcription factors (i.e., genes appeared in the same module) can be predicted by the same regulatory program and mRNA levels of transcription factors. For computational convenience, we also add the following assumptions. We model the relative changes of mRNA levels with respect to a reference condition and quantize those changes into three states: up-regulation, down-regulation, no change.

The key idea of our model is to characterize a regulatory program in terms of two properties of individual transcription factors. First, a transcription factor possesses a consistent function as an activator or a repressor. This function is not inverted in the context of combinatorial control. Second, a transcription factor may take effect only if its expression changes in certain direction. We categorize the direction of effectiveness into four types. A regulator is necessary if decreasing

Table 1. Responses of regulated genes in each combinatorial category

	necessary	sufficient	both	neither
activator	$f \downarrow \Rightarrow g \downarrow$	$f \uparrow \Rightarrow g \uparrow$	$f \downarrow \Rightarrow g \downarrow, f \uparrow \Rightarrow g \uparrow$	g any value
repressor	$f \downarrow \Rightarrow g \uparrow$	$f \uparrow \Rightarrow g \downarrow$	$f \downarrow \Rightarrow g \uparrow, f \uparrow \Rightarrow g \downarrow$	g any value

its expression level leads to the responses opposite to its function. A regulator is sufficient if increasing its expression level leads to the responses consistent with its function. A regulator can be both necessary and sufficient or neither necessary nor sufficient. Unlike the function of a single regulator, we allow the direction of effectiveness of a transcription factor varies when it participates in different regulatory models. The predicted response of a regulatory program of a single regulator is uniquely determined by these two properties. Table 1 lists the predicted responses from different states of a single transcription factor.

A combinatorial function of multiple regulators gives predicted responses under each possible input state. By assuming the function and the direction of effectiveness of each regulator are preserved in all input states, we can construct the combinatorial function from the predicted response corresponding to each regulator. Briefly, each joint input state is the concatenation of the input states of single regulators. For each joint input state, the combinatorial function reports the consensus of predictions according to the input state and the two properties of each regulator. If contradiction occurs then the function reports an uncertain output. The rules of generating the output of the combinatorial function from predictions of individual regulators are described in Section 2.2.

The functional class generated by this characterization represents only a small subset of all possible combinatorial functions: the number of possible combinations of these two properties for n inputs is 8^n, whereas the number of all possible tri-state Boolean functions with n inputs is 3^{3^n}. With drastic reduction of the possible functions we obtain a more tractable class that is possible to estimate from limited data. While the number of possible functions is still exponential in n, we can enumerate the possibilities for small n.

Despite its simplification, characterization of a regulatory program with properties of single regulators still retains some combinatorial interactions between regulators. Some of these combinatorial effects have clear mechanistic interpretations. For example, if all regulators in a model are necessary, then they are likely to form a complex or cooperatively bind together on promoters. In contrast, if all regulators are sufficient, then they may independently act on promoters. In general, we can view a necessary regulator as essential for maintaining a basal transcription level under the reference condition, and a sufficient regulator as providing an additive enhancement or reduction of gene expression.

2.2   Definition of a Regulatory Model

We define a model of transcription regulation to have three components: a set of transcription factors, a set of genes controlled by these transcription factors, and a regulatory program specifying the relation between the expression data of regulators and regulated genes. We first define a deterministic regulatory program as a function which maps the mRNA state of transcription factors into the mRNA state of a "typical" response of regulated genes.

$$f : S^n \to S. \tag{1}$$

where $S = \{-1, 0, +1\}$ is the quantized state expression changes and n the input size. According to the module assumption, all regulated genes in a model are controlled by the same regulatory program.

The function of a single regulator is uniquely determined by the function and direction of effectiveness of the regulator, as shown in Table 1. Thus at each state of multiple regulators, we can predict the output response according to the input state of each regulator. We adopt the following rules to synthesize the predicted responses from single regulators. If the predicted responses are all +1s or 0s, then the output is +1. If the predicted responses are all -1s or 0s, then the output is -1. If the predicted responses contain both +1s and -1s, or are all 0s, then the output is 0. These rules simply report the consensus of predicted responses and output 0 if contradiction occurs. Notice we do not distinguish between the uncertain state and the state of an insignificant change under these rules. We can thus construct the combinatorial function f from Table 1 and the synthesis rules. An example of a deterministic combinatorial function of two necessary activators is shown in Table 2.

Table 2. The combinatorial function of two necessary activators

f_1	f_2	g
-1	-1	-1
-1	0	-1
-1	+1	-1
0	-1	-1
+1	-1	-1
o.w.	o.w.	0

The deterministic function is too rigid and does not consider the uncertainty of the regulatory program. To take uncertainty into account, we construct a probabilistic regulatory program as a conditional probability function:

$$P : S^n \times S \to [0, 1]. \tag{2}$$

The conditional probability is related to the deterministic function in the following way. Denote c_{ge} as the expression state of regulated gene g in experiment e, and c_{Re} as the expression state of regulator set R in experiment e. The conditional probability $P(c_{ge}|c_{Re}, f) \equiv P(c_{ge}|f(c_{Re}))$ depends on the regulated gene expression c_{ge} and the output of the deterministic function $f(c_{Re})$. The c_{ge} that agrees with $f(c_{Re})$ is assigned a high probability. However, when $f(c_{Re}) = 0$ each c_{ge} state is assigned an equal probability. Table 3 shows the conditional probability table, where α is a free parameter.

Table 3. The table of $P(c_{ge}|f(c_{Re}))$

$f(c_{Re})$	$P(c_{ge} = -1|f(c_{Re}))$	$P(c_{ge} = 0|f(c_{Re}))$	$P(c_{ge} = +1|f(c_{Re}))$
-1	$1 - \alpha$	α	0
0	$\frac{1}{3}$	$\frac{1}{3}$	$\frac{1}{3}$
+1	0	α	$1 - \alpha$

3 Identifying Regulatory Models

In this section we describe a method of identifying regulatory models from protein-DNA binding and gene expression data. We first define a scoring function (log likelihood function) of binding and expression data according to the model. Next, we adopt a greedy algorithm to identify the models which optimize the scoring function, and evaluate the significance of the inferred model.

3.1 Likelihood Function of a Regulatory Model

We define a log likelihood function of a regulatory model in terms of how well it fits binding and expression data. It contains two terms. The term corresponding to binding data is the log likelihood ratio between the regulatory model that each regulator binds to each regulated gene, versus the null model that the binding of each (protein,promoter) pair occurs with probability $\frac{1}{2}$. The term corresponding to expression data is the log likelihood ratio between the regulatory model that the expression states of the regulators and regulated genes conform with the regulatory program, versus the null model that there is no relation between the expression states of regulators and regulated genes. The joint scoring function is the weighted sum of these two terms.

We define the following notations for the log likelihood function of binding and expression data. Denote $M = (R, G, f)$ as a regulatory model, where R and G are regulators and regulated gene sets and f the (deterministic) regulatory program. For each $r \in R$ and $g \in G$, define b_{rg} as a binary variable indicating whether r binds to g. b_{rg} is not directly observed but through a noisy measurement outcome x_{rg} from binding data. Denote E as a collection of expression experiments. For each $r \in R$ and $e \in E$, define c_{re} as the expression change of regulator r in experiment e. c_{re} is linked with a noisy measurement outcome x_{re} from microarray data. For each $g \in G$ and $e \in E$, c_{ge} and x_{ge} are defined analogously. Furthermore, denote $\{b_{rg}\}$ as a state of all indicator variables $b_{rg} : r \in R, g \in G$. $\{c_{re}\}$ and $\{c_{ge}\}$ are defined analogously. Also denote c_{Re} as a state of all $c_{re} : r \in R$ in a specific experiment e.

The marginal likelihood function of binding data under a hypothesis H is

$$P(\{x_{rg}\}|H) = \sum_{\{b_{rg}\}} P(\{b_{rg}\}|H)P(\{x_{rg}\}|\{b_{rg}\}). \tag{3}$$

The conditional probability $P(x_{rg}|b_{rg})$ of each pair-wise interaction reflects the confidence of binding (for example, CHIP-chip) experiments. We use an asymptotic statistic and model selection criterion to calculate the ratio $\frac{P(x_{rg}|b_{rg}=1)}{P(x_{rg}|b_{rg}=0)}$ from the measurement p-value. Details are described in [12].

We are interested in two $P(\{b_{rg}\}|H)$ priors. The only $\{b_{rg}\}$ state consistent with the regulatory model M is each factor binds to each regulated gene. Denote this hypothesis of binding states as H_1:

$$H_1 : P(\{b_{rg}\}|H_1) = \prod_{r \in R, g \in G} \delta(b_{rg} = 1). \tag{4}$$

where $\delta(.)$ is the indicator function. In contrast, for a null model H_0 under which the regulators do not have any specific relation to the genes, the prior probability of $\{b_{rg}\}$ is given by

$$H_0 : P(\{b_{rg}\}|H_0) = \frac{1}{2^{|R||G|}}. \tag{5}$$

By applying both priors and the independence of each x_{rg}, the log likelihood ratio becomes:

$$L^b(R, G) = \log P(\{x_{rg}\}|H_1) - \log P(\{x_{rg}\}|H_0)$$
$$= |R||G| \log 2 + \sum_{(r,g)} [\log P(x_{rg}|b_{rg} = 1) - \log(P(x_{rg}|b_{rg}=1) + P(x_{rg}|b_{rg}=0))]. \tag{6}$$

The log likelihood ratio of expression data can be similarly constructed. The marginal likelihood function of expression data under a hypothesis H is

$$P(\{x_{re}\}, \{x_{ge}\}|H) = \sum_{\{c_{re}\}, \{c_{ge}\}} P(\{c_{re}\}, \{c_{ge}\}|H)P(\{x_{re}\}|\{c_{re}\})P(\{x_{ge}\}|\{c_{ge}\}). \tag{7}$$

Similar to binding data, the null hypothesis of expression data assigns a uniform probability to each possible expression state $\{c_{re}\}$ and $\{c_{ge}\}$:

$$H_0 : P(\{c_{re}\}\{c_{ge}\}|H_0) = \frac{1}{3^{|E|(|R|+|G|)}}. \tag{8}$$

The alternative model H_1 specifies the relation between c_{ge} and c_{Re} in each experiment e. It is specified by function f and Table 3. Each input state c_{Re} is assigned a uniform probability as in H_0.

$$H_1 : P(\{c_{re}\}\{c_{ge}\}|H_1) = \prod_{e \in E} [\frac{1}{3^{|R|}} \prod_{g \in G} P(c_{ge}|f(c_{Re}))]. \tag{9}$$

The conditional probabilities $P(\{x_{re}\}|\{c_{re}\})$ and $P(\{x_{ge}\}|\{c_{ge}\})$ are again dataset-specific and independent of the regulatory model. We will discuss the choice of error models in Section 4.

Combining equations 7, 8, 9, we evaluate the log likelihood ratio of expression data. Skipping intermediate steps,

$$
\begin{aligned}
L^e(R, G, f) &= \log P(\{x_{re}\}, \{x_{ge}\}|H_1) - \log P(\{x_{re}\}, \{x_{ge}\}|H_0) \\
&= -|E||R|\log 3 + \sum_{e \in E}[\log(\sum_{v \in \{-1,0,+1\}} P_v(e) \cdot \prod_{g \in G} \sum_{c_{ge}} P(c_{ge}|v)P(x_{ge}|c_{ge}))] \\
&+ |E|(|R| + |G|)\log 3 - \sum_{e \in E}[\sum_{r \in R}\log(P(x_{re}|c_{re} = +1) + P(x_{re}|c_{re} = -1) \\
&+ P(x_{re}|c_{re} = 0)) + \sum_{g \in G}\log(P(x_{ge}|c_{ge} = +1) + P(x_{ge}|c_{ge} = -1) + P(x_{ge}|c_{ge} = 0))].
\end{aligned}
$$

$$(10)$$

where $P_v(e)$ denotes the probability of the regulator states in experiment e which generate deterministic output v:

$$
P_v(e) = \sum_{\{c_{Re}\}} \delta(f(c_{Re}) = v) \cdot P(x_{Re}|c_{Re}).
$$

$$(11)$$

We define the joint log likelihood ratio as the weighted sum of the log likelihood functions of binding and expression data:

$$
L(R, G, f) = L^b(R, G) + \lambda L^e(R, G, f).
$$

$$(12)$$

λ is a free parameter specifying the relative importance of expression data with respect to binding data. Since the number of expression experiments far exceeds the number of binding experiments, we have to degrade the importance of expression data in order to make binding data relevant.

3.2   Algorithm of Identifying Regulatory Models

We want to identify the regulatory models which optimize the joint scoring function in equation 12. This problem is difficult due to the enormous number of combinations of regulators, regulated genes and regulatory programs. We use a greedy algorithm which incrementally incorporates regulated genes and identifies the optimal regulatory program. The key steps in the algorithm are as follows.

1. Find a collection of regulator sets which co-bind to a certain number of genes according to the CHIP-chip data. The thresholds of determining significant binding events (the p-value threshold of binding data) and the number of co-bound genes are free parameters. We set $p \leq 0.005$ and regulators co-bind to ≥ 10 genes. Furthermore, we only consider the sets of ≤ 3 regulators.
2. For each candidate regulator set, identify the optimal regulated genes and regulatory programs. We are able to exhaust all possible regulatory programs due to the simplifications discussed earlier. For each regulatory program, we incrementally add genes into the regulated set, such that the log likelihood score is maximized. Since equation 12 increases with the number of regulated genes in the model, we have to specify a criterion for stopping adding genes

in the set. We will describe a p-value of calculation adding a new gene in Section 3.3. We allow each gene to be assigned to multiple regulatory models. We then compare the scores of regulatory programs (each has a different gene set). Because the log likelihood score grows with the number of genes, we compare the scores of fixed sized gene sets by choosing top n (n is the fixed size) genes according to the order of adding genes. The fixed size is the size of the smallest gene set among all regulatory programs. The result of step 2 is a regulatory program and a regulated gene set for each regulator set.

3. Some of the regulatory programs may be spurious or do not have functional roles. We evaluate the p-value of a regulatory program log likelihood score by using a permutation test. Details will be discussed in Section 3.3.

4. Due to insufficient data there are many regulatory programs which fit the data equally or nearly equally well. Thus reporting one regulatory program may not be very informative. We report the direction of effectiveness for each regulator which is the consensus among the optimal regulatory programs. We also evaluate the p-value of each reported direction of effectiveness. Details will be discussed in Section 3.3.

Step 2 has to be elaborated. Each regulatory program induces a different set of regulated genes. Because the log likelihood score in equation 12 grows with the number of regulated genes, the regulatory program with the largest set of regulated genes will always be chosen if we maximize the joint log likelihood score. To remove the effect of different regulated gene set sizes, we fix the size of regulated gene sets in the following way. Recall each gene is incorporated in the model in a greedy fashion, so the first n genes of a regulatory program are the top n genes which best conform with the regulatory program. We discard the regulatory programs with small regulated gene sets (< 5 genes) and identify the minimum size among the remaining regulated gene sets. We then compare the log likelihood scores of regulatory programs on the fixed-sized regulated gene sets. This procedure is a tentative solution to alleviate the effect of gene set size on the log likelihood score. In the long run a more principled way of normalizing equation 12 in terms of regulated gene set size is needed.

3.3   Evaluating the Significance of Regulatory Models

We have used three significant measures (p-values) in the algorithm procedures. The first p-value evaluates the significance of adding a new gene in the regulated gene set. This p-value is calculated by comparing the increment of the log likelihood score generated from empirical data to the increment from random expression data. We consider a randomization scenario that $P(x_{ge}|c_{ge} = 0), P(x_{ge}|c_{ge} = \pm 1)$ of the newly added gene are uniformly sampled from the simplex $P(x_{ge}|c_{ge} = 0) + P(x_{ge}|c_{ge} = -1) + P(x_{ge}|c_{ge} = +1) = 1$. Rather than random samplings, the p-value under this scenario can be analytically approximated. Details about the approximation are described in the Supplementary Webpage.

The second p-value evaluates the significance of a specific regulatory model. It is calculated from the following permutation test procedure. The expression data of regulated genes are randomly permuted (over genes and experiments). The optimal regulatory program and its log likelihood score from each permuted data are calculated, and the p-value is the fraction of optimal log likelihood scores from random data that exceed the empirical score. Details about the procedure are reported in the Supplementary Webpage.

The third p-value calculates the significance of the combinatorial property of a regulator. It is calculated according to the gap of log likelihood scores between the best model where this property holds and the best model where this property does not hold. For example, to evaluate the significance of "r_1 is a necessary activator", we find the optimal model M_1 among the models where r_1 is a necessary activator and the optimal model M_0 among the models where r_1 is not a necessary activator. We compare the empirical gap score with the gap scores obtained by randomly permuting gene expression data. Notice the gap score of each permuted data is obtained by re-optimizing the regulatory models to fit the permuted data. The p-value is the fraction of the random gap scores exceeding the empirical gap. Details about the procedure also can be seen in the Supplementary Webpage.

4   Empirical Analysis

We applied the algorithm of identifying regulatory models to the protein-DNA interaction data of 106 transcription factors [11] and two sets of large-scale gene expression data: Rosetta Compendium data of gene knock-outs [13] and stress response gene expression data published by Gasch et al. [14]. Rosetta data contains the log ratios and p-values of steady-state measurements, whereas Gasch data provides log ratios of time-course measurements. For simplicity we fix the regulatory functions (activators or repressors) of single regulators according to previous studies.

The conditional probabilities $P(\{x_{rg}\}|\{b_{rg}\})$ of binding data and $P(\{x_{re}\}|\{c_{re}\})$ and $P(\{x_{ge}\}|\{c_{ge}\})$ of Rosetta gene expression data were evaluated using the approximation described in [12]. The conditional probabilities $P(\{x_{re}\}|\{c_{re}\})$ and $P(\{x_{ge}\}|\{c_{ge}\})$ of the Gasch data were evaluated from Gaussian and exponential distributions of the time-course responses of perturbations. Details are described in the Appendix.

We summarize and analyze the inferred models in the following aspects. We first visualize the regulatory models inferred from two expression datasets and discuss their inferred combinatorial properties. We then validate the inferred models with gene function ontology, literature survey, and sensitivity analysis.

4.1   Models Inferred from Rosetta and Gasch Data

Figure 1 summarizes the information about regulatory models inferred from Rosetta and Gasch data. We only consider the regulatory models with up to three regulators. We represent a regulatory model as a bi-partite graph between regulators (circles) and a regulated gene set (a square). The color of a regulator

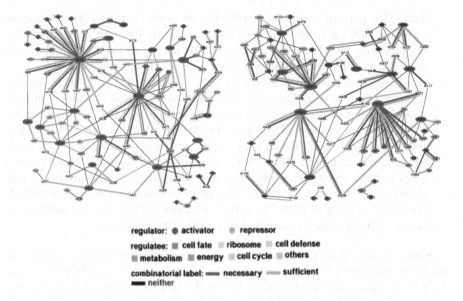

regulator: ● activator ○ repressor
regulatee: ▦ cell fate ▦ ribosome ▦ cell defense
▦ metabolism ▦ energy ▦ cell cycle ▦ others
combinatorial label: ━━ necessary ━━ sufficient
━━ neither

Fig. 1. Models inferred from Rosetta (left) and Gasch (right) data

indicates its regulatory function as an activator (red) or a repressor (green). The color of a regulated gene set indicates the MIPS functional categories enriched in the regulated gene set ($p \leq 0.06$ according to hyper-geometric test with Bonferroni correction). The color of an edge indicates the direction of effectiveness of a regulator in a model: red for necessary, green for sufficient, and black for neither. Two edges can exist between two nodes since a regulator can be both necessary and sufficient. The width of an edge indicates the confidence about about necessity or sufficiency as described in Section 4. We use the visualization software Cytoscape (www.cytoscape.org) to draw the graphs.

We found the combinatorial properties of many inferred regulatory models are consistent with the knowledge about the combinatorial interactions of these transcription factors. We summarize these interactions into three categories and draw a number of illustrative examples for each category.

– Each regulator is necessary for a regulated gene set. This pattern appears in regulator pairs such as (Ino2,Ino4), (Swi4,Swi6), (Swi6,Mbp1), (Fkh1,Fkh2) in Rosetta models. These regulator pairs are known to be components of protein complexes for transcriptional activation. Ino2-Ino4 complex regulates genes involved in phospholipid synthesis [15]. Protein complexes Swi6-Swi4, Swi6-Mbp1 and Fkh1-Fkh2 activate genes expressed during G1/S, S/G2 or G2/M phases of yeast cell cycle [16].
– Each regulator is sufficient for a regulated gene set. This pattern is common for stress response regulators, for example, (Msn4,Yap1), (Msn2,Yap1), (Msn2,Hsf1) pairs in Rosetta data and (Msn4,Hsf1), (Msn4,Yap1) in Gasch

data. This pattern is consistent with the property that each stress response regulator either activates the gene under a slightly different stress condition (for example, Hsf1 for heat shock and Yap1 for hyperoxia) or contributes in an additive or redundant fashion (for example, Msn2 and Msn4) [14].

– Some regulators are both necessary and sufficient, and the others are not strongly effective in either direction. Examples in Rosetta models include several small modules co-regulated by Gcn4 and one of the following regulators involved in amino acid synthesis: Leu3, Cbf1, Abf1, and several ribosome gene sets regulated by Rap1, Fhl1 and several other factors in Gasch models. In these examples, there exist some "master regulators" which control genes in both directions, while other regulators are not correlated with regulated genes at expression levels. This property does not necessarily exclude the functional role of these "inactive" regulators. They may be possible cofactors which regulate transcription via other mechanisms.

Since our regulatory models are based on simplifying assumptions, many true combinatorial interactions of regulators are not retrieved. It is difficult to assess the false negatives of the algorithm due to the lack of the complete knowledge about combinatorial gene regulation. Instead, we draw several illustrative examples from known combinatorial interactions of yeast genes.

– The well-known interaction of Gal4-Gal80 complex on galactose metabolic genes does not appear in Figure 1. The Rosetta module regulated by Gal4 (m473) is not enriched with galactose metabolic genes, and Gal80 does not appear in Figure 1. This is because the expression level of Gal4 is low even under active state [3]. Hence its regulatory function on galactose metabolic genes cannot be revealed by expression data alone. Although Gal80 expression level is known to modulate in certain datasets (e.g., [3]), it does not vary significantly in both Rosetta and Gasch data.

– The combinatorial interaction of Ste12 and Dig1 on pheromone response genes is only partially retrieved. Dig1 inhibits the phosphorylation of Ste12 [17], hence the inhibitory function of Dig1 is valid only when Ste12 is present. This combinatorial function cannot be captured by our models since the effectiveness of a regulator depends on the state of other regulators.

– Sok2 is known to be both activator and repressor for different genes [18]. We assign it as an repressor since it represses more genes. However, this assignment also excludes the regulatory models where Sok2 is an activator.

4.2   Validation of Inferred Models

In addition to the qualitative properties described in Section 4.1, we performed three quantitative validations on the inferred models. First, we investigated the enrichment of functional categories in the regulated gene set according to Munich Information Center for Protein Sequences (MIPS) database (http://mips.gsf.de/) in the regulated gene sets. Second, we checked from previous works whether regulators participating in the same model were known to have functional interactions. Third, we demonstrated that the inferred models were robust against the variation of free parameter values.

For each regulatory model, we evaluated the hyper-geometric p-values of the enrichment of MIPS categories with Bonferroni correction. We considered the models with significant log likelihood values (permutation p-value ≤ 0.02 for Rosetta models and p-value ≤ 0.001 for Gasch models, including the models of single regulators). Overall, about half of the inferred models are enriched with at least one MIPS category ($p \leq 0.06$): 46% of the Rosetta models (51 out of 110) and 45% of the Gasch models (65 out of 144) are enriched. Due to the incompleteness of the MIPS database and the conservative estimation of Bonferroni correction, more inferred models are expected to be involved in specific cellular processes.

We also searched PubMed and Incyte Yeast Proteome Databases (http://www.incyte.com/login.html) to check whether regulators participating in the same model were known to jointly control one or multiple genes. More than two thirds of the regulator sets in the significant models were verified in previous works: 60% of the significant Rosetta models with multiple regulators (46 out of 77) and 67% of the significant Gasch models with multiple regulators (46 out of 69) contain regulators whose interactions were reported in previous works. The complete list of all regulatory models and their validations are reported in the Supplementary Webpage.

We further demonstrated the inferred models were robust against the variations of three free parameters: λ appeared in the joint log likelihood function (equation 12) is the relative weight between expression and binding data, α in Table 3 relates the the prediction of a regulatory program to the hidden states of expression changes, p^{stop} in the greedy algorithm specifies the stopping criterion of the p-values of adding genes (Section 3.2). The default setting of these param-

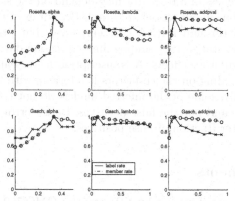

Top: Rosetta data. Bottom: Gasch data.
Solid line: overlap of combinatorial labels.
Dash line: overlap of regulated gene sets.

Fig. 2. Robustness tests on parameters

eters is $\lambda = 0.1, \alpha = \frac{1}{3}, p^{stop} = 0.1$. We performed robustness tests by varying each parameter while fixing the other two as the default values. Inferred models generated from the new parameter settings were compared to the default models in two aspects. First, we calculated the average overlap rate of regulated gene sets (with respect to the default models) over all models. Second, we counted the fraction of new models which had identical inferred directions of effectiveness to the default models. Figure 2 shows the sensitivity of parameters in Rosetta and Gasch models. Both sensitivity measures are very robust against each parameter in each dataset except α on Rosetta data. For example, when varying λ from 0.01 to 0.9, the average overlap rate of Gasch models ranges between 90% and 100% and more than 85% of inferred models agree on directions of effectiveness. In contrast, models inferred from Rosetta data are sensitive to α: the average overlap rate drops to 50% when α varies from $\frac{1}{3}$ to 0.1.

5   Discussion

We have described a simple computational approach to capture combinatorial effects of multiple transcription factors in transcription control. We identify regulatory models – including subsets of regulators and genes together with a regulatory program – from binding and expression data. We define regulatory programs with multiple regulators according to two properties of single transcription factors: 1) the function of a regulator and 2) its direction of effectiveness. The inferred models agree substantially with known functions and interactions. Moreover, the inferred models are robust against specific parameter values.

There are, however, many unresolved issues. Most combinatorial functions cannot be reduced to the properties of individual regulators. For example, the direction of effectiveness of a regulator may depend on the state of other regulators. The assumptions in our model are simplistic. For example, some regulators are not modulated through mRNA (protein) levels but primarily by altering protein modification states [19]. Binding and expression data alone are unlikely to capture such regulatory effects. A transcription factor can be both activator and repressor, depending on the co-factors it interacts with and the sets of regulated genes. Finally, some of the inferred models do not correspond to known biological functions and may be false positives. Better error models are needed to weed out a greater fraction of false positives.

Acknowledgements

This work was supported in part by NIH grant(s) GM68762 and GM69676. We thank Julia Zeitlinger and Ernst Fraenkel from MIT Whitehead Institute, John Barnett, Georg Gerber, Karen Sachs, Jason Rennie, David Gifford from MIT Computer Science and Artificial Intelligence Laboratory for helpful commments and discussions.

References

1. McNabb, D. et al. Cloning of yeast HAP5: a novel subunit of a heterotrimeric complex required for CCAAT binding. Genes Development. **9(1)** (1995) 47-58

2. Friedman, N. et al. Using Bayesian networks to analyze expression data. Journal of Computational Biology. **7** (2000) 601-620

3. Hartemink, A. et al. Using graphical models and genomic expression data to statistically validate models of genetic regulatory networks. Pacific Symposium of Biocomputing. (2001) 422-433

4. Segal, E. et al. From promoter sequence to expression: a probabilistic framework. Proceedings of the 6th International Conference on Research in Computational Molecular Biology. (2002) 263-272

5. Bar-Joseph, Z. et al. Computational discovery of gene modules and regulatory networks. Nature Biotechnology. **21** (2003) 1337-1342

6. Pilpel, Y. et al. Identifying regulatory networks by combinatorial analysis of promoter elements. Nature Genetics. **29** (2001) 153-159

7. Tong, A. et al. Global mapping of the yeast genetic interaction network. Science. **303** (2004) 808-813

8. Tanay, A. et al. Computational expansion of genetic networks. Bioinformatics. **17 Suppl 1** (2001) S270-S278

9. Segal, E. et al. Module networks: identifying regulatory modules and their condition-specific regulators from gene expression data. Nature Genetics. **34(2)** (2003) 166-176

10. Yuh, C.h. et al. Genomic cis-regulatory logic: experimental and computational analysis of a sea urchin gene. Science. **279** (1998) 1896-1902

11. Lee, T. et al. A transcriptional regulatory network map for *Saccharomyces cerevisiae*. Science. **298** (2002) 799-804

12. Yeang, C.H. et al. Physical network models. Journal of Computational Biology. **11(2-3)** (2004) 243-262

13. Hughes, T. et al. Functional discovery via a compendium of expression profiles. Cell. **102** (2000) 109-126

14. Gasch, A. et al. Genomic expression programs in the response of yeast cells to environmental changes. Molecular Biology of Cell. **11(12)** (2000) 4241-4257

15. Ambroziak, J. et al. INO2 and INO4 gene products, positive regulators of phospholipid biosynthesis in *Saccharomyces cerevisiae*, form a complex that binds to the INO1 promoter. Journal of Biological Chemistry. **269(21)** (1994) 15344-15349

16. Simon, I. et al. Serial regulation of transcriptional regulators in the yeast cell cycle. Cell. **106** (2001) 697-708

17. Bardwell, L. et al. Differential regulation of transcription: repression by unactivated mitogen-activated protein kinase Kss1 requires Dig1 and Dig2 proteins. PNAS. **95(26)** (1998) 15400-15405

18. Shenhar, G. et al. A positive regulator of mitosis, Sok2, functions as a negative regulator of meiosis in *Saccharomyces Cerevisiae*. Cellular Biology. **21(5)** (2001) 1603-1612

19. Lee, J. et al. YAP1 and SKN7 control two specialized oxidative stress response regulons in yeast. Journal of Biological Chemistry. **274(23)** (1999) 16040-16046

Appendix: Quantization of Time-Course Expression Data

In the Appendix we will show a method of evaluating the conditional probabilities $P(x_{re}|c_{re})$ and $P(x_{ge}|c_{ge})$ from time-course gene expression data. In the stress response dataset, x_{re} and x_{ge} are time-course measurements of expression responses under a stress condition. The goal is to convert x_{re} into conditional probabilities $P(x_{re}|c_{re} = +1), P(x_{re}|c_{re} = -1), P(x_{re}|c_{re} = 0)$.

Denote $y \in \{-1, 0, +1\}$ as the actual, quantized expression change of a gene under one experimental condition, and $x(t_1), \cdots, x(t_n)$ are its n time-course measurements. We relate the discrete state y to measurements $x(t_1), \cdots, x(t_n)$ with a two-level process. The discrete state y generates a continuous time-course expression profile $m(t_1), \cdots, m(t_n)$; and $x(t_1), \cdots, x(t_n)$ are noisy measurements of $m(t_1), \cdots, m(t_n)$. We model measurement errors $x(t_1) - m(t_1), \cdots, x(t_n) - m(t_n)$ as iid Gaussian random variables with zero mean and variance σ^2.

The actual expression profile $m(t_1), \cdots, m(t_n)$ is a zero vector given $y = 0$. Thus $P(x(t_1), \cdots, x(t_n)|y = 0)$ is the product of normal densities:

$$P(x(t_1), \cdots, x(t_n)|y = 0) = \left(\frac{1}{2\pi\sigma^2}\right)^{\frac{n}{2}} \prod_{i=1}^{n} e^{-\frac{x(t_i)^2}{2\sigma^2}}. \tag{13}$$

We model the prior probabilities $P(m(t_1), \cdots, m(t_n)|y = \pm 1)$ with an iid exponential distribution:

$$P(m(t_1), \cdots, m(t_n)|y = +1) = \prod_{t_i=1}^{n} P(m(t_i)|y = +1).$$
$$P(m(t_i)|y = +1) = \begin{cases} \gamma e^{-\gamma m(t_i)} & \text{if } m(t_i) \geq 0, \\ 0 & \text{otherwise.} \end{cases} \tag{14}$$

$P(m(t_1), \cdots, m(t_n)|y = +1)$ assigns a non-zero probability to each non-negative expression profile, and penalizes the expression profiles deviating from 0. $P(m(t_1), \cdots, m(t_n)|y = -1)$ is defined analogously. By marginalizing over $m(t_i)$, the conditional probability $P(x(t_1), \cdots, x(t_n)|y = +1)$ becomes

$$P(x(t_1), \cdots, x(t_n)|y = +1) = \prod_{i=1}^{n} \int_0^\infty P(m(t_i)|y = +1)P(x(t_i)|m(t_i))dm(t_i)$$
$$= \prod_{i=1}^{n} \gamma e^{(-\gamma x(t_i)+\frac{1}{2}\gamma^2\sigma^2)}(1 - \Phi(\frac{-(x(t_i)-\gamma\sigma^2)}{\sigma})). \tag{15}$$

where $\Phi(.)$ is the standard normal cumulative distribution function. Similarly,

$$P(x(t_1), \cdots, x(t_n)|y = -1) = \prod_{i=1}^{n} \int_{-\infty}^0 P(m(t_i)|y = -1)P(x(t_i)|m(t_i))dm(t_i)$$
$$= \prod_{i=1}^{n} \gamma e^{(\gamma x(t_i)+\frac{1}{2}\gamma^2\sigma^2)}(\Phi(\frac{-(x(t_i)+\gamma\sigma^2)}{\sigma})). \tag{16}$$

σ and γ are free parameters. In the empirical analysis we set $\sigma = \gamma = 0.5$ for they are close to the variance of the entire Gasch data.

Supplementary Webpage

Details about the calculations of p-values and inferred regulatory models can be found in the Supplementary Webpage http://www.csail.mit.edu/ tommi/suppl/ recomb05/.

Predicting Transcription Factor Binding Sites Using Structural Knowledge

Tommy Kaplan[1,2], Nir Friedman[1,*], and Hanah Margalit[2,*]

[1] School of Computer Science,
The Hebrew University, Jerusalem 91904, Israel
{tommy, nir}@cs.huji.ac.il
[2] Dept. of Molecular Genetics and Biotechnology,
Hadassah Medical School, The Hebrew University,
Jerusalem 91120, Israel
hanah@md.huji.ac.il

Abstract. Current approaches for identification and detection of transcription factor binding sites rely on an extensive set of known target genes. Here we describe a novel structure-based approach applicable to transcription factors with no prior binding data. Our approach combines sequence data and structural information to infer context-specific amino acid-nucleotide recognition preferences. These are used to predict binding sites for novel transcription factors from the same structural family. We apply our approach to the Cys_2His_2 Zinc Finger protein family, and show that the learned DNA-recognition preferences are compatible with various experimental results. To demonstrate the potential of our algorithm, we use the learned preferences to predict binding site models for novel proteins from the same family. These models are then used in genomic scans to find putative binding sites of the novel proteins.

1 Introduction

Specific binding of transcription factors to cis-regulatory elements is a crucial component of transcriptional regulation. Previous studies have used both experimental and computational approaches to determine the relationships between transcription factors and their targets. In particular, probabilistic models were employed to characterize the binding preferences of transcription factors, and to identify their putative sites in genomic sequences [24, 27]. This approach is useful when massive binding data are available, but it cannot be applied to proteins without extensive experimental binding studies. This difficulty is particularly emphasized in view of the genome projects, where new proteins are classified as DNA-binding according to their sequence, yet there is no information about the genes they regulate.

To address the challenge of profiling the binding sites of novel proteins, we propose a family-wise approach that builds on structural information and on

---

* Correspondence authors.

S. Miyano et al. (Eds.): RECOMB 2005, LNBI 3500, pp. 522–537, 2005.
© Springer-Verlag Berlin Heidelberg 2005

the known binding sites of other proteins from the same family. We use solved protein-DNA complexes [16, 18] to determine the exact architecture of interactions between nucleotides and amino acids at the DNA-binding domain. Although sharing the same structure, different proteins from a structural family obtain different binding specificities due to the presence of different residues at the DNA-binding positions. To predict their binding site motifs, we need to identify the residues at these key positions and understand their DNA-binding preferences.

In previous studies, we used the empirical frequencies of amino acid-nucleotide interactions [17–19] in solved protein-DNA complexes (from various structural families) to build a set of *DNA-recognition preferences*. This approach assumed that an amino acid has common nucleotide-binding preferences for all structural domains and at all binding positions. However, there are clear experimental indications that this assumption is not always valid: a particular amino acid may have different binding preferences depending on its positional context [9, 10, 14]. To estimate these context-specific DNA-recognition preferences, we need to determine the appropriate context of each residue, which may depend on its relative position and orientation with respect to the nucleotide. For this, we need to collect statistics about the DNA-binding preferences at this context. Naively, this can be achieved from a large ensemble of solved protein-DNA complexes from the same family. Unfortunately, sufficient data of this type are currently unavailable.

To overcome this obstacle, we propose to estimate context-specific DNA-recognition preferences from available sequence data using statistical estimation procedures. The input of our learning algorithm are pairs of transcription factors and their target DNA sequences [27]. We then recognize the specific residues and nucleotides that participate in protein-DNA interaction, and collect statistics about the DNA-binding preferences of residues at different contexts of the binding domain. These preferences can then be used to predict binding sites of other transcription factors from the same family, for which no known targets are available.

2   The Canonical Cys$_2$His$_2$ Zinc Finger DNA-Binding Family

We apply our approach to the Cys$_2$His$_2$ Zinc Finger DNA-binding family. This family is the largest known DNA-binding family in multi-cellular organisms [26] and has been studied extensively [28]. Many members of this family bind DNA targets according to a stringent binding model [13, 20], which maps the exact interactions between specific residues in the DNA-binding domain along with nucleotides at the DNA site (Figure 1). We term this the *canonical binding model*. In addition, Zinc Finger proteins whose DNA-binding domains are similar to those that bind through this model are termed canonical. According to the canonical binding model the residues involved in DNA binding are located at positions 6, 3, 2, and -1 relatively to the beginning of the α-helix in each fin-

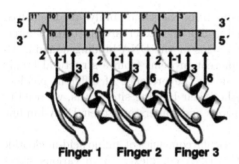

Fig. 1. The canonical Cys2His2 Zinc Finger DNA-binding model, based on solved protein-DNA complexes [13, 20]. A protein with three fingers is shown. In each finger, residues at positions 6, 3, 2, and -1 (relatively to the beginning of the α-helix) interact with adjacent nucleotides in the DNA molecule. (Figure adapted from Prof. Aaron Klug, with permission)

ger (Figure 1). Our goal is to extract position-specific amino acid-base binding preferences for each of these positions.

2.1 Sequences of Zinc Finger Proteins and Their DNA Binding Sites

To estimate the recognition preferences, we use the sequences of many Zinc Finger proteins together with their native DNA targets (extracted from the TRANS-FAC database [27]). To identify the canonical Cys2His2 Zinc Fingers based on their sequence, we trained a profile HMM [12] on 31 experimentally determined canonical domains [28], and used it to classify the remaining Cys2His2 Zinc Finger domains in TRANSFAC [27]. From the canonical ones, we only selected proteins with two to four properly spaced fingers. This resulted in 61 canonical proteins, and 455 protein-DNA pairs. We use these as our training data in subsequent steps. The total number of fingers in this dataset was 1320, and the total length of all binding sites was 9761bp (average length of 21bp per site).

2.2 Identification of DNA-Binding Residues

Next, we conceptually "thread" each protein-DNA pair onto the canonical binding model, to obtain an ensemble of residue-nucleotide interactions, from which we can estimate the recognition preferences. To do so, we must first identify the DNA-binding residues. We identify these positions using their relative positioning in the Cys2His2 conserved pattern: CX(2-4)CX(11-13)HX(3-5)H. Although theoretically there can be 20^4 different combinations of amino-acids at the four interacting positions, we found only 80 different combinations among the 1320 fingers in our database.

2.3 Identification of DNA Binding Sites

Now that we can identify the interacting residues, we face the problem of identifying the stretch of nucleotides they interact with. Unfortunately, the exact binding locations of the transcription factors are not pinpointed in TRANS-FAC, and thus we must employ statistical tools to infer them. In short, we wish to enumerate over all possible alignments of the DNA, and consider the likelihood of each DNA site given the interacting residues. For this we need to use the position-specific amino acid-base recognition preferences that we aim to estimate. We demonstrate in Section 3 how this is achieved, but first let us describe the probabilistic model of the DNA binding site, given that such preferences are available.

2.4   Probabilistic Model for Protein-DNA Interactions

We now consider how to model the DNA binding preferences of a protein given its amino acid sequence. In a probabilistic framework, we describe a model that assigns probabilities for any sequence of nucleotides at the binding site, given the residues they interact with. For a canonical Zinc Finger protein, we denote by $A = \{A_{i,p} : i = \{1, \ldots, k\}, p \in \{-1, 2, 3, 6\}\}$ the set of interacting residues in the different four positions of the k fingers (ordered from the N- to the C-terminus). Let $N_1, \ldots, N_L$ be a target DNA sequence. The conditional probability of an interaction with a DNA subsequence, starting from the j'th position in the DNA, is:

$$P(N_j, \ldots, N_{j+3k-1}|A) = \prod_{i=1}^{k} P_6(N_{j+3(i-1)}|A_{k+1-i,6})P_3(N_{j+3(i-1)+1}|A_{k+1-i,3})$$
$$P_{-1}(N_{j+3(i-1)+2}|A_{k+1-i,-1}) \tag{1}$$

where $P_p(N|A)$ is the conditional probability of nucleotide N given amino acid A at position p. These probabilities are the parameters of the model. For each of the four interacting positions, there should be a matrix of the conditional probabilities of the four nucleotides given all 20 residues. We call these matrices the *DNA-recognition preferences*.

We model the non-interacting nucleotides in the sequence using a background model P_{BG} (e.g. a uniform mononucleotide model), thus the probability of a sequence of length L that contains a binding site at position j is:

$$P(N|A, j) = P_{BG}(N_1, \ldots, N_{j-1})P(N_j, \ldots, N_{j+3k-1}|A)P_{BG}(N_{j+3k}, \ldots, N_L) \tag{2}$$

Since the exact positioning j of the binding site is not known, we enumerate over all possible values:

$$P(N|A) = \sum_j P(j)P(N|A, j) \tag{3}$$

where $P(j)$ is the prior probability of binding at position j. To handle truncated sites in the TRANSFAC database, we allow j to range between $-3/4 * 3k$ (when only the last quarter of the binding site is present) and $L - 3/4 * 3k + 1$ (only first quarter is present). We use a uniform prior over the j's, with the exception of missing nucleotides, which are penalized exponentially.

The model, as described above, does not account for interactions by the amino acids in positions 2 in each finger nor reverse complement binding. The latter requires only minimal adjustments, and is handled using an additional orientation variable. Handling the residues at position 2 is a bit trickier. According to the canonical binding model (Figure 1), the amino acid at position 2 interacts with the nucleotide that is complementary to the nucleotide interacting with position 6 of the previous finger. Thus, when we have a base pair interacting with two amino acids, we replace the term $P_6(N_{j+3(i-1)}|A_{k+1-i,6})$ by a term:

$$\alpha P_6(N_{j+3(i-1)}|A_{k+1-i,6}) + (1 - \alpha)P_2(N_{j+3(i-1)}|A_{k+2-i,2}) \qquad (4)$$

for $i > 1$, where α is a weighting coefficient that depends on the number of samples seen while estimating the recognition preferences at each position. Moreover, we add the term $P_2(N_{j+3(i-1)}|A_{k+2-i,2})$, for $i = k + 1$, to capture the last nucleotide, that is in interaction with position 2 of the first finger.

3   Learning DNA-Recognition Preferences from Sequence Data

We use the sequences of the proteins and their target DNA sites, and estimate four sets of position-specific DNA-recognition preferences that maximize the likelihood of the DNA given the binding proteins. As stated above, although the DNA sequences in our database were reported as bound by their corresponding proteins [27], the exact binding locations are not documented. Thus, we need to simultaneously identify the exact binding locations and optimize the parameters of DNA-recognition. For this, we use the iterative *Expectation Maximization* (EM) algorithm [11]. We start with some initial choice of DNA-recognition preferences (possible choices are discussed below). The algorithm proceeds iteratively, by carrying out these two steps, as illustrated in Figure 2.

E-step: For every protein-DNA pair, we compute the expected posterior probability that the binding begins in position j, using the current sets of DNA-recognition preferences θ_t.

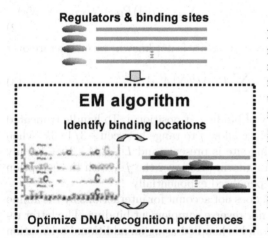

Fig. 2. Estimating the DNA-recognition preferences. The recognition preferences are estimated from unaligned pairs of transcription factors and their DNA targets from the TRANSFAC database [27] (shown on top). The EM algorithm is used to simultaneously assess the exact binding positions of each protein-DNA pair (bottom-right), and to estimate four sets of position-specific DNA-recognition preferences (bottom-left)

$$P(j|A, N) = \frac{P_{\theta_t}(N|A, j)}{\sum\limits_{j'} P_{\theta_t}(N|A, j')} \tag{5}$$

M-step: Next, we update the sets of DNA-recognition preferences θ_{t+1} to maximize the likelihood of the current binding positions j's for all protein-DNA pairs. This is based on the posterior probabilities that were computed in the E-step. Specifically, the conditional probability $P_p(n|a)$ of each nucleotide n given amino acid a at position p of the Zinc finger domain is estimated in θ_{t+1} using the expected number of interactions between n and a at p over all protein pairs, given the posterior probabilities j's.

The EM algorithm is proved to converge, since each of these two steps increases the likelihood of the data [11]. Obviously, this does not ensure that the final sets of DNA-recognition preferences θ_T are the *optimal* ones, due to sub-optimal local maxima of the likelihood function. This can be overcome by applying the EM procedure with multiple random starting points or by using prior knowledge starting points. An additional potential pitfall is over-fitting the recognition preferences of rare residues. To address this problem and ensure that the estimated parameters for rare amino-acids are close to uniform distribution (i.e., uninformative), we use a standard method of regularization by *pseudo-counts*. By applying a uniform *Dirichlet* prior, we add a constant (0.7 in the results below) to each amino acid-nucleotide count computed at the end of the E-step. We then perform a maximum *a-posteriori* estimation rather than maximum likelihood estimation.

We evaluate the robustness and convergence rate of the EM algorithm using a 10-fold cross validation procedure. In each round, we remove part of the data, train on the remaining pairs, and test the likelihood of the held-out protein-DNA pairs. We use this procedure to test various initialization options, including random starting points, and the general protein-DNA recognition preferences that were learned from all protein-DNA families [17]. Figure 3 shows the average likelihood per interaction on held-out test data, for various starting points. As shown, the EM algorithm performs best when initialized with the general recognition preferences, converging within few iterations. Similar likelihood results were ob-

Fig. 3. 10-fold cross-validation tests show the average likelihood per interaction (in bits) of held-out test data. We show the likelihood along 14 EM iterations using various starting points. The thick red line marks the likelihood obtained when starting from the general set of DNA-recognition preferences [18]. Random starting points are plotted using thin blue lines. The likelihood of the data according to the general set of DNA-recognition preferences [17] is shown with the horizontal dashed line

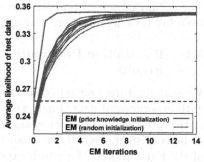

Fig. 4. Four sets of position-specific DNA-recognition preferences for canonical Cys$_2$His$_2$ Zinc Fingers. The estimated sets of DNA-recognition preferences for the DNA-binding residues at positions 6, 3, 2, and -1 of the Zinc Finger domain are displayed as sequence logos. At each position, the associated distribution of nucleotides is displayed for each amino acid. The total height of letters represents the information content (in bits) of the position, and the relative height of each letter represents its probability. Color intensity indicates the level of confidence for the preferences of a given amino acid at a certain position (where pale colors indicate low confidence positions due to a small number of occurrences of the residue at the specific position within the training data). Some of the DNA-binding preferences are general, regardless of the residue's position within the Zinc Finger domain (e.g. the tendency of lysine to bind guanine (G)), while others are position-dependent (e.g. the tendency of phenylalanine to bind cytosine only when in position 2)

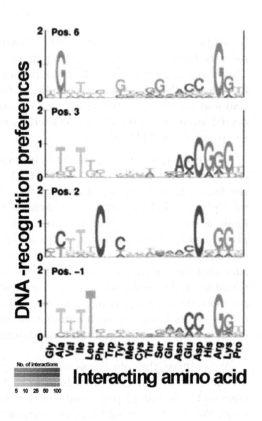

tained using random starting points, although the convergence rate is somewhat slower. Figure 3 also shows that the algorithm does not over-fit the training data, as this would have led to deteriorated performances over the held-out test data. The optimized sets of position-specific DNA-recognition preferences (estimated from full training data) are shown in Figure 4.

3.1 Recognition Preferences Are Consistent with Experimental Results

We evaluated the four sets of DNA-recognition preferences by comparing them with experimental data. First, we compared the derived preferences with qualitative preferences based on phage-display experiments [28] and found the two to be consistent. Second, we predicted binding site models for various variants of the Egr-1 protein, for which experimental binding data were available using DNA-microarrays [6]. We then used the predicted binding models to score each

Table 1. Correlation with experimentally measured binding affinities. We compare the ranking of the predicted binding sites to the experimental binding results of Bulyk et al. [6,7], using Spearman rank correlation. Oligonucleotides with low binding affinities (baseline noise value) were considered as non-viable and not taken into account

Variant of Egr-1	Spearman correlation coefficient	number of viable oligos	p-value
wt	0.73	15	<0.0025
LRHN	0.60	12	<0.025
REDV	0.83	15	<0.0005
RGPD	0.67	17	<0.0025

of the possible DNA binding sites that were tested in the experimental study. We found that our predictions were significantly correlated with the experimentally measured binding affinities (Table 1).

3.2 Predicting Binding Sites of Novel Proteins Within Genomic Sequences

We now turn to evaluate the ability of the estimated DNA-recognition preferences to identify the binding sites of novel proteins within genomic sequences.

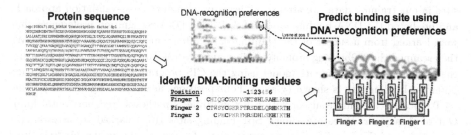

Fig. 5. Predicting the DNA binding site motifs of novel transcription factors. Given the sequence of a novel protein (shown on left), its DNA-binding domains (in blue) are identified using the Cys$_2$His$_2$ conserved pattern. The residues at the key positions (6, 3, 2 and -1) of each finger (marked in red in the middle-bottom panel) are then assigned onto the canonical binding model (on right), and the sets of position-specific DNA-recognition preferences (middle-top panel) are used to construct a probabilistic model of the DNA binding site (right). For example, position 1 in the binding site is determined by the binding preferences of the lysine (K) at the sixth position of the third finger (dotted red and black lines). We predict the nucleotide probabilities at this position using the appropriate recognition preferences (dotted black line). A web-server for predicting the binding sites of Cys$_2$His$_2$ Zinc Finger proteins can be accessed at http://compbio.cs.huji.ac.il/predictBS

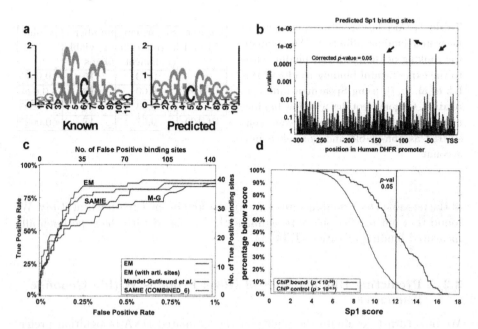

Fig. 6. Validation of the DNA-recognition preferences. (a) The predicted binding site model of human Sp1 protein is compared to its known site (matrix V$SP1_Q6 from TRANSFAC [27], based on 108 aligned binding sites). To prevent bias by known Sp1 sites in our training data, we applied a "leave-protein-out" cross-validation approach, and predicted the DNA-binding model of Sp1 using DNA-recognition preferences that were learned from a subset of the training data, after removing all Sp1 sites. (b) We then scanned the 300bp-long promoter of human dihydrofolate reductase (DHFR) using the predicted Sp1 binding site model using the CIS program [2]. The p-value of each potential binding site is shown (y-axis). Four positions achieved a significant p-value (≤ 0.05) after a Bonferroni correction for multiple hypotheses (red horizontal line). Out of these four, three are known Sp1 binding sites [15] (marked by arrows). (c) A summary of the *in silico* binding experiments for natural 43 known binding sites. Shown is the tradeoff between False Positive rate (x-axis) and True Positive rate (y-axis) as the threshold for putative binding sites is changed, using an ROC curve. For every threshold point, our sets of recognition preferences (marked EM) achieve higher accuracy than the preferences of Mandel-Gutfreund et al. [17] (marked M-G) and Benos et al. [4] (marked SAMIE). Interestingly, when the DNA-recognition preferences were estimated from training data that were expanded to include artificial sequences, from TRANSFAC we obtained inferior results (dotted red line). (d) Comparison of cumulative distributions of Sp1 logodd scores within genomic sequences of Sp1 targets and non-targets determined by unbiased chromatin immunoprecipitation (*ChIP*) scans of human Chromosomes 21 and 22 [8]. The predicted Sp1 motif appears in a significant manner ($p \leq 0.05$) in 45% of the target sequences but only in 5% of the control sequences

3.3   Predicting the Binding Site Models of Novel Proteins

Given a novel Cys_2His_2 Zinc Finger protein, we first need to analyze its sequence and predict a binding site model. We first identify the four key residues at each DNA-binding domain, and then we utilize the learned sets of DNA-recognition preferences by assigning the appropriate probabilities and constructing a probabilistic model of the binding site. This is illustrated in Figure 5.

For example, Figure 6a compares the known binding site model of Sp1, to the one predicted using our approach. To prevent bias by known Sp1 sites in our training data, we apply a *"leave-protein-out"* cross-validation analysis, and predict the DNA-binding model of Sp1 using DNA-recognition preferences that were learned from a reduced dataset without Sp1 binding sequences.

3.4   *In Silico* Binding Experiments

We now use the predicted binding site models to scan genomic sequences for putative binding sites. Using the CIS algorithm [2], we score each possible binding position on the two DNA strands using a log-odds score (the log of the ratio between the probability of the binding site given the predicted model, and its probability given a 3rd-order Markov model trained on genomic sequences). We then estimate the p-value of these scores and apply a *Bonferroni* correction to account for multiple tests within the same promoter region. Sites with a significant p-value (≤ 0.05 after Bonferroni correction) were marked as putative binding sites. For example, Figure 6b demonstrates such an *in silico* binding experiment for the human dihydrofolate reductase (DHFR) promoter, using the predicted binding site model of Sp1.

3.5   Quantitative Validation of Binding Site Predictions for Novel Proteins

To further evaluate the sets of recognition preferences, we mined the literature for experimentally verified binding sites of canonical Cys_2His_2 Zinc finger proteins. These include 43 binding sites, from 21 pairs of transcription factors and the natural genomic promoter regions of their target genes (some proteins have multiple binding sites per promoter). As described above, we utilized the learned DNA-recognition preferences to predict binding site models for the involved transcription factors, and used them to scan the respective promoter regions for putative binding sites. To ensure the validity of the test, we applied a "leave-protein-out" cross-validation test as specified above. Figure 6c summarizes these 21 *in silico* binding experiments using an ROC curve. Using $p = 0.05$ (with Bonferroni correction), our method marked 30 locations as putative binding sites, out of which 21 match experimental knowledge (sensitivity of 49% and specificity of 70%, hyper-geometric p-value $< 10^{-48}$).

3.6   Comparison with Other Computational Approaches

In a similar manner, we generated probabilistic binding site models for these transcription factors using the recognition preferences of Mandel-Gutfreund et al. [17] (that were used as a starting point in our learning algorithm), and repeated the quantitative analysis. As we show in Figure 6c, predictions based on these preferences have inferior accuracy.

In a recent study, Benos et al. [4] used *in vitro* specialized experimental data (such as SELEX and phage display) to assign position-specific DNA-recognition preferences for the Cys_2His_2 Zinc Finger family (see detailed comparison of the two approaches in the Discussion section). As before, we used their preferences to generate probabilistic binding site models for these transcription factors, and then used them to scan the corresponding promoter regions. Once again, Figure 6c shows that predictions based on our sets of DNA-recognition preferences are more accurate.

3.7   Predictions Based on Genomic Data

To further evaluate our predictions on long genomics sequences, we used the binding locations of Sp1 along human Chromosomes 21 and 22, as mapped by an unbiased genome-wide chromatin immunoprecipitation ($ChIP$) assay [8]. We compiled two datasets of 1Kb-long sequences: one dataset included sequences that exhibited highly significant binding in the $ChIP$ assay, while the other dataset included sequences that showed no binding at all (to be used as a control). We then performed *in silico* binding experiments using CIS [2], searching the sequences by the predicted binding site model of Sp1. Figure 6d compares the abundance of putative hits in both datasets. As can be seen, using a Bonferroni corrected threshold of 0.05, putative Sp1 binding sites were found in 45% of the experimentally-bound sequences, while only in 5% of the control sequences.

4   Discussion

In this paper we propose a general framework for predicting the DNA binding site models of novel transcription factors from known families. Our framework combines structural information about a DNA-binding family, with sequence data about binding sites for other proteins in the same family. We apply our approach to the canonical Cys_2His_2 Zinc Finger DNA-binding family, and use a statistical estimation algorithm to derive a set of amino acid-nucleotide recognition preferences for each key position in the Zinc Finger DNA-binding domain. These recognition preferences can then be used to predict the binding site models of novel proteins from the same family. Finally, we use the predicted models to scan regulatory genomic regions of target genes, and identify their putative binding sites.

5   Prediction of Binding Sites Using Structure-Based Approaches

Structure-based approaches for prediction of transcription factor binding sites have recently gained much interest [4, 14, 17, 23, 25]. Most of the structural approaches define a protein-DNA binding model based on solved protein-DNA complexes, and attempt to identify DNA subsequences that fit best the amino acids that are determined as interacting with the DNA. While some of these studies [14, 19] used ensembles of solved protein-DNA complexes from all DNA-binding domains to extract general preferences for amino acid-base recognition, we and others focus on a single DNA-binding domain. Although less general, we hope that such an approach will lead to more fine-grained definitions of the binding preferences.

In a recent study, Benos et al. [4] assigned position-specific DNA-recognition potentials for the Cys_2His_2 Zinc Finger family. Although the model they used is quite similar to ours, there are significant differences between the two. First, they relied only on aligned binding sites from *in vitro* specialized experiments, such as SELEX and phage display, to train their recognition preferences. Second, their assays screened artificial sequences of both artificial proteins and artificial DNA targets. In contrast, we rely on longer, unaligned natural binding data. Previous studies showed that there are discrepancies between SELEX-derived motifs and those derived from natural binding sites [21, 22]. As we showed, our sets of estimated DNA-recognition preferences are more consistent with independent experimental results [6, 9, 10, 28] and are superior to similar preferences derived by the other computational methods [4, 17]. To further illustrate this point, we returned the artificial binding sequences from TRANS-FAC back into our training data, and obtained inferior predictions. Figure 6c summarizes a quantitative comparison between all models in identifying binding sites of novel proteins within genomic sequences. It should be stressed out that in order to prevent unfair bias, we use a "leave-protein-out" cross validation, hence removing all binding sites of a protein from the training data before testing it.

5.1   Analysis of the Estimated DNA-Recognition Preferences

A close examination of the learned sets of DNA-recognition preferences suggests that the protein-DNA recognition code is not deterministic, but rather spans a range of preferences. Moreover, our analyses show that a residue may have different nucleotide preferences depending on its context. For some amino acids, the qualitative preferences remain the same across various positions, while the quantitative preferences vary (e.g. arginine, see Figure 4). The DNA-binding preferences of other residues change across various positions. For example, histidine at position 3 tends to interact with guanine, while it shows no preference to any nucleotide at all other positions. Another example is the tendency of alanine at position 6 to face guanine. This preference, which was revealed automatically by our analysis, does not comply with both the chemical nature of alanine's side chain, nor with general examinations of amino acid-nucleotide interactions [14, 17]. We suspect that it is affected by the large fraction of Sp1 targets in our dataset. This potential interaction was implied before in Sp1 binding sites [5] and may reflect an interaction between the residue at position 2 with the complementary cytosine.

5.2   Inter-position Dependencies in the Binding Site

The Cys_2His_2 binding model inherently assumes that all positions within the binding site are independent of each other. This assumption is used in most computational approaches that model binding sites. Two papers [3, 7] discuss this issue in the context of the Cys_2His_2 Zinc Finger domain. Their analyses of binding affinity measurements suggest that some weak dependencies do exist among some positions of the binding sites of Egr-1. Nonetheless, a reasonable approximation of the binding specificities is obtained even when ignoring these dependencies. In another recent study [1], we evaluated probabilistic models that are capable of capturing such inter-position dependencies within binding sites. Our results showed that dependencies can be found in the binding sites of many proteins from various DNA-binding domains (especially from the helix-turn-helix and the homeo domains). However, our results also implied that using such models of dependencies in modeling the binding sites of Zinc Finger proteins does not lead to significant improvements [1]. Thus, we believe that the Cys_2His_2 binding model we use here is indeed a reasonable approximation of the actual binding.

5.3   Genome-wide Predictions of Binding Sites and Target Genes

In the current era, there is a growing gap between the number of known protein sequences and the number of experimentally verified binding sites. To better understand regulatory mechanisms in newly solved genomes, it is crucial to identify the direct target genes of novel DNA-binding proteins. Our method opens the way for such genome-wide assays. By predicting the binding site models of regulatory proteins, one might attempt to also classify the genes to those that contain significant binding sites at their regulatory promoter regions (hence, putative target genes) and those that do not. As we showed, our approach can scale up to such genome-wide scans.

5.4   Applications to Other DNA-Binding Domains

Theoretically, our approach can be extended to handle other structural families. In Figure 7, we analyze the number of binding sites needed for estimating the DNA-recognition preferences. We show that ~200 sites are sufficient for achieving similar likelihood values. Other possible families of DNA-binding domains, such as the leucine zipper, the homeodomain and the helix-turn-helix domain, have enough sites in TRANSFAC to allow similar analyses (1191, 505 and 201 sites, respectively). Unfortunately, this move requires that the various proteins

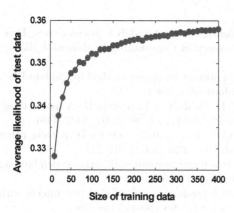

Fig. 7. Likelihood of held-out test data given different sizes of training datasets. The original data (455 canonical Cys_2His_2 zinc finger sites from TRANSFAC 7.3) were split into 10 equally-sized sets. We used each set as held-out test data, while applying the following procedure 10 times: various portions of different sizes (from 10 to 400 binding sites) were sampled from the remaining 90% of the data, and were used as training data for the EM algorithm. We then calculated the average likelihood of the held-out 10%

of the target DNA-binding domain will follow a common simple DNA-binding model. This is not the case for some families, where the binding models are far more flexible and complex. To handle these cases, more advanced models and learning techniques will be needed. Furthermore, for some families there is no simple way of inferring the interacting residues, based on the sequence of the protein (unlike the conserved Cys_2His_2 pattern in the Zinc Finger domain), and so the possible search space grows even further.

In spite of these drawbacks, we believe that structural approaches, as the one we show here, will lead to successful predictions of binding site models, and following that, to accurate identification of the target genes of novel proteins, even on genome-wide scales. Eventually, such approaches will be utilized to reconstruct larger and larger portions of the transcriptional regulatory networks that control the living cell.

Acknowledgments

The authors wish to thank Yael Altuvia, Ernest Fraenkel, Benjamin Gordon, Ruth Hershberg, Dalit May, Lena Nekludova, Aviv Regev, and Eran Segal for helpful discussions. T.K. is supported by the Horowitz Fellowship. This work was supported in part by grants from the Israeli Ministry of Science and the Israeli Science Foundation (administered by the Israeli Academy of Sciences and Humanities).

Availability

A web-server for predicting the binding sites of Cys_2His_2 Zinc Finger proteins, based on their sequences and on the estimated recognition preferences, can be accessed at http://compbio.cs.huji.ac.il/predictBS.

References

1. Barash, Y., *et al.*: Modeling dependencies in Protein-DNA binding sites. Proc. of the 7th International Conf. on Research in Computational Molecular Biology (2003) 28–37
2. Barash, Y., *et al.*: CIS: Compound Importance Sampling method for protein-DNA binding site *p*-value estimation. Bioinformatics (2004)
3. Benos, P.V., Bulyk, M.L., Stormo, G.D.: Additivity in protein-DNA interactions: how good an approximation is it? Nucleic Acids Res. **30** (2002) 4442–4451
4. Benos, P.V., Lapedes, A.S., Stormo, G.D.: Probabilistic code for DNA recognition by proteins of the EGR family. J. Mol. Biol. **323** (2002) 701–727
5. Berg, J.M.: Sp1 and the subfamily of zinc finger proteins with guanine-rich binding sites. Proc. Natl. Acad. Sci. USA **89** (1992) 11109–11110
6. Bulyk, M.L., *et al.*: Exploring the DNA-binding specificities of zinc fingers with DNA microarrays. Proc. Natl. Acad. Sci. USA **98** (2001) 7158–7163
7. Bulyk, M.L., Johnson, P.L.F., Church, G.M.: Nucleotides of transcription factor binding sites exert interdependent effects on the binding affinities of transcription factors. Nucleic Acids Res. **30** (2002) 1255–1261
8. Cawley, S., *et al.*: Unbiased mapping of transcription factor binding sites along human chromosomes 21 and 22 points to widespread regulation of noncoding RNAs. Cell **116**(4) (2004) 499–509
9. Choo, Y., Klug, A.: Selection of DNA binding sites for zinc fingers using rationally randomized DNA reveals coded interactions. Proc. Natl. Acad. Sci. USA **91** (1994) 11168–11172
10. Choo, Y., Klug, A.: Toward a code for the interactions of zinc fingers with DNA: selection of randomized fingers displayed on phage. Proc. Natl. Acad. Sci. USA **91** (1994) 11163–11167
11. Dempster, A.P., Laird, N.M., Rubin, D.B.: Maximum Likelihood form incomplete data via the EM algorithm. J. Royal Stat. Soc. B. **39** (1977) 1–38
12. Eddy, S.R.: Profile hidden Markov models. Bioinformatics **14** (1998) 755–763
13. Elrod-Erickson, M., Benson, T.E., Pabo, C.O.: High-resolution structures of variant Zif268-DNA complexes: implications for understanding zinc finger-DNA recognition. Structure **6** (1998) 451–464
14. Kono, H., Sarai, A.: Structure-based prediction of DNA target sites by regulatory proteins. Proteins **35** (1999) 114–131
15. Kriwacki, R.W., *et al.*: Sequence-specific recognition of DNA by zinc-finger peptides derived from the transcription factor Sp1. Proc. Natl. Acad. Sci. USA **89** (1992) 9759–9763
16. Luscombe, N.M., Laskowski, R.A., Thornton, J.M.: Amino acid-base interactions: a three-dimensional analysis of protein-DNA interactions at an atomic level. Nucleic Acids Res. **29** (2001) 2860–2874
17. Mandel-Gutfreund, Y., Baron, A., Margalit, H.: A structure-based approach for prediction of protein binding sites in gene upstream regions. Proc. of the Pac. Symp. Biocomput. (2001) 139–150
18. Mandel-Gutfreund, Y., Schueler, O., Margalit, H.: Comprehensive analysis of hydrogen bonds in regulatory protein DNA-complexes: in search of common principles. J Mol Biol. 253 (1995) 370–382
19. Mandel-Gutfreund, Y., Margalit, H.: Quantitative parameters for amino acid-base interaction: implications for prediction of protein-DNA binding sites. Nucleic Acids Res. **26** (1998) 2306–2312

20. Pavletich, N.P., Pabo, C.O.: Zinc finger-DNA recognition: crystal structure of a Zif268-DNA complex at 2.1 Å. Science **252** (1991) 809–817
21. Robison, K., McGuire, A.M., Church, G.M.: A comprehensive library of DNA-binding site matrices for 55 proteins applied to the complete Escherichia coli K-12 genome. J. Mol. Biol. **284** (1998) 241–254
22. Shultzaberger, R.K., Schneider, T.D.: Using sequence logos and information analysis of Lrp DNA binding sites to investigate discrepancies between natural selection and SELEX. Nucleic Acids Res. **27** (1999) 882–887
23. Steffen, N.R., et al.: DNA sequence and structure: direct and indirect recognition in protein-DNA binding. Bioinformatics **18 Suppl 1** (2002) S22–S30
24. Stormo, G.D.: DNA binding sites: representation and discovery. Bioinformatics **16**(1) (2000) 16–23
25. Suzuki, M., Gerstein, M., Yagi, N.: Stereochemical basis of DNA recognition by Zn fingers. Nucleic Acids Res. **22** (1994) 3397–3405
26. Tupler, R., Perini, G., Green, M.R.: Expressing the human genome. Nature **409**(6822) (2001) 832–833
27. Wingender, E., et al.: The TRANSFAC system on gene expression regulation. Nucleic Acids Res. **29** (2001) 281–283
28. Wolfe, S.A., et al.: Analysis of zinc fingers optimized via phage display: evaluating the utility of a recognition code. J. Mol. Biol. **285** (1999) 1917–1934

Motif Discovery Through Predictive Modeling of Gene Regulation

Manuel Middendorf[1], Anshul Kundaje[2], Mihir Shah[2], Yoav Freund[2,4,5],
Chris H. Wiggins[3,4], and Christina Leslie[2,4,5]

[1] Department of Physics
[2] Department of Computer Science
[3] Department of Applied Mathematics
[4] Center for Computational Biology and Bioinformatics
[5] Center for Computational Learning Systems,
Columbia University, New York, NY 10027
cleslie@cs.columbia.edu
http://www.cs.columbia.edu/compbio/medusa

Abstract. We present MEDUSA, an integrative method for learning
motif models of transcription factor binding sites by incorporating promoter sequence and gene expression data. We use a modern large-margin
machine learning approach, based on boosting, to enable feature selection
from the high-dimensional search space of candidate binding sequences
while avoiding overfitting. At each iteration of the algorithm, MEDUSA
builds a motif model whose presence in the promoter region of a gene,
coupled with activity of a regulator in an experiment, is predictive of
differential expression. In this way, we learn motifs that are functional
and predictive of regulatory response rather than motifs that are simply
overrepresented in promoter sequences. Moreover, MEDUSA produces a
model of the transcriptional control logic that can predict the expression
of any gene in the organism, given the sequence of the promoter region
of the target gene and the expression state of a set of known or putative
transcription factors and signaling molecules. Each motif model is either
a k-length sequence, a dimer, or a PSSM that is built by agglomerative
probabilistic clustering of sequences with similar boosting loss. By applying MEDUSA to a set of environmental stress response expression data
in yeast, we learn motifs whose ability to predict differential expression
of target genes outperforms motifs from the TRANSFAC dataset and
from a previously published candidate set of PSSMs. We also show that
MEDUSA retrieves many experimentally confirmed binding sites associated with environmental stress response from the literature.

1 Introduction

One of the central challenges in computational biology is the elucidation of mechanisms for gene transcriptional regulation using functional genomic data. The
problem of identifying binding sites for transcription factors in the regulatory sequences of genes is a key component in these computational efforts. While there

S. Miyano et al. (Eds.): RECOMB 2005, LNBI 3500, pp. 538–552, 2005.
© Springer-Verlag Berlin Heidelberg 2005

is a vast literature on this subject, only a few different conceptual approaches have been tried, and each of these standard approaches has its limitations.

The most widely-used methodology for computational discovery of putative binding sites is based on clustering genes—usually by similarity of gene expression profiles, sometimes combined with annotation data—and searching for motif patterns that are overrepresented in the promoter sequences of these genes in the belief that they may be coregulated. Popular motif discovery programs in this paradigm include MEME [1], Consensus [2], Gibbs Sampler [3], AlignACE [4] and many others. The cluster-first methodology has several drawbacks. First, it is not always true that genes with correlated gene expression profiles are in fact coregulated genes whose regulatory regions contain common binding sites. Moreover, by focusing on coregulated genes, one fails to consider more complicated combinatorial regulatory programs and the overlapping regulatory pathways that can affect different sets of genes under different conditions. Recently, more sophisticated graphical models for gene expression data have been introduced to try to partition genes into "transcriptional modules" [5]—clusters of genes that obey a common transcriptional program depending on a small number of regulators—or to learn overlapping clusters of this kind [6]. These graphical model approaches use the abstraction of modules to give an interpretable representation of putative relationships between genes and to suggest biological hypotheses. One expects that using these more complex clustering algorithms as a preprocessing step for motif discovery would lead to improved identification of true binding sites; however, it is difficult to assess how much of an advantage one might obtain.

Another well-established motif discovery approach is the innovative RE-DUCE method [7] and related algorithms [8, 9]. REDUCE avoids the cluster-first methodology by considering the genome-wide expression levels given by a single microarray experiment, and it discovers sequences whose presence in promoter sequences correlates with differential expression. Since REDUCE uses linear regression to iteratively identify putative binding sites, it must enforce strict tests of statistical significance to avoid overfitting in a large parameter space corresponding to the set of all possible sequence candidates. Therefore, REDUCE can find the strongest signals in a dataset but will not attempt to find more subtle sites that affect fewer genes. Since the algorithm fits parameters independently for each microarray experiment, the issue of condition-specific regulation enters the analysis only as post-processing step rather than through simultaneous training from multiple conditions.

In this paper, we introduce a new motif discovery algorithm called MEDUSA (Motif Element Discrimination Using Sequence Agglomeration) that learns putative binding sites associated with condition-specific regulation in a large gene expression dataset. MEDUSA works by extracting binding site motifs that contribute to a *predictive model* of gene regulation. More specifically, MEDUSA builds motif models whose presence in the promoter region of a gene, together with the activity of regulators in an experiment, is predictive of differential expression. Like REDUCE, MEDUSA avoids the cluster-first methodology and builds a single regulatory model to explain the response of all target genes.

However, unlike REDUCE, MEDUSA learns from multiple and diverse gene expression experiments, using the expression states of a set of known regulatory to represent condition-specific regulatory conditions. Moreover, MEDUSA is based on a classification approach (using large-margin machine learning) rather than linear regression, to avoid overfitting in the high-dimensional search space of candidate binding sequences. In addition to discovering binding site motifs, MEDUSA produces a model of the condition-specific transcriptional control logic that can predict the expression of any gene, given the gene's promoter sequence and the expression state of a set of known transcription factors and signaling molecules.

The core of MEDUSA is a boosting algorithm that adds a binding site motif (coupled with a regulator whose activity helps predict up/down regulation of genes whose promoters contain the motif) to an overall gene regulation model at each boosting iteration. Each motif model is either a k-length sequence (or "k-mer"), a dimer, or a PSSM. The PSSMs are generated by considering the most predictive k-mer features (Fig. 2) selected at a given round of boosting that are associated with a common regulator; we then perform agglomerative probabilistic clustering of these k-mers into PSSMs, and we select from all the candidate PSSMs seen during clustering the one that minimizes boosting loss (Fig. 2). In experiments on a set of environmental stress response expression data in yeast, we learn motifs together with regulation models that achieve accurate prediction of up/down regulation of target genes in held-out experiments. In fact, we show that the performance of the learned motifs for prediction of differential expression in test data is stronger than the performance of motifs from the TRANSFAC dataset or from a previously published candidate set of PSSMs. For these environmental stress response experiments, we also show that MEDUSA retrieves many experimentally confirmed binding sites from the literature.

We first introduced the idea of *predictive modeling* of gene regulation with the GeneClass algorithm [10]. However, GeneClass uses a fixed set of candidate motifs as an input to the algorithm and cannot perform motif discovery. We note also that there have been previous efforts to incorporate motif discovery in an integrative model for sequence and expression data using the probabilistic graphical model framework [11]. This graphical model approach again uses the abstraction of "modules" to learn sets of motifs associated with clusters of genes, giving a high-level modular representation of gene regulation. As explained above, MEDUSA does not produce an abstract module representation. However, it has two advantages over graphical model methods. First, MEDUSA uses a large-margin learning approach that helps to improve the *generalization* of the learned motifs and regulation model, and we can evaluate prediction accuracy on held-out experiments to assess our confidence in the model. Second, training graphical models requires special expertise to avoid poor local minima in a complex optimization problem, while MEDUSA can be run "out-of-the-box". Code for MEDUSA is publicly available and can be downloaded from the supplementary website for the paper, http://www.cs.columbia.edu/compbio/medusa.

2   Methods

2.1   Learning Algorithm

MEDUSA learns binding site motifs together with a predictive gene regulation model using a specific implementation of Adaboost, a general discriminative learning algorithm proposed by Freund and Schapire [12]. Adaboost's basic idea is to iteratively apply variants of a simple, weakly discriminative learning algorithm, called the *weak* learner, to different weightings of the same training set. The only requirement of the weak learner is that it predicts the class label of interest with greater than 50% accuracy. At each iteration, weights are recalculated so that examples which were misclassified at the previous iteration are more highly weighted. Finally, all of the weak prediction rules are combined into a single *strong* rule using a weighted majority vote. As discussed in [13], boosting is a large-margin classification algorithm, able to learn from a potentially large number of candidate features while maintaining good generalization error (that is, without over-fitting the training data).

The discretization of expression data (see Sect. 3.2) into up- and down-regulated expression levels allows us to formulate the problem of predicting regulatory response of target genes as the *binary classification* task of learning to predict up and down examples. Rather than viewing each microarray experiment as a training example, MEDUSA considers all genes and experiments simultaneously and treats every gene-experiment pair as a separate instance, dramatically increasing the number of training examples available. For every gene-experiment example, the gene's expression state in the experiment (up- or down-regulation) gives the output label $y_{ge} = \pm$. As we explain below (see Sect. 3.2), positive and negative examples correspond to statistically significant up- and down-regulated expression levels; examples with baseline expression levels are omitted from training.

The inputs to the learner are (i) the promoter sequences of the target genes and (ii) the discretized expression levels of a set of putative regulator genes. The sequence data is represented only via occurrence or non-occurrence of a sequence element or motif. A full discussion of how MEDUSA determines a set of sequence and motif candidates to be considered at each round of boosting is given in Sect. 2.2. Let the binary matrix $M_{\mu g}$ indicate the presence ($M_{\mu g} = 1$) or absence ($M_{\mu g} = 0$) of a motif μ in the promoter sequence of gene g, and let the binary matrices $P_{\pi e}^{\sigma}$ indicate the up-regulation ($\sigma = +$) or down-regulation ($\sigma = -$) of a regulator π in experiment e ($P_{\pi e}^{\sigma} = 1$, if regulator π is in state σ in experiment e, and $P_{\pi e}^{\sigma} = 0$, otherwise). Our weak rules split the gene-experiment examples in the training data by asking questions of the form '$M_{\mu g} P_{\pi e}^{\sigma} = 1$?'; i.e., 'Is motif μ present, and is regulator π in state σ?'. In this way, each rule introduced corresponds to a putative interaction between a regulator and some sequence element in the promoter of the target gene that it regulates.

The weak rules are combined by weighted majority vote using the structure of an alternating decision tree [14, 10]. An example is given in Fig. 1. The weak rules are shown in rectangles. Their associated weights, indicating the strength

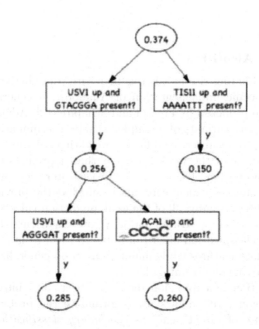

Fig. 1. Example of an alternating decision tree: The rectangles represent weak rules, learned by MEDUSA, Examples for which the condition holds follow the path further down the tree ('y') and have their scores incremented by the prediction score given in the ovals. The final prediction is the sum of all scores that the example reaches

of their contribution to the majority vote, are shown in ovals. If the {motif presence, regulator state} condition for a particular rule holds in the example considered, the weight of the rule is added to the final prediction score. The weight can be either positive or negative, contributing to up- or down-regulation respectively. Rules that appear lower in the tree are conditionally dependent on the rules in ancestor nodes. For example, in Fig. 1, only if USV1 is up-regulated and both motifs GTACGGA and AGGGAT are present is the score 0.285 added to the prediction score. The tree structure is thus able to reveal combinatorial interactions between regulators and/or motifs. The sign of the final prediction score gives the prediction, and the absolute value of the score indicates the level of confidence. In this work, we consider both sequences and position-specific scoring matrices (PSSM) (an example is shown in the lower right node of Fig. 1) as putative motifs (see Sect. 2.2).

Each iteration of the boosting algorithm results in the addition of a new node (corresponding to a new weak rule) to the tree. The weak rule and its position in the tree at which it is added are chosen by minimizing the boosting loss over all possible combinations of motifs, regulators, and regulator-states, and over all possible positions ("preconditions") in the current tree. A pseudo-code description is given in Fig. 2.

Definitions:

$\hat{c}$	= precondition associated with a specific position in the tree
$c_{\mu\pi\sigma}$	= weak rule associated with motif μ and regulator π in state σ
w_{ge}	= weight of example (g, e)
$W[c(g, e)]$	= $\sum_{c(g,e)=1} w_{ge}$, for a given condition c
$\neg c$	= not c
$Z(\hat{c}, \mu, \pi, \sigma)$	= boosting loss
	= $W[\neg\hat{c}] + 2\sqrt{W[\hat{c} \wedge c_{\mu\pi\sigma}]W[\hat{c} \wedge \neg c_{\mu\pi\sigma}]}$
y_{ge}	= label of example (g, e)
T	= total number of boosting iterations
$F_t(g, e)$	= prediction function at iteration t
α_t	= weight of weak rule t contributing to the final prediction score

Initialization:
$F_0(g, e) = 0$, for all (g, e)

Main loop:
for $t = 1 \ldots T$
   $w_{ge} = e^{-y_{ge} F_{t-1}(g,e)}$
   call Hierarchical Motif Clustering (Sec. 2.2).
   get a set of proposed PSSMs.
   minimize boosting loss:
   $\mathbf{c}^* = \mathrm{argmin}_{\hat{c},\mu,\pi,\sigma} Z(\hat{c}, \mu, \pi, \sigma)$
   calculate weight of the new weak rule $\mathbf{c}^*$:
   $\alpha_t = \frac{1}{2} \ln \frac{W[\mathbf{c}^* \wedge (y_{ge}=+)]}{W[\mathbf{c}^* \wedge (y_{ge}=-)]}$
   add new node $\mathbf{c}^*$ with weight α_t to the tree
   $F_t(g, e) = F_{t-1}(g, e) + \alpha_t \mathbf{c}^*(g, e)$
end for
$\mathrm{sign}(F_T(g, e))$ = prediction for example (g, e)
$|F_T(g, e)|$ = prediction confidence for (g, e)

Fig. 2. Pseudo-code description of the learning algorithm

The implementation uses efficient sparse matrix multiplication in MATLAB, exploiting the fact that our motif-regulator features are outer products of motif occurrence vectors and regulator expression vectors, and allows us to scale up to significantly larger datasets than in [10].

2.2   Hierarchical Motif Clustering

At each boosting iteration, MEDUSA considers all occurrences of k-mers ($k = 2, 3, \ldots, 7$) and dimers with a gap of up to 15 bp (see Sect. 3.4) in the promoter sequence of each gene as candidate motifs. Since slightly different sequences might in fact be instances of binding sites for the same regulator, MEDUSA performs a hierarchical motif clustering algorithm to generate more general candidate PSSMs as binding site models. The motif clustering uses k-mers and dimers associated with low boosting loss as a starting point to build PSSMs: these sequences are viewed seed PSSMs, and then the algorithm proceeds by iteratively merging similar PSSMs, as described below. The generated PSSMs are then considered as additional putative motifs for the learning algorithm.

A position-specific scoring matrix (PSSM) is represented by a probability distribution $p(x_1, x_2, \ldots, x_n)$ over sequences $x_1 x_2 \ldots x_n$, where $x_i \in \{A, C, G, T\}$. The emission probabilities are assumed to be independent at every position such that $p(x_1, \ldots, x_n) = \prod_{i=1}^{n} p_i(x_i)$. For a given input sequence the PSSM returns a log-odds score $S = \sum_{i=1}^{n} \ln (p_i(x_i)/p^{bg}(x_i))$ with respect to background probabilities p^{bg}. A score threshold can then be chosen to define whether the input sequence is a hit or not.

When comparing two PSSMs, we allow possible offsets between the two starting positions. In order to give them the same lengths, we pad either the left or right ends with the background distribution. We then define a distance measure $d(p, q)$ as the minimum over all possible position offsets of the JS entropy [15] between two PSSMs p and q.

$$d(p, q) \equiv \min_{\text{offsets}} \left[w_1 D_{KL}(p||w_1 p + w_2 q) + w_2 D_{KL}(q||w_1 p + w_2 q) \right],$$

where D_{KL} is the Kullback-Leibler divergence [15]. By using $p(x_1 \ldots x_n) = \prod_{i=1}^{n} p_i(x_i)$ and $\sum_{x_i} p_i(x_i) = 1$ (and the analogous equations for q) one can easily show that $D_{KL}(p||q) = \sum_{i=1}^{n} D_{KL}(p_i||q_i)$. The relative weights of the two PSSMs, w_1 and w_2, are here defined as $w_{1,2} = N_{1,2}/(N_1 + N_2)$, where N_1, N_2 are the numbers of target genes for the given PSSM. Note that this distortion measure is not affected by adding more "padded" background elements either before or after the PSSM. Our merge criterion is similar to the one used in the agglomerative information bottleneck algorithm [16], though we also consider offsets in our merges.

At every boosting iteration, we first find the weak rule c_{tmp} among all possible combinations of regulators, regulator-states and sequence motifs (k-mers and dimers), that minimizes boosting loss. The 100 motifs with lowest loss appearing with the same regulator, regulator-state, and precondition as in c_{tmp} are then input to the hierarchical clustering algorithm. Sequence motifs can be regarded as PSSMs with 0/1 emission probabilities, smoothed by background probabilities. By iteratively joining the PSSMs with smallest $d(p, q)$, the clustering proposes a set of 99 PSSMs from various stages of the hierarchy. At every merge of two PSSMs, the score threshold associated with the new PSSM is found by optimizing the boosting loss. Note also that the new PSSM can be longer than either of the two PSSMs used in the merge, due to the procedure of merging with offsets; in this way, we can obtain candidate PSSMs longer the maximum seed k-mer length of 7. The number of target genes, which determines the weight of the PSSM for further clustering, is calculated by counting the number of promoter sequences which score above the threshold. The new node that is then added to the alternating decision tree is the weak rule that minimizes boosting loss considering all sequence motifs and PSSMs.

3   Statistical Validation

3.1   Dataset

We use the environmental stress response (ESR) dataset of Gasch *et al.* [17], which consists of 173 cDNA microarray experiments measuring the expression of 6152 *S. cerevisiae* genes in response to diverse environmental perturbations. All measurements are given as $\log_2$ expression values (fold-change with respect to an unstimulated reference condition). Note that our analysis does not require a normalization to a zero-mean, unit-variance distribution, as is often employed; instead we wish to retain the meaning of the true zero (that is, the reference state).

3.2   Discretization

We discretize expression data by using a noise model that accounts for intensity specific effects in the raw data from both the Cy3 (R) and Cy5 (G) channels. In order to estimate the null model, we use the three replicate unstimulated experiments published with the same dataset [17]. Plots of $M = \log_2(R/G)$ versus $A = \log_2(\sqrt{RG})$ (Fig. 3) show the intensity specific distribution of the noise in the expression values. We compute the cumulative empirical null distribution of M conditioned on A by binning the A variable into small bin sizes, maintaining a good resolution while having sufficient data points per bin. For any expression value (M, A) of a gene in an experiment, we estimate a p-value based on the

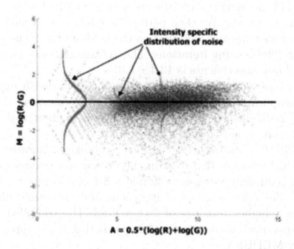

Fig. 3. Expression discretization. A noise distribution is empirically estimated using data from three unstimulated reference experiments. The noise model takes into account intensity-specific effects. By choosing a p-value cutoff of 0.05 we discretize differential expression into up-regulated, down-regulated, and baseline levels

null distribution conditioned on A, and we use a p-value cutoff of 0.05 to discretize the expression values into +1, -1 or 0 (up-regulation, down-regulation, or baseline). The discretization allows us to formulate the prediction problem as a classification task.

3.3   Candidate Regulators

The regulator set consists of 475 genes (transcription factors, signaling molecules, kinases and phosphatases), including 466 which are used in Segal *et al.* [5] and 9 generic (global) regulators obtained from Lee *et al.* [18].

3.4   Motif Set

We scan the 500 bp 5'-UTR promoter sequences of all *S. cerevisiae* genes from the Saccharomyces genome Database (SGD) for all occurring k-mer motifs ($k = 2, 3, \ldots, 7$). We also include 3-3 and 4-4 dimer motifs allowing a middle gap of up to 15 bp. We restrict the set of all dimers to those whose two components have specific relationships, consistent with most known dimer motifs: equal, reversed, complements, or reverse-complements. As described in Sect. 2.2, we use an information-theoretic, hierarchical clustering scheme to infer a set of PSSMs at each boosting iteration. The complete candidate motif set is then the union of all k-mers, dimers, and PSSMs, with a cardinality of $10962 + 1184 + 99 = 12245$.

3.5   Cross-Validation

We divide the 173 microarray experiments into five folds, keeping replicate experiments in the same fold. We then perform five-fold cross-validation, training the classifier on four folds and testing it on the held-out fold. The learning algorithm is run for 700 boosting iterations. The average test-loss for prediction on all genes in held-out experiments is $13.4 \pm 3.9\%$.

For comparison, we run the same learning algorithm with experimentally-confirmed or computationally-predicted motifs in the literature. In these runs, the hierarchical motif clustering is left out, and the set of putative motifs contains only those that were proposed in the literature.

The TRANSFAC database [19] contains a library of known and putative binding sites which can be used to scan the promoter sequence of every gene. After removing redundant sites, we compile a list of 354 motifs. The boosting algorithm with the same number of iterations and the same folds for cross-validation gives a higher test-loss of $20.8 \pm 2.8\%$ The compiled TRANSFAC motifs thus have a much weaker strength in predicting gene expression than the motifs found by MEDUSA.

The same comparison was performed with a list of 356 motifs found in [20] by using a state-of-the-art Gibbs sampling algorithm on groups of genes clustered by expression data and annotation information. These motifs also gave weaker predictive strength than those discovered by MEDUSA with an average test-loss of $16.1 \pm 3.5\%$.

We are thus able to identify motifs which have a significantly stronger prediction accuracy (on independent held-out experiments) than motifs previously identified in the literature.

4   Biological Validation

To confirm that MEDUSA can retrieve biologically meaningful motifs, we run additional experiments, randomly holding out 10% of the (gene,experiment) examples and training MEDUSA on the remaining examples. We learn ungapped k-mers and dimers simultaneously. After 1000 iterations, we obtain a test loss of 11% and a set of 1000 PSSMs. We then compare to several known and putative binding sites, consensus sequences and PSSMs from five databases: TRANS-FAC [19], TFD, SCPD, YPD and a set of PSSMs found by AlignACE [20]. After converting the sequences and consensus patterns to PSSMs, smoothed by background probabilities, we compare all PSSMs with the ones found by MEDUSA using $d(p, q)$ (see Sect. 2.2) as a distance measure. We define the best match for each of MEDUSA's PSSMs as the PSSM that is closest to it in terms of $d(p, q)$.

Each node in the alternating decision tree defines a particular subset of genes, namely those having at least one example that passes through the particular node. In this way, we can associate motifs with Gene Ontology (GO) annotations by looking for enriched GO annotations in the gene subsets, and we can estimate the putative functions of the targets of a transcription factor that might bind to the PSSM in each node. We see matches to variants of the STRE element, the binding site for the MSN2 and MSN4 general stress response transcription factors. The genes passing through nodes containing these PSSMs are significantly enriched for the GO terms carbohydrate metabolism, response to stress and energy pathways, consistent with the known functions of MSN2/4. GCR1 and RAP1 are known to transcriptionally regulate ribosomal genes, consistent with enriched GO annotations associated with the nodes of the specific PSSMs. The heat shock factor HSF1—which binds to the heat shock element (HSE)—plays a primary role in stress response to heat as well as several other stresses. The heat shock element exists as a palindromic sequence of the form *NGAANNTTCN*. We find almost an exact HSE in the tree. In *S. cerevisiae*, several important responses to oxidative and redox stresses are regulated by Yap1p, which binds to the YRE element. We find several strongly matching variants of the YRE. It is interesting to note that comparison of PSSMs from AlignACE with our PSSMs revealed the PAC and RRPE motifs to be among the top three matches. These PSSMs also appear in the top 10 iterations in the tree, indicating they are also strongly predictive of the target gene expression. Both these putative regulatory motifs have been studied in great depth with respect to their roles in rRNA processing and transcription as well their combinatorial interactions. The enriched GO annotations of these nodes are the same as their putative functions. The tree contains 122 dimer motifs with variable gaps. These include the HSE motif (*GAANNTTC*), HAP1 motif (*CCGN*CCG*), GIS1 motif (*AGGGGCCCCT*)

TFNAME	DB-MOTIF	MOTIF	DBNAME	d(p,q)
CBF1	CACGTG		YPD	0.032635
CGG everted repeat	CGGN*CCG		YPD	0.032821
MSN2			TRANSFAC	0.085626
HSF1	TTCNNNGAA		SCPD	0.102410
XBP1			TRANSFAC	0.140561
STE12			TRANSFAC	0.256750
GCN4			SCPD	0.292221
TBP			TRANSFAC	0.376601
HAP1	CGGNNNTWNCGG		YPD	0.423004
RAP1	RMACCCA		SCPD	0.523059
mPAC			AlignACE	0.552493
mRRPE	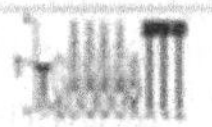	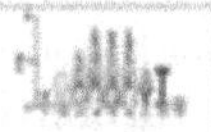	AlignACE	0.630740
PHO4			TRANSFAC	0.672961
YAP1			TRANSFAC	0.777816
MIG1	CCCCACAAA		YPD	0.799412
MET31,32	AAACTGTGG		YPD	0.84893
HAP2,3,4			TRANSFAC	1.070837

Fig. 4. Matching MEDUSA's PSSMs to motifs known in the literature: By using $d(p,q)$ (see Sect. 2.2) as a distance measure, we match PSSMs identified by MEDUSA's to motifs known in the literature. The table shows the logos of MEDUSA's PSSMs (column 3), the matching motif of the database (column 2), the corresponding transcription factor (column 1), the name of the database (column 4) and the distance $d(p,q)$ (column 5)

as well as variants of the *CCG* everted repeat. Several important biologically verified PSSMs learned by MEDUSA are given in Fig. 4. A complete comparison study of MEDUSA's PSSMs with each of the above mentioned databases as well as Gene Ontology analysis is available on the online supplementary website.

An added advantage of MEDUSA is that we can study the regulators whose mRNA expression is predictive of the expression of targets. These regulators are paired with the learned PSSMs. Of the 475 regulators (transcription factors, kinases, phosphatases and signaling molecules) used in the study, 234 are present in the tree. We can rank these regulators by abundance score (AS), namely the number of times a regulator appears in the tree in different nodes. If a regulator has a large AS, then it affects the prediction of several target genes through several nodes. The top 10 regulators include TPK1, USV1, AFR1, XBP1, ATG1, ETR1, SDS22, YAP4, PDR3. TPK1 is the kinase that affects the cellular localization of the general stress response factors MSN2/4. XBP1 is an important stress related repressor. USV1 was also identified by Segal *et al.* [5] to be a very important stress response regulator. A complete analysis of the regulators as well their association with specific motifs is available on the supplementary website.

5   Discussion

We have proposed a new algorithm called MEDUSA for learning binding site motifs together with a predictive model for gene regulation. MEDUSA jointly learns from promoter sequence data and multiple gene expression experiments, together with a candidate list of putative regulators (transcription factors and signaling molecules), and builds motif models whose presence in the promoter region of a target gene, together with the activity of regulators in an experiment, is predictive of up/down regulation of the gene. We can readily evaluate the predictive accuracy of the learned motifs and regulation model on test data, and we present results for a yeast environmental stress response dataset that demonstrate that MEDUSA's binding site motifs are better able to predict regulatory response on held-out experiments than binding site sequences taken from TRANSFAC or previously published computationally-derived PSSMs.

Popular cluster-first motif discovery strategies often require complex or even manual preprocessing to determine suitable putative clusters of coregulated genes. In practice, in addition to using gene expression profiles in the clustering algorithm, one might need to incorporate annotation data or even use hand curation to properly refine the putative clusters [4]. One must then carefully apply a standard motif discovery algorithm to find overrepresented motifs in the promoter sequences of genes in each cluster, which may involve optimizing parameters in the algorithm and thresholds for each of the extracted motif models. By contrast, MEDUSA avoids clustering and manual preprocessing altogether, and automatically determines PSSMs together with thresholds used for determining PSSM hits by optimizing boosting loss. In our experiments, MEDUSA learned many of the binding site motifs associated with various environmental stress responses in the literature.

Recent work using the framework of probabilistic graphical models has also presented an algorithm for learning putative binding site motifs in the context of building an integrated regulation model [11]. The graphical modeling approach is appealing due to its descriptive nature: since the graph structure encodes how different variables are meant to be related, it is clear how to try to interpret the results. The MEDUSA algorithm builds binding site motifs while producing a single regulation model for all target genes without introducing conceptual subunits like "clusters" or "transcriptional modules". This single regulation model is arguably more biologically realistic and can capture combinatorial regulatory effects on overlapping sets of targets. The regulation model can also be interpreted as a gene regulatory network, since the activity of regulators predicts differential expression of targets via binding sites, although necessarily this network is large and contains many nodes. Nonetheless, we can use this model to address specific biological questions, for example by restricting attention to particular target genes or experiments [21], allowing meaningful interpretation.

One difficulty of using complex graphical models is that they require careful training methodologies to avoid poor local optima and severe overfitting. MEDUSA can be run "out-of-the-box", making it easy to reproduce results and allowing non-specialists to apply the algorithm to new datasets. Moreover, it is difficult to statistically validate the full structure or the components of complex graphical models; in the literature, most work using these models for gene regulation has focused on biological validation of particular features in the graph rather than generalization measures like test loss. MEDUSA's predictive methodology—using large-margin learning strategies to focus on improving generalization—produces binding site motifs that achieve good accuracy for prediction of regulatory response on held-out experiments. The fact that we can easily evaluate the predictive performance of our learned motifs and regulation model gives us a simple statistical test of confidence in our results. The superior performance of MEDUSA in discovering predictive motifs is very encouraging for applying such large-margin techniques to analysis of expression data for as-yet unannotated genomes and for elucidating the transcriptional regulatory mechanisms of more complex organisms.

Acknowledgments

AK is supported by NSF EEC-00-88001. CW and MM are partially supported by NSF ECS-0332479, ECS-0425850 and NIH GM36277. CL and CW are supported by NIH grant LM07276-02, and CL is supported by an Award in Informatics from the PhRMA Foundation.

References

1. Bailey, T.L., Elkan, C.P.: Fitting a mixture model by expectation-maximization to discover motifs in biopolymers. In Altman, R., Brutlag, D., Karp, P., Lathrop, R., Searls, D., eds.: Proceedings of the Second International Conference on Intelligent Systems for Molecular Biology, AAAI Press (1994) 28–36
2. Hertz, G.Z., Stormo, G.D.: Identifying DNA and protein patterns with statistically significant alignments of multiple sequences. Bioinformatics **15** (1999) 563–577
3. Lawrence, C.E., Altschul, S.F., Boguski, M.S., Liu, J.S., Neuwald, A.F., Wootton, J.C.: Detecting subtle sequence signals: A Gibbs sampling strategy for multiple alignment. Science **262** (1993) 208–214
4. Hughes, J.D., Estep, P.W., Tavazoie, S., Church, G.M.: Computational identification of cis-regulatory elements associated with groups of functionally related genes in Saccharomyces cerevisiae. J. Mol. Biol. **296** (2000) 1205–14
5. Segal, E., Shapira, M., Regev, A., Pe'er, D., Botstein, D., Koller, D., Friedman, N.: Module networks: Identifying regulatory modules and their condition specific regulators from gene expression data. Nature Genetics **34** (2003) 166–176
6. Battle, A., Segal, E., Koller, D.: Probabilistic discovery of overlapping cellular processes and their regulation. In: Proceedings of the eighth annual international conference on Computational molecular biology, ACM Press (2004) 167–176
7. Bussemaker, H.J., Li, H., Siggia, E.D.: Regulatory element detection using correlation with expression. Nature Genetics **27** (2001) 167–171
8. Conlon, E.M., Liu, X.S., Lieb, J.D., Liu, J.S.: Integrating regulatory motif discovery and genome-wide expression analysis. Proceedings of the National Academy of Sciences USA **100** (2003) 3339–3344
9. Zilberstein, C.B.Z., Eskin, E., Yakhini, Z.: Sequence motifs in ranked expression data. In: Proceedings of the First RECOMB Satellite Workshop on Regulatory Genomics. (2004)
10. Middendorf, M., Kundaje, A., Wiggins, C., Freund, Y., Leslie, C.: Predicting genetic regulatory response using classification. Proceedings of the Twelfth International Conference on Intelligent Systems for Molecular Biology (ISMB 2004) (2004)
11. Segal, E., Yelensky, R., Koller, D.: Genome-wide discovery of transcriptional modules from DNA sequence and gene expression. Bioinformatics **19** (2003) 273–282
12. Schapire, R.E.: The boosting approach to machine learning: An overview. In: MSRI Workshop on Nonlinear Estimation and Classification. (2002)
13. Schapire, R.E., Freund, Y., Bartlett, P., Lee, W.S.: Boosting the margin: A new explanation for the effectiveness of voting methods. The Annals of Statistics **26** (1998) 1651–1686
14. Freund, Y., Mason, L.: The alternating decision tree learning algorithm. Proceedings of the Sixteenth International Conference on Machine Learning (1999) 124–133
15. Cover, T., Thomas, J.: Elements of Information Theory. John Wiley, New York (1990)
16. Slonim, N., Friedman, N., Tishby, N.: Unsupervised document classification using sequential information maximization. In: Proceedings of the 25th annual international ACM SIGIR conference on Research and development in information retrieval, ACM Press (2002) 129–136
17. Gasch, A.P., Spellman, P.T., Kao, C.M., Carmel-Harel, O., Eisen, M.B., Storz, G., Botstein, D., Brown, P.O.: Genomic expression programs in the response of yeast cells to environmental changes. Molecular Biology of the Cell **11** (2000) 4241–4257

18. Lee, T.I., Rinaldi, N.J., Robert, F., Odom, D.T., Bar-Joseph, Z., Gerber, G.K., Hannett, N.M., Harbison, C.R., Thompson, C.M., Simon, I., Zeitlinger, J., Jennings, E.G., Murray, H.L., Gordon, D.B., Ren, B., Wyrick, J.J., Tagne, J., Volkert, T.L., Fraenkel, E., Gifford, D.K., Young, R.A.: Transcriptional regulatory networks in Saccharomyces cerevisiae. Science **298** (2002) 799–804

19. Wingender, E., Chen, X., Hehl, R., Karas, H., Liebich, I., Matys, V., Meinhardt, T.., Prüss, M., Reuter, I., Schacherer, F.: TRANSFAC: an integrated system for gene expression regulation. Nucleic Acids Research **28** (2000) 316–319

20. Pilpel, Y., Sudarsanam, P., Church, G.M.: Identifying regulatory networks by combinatorial analysis of promoter elements. Nature Genetics **2** (2001) 153–159

21. Kundaje, A., Middendorf, M., Shah, M., Wiggins, C., Freund, Y., Leslie, C.: A classification-based framework for predicting and analyzing gene regulatory response. (Web supplement: http://www.cs.columbia.edu/compbio/robust-geneclass)

HAPLOFREQ - Estimating Haplotype Frequencies Efficiently

Eran Halperin[1],* and Elad Hazan[2]

[1] International Computer Science Institute, Berkeley, CA, 94704 USA
http://www.icsi.berkeley.edu/~heran
[2] Department of Computer Science, Princeton University, Princeton NJ, 08544 USA
http://www.cs.princeton.edu/~ehazan

Abstract. A commonly used tool in disease association studies is the search for discrepancies between the haplotype distribution in the case and control populations. In order to find this discrepancy, the haplotypes frequency in each of the populations is estimated from the genotypes.

We present a new method HAPLOFREQ to estimate haplotype frequencies over a short genomic region given the genotypes or haplotypes with missing data or sequencing errors. Our approach incorporates a maximum likelihood model based on a simple random generative model which assumes that the genotypes are independently sampled from the population. We first show that if the phased haplotypes are given, possibly with missing data, we can estimate the frequency of the haplotypes in the population by finding the *global* optimum of the likelihood function in *polynomial time*. If the haplotypes are not phased, finding the maximum value of the likelihood function is NP-hard. In this case we define an alternative likelihood function which can be thought of as a relaxed likelihood function. We show that the maximum relaxed likelihood can be found in polynomial time, and that the optimal solution of the relaxed likelihood approaches asymptotically to the haplotype frequencies in the population.

In contrast to previous approaches, our algorithms are guaranteed to converge in polynomial time to a global maximum of the different likelihood functions. We compared the performance of our algorithm to the widely used program PHASE, and we found that our estimates are at least 10% more accurate than PHASE and about ten times faster than PHASE.

Our techniques involve new algorithms in convex optimization. These algorithms may be of independent interest. Particularly, they may be helpful in other maximum likelihood problems arising from survey sampling.

1 Introduction

Most of the genetic variation among different people can be characterized by single nucleotide polymorphisms (SNPs) which are mutations at a single nucleotide

---

* Most of this work was done while the author was in a Research Associate in the Computer Science department of Princeton University

S. Miyano et al. (Eds.): RECOMB 2005, LNBI 3500, pp. 553–568, 2005.

position that occurred once in human history and were passed on through heredity. Characterization of the generic variation is an important research tool for trait association and disease association in particular. In order to understand the structure of this variation, we need to be able to determine the *haplotypes* of individuals, or which nucleotide base occurs at each position for each chromosome. The effort to characterize human variation, currently a major focus for the international community, will be a tremendous undertaking requiring obtaining the haplotype information from a large collection of individuals from diverse populations [16].

As opposed to haplotypes, the genotype gives the bases at each SNP for both copies of the chromosome, but loses the information as to the chromosome on which each base appears. Unfortunately, many sequencing techniques provide the genotypes and not the haplotypes. Haplotype analysis has become increasingly common in genetic studies of human disease. However, many of these methods rely on phase information, that is, the haplotype information vs. the genotype information. Phase can be inferred by genotyping family members of each subject, but this has its downsides because of logistic and budget issues. Alternatively, laboratory techniques such as long range PCR or chromosomal isolation have been also used [18, 15] but these are often costly and are not suitable for large scale polymorphism screening. As an alternative to those technologies, many computational methods have been developed for phasing the genotypes (e.g. [1, 6, 7, 13, 19, 17, 5, 8, 12]).

Even though much of the attention was aimed at finding the haplotype phase, it is usually crucial to estimate correctly the haplotype frequencies in the population and not necessarily to phase the individual genotypes. For instance, in disease association studies, it is usually more informative to find the discrepancies between the control haplotype distribution and the cases haplotype distribution, than to find the phase of the haplotypes. The most likely estimation for the haplotype distribution in a population can be viewed as a weighted average over all possible phasing options. Therefore, finding the most likely phase and counting the number of occurrences of each haplotype could be used as a crude estimate for the haplotype distribution. On the other hand, in some cases this crude estimate may be inaccurate and more accurate frequency estimators are needed.

There are however a few EM-based (Expectation Maximization) algorithms that directly estimate the haplotype frequencies [3, 4, 10, 14]. These methods use a likelihood function based on the underlying assumption that the Hardy-Weinberg equilibrium holds (that the two haplotypes of an individual are independently drawn from the haplotype distribution in the population). In particular, those methods try to find a *haplotype* distribution which maximizes the probability of observing genotypes in the given sample, under the assumption of Hardy-Weinberg equilibrium.

One of the main drawbacks in all previous methods is that there is no guarantee that the algorithm converges to a global maximum, or that the algorithm converges in polynomial time. Both the convergence of the EM algorithm to a global optimum and its running time are heavily affected by the starting point of the algorithm which is usually a 'reasonable' guess or a random point.

We present a method called HAPLOFREQ which aims in overcoming the above limitations of previous approaches. Similarly to previous approaches, we use a likelihood function model. Our approach is different from previous approaches in the following aspects. First, we use an algorithm which is provably guaranteed to run efficiently and to find the haplotype distribution assuming that the number of samples is large enough and assuming a uniform error model. Second, we consider two different likelihood functions, one that assumes Hardy-Weinberg equilibrium and another that does not. The latter is used in order to find the *genotype* distribution given missing data, or the *haplotype* distribution given phased haplotypes with missing data. For instance, the phased haplotypes are given when sequencing chromosome X in men, or when sequencing the genome of certain organisms that are either haploid or have a short life span[1]

In the case where the Hardy-Weinberg equilibrium holds, the maximum likelihood function is a multinomial of very high degree. In order to find the maximum value of this multinomial we relax the problem by allowing the variables to be n-dimensional vectors instead of real numbers. We then use convex programming methods which involve linear constraints, multinomial functions and positive semidefinite constraints in order to find the maximum value of the relaxed problem. This relaxed objective function can be thought of as an alternative likelihood function since we show that the maximum value of the relaxed function approaches asymptotically to the haplotypes frequencies in the population.

We measured the performance of our algorithm over various data sets and compared it to the widely used program PHASE [19]. We found that our algorithm is consistently more accurate and much faster than PHASE. In Section 5 we describe our experiments and their results.

2 Estimating Haplotype Frequencies

One of the most natural tools in disease association studies is the search for discrepancies in the allele distribution between the cases and the controls. A natural extension of this tool is the search for discrepancies between the haplotype distribution in each of the populations. In particular, the haplotype frequency is calculated from the samples of each of the populations, and a statistical test (e.g. chi squared) is performed in order to assess whether the haplotype distributions of the two populations are identical. If the distributions are significantly different, then the region is likely to be correlated with the disease.

In order to estimate the haplotype frequencies in a population, a geneticist would sample a set of n individuals from the population. Throughout the paper we assume that these n individuals are independently sampled from a large population. Each sample consists of a genotype, which is the information of the two copies of the chromosome in each base. The haplotype information therefore

---

[1] For example in Drosophila, the phased haplotypes can be obtained by breeding.

has to be derived from the genotype information. Furthermore, the sequenced data usually contains some missing data, which adds another complexity to the problem. In this paper we focus on estimating the haplotype frequencies from the genotype data, with missing data. Our model and algorithms have natural extensions for other types of noise, such as sequencing errors. More detail on this issue in the full version.

2.1   A Maximum Likelihood Approach

In order to formalize the above scenario, we first need to set some notations and definitions. A *complete haplotype* is a binary string of length k. The values 0 and 1 correspond to the mutation and the wild type alleles. A *partial haplotype* is a string over $\{0, 1, *\}^k$. The character '*' corresponds to an unknown value (missing data).

We denote a genotype by a string over $\{0, 1, 2, *\}^k$, where 0,1 correspond to homozygous sites (i.e. the bases of the mother's chromosome and the father's chromosomes are the same), the value '2' corresponds to a heterozygous position, that is, a position where the mother chromosome carries a different base than the father chromosome and '*' corresponds to unknown values for both haplotypes. For a given genotype g or haplotype h, we denote by $g(i)$ ($h(i)$ respectively) its value in the i-th coordinate.

We say that a genotype $g \in \{0, 1, 2, *\}^k$, and a pair of complete haplotypes $h^1, h^2 \in \{0, 1\}^k$ are **compatible** if for every position i, if $g(i) \in \{0, 1\}$ then $h^1(i) = h^2(i) = g(i)$ and if $g(i) = 2$ then $h^1(i) \neq h^2(i)$.

For a genotype g, we define $\mathcal{C}(g)$ to be the set of pairs of haplotypes that are compatible with g. We assume that the genotypes admit a Hardy-Weinberg equilibrium, that is, that the two haplotypes of each individual are independently picked from the distribution of haplotypes in the population.

Let $\mathcal{P}$ be a distribution over the set of all possible complete haplotypes of length k. We denote by $p(h)$ the probability assigned to the haplotype h by $\mathcal{P}$. We consider the following likelihood function [3] of a set of partial genotypes $\mathcal{G}$ and a distribution $\mathcal{P}$:

$$\mathcal{L}(\mathcal{G}, \mathcal{P}) = \prod_{g \in \mathcal{G}} \sum_{(h_1, h_2) \in \mathcal{C}(g)} p(h_1) p(h_2). \tag{1}$$

The function $\mathcal{L}(\mathcal{G}, \mathcal{P})$ is simply the probability of observing the genotypes $\mathcal{G}$ in a random sample of the population under Hardy-Weinberg equilibrium, given that the distribution of complete haplotypes in the population is $\mathcal{P}$ and that the distribution of missing data in a genotype g does not depend on the contents of g.

When the sample size approaches infinity, the maximum likelihood is attained when $\mathcal{P}$ is the actual distribution of haplotypes in the population[2]. Therefore,

---

[2] This is true under some reasonable assumptions on the distribution of the missing data.

it is only natural to aim in finding the distribution $\mathcal{P}$ which maximizes the likelihood and to estimate the distribution of haplotypes in the population as $\mathcal{P}$. Previous methods [3, 4, 19] use Expectation Maximization (EM) in order to find the maximum likelihood. When using EM, both the running time and the convergence to a global maximum depend on the starting point. In particular, these algorithms may be exponential, and they may give a non-optimal solution, even when the number of samples is large. In Section 4 we introduce an alternative approach which is guaranteed to converge to a global optimum of another likelihood function $\mathcal{L}_2(\mathcal{G}, \mathcal{P})$. We further show in Section 4.1 that $\mathcal{L}$ and $\mathcal{L}_2$ have the same asymptotic behavior under Hardy-Weinberg.

2.2   Working with Phased Data

In some cases we are given the phased genotypes, possibly with missing data. For instance, some sequencing techniques provide the haplotypes and not the genotypes. In haploid organisms, or in diploid organisms with short life span such as drosophila[3] we can get the phased haplotypes. It is therefore interesting to estimate the haplotype frequencies given a sample of haplotypes with missing data. This approach may also be useful to estimate the *genotype* frequencies, given a sample of genotypes with missing data. The latter may be particularly important when there are departures from Hardy-Weinberg equilibrium in the underlying genotype distribution.

In order to formalize the above scenario, we need to introduce a few more notations and definitions. We say that a partial haplotype $h_1 \in \{0, 1, *\}^k$ is **consistent** with a complete haplotype $h_2 \in \{0, 1\}^k$ if they share the same values whenever $h_1(i) \neq *$. Given a partial haplotype h, we define $\mathcal{C}(h)$ to be the set of complete haplotypes that are consistent with h.

As before, let $\mathcal{P}$ be a distribution over the set of all possible complete haplotypes of length k. Given the set of partial haplotypes $\mathcal{H}$, the likelihood of $\mathcal{P}$ is given by

$$\mathcal{L}(\mathcal{H}, \mathcal{P}) = \prod_{h \in \mathcal{H}} \sum_{h' \in \mathcal{C}(h)} p(h').$$

The function $\mathcal{L}(\mathcal{H}, \mathcal{P})$ is simply the probability of observing the partial haplotypes $\mathcal{H}$ in a random sample of the population, given that the distribution of complete haplotypes in the population is $\mathcal{P}$ and that the distribution of missing data in a haplotype h does not depend on the contents of h.

Again, in order to estimate the haplotype frequencies, we find the distribution $\mathcal{P}$ that maximizes the likelihood $(\mathcal{H}, \mathcal{P})$. In Section 3 we introduced an efficient polynomial time algorithm that finds the global maximum of $(\mathcal{H}, \mathcal{P})$. This is quite surprising given that we essentially find a maximum point of a polynomial of potentially high degree. In general, finding an extremum of a polynomial is an intractable problem.

---

[3] In diploid organisms the haplotype data is found through breeding.

3 Estimating Haplotype Frequencies from a Phased Sample

In this section we introduce an algorithm which estimates the haplotype frequencies in a population given a sample of phased haplotypes with missing data.

Formally, given a set $\mathcal{H}$ of n partial haplotypes, we are interested in finding a distribution $\mathcal{P}$ which maximizes the function $\mathcal{L}(\mathcal{H}, \mathcal{P})$ which is given in the previous section. Thus, finding the distribution of maximum likelihood can be done by solving the following mathematical programming problem:

$$\text{Maximize } \prod_{h \in \mathcal{H}} \sum_{h' \in \mathcal{C}(h)} p(h')$$
$$\text{s.t. } \sum_{h \in \{0,1\}^k} p(h) = 1$$
$$p(h) \geq 0, \qquad h \in \{0,1\}^k$$

We will use the following definition in order to simplify the notations.

Definition 1. *Given a partial haplotype $h \in \{0, 1, *\}^k$ and a set of haplotypes $S = \{h_1, ..., h_m\} \subseteq \{0,1\}^k$, define the **compatibility vector** of h with respect to S as a vector $A_h \in \{0,1\}^m$ such that $A_h(i) = 1$ if $h_i \in \mathcal{C}(h)$ and $A_h(i) = 0$ otherwise.*

Note that in practice the values of k are relatively small, and the set of possible haplotypes is limited to a reasonable size. A typical value for k is in the range of $10 - 50$, and there are typically at most a few hundreds of possible haplotypes, that is, haplotypes that are compatible with one of the genotypes.

Using this definition, the maximum likelihood formulation above is equivalent to solving the following problem:

Definition 2 (FREQUENCY ESTIMATION OF PHASED GENOTYPES).
Input: *A matrix $A \in \{0,1\}^{n \times m}$ consisting of n row vectors $\{A_1, ..., A_n\} \in \{0,1\}^m$*
Goal: *Find a vector $p \in \Re_+^n$, such that:*

1. *$\sum_{i=1}^m p_i = 1$; $\forall i \ p_i \geq 0$*
2. *Let $q \stackrel{def}{=} A \cdot p$. Then the following quantity is maximized: $f(p) = \prod_{i=1}^n q_i$*

3.1 Algorithms for FREQUENCY ESTIMATION OF PHASED GENOTYPES

In the full version of the paper we prove the following theorem using a variant of the Ellipsoid algorithm[11].

Theorem 1. FREQUENCY ESTIMATION OF PHASED GENOTYPES *is solvable in polynomial time.*

We will now describe a much more efficient algorithm which, for every $\varepsilon > 0$, finds a $(1 + \epsilon)$-approximate solution in time polynomial in the input size and $\frac{1}{\epsilon}$.

Let the input matrix be $A \in \{0,1\}^{n \times m}$. Denote $q = A \cdot p$ and denote the solution vector (optimal probabilities assigned to the haplotypes) by o. Also define $w = A \cdot o$ as the "weights" vector of the optimum solution. Let f be the objective function, that is $f(x) = \prod_i (A_i x)$.

It is easy to see that we can always obtain an objective value of $f(p) \geq \left(\frac{1}{n}\right)^n$ by picking a non-empty column from each row and then assigning the uniform distribution over all columns picked. Alternatively, we can obtain an initial value of $f(p) \geq \left(\frac{1}{m}\right)^n$ by the uniform distribution over all columns (assigning all probabilities to be $\frac{1}{m}$).

Let $\tau = \min_{p_i > 0} p_i$. In particular, we have that every non-zero p_i or q_i value is at least τ. In the full version of the paper we show that one can assume that $\tau \geq 1/m^2$. In practice τ is normally a constant in the order of 0.001.

Our algorithm is a hill climbing algorithm. We start from one of the trivial solutions above, and then make a series of improvement steps until required performance guarantee is reached. In each improvement step we amend the current vector of probabilities p to $p' = p + \delta$, such that to improve the overall value. The algorithm, called HAPLOFREQ which takes as an input a precision parameter ε is given in Figure 1.

Procedure HAPLOFREQ(ε)
$p \leftarrow 1 \cdot \frac{1}{m}$
$\forall i$ set $q_i \leftarrow A_i p$
$\delta \leftarrow$ FINDDELTA(p, q, A)
while $\sum_i \frac{A_i \delta}{q_i} \geq \varepsilon$ **do**

   Update p to be: $p \leftarrow p + \frac{\tau^2 \varepsilon}{8n} \delta$
   $\delta \leftarrow$ FINDDELTA(p, q, A)
return p

Fig. 1. Algorithm HAPLOFREQ

The only missing part in the description of HAPLOFREQ is the way we find the vector δ. This part is done in the procedure FINDDELTA. We need to make sure that the procedure FINDDELTA finds a vector δ such that $p + \frac{\tau^2 \varepsilon}{8n} \delta$ is an improved solution to the problem.

Definition 3. *Define a ε-good vector with respect to a current solution p as a vector δ that satisfies:*

$$\sum_{i=1}^{m} \delta_i = 0 \ ; \quad 0 \leq \delta_i + p_i \leq 1 \ ; \quad \sum_{i=1}^{n} \frac{A_i \delta}{q_i} \geq \varepsilon$$

The procedure FINDDELTA returns an ε-good vector if one exists. Note that FINDDELTA can be implemented using linear programming, but this is not very efficient. The proof for the following lemma is given in the full version of the paper.

Lemma 1. *The procedure FINDDELTA finds a ε-good vector if one exists, and it can be implemented in time $\tilde{O}(nm)$. Furthermore, it returns a vector δ such that $|A_i \cdot \delta| \leq 2$.*

(Here an in the rest of the paper the $\tilde{O}$ notation is used to suppress polylog-arithmic factors.) In the rest of this section we prove the following theorem:

Theorem 2 (Main). *For any constant $\varepsilon > 0$, the algorithm HAPLOFREQ(ε) finds a $(1 + \varepsilon)$-approximate solution in polynomial time.*

To prove this theorem, we first prove that we can always find an ε-good vector if our current solution is not a e^ε-approximate solution. We then show that using a ε-good vector we can improve our current solution, and that polynomially many improvements suffice to obtain a $(1 + \varepsilon)$-approximate solution.

Lemma 2. *If $\frac{OPT}{ALG} = \frac{f(o)}{f(p)} \geq e^\varepsilon$, then there exists an ε-good vector δ.*

Proof. The optimal solution gives rise to a natural vector $\delta := o - p$. It obviously satisfies the first two conditions above, and as for the last:

$$\sum_{i=1}^{n} \frac{A_i \delta}{q_i} = \sum_{i=1}^{n} \frac{w_i - q_i}{q_i} \geq n \cdot \sqrt[n]{\prod_{i=1}^{n} \frac{w_i}{q_i}} - n \geq n \cdot \sqrt[n]{e^\varepsilon} - n = n \cdot (e^{\varepsilon/n} - 1) \geq \varepsilon$$

where the first inequality follows from the arithmetic and geometric mean inequality and the last inequality is true since $e^x > x + 1$.

Lemma 2 shows that if we are still far from the optimal solution then there is at least one ε-good vector. Since one such vector exists, FINDDELTA is guarantied to provide such a vector. We now show that the resulting improvement step brings us closer to the optimum.

Lemma 3. *Let $\varepsilon > 0$ and let δ be a ε-good vector with respect to p. As before, let $\tau = \min_{p_i > 0} p_i$. Let $p' := p + \sigma \delta$ where $\sigma = \frac{\tau^2 \varepsilon}{8n}$. Then $\frac{f(p')}{f(p)} \geq e^{\varepsilon \sigma / 2}$.*

Proof. Denote $c_i := \frac{A_i \delta}{q_i}$. By Lemma 1 we know that $|A_i \delta| \leq 2$, and therefore $|c_i| \leq \frac{2}{\tau}$ (since $q_i \geq \tau$). Hence:

$$\log \left(\frac{f(p')}{f(p)} \right) = \log \left(\prod_{i=1}^{n} \frac{A_i(p + \sigma \delta)}{q_i} \right) = \sum_{i=1}^{n} \log \frac{A_i(p + \sigma \delta)}{q_i} = \sum_{i=1}^{n} \log(1 + \sigma c_i)$$

$$\geq \sum_{i=1}^{n} \left[(\sigma c_i) - (\sigma c_i)^2 \right] \geq \sigma \varepsilon - \sigma^2 \sum_{i=1}^{n} c_i^2 \geq \sigma \varepsilon - \frac{4n\sigma^2}{\tau^2} \geq \frac{\varepsilon \sigma}{2},$$

where the approximation to the logarithm holds since $|\sigma c_i| \leq \frac{1}{2}$.

Now we can prove Theorem 2:

Proof (Theorem 2). Let p_i be the vector obtained by the algorithm after i improvement steps, and let $a_i = \log(f(p_i))$. Let o be an optimal solution and let

$opt = \log(f(o))$. By Lemma 3, $a_{i+1} \geq a_i + \sigma(opt - a_i)/2$, where $\sigma = \frac{\tau^2 \epsilon}{8n}$, as long as $opt - a_i \geq \epsilon$. Therefore, as long as $opt - a_i \geq \epsilon$, we get that

$$opt - a_{i+1} \leq (1 - \frac{\tau^2 \epsilon}{8n})(opt - a_i).$$

Since we can start from a solution of value at least n^{-n} and the optimal solution is bounded by 1, we have that $opt - a_0 \leq n \log n$. Therefore, after $r = \tilde{O}(\frac{n}{\tau^2 \epsilon})$ iterations we find a solution such that $a_r \geq opt - \epsilon$.

Note that our analysis of the running time of HAPLOFREQ so far had a polynomial dependence on the precision parameter τ. In the full version of the paper we prove the following lemma, which shows that the running time is indeed polynomial in the size of the problem (or rather that we may assume that τ is large).

Lemma 4. *For each solution* p *throughout the algorithm it holds that* $\min_i q_i \geq \frac{1}{m^2}$.

In practical instances, the size of τ is a constant, and therefore the algorithm perform $\tilde{O}(n)$ iterations. In the worst case, the algorithm performs $\tilde{O}(nm^4)$ iterations.

4 Estimating Haplotype Frequencies from Unphased Genotypes

We now turn to the case where we have a set of genotypes and our goal is to find the frequencies of the underlying haplotypes. Recall that under the Hardy-Weinberg equilibrium, the likelihood function of a set of genotypes $\mathcal{G}$ and a distribution $\mathcal{P}$ is given by equation (1). Thus, finding the haplotype distribution with the maximum likelihood can be done by solving the following mathematical programming problem:

$$\text{Maximize } \prod_{g \in \mathcal{G}} \sum_{(h_1,h_2) \in \mathcal{C}(g)} p(h_1)p(h_2)$$
$$\text{s.t.} \quad \sum_{h \in \{0,1\}^k} p(h) = 1$$
$$p(h) \geq 0, \quad\quad h \in \{0,1\}^k$$

We follow the approach used for phased data, and try to solve this mathematical program by first abstracting it out. We first need the following definition, which is analogous to Definition 1.

Definition 4. *Given a genotype* $g \in \{0,1,2,*\}^k$ *and a set of haplotypes* $S = \{h_1, ..., h_m\} \subseteq \{0,1\}^k$, *define the (symmetric)* **compatibility matrix** *of* g *with respect to* S *as a matrix* $A^g \in \{0,1\}^{m \times m}$ *such that* $A^g_{ij} = 1$ *if* $(h_i, h_j) \in \mathcal{C}(g)$ *and* $A^g_{ij} = 0$ *otherwise.*

It is easy to verify that the maximum likelihood formulation given above can be solved if the following problem can be solved:

Definition 5 (Frequency Estimation of Unphased Genotypes).
Input: *A set of matrices* $\{A_1, ..., A_n\} \in \{0,1\}^{m \times m}$
Goal: *Let* $\mathcal{P} \subseteq [0,1]^m$ *be the polytope of all probability distribution vectors over* m *elements* $\boldsymbol{p} \in \Re^m$ *(that is, the set of all vectors* $\boldsymbol{p}$ *such that* $\forall i\ p_i \geq 0$ *and* $\sum_i p_i = 1$*). Find the vector in* $\mathcal{P}$ *that maximizes the product* $\prod_i \boldsymbol{p}^T A_i \boldsymbol{p}$*. Formally:*

$$\max_{\boldsymbol{p} \in \mathcal{P}} f(\boldsymbol{p}) = \max_{\boldsymbol{p} \in \mathcal{P}} \prod_{i=1}^{n} \boldsymbol{p}^T A_i \boldsymbol{p}$$

Unfortunately, the above mathematical program is NP-hard (details are omitted from this version). We therefore suggest to use a different likelihood function $\mathcal{L}_2$ which can be thought of as a relaxation of $\mathcal{L}$. Instead of having a distribution $\mathcal{P}$ over the haplotypes, we assign to each haplotype h_i a k-dimensional vector $\boldsymbol{v}_i$, such that $\sum_{i=1}^{k} \boldsymbol{v}_i = v_0$ where $\|v_0\| = 1$. The likelihood $\mathcal{L}_2$ is now defined as a function of $\mathcal{G}$ and of $\mathcal{V} = \{\boldsymbol{v_1}, \ldots, \boldsymbol{v_m}\}$:

$$\mathcal{L}_2(\mathcal{G}, \mathcal{V}) = \prod_{g \in \mathcal{G}} \sum_{(h_i, h_j) \in \mathcal{C}(g)} \boldsymbol{v}_i \cdot \boldsymbol{v}_j.$$

We call $\mathcal{V}$ a vector distribution of the haplotypes. Note that if we restrict the vectors of $\mathcal{V}$ to be in one dimensional space then $\mathcal{V}$ is a probability distribution and $\mathcal{L}_2(\mathcal{G}, \mathcal{V}) = \mathcal{L}(\mathcal{G}, \mathcal{V})$. The vectors $\mathcal{V}$ can be represented by a matrix $\mathcal{P}$ such that $P_{ij} = \boldsymbol{v}_i \cdot \boldsymbol{v}_j$, i.e., P_{ij} is the scalar product of $\boldsymbol{v}_i$ and $\boldsymbol{v}_j$. Such a matrix P is called a positive semidefinite (PSD) matrix (see [20]). Therefore, an analogous problem to Frequency Estimation of Unphased Genotypes is the following:

Definition 6 (Maximum Relaxed Unphased Likelihood).
Input: *A set of matrices* $\{A_1, ..., A_n\} \in \{0,1\}^{m \times m}$
Goal: *Let* $\mathcal{Q}$ *be the cone of all positive-semi-definite matrices* $P \in \Re^{m \times m}$ *that satisfy* $\sum_{i,j} P_{ij} = 1$ *,* $\forall i, j\ .P_{ij} \geq 0$*. Find the PSD matrix in* $P \in \mathcal{Q}$ *that maximizes the product* $\prod_i A_i \bullet P$ *(where* $\bullet$ *stands for the Frobenius inner product[4]) Formally:*

$$\max_{p \in \mathcal{Q}} f(P) = \max_{P \in \mathcal{Q}} \prod_{i=1}^{n} A_i \bullet P$$

Clearly, if we can solve Maximum Relaxed Unphased Likelihood we could find the vector distribution $\boldsymbol{V}$ which maximizes $\mathcal{L}_2(G, V)$. In Section 4.2 we introduce an efficient algorithm which solves Maximum Relaxed Unphased Likelihood in polynomial time.

---

[4] The Frobenius inner product of matrices X and Y is $\sum_{i,j} X_{ij} Y_{ij}$.

4.1 Asymptotic Behavior of the Likelihood Function

Finding the maximum likelihood of $\mathcal{L}_2$ does not ensure us that we will converge to the correct haplotype distribution when the number of samples is sufficiently large. We now turn to show that under Hardy-Weinberg equilibrium, and under the assumption that there is no missing data, if the sample size is large enough, the maximum of $\mathcal{L}_2(\mathcal{G}, \mathcal{V})$ is attained in a point which converges to the correct distribution.

Lemma 5. *Under the Hardy-Weinberg, the solution to* MAXIMUM RELAXED UNPHASED LIKELIHOOD *converges to the haplotype frequencies in the population.*

Proof. Let the set of sampled genotypes be $\mathcal{G}$. Denote by $p(g)$ the frequency of genotype g in the population. Therefore, $p(g)$ is the probability to sample a genotype $g \in \mathcal{G}$. When the number of samples goes to infinity, the ratio of sampled genotypes g approaches $p(g)$. If the ratio is exactly $p(g)$, then maximizing $\mathcal{L}_2(\mathcal{G}, \mathcal{V})$ is equivalent to maximizing the function

$$\prod_{g \in \mathcal{G}} \left(\sum_{h_i, h_j \in \mathcal{C}(g)} v_i \cdot v_j \right)^{p_g}$$

It is easy to see that this objective is maximized when:

$$\forall_{g \in \mathcal{G}} \sum_{h_i, h_j \in \mathcal{C}(g)} v_i \cdot v_j = p_g.$$

As $|\mathcal{G}| \mapsto \infty$, we know that $p_g \mapsto \sum_{i,j \in \mathcal{C}(g)} p_i p_j$. Therefore, one optimal solution to this equation system is the solution $v_i = p_i$. Observe that equations above imply that the homozygous genotype g_{ii} with haplotype h_i satisfies that $p_i^2 = \|v_i\|^2$. These restrictions, together with the rest of the constraints, determine $\mathcal{V}$ uniquely. We can now use the fact that the function $\max_{\mathcal{V}} \mathcal{L}_2(\mathcal{G}, \mathcal{V})$ is a continuous function G in order to complete the proof.

Now that we know that the solution of MAXIMUM RELAXED UNPHASED LIKELIHOOD converges to the correct solution, in particular we know that for large enough sample the vectors v_i should be closed to one dimensional. We therefore define $p_i = v_i \cdot 1$ as the suggested probability distribution. By Lemma 5, as the number of samples grow, the probabilities p_i get closer to to the true frequencies in the population.

4.2 A Polynomial Time Approximation Algorithm for MAXIMUM RELAXED UNPHASED LIKELIHOOD

In this subsection we give a hint regarding the polynomial time algorithm for the MAXIMUM RELAXED UNPHASED LIKELIHOOD. The complete version of this paper will fill in the remaining details.

The general framework for our algorithm is identical to the algorithm for the linear case. An initial solution can easily be obtained by assigning all probabilities

---

Procedure HAPLOFREQ2(ε)
$P \leftarrow J \cdot m^{-2}$
set $q_i \leftarrow A_i \bullet P$
$\Delta \leftarrow$ FINDPSDDELTA($P, q, \{A_i\}$)
while $\sum_i \frac{A_i \bullet \Delta}{q_i} \geq \varepsilon$ **do**
    Update P to be: $P \leftarrow P + \frac{\tau^2 \varepsilon}{2m} \Delta$
    $\Delta \leftarrow$ FINDPSDDELTA($P, q, \{A_i\}$)
return P

Fig. 2. Algorithm HAPLOFREQ2 to estimate the haplotype frequencies given unphased data. The procedure FINDPSDDELTA is similar to the procedure FINDDELTA and is given in the full version of the paper

to be $\frac{1}{m}$ (that is, a positive semidefinite matrix P where $p_{ij} = \frac{1}{m^2}$). Starting from this trivial solution, the algorithm makes a series of local improvements until the required performance guarantee is reached. The algorithm, called HAPLOFREQ2, is given in Figure 2.

In the full version of the paper, we prove the following

Theorem 3. *For any constant $\varepsilon > 0$, the algorithm* HAPLOFREQ2*(ε) finds a $(1 + \varepsilon)$-approximate solution in polynomial time.*

The proof is similar in nature to the proof of Theorem 2, with several technical points that need attention.

One technically concerns the amendment matrix Δ. At every improvement step the resulting matrix should again be positive semidefinite. The optimization routine used to find this amendment matrix, called FINDPSDDELTA, uses semidefinite programming for this task.

A semidefinite program is similar to a linear program with additional constraints that some of the variables form a positive semidefinite matrix, and is known to be solvable in polynomial time [20].

5   Experimental Results

We implemented the algorithms HAPLOFREQ and HAPLOFREQ2 (described in Sections 3.1 and 4.2 respectively) and compared them to the widely used software PHASE [19].

The data sets. We applied our algorithm to the data set studied in [2]. This data set is a 500 kilobase region of chromosome 5q31 containing 103 SNPs. In this study, genotypes for the 103 SNPs are collected from 129 mother, father and child trios from European derived population in an attempt to identify a genetic risk for Crohn's disease. In order to evaluate the performance of HAPLOFREQ we need to know what the underlying distribution in the population is. We therefore simulated data in the following way: We first partitioned the data into nonoverlapping contigous regions containing $5, 12$ and 19 SNPs (thus resulting in twenty regions with five SNPs, eight regions with twenty, and five regions with nineteen SNPs). We phased each of these regions using PHASE [19]. The resulting phased haplotypes induce a distribution. We then generated more data sets by picking haplotypes randomly and independently from that distribution, and by adding randomly scattered missing data and randomly scattered sequencing errors. We generated this way sets of $25, 100$ and 500 genotypes. This resulted in approximately 700 sets of genotypes, with varying amount of missing data (up to 10%) and varying amount of sequencing errors (up to 1%). Furthermore, the different number of SNPs per dataset and the different number of genotypes per dataset allow us to measure the performance of the algorithms in different scenarios. Note that these simulations implicitly assume that the underlying genotype distribution in the population satisfies the Hardy-Weinberg Equilibrium. On the other hand, when we sample from that distribution, the sampling deviations result in departures from Hardy-Weinberg.

In the study of [2], they partitioned the data into 11 haplotype blocks. We use these blocks in order to generate more data sets. For each of the blocks we use the trios in order to deduce the haplotypes of the children. In order to estimate the haplotype distribution we simply count the non-ambiguous haplotypes.

Implementation details. Both HAPLOFREQ and HAPLOFREQ2 assume that the number of possible haplotypes is limited - and usually small. In many scenarios this is actually the case, but in order to avoid extensive running time in the other cases, we use a preprocessing mechanism which filters out haplotypes that seem to have an extremely small frequency. The preprocessing mechanism is based on a greedy procedure, similar to the one given in [9]. After the preprocessing we are typically left with about 50 possible haplotypes on which we run our algorithms.

In each iteration of HAPLOFREQ2 we have to solve a semidefinite program. Even though semidefinite programs can be solved in polynomial time, existing software packages are relatively slow. We therefore implemented a semidefinite programming solver which is specifically tailored for our needs. This implementation takes advantage of some properties of our algorithm in order to speed up the semidefinite solver.

The resulting implementation of both algorithms is very efficient. In particular, HAPLOFREQ (on haplotypes) typically runs 15 to 25 times faster than PHASE and HAPLOFREQ2 (on genotypes) typically runs 3 to 10 times faster. Figure 3 gives a concise comparison of measured running times of PHASE, HAPLOFREQ and HAPLOFREQ2.

Distance measures. We use two measures for the distance between two distributions. The first measure is the l_1 norm of the difference between the two distributions. Given two distributions, $\{p_1, \ldots, p_k\}$ and $\{q_1, \ldots, q_k\}$, the l_1 norm of their difference is defined as $\sum_{i=1}^{k} |p_i - q_i|$. We also used the chi-square difference, that is, $\sum_{i=1}^{k} \frac{(p_i - q_i)^2}{q_i}$. The chi-squared distance is particularly interesting since when an association study is performed, one uses the chi-squared test in order to test the hypothesis that the two underlying distributions are the same.

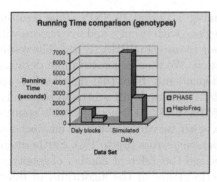 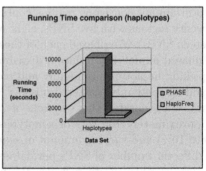

Fig. 3. The running times of PHASE vs. HAPLOFREQ and HAPLOFREQ2 on the data sets taken from [2]

For the first metric, we sum only over the probabilities q_i that are greater than 0.01, and for the second over those that are greater then 0.05.

Accuracy of estimations. We compared the accuracy of the frequency estimations of our HAPLOFREQ2 to PHASE [19]. For each of the simulated data sets and the blocks of [2], we ran both PHASE and HAPLOFREQ2. We found that HAPLOFREQ2 is typically $10 - 50\%$ more accurate than PHASE. In general, we found that the longer the regions are the bigger the difference between the two algorithms. Furthermore, adding missing data and sequencing errors also increases the difference in the performance of the two algorithms. On the other hand, increasing the number of genotypes reduces the difference between the two methods.

We also measured the performance of the algorithm HAPLOFREQ with the performance of PHASE. For this comparison, we used the phased haplotype information as the input (since the data sets are simulated we can have the

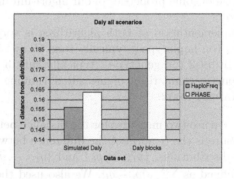

Fig. 4. Average l_1 distance from the actual distribution on the Daly data sets (both blocks and simulated)

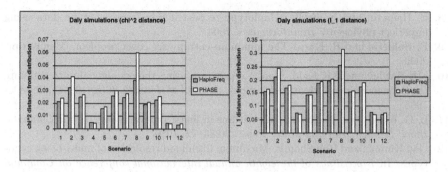

Fig. 5. Average chi^2 and l_1 distances from the actual distribution on the simulated Daly data, with various simulation parameters (i.e amount of missing data, number of genotypes, etc.)

phased haplotypes). Since PHASE assumes that the haplotypes are not phased, we gave PHASE as an input a homozygous genotype for each of the haplotypes in our data sets. In order to get the haplotypes in the Daly blocks, we used the trios information and set as missing each of the ambiguous SNPs. We found again that HAPLOFREQ is typically $10 - 50\%$ more accurate than PHASE.

We note that the deviations caused by sampling have an enormous affect on all three algorithms. In particular, the distributions found by all three algorithms (PHASE, HAPLOFREQ and HAPLOFREQ2) are much closer to the sampled distribution than to the underlying population distribution. A complete summary of the comparison can be found in Figures 4 and 5.

References

1. AG Clark. Inference of haplotypes from pcr-amplified samples of diploid populations. *Journal of Molecular Biology and Evolution*, 7(2):111–22, Mar 1990.
2. MJ Daly, JD Rioux, SF Schaffner, TJ Hudson, and ES Lander. High-resolution haplotype structure in the human genome. *Nature Genetics*, 29(2):229–32, Oct 2001.
3. L Excoffier and M Slatkin. Maximum-likelihood estimation of molecular haplotype frequencies in a diploid population. *Molecular Biology and Evolution*, 12(5):921–7, Sept 1995.
4. D. Fallin and NJ. Schork. Accuracy of haplotype frequency estimation for biallelic loci, via the expectation-maximization algorithm for unphased diploid genotype data. *American Journal of Human Genetics*, 67:947–959, 2000.
5. D Gusfield. Haplotyping as perfect phylogeny: Conceptual framework and efficient solutions. In *Proceedings of the 6th Annual International Conference on (Research in) Computational (Molecular) Biology*, 2002.
6. D Gusfield. A practical algorithm for optimal inference of haplotypes from diploid populations. In *Proceedings of the 8th International Conference on Intelligent Systems for Molecular Biology (ISMB)*, 2000.
7. D Gusfield. Inference of haplotypes from samples of diploid populations: complexity and algorithms. *Journal of Computational Biology*, 8(3):305–23, 2001.

8. E. Halperin and E. Eskin. Haplotype reconstruction from genotype data using imperfect phylogeny. *Bioinformatics*, 2004.

9. E. Halperin and R. Karp. The minimum-entropy set cover problem. Manuscript, 2003.

10. ME Hawley and KK Kidd. Haplo: a program using the em algorithm to estimate the frequencies of multi-site haplotypes. *Journal of Heredity*, 86(5):409–11, Sep-Oct 1995.

11. L. G. Khachiyan. Polynomial algorithms in linear programming. *USSR Computational Mathematics and Math. Phys.*, 20:53–72, 1980.

12. Gad Kimmel and Ron Shamir. Maximum likelihood resolution of multi-block genotypes. In *Proceedings of the eighth annual international conference on Computational molecular biology*, pages 2–9. ACM Press, 2004.

13. G. Lancia, V. Bafna, S. Istrail, R. Lippert, and R. Schwartz. Snps problems, algorithms and complexity, european symposium on algorithms. In Springer-Verlag, editor, *Proceedings of the European Symposium on Algorithms (ESA-2001), Lecture Notes in Computer Science*, volume 2161, pages 182–193, 2001.

14. JC Long, RC Williams, and M Urbanek. An e-m algorithm and testing strategy for multiple-locus haplotypes. *American Journal of Human Genetics*, 56(3):799–810, Mar 1995.

15. S. Michalatos-Beloin, SA. Tishkoff, KL. Bently, KK. Kidd, and G. Ruano. Molecular haplotyping of genetic markers 10 kb apart by allele-specific long-range pcr. *Nucleic Acids Res*, 24:4841–4843, 1996.

16. NIH. Large-scale genotyping for the haplotype map of the human genome. RFA: HG-02-005, 2002.

17. Niu, Qin, Xu, and Liu. In silico haplotype determination of a vast set of single nucleotide polymorphisms. Technical report, Department of Statistics, Harvard University, 2001.

18. N Patil, AJ Berno, DA Hinds, WA Barrett, JM Doshi, CR Hacker, CR Kautzer, DH Lee, C Marjoribanks, DP McDonough, BT Nguyen, MC Norris, JB Sheehan, N Shen, D Stern, RP Stokowski, DJ Thomas, MO Trulson, KR Vyas, KA Frazer, SP Fodor, and DR Cox. Blocks of limited haplotype diversity revealed by high-resolution scanning of human chromosome 21. *Science*, 294(5547):1719–23, Nov 23 2001.

19. M. Stephens, N. Smith, and P. Donnelly. A new statistical method for haplotype reconstruction from population data. *American Journal of Human Genetics*, 68:978–989, 2001.

20. Henry Wolkowicz, Romesh Saigala, and Lieven Vandenberghe. Handbook of semidefinite programming. *International Series in Operations Research and Management Science*, 27, 2000.

Improved Recombination Lower Bounds for Haplotype Data

Vineet Bafna and Vikas Bansal

Department of Computer Science and Engineering,
University of California at San Diego, La Jolla, CA 92093-0114, USA
{vbafna, vibansal}@cs.ucsd.edu

Abstract. Recombination is an important evolutionary mechanism responsible for the genetic diversity in humans and other organisms. Recently, there has been extensive research on understanding the fine scale variation in recombination rates across the human genome using DNA polymorphism data. A combinatorial approach toward this is to estimate the minimum number of recombination events in any history of the sample. Recently, Myers and Griffiths [1] proposed two measures, R_h and R_s, that give lower bounds on the minimum number of recombination events. In this paper, we provide new and improved methods (both in terms of running time and ability to detect past recombination events) for computing recombination lower bounds. Our principal results include:

- We show that computing the lower bound R_h is NP-hard and adapt the greedy algorithm for the set cover problem [2] to obtain a polynomial time algorithm for computing a diversity based bound R_g. This algorithm is several orders of magnitude faster than the Recmin program [1] and the bound R_g matches the bound R_h almost always.
- We also show that computing the lower bound R_s is also NP-hard using a reduction from MAX-2SAT. We give a $O(m2^n)$ time algorithm for computing R_s for a dataset with n haplotypes and m SNP's. We propose a new bound R_I which extends the history based bound R_s using the notion of intermediate haplotypes. This bound detects more recombination events than both R_h and R_s bounds on many real datasets.
- We extend our algorithms for computing R_g and R_s to obtain lower bounds for haplotypes with missing data. These methods can detect more recombination events for the LPL dataset [3] than previous bounds and provide stronger evidence for the presence of a recombination hotspot.
- We apply our lower bounds to a real dataset [4] and demonstrate that these can provide a good indication for the presence and the location of *recombination hotspots*.

1   Introduction

Recombination is one of the major evolutionary mechanisms responsible for genetic diversity in many organisms. Although all genetic variation starts from mu-

S. Miyano et al. (Eds.): RECOMB 2005, LNBI 3500, pp. 569–584, 2005.

tation, recombination can give rise to new variants by combining types already present in the population. Recombination tends to break the dependence among alleles on either side of the crossover and hence reduce the Linkage Disequilibrium (LD). Recent studies of human polymorphism data ([5], [6], [7]) suggested an interesting block like structure of the genome, where long stretches of the human genome known as *LD blocks* (with high LD) show signs of little or no recombination and the recombination events occur in so called *recombination hot-spots*. Jeffreys et al. [4] analyzed a 216-kb region from the MHC region using sperm typing and identified clusters of recombination hotspots separated by long regions (60-90 kbs) of low diversity. However, the experimental determination of recombination rates at high resolution is technically difficult and costly. An alternative approach is to use population genetics data to infer the fine-scale variations in recombination rates. A variety of statistical methods based on different population genetics models have been proposed to estimate recombination rates from polymorphism data (see e.g. [8], [9], [10], [11]). The emergence of genome-wide diversity studies, such as the HapMap project[12], has accelerated efforts towards constructing a fine-scale recombination map of the human genome. More recently, two large scale studies [13, 14] have shown fine-scale recombination rate variation and recombination hotspots to be a ubiquitous feature of the human genome.

In contrast to model based methods to infer recombination rates, an alternative approach for characterizing the heterogeneity in recombination is to obtain a direct count of past recombination events from population genetics data. Population genetics data, in particular haplotype data contains signature patterns left behind by past recombination events. A parsimonious approach to counting recombination events is to compute the minimum number of recombination events required to explain any evolutionary history of the sample assuming that each segregating sites mutates only once. This may be achieved by trying to reconstruct the underlying graph or phylogenetic network that uses the minimum number of recombination events. This problem is computationally challenging and has resisted efforts for even an exponential time algorithm [15, 16, 17] (see [18] and [19] for some recent work on efficient algorithms for restricted versions of this problem). Therefore, research in this area has focused on computing lower bounds on the minimum number of recombination events. Although most historical recombination cannot be recovered, one expects that the number of recombination events detected for a particular genomic region is a good indicator of the underlying recombination rate for that region. Myers and Griffiths [1] demonstrated the R_h lower bound to be much more powerful than previous lower bounds in detecting recombination events through simulation studies and found a strong clustering of recombination events in the center of the lipoprotein lipase gene [3].This region has previously been characterized to be a putative recombination hotspot [20]. Fearnhead et al. [21] applied the R_h method of Myers and Griffiths [1] to detect recombination events in the β-globin gene cluster which has a well-characterized recombination hotspot. They found that the results obtained using this method were consistent with their estimates obtained using a full likelihood method.

1.1   Our Contribution

In this paper, our objective is to explore the problem of computing lower bounds on the number of recombination events both from an algorithmic and application perspective.

We provide a theoretical formulation for the lower bound, R_h and show that it is NP-hard to compute this bound. However, on the positive side, using the greedy algorithm for the set cover problem [2], we present a $O(mn^2)$ time algorithm which computes a lower bound R_g for a dataset with n rows and m segregating sites. This algorithm outperforms the Recmin program [1] by several orders of magnitude on large datasets (e.g. the Daly dataset [6]) and finds almost identical lower bounds.

We also show that computing the lower bound R_s is NP-hard using a reduction from MAX-2SAT. We give an $O(m2^n)$ time algorithm for computing R_s which enables us to apply it to real datasets. The previous implementation of Myers and Griffiths [1] had only an $\Omega(m \cdot n!)$ bound and is intractable for more than 10-15 haplotypes. Next, we show that the lower bound R_s can underestimate the true number of recombination events since it does not consider missing haplotypes. We propose a new bound R_I which extends R_s using the notion of intermediate haplotypes. This bound finds the optimal bound of 7 for the haplotypes from the ADH locus of *Drosophila Melanogaster* [22] and detects more recombination events than the R_s method on several datasets from the SeattleSNP database [23].

Most real haplotype datasets have some amount of missing data. A simple way of handling missing data is to not consider SNP's which have missing alleles for some haplotypes. We provide extensions of the bounds R_g and R_s for efficiently computing bounds for haplotype datasets with missing data. These bounds applied to the LPL dataset [3] detect many more recombination events (in comparison to the number detected by ignoring the sites with missing data) which provide strong support for the presence of a recombination hotspot [20]. Finally, we apply our methods to the polymorphism data from the MHC region and show plots which clearly indicate the presence of recombination hotspots that were detected by Jeffreys et al. [4] through sperm typing. We also find that the location of the hotspots (determined using sperm typing) are in good agreement with the values obtained using recombination lower bounds.

2   Basic Definitions and Previous Work

A single nucleotide polymorphism (commonly known as a SNP) is a position in the genome where multiple (predominantly two) bases are observed in the population. Very few polymorphic sites (about 0.1%) in humans have been found to be tri-allelic, i.e. having more than two different bases at the given site. Therefore, it is reasonable to make the *infinite-sites* or no-homoplasy assumption while dealing with human polymorphism data. As there are only two alleles at every site (the ancestral and the mutant), the extant data is represented by a binary matrix M with n rows and m columns, with the two nucleotides arbitrarily renamed 0 and 1. Hence, all our results on binary character data are applicable to real haplotype data.

2.1   Phylogenetic Networks and Recombination Lower Bounds

A recombination event at site p, between two haplotypes A and B, produces a recombinant sequence C, which is either a concatenation of sites $A[1 \ldots p]$ with $B[p + 1 \ldots m]$ or $B[1 \ldots p]$ with $A[p + 1 \ldots m]$. A phylogenetic network G for a set M of n sequences is a directed acyclic graph with a root. The root has no incoming edges. Each node in G is labeled by a m-length binary sequence where m is the number of sites. Each leaf of this graph is labeled by a sequence in M. Each node other than the root has either one or two incoming edges. A node with two incoming edges is called a *recombination* node. Some of the edges are labeled by the columns (sites) of M which correspond to a mutation event at that site. For a non-recombination node v, let e be the single incoming edge into v. The sequence labeling v can be obtained from the sequence labeling v's parent by changing the value at the sites which label the edge e from 0 to 1 (assuming that the root sequence is all-0). Each recombination node v is associated with an integer r_v (in the range $[2, m]$), called the recombination point for v. Corresponding to the recombination at node v, one of the two sequences labeling the parents of v is denoted as P and the other one as S. The sequence labeling node v is a concatenation of the first $r_v - 1$ characters of P with the last $m - r_v + 1$ characters of P. The sequences labeling the leaves of the phylogenetic network are referred to as *extant* sequences. A phylogenetic network G explains a set M of n haplotypes iff each sequence labels exactly one of the leaves of G. For a given set of haplotypes, there can be many possible phylogenetic networks with varying number of recombination events which explain the set. We define m_M to be the *minimum number of recombinations required to explain* M, i.e. there exists a phylogenetic network with m_M number of recombinations which explains M and there is no phylogenetic network with fewer number of recombination events that explains M.

The lower bound R_m, introduced by Hudson and Kaplan [24] is based on the *four-gamete test*; if for a pair of SNP's with ancestral and mutant alleles a/b and c/d respectively, all four possible gametes (ac, ad, bc, bd) are present, then at least one recombination event must have happened between the pair of loci under the assumption that no site mutates more than once. Based on this idea, one can find all intervals in which recombination must have occurred and choose the largest set of non-overlapping intervals from this collection. The bound R_m is the number of intervals in this set. However, R_m is a conservative estimate of the actual number of recombination events [24]. One can use haplotype diversity to infer more than one recombination event in an interval. Consider an interval with m segregating sites. If $n(> m + 1)$ distinct haplotypes are observed in this interval, then at most m haplotypes can be explained using mutation events. Assuming that the ancestral haplotype is present in the sample, the remaining $n - m - 1$ haplotypes must arise due to recombination events. Hence, one can infer a lower bound of $n - m - 1$ for the interval. Moreover, one can choose any subset of segregating sites for an interval and compute this difference to obtain another lower bound for that region. Taking the maximum bound over all subsets of segregating sites in a particular region, gives the best lower bound, R_h [1].

The bounds R_m and R_h do not explicitly consider possible histories of the sample. The lower bound R_s [1], computes for every history (an ordering of the haplotypes), a simplified number of recombination events, such that any a phylogenetic network that is consistent with this history, requires more recombination events than this number. By minimizing over all possible histories, one obtains a lower bound on the minimum number of recombination events. Myers and Griffiths [1] provide an algorithmic definition for the bound R_s. Their algorithm performs three kinds of operations on a given matrix: row deletion, column deletion and non-redundant row removal. A *row deletion* can be performed if the given row is identical to another row in the matrix. Such a row is also referred to as a *redundant* row. A *column deletion* can be done if the column (site) is *non-informative* (all but one rows have the same allele at this site). A *non-redundant row removal* is a row removal when there are no non-informative sites in the matrix and no redundant rows. Given an ordering of the n rows, the algorithm performs a sequence of column deletions, row deletions and non-redundant row removals until there is no row left in the matrix M. The minimum number of non-redundant row removal events over all possible histories gives the bound R_s. Since, the procedure considers all $n!$ histories, the worst case complexity of this procedure is $\Omega(m.n!)$. For some recent work on new methods for obtaining computing lower bounds, the interested reader is referred to [1, 25].

2.2   Combining Local Recombination Bounds

Myers and Griffiths [1] presented a general framework for computing recombination lower bounds from haplotype data. This framework can combine local recombination bounds on continuous subregions of a larger region to obtain recombination bounds for the larger parent region. Consider a matrix M with m segregating sites labeled 1 to m. Suppose that one has computed, for every interval (i, j) $(1 \leq i < j \leq m)$, a lower bound b_{ij} on the number of recombination events between the sites i and j. Each local lower bound b_{ij} can be computed by any lower bound method described previously and bounds for different intervals may be obtained by different methods.

In the second step, which is essentially a dynamic programming algorithm, one computes a new lower bound B_{ij} on the minimum number of recombination events between the sites i and j using the local bounds $b_{i'j'}, i' \leq i < j \leq j'$. The local bound B_{ij} can be computed as $B_{ij} = \max_{k=i+1}^{j-1} (B_{ik} + b_{kj})$. Note that the combined lower bound B_{ij} can be substantially better than the corresponding local bound b_{ij} for an interval (i, j). It is important to note that all the practical results in this paper are obtained by computing lower bounds (by using the corresponding lower bound method) for all intervals of length w (specified as a parameter) for the given dataset, and combining them using the dynamic programming algorithm.

3   Bounds Based on Haplotype Diversity

Consider a matrix M and let $S' \subseteq S$ be a subset of sites in M. For a subset S' of segregating sites, we denote the set of distinct haplotypes induced by S' as $H(S')$. The R_h bound of Myers and Griffiths[1] is based on the observation that $|H(S')| - |S'| - 1$ is a lower bound on the number of recombinations for every subset S'. Since the number of subsets is 2^w for a region of width w, Myers and Griffiths [1] use the approach of computing this difference for subsets of size at most s where $s < w$ is a specified parameter. Increasing s can provide better bounds with an increase in computation time since the running time is exponential in s. We define the algorithmic problem associated with the computation of the bound R_h as follows:

MDS: Most Discriminative SNP subset problem

Input: A binary matrix M and an integer k, where S is the set of columns of M.

Output: Is there a subset S' of S, such that $|H(S')| - |S'| - 1 \geq k$.

Computing the R_h bound is equivalent to finding the largest value of k for which the MDS problem has a solution. We show that MDS problem is NP-complete by using a reduction from the *Test Collection Problem*[27]. An instance of the test collection problem consists of a collection $\mathcal{C}$ of subsets of a finite set $\mathcal{S}$ and an integer k, and the objective is to decide if there is a sub-collection $\mathcal{C}' \subseteq \mathcal{C}$ such that for each $x, y \in \mathcal{S}$ there exists $c \in \mathcal{C}'$ that contains exactly one of x and y and $|\mathcal{C}'| \leq k$. An instance of the test collection problem can be encoded as a binary matrix M of size $|\mathcal{S}| \times |\mathcal{C}|$. Each row of the matrix corresponds to an element of the finite set $\mathcal{S}$ and $M[x, c] = 1$ if the subset c contains the element x and 0 otherwise. Here, the objective is to find a subset S' of the columns of M of size at most k such that for every pair of rows in M, there is a column in S' that can distinguish between them, i.e. $|S'| \leq k$ and $H(S') = |\mathcal{S}|$. Using this encoding we show that the MDS problem is NP-complete (the proof is omitted for lack of space).

Lemma 1. *The MDS problem is NP-complete.*

3.1   The Lower Bound R_g

From the above encoding, it is easy to see that computing the bound R_h is equivalent to finding a a smallest subset of columns C such that for every pair of haplotypes (rows) (x, y) in M, there is at least one column $c \in C$ such that $M[x, c] \neq M[y, c]$.

We adapt the standard greedy algorithm for the set cover problem [2] to devise an algorithm (described in Figure 1) for computing a lower bound; denoted as R_g. It is well known that the greedy algorithm gives a $1 + 2 \ln n$ approximation for the test collection problem where $n = |\mathcal{S}|$, the size of the ground set. However, this approximation ratio does not apply to the MDS problem. Although, in general $R_g \leq R_h$, we found that the overall bound (obtained by combining the local bounds computed using R_g) was equal to the corresponding bound returned by the Recmin program [1] for almost all datasets we did the comparison for

COMPUTE_$R_g(M)$
1. Repeat until possible
     If two rows in M are identical, coalesce them.
     If a site s is non-informative, remove the site s.
2. Let M' be the reduced matrix with n rows and m sites
3. Initialize $d(x,y) = 0$ for all pairs of rows and $I = \phi$
4. while $d(x,y) = 0$ for some pair
5.   Let s' be the column that can distinguish between the maximum pairs
          for which $d(x,y) = 0$
7.   set $d(x,y) = 1$ for all (x,y) s.t. $M'[x,s'] \neq M'[y,s']$
8. $I = I \cup \{s'\}$
9. Return $|H(I)| - |I| - 1$

Fig. 1. The greedy algorithm for computing the bound R_g

Table 1. Comparison of the performance of the Recmin program [1] and our R_g bound for the Daly haplotypes with different values of the parameters; maximum subset size (s) and maximum width (w). Note that the R_g bound requires only the parameter w

	Recmin program [1]		R_g bound	
Parameters	Bound	time	Bound	time
w=15 s=6	134	4 secs	180	01 secs
w=20 s=10	183	2.5 mins	188	03 secs
w=25 s=10	186	31 mins	198	06 secs
w=30 s=15	200	29 hrs	199	11 secs
w=35	-	-	203	15 secs

(we believe that this is due to the effect of combining the local bounds). The running time of the Recmin program [1] is proportional to $\sum_{i=2}^{s} \binom{w}{i}$ where w is the maximum number of segregating sites in a region for which the local bound is computed and s is the maximum subset size used for computing the R_h bound. In contrast, in order to compute the best bound by combining the local R_g bounds, we require only one parameter, i.e maximum width and the overall running time is $O(n^2 m w^2)$. To illustrate the kind of improvements we obtain using R_g, we compare the bounds (Recmin and R_g) for the phased haplotypes (258 haplotypes on 103 SNP's) of the Daly [6] dataset obtained from the Hap Webserver [28] (see Table 3.1).

4   History Based Lower Bounds

Myers and Griffiths[1] only give a procedural definition of the bound R_s, and their description is somewhat informal. The time complexity of their procedure (as described in Algorithm 3 in [1]) is $O(m\,n!)$, where n is the number of rows, and m the number of columns. We give a theoretical formulation of the bound

Compute_R_s(M)
1. **for all** row subsets r: $R_S[r] = 0$
2. **for all** subsets r picked in an increasing order
3.   **if** $\exists$ a redundant row in r
4. $R_S[r] = \min_i \{R_S[r_{-i}]\}$ (* for all rows i s.t. i is redundant *)
5.   **else**
6. $R_S[r] = \min_i \{1 + R_S[r_{-i}]\}$ (* for all rows i s.t. $r_i = 1$ *)
7. return $R_S(\mathbf{1})$

Fig. 2. An $O(m2^n)$ algorithm for computing R_s (**1** refers to all-ones vector of length n)

R_s which allows us to develop an exponential time algorithm for computing it and also show that computing R_s is NP-hard.

We define a history for a set of n rows as simply an ordering of the rows. We start by redefining R_s in terms of appropriate cost of a row in a given history. Consider a history $H = r_1 \rightarrow r_2 \ldots \rightarrow r_n$. The cost of row r_i in the history, denoted by $C_s(r_i)$, is 0 if after removing non-informative columns from $r_1, r_2, \ldots, r_i$, the row r_i turns out to be identical to one of the rows $r_1, \ldots, r_{i-1}$ and 1 otherwise. Then we have

$$C_s(H) = \sum_i C_s(r_i) \text{ and } R_s(M) = \min_{\text{history } H} C_s(H)$$

We defer the discussion of why R_s is a lower bound to Theorem 2 (where we prove that R_I is a lower bound). Consider a bit vector r of lengths n. Let M_r denote a submatrix of M which contains only rows i such that $r_i = 1$. Define a partial order on the vectors as follows: $v_1 \leq v_2$ if $v_2[i] = 1$ whenever $v_1[i] = 1$. Define the vector v_{-i} as the v with the $i-$th bit set to 0. Let $R_S[v]$ denote the R_s bound for the corresponding sub-matrix. The procedure in Figure 2 gives an $O(m2^n)$ algorithm for computing R_s. This dynamic programming algorithm can bring significant improvements in running time. Note that in order to compute R_s, step 4 of the algorithm in Figure 2 can be replaced by $R_S[r] = R_S[r_{-i}]$ for any row i that is redundant. Using a non-trivial reduction from the MAX-2SAT problem we show that computing the bound R_s for a matrix is NP-hard (proof omitted).

Theorem 1. *Computing $R_s(M)$ is NP-hard.*

The R_s bounds searches over possible histories of the set of haplotypes and one would expect the bound to be better than the diversity based bound R_h. In practice, however, both R_h and R_s underestimate the true bound in many instances.

4.1 Recombinant Intermediates and the Bound R_I

We use an example to demonstrate how R_s can be improved. Consider the set of $n + 2$ haplotypes with n sites shown in Figure 3. For illustration $n = 7$.

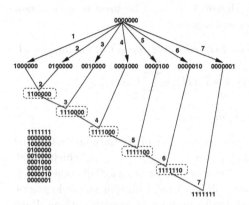

Dataset	Size	R_s	R_I
CSF3	15 x 17	3	4
MMP3	21 x 41	6	9
ABO	68 x 197	70	73
DCN	31 x 117	16	18
HMOX1	34 x 53	14	16
F2RL3	28 x 29	10	11
EPHB6	31 x 62	23	25

Fig. 3. On the left is a set of 9 haplotypes for which R_s is 1 and a phylogenetic network for the set of haplotypes with 6 recombination events $R(_I = 6)$. On the right is a table which compares the number of detected recombination events using R_s and R_I for the phased haplotype datasets for various genes obtained from the SeattleSNP project [23]

Note that if the history was forced to start with the first two haplotypes, each of the following n rows could only be removed through a non-redundant row removal, and we would have a recombination bound of n. However, if we choose 1111111 to be the last haplotype in the history, then removing it makes every column non-informative. As R_s is the minimum over all histories, $R_s(M) = 1$. However, at least 6 recombinations are needed. Note that for this particular example, we can boost the R_s bound to the correct value by applying the dynamic programming algorithm [1] for combining local bounds. However, the example illustrates a problem with R_s, which is that in explaining a non-redundant row-removal, we only charge a *SINGLE* recombination event. Therefore, if 1111111 was indeed the last haplotype in the true history, then adding it would require 5 recombinants (the haplotypes in dashed boxes) NOT from the current set (as explained in Figure 3).

We use this idea to improve the R_S bound. Consider a history $H = r_1 \rightarrow r_2 \ldots \rightarrow r_n$. Let $\mathcal{I}_j(H)$ denote the minimum number of recombination events in obtaining r_j, given any phylogenetic network for $r_1, \ldots, r_{j-1}$. We allow the use of recombinant intermediates, and so $\mathcal{I}_j(H)$ can be greater than one. In general, the use of recombinant intermediates is tricky because the intermediates may help explain some of the existing haplotypes by simple mutations. In order to prove a lower bound, we introduce the concept of a *direct recombination*. We define $C_d(r_i)$ for a haplotype r_i in a given history H as follows:

$$C_d(r_i) = \begin{cases} 0 & r_i \text{ is different from all } r_{j<i} \text{ in a non-informative column.} \\ 0 & r_i \text{ is identical to } r_{j<i} \text{ after removing non-informative columns} \\ 1 & \text{Otherwise} \end{cases} \quad (1)$$

We observe that the definition of $C_d(r_i)$ holds for a set of haplotypes $\{r_1, r_2, \ldots, r_{i-1}, r_i\}$ and denote this generic definition as $C_d(r_i, \{r_1, r_2, \ldots, r_{i-1}\})$. Note that

$C_d(r_i) \leq C_s(r_i)$ for all i in a history. However, C_d can be used to give a new lower bound on the total number of recombinations.

Theorem 2. *Let $\mathcal{H}$ denote the set of all histories over the set of haplotypes M. Then*

$$R_I = \min_{H \in \mathcal{H}} \max_j \{\sum_{i<j} C_d(r_i) + \mathcal{I}_j(H) + \sum_{i \geq j} C_s(r_i)\}$$

is a lower bound on the number of recombinations.

Proof. Recall that m_M denotes the minimum number of recombinations in any history of M. We construct one history $H = r_1 \rightarrow r_2 \ldots \rightarrow r_n$ in which which $\sum_{i<j} C_d(r_i) + \mathcal{I}_j(H) + \sum_{i>j} C_s(r_i)$ is a lower bound on m_M for all choices of j. This is sufficient because we minimize over all histories. Consider an phylogenetic network $\mathcal{A}$ that explains m_M with a minimum number of recombinations. Each node v in the phylogenetic network corresponds to a haplotype r_v, which may or may not be in M. Haplotype $r \in M$ is a *direct witness* for a recombinant node v if $r = r_v$. It is an *indirect witness* if it can be derived from r_v solely by mutation events. A predecessor relationship $<_P$ is defined for some haplotypes $r_i, r_j \in M$. Specifically $r_i <_P r_j$ if r_i is a (direct or indirect) witness to a recombinant node on a path from the root to r_j. Note that $<_P$ is a partial order. Next, choose a history H (a total ordering) that is consistent with $<_P$. Note that $C_s(r_i) = 1$ if and only if r_i is the first witness to a recombination node in A to appear in H (thereby proving that $R_s(M)$ is a lower bound). Likewise $C_d(r_i) = 1$ if and only if r_i is the first direct witness to a recombination node in A to appear in H. As each recombination node contributes at most 1, $R_s = \sum_i C_s(r_i)$ is a valid lower bound on the number of recombinations. Consider an arbitrary r_j with $C_s(r_j) = 1$. Instead of charging 1 to the number of recombination events, we charge a value $\mathcal{I}_j(H)$ equal to the minimum number of recombinations needed to obtain r_j from $r_1, r_2, \ldots, r_{j-1}$. Consider the sequence of intermediate recombination events that were used to obtain r_j. None of these nodes have a direct witness. Therefore the nodes in $r_1, r_2, \ldots, r_{j-1}$ that had a C_d value of 1 correspond to other recombination nodes.

Next, the haplotypes $r_{i>j}$ that follow r_i are charged $C_s(r_i)$. Whenever, $C_s(r_i) = 1$, it is because r_i is the first witness to a recombination node in A to appear in H. By construction, this recombination node is not on any path from root to r_j, and therefore wasn't charged when considering intermediates for r_j. Therefore, each recombination node is charged at most once and the bound holds.

It is easy to see that $R_I \geq R_s$. In order to compute R_I, we need to compute $\mathcal{I}_j(H)$ for all haplotypes j, and all histories H. To do this more efficiently, we define $\mathcal{I}_j$ over subsets, instead of histories. We denote a subset of haplotypes by the bit-vector r of size n where $r_i = 1$ iff $r_i \in r$ and define $\mathcal{I}_j[r]$ as minimum number of recombination events needed to obtain r_j, over any history of the haplotypes in r. Likewise, define $R_d(r)$ as the minimum number of direct recombinations in any history of the haplotype subset r. The algorithm in Figure 4 describes how to compute R_I in time $O(n2^n I(m.n))$ time, where $I(m,n)$ is the time to compute $\mathcal{I}_j[r]$ for any subset r.

Compute_R_I(M)
1. **for all** row subsets r : $R_d[r] = 0$; $R_I[r] = 0$
2. **for all** subsets r chosen in an increasing order
3.   **if** $\exists i$ s.t. $r_i = 1$ and row i is redundant
4. $R_d[r] = R_d[r_{-i}]$; $R_I[r] = R_I[r_{-i}]$
5.   **else**
6.   **for all** rows i s.t. $r_i = 1$
7. $R_{d,i} = \min_i\{C_d(r_i, r_{-i}) + R_d[r_{-i}]\}$
8. $R_{I,i} = \min_i\{\max\{1 + R_I[r_{-i}], R_d[r_{-i}] + \mathcal{I}_i[r_{-i}]\}\}$
9.   **end for**
10. $R_d[r] = \min_i\{R_{d,i}\}$; $R_I[r] = \min_i\{R_{I,i}\}$
11.   **end if**
12. **end for**
13. return $R_I(M)$

Fig. 4. An $O(2^n I(m,n))$ algorithm for computing R_I. $\mathcal{I}_i[r_{-i}]$ denotes the minimum number of recombinant intermediates needed to compute haplotype r_i given the subset r with r_i removed

4.2   Computing Recombinant Intermediates

Our goal is to compute $\mathcal{I}_i[r]$ efficiently. Haplotype i is assumed to arise later in history the in r and is therefore a mosaic of sub-intervals of the haplotypes in r. The mosaic can be expressed by a sequence of pairs $M = (h_1, j_1), (h_2, j_2), \ldots,$ (h_k, j_k) interpreted as follows: In h_i, columns $1, \ldots, j_1$ came from haplotype h_1, columns $j_1 + 1, \ldots, j_2 + 1$ from h_2, and so on. If M were the true mosaic, then h_i would need $k - 1$ recombinant intermediates. Thus, we need to minimize this.

First, we can ignore all columns that are identical for all haplotypes in r. If h_i has a different value in any of these columns, it can be explained by a mutation. If it has the identical value, the column can be explained using any haplotype and will not contribute to recombination. Ignoring these columns, the following is true: if columns $j_1, \ldots, j_2$ of h_i arise from haplotype h, then the values of h and h_i must be identical in columns j_1 through j_2. If any columns c was different ($h_i[c] \neq h[c]$), to explain it by a mutation would violate the infinite-sites assumption. This observation allows us to solve the problem of computing $\mathcal{I}_i[r]$ efficiently.

For column $c, 1 \leq c \leq m$ and haplotype h, let $I[c, h]$ denote the minimum number of recombinations needed to explain the first c columns of haplotype h_i such that the c-th column arose from haplotype h. This is sufficient because $\mathcal{I}_i[r] = \min_h\{I[m, h]\}$. $I[c, h]$ can be computed using the following recurrence:

$$I[c, h] = \begin{cases} 0 & c = 0 \\ \infty & h_i[c] \neq h[c] \\ \min\{I[c - 1, h], \min_{h' \neq h}\{1 + I[c - 1, h']\}\} & o/w \end{cases}$$

4.3   Results for R_I Bound

Besides the simulated example (in Figure 3), real datasets are known where R_s and R_h are sub-optimal. As an example, the R_h and R_s bounds for Kreitman's data [22] from the ADH locus of *Drosophila Melanogaster* are both 6. Song and Hein [25] showed that their set theoretic lower bound gave a bound of 7 and proved this to be optimal by actually constructing an phylogenetic network which requires 7 recombination events. Our new lower bound R_I also returns the optimal bound of 7. However, the set theoretic-bound [25] does not have an explicit algorithmic description. On the other hand, the R_I bound can be computed for large datasets (100×500 matrix can be analyzed in about an hour on a standard PC) and gives improved bounds for a number of real datasets (see the table in figure 3 for a partial list).

5   Bounds for Haplotypes with Missing Data

A complete haplotype is an element of $\{0,1\}^m$ where m is the number of SNP's and the j-th component indicates the nucleotide at that position. However, due to errors or other reasons, the allele at a particular position for a individual is sometimes not available. In such a scenario, some of the haplotypes are partial or incomplete. A partial haplotype is an element of $\{0,1,?\}^m$ where ? represents the positions where the allele is unknown. Since most of the real haplotype data has missing entries, it is important to find efficient methods to find recombination lower bounds for haplotypes with missing data. The lower bounds R_h and R_s do not naturally extend for a incomplete haplotype matrix. However, in this section, we show how both the greedy algorithm for computing R_g and the exponential algorithm for computing R_s can be extended for an incomplete matrix without much increase in the computational complexity. We first need to modify the definitions of non-informative site and redundant row. A site is non-informative if it has all but one alleles of one type (ignoring the missing alleles). For comparing two rows, we define $M[x,a] \neq M[y,b]$ if and only if $M[x,a] \neq '?'$ and $M[y,b] \neq '?'$ and $M[x,a] \neq M[y,b]$.

In the last step of the greedy algorithm (Figure 1), the algorithm returns the bound $H(I) - I - 1$. For a matrix with missing entries, it is not straightforward to compute $H(I)$. However, consider an assignment to the ?'s that minimizes $H(I)$. Then the difference $H(I) - I - 1$ gives a valid lower bound, i.e. a bound which is valid for all possible assignments to the missing entries. However, one then needs to solve the minimum haplotype completion problem; where given an haplotype matrix with missing entries, the objective is to complete the missing entries so as to minimize the number of distinct haplotypes. This problem was shown to be NP-hard by Kimmel et al. [29]. However, for our purposes, we use a simple heuristic to find the minimum number of rows that can be distinguished using the non-missing entries. This gives a valid lower recombination lower bound that is easily computable.

5.1 R_{sm}: History Based Bound for Missing Data

Consider a set of haplotypes M where some of the haplotypes are incomplete. We define a completion to be assignment of 0 or 1 to every missing allele in M. Clearly, there exists a completion M' of M and a corresponding ordering H for that complete matrix M', such that the number of row removal operations is minimum over all completions and all orderings. In other words, the definition of the R_s lower bound has to be modified as: $R_s(M) = \min_{M'}[R_s(M')]$ where M' is a completion of M.

Since a complete matrix is a special case of a incomplete matrix, it follows that it is also NP-hard to compute the modified version of R_s for an incomplete matrix. The $O(m \cdot 2^n)$ algorithm for computing R_s (described in Figure 2) can be used for computing the bound R_{sm} (this bound may not exactly equal $R_s(M)$) for an incomplete matrix with the modified definitions of redundant row and non-informative site. The next lemma shows that $R_{sm}(M)$ is less than $R_s(M)$ and is hence a valid lower bound.

Lemma 2. For a incomplete haplotype matrix M, $R_{sm}(M) \le R_s(M)$.

5.2 Application to Haplotype Data from LPL Locus

A 9.7-kb region in the human LPL gene was sequenced by Nickerson et al. [3] in 71 individuals from three different populations. The haplotype data comprised of 88 haplotypes defined by 69 variable sites with about 1.2% missing data. This data has previously been analyzed for haplotype diversity and recombination by Clark et al. [30], Templeton et al. [20] and Myers and Griffiths [1]. In table 5.2, we compare the bounds obtained for different sub-regions of the LPL region for various populations. The overall bound for the whole region is 70 if one ignores the sites with missing data (see [1]), while we obtain a much improved bound of 87 by applying our R_h/R_{sm} bounds along with the dynamic programming framework. Templeton et al. [20] had found the 29 recombination events detected using their method to be clustered near the center of the region (approximately between the sites 2987 and 4872). It is interesting to note that number of detected recombination events (37) in this region increases sig-

Table 2. The number of detected recombination events using methods for missing data for the LPL datasets. The number in bracket indicates the corresponding lower bound obtained by ignoring sites with missing alleles [1]. The region (2987-4872) corresponds to the suggested hotspot [20]

	Site Range			
Region	106-2987	2987-4872	4872-9721	Full
Jackson	10(10)	11(9)	17(13)	39(36)
Finland	2(2)	13(13)	13(11)	31(27)
Rochester	1(1)	13(13)	7(7)	22(21)
Combined	13(12)	37(22)	36(28)	87(70)

nificantly (from 22) when one takes into account the sites with missing alleles. Thus, the bounds obtained using our improved methods which can handle missing data, seem to provide strong support for the presence of a recombination hotspot suggested by Templeton et al. [20]. This demonstrates that the ability to extract past recombination events can be crucial to detecting regions with elevated recombination rates.

6   Lower Bounds and Recombination Hotspots

In humans, pedigree studies have shown variation in recombination rates on a megabase scale, and analyses of sperm crossovers in males [7,4] have identified hotspots of length 1-2kbs where recombination events cluster. However, characterizing fine-scale variation in recombination rates using pedigree studies (at the kb scale) is difficult and experimental difficulties limit the large-scale application of sperm analyses. After several studies [6,5] observed a block-like structure in patterns of linkage disequilibrium in the human genome, it has been speculated that most or all recombination occurs in recombination hotspots [31]. The problem of detecting recombination hotspots (roughly defined as a region in which the recombination rate is much higher than the average recombination rate) using DNA polymorphism data has been considered by several studies [21,11] which proposed statistical based methods to give quantitative estimates of recombination rates.

Here, we apply our lower bounds to the population data from a 216-kb segment of the class II region of the Major histocompatibility complex (MHC). Jeffreys et al. [4] sequenced 50 individuals from UK in this region and identified

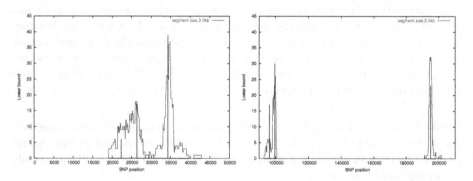

Fig. 5. Sliding window plot of recombination lower bounds (window of size 2 kb incremented 0.1 kb at each step) for the 216-kb segment of the class II region of the major histocompatibility complex (MHC). The vertical black lines (height scaled by logarithm of the mean recombination rate obtained from sperm typing for that hotspot) show the approximate locations of the center of the six hotspots inferred using sperm crossover analysis by Jeffreys et al. [4]. The TAP2 hotspot [7] is the last hotspot near the 200-kb region

six recombination hotspots using sperm crossover analysis. Since the available data is unphased, we applied our lower bounds to the haplotypes estimated by the PHASE program [32]. Three separate studies [14, 21, 13] have applied their methods to infer recombination hotspots for this dataset. Although, sperm typing and recombination lower bounds measure very different things, we find that the lower bounds are able to locate most of the recombination hotspots with high accuracy (see Figure 5). Five regions show very good evidence of elevated recombination with excellent agreement with the center of the corresponding hotspots (as found by Jeffreys et al. [4]), with only one of the characterized hotspots (DMB1 near the 96kb region) showing a weak signal. This clearly demonstrates the ability of recombination lower bound methods to provide first hand indication of the presence and the location of hotspots. One criticism of lower bound methods is that we do not model events such as repeat mutations and gene conversion. However, such events are rare and our results (see also Myers and Griffiths [1]) suggest that this has only moderate effects on the bounds.

References

1. Myers, S., Griffiths, R.: Bounds on the Minimum Number of Recombination Events in a Sample History. Genetics **163** (2003) 375–394
2. Johnson, D.: Approximation algorithms for combinatorial problems. Journal of Comput. System Sci. **9** (1972) 256–278
3. Nickerson, D. et al.: DNA sequence diversity in a 9.7-kb region of the human lipoprotein lipase gene. Nature Genetics **19** (1998) 233–240
4. Jeffreys, A.J., Kauppi, L., Neumann, R.: Intensely punctate meiotic recombination in the class II region of the major histocompatibility complex. Nature Genetics **29** (2001) 217–222
5. Gabriel, S.B. et al.: The structure of haplotype blocks in the human genome. Science **296** (2002) 2225–2229
6. Daly, M.J., Rioux, J.D., Schaffner, S.F., Hudson, T.J., Lander, E.S.: High-resolution haplotype structure in the human genome. Nature Genetics **29** (2001) 229–232
7. Jeffreys, A., Ritchie, A., Neumann, R.: High resolution analysis of haplotype diversity and meiotic crossover in the human tap2 recombination hotspot. Hum. Mol. Genet. **9** (2000) 725–733
8. Griffiths, R.C., Marjoram, P.: Ancestral inference from samples of DNA sequences with recombination. Journal of Computational Biology **3** (1996) 479–502
9. Fearnhead, P., Donnelly, P.: Estimating recombination rates from population genetic data. Genetics **159** (2001) 1299–1318
10. Hudson, R.R.: Two-locus sampling distributions and their applications. Genetics **159** (2001) 1805–1817
11. Li, N., Stephens, M.: Modeling linkage disequilibrium and identifying recombination hotspots using single-nucleotide polymorphism data. Genetics **165** (2003) 2213–2233
12. The International HapMap Consortium: The international hapmap project. Nature **426** (2003) 789–796
13. McVean, G. et al.: The fine-scale structure of recombination rate variation in the human genome. Science **304** (2004) 581–584

14. Crawford, D. et al.: Evidence for substantial fine-scale variation in recombination rates across the human genome. Nature Genetics **36** (2004) 700–706
15. Hein, J.: Reconstructing Evolution of sequences subject to recombination using parsimony. Math. Biosci. **98** (1990) 185–200
16. Hein, J.: A Heuristic Method to Reconstruct the History of Sequences Subject to Recombination. J. Mol. Evol. **20** (1993) 402–411
17. Song, Y., Hein, J.: Parsimonious Reconstruction of Sequence Evolution and Haplotype Blocks: Finding the Minimum Number of Recombination Events. WABI (2003) 287–302
18. Wang, L., Zhang, K., Zhang, L.: Perfect phylogenetic networks with recombination. Journal of Computational Biology **8** (2001) 69–78
19. D.Gusfield, Eddhu, S., C.Langley: Efficient reconstruction of phylogenetic networks with constrained recombination. In Proc. of IEEE CSB Conference. (2003) 363–374
20. Templeton, A. et al.: Recombinational and mutational hotspots within the human lipoprotein lipase gene. American Journal of Human Genetics **66** (2000) 69–83
21. Fearnhead, P. et al.: Application of coalescent methods to reveal fine-scale rate variation and recombination hotspots. Genetics **167** (2004) 2067–2081
22. Kreitman, M.: Nucleotide Polymorphism at the Alcohol Dehydrogenase Locus of Drosophila Melanogaster. Nature **304** (1983) 412–417
23. SeattleSNPs. NHLBI Program for Genomic Applications, UW-FHCRC, Seattle, WA. http://pga.gs.washington.edu (2004)
24. Hudson, R.R., Kaplan, N.L.: Statistical properties of the number of recombination events in the history of a sample of DNA sequences. Genetics **111** (1985) 147–164
25. Song, Y., Hein, J.: On the minimum number of recombination events in the evolutionary history of dna sequences. Journal of Mathematical Biology **48** (2004) 160–186
26. Bafna, V., Bansal, V.: The number of recombination events in a sample history: Conflict graph and lower bounds. IEEE Trans. on Comp. Biology and Bioinformatics **1** (2004) 78–90
27. Garey, M.R., Johnson, D.S.: Computers and Intractability: A Guide to the Theory of NP-completeness. W.H. Freeman and Company (1979)
28. Eskin, E., Halperin, E.: Haplotype reconstruction from genotype data using imperfect phylogeny. Bioinformatics **20** (2003) 1842–9
29. Kimmel, G., Shamir, R.: The incomplete perfect phylogeny haplotype problem. Second RECOMB Satellite Workshop on Computational Methods for SNPs and Haplotypes (2004)
30. Clark, A. et al.: Haplotype structure and population genetic inferences from nucleotide-sequence variation in human lipoprotein lipase. American Journal of Human Genetics **63** (1998) 595–612
31. Goldstein, D.B.: Islands of linkage disequilibrium. Nature Genetics **29** (2001) 109–111
32. Stephens, M., Smith, N.J., Donnelly, P.: A new statistical method for haplotype reconstruction from population data. American Journal of Human Genetics **68** (2001) 978–989

A Linear-Time Algorithm for the Perfect Phylogeny Haplotyping (PPH) Problem⋆

Zhihong Ding, Vladimir Filkov, and Dan Gusfield

Department of Computer Science, University of California, Davis
{dingz, filkov, gusfield}@cs.ucdavis.edu

Abstract. Since the introduction of the Perfect Phylogeny Haplotyping (PPH) Problem in Recomb 2002 [15], the problem of finding a linear-time (deterministic, worst-case) solution for it has remained open, despite broad interest in the PPH problem and a series of papers on various aspects of it. In this paper we solve the open problem, giving a practical, deterministic linear-time algorithm based on a simple data-structure and simple operations on it. The method is straightforward to program and has been fully implemented. Simulations show that it is much faster in practice than prior methods. The value of a linear-time solution to the PPH problem is partly conceptual and partly for use in the inner-loop of algorithms for more complex problems, where the PPH problem must be solved repeatedly.

1 Introduction

In diploid organisms (such as humans) there are two (not completely identical) "copies" of each chromosome, and hence of each region of interest. A description of the data from a single copy is called a *haplotype*, while a description of the conflated (mixed) data on the two copies is called a *genotype*. In complex diseases (those affected by more than a single gene) it is often much more informative to have haplotype data (identifying a set of gene alleles inherited together) than to have only genotype data. The international Haplotype Map Project [21] is focussed on determining the common SNP haplotypes in several diverse human populations.

Today, the underlying data that forms a haplotype is usually a vector of values of m *single nucleotide polymorphisms (SNP's)*. A SNP is a single nucleotide site where exactly two (of four) different nucleotides occur in a large percentage of the population. In general, it is not feasible to examine the two haplotypes separately, and *genotype* data rather than haplotype data is usually obtained. Then one tries to infer the original haplotype pairs from the observed genotype data. We represent each of the n input *genotypes* as vectors, each with m sites, where each site in a vector has value 0, 1, or 2. A site i in the genotype vector g

---

⋆ Research partially supported by grant EIA-0220154 from the National Science Foundation. Thanks to Chuck Langley for helpful discussions.

S. Miyano et al. (Eds.): RECOMB 2005, LNBI 3500, pp. 585–600, 2005.

has a value of 0 (respectively 1) if site i has value 0 (or 1) on both the underlying haplotypes that generate g. Otherwise, site i in g has value 2. Note that we do not know the underlying haplotype pair that generates g, but we do know g.

Given an input set of n genotype vectors of length m, the *Haplotype Inference (HI) Problem* is to find a set of n pairs of binary vectors (with values 0 and 1), one pair for each genotype vector, such that each genotype vector is explained (can be generated by the associated pair of haplotype vectors). The ultimate goal is to computationally infer the true haplotype pairs that generated the genotypes. This would be impossible without the implicit or explicit use of some genetic model, either to assess the biological fidelity of any proposed solution, or to guide the algorithm in constructing a solution. The most powerful such genetic model is the population-genetic concept of a *coalescent* [25, 22]. The coalescent model of SNP haplotype evolution says that without recombination the evolutionary history of $2n$ haplotypes, one from each of $2n$ individuals, can be displayed as a rooted tree with $2n$ leaves, where some ancestral sequence labels the root of the tree, and where each of the m sites labels exactly one edge of the tree. A label i on an edge indicates the (unique) point in history where a mutation at site i occurred. Sequences evolve down the tree, starting from the ancestral sequence, changing along a branch e by changing the state of any site that labels edge e. The tree "generates" the resulting sequences that appear at its leaves. In more computer science terminology, the coalescent model says that the $2n$ haplotype (binary) sequences fit a *perfect phylogeny*. See [15] for further explanation and justification of the perfect phylogeny haplotype model. Generally, most solutions to the HI problem will not fit a perfect phylogeny, and this leads to

The Perfect Phylogeny Haplotyping (PPH) Problem: Given an n by m matrix S that holds n genotypes from m sites, find n pairs of haplotypes that generate S and fit a perfect phylogeny.

It is the requirement that the haplotypes fit a perfect phylogeny, and the fact that most solutions to the HI problem will not, that enforce the coalescent model of haplotype evolution, and make it plausible that a solution to the PPH problem (when there is one) is biologically meaningful.

The PPH problem was introduced in [15] along with a solution whose worst-case running time is $O(nm\alpha(nm))$, where α is the extremely slowly growing inverse Ackerman function. This nearly-linear-time solution is based on a linear-time reduction of the PPH problem to the *graph realization* problem, a problem for which a near-linear-time method [4] was known for over fifteen years. However, the near-linear-time solution to the graph realization problem is very complex (only recently implemented), and is based on other complex papers and methods, and so taken as a whole, this approach to the PPH problem is hard to understand, to build on, and to program. Further, it was conjectured in [15] that a truly linear-time ($O(nm)$) solution to the PPH problem should be possible.

After the introduction of the PPH problem, a slower variation of graph-realization approach was implemented [6], and two simpler, but also slower methods (based on "conflict-pairs" rather than graph theory) were later introduced [2, 10]. All three of these approaches have best and worst-case running times of

$\theta(nm^2)$. Another paper [26] developed similar insights about conflict-pairs without presenting an algorithm to solve the PPH problem. The PPH problem is now well-known (for example discussed in several surveys on haplotyping methods [5, 17, 18, 16]). Related research has examined extensions, modifications or specializations of the PPH problem [24, 19, 11, 8, 9, 3], or examined the problem when the data or solutions are assumed to have some special form [20, 12, 13]. Some of those methods run in linear time, but only work for specializations of the full PPH problem [12, 13], or are only correct with high probability (with some model) [8, 9]. The problem of finding a deterministic, linear-time algorithm for all data has remained open, and a recent paper [1] shows that conflict-pairs methods are unlikely to be implementable in linear time.

In this paper, we completely solve the open problem, giving a deterministic, linear-time (worst-case) algorithm for the PPH problem, making no assumptions about the form of the data or the solution. The algorithm is graph-theoretic, based on a simple data-structure and standard operations on it. The linear-time bound is trivially verified, and the correctness proofs are of moderate difficulty. The algorithm is straightforward to implement, and has been fully implemented. Tests show it to be much faster in practice as well as in theory, compared to the other existing programs. As in some prior solutions, the method provides an implicit representation of all the PPH solutions.

In addition to the conceptual value of our solution, its practical value can be significant. Currently, the full structure of haplotypes in human populations and subpopulations is not known, and there are some genes with high linkage disequilibrium that extends over several hundred kilobases (suggesting very long haplotype blocks with a perfect or near-perfect phylogeny structure). So it is too early to know the full range of *direct* application of this algorithm to long sequences (see [7] for a more complete discussion). Moreover, faster algorithms are of practical value when the PPH problem is repeatedly solved in the inner-loop of an algorithm. For example, in [7] and [26], one finds, from every SNP site, the longest interval starting at that site for which there is a PPH solution. Moreover, there are applications where one may examine *subsets* of sites to find subsets for which there is a PPH solution. In such applications, efficiencies in the inner loop will be significant, even if each subset is relatively small.

2   The Shadow Tree

Our algorithm builds and uses a directed, rooted graph, called a *"shadow tree"*, as its primary data structure. There are two types of *"edges"* in the shadow tree: *tree edges* and *shadow edges*, which are both directed towards the root. Tree and shadow edges are labeled by column numbers from S (with shadow edges having bars over the labels). For each tree edge, i, there is a shadow edge, $\bar{i}$, in the shadow tree. The end points of each tree and shadow edge are called *connectors*, and can be of two types: H or T connectors, corresponding to the head (arrow) or tail of the edge.

The shadow tree also contains directed *"links"*. From a graph theory standpoint these are also edges, but we reserve the word *"edge"* for tree and shadow edges. Links are used to connect certain tree and shadow edges, and are needed for linear-time manipulation of the shadow tree. Each link is either *free* or *fixed*, and always points away from an H connector. When we say edge E *"links to"* E', we mean there is a link from the H connector of E to a connector of E'.

Since links can point to either an H or a T connector, the "parent of" relationship between edges is not the same as the "links to" relationship, and is defined recursively: If an edge links to the root, then its parent is the root. If an edge E links to the T connector of an edge E_p, then the parent of E, $p(E)$, is defined as E_p. However, if E links to the H connector of an edge E', $p(E)$ is defined to be the same as $p(E')$. For convenience, we define the parent of a connector as the parent of the edge that contains the connector. See Fig. 1 for an illustration of all these elements.

Tree edges, shadow edges, and *fixed* links are organized into *classes*, which are subgraphs of the shadow tree. Every free link connects two classes, while each fixed link is contained in a single class. We will see later that each class in the shadow tree encodes a subgraph that must be contained in **all** solutions to the PPH problem. In each class, if the links are contracted, then the remaining edges form two rooted trees (except for the root class which has only one rooted tree), where if one subtree contains a tree edge the other contains its shadow edge. The roots of the two subtrees are called the *"class roots"* of this class, and every class root is an H connector. Each class X (except for the root class) attaches to one other unique *"parent"* class $p(X)$ by using two free links. Each link goes from a class root of X to a distinct connector in $p(X)$. The connectors in $p(X)$ that are linked to are called *"join points"*. As an example, see Fig. 1.

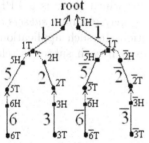
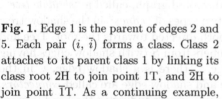
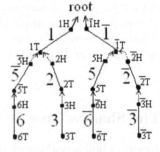

Fig. 1. Edge 1 is the parent of edges 2 and 5. Each pair $(i, \bar{i})$ forms a class. Class 2 attaches to its parent class 1 by linking its class root 2H to join point 1T, and $\overline{2}$H to join point $\overline{1}$T. As a continuing example, edges 4 and $\overline{4}$ will be added later

Fig. 2. The result of flipping the class of edges 5 and $\overline{5}$, and flipping the class of edges 6 and $\overline{6}$ in Fig. 1, followed by merging these two classes. Free links are drawn as dotted lines with arrows, while fixed links as solid lines with arrows

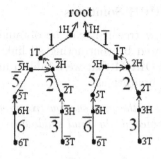

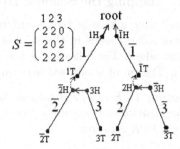

$$S = \begin{pmatrix} 1 & 2 & 3 \\ 2 & 2 & 0 \\ 2 & 0 & 2 \\ 2 & 2 & 2 \end{pmatrix}$$

Fig. 3. The result of flipping the class of edges 2 and $\overline{2}$ in Fig. 2, followed by merging it with the class of edges 5 and $\overline{5}$. The class roots of the merged class are 2H and $\overline{2}$H

Fig. 4. The final shadow tree after processing the given genotype matrix. It's an implicit representation of all PPH solutions for S

2.1   Operations on the Shadow Tree

As the algorithm processes the matrix S, new edges are added and information about old edges is updated. Three operations are used to modify the shadow tree, *edge addition*, *class flipping*, and *class merging*.

An edge is *added* to the shadow tree by creating a single edge class, consisting of the edge and its shadow edge, and then linking both edges to certain connectors in the shadow tree. Both edges of the first class created in the algorithm are linked to the root.

A class X can *flip* relative to its parent class $p(X)$ by switching the links that connect X to $p(X)$. A flip does not change any class roots or any join points, but simply switches which of the two class roots links to which of the two join points. See Fig. 2 for an example.

The algorithm may choose to *merge* two classes resulting in a larger class. A class X may merge with its parent class $p(X)$, or two classes with the same parent may merge. No other merges are possible. In the first case, the free links connecting X to $p(X)$ are changed to fixed links, and the class roots of $p(X)$ become the class roots of the new class. See Fig. 2 for an example. In the second case, when two classes X and X' have same parent, the links from the class roots of X become fixed, and are changed to point to the class roots of X' (assuming that column numbers of edges that contain class roots of X' are smaller than those of the class roots of X). After merging, the class roots of X' become the class roots of the new class. See Fig. 3 for an example of this cases. Three or more classes can be merged by executing consecutive merges.

The algorithm can "*walk up*" in the shadow tree by following links from H connectors of tree or shadow edges, until the walk reaches the root. The algorithm can efficiently find class roots and join points of a class by walking up in the shadow tree and checking if a link encountered is fixed or free.

2.2   Mapping the Shadow Tree to All PPH Solutions

We say that a tree is "*contained in*" a shadow tree if it can be obtained by flipping some classes in the shadow tree followed by contracting all links and shadow edges. The following is the KEY THEOREM that we establish in this paper. The proof is omitted here due to space limitations.

Theorem 1. *Every PPH solution is contained in the final shadow tree produced by the algorithm. Conversely, every tree contained in the final shadow tree is a distinct PPH solution.*

For example (Fig. 4), by flipping the class of 2, $\bar{2}$, 3, and $\bar{3}$, and then doing the required contractions, we get all PPH solutions for S, which are root(1(2), 3) and root(1(3), 2). Note that flipping the root class results in the same tree. Thus a final shadow tree with p classes implicitly represents 2^{p-1} PPH solutions.

2.3   Invariant Properties

The linear time PPH algorithm processes the input matrix S one row at a time, starting at the first row. At every step, the algorithm maintains certain properties of the shadow tree which are necessary for the correctness and the running time.

Theorem 2. *The shadow tree has the following invariant properties:*

Property 1: *For any column i in S, the edge labeled by i is in the shadow tree if and only if the shadow edge $\bar{i}$ is; i and $\bar{i}$ are in the same class, and are in different subtrees of the class (except for the root class).*

Property 2: *Each class (except for the root class) attaches to exactly one other class, and the two join points are in different subtrees of the parent class unless it is the root class.*

Property 3: *Along any directed path towards the root the column numbers of the edges (tree or shadow edges) strictly decrease. Also, for any two edges E and E', if E was added to the shadow tree while processing a row k, and E' was added when processing a row greater than k, then E' can never be above E on a path to the root in the shadow tree.*

When each edge is added to the growing shadow tree, the algorithm ensures that these properties are satisfied. None of the operations to the shadow tree later changes these properties.

3   Some Definitions

We use E_i to denote an edge, and C_i to denote a column number (i could be any integer between 1 and m). The "*class of edge E_i*" is defined as the class that contains E_i. The "*class root of E_i*" is defined as the root of the subtree that contains E_i, in the class of E_i.

We define three functions *col*, *te*, and *se*. Function *col* takes an edge or a connector as input and returns the column number of that edge or the column number of the edge which the connector belongs to. Function te (or se) takes a column number or an edge as input and returns the tree edge (or shadow edge respectively) of that column number or edge. If the input is the root of the shadow tree, then function col, te, and se each returns the root. The class of C_i is defined as the class that contains $te(C_i)$.

For two columns C_i and C_j, $C_j < C_i$ means that column C_j is to the left of column C_i in S. The root is defined as smaller than any column number.

A "*2 entry C_i in row k*" means that the entry at column C_i and row k in S has a value 2. A "*new 2 entry C_i in row k*" means that there is no 2 entry at C_i in rows 1 through $k-1$. An "*old 2 entry C_i in row k*" means there is at least one 2 entry at C_i in rows 1 through $k-1$.

We say that a tree T contained in shadow tree ST is "*in*" a PPH solution if T can be obtained from a PPH solution after contracting all edges corresponding to columns not in ST.

For any column C_i in S we define the "*leaf count*" of column C_i as the number of 2's in column C_i plus twice the number of 1's in column C_i.

4   Algorithm

We assume throughout the paper that the columns of S are arranged by decreasing leaf count, with the column containing the largest leaf count on the left. For ease of exposition, in this section we first describe a linear time algorithm for the PPH problem where S is assumed to only contain entries of value 0 and 2, and the all-zero sequence is the ancestral sequence in any solution. We will relax these assumptions, and solve the general PPH problem in Sec. 5.

The algorithm processes the input matrix S one row at at time, starting at the first row. We let $T(k)$ denote the shadow tree produced after processing the first k rows of S. For row $k+1$, the algorithm puts the column numbers of all old 2 entries in row $k+1$ into a list $OldEntryList$, and puts column numbers of all new 2 entries in row $k+1$ into a list $NewEntryList$.

The algorithm needs two observations. First, all edges labeled with columns that have 2 entries in row $k+1$ must form two paths to the root in any PPH solution, and no edges labeled with columns that have 0 entries in row $k+1$ can be on either of these two paths. Second, along any path to the root in any PPH solution, the successive edges are labeled by columns with strictly increasing leaf counts. These two observations are simple, but powerful and are intuitively why we can achieve linear time, while no such solution exists for the general graph realization problem.

The algorithm processes a row $k+1$ using three procedures. The first procedure, $OldEntries$, tries to create two directed paths to the root of $T(k)$ that contain all the tree edges in $T(k)$ corresponding to columns in OldEntryList. Those two paths may also contain some shadow edges. The subgraph defined by those two directed paths is called a "*hyperpath*". The process of creating a hyperpath may involve flipping some classes, and may also identify classes that need to be merged, fixing the relative position of the edges in the merged class in all PPH solutions. In the second procedure, *FixTree*, the algorithm locates any additional class merges that are required. In the third procedure, *NewEntries*, the algorithm adds the tree and shadow edges corresponding to the columns in NewEntryList, and may do additional class merges. The resulting shadow tree is $T(k + 1)$.

4.1   Procedure OldEntries

Proc. OldEntries is divided into two procedures, FirstPath followed by Second-Path. Proc. FirstPath constructs a path (called FirstPath) to the root that consists of tree edges of some column numbers in OldEntryList. The shadow tree produced after this procedure is denoted by $T_{FP}(k)$.

Procedure FirstPath: Assume that column numbers in OldEntryList (and lists used later) are ordered decreasingly, with the largest one, C_i, at the head of the list. The algorithm does a front to back scan of OldEntryList, starting from C_i, and a parallel walk up in $T(k)$, starting from edge $te(C_i)$. Let C_j denote the next entry in OldEntryList, and let E_p be the parent of $te(C_i)$ in $T(k)$. If E_p and $te(C_i)$ are not in the same class, then let E'_p denote the resulting parent of $te(C_i)$ if we flip the class of C_i. If E'_p is a tree edge and $col(E_p) \leq col(E'_p)$, then the algorithm will flip the class of C_i and set E_p to E'_p (it can be proven that if E'_p is a shadow edge, then $col(E_p) \geq col(E'_p)$). This class flipping is done to simplify the exposition in the paper and proofs of correctness of the algorithm.

The ideal case is that E_p is the tree edge $te(C_j)$, in which case we can move to the next entry in OldEntryList, and simultaneously move up one edge in $T(k)$. The ideal case continues as long as the next entries in OldEntryList correspond to the parent edges encountered in the shadow tree, and those edges are tree edges. The procedure ends when there is no entry left in OldEntryList, and we move to the root of the shadow tree.

However, there are three cases, besides the ideal case, that can happen. One case is that E_p is a shadow edge, which can only happen when $te(C_i)$ and E_p are in the same class. Then we simply walk past E_p (i.e. let $E_p = p(E_p)$), without moving past entry C_i in OldEntryList. A second case is that E_p is a tree edge, but $col(E_p) < C_j$. This indicates that $te(C_i)$ and $te(C_j)$ can never be on the same path to the root, and the algorithm adds C_j to the head of a list called *CheckList*, to be processed in Proc. SecondPath. The third case is that E_p is a tree edge, but $col(E_p) > C_j$ (and hence $col(E_p)$ has a 0 entry in row $k + 1$). This indicates that edges $te(C_i)$ and E_p must be on different paths to the root of $T(k + 1)$, and the algorithm flips the class that contains $te(C_i)$ to avoid edge E_p. In that case, the algorithm will also merge the classes containing $te(C_i)$ and E_p to fix the relative position of those edges in any PPH solution. However, if $te(C_i)$ and E_p are in the same class when this case occurs, then even flipping the class of $te(C_i)$ won't avoid the problem, and hence the algorithm reports that no PPH solution exists.   □

$$\begin{matrix} 1\,2\,3\,4\,5\,6\,7 & \text{OldEntryList:} \\ \left(\begin{array}{c} 2\,2\,2\,0\,0\,0\,0 \\ 2\,0\,0\,0\,2\,2\,0 \\ \!\!\!\!\gg\! 2\,2\,2\,2\,0\,2\,2 \\ 2\,2\,2\,2\,0\,0\,0 \\ 2\,0\,0\,0\,2\,0\,0 \end{array} \right) & \begin{array}{c} 1,2,3,6 \\ \text{NewEntryList:} \\ 4,7 \\ \text{CheckList:} \\ 2,3 \end{array} \end{matrix}$$

Fig. 5. The shadow tree after processing the first two rows of this genotype matrix is shown in Fig. 1. The shadow tree at the end of Proc. FirstPath for row 3 is shown in Fig. 2. Lists shown are for row 3

As an example, see Fig. 5. Proc. FirstPath performs a flip/merge-class in the third case. Let $T'(k)$ denote the shadow tree $T(k)$ after that flip/merge, and note that as a result, $T'(k)$ contains some, but not all, trees contained in $T(k)$. Then, it can be proven that a tree T contained in $T(k)$ is not contained in $T'(k)$ only if T is not in any PPH solution. The algorithm also puts column number C_j into CheckList in the second case during processing C_i, which indicates that $te(C_j)$ and $te(C_i)$ cannot be on the same path to the root in any PPH solution (the proof is omitted due to space limitations). The claims above essentially say that when Proc. FirstPath takes any "non-obvious" action, either flipping and merging classes or putting a column number into Checklist, it is "forced" to do so. The algorithm may perform other class flips and merges in the procedures described later.

At the end of Proc. FirstPath, any columns in OldEntryList, whose corresponding tree edges are not on FirstPath, have been placed into CheckList. Proc. SecondPath tries to construct a second path (called SecondPath) to the root that contains all the tree edges in $T(k)$ corresponding to columns in CheckList. The shadow tree produced after this procedure is denoted by $T_{SP}(k)$, and it contains a hyperpath for row $k+1$.

Procedure SecondPath: Let C_i be the largest column number in CheckList, and let C_j denote the next entry in CheckList. The algorithm does a front to back scan of CheckList, starting from column C_i, and a parallel walk up in $T_{FP}(k)$, starting from edge $te(C_i)$. The parent of $te(C_i)$ in $T_{FP}(k)$, denoted as E_p, is obtained in the same way as in Proc. FirstPath.

The rest of the algorithm is similar to Proc. FirstPath, with two major differences. First, the second case in Proc. FirstPath (when E_p is a tree edge, $col(E_p) < C_j$) now causes the algorithm to determine that no PPH solution exists. Second, the third case in Proc. FirstPath (when E_p is a tree edge, $col(E_p) > C_j$), now indicates two possible subcases. In the first subcase, if $col(E_p)$ has a 0 entry in row $k+1$, then as in Proc. FirstPath, the algorithm determines that edges $te(C_i)$ and E_p must be on different paths to the root of $T(k+1)$, and it does a flip/merge-class as in Proc. FirstPath. In the second subcase, if $col(E_p)$ is in OldEntryList, but not in CheckList, then it must be that E_p is on FirstPath. SecondPath is about to use a tree edge that is already on FirstPath, and hence some action must be taken to avoid this conflict. In this case, there is a direct way to complete the construction of SecondPath. The algorithm calls Proc. DirectSecondPath, and ends Proc. SecondPath.   □

When SecondPath is about to use a tree edge, E_p, that is on FirstPath, Proc. DirectSecondPath is called to decide whether E_p must stay on FirstPath, or whether it must be on SecondPath, or if it can be on either path to root (it can be shown that only these three cases yield valid PPH solutions). The algorithm also performs the appropriate class flips and merges to ensure that E_p stays on the path chosen by the algorithm regardless of later class flips, in the first two cases, or that FirstPath and SecondPath have no tree edge in common, in the third case.

Procedure DirectSecondPath: Recall that $te(C_i)$ is the tree edge on Second-Path whose parent edge is E_p. Let E_k denote the tree edge on FirstPath whose parent edge is E_p at the end of Proc. FirstPath. The following tests determine which path to put E_p on.

Test1: If after flipping the class of C_i and the class of E_k, E_p is either on both FirstPath and SecondPath, or on none of them, then no hyperpath exists for row $k + 1$, and hence no solution exists for the PPH problem.

Test2: If E_p is in the same class as E_k (respectively $te(C_i)$), then E_p must be on FirstPath (respectively SecondPath).

Test3: If after flipping the class of C_i and the class of E_k so that E_p is on FirstPath (respectively SecondPath), but not on SecondPath (respectively First-Path), and there doesn't exist a hyperpath in the shadow tree after the flip, then E_p must be on SecondPath (respectively FirstPath).

If the test results indicate that E_p must be on both FirstPath and Second-Path, then no hyperpath exists for row $k + 1$, and hence no solution exists for the PPH problem.

If the test results indicate that E_p must be on FirstPath (respectively Sec-ondPath), then flip the class of C_i and the class of E_k so that E_p is on FirstPath (respectively SecondPath), but not on SecondPath (respectively FirstPath), and merge the classes of E_p, C_i, and E_k.

If the test results show that E_p can be on either path, then for concreteness, flip one of the class of C_i and the class of E_k so that E_p is on FirstPath, but not on SecondPath, and merge the class of C_i with the class of E_k.   □

As an example, the shadow tree at the end of Proc. SecondPath for row 3 of the matrix in Fig. 5 is shown in Fig. 3. In this example the algorithm determines that tree edge 1 can be on either FirstPath or SecondPath.

4.2   Procedure FixTree

Proc. FixTree finds and merges more classes, if necessary, to remove trees contained in $T_{SP}(k)$ that are not in any PPH solutions. It first extends Second-Path with shadow edges whose column numbers are in OldEntryList of row $k + 1$. The subgraph defined by FirstPath and the extended SecondPath is call

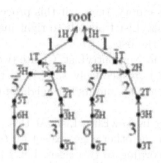 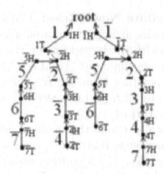

Fig. 6. The shadow tree at the end of Proc. FixTree for row 3 of the matrix in Fig. 5. In Proc. FixTree for this example, $E_1 = 6$, $E'_2 = E_2 = 3$, the class of edge 3 is merged with the class of edge 2. The class roots of the merged class are 2H and $\overline{2}$H

Fig. 7. The shadow tree at the end of Proc. NewEntries for row 3 of the matrix in Fig. 5. In Proc. NewEntries for this example, $E_1 = 6$, $E'_2 = E_2 = 3$, $C_h = 4$, $C_t = 7$, $E_{h1} = 3$, $E_{h2} = 1$, $E_{h2'} = \overline{3}$. Note that edge $\overline{7}$ links to edge 6 instead of $\overline{4}$

an "*extended hyperpath*"; it contains the hyperpath found earlier. By utilizing the extended hyperpath the algorithm can determine which additional classes need to be merged. The shadow tree produced after this procedure is denoted by $T_{FT}(k)$.

Procedure FixTree: (See Fig. 6 for an example) Let E_1 denote the tree edge of the largest column number in OldEntryList, i.e., the lowest edge of FirstPath. Let E_2 denote the tree edge of the largest column number in OldEntryList whose tree edge is not on FirstPath, i.e., the lowest edge of SecondPath.

Find a maximal path from E_2, towards the leaves in $T_{SP}(k)$, consisting of shadow edges whose column numbers are in OldEntryList. It can proven that such a maximal path is unique. Let E'_2 denote the edge that is the lower end of the maximal path; if the path doesn't contain any edge, then let $E'_2 = E_2$.

Repeat the following process until either E_1 and E'_2 are in the same class, or E'_2 is the parent of the class root of $se(E_1)$: if the column number of the edge containing the class root of E_1 is larger than the column number of the edge containing the class root of E_2, then merge the class of E_1 with its attaching class, otherwise merge the class of E_2 with its attaching class. □

4.3   Procedure NewEntries

Proc. NewEntries creates and adds edges corresponding to columns in NewEntryList of row $k + 1$ to $T_{FT}(k)$. Ideally it tries to attach new edges to the two ends of the extended hyperpath constructed in Proc. FixTree. If some new edges cannot be added in this way, the algorithm finds places to attach them. It then merges more classes, if necessary, so that there are two directed paths to the root in $T(k + 1)$ containing all the tree edges corresponding to the columns that have 2 entries in row $k + 1$, no matter how any classes are flipped.

Procedure NewEntries: If NewEntryList is empty, then exit this procedure. Otherwise arrange column numbers in NewEntryList from left to right increasingly, with the largest one on the right end of the list.

Create edges $te(C_i)$ and $se(C_i)$ for each C_i in NewEntryList. Create two free links pointing from the H connector of $te(C_i)$ (respectively $se(C_i)$) to the T connector of $te(C_j)$ (respectively $se(C_j)$), for each C_i and its left neighbor C_j in NewEntryList.

Let C_h denote the smallest column number in NewEntryList. At this point, each new edge is attached, using a free link, to one other edge, except for $te(C_h)$ and $se(C_h)$. The algorithm attaches them according to two cases. Let E_1, E_2, and E_2' be the same as in Proc. FixTree.

In the first case, when $col(E_1) < C_h$, $te(C_h)$ and $se(C_h)$ are attached to the two ends of the extended hyperpath. It creates a free link pointing from the H connector of $te(C_h)$ to the T connector of E_1. It creates a free link pointing from the H connector of $se(C_h)$ to the T connector of E_2', if E_2' is in the class of E_1, and otherwise to a connector in the class of E_1 whose parent is E_2'.

In the second case, when $col(E_1) > C_h$, by Property 3 of Theorem 2, none of $te(C_h)$ and $se(C_h)$ can attach to E_1. If $col(E_2) > C_h$, then no PPH solution exists no matter where new edges are attached; otherwise the algorithm finds two edges (E_{h1} and $E_{h2'}$) to attach $te(C_h)$ and $se(C_h)$, as follows.

Let E_{h1} denote the tree edge of the largest column number in OldEntryList that is less than C_h. Let E_{h2} denote the tree edge of the largest column number in OldEntryList that is less than C_h, and not on the path from E_{h1} to the root. If E_{h1} or E_{h2} doesn't exist, then let it be the root.

Similar to Proc. FixTree, the algorithm finds a maximal path from E_{h2} toward the leaves in $T_{FT}(k)$, consisting of shadow edges whose column numbers are in OldEntryList and less than C_h. Let $E_{h2'}$ denote the edge that is at the lower end of the maximal path.

If E_{h1} is on the path from E_1 (respectively E_2') to the root, then create a free link pointing from the H connector of $se(C_h)$ (respectively $te(C_h)$) to the T connector of E_{h1}, and create a free link pointing from the H connector of $te(C_h)$ (respectively $se(C_h)$) to the T connector of $E_{h2'}$ if $E_{h2'}$ is in the class of E_{h1}, otherwise to a connector in the class of E_{h1} whose parent is $E_{h2'}$.

If there are column numbers in NewEntryList that are larger than $col(E_1)$, then let C_t denote the smallest one among them ($C_h < col(E_1) < C_t$). $se(C_t)$ is a new edge that has been attached to an edge by the algorithm. As a special case, the algorithm changes the link from the H connector of $se(C_t)$ to point to the T connector of E_1.

All new edges are added to $T_{FT}(k)$ according to case 1 and 2. The algorithm then merges the class of C_h with the classes of column numbers in NewEntryList that are less than $col(E_1)$, and merges the class of C_h with the classes of column numbers in OldEntryList that are larger than C_h. $\qquad\square$

As an example, Fig. 7 shows the shadow tree T produced by the algorithm after processing the first three rows of the matrix S in Fig. 5. T is also the final shadow tree for S. It can be verified that Theorem 1 holds for S and T.

4.4   Correctness and Efficiency

Due to space limitations we cannot include proofs for the theorems in this paper. Proofs are given in the full version of the paper (available at [14]). For each row of S, the algorithm does a fixed number of scans of the entries in that row, and a fixed number of parallel walk ups in the shadow tree. There are some steps in the algorithm that require a traversal of the shadow tree (finding a maximal path from an edge, for example), but such operations happen at most once in each procedure, and hence at most once in the processing of each row. However, they can actually be implemented efficiently without traversing the shadow tree (we omit the details). It takes constant time to scan one entry in S, or to walk up one edge in the shadow tree. Each flip and/or merge is associated with an edge in a walk, and each flip or merge is implemented in constant time. Hence, the time for each row is $O(m)$, and the total time bound is $O(nm)$, where n and m are the number of rows and the number of columns in S.

5   General PPH Problem

Now we solve the general PPH problem for S with entries of value 0, 1, and 2. We assume that the rows of S are arranged by the position of rightmost 1 entry in each row decreasingly, with the first row containing the rightmost 1 entry in S. It is easy to prove that if there exists PPH solution(s) for S, then entries of value 1 are to the left of entries of value 2 in each row of S.

To solve the general PPH problem, we need to first build an "*initial perfect phylogeny*" T_i for S. The initial perfect phylogeny is described in detail in [15], and is built as follows. Let C_1 (respectively R_1) denotes the set of columns (respectively rows) in S that each contain at least one entry of value 1. We build T_i by first creating, for each row i in R_1, an ordered path to the root consisting of edges labeled by columns that have entries of value 1 in row i, with the edge of the smallest column label attaching to the root. We can then simply merge the identical initial segments of all these paths to create T_i. As shown in [15], T_i can be built in linear time, and must be in every PPH solution for S.

We build an "*initial shadow tree*" ST_i based on T_i by changing each edge in T_i into a tree edge in ST_i, creating an H connector and a T connector for each tree edge in ST_i, and creating a fixed link pointing from the H connector of each tree edge, corresponding to an edge E in T_i, to the T connector of the tree edge whose corresponding edge in T_i is the parent of E. There is no shadow edges in ST_i, and the tree edges in ST_i form one class.

5.1   Algorithm with Entries of Value 1

The underlying idea of the algorithm is that in any PPH solution for S, all the edges labeled with columns that have entries of value 2 in row $k + 1$ must form two paths toward an edge in the initial tree. From that edge, there is a path to the root consisting of edges labeled with columns that have entries of value 1 in row $k + 1$.

The algorithm for the PPH problem with entries of value 1, denoted as the algorithm with 1 entries, is very similar to the algorithm in Sec. 4. There are three differences. First, the algorithm with 1 entries builds and uses an initial shadow tree ST_i. Second, we now call an entry C_i an "*old 2 entry C_i in row $k+1$*" if there is at least one entry of either value 2 or 1 at C_i in rows 1 through k. The third difference is the most important one. In the algorithm with 1 entries, whenever we use the term "*root*" during the processing of row $k + 1$, we mean the *root for row $k+1$*. The root for row $k + 1$ is defined as the T connector of the tree edge in the initial shadow tree ST_i whose column number has the rightmost 1 entry in row $k + 1$. If there is no entry with value 1 in row $k + 1$, then the root for row $k + 1$ is defined as the root of ST_i. Every new edge attached to the root for row $k + 1$ becomes part of the same class as the root of ST_i. This is a simple generalization of the earlier algorithm, since earlier, the root for each row is the root of the whole shadow tree.

5.2   Remaining Issues

Identical Columns: We use an example to demonstrate how to deal with identical columns. Suppose that after arranging columns of the matrix S by decreasing leaf count, columns 5, 6, 7 are identical. We first remove columns 6, 7 from S, and obtain a new matrix S' with distinct columns. Note that we use the same column indices of S to label columns in S', i.e., column 5 of S' has a column label 5, but column 6 of S' has a column label 8. Then we solve the PPH problem on S' by using our previous algorithm. Once a final shadow tree T' for S' is constructed, we can get a final shadow tree T for S according to two cases.

In the first case the class of column 5 in T' consists of just edge 5 and $\bar{5}$. We then split tree edge 5 into three tree edges 5, 6, 7, and split shadow edge $\bar{5}$ into $\bar{5}$, $\bar{6}$, $\bar{7}$ in T. The result is equivalent to saying that 7T free links to 6H, 6T free links to 5H, and the links that link to 5H in T' now link to 7H in T. The same idea holds for shadow edges. In the second case, the class of column 5 in T' consists of edges other than 5 and $\bar{5}$. Then we want 7T to link to 6H with a fixed link, and 6T to fix link to 5H, and the links that link to 5H in T' now link to 7H in T. The same idea holds for shadow edges.

Unknown Ancestral Sequence: As mentioned in [15], the PPH problem with unknown ancestral sequence can be solved by using *majority sequence* as the root sequence, and then applying our algorithm. See [15] for more details.

6   Results

We have implemented our algorithm for the general PPH problem in C, and compared it with existing programs for the PPH problems. Program DPPH [2] was previously established as the fastest of the existing programs [7]. Some representative examples are shown in the table below. In the case of $m = 2000, n = 1000$, our program is about 250 times faster than DPPH, and the linear behavior of its running time is clear. This result is an average of 10 test cases. As in [7], our test data is generated by the program ms [23]. The cases of 50 and 100 sites and 1000 individuals are included because they reflect the sizes of subproblems that are of current interest in larger genomic scans. In those applications, there may be a huge number of such subproblems that will be examined. Our program and pseudocode can be downloaded at [14].

Sites (m)	Individuals (n)	# test cases	Average Running Time (seconds)	
			DPPH	Our program
5	1000	20	0.01	0.006
50	1000	20	0.20	0.07
100	1000	20	1.06	0.11
500	250	30	5.72	0.13
1000	500	30	45.85	0.48
2000	1000	10	467.18	1.89

References

1. V. Bafna, D. Gusfield, S. Hannenhalli, and S. Yooseph. A note on efficient computation of haplotypes via perfect phylogeny. *J. Comp. Bio.*, 11(5):858–866, 2004.
2. V. Bafna, D. Gusfield, G. Lancia, and S. Yooseph. Haplotyping as perfect phylogeny: A direct approach. *J. Computational Biology*, 10:323–340, 2003.
3. T. Barzuza, J.S. Beckmann, R. Shamir, and I. Pe'er. Computational Problems in Perfect Phylogeny Haplotyping: Xor-Genotypes and Tag SNP's. In *Proc. of CPM 2004*.
4. R.E. Bixby and D.K. Wagner. An almost linear-time algorithm for graph realization. *Mathematics of Operations Research*, 13:99–123, 1988.
5. P. Bonizzoni, G. D. Vedova, R. Dondi, and J. Li. The haplotyping problem: Models and solutions. *J. Computer Science and Technology*, 18:675–688, 2003.
6. R.H. Chung and D. Gusfield. Perfect phylogeny haplotyper: Haplotye inferral using a tree model. *Bioinformatics*, 19(6):780–781, 2003.
7. R.H. Chung and D. Gusfield. Empirical Exploration of Perfect Phylogeny Haplotyping and Haplotypers. In *Proc. the 9'th International Conference on Computing and Combinatorics*, volume 2697 of *LNCS*, pages 5–9, 2003.
8. P. Damaschke. Fast perfect phylogeny haplotype inference. *14th Symp. on Fundamentals of Comp. Theory FCT'2003*, LNCS 2751, 183–194, 2003.
9. P. Damaschke. Incremental haplotype inference, phylogeny and almost bipartite graphs. *2nd RECOMB Satellite Workshop on Computational Methods for SNPs and Haplotypes*, pre-proceedings, 1–11, 2004.
10. E. Eskin, E. Halperin, and R.M. Karp. Efficient Reconstruction of Haplotype Structure via Perfect Phylogeny. *J. Bioinformatics and Computational Biology*, 1(1):1–20, 2003.
11. E. Eskin, E. Halperin, and R. Sharan. Optimally Phasing Long Genomic Regions using Local Haplotype Predictions. In *Proc. of the Second RECOMB Satellite Workshop on Computational Methods for SNPs and Haplotypes*, Pittsburg, USA, Feburary 20–21, 2004.

12. J. Gramm, T. Nierhoff, T. Tantau, and R. Sharan. On the Complexity of Haplotyping Via Perfect Phylogeny. Presented at the *Second RECOMB Satellite Workshop on Computational Methods for SNPs and Haplotypes*, February 20–21, Pittsburgh, USA. Proceedings to appear in LNBI, Springer, 2004.

13. J. Gramm, T. Nierhoff, and T. Tantau. Perfect Path Phylogeny Haplotyping with Missing Data is Fixed-Parameter Tractable. Accepted for the *First International Workshop on Parametrized and Exact Computation (IWPEC 2004)*, Bergen, Norway, September 2004. Proceedings to appear in LNCS, Springer.

14. D. Gusfield. http://wwwcsif.cs.ucdavis.edu/~gusfield/lpph/.

15. D. Gusfield. Haplotyping as perfect phylogeny: Conceptual framework and efficient solutions (extended abstract). In *Proc. of RECOMB 2002*, pages 166–175, 2002.

16. D. Gusfield. An overview of combinatorial methods for haplotype inference. In S. Istrail, M. Waterman, and A. Clark, editors, *Computational Methods for SNPs and Haplotype Inference*, volume 2983 of *LNCS*, pages 9–25. Springer, 2004.

17. B.V. Halldórsson, V. Bafna, N. Edwards, R. Lippert, S. Yooseph, and S. Istrail. A survey of computational methods for determining haplotypes. In *Proc. of the First RECOMB Satellite on Computational Methods for SNPs and Haplotype Inference*, Springer LNBI 2983, pages 26–47, 2003.

18. B. Halldórsson, V. Bafna, N. Edwards, R. Lipert, S. Yooseph, and S. Istrail. Combinatorial problems arising in SNP and haplotype analysis. In C. Calude, M. Dinneen, and V. Vajnovski, editors, *Discrete Mathematics and Theoretical Computer Science. Proc. of DMTCS 2003*, volume 2731 of *Springer LNCS*.

19. E. Halperin and E. Eskin. Haplotype reconstruction from genotype data using Imperfect Phylogeny. *Bioinformatics*, 20:1842–1849, 2004.

20. E. Halperin and R.M. Karp. Perfect Phylogeny and Haplotype Assignment. In *Proc. of RECOMB 2004*, pages 10–19, 2004.

21. L. Helmuth. Genome research: Map of the human genome 3.0. *Science*, 293(5530):583–585, 2001.

22. R. Hudson. Gene genealogies and the coalescent process. *Oxford Survey of Evolutionary Biology*, 7:1–44, 1990.

23. R. Hudson. Generating samples under the Wright-Fisher neutral model of genetic variation. *Bioinformatics*, 18(2):337–338, 2002.

24. G. Kimmel, and R. Shamir. The Incomplete Perfect Phylogeny Haplotype Problem. Presented at the *Second RECOMB Satellite Workshop on Computational Methods for SNPs and Haplotypes*, February 20–21, 2004, Pittsburgh, USA. To appear in *J. Bioinformatics and Computational Biology*.

25. S. Tavare. Calibrating the clock: Using stochastic processes to measure the rate of evolution. In E. Lander and M. Waterman, editors, *Calculating the Secretes of Life*. National Academy Press, 1995.

26. C. Wiuf. Inference on Recombination and Block Structure Using Unphased Data. *Genetics*, 166(1):537–545, January 2004.

Human Genome Sequence Variation and the Inherited Basis of Common Disease

David Altshuler

Broad Institute of MIT and Harvard and
Massachusetts General Hospital, Department of Molecular Biology,
50 Blossom Street, Wellman 831, Boston, MA 01224, USA

A central goal in medical research is to understand how variation in DNA sequence, environment and behavior combine to cause common human diseases. To date, most progress has been limited to rare conditions in which mutation of a single gene is both necessary and sufficient to cause disease. In common diseases, in contrast, multiple genes combine with environmental influences, meaning that the contribution of any single gene variant is probabilistic rather than deterministic. Where gene effects are partial, rather than complete, their effects can only be demonstrated by measuring frequencies of variants in collections of patients and controls much larger than a single family.

A comprehensive search over common variation is increasingly practical due to dramatic advances characterizing and cataloguing human genome sequence variation. In each individual, approximately 90% of heterozygous sites are due to variants with a worldwide frequency $> 1\%$, with the remaining 10% due to sites that are individually rare and heterogeneous. The large influence of common variation is due to the historically small size of the human population, with most heterozygosity due to variants that are ancestral and shared, rather than de novo and recent. The relative contribution of rare and common variation to disease is based on the historical evolutionary selection that acted on each causal variant.

The shared ancestry of human variation also explains the correlations among nearby variants, known as linkage disequilibrium, which has the practical implication that a subset of variants capture information about a much more complete collection. To date, 8.8M human variants have been discovered and deposited in the public SNP database dbSNP, and over 1M genotyped in 270 individuals as part of the International Haplotype Map Project. Analysis of these data reveals the general patterns of linkage disequilibrium in the human genome, as well as outliers that may indicate evolutionary selection or unusual chromosomal alterations such as inversions, duplications and deletions.

A practical question is to understand the tradeoff of efficiency and power in selecting subsets of markers based on patterns of linkage disequilibrium, known as "tag SNPs." Evaluation of such methods must include explicit description of the underlying disease model, the method used to select tags, how they will be employed in downstream disease studies, the dependence on the dataset from which they are selected, and the multiple comparisons implicit in each approach.

S. Miyano et al. (Eds.): RECOMB 2005, LNBI 3500, pp. 601–602, 2005.
© Springer-Verlag Berlin Heidelberg 2005

We have implemented and will describe a simulation strategy based on empirical genotype data, in which each parameter above is evaluated. As expected, there is no single "optimal" tag SNP approach, but rather a series of tradeoffs based on assumptions about the likelihood of encountering and allele frequency of true causal variants, their representation in the dataset from which tags are selected, the patterns of linkage disequilibrium in each population.

Stability of Rearrangement Measures
in the Comparison of Genome Sequences

David Sankoff and Matthew Mazowita

Department of Mathematics and Statistics,
University of Ottawa, 585 King Edward Avenue,
Ottawa, Canada K1N 6N5
{sankoff, mmazo039}@uottawa.ca

Abstract. We present data-analytic and statistical tools for studying
rates of rearrangement of whole genomes and to assess the stability of
these methods with changes in the level of resolution of the genomic
data. We construct data sets on the numbers of conserved syntenies
and conserved segments shared by pairs of animal genomes at different
levels of resolution. We fit these data to an evolutionary tree and find
the rates of rearrangement on various evolutionary lineages. We document the lack of clocklike behaviour of rearrangement processes, the
independence of translocation and inversion rates, and the level of resolution beyond which translocations rates are lost in noise due to other
processes.

1   Introduction

The goal of this paper is to present data-analytic and statistical tools for studying rates of rearrangement of whole genomes and to assess the stability of these
methods with changes in the level of resolution of the genomic data. From secondary data provided by the UCSC Genome Brower, we construct data sets on
the number of *conserved syntenies* (pairs of chromosomes, one from each species,
containing at least one sufficiently long stretch of homologous sequence) and the
number of *conserved segments* (i.e., the total number of such stretches of homologous sequence) shared by pairs of animal genomes at levels of resolution
from 30 Kb to 1Mb. For each lineage, we calculate rates of interchromosomal
and intrachromosomal rearrangement, with and without the assumption these
are due to translocations and inversions of specific kinds.

   The key to using whole genome sequences to study evolutionary rearrangements is being able to partition each genome into segments conserved by two
genomes since their divergence. In the higher animals and many other eukaryotes this can be extremely difficult, given the high levels of occurrence of trans-
posable elements, retrotranspositions, paralogy and other repetitive and/or inserted sequences, deletions, conversion and uneven sequence divergence. The
inference of conserved segments becomes a multistage procedure parametrized

S. Miyano et al. (Eds.): RECOMB 2005, LNBI 3500, pp. 603–614, 2005.

by repeat-masking sensitivities, alignment scores, penalties and thresholds and various numerical criteria for linking smaller segments into large ones. Successful protocols have been developed independently by two research groups [5, 7] using somewhat different strategies to combine short regions of elevated similarity to construct the conserved segments, bridging singly gapped or doubly gapped regions where similarity does not attain a threshold criterion and ignoring short inversions and transpositions that have rearranged one sequence or the other. We develop our data sets on syntenies and segments from the continuously updated output of the UCSC protocol made available on the genome browser.

Building on the ideas in [11–13], we derive an estimator for the number of reciprocal translocations responsible for the number of conserved syntenies between two genomes, and use simulations to show that the bias and standard deviation of this estimator are less than 5 %, under two models of random translocation, with and without strict conservation of the centromere. By contrasting the number of conserved segments with the number of conserved syntenies, we can also estimate the number inversions or other intra-chromosomal events.

Our results include:

- A loss of stability of the data at resolutions starting at about 100 Kb.
- A highly variable proportion of translocations relative to inversions across lineages, for a fixed level of resolution.
- The relative stability of translocation rates compared to inversion rates as resolution is refined.
- The absence of correlation between accumulated rearrangements and chronological time elapsed, especially beyond 20 Myr.

2   The Data

We examined the UCSC browsers for six mammalian species and the chicken and constructed our sets of segments and syntenies for the pairs shown in Table 1. We did not use other browsers, or other nets on the four browsers in the table, because either the browser-nets pair are not posted, or because only a build older than the one used in our table was posted.

For each of the pairs in Table 1, four data sets were constructed, one at each of the 1 Mb, 300 Kb, 100 Kb and 30 Kb levels. This contains all segments larger than the resolution level as measured by the segment starting and ending points. We only counted segments in autosomes, except for the comparisons with chicken where the sex chromosomes were also included.

One complication in identifying the segment stems from the key technique of the net construction in [5], which allows very long double gaps in alignments. These gaps are often so long that they contain nested large alignments with chromosomes other than that of the main alignment. Whenever a gap in an alignment contained a nested alignment larger than the level of resolution, we

Table 1. Browsers and nets providing segments

Browser Species↓	Net Species and Build Number					
	human Hg17	chimp PanTro1	mouse Mm5	rat rn3	dog canFam1	chicken GalGal2
human		√	√	√	√	√
mouse	√			√	√	√
rat	√		√			
dog	√		√			

Table 2. Data on conserved syntenies c', conserved segments n, inferred translocations $\hat{t}$ and inferred inversions $\hat{i}$

Net	Human Browser				Mouse Browser				Rat Browser				Dog Browser			
	c'	n	$\hat{t}$	$\hat{i}$	c'	n	$\hat{t}$	$\hat{i}$	c'	n	$\hat{t}$	$\hat{i}$	c'	n	$\hat{t}$	$\hat{i}$
Human																
					143	1099	77.0	463.0	134	1484	71.2	660.8	106	636	34.5	264.5
					120	496	59.4	179.1	114	711	55.8	289.7	89	356	25.7	133.3
					107	328	50.3	104.2	102	358	47.3	121.7	83	256	22.6	86.4
					100	241	45.6	65.4	94	241	41.9	68.6	79	180	20.6	50.4
Mouse																
30Kb	181	1186	103.9	478.1					98	1833	44.6	861.9	189	731	79.8	266.7
100Kb	121	533	57.0	198.5					65	741	23.9	336.6	172	428	70.2	124.8
300Kb	108	351	48.3	116.2					48	189	14.3	70.2	158	321	62.5	79.0
1Mb	103	256	45.1	71.9					42	88	11.1	22.9	144	246	54.8	49.2
Rat																
30Kb	179	1806	102.1	789.9	97	2008	43.7	950.8								
100Kb	119	822	55.7	344.3	60	699	21.4	318.6								
300Kb	104	393	45.7	139.8	43	172	12.1	64.4								
1Mb	95	249	40.1	73.4	41	87	11.1	22.9								
Dog																
30Kb	210	1206	131.6	460.4	197	1028	128.0	376.5								
100Kb	108	465	48.3	173.2	178	502	108.1	133.4								
300Kb	85	274	34.0	92.0	162	341	93.1	67.9								
1Mb	80	194	31.0	55.0	148	251	81.0	35.0								
Chicken																
30Kb	183	1834	101.2	804.3	199	1773	122.5	754.0								
100Kb	114	992	50.8	433.7	159	1025	86.6	415.9								
300Kb	84	598	32.5	255.0	134	601	67.4	223.1								
1Mb	71	330	25.0	128.5	108	335	49.5	108.0								
Chimp																
30Kb	79	2824	30.4	1370.6												
100Kb	33	543	5.4	255.1												
300Kb	25	143	1.5	59.0												
1Mb	24	65	1.0	20.5												

broke the main alignment in two and counted the segments before and after the gap separately, assuming they remained long enough, as well as the segment in the gap.

Table 2 shows the results of our data extraction procedure. Entries for $\hat{t}$ and $\hat{i}$ calculated according to equations (7) and (1), respectively.

3   Models of Translocation

In order to derive and validate our estimator of translocation rates, we model the autosomes of a genome as c linear segments with lengths $p(1), \cdots, p(c)$, proportional to the number of base pairs they contain, where $\sum_{i=1}^{c} p(i) = 1$. We assume the two breakpoints of a translocation are chosen independently according to a uniform distribution over all autosomes, conditioned on their not being on the same chromosome. (There is no statistical evidence [9] that translocational breakpoints cluster in a non-random way on chromosomes, except in a small region immediately proximal – within 50-300Kb – to the telomere in a wide spectrum of eukaryote lineages [6].)

A reciprocal translocation between two chromosomes h and k consists of breaking each one, at some interior point, into two segments, and rejoining the four resulting segments such that two new chromosomes are produced.

In one version of our model, we impose a left-right orientation on each chromosome, such that a left-hand fragment must always rejoin a right-hand fragment. This ensures that each chromosome always retains a segment, however small it may become, containing its original left-hand extremity. This restriction models the conservation of the centromere without introducing complications such as trends towards or away from acrocentricity. With further translocations, if a breakpoint falls into a previously created segment on chromosome i, it divides that segment into two new segments, the left-hand one remaining in chromosome i, while the right-hand one, and all the other segments to the right of the breakpoint, are transferred to the other chromosome involved in the translocation. It is for this version of the model that we will derive an estimator of the number of translocations, and that we will simulate to test the estimator.

In another version of the model, an inverted left-hand fragment may rejoin another left-hand fragment and similarly for right-hand fragments. This models a high level of neocentromeric activity. We will also simulate this model to see how our estimator (derived from the previous model) fares.

We do not consider chromosome fusion and fission, so that the number of chromosomes is constant throughout the time period governed by the model. Later, in our analysis of animal genomes, we simply assume that the case where fusions or fissions occur will be well approximated by interpolating two models (with fixed chromosome number) corresponding to the two genomes being compared.

Moreover, in our simulations, we do not consider the effects of inversions on the accuracy of estimator. In previous work [12], we showed that high rates of long inversions would severely bias the estimator upwards, but that the rates and distribution of inversion lengths documented for mammalian genomes [5, 7] had no perceptible biasing effect.

In our simulations we impose a threshold and a cap on chromosome size, rejecting any translocation that results in a chromosome too small or too large. Theories about meosis, e.g. [15], can be adduced for these constraints, though there are clear exceptions, such as the "dot" chromosomes of avian and some reptilian and other vertebrate genomes [1, 3].

The total number of segments on a human chromosome i is

$$n^{(i)} = t^{(i)} + 2u^{(i)} + 1, \tag{1}$$

where $t^{(i)}$ is the number of translocational breakpoints on the chromosome, and $2u^{(i)}$ is the number of inversion breakpoints.

4   Prediction and Estimation

We assume that our random translocation process is temporally reversible, and to this effect we show in Figure 1 and Section 5.1 that the equilibrium state of our process well approximates the observed distribution of chromosome lengths in the human genome. In comparing two genomes, this assumption allows us to treat either one as ancestral and the other as derived, instead of having to consider them as diverging independently from a common ancestor.

At the outset, assume the first translocation on the lineage from genome A to genome B involves chromosome i. The assumption of a uniform density of breakpoints across the genome implies that the "partner" of i in the translocation will be chromosome j with probability $p_i(j) = \frac{p(j)}{1-p(i)}$. Thus the probability that the new chromosome labelled i contains no fragment of genome A chromosome j, where $j \neq i$, is $1 - p_i(j)$. For small $t^{(i)}$, after chromosome i has undergone $t^{(i)}$ translocations, the probability that it contains no fragment of the genome A chromosome j is approximately $(1 - p_i(j))^{t^{(i)}}$, neglecting second-order events, for example, the event that j previously translocated with one or more of the $t^{(i)}$ chromosomes that then translocated with i, and that a secondary transfer to i of material originally from j thereby occurred.

Then the probability that the genome B chromosome i contains at least one fragment from j is approximately $1 - (1 - p_i(j))^{t^{(i)}}$ and the expected number of genome A chromosomes with at least one fragment showing up on genome B chromosome i is

$$E(c^{(i)}) \approx 1 + \sum_{j \neq i} [1 - (1 - p_i(j))^{t^{(i)}}] \tag{2}$$

so that

$$c - E(c^{(i)}) \approx \sum_{j \neq i} (1 - p_i(j))^{t^{(i)}}, \tag{3}$$

where the leading 1 in (2) counts the fragment containing the left-hand endpoint of the genome A chromosome i itself. We term $c^{(i)}$ the number of *conserved syntenies* on chromosome i.

Suppose there have been a total of t translocations in the evolutionary history. Then

$$\sum_i t^{(i)} = 2t. \tag{4}$$

We can expect these to have been distributed among the chromosomes approximately as

$$t^{(i)} = 2tp(i), \tag{5}$$

so that

$$c^2 - \sum_i E(c^{(i)}) \approx \sum_i \sum_{j \neq i} (1 - p_i(j))^{2tp(i)}. \qquad (6)$$

Substituting the $c^{(i)}$ for the $E(c^{(i)})$ in eqn (6) suggests solving

$$c^2 - \sum c^{(i)} = \sum_i \sum_{j \neq i} [1 - p_i(j)]^{2\hat{t}p(i)}, \qquad (7)$$

for $\hat{t}$ to provide an estimator of t. Newton's method converges rapidly for the range of parameters used in our studies, as long as not all $c^{(i)} = c$. (We know of no comparative map where even one chromosome of one genome shares a significant syntenic segment with every autosome of the other genome, much less a map where every chromosome is thus scrambled.)

5   Simulations

5.1   Equilibrium Distribution of Chromosome Size

Models of accumulated reciprocal translocations for explaining the observed range of chromosome sizes in a genome date from the 1996 study of Sankoff and Ferretti [10]. They proposed a lower threshold on chromosome size in order to reproduce the appropriate size range in plant and animal genomes containing from two to 22 autosomes. A cap on largest chromosome size has also been proposed [15] and shown to be effective [4]. Economy and elegance in explaining chromosome size being less important in the present context than simulating a realistic equilibrium distribution of these sizes, we imposed both a threshold of 50 Mb and a cap of 250 Mb on the process described in Section 3, simply rejecting any translocations that produced chromosomes out of the range. These values were inspired by the relative stability across primates and rodents evident in the data in Table 3, though they are less pertinent for the dog, with a larger number of correspondingly smaller chromosomes, and chicken, which has several very small chromosomes.

Table 3. Shortest and longest chromosome, in Mb

genome	shortest	longest
mouse	61	199
human	47	246
rat	47	268
chimp	47	230
dog	26	125
chicken	0.24	188

Simulating the translocation process 100 times up to 10,000 translocations each produced the equilibrium distribution of chromosome sizes in Figure 1. The superimposed distribution of human autosome sizes is very close to the equilibrium distribution.

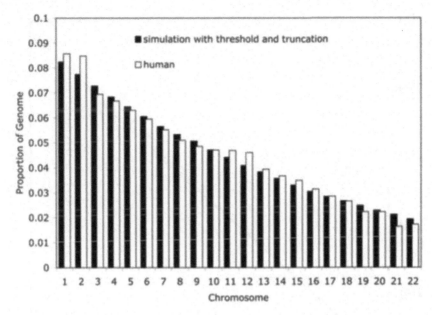

Fig. 1. Comparison of equilibrium distribution of simulated chromosome sizes with human autosome sizes

5.2   Performance of the Estimator

Figure 2 depicts the estimated number of translocations as a function of the true number t in the simulation. The estimator $\hat{t}$ appears very accurate, only lightly biased (less than 5 % for $t < 200$), with small error rates (s.d. less than 5 % for $t < 200$).

6   Fitting the Data to Animal Phylogeny

To infer the rates of rearrangement on evolutionary lineages, we assumed the phylogenetic tree in Figure 3. Because of the limited number of genome pairs for which we have data, we artificially attributed all the chimp-human divergence to the chimp lineage, and could not estimate the translocational divergence during the mammalian radiation, i.e. between the divergence of the dog lineage and the common primate/rodent lineage.

We fit the data in Table 2 to the tree by solving the system of linear equations between the additive path lengths in the tree and the inferred rearrangement distances, namely $c' - c$ the number of new syntenies created on a path, $\hat{t}$ the number of translocations inferred to have occurred, $n - c$ the number of segments created, and $\hat{i}$ the inferred number of inversions. Where browser-net pairs were available in both directions, we averaged the two results in Table 2 to produce a single equation. As there are more pairs than edges, we also used an averaging procedure in solving the equations to produce the results in Table 4.

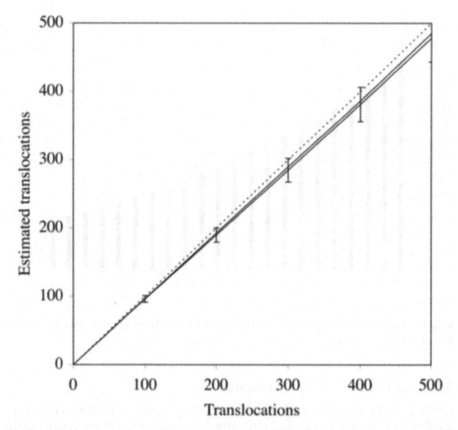

Fig. 2. Mean value, over 100 runs, of $\hat{t}$ as a function of t . Dotted line: $\hat{t} = t$. Lower line with ± 1 s.d. error bars: model with centromere. Upper line: model without centromere

The very approximate temporal edge lengths given in Table 4 were based, for p, on twice the usual estimate (6 Myr) of chimp-human divergence to account for both human and chimp evolution; for m and r the date for the rat-mouse divergence (≈ 20 Myr); for h and d the date for mammalian radiation; for a the same date less the 20 Myr of murid evolution; and for c twice the mammalian-reptile divergence time (310 Myr) less the 85 Myr since the mammalian radiation.

7   Observations

The most striking trend in Table 4 is the dramatic increase in both conserved syntenies and conserved segments in almost every lineage (except a) as the level of resolution is refined, starting at 100 Kb but accelerating rapidly at 30 Kb. It seems likely that the increased level of translocations inferred is artifactual, the apparent level of conserved syntenies reflecting retrotransposition and other interchromosomal process and not reciprocal translocation [16].

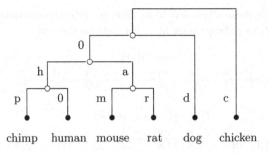

Fig. 3. Unrooted phylogeny for fitting translocation measures. Edges labelled "0" indicate the two endpoints are collapsed; none of the pairwise measures bear on the location of the vertices between chicken and dog, and between human and chimp. These two lengths can be assumed to be short in any case (period of rapid mammalian radiation and human-chimp divergence, respectively)

That this increase does not reflect translocational distance is further evidenced by the loss at 30 Kb of clear trends among the lineages visible at less refined resolutions, such as the very low values for human, mouse and rat compared to the other lineages. Thus we can conclude that below the 100 Kb level, the study of translocational rearrangement by our statistical approach is no longer feasible.

Even at the less refined levels of resolution, any correlation between chronological time and translocational distance breaks down somewhere between 20 and 65 Myr. As has been remarked previously [2], the chicken evidences a low rate of translocation. The dog on the other hand, shows a high rate.

Turning to the results on inversions, the rapid increase in segments and inferred inversions at refined resolutions may, in contrast to translocations, be a real effect. It is known that the inversions of small size are very frequent in these genomes [5], with a mean size less than 1 Kb, so that it can be expected that the number of inversions inferred will continue to accelerate with increased resolution. Indeed, even between 1 Mb and 300 Kb, while translocation rates are relatively stable, inversion rates increase substantially.

The pattern of inversion rates among the lineages is very different from that of translocations. Here, chicken has a very high rate while dog has a low one, the opposite of what was seen with translocations. Perhaps most startling is the high rate recovered for the chimp lineage. This disproportion is likely an artifact; whereas in more distant comparisons the alignments of many inverted segments may not be detected due to sequence divergence, the proximity of the human and chimpanzee genomes allows for a very high recovery rate.

Again with inversions, as with translocations, there is little correlation of lineage-specific rates and the chronological span of the lineage.

Finally, when inversion and translocation rates are compared at a fixed level of resolution, no systematic association can be seen.

Table 4. Tree edge-lengths estimated from pairwise interchromosomal measures in Table 2. Negative entries indicate poor fit of the data to this tree

Time (Myr)

	h	a	m	r	c	d	p
	85	65	20	20	535	85	12

New syntenies created

	h	a	m	r	c	d	p
1Mb	13.5	52.8	14.8	7.3	28.5	43.0	2.0
300Kb	10.9	60.6	15.5	10.5	45.0	49.3	3.0
100Kb	18.5	57.8	23.8	19.3	66.0	57.5	11.0
30Kb	56.9	42.6	42.0	36.0	99.8	75.5	57.0

Translocations

	h	a	m	r	c	d	p
1Mb	6.0	31.6	7.8	3.3	14.6	24.2	1.0
300Kb	3.6	37.8	8.0	5.2	25.3	28.4	1.5
100Kb	7.1	38.5	12.6	10.1	39.6	34.0	5.4
30Kb	34.7	31.8	23.9	20.2	66.7	48.3	30.4

New segments created

	h	a	m	r	c	d	p
1Mb	96.3	95.8	36.0	32.0	198.0	74.5	43.0
300Kb	141.1	115.1	62.8	98.3	419.5	109.3	121.0
100Kb	224.0	45.5	224.5	476.0	741.0	161.5	521.0
30Kb	585.5	-163.0	699.5	1201.5	1222.0	310.0	2802.0

Inversions

	h	a	m	r	c	d	p
1Mb	42.1	16.3	10.2	12.7	83.9	13.1	20.5
300Kb	67.0	19.8	23.4	43.9	184.0	26.3	59.0
100Kb	104.9	-15.8	99.7	227.9	330.4	46.8	255.1
30Kb	188.2	-43.5	325.8	580.6	543.8	-33.0	1370.6

8   Discussion

We have proposed an estimator of the number of translocations intervening between two rearranged genomes, based only on the numbers of conserved syntenies on each chromosome, the lengths of the chromosomes and a simplified random model of interchromosomal exchange. This estimator proves to be very accurate in simulations, which is remarkable given that it only explicitly takes into account the first-order effects of interchromosomal exchange.

In this paper, we applied our estimator to animal genome comparisons at various levels of resolution. This showed that translocation estimates are stable at coarse resolutions, while inversions increased markedly. This reflects the discovery of high numbers of smaller-scale local arrangements recognizable from genomic sequence [5]. At very detailed levels of resolution, inferred translocations numbers probably reflect processes other than translocation, though increased inversion inferences are more likely to reflect the inversion process.

Our estimates of the number of translocations and inversions in the evolutionary divergence of animals are only about a half of what has been published by Pevzner and colleagues [7, 8, 2] for corresponding level of resolution. Their estimates are based on an algorithmic reconstruction of the details of evolutionary history. Our model assumes each translocation and inversion creates two new segments, but the algorithms require a number of rearrangements almost equal to the number of segments to account for how the segments are ordered on the chromosomes. This accounts for the difference between the two sets of results.

Acknowledgements

Thanks to Phil Trinh for help with the genome browsers and other tools. Research supported in part by a Discovery grant from the Natural Sciences and Engineering Research Council of Canada (NSERC). DS holds the Canada Research Chair in Mathematical Genomics and is a Fellow in the Evolutionary Biology Program of the Canadian Institute for Advanced Research.

References

1. Bed'hom, B. (2000). Evolution of karyotype organization in *Accipitridae*: A translocation model. In Sankoff, D. and Nadeau, J. H. (eds) *Comparative Genomics: Empirical and Analytical Approaches to Gene Order Dynamics, Map Alignment and Evolution of Gene Families*. Dordrecht, NL, Kluwer, 347–56.
2. Bourque, G., Zdobnov, E., Bork, P., Pevzner, P.A. and Tesler, G. (2005) Comparative architectures of mammalian and chicken genomes reveal highly variable rates of genomic rearrangements across different lineages. *Genome Research*, **15**, 98–110.
3. Burt, D.W. (2002). Origin and evolution of avian microchromosomes. *Cytogenetic and Genome Research*, **96**, 97–112.
4. De, A., Ferguson, M., Sindi, S. and Durrett, R. (2001). The equilibrium distribution for a generalized Sankoff-Ferretti model accurately predicts chromosome size distributions in a wide variety of species. *Journal of Applied Probability*, **38**, 324–34.
5. Kent, W. J., Baertsch, R., Hinrichs, A., Miller, W. and Haussler, D. (2003). Evolution's cauldron: Duplication, deletion, and rearrangement in the mouse and human genomes. *Proceedings of the National Academy of Sciences, USA*, **100**, 11484–9.
6. Mefford, H.C. and Trask, B.J. (2002). The complex structure and dynamic evolution of human subtelomeres. *Nature Reviews in Genetics*, **3**, 91-102; 229.
7. Pevzner, P. A. and Tesler, G. (2003). Genome rearrangements in mammalian genomes: Lessons from human and mouse genomic sequences. *Genome Research*, **13**, 37-45
8. Pevzner, P. A. and Tesler, G. (2003). Human and mouse genomic sequences reveal extensive breakpoint reuse in mammalian evolution. *Proceedings of the National Academy of Sciences, USA*, **100**, 7672-7

9. Sankoff, D., Deneault, M., Turbis, P. and Allen, C.P. (2002) Chromosomal distributions of breakpoints in cancer, infertility and evolution. *Theoretical Population Biology*, **61**, 497–501.
10. Sankoff, D. and Ferretti, V. (1996). Karotype distributions in a stochastic model of reciprocal translocation. *Genome Research*, **6**, 1–9.
11. Sankoff, D., Ferretti, V, and Nadeau, J.H. (1997) Conserved segment identification. *Journal of Computational Biology*, **4**, 559–65.
12. Sankoff, D. and Mazowita, M. (2005) Estimators of translocations and inversions in comparative maps. In Lagergren, J. (ed.) *Proceedings of the RECOMB 2004 Satellite Workshop on Comparative Genomics, RCG 2004*, Lecture Notes in Bioinformatics, **3388**, Springer, Heidelberg, 109–122
13. Sankoff, D., Parent, M.-N. and Bryant, D. (2000). Accuracy and robustness of analyses based on numbers of genes in observed segments. In Sankoff, D. and Nadeau, J. H. (eds) *Comparative Genomics: Empirical and Analytical Approaches to Gene Order Dynamics, Map Alignment and Evolution of Gene Families.* Dordrecht, NL, Kluwer, 299–306.
14. Sankoff, D. and Trinh, P. (2004). Chromosomal breakpoint re-use in the inference of genome sequence rearrangement. *Proceedings of RECOMB 04, Eighth International Conference on Computational Molecular Biology.* New York: ACM Press, 30–5.
15. Schubert, I. and Oud, J.L. (1997). There is an upper limit of chromosome size for normal development of an organism. *Cell*, **88**, 515–20.
16. Trinh, P., McLysaght, A. and Sankoff, D. (2004) Genomic features in the breakpoint regions between syntenic blocks. *Bioinformatics*, **20**, I318–I325.

On Sorting by Translocations

Anne Bergeron[1], Julia Mixtacki[2], and Jens Stoye[3]

[1] LaCIM, Université du Québec à Montréal, Canada
anne@lacim.uqam.ca
[2] Fakultät für Mathematik, Universität Bielefeld, 33594 Bielefeld, Germany
julia.mixtacki@uni-bielefeld.de
[3] Technische Fakultät, Universität Bielefeld, 33594 Bielefeld, Germany
stoye@techfak.uni-bielefeld.de

Abstract. The study of genome rearrangements is an important tool in comparative genomics. This paper revisits the problem of sorting a multichromosomal genome by translocations, i.e. exchanges of chromosome ends. We give an elementary proof of the formula for computing the translocation distance in linear time, and we give a new algorithm for sorting by translocations, correcting an error in a previous algorithm by Hannenhalli.

1 Introduction

We revisit the problem of sorting multichromosomal genomes by translocations that was introduced by Kececioglu and Ravi [7] and Hannenhalli [5]: Given two genomes A and B, the goal is to find a shortest sequence of exchanges of nonempty chromosome ends that transforms A into B. The length of such a shortest sequence is the translocation distance between A and B, and the problem of computing this distance is called the translocation distance problem.

The study of genome rearrangements allows to better understand the processes of evolution and is an important tool in comparative genomics. However, the combinatorial theories that underly rearrangement algorithms are complex and prone to human errors [9, 10].

Given their prevalence in eukaryotic genomes [4], a good understanding of translocations is necessary. Using tools developed in the context of sorting two signed genomes by inversions, we establish on solid grounds Hannenhalli's equation for the translocation distance, and give a new algorithm for sorting by translocations.

Restricting genome rearrangements to translocations only might look, at first glance, a severe constraint. However, mastering the combinatorial knowledge of a single operation is always a step towards a better understanding of the global picture. As more and more genomes are decoded, sound mathematical models, and correct algorithms will play a crucial role in analyzing them.

The next section introduces the basic background needed in the following. The third section gives a counter-example to Hannenhalli's algorithm. Section 4

S. Miyano et al. (Eds.): RECOMB 2005, LNBI 3500, pp. 615–629, 2005.

presents a new proof and formula for the translocation distance formula, and Section 5 discusses the algorithms.

2   Definitions and Examples

2.1   Genes, Chromosomes and Genomes

As usual, we represent a *gene* by a signed integer where the sign represents its orientation. A *chromosome* is a sequence of genes and does not have an orientation. A *genome* is a set of chromosomes. We assume that each gene appears exactly once in a genome. If the k-th chromosome in a genome A of N chromosomes contains m_k genes, then the genes in A are represented by the integers $\{1, \ldots, n\}$ where $n = \sum_{k=1}^{N} m_k$:

$$A = \{(a_{11}\, a_{12}\, \ldots\, a_{1m_1}), (a_{21}\, a_{22}\, \ldots\, a_{2m_2}), \ldots, (a_{N1}\, a_{N2}\, \ldots\, a_{Nm_N})\}.$$

For example, the following genome consists of three chromosomes and nine genes:

$$A_1 = \{(4\quad 3),\quad (1\quad 2\quad -7\quad 5),\quad (6\quad -8\quad 9)\}.$$

For an interval $I = a_i\, \ldots\, a_j$ of elements inside a chromosome we denote by $-I$ the reversed interval where the sign of each element is changed, i.e. $-I = -a_j\, \ldots - a_i$. Since a chromosome does not have an orientation, we can *flip* the chromosome $X = (x_1, x_2, \ldots, x_k)$ into $-X = (-x_k, \ldots, -x_1)$ and still have the same chromosome. More precisely, let us consider two chromosomes X and Y. We say that a chromosome X is *identical* to a chromosome Y if either $X = Y$ or $X = -Y$. Genomes A and B are *identical* if for each chromosome contained in A there is an identical chromosome in B and vice versa.

A *translocation* transforms the chromosomes $X = (x_1, \ldots, x_i, x_{i+1}, \ldots, x_k)$ and $Y = (y_1, \ldots, y_j, y_{j+1}, \ldots, y_l)$ into new chromosomes $(x_1, \ldots, x_i, y_{j+1}, \ldots, y_l)$ and $(y_1, \ldots, y_j, x_{i+1}, \ldots, x_k)$. It is called *internal* if all exchanged chromosome ends are non-empty, i.e. $1 < i < k$ and $1 < j < l$.

Given a chromosome $X = (x_1, x_2, \ldots, x_k)$, the elements x_1 and $-x_k$ are called its *tails*. Two genomes are *co-tailed* if their sets of tails are equal. Note that an internal translocation does not change the set of tails of a genome.

In the following, we assume that the elements of each chromosome of the target genome B are positive and in increasing order. For example, we have that

$$A_1 = \{(\,4\,\,3),\,(1\,\,2\,-7\,\,5),\,(6\,-8\,\,9)\}$$

$$B_1 = \{(1\quad 2\quad 3),\,(4\,\,5),\,(6\quad 7\quad 8\quad 9)\}.$$

The *sorting by translocations problem* is to find a shortest sequence of translocations that transforms one given genome A into the genome B. We call the length of such a shortest sequence the *translocation distance* of A, and denote this number by $d(A)$. The problem of computing $d(A)$ is called the *translocation distance problem*.

In the following, we will always assume that translocations are internal. Therefore, in the sorting by translocations problem, genomes A and B must be co-tailed.

Translocations on a genome can be simulated by inversions of intervals of signed permutations, see [6, 9, 10]. For a genome A with N chromosomes, there are $2^N N!$ possible ways to chain the N chromosomes, each of these is called a *concatenation*. Given a concatenation, we extend it by adding a first element 0 and a last element $n + 1$. This results in a signed permutation P_A on the set $\{0, \ldots, n + 1\}$:

$$P_A = (0\ a_{11}\ a_{12}\ \ldots\ a_{1m_1}\ a_{21}\ a_{22}\ \ldots\ a_{2m_2}\ \ldots\ a_{N1}\ a_{N2}\ \ldots\ a_{Nm_N}\ n + 1).$$

An *inversion* of an interval reverses the order of the interval while changing the sign of all its elements. We can model translocations on the genome A by inversions on the signed permutation P_A. Sometimes it is necessary to flip a chromosome. This can also be modeled by the inversion of a chromosome, but does not count as an operation in computing the translocation distance since the represented genomes are identical. See Fig. 1 for an example.

In the following sections we consider several concepts such as elementary intervals, cycles and components that are central to the analysis of the sorting by translocation problem. These concepts were originally developed for the analysis of the inversion distance problem. The notation follows [2].

$P_{A_1} = (0\quad 4\quad 3\quad 1\quad 2\quad \text{-}7\quad 5\quad 6\quad \text{-}8\quad 9\quad 10)$

$A_1 = \{(4\quad 3), (1\quad 2\quad \text{-}7\quad 5), (6\quad \text{-}8\quad 9)\}$

$\quad\quad (0\quad 4\quad 3\quad \text{-}5\quad 7\quad \text{-}2\quad \text{-}1\quad 6\quad \text{-}8\quad 9\quad 10)$

$\{(4\quad \text{-}7\quad 5), (1\quad 2\quad 3), (\text{-}9\quad 8\quad \text{-}6)\}$

$\quad\quad (0\quad 4\quad \text{-}7\quad 5\quad \text{-}3\quad \text{-}2\quad \text{-}1\quad 6\quad \text{-}8\quad 9\quad 10)$

$\quad\quad (0\quad 4\quad \text{-}7\quad \text{-}6\quad 1\quad 2\quad 3\quad \text{-}5\quad \text{-}8\quad 9\quad 10)$

$\{(4\quad \text{-}7\quad \text{-}6), (1\quad 2\quad 3), (\text{-}5\quad \text{-}8\quad 9)\}$

$\quad\quad (0\quad 4\quad \text{-}7\quad \text{-}6\quad 1\quad 2\quad 3\quad \text{-}9\quad 8\quad 5\quad 10)$

$\quad\quad (0\quad 4\quad 9\quad \text{-}3\quad \text{-}2\quad \text{-}1\quad 6\quad 7\quad 8\quad 5\quad 10)$

$\{(\text{-}9\quad \text{-}4), (1\quad 2\quad 3), (\text{-}5\quad \text{-}8\quad \text{-}7\quad \text{-}6)\}$

$\quad\quad (0\quad \text{-}9\quad \text{-}4\quad \text{-}3\quad \text{-}2\quad \text{-}1\quad 6\quad 7\quad 8\quad 5\quad 10)$

$\quad\quad (0\quad \text{-}9\quad \text{-}8\quad \text{-}7\quad \text{-}6\quad 1\quad 2\quad 3\quad 4\quad 5\quad 10)$

$\quad\quad (0\quad \text{-}5\quad \text{-}4\quad \text{-}3\quad \text{-}2\quad \text{-}1\quad 6\quad 7\quad 8\quad 9\quad 10)$

$B_1 = \{(1\quad 2\quad 3), (4\quad 5), (6\quad 7\quad 8\quad 9)\}\quad Id = (0\quad 1\quad 2\quad 3\quad 4\quad 5\quad 6\quad 7\quad 8\quad 9\quad 10)$

Fig. 1. *Left:* An optimal sorting scenario for the translocation distance problem for the genomes A and B; the exchanged chromosome ends are underlined. *Right:* Given an arbitrary concatenation, the problem can be modeled by sorting the signed permutation P_A by inversions; solid lines denote inversions that represent translocations, dashed lines denote inversions that flip chromosomes

2.2   Elementary Intervals and Cycles

Let A be a genome on the set $\{1, \ldots, n\}$. We consider the extended signed permutation P_A defined by an arbitrary concatenation of the chromosomes of A.

Definition 1. *A pair $p \cdot q$ of consecutive elements in a signed permutation is called a* point. *A point is called an* adjacency *if it is a point of the form $i \cdot i + 1$ or $-(i + 1) \cdot -i$, $0 \leq i \leq n$, otherwise it is called a* breakpoint.

The signed permutation P_A has $n + 1$ points, $N - 1$ of them are between tails, and two other points are between 0 and a tail and between a tail and $n+1$. Those $N + 1$ points define the concatenation of the genome A, and are called *white* points. The points inside chromosomes are *black* points.

For example, the signed permutation P_{A_1} has ten points; three of them are adjacencies, and all the other points are breakpoints. The points $0 \cdot 4$, $3 \cdot 1$, $5 \cdot 6$ and $9 \cdot 10$ are white.

$$P_{A_1} = (0 \quad 4 \quad 3 \quad 1 \quad 2 \quad -7 \quad 5 \quad 6 \quad -8 \quad 9 \quad 10)$$

When sorting, eventually all black points must become adjacencies. A translocation acts on two black points inside different chromosomes. We can flip chromosomes by performing an inversion between two white points.

Definition 2. *For each pair of unsigned elements $(k, k + 1)$, $0 \leq k < n + 1$, define the* elementary interval I_k *associated to the pair $k \cdot k + 1$ of unsigned elements to be the interval whose endpoints are:*

1. *the right point of k, if k is positive, otherwise its left point;*
2. *the left point of $k + 1$, if $k + 1$ is positive, otherwise its right point.*

Since we assume that genomes are co-tailed and that the elements of the target genome are positive and in sorted order, the two endpoints of an elementary interval will always be either both black or both white. From Definition 2 it follows that exactly two elementary intervals of the same color meet at each breakpoint.

Definition 3. *A black (or white) cycle is a sequence of breakpoints that are linked by black (respectively white) elementary intervals. Adjacencies define trivial cycles.*

The elementary intervals and cycles of our example permutation P_{A_1} are shown in Fig. 2.

The white cycles formed by the $N + 1$ white points depend on the concatenation. Since the order and the orientation of the chromosomes are irrelevant for the sorting by translocation problem, we focus on the black cycles that are formed by the $n-N$ black points. The number of black cycles of P_A is maximized, and equals $n - N$, if and only if genome A is sorted.

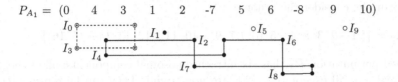

Fig. 2. Elementary intervals and cycles of the signed permutation P_{A_1}

2.3   Effects of a Translocation on Elementary Intervals and Cycles

In the previous section we have seen that we have to reduce the number of black breakpoints or increase the number of black cycles of P_A in order to sort a genome A by translocations. Thus, we are interested in how a translocation changes the number of breakpoints, as well as the number of cycles.

Lemma 1 (Kececioglu and Ravi [7]). *A translocation in genome A modifies the number of black cycles of P_A by 1, 0 or -1.*

Following the terminology of Hannenhalli [5], a translocation is called *proper* if it increases the number of black cycles by 1, *improper* if it leaves the number of black cycles unchanged and *bad* if it decreases the number of black cycles by 1. As a consequence of Lemma 1 we get the lower bound $d(A) \geq n - N - c$, where c is the number of black cycles of a genome A.

An elementary interval whose endpoints belong to different chromosomes is called *interchromosomal*, otherwise it is called *intrachromosomal*. Given an interchromosomal elementary interval I_k of P_A, we can always assume that elements k and $k+1$ have different signs, since we can always flip a chromosome. This implies that the corresponding translocation creates a new adjacency: either $k \cdot k+1$ or $-(k+1) \cdot -k$. Hence we have:

Lemma 2. *For each interchromosomal elementary interval in P_A, there exists a proper translocation in the genome A.*

2.4   Intrachromosomal Components

As discussed in [1] for the inversion distance problem, elementary intervals and cycles can be grouped into higher structures:

Definition 4. *A component of a signed permutation is an interval from i to $i+j$ or from $-(i+j)$ to $-i$, where $j > 0$, whose set of elements is $\{i, \ldots, i+j\}$, and that is not the union of smaller such intervals.*

We refer to a component by giving its first and last element such as $[i \ldots j]$. When the elements of a component belong to the same chromosome, then the component is said to be *intrachromosomal*. An intrachromosomal component is called *minimal* if it does not contain any other intrachromosomal component. An intrachromosomal component that is an adjacency is called *trivial*, otherwise *non-trivial*.

For example, consider the genome

$$A_2 = \{(\ 1\ -2\ 3\ 8\ 4\ -5\ 6),(\ 7\ 9\ -10\ 11\ -12\ 13\ 14\ -15\ 16)\}.$$

The signed permutation P_{A_2} has six intrachromosomal components; all of them are minimal and all except $[13\ldots 14]$ are non-trivial. They can be represented by a boxed diagram such as in Fig. 3. Note that $[3\ldots 9]$ is a component that is not intrachromosomal.

The relationship between intrachromosomal components plays an important role in the sorting by translocations problem. As shown in [3], two different intrachromosomal components of a chromosome are either disjoint, nested with different endpoints, or overlapping on one element.

When two intrachromosomal components overlap on one element, we say that they are *linked*. Successive linked intrachromosomal components form a *chain*. A chain that cannot be extended to the left or right is called *maximal*. We represent the nesting and linking relation of intrachromosomal components of a chromosome in the following way:

Definition 5. *Given a chromosome X and its intrachromosomal components, define the forest F_X by the following construction:*

1. *Each non-trivial intrachromosomal component is represented by a round node.*
2. *Each maximal chain that contains non-trivial components is represented by a square node whose (ordered) children are the round nodes that represent the non-trivial intrachromosomal components of this chain.*
3. *A square node is the child of the smallest intrachromosomal component that contains this chain.*

We extend the above definition to a forest of a genome by combining the forests of all chromosomes:

Definition 6. *Given a genome A consisting of chromosomes $\{X_1, X_2, \ldots, X_N\}$. The forest F_A is the set of forests $\{F_{X_1}, F_{X_2}, \ldots, F_{X_N}\}$.*

Note that the forest F_A can consist of more than one tree in contrast to the unichromosomal case [1]. Figure 3 shows the forest F_{A_2} that consists of three trees.

2.5   Effects of a Translocation on Intrachromosomal Components

We say that a translocation *destroys* an intrachromosomal component C if C is not an intrachromosomal component in the resulting genome. When sorting a genome, eventually all its non-trivial intrachromosomal components, and hence all its trees, are destroyed.

The only way to destroy an intrachromosomal component with translocations is to apply a translocation with one endpoint in the component, and one endpoint in another chromosome. Such translocations always merge cycles and thus

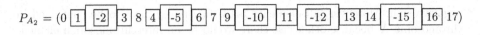

$$P_{A_2} = (0\ \boxed{1}\ \boxed{-2}\ \boxed{3}\ 8\ \boxed{4}\ \boxed{-5}\ \boxed{6}\ 7\ \boxed{9}\ \boxed{-10}\ \boxed{11}\ \boxed{-12}\ \boxed{13}\ \boxed{14}\ \boxed{-15}\ \boxed{16}\ 17)$$

F_{A_2} :

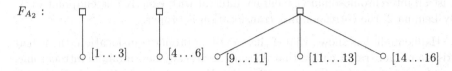

Fig. 3. The intrachromosomal components of the signed permutation P_{A_2} of the genome $A_2 = \{(1 - 2\ \ 3\ \ 8\ \ 4 - 5\ \ 6), (7\ \ 9 - 10\ \ 11 - 12\ \ 13\ \ 14 - 15\ \ 16)\}$ and the forest F_{A_2}

are always bad. Yet, a translocation may destroy more than one component at the same time. In fact, a translocation that acts on one point of an intrachromosomal component C destroys C and all the intrachromosomal components that contain C. Thus, at most two minimal intrachromosomal components on two different chromosomes, plus all intrachromosomal components containing these two components, can be destroyed by a single translocation.

It is also possible to eventually destroy by a single translocation two intrachromosomal components that initially belong to two different trees of the same chromosome. The next results show how.

Lemma 3. *If a chromosome X of genome A contains more than one tree, then there exists a proper translocation involving chromosome X.*

Proof. Consider the chromosome $X = (x_1 \ldots x_m)$. We assume that all elementary intervals involving chromosome X are intrachromosomal. The first step is to show that then the whole chromosome is an intrachromosomal component. We have to show that the first element of the chromosome is the smallest element and the last element is the greatest, if both are positive, and the reverse, if both are negative, and that all elements between the smallest and the greatest are contained in the chromosome.

Let i be the smallest unsigned element contained in chromosome X. Suppose that i has positive sign and $x_1 \neq i$. The left point of i is an endpoint of the elementary interval I_{i-1}. Since i is the smallest element, the unsigned element $i-1$ belongs to a chromosome different from X. Therefore the elementary interval I_{i-1} is interchromosomal. This contradicts our assumption that all elementary intervals involving the chromosome X are intrachromosomal.

Let j be the greatest unsigned element contained in chromosome X. Suppose that j has positive sign and $x_m \neq j$. Then the right point of j is an endpoint of the elementary interval I_j, and the element $j + 1$ belongs to another chromosome. Thus, the elementary interval I_j is interchromosomal contradicting our assumption.

By a similar argumentation, we can show that $x_1 = -j$, if j is the greatest element and has negative sign, and $x_m = -i$, if i the smallest element and has negative sign. Moreover, all elements between i and j have to be contained in chromosome X because otherwise there would be an interchromosomal elementary interval. Thus, chromosome X itself is an intrachromosomal component, and contains a single tree. This leads to a contradiction. Therefore, there must exist an interchromosomal elementary interval with exactly one endpoint in X. By Lemma 2, the corresponding translocation is proper. $\qquad\square$

Hannenhalli has shown that if there exists a proper translocation, then there exists a proper translocation that does not create any new minimal intrachromosomal components (see Theorem 10 in [5]). However, as we will see in Section 3, Hannenhalli's result is not sufficient to prove his claims, and leads to an incorrect algorithm. The following theorem states a stronger result, which is necessary to prove the distance formula and to develop sound algorithms.

Theorem 1. *If a chromosome X of genome A contains more than one tree, and no other chromosome of A contains any non-trivial intrachromosomal component, then there exists a proper translocation involving chromosome X that does not modify F_A.*

Proof. A proper translocation can modify F_A either by linking two existing non-trivial intrachromosomal components, or by creating new ones. In the first case, the two existing components must be in separate chromosomes, contrary to the hypothesis.

By Lemma 3, there exists at least one proper translocation involving chromosome X. Assume that they all create new non-trivial components, and consider a proper translocation T that creates a component $[i \ldots j]$ of minimal length, where $i < j - 1$. We will show that then there must exist another proper translocation that either creates smaller components, or does non create non-trivial components.

Since T creates the component $[i \ldots j]$, by flipping chromosomes as necessary, the signed permutation P_A can be written as:

$$P_A = (\ldots \; i \; \ldots \; x \; \ldots -j \ldots -y \; \ldots \;).$$
$$\underset{T}{\bullet\!\!-\!\!-\!\!-\!\!-\!\!-\!\!-\!\!\bullet}$$

where i and x are on the same chromosome and j and y on a different chromosome. Translocation T transforms P_A into $P_{A'}$:

$$P_{A'} = (\ldots \; i \; \ldots \; x \; y \; \ldots \; j \; \ldots \;).$$

Since the interval $(i \ldots x \; y \ldots j)$ is a component, neither $(i \ldots x)$ nor $(y \ldots j)$ can be a component, otherwise we would have nested components with the same endpoints. Moreover, since $i < j - 1$, we have that $i \neq x$ or $j \neq y$ (or both).

Suppose $i \neq x$, this means that there exists an elementary interval J that has one endpoint between i and x and the other endpoint between j and y, otherwise $(i \ldots x)$ would be a component. Thus J is an interchromosomal elementary interval of P_A.

$$P_A = (\ldots \; i \; \ldots \; x \; \ldots -j \ldots -y \; \ldots \;)$$
$$\underset{J}{\bullet\!\!-\!\!-\!\!-\!\!-\!\!-\!\!-\!\!\bullet}$$

Applying the corresponding translocation to A yields:

$$P_{A''} = (\ldots \ i \ \ldots \ j \ \ldots -x \ldots -y \ \ldots).$$

where i and j are on the same chromosome, and x and y on a different one.

If x or y belong to a new non-trivial component in $P_{A''}$, then this component must by strictly shorter than $[i \ldots j]$, since both x and y are in $\{i \ldots j\}$.

A new non-trivial component cannot contain both i and j, since the element $x \in \{i \ldots j\}$ is on a different chromosome. If it contains i and is longer than $[i \ldots j]$, then it must be an interval of the form: $(i' \ldots i \ldots j')$, where $i' < i < j' < j$. But all the elements at the right of i are greater than i, and all the elements at the left of i are smaller than i, implying that either $i' = i$ or $i = j'$, which is a contradiction. Similar arguments hold if the new non-trivial component contains j and is longer than $[i \ldots j]$.

The case where $j \neq y$ can be treated similarly. □

Efficient sorting by translocations will use the fact that trees belonging to different chromosomes can be easily dealt with. When all the trees are in one chromosome, we want to *separate* them, that means move them to different chromosomes. The next result states that such a separation is always possible with translocations that do not modify the topology of the forest.

Corollary 1. *If a chromosome X of genome A contains more than one tree, and no other chromosome of A contains any non-trivial intrachromosomal component, then the trees can be separated by proper translocations without modifying F_A.*

Proof. By Theorem 1, there exists a proper translocation that does not change F_A. Such a proper translocation either separates the trees or not. If all the trees are still contained in the same chromosome, then, by the same argument, there exists another proper translocation that does not change the number of trees. Thus, there always exists either a separating or a non-separating proper translocation. Since the number of successive proper translocations is finite, there always exists a sequence of proper translocations that separates the trees. □

3   A Discussion of Hannenhalli's Algorithm

In order to compute the translocation distance, Hannenhalli [5] introduced the notions of *subpermutations* and *even-isolation*. *Subpermutations* are equivalent to the non-trivial intrachromosomal components defined in the previous section. A genome A has an *even-isolation* if all the minimal subpermutations of A reside on a single chromosome, the number of minimal subpermutations is even, and all the minimal subpermutations are contained within a single subpermutation. Hannenhalli showed that

$$d(A) = n - N - c + s + o + 2i$$

where s denotes the number of minimal subpermutations, $o = 1$ if the number of minimal subpermutations is odd and $o = 0$ otherwise, and $i = 1$ if P has an even-isolation and $i = 0$ otherwise.

Based on the above equation, Hannenhalli gave a polynomial time algorithm for the sorting by translocations problem (Algorithm 1) where a translocation is called *valid* if it decreases the translocation distance.

Algorithm 1 (Hannenhalli's algorithm, from [5])

1: **while** A is not identical to the target genome **do**
2:    **if** there is a proper translocation in A **then**
3:       select a valid proper translocation ρ
4:    **else**
5:       select a valid bad translocation ρ
6:    **end if**
7:    $A \leftarrow A\rho$
8: **end while**

The main assumption behind the algorithm is that if there exists a proper translocation, then there always exists a valid proper translocation (Theorem 12 in [5]). This is based on the argument that there exists a proper translocation that increases the number of cycles by 1 and does not change the number of minimal subpermutations. Hannenhalli wrongly concludes that such a proper translocation cannot create an even-isolation. The following genome shows that, apart from the obvious way to create an even-isolation by creating new subpermutations, there is a second way:

$$A_3 = \{(1 \quad 2 \quad 4 \quad 3 \quad 5 \quad 12), (11 \quad 6 \quad 8 \quad 7 \quad 9 \quad 10)\}.$$

Genome A_3 has exactly one proper translocation, yielding

$$A_3' = \{(1 \quad 2 \quad 4 \quad 3 \quad 5 \quad 6 \quad 8 \quad 7 \quad 9 \quad 10), (11 \quad 12)\}.$$

This translocation creates an even-isolation by chaining the two existing subpermutations $[2 \ldots 5]$ and $[6 \ldots 9]$. Therefore the translocation is not valid.

In order to prove the translocation formula, Hannenhalli first shows that if there exists a proper translocation, then there exists an alternative proper translocation that does not create new minimal subpermutations (Theorem 10 in [5]). Then Hannenhalli assumes that there is no proper translocation and follows by indicating how to destroy subpermutations (Theorem 13 in [5]). These results lead to an algorithm based on the false impression that the subpermutations can be destroyed independently of the sorting procedure.

Sometimes, in an optimal sorting scenario, we first have to destroy the subpermutations as it is the case of genome A_3. But in other cases, we first have to separate the subpermutations before destroying them. For example, consider the following genome:

$$A_4 = \{(-9 \quad 8 \quad -7 \quad 4 \quad -3 \quad 2 \quad -1), (10 \quad 6 \quad 5 \quad 11)\}.$$

In order to sort genome A_4 optimally, we first have to apply a proper translocation separating the subpermutations $[-9 \ldots -7]$ and $[-3 \ldots -1]$.

$$A_4' = \{(-9 \quad 8 \quad -7 \quad 4 \quad 5 \quad 11), (10 \quad 6 \quad -3 \quad 2 \quad -1)\}$$

In the resulting genome A_4', the two subpermutations belong to different chromosomes so that we can destroy them by a single bad translocation.

However, in the next section we will show that Hannenhalli's equation for the translocation distance holds, but that any sorting strategy should deal with destroying intrachromosomal components at each iteration step.

4   Computing the Translocation Distance

Given a genome A and the forest F_A, let L be the number of leaves, and T the number of trees of the forest. The following lemma will be central in proving the distance formula and establishing an invariant for the sorting algorithm.

Lemma 4. *Let A be a genome whose forest has L leaves and T trees. If L is even, and $T > 1$, then there always exists a sequence of proper translocations, followed by a bad translocation, such that the resulting genome A' has $L' = L - 2$ leaves and $T' \neq 1$ trees.*

Proof. If all the trees are on the same chromosome then, by Corollary 1, we can separate the forest with proper translocations without modifying T or L.

Assume that there exist trees on different chromosomes. In the following, we show how to pair two leaves such that the bad translocation destroying the corresponding intrachromosomal components reduces the number of leaves by two. We have to show that $T' > 1$ or $T' = 0$. Therefore, we consider the following cases.

If $T = 2$, then either both trees have an even number of leaves or both have an odd number of leaves since the total number of leaves is even. If both trees have an even number of leaves, we pair any two leaves belonging to different trees and destroy them. In this case, the number of trees can only be increased. If both trees have an odd number of leaves, then we choose the middle leaves of both trees. In the best case, if both trees consist of a single leaf each, we get $T' = 0$, or otherwise $T' > 1$.

If $T > 2$ and one of the trees has an even number of leaves, we pair one of its leaves with any other leaf of a tree that belongs to a different chromosome. Since at most one tree will be destroyed, it follows that $T' > 1$.

If $T > 2$ and all the trees have an odd number of leaves, then T must be even since the total number of leaves is even. Hence the number of trees is at least four and we can choose any two leaves of the trees that belong to different chromosomes. It follows immediately that $T' > 1$. □

Lemma 4 implies that when the number of leaves is even, and $T > 1$, we can always destroy the forest optimally: we can use proper translocations to separate the forest, and then remove two leaves with a bad translocation. Eventually, all trees are destroyed, i.e. $T = 0$. The basic idea is to reduce all other cases to the simple case of Lemma 4.

Theorem 2. *Let A be a genome with c black cycles and F_A be the forest associated to A. Then*

$$d(A) = n - N - c + t$$

where

$$t = \begin{cases} L + 2 & \text{if } L \text{ is even and } T = 1 & (1) \\ L + 1 & \text{if } L \text{ is odd} & (2) \\ L & \text{if } L \text{ is even and } T \neq 1. & (3) \end{cases}$$

Proof. We first show $d(A) \geq n - N - c + t$. Consider an optimal sorting of length d containing p proper translocations and b bad translocations, thus $d = p + b$. Since b translocations remove b cycles, and p translocations add p cycles, we must have:

$$c - b + p = n - N, \text{ implying } d = n - N - c + 2b.$$

We will show that $2b \geq t$, implying $d \geq n - N - c + t$.

Since a bad translocation removes at most two leaves, we have that $b \geq L/2$, if L is even, and $b \geq (L + 1)/2$, if L is odd. Therefore, in cases (2) and (3), it follows that $b \geq t/2$.

If there is only one tree with an even number of leaves, then there must be a bad translocation B in the optimal sorting that has one endpoint in a tree and the other not contained in a tree. If this translocation does not destroy any leaves, then $b \geq 1 + L/2$. If translocation B destroys a minimal component, it destroys exactly one, and the minimal number of bad translocations needed to get rid of the remaining ones is $((L - 1) + 1)/2$, implying again that $b \geq 1 + L/2$. Thus, in case (1), we also have $b \geq t/2$.

In order to show that $d(A) \leq n - N - c + t$, we will exhibit a sequence of proper and bad translocations that achieve the bound $n - N - c + t$.

In case (2), if L is odd and $T = 1$, we destroy the middle leaf of the tree. Then $L - 1$ is even, and $T > 1$ or $T = 0$. If $T > 1$, then the preconditions of Lemma 4 apply, and the total number of bad translocations will be $1 + (L - 1)/2$.

If L is odd and $T > 1$, we destroy a single leaf of some tree with more than one leaf, if such a tree exists. Otherwise, we must have $T > 2$, since the number of leaves is odd, and we destroy any leaf. In both cases, we have $T' > 1$. Again, the total number of bad translocations will be $1 + (L - 1)/2$.

In case (3), if L is even and $T \neq 1$, then the preconditions of Lemma 4 apply, and the total number of bad translocations will be $L/2$.

In case (1), if L is even and $T = 1$, destroy any leaf and apply case (2), the total number of bad translocations will be $1 + L/2$.   □

For example, the genome

$$A_2 = \{(\,1\,-2\,\,3\,\,8\,\,4\,-5\,\,6\,), (\,7\,\,9\,\,-10\,\,11\,-12\,\,13\,\,14\,-15\,\,16\,)\}$$

of Section 2.4 consists of two chromosomes and 16 elements. The signed permutation P_{A_2} has seven black cycles. The forest F_{A_2} has three trees and five leaves (see Fig. 3). Therefore, we have

$$d(A_2) = n - N - c + t = 16 - 2 - 7 + 6 = 13.$$

5   Algorithms

In this section we present two algorithms. The first algorithm allows to compute the translocation distance between two genomes in linear time, a result previously given by Li *et al.* [8], although we believe that our algorithm is simpler than theirs. The second algorithm is the first correct polynomial time algorithm for sorting a genome by translocations.

The algorithm to compute the translocation distance is similar to the one to compute the reversal distance presented in [1]. We only sketch the algorithm here and discuss those parts that need to be modified.

Assume that a genome A and an extended signed permutation P_A are given. The algorithm consists of three parts. In the first part, the cycles of P_A are computed by a left-to-right scan of P_A without taking into account the points between tails. The second part is the computation of the intrachromosomal components. We apply to each chromosome the linear-time algorithm of [1] to compute the direct and reversed components of a permutation. Note that the intrachromosomal components are equivalent to the direct and reversed components. Finally, in the third part of the algorithm the forest F_A is constructed by a single pass over the intrachromosomal components, and the distance can then easily be computed using the formula of Theorem 2.

Altogether, we can state the following theorem, previously established in [8].

Theorem 3. *The translocation distance $d(A)$ of a genome A can be computed in linear time.*

We now turn to the sorting by translocations problem. An algorithm that sorts a genome optimally is shown in Algorithm 2. Assume that the forest F_A of the genome A is given. We denote by L the number of leaves and by T the number of trees of the forest.

Initially, we apply one or two translocations in order to arrive at the preconditions of Lemma 4. If the forest consists of a single tree with an even number of leaves (line 2), we destroy any leaf. In the resulting genome, if the number of leaves is odd and in a single tree, we destroy its middle leaf, if there is more than one tree, we apply a translocation that destroys one leaf of the greatest tree. In all cases, we get a genome A' with $T' = 0$, or $T' > 1$ and L' even.

Then, as long as there exist intrachromosomal components (i.e. $T > 1$ and L is even), we can destroy the forest optimally as described in Lemma 4: we use proper translocations to separate the forest, and remove two leaves with each bad translocation. Once all intrachromosomal components are destroyed (i.e. $T = 0$), we can sort the genome using proper translocations that do not create

---

Algorithm 2 (Sorting by translocations algorithm)

1: L is the number of leaves, and T the number of trees in the forest F_A associated
   to the genome A
2: **if** L is even **and** $T = 1$ **then**
3:    destroy one leaf such that $L' = L - 1$
4: **end if**
5: **if** L is odd **then**
6:    perform a bad translocation such that $T' = 0$, or $T' > 1$ and $L' = L - 1$
7: **end if**
8: **while** A is not sorted **do**
9:    **if** there exist intrachromosomal components on different chromosomes **then**
10:      perform a bad translocation such that $T' = 0$, or $T' > 1$ and L' is even
11:    **else**
12:      perform a proper translocation such that T and L remain unchanged
13:    **end if**
14: **end while**

---

new non-trivial intrachromosomal components. Such proper translocations exist as we have shown in the proof of Theorem 1. Thus, there always exists either a proper translocation that does not modify the topology of the forest, or a bad translocation that maintains the preconditions of Lemma 4. This establishes the correctness of the algorithm and we have:

Theorem 4. *Algorithm 2 solves the sorting by translocations problem in $O(n^3)$ time.*

Initially, the forest F_A associated to a genome A is constructed. This can be done in $\mathcal{O}(n)$ time as discussed above. The algorithm requires at most $\mathcal{O}(n)$ iterations. The bad translocations of line 9 can be found in constant time as described in the proof of Lemma 4. Since there are $\mathcal{O}(n)$ proper translocations and each translocation requires the construction of the forest to verify the condition $T' = T$ and $L' = L$, the search for a proper translocation in line 11 takes $\mathcal{O}(n^2)$ time. Hence, the total time complexity of Algorithm 2 is $\mathcal{O}(n^3)$.

6   Conclusion

The real challenge in developing genome rearrangement algorithms is to propose algorithms whose validity can be checked, both mathematically and biologically. The most useful set of rearrangements operations is currently including translocations, fusions, fissions and inversions [10]. Unfortunately, we think that few people are able to assess the mathematical validity of the current algorithms. The work we have done in this paper opens the way to simpler description and implementation of such algorithms.

References

1. A. Bergeron, J. Mixtacki, and J. Stoye. Reversal distance without hurdles and fortresses. In *CPM 2004 Proceedings*, volume 3109 of *LNCS*, pages 388–399. Springer Verlag, 2004.

2. A. Bergeron, J. Mixtacki, and J. Stoye. The inversion distance problem. In O. Gascuel, editor, *Mathematics of Evolution and Phylogeny*, chapter 10, pages 262–290. Oxford University Press, Oxford, UK, 2005.

3. A. Bergeron and J. Stoye. On the similarity of sets of permutations and its applications to genome comparison. In *Proceedings of COCOON 03*, volume 2697 of *LNCS*, pages 68–79. Springer Verlag, 2003.

4. Mouse Genome Sequencing Consortium. Initial sequencing and comparative analysis of the mouse genome. *Nature*, 420:520–562, 2002.

5. S. Hannenhalli. Polynomial-time algorithm for computing translocation distance between genomes. *Discrete Appl. Math.*, 71(1-3):137–151, 1996.

6. S. Hannenhalli and P. A. Pevzner. Transforming men into mice (polynomial algorithm for genomic distance problem). In *FOCS 1995 Proceedings*, pages 581–592. IEEE Press, 1995.

7. J. D. Kececioglu and R. Ravi. Of mice and men: Algorithms for evolutionary distances between genomes with translocation. In *Proceedings of the sixth ACM-SIAM Symposium on Discrete Algorithm*, pages 604–613. Society of Industrial and Applied Mathematics, 1995.

8. G. Li, X Qi, X. Wang, and B. Zhu. A linear-time algorithm for computing translocation distance between signed genomes. In *CPM 2004 Proceedings*, volume 3109 of *LNCS*, pages 323–332. Springer Verlag, 2004.

9. M. Ozery-Flato and R. Shamir. Two notes on genome rearrangements. *J. Bioinf. Comput. Biol.*, 1(1):71–94, 2003.

10. G. Tesler. Efficient algorithms for multichromosomal genome rearrangements. *J. Comput. Syst. Sci.*, 65(3):587–609, 2002.

Author Index

Lecture Notes in Bioinformatics